D0468561

Electronic Communications

Electronic Communications

William Sinnema
Robert McPherson

Prentice-Hall Canada Inc., Scarborough, Ontario

Canadian Cataloguing in Publication Data

Sinnema, William, 1937-1988
Electronic communications

Includes index.
ISBN 0-13-251455-9

1. Telecommunication. I. McPherson, Robert Gordon. II. Title.

TK5101.S56 1989 621.38'0413 C89-094780-5

Prentice Hall, Inc., Englewood Cliffs, New Jersey
Prentice-Hall International, Inc., London
Prentice-Hall of Australia, Pty., Ltd., Sydney
Prentice-Hall of India Pvt., Ltd., New Delhi
Prentice-Hall of Japan, Inc., Tokyo
Prentice-Hall of Southeast Asia (Pte.) Ltd., Singapore
Editora Prentice-Hall do Brasil Ltda., Rio de Janeiro
Prentice-Hall Hispanoamericana, S.A., Mexico

ISBN 0-13-251455-9

Production Editors: Peter Buck, Elizabeth Long
Copy Editor: Barbara Zeiders
Proofreader: Martha Williams
Interior Design: Joe Chin
Original Art: Agnes Szilagyi
Cover Design: Marjorie Pearson
Production Coordinators: Anna Orodi, Sandra Paige
Typesetting: EPS Group

1 2 3 4 5 THB 95 94 93 92 91

Printed and bound in Canada by T.H. Best Printing Company

Dedication

This text is dedicated to the book's principal editor Bill Sinnema. Mr Sinnema was an electronics technology instructor and the author of two other textbooks. He was widely respected by both his students and colleagues not only for his technical expertise and teaching abilities, but also for his quiet human qualities. His death in an accident was sorrowfully felt by his colleagues and co-authors of this text.

Contents

14 Wave Propagation 569

Dennis Morland

15 Antennas 596

Dennis Morland

16 An Introduction to Fiber Optic Communications 643

Robert McPherson

17 Introduction to Cellular Radio 663

Wayne Wolinski

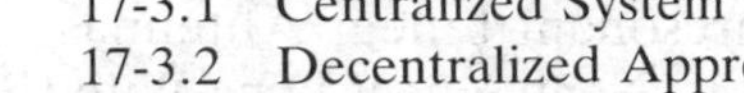

18 Satellite Communications 673

Walter Kalin

Preface

This text is a comprehensive study of electronic communication circuits and systems, aimed primarily at the technologist student. A general knowledge of basic AC circuit theory and solid state fundamentals is assumed, and although calculus is occasionally employed, an effort has been made to avoid the use of advanced mathematics. The book is intended for use in a two-year electronics technology program.

The first chapter of this text introduces the concepts and terms used in electronic communications. A number of simplified techniques are presented that help to describe how information-bearing signals are processed by a communication system. The basic operation of a superhet receiver is examined using these techniques.

Chapters 2 to 5 undertake to examine the operation of the fundamental functional blocks (tuned circuits, amplifiers, filters, oscillators, mixers, modulators, etc.) that make up the parts of a radio communication system. With this as a foundation, chapters 6, 7, and 8 discuss the generation and detection of signals in analog modulation systems (AM, SSB, FM/PM).

Noise, the fundamental limitation of all communication systems is discussed at length in Chapter 9.

Chapter 10 examines the telephone system from the subscriber's premises to the Central Office and the Toll Network. Transmission, switching, and signaling concepts are covered in a practical manner for each part of the network.

Chapter 11 considers the digital communication field, particularly as it relates to the telecommunications industry. This is followed in Chapter 12 by a general introduction to data communication networks, their form of interconnection, and protocols.

In Chapter 13, transmission line and waveguide structures are introduced and the propagation of electromagnetic waves on these structures is examined. The use of a Smith chart to predict the operation of the structures is demonstrated.

Chapters 14 and 15 investigate the launching, travel, and reception of free space electromagnetic waves in a radio communication system. Chapter 15 presents fundamental antenna properties, while Chapter 14 examines how a wave leaving the transmitter is affected by the path it follows to the receiver.

The operation and relative merits of fibre optic communication are examined in Chapter 16.

The rapidly developing field of cellular radio is introduced in Chapter 17.

An extensive introduction to the use of satellite-based communication systems is undertaken in Chapter 18.

The authors would like to thank our many colleagues and students whose input and responses have helped to shape this text and sharpen its effectiveness. In particular, we would like to thank the following technical specialists who were good enough to review early drafts of the text and provide input for its improvement:

R.J. Angell, Sr., *Confederation College*
Trevor Glave, *British Columbia Institute of Technology*
R.E. Greenwood, *Ryerson Polytechnical Institute*
Bruce Horin, *Humber College*
H.H. Manoochehri, *Humber College*
Peter H. Risdahl, *Southern Alberta Institute of Technology*
Peter C. Szilagyi, *Sault College*

Introduction

Robert McPherson

1-1 OBJECTIVES

This initial chapter presents some fundamental concepts of electronic communication. These are intended to provide readers with sufficient background to serve as an overview of communication *systems*. Subsequent chapters deal in much greater detail with the specific operation of the various functional blocks within a communication system.

The opening sections of this chapter present a model for representing communication systems and then examine the concepts of signal modulation, channel multiplexing, and the allocation of frequency spectra to potential broadcasters. Subsequent sections of the chapter examine the relative merits of using frequency—rather than time—domain respresentations of communication signals and present some simplified techniques for describing the effects that devices such as mixers have on communication signals. The final section undertakes a block diagram description of superheterodyne receivers and examines the advantages and limitations of a typical superheterodyne receiver.

1-2 COMMUNICATION SYSTEM MODEL

The purpose of a communication system is to transfer information from one point to another. There exist, of course, many methods to accomplish this end, each with its own relative advantages and disadvantages. We may, however, view most communication systems within the framework of the general model shown in Fig. 1-1. Consider as an example the sequence that occurs when you speak to someone. You (the information source) put your thought into words (encode) and speak (modulate an acoustic wave). The voice travels through the air (channel) to the listener (receiver). The listener's ear converts the sound to nerve impulses (demodulates), which travel to the brain to be interpreted (decoded) as to meaning (information recovered).

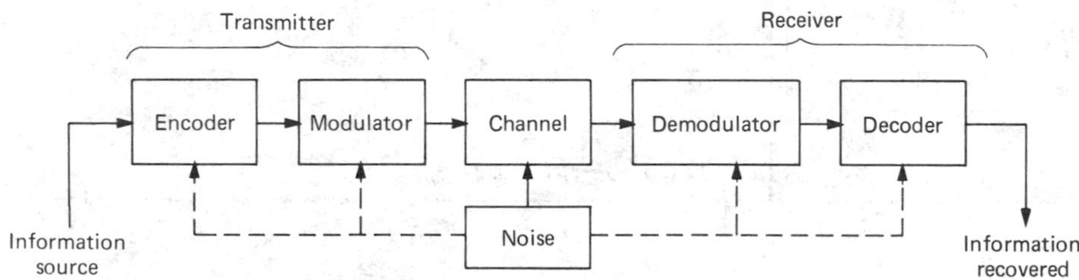

Figure 1-1 General communication system.

In addition, consider what occurs if the process is performed in a room with a high level of background noise. Unless the speaker and listener are sufficiently close together that the speaker's voice (signal) heard by the listener (receiver) is significantly above the ambient noise level, the message received will be distorted if not lost altogether. We must therefore include noise sources in our model as these ultimately limit the range and usefulness of any communication system.

1-3 ELECTRONIC COMMUNICATION, MODULATION, AND MULTIPLEXING

We will restrict our present examinations to electronic communication systems wherein the information to be transmitted is impressed on a high-frequency electromagnetic wave. The wave is then delivered to the receiver by a suitable transmission line or by direct broadcast through the air. The effects of noise signals on this signal transmission are not addressed at this time but are examined in some detail in Chapter 9.

The wave used to carry the information is typically a high-frequency sinusoid of the form

$$v(t) = A\cos(2\pi ft + \theta) \tag{1-1}$$

The information to be transmitted is impressed on the carrier wave by "modulating" (varying) one of the wave characteristics in direct proportion to the information signal. If the carrier wave amplitude (A) is adjusted in proportion to the information signal, the result is *amplitude modulation* (AM). Alternatively, if the frequency (f) of the carrier wave is adjusted in proportion to the information signal, a frequency modulation (FM) signal is generated. If the phase (θ) is adjusted, it is called *phase modulation* (PM).

Many modulation schemes exist. The one effect common to all modulation schemes is that the process of impressing an information signal onto a carrier *always* results in the generation of signal components which are at frequencies somewhat different from the carrier frequency. In order to deliver the modulated signal with its impressed information to the receiver, the communication system must be capable not only of transferring the carrier frequency signal but also of passing a *band* of frequencies around the carrier frequency. How wide a band of frequencies must be passed depends on the specific modulation technique and the amount of information to be transferred. In general, the higher the information rate, the greater the bandwidth required.

To transmit an information signal through a communication system it is not absolutely necessary to impress the information onto a high-frequency carrier. It is possible simply to send the information waveform directly. This, for example, is what is done with local telephone calls where both telephones are connected to the same local exchange. The voice is converted into an electrical signal and the signal is delivered to the receiver via a twisted copper wire pair.

Direct transmission of the information waveform will, however, typically prevent multiplexing of the transmission facility. Consider, for example, a cable TV distribution system. It would be possible to transmit the video signal for one channel through a cable system directly to the various receivers. A second video signal from another channel could not, however, be added to the system since the receivers would not have any basis on which to separate the two signals. If, on the other hand, the video signal is used to modulate a high-frequency carrier, the information of the channel is contained within a finite band of frequencies around the carrier frequency. Second and subsequent channels may be added to the system by selecting different carrier frequencies and ensuring that a sufficient spread exists between the carrier frequencies to avoid overlap of the individual channel bandwidths. This arrangement is outlined in Fig. 1-2. Each receiver on the system can selectively filter out its desired channel and then recover the desired video information.

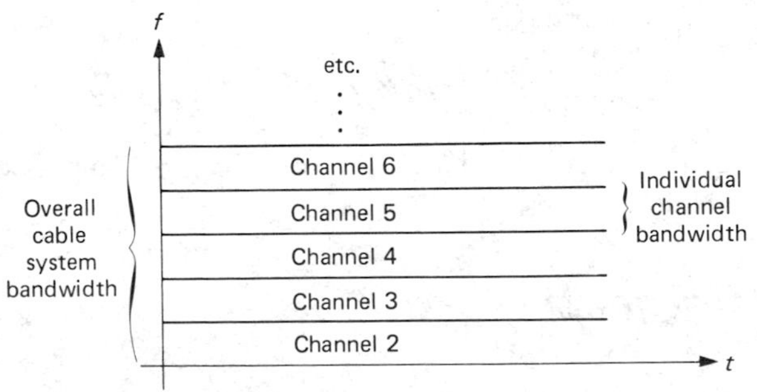

Figure 1-2 Frequency-division multiplexing.

A second limitation to direct transmission of the information signal occurs if the information is to be broadcast as a radio wave rather than being carried on a dedicated cable. To launch a radio wave into space, an antenna structure is required. For reasons that are beyond the scope of the present discussion (see Chapter 15 instead), the structure must be at least a tenth of a wavelength in extent to launch a radio wave with any degree of efficiency. The wavelength of a sinusoidally varying radio wave is defined as the distance a wave will travel in one cycle. Since radio waves travel at the speed of light ($c = 3 \times 10^8$ m/s) the wavelength is

$$\lambda = \frac{c}{f} \tag{1-2}$$

Consider, for example, the wavelength of a 1-kHz audio tone. One wavelength is 300 km. Clearly, an antenna to launch this wave would have to be unrealistically large. If, however, the audio signal is used to modulate a 300-MHz carrier, the wavelength of the wave to be launched is only 1 m in extent and construction of an antenna is feasible.

In Fig. 1-2, the transmission system's information-carrying capacity has been multiplexed by giving each channel continuous access to a limited portion of the overall channel bandwidth. An alternative form of multiplexing that is widely used in transmitting digital information is *time-division multiplexing*. The format of this multiplexing scheme is shown in Fig. 1-3. In time-division multiplexing (TDM) each channel is allowed access to the entire channel bandwidth, but only within the designated time slots. Comparing this to *frequency-division multiplexing* (FDM), we may note that in TDM each channel is "on" for only a fraction of the time. In each time slot, however, the wide bandwidth allows a large information rate to be used. Channel information in TDM is thus sent in brief high-rate bursts. Channel information in FDM is sent continuously but at a lower rate (less bandwidth). The two schemes are thus theoretically similar in their information-handling capabilities.

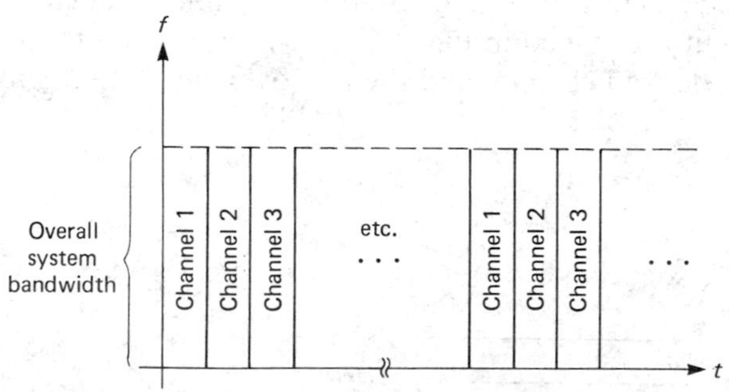

Figure 1-3 Time-division multiplexing.

1-4 RADIO-FREQUENCY SPECTRUM ALLOCATIONS

When information is to be transferred between two widely separated points, one of the most attractive techniques is to broadcast a radio wave modulated with the information signal. The advantage of direct broadcast is that no physical construct (coaxial cable, fiber optic cable, twisted wire pair) must be built between the source and the receiver. Since phenomenal cost savings are possible with such a system, there is a huge demand to broadcast signals over the spectrum of possible broadcast frequencies. The broadcast spectrum is therefore treated as a very valuable and finite resource. Permission to broadcast is carefully controlled and regulated by various governmental and international organizations. The International Telecommunications Union (ITU), an agency of the United Nations, has divided the radio-frequency spectrum into several bands. Beginning with the very low frequency band (VLF), Table 1-1 gives the names and designations for the various frequency (or wavelength) bands.

TABLE 1-1 Standard Frequency Ranges[a]

Band	Designation	Frequency Range	Wavelength (λ)
VLF	Very low frequency	$3 < f < 30$ kHz	$100 > \lambda > 10$ km
LF	Low frequency	$30 < f < 300$ kHz	$10{,}000 > \lambda > 1000$ m
MF	Medium frequency	$300 < f < 3000$ kHz	$1000 > \lambda > 100$ m
HF	High frequency	$3 < f < 30$ MHz	$100 > \lambda > 10$ m
VHF	Very high frequency	$30 < f < 300$ MHz	$10 > \lambda > 1$ m
UHF	Ultra high frequency	$300 < f < 3000$ MHz	$1 > \lambda > 0.1$ m
SHF	Super high frequency	$3 < f < 30$ GHz	$10 > \lambda > 1$ cm
EHF	Extremely high frequency	$30 < f < 300$ GHz	$10 > \lambda > 1$ mm
	Submillimeter	$300 < f < 3000$ GHz	$1 > \lambda > 0.1$ mm
	Far infrared	$3000 < f < 30{,}000$ GHz	$100 > \lambda > 10$ μm
	Near infrared	$30{,}000 < f < 394{,}737$ GHz	$10 > \lambda > 0.76$ μm
	Visible light	$394{,}737 < f < 769{,}231$ GHz	$0.76 > \lambda > 0.39$ μm
UV	Ultraviolet light	$7.69 \times 10^{14} < f < 3 \times 10^{16}$ Hz	$0.39 > \lambda > 0.01$ μm
	X-rays	$3 \times 10^{16} < f < 3 \times 10^{18}$ Hz	$100 > \lambda > 1$ Å
	Gamma rays	$3 \times 10^{18} < f < 3 \times 10^{20}$ Hz	$1 > \lambda > 0.01$ Å
	Cosmic rays	$3 \times 10^{20} < f < 3 \times 10^{24}$ Hz	$0.01 > \lambda > 1 \times 10$ Å

[a]kHz = 1000 hertz;
MHz = 1000 kHz;
GHz = 1000 MHz;
micrometer (μm) = 10^{-6} meter (m);
angstrom (Å) = 10^{-10} m = 1Å.

Under the sponsorship of the ITU, the World Administrative Radio Conferences (WARC) allocates specific frequency bands to various services. The radio spectrum is used by broadcasters, police communications and radar, navigational aids in the air and on the sea, radio amateurs, telecommunication carriers, CB operators, microwave ovens, and so on. Table 1-2 lists a few of the spectrum allocations. Although only a small fraction of the allocations are shown, the diversity of the applications is noteworthy.

TABLE 1-2 Some Spectrum Allocations

Frequency Band	Allocation (type of service)	Description
10–14 kHz	Omega	Long-range marine and aviation navigation aid Earth–ionosphere appears as a waveguide
100 kHz	Loran C	Long-range marine radio navigation aid
535–1605 kHz	AM radio	AM radio broadcasting
54–72 MHz	TV	Channels 2–4
76–88 MHz	TV	Channels 5 and 6
88–108 MHz	FM radio	100 FM channels spaced 200 kHz apart
174–216 MHz	TV	Channels 7–13
470–608 MHz	TV	Channels 14–36
614–806 MHz	TV	Channels 38–69
2450 MHz		Microwave ovens
3.7–4.2 GHz		*Anik 1* satellite downlink
5.975–6.425 GHz		*Anik 1* satellite uplink
10.525 GHz		Police radar
3.5–4.0 MHz	80-m band	Amateur radio
7.0–7.3 MHz	40-m band	Amateur radio
14.0–14.35 MHz	20-m band	Amateur radio
21–21.45 MHz	15-m band	Amateur radio
28–29.7 MHz	10-m band	Amateur radio
50–54 MHz	6-m band	Amateur radio
144–148 MHz	2-m band	Amateur radio
2.5 MHz	WWV (Fort Collins, Colorado) WWVH (Kekaha, Hawaii)	Broadcasting of high-accuracy standard radio frequencies and time signals
5 MHz	"	"
10 MHz	"	"
15 MHz	"	"

1-5 TELECOMMUNICATION STANDARDS ORGANIZATIONS

The International Telecommunication Union (ITU) was founded to promote and foster international telecommunications on a worldwide basis, to carry on studies of technical problems relating to the telecommunications network, and to assure proper interworking of the various national systems. The ITU is a specialized agency of the United Nations, but predates it, having come into existence in 1932 with the merger of two older organizations, the International Telegraph Union (1865–1932) and the Radio Telegraph Union (1903–1932). Within the ITU there are three major divisions:

1. *The Consultative Committee on International Telegraphy and Telephony* (*CCITT*). This committee is responsible for making technical recommendations and studying tariff questions relating to telegraphy and telephony.

2. *The Consultative Committee on International Radio* (*CCIR*). This committee makes technical and operational recommendations relating to radio communications.
3. *The International Frequency Registration Board* (*IFRB*). This group registers and examines assignments for space and terrestrial services, records frequency assignments for orbital positions, and advises on the maximum number of radio channels in those portions of the spectrum where interference would occur.

The ITU also holds special conferences to consider new opportunities and particular issues in the light of technological advances. One such conference was the 1979 World Administrative Conference (WARC-79), where the international regulations for spectrum use were reviewed and modified. An earlier WARC-63 conference was devoted to space communication, and large frequency bands were allocated to space services. The ITU operates under the ITU convention. This convention is like a treaty, to which the signatories are bound. It states the structure, goals, and responsibilities of the ITU and its members.

Each country has its own communication authorities that implement the decision of the ITU conference. In Canada the Federal Minister of Communications is responsible for all matters over which the Parliament of Canada has jurisdiction relating to telecommunications. The Department of Communications (DOC), the Canadian Radio Television and Telecommunications Commission (CRTC), the Canadian Broadcasting Corporation (CBC), and Teleglobe Canada are all independent organizations that report directly to the Minister of Communications (Fig. 1-4).

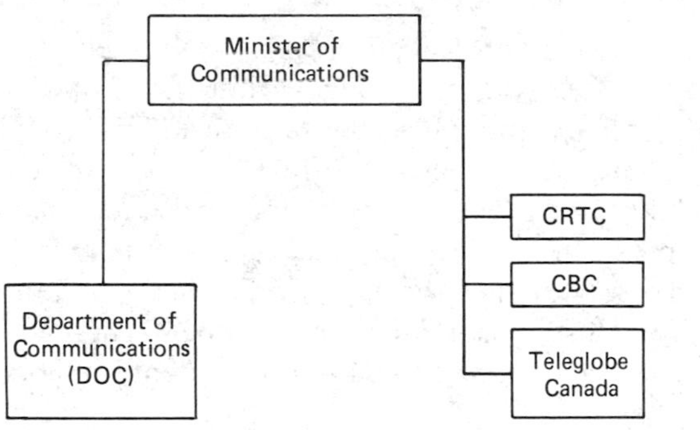

Figure 1-4 Canadian communication authorities.

The Department of Communications itself is comprised of four sections:

1. The policy sector, which formulates and implements policies relating to domestic and international telecommunications.
2. The research sector, which develops new communication systems and provides scientific advice for policymakers. Its principal in-house research facility is located at Shirley Bay (near Ottawa), Ontario. It also distributes research contracts to various Canadian universities.
3. The space section, which is reponsible for satellite communications operations, such as the *Anik* series.

4. The spectrum management and government telecommunications sector, of which the spectrum management service is responsible for the planning, allocating, and regulation of the frequency spectrum; and the government telecommunications agency plans and provides telecommunication services for the federal government.

The CRTC is the licensing and regulating authority for all broadcasting in Canada. It also has responsibilities in telecommunications in such areas as tariffs, regulations, and service complaints.

The CBC is responsible for the national broadcasting service, operating French and English TV, AM, and FM radio networks. It must provide a balanced service of information and contributes to national unity and a continuing expression of Canadian identity.

Teleglobe Canada provides for all the external or overseas telecommunication services. It also participates in the deliberations of such organizations as the Commonwealth Telecommunications Organization (CTO), the International Telecommunications Satellite Organization (Intelsat), the ITU, and the Canadian Telecommunications Carriers Association (CTCA). It operates satellite earth stations at Lake Cowichan, British Columbia; Mill Village 1 and 2, Nova Scotia; and Des Laurentides, Weir, Quebec.

To supply the many communication services required across Canada, the Canadian Telecommunications Carriers and the federal government have formed Telesat Canada, whose function is to establish a system of domestic satellite communications. Its aim is to give people anywhere in Canada an opportunity to communicate with each other instantly and reliably. It presently operates a series of *Anik* (meaning "brother" in Inuktitut) domestic satellites operating in the 6/4 and 14/12 GHz bands.

In the United States, the 1934 Communications Act established the Federal Communications Commission (FCC) to regulate nongovernment use of frequencies and to regulate interstate and foreign communications by wire and radio. It licenses transmitters within the United States and aboard U.S.-registered ships and aircraft.

The regulating reponsibility for frequency spectrum use by federal agencies has been delegated by the president to the National Telecommunications and Information Administration (NTIA) within the U.S. Department of Commerce. NTIA and the FCC maintain a close liaison with one another. The decisions of the NTIA are based on the recommendations of the Interdependent Radio Advisory Committee (IRAC).

1-6 MATHEMATICAL TOOLS

As a signal passes through a communication system the waveform of the signal undergoes many substantial changes as filters, mixers, demodulators, and other communication circuits operate on the signal desired. The effect that these various circuits have on the signal can be examined using various mathematical techniques. A series of relatively simple techniques are described in this section and the examples are, by and large, related directly to communication circuit applications.

1-6.1 Time-Domain and Frequency-Domain Representations

Consider the arbitrary periodic waveform with period T shown in Fig. 1-5. This representation of the waveform is termed the *time-domain representation* and is simply a picture of the voltage amplitude variations with time. This is a format with which we are very familiar, because it is what we see when using an ordinary oscilloscope. There is, however, an alternative (and in communciations more valuable) format that can be used to represent the waveform, called the *frequency-domain representation*.

The basis for this format is the Fourier series expansion of a function. It can be shown that any periodic function of time, with period T, can be fitted to an expansion of the form

$$V(t) = a_0 + C_1 \cos(\omega t + \phi_1) + C_2 \cos(2\omega t + \phi_2) + C_3 \cos(3\omega t + \phi_3) + C_4 \cos(4\omega t + \phi_4) + \cdots \tag{1-3}$$

where $\omega = 2\pi f_1 = 2\pi/T$, and $C_1, C_2, C_3, \ldots$ and $\phi_1, \phi_2, \phi_3, \ldots$ are all constants which may be calculated for a particular waveform shape. Thus any periodic waveform may be considered to be the summation of a number of purely sinusoidal waveforms, the frequencies of which are different but harmonically related (i.e., integral multiples of the "fundamental" frequency f_1; see Fig. 1-5).

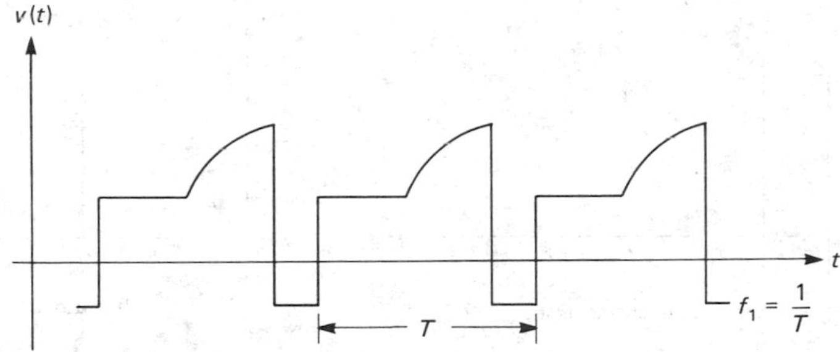

Figure 1-5 Periodic waveform.

The frequency-domain representation of a function is simply a line graph of the magnitude constants ($C_1, C_2, \ldots$) versus the frequency of the sinusoidal terms. This is illustrated in Fig. 1-6. The main advantage of working with the frequency-domain representation of a function rather than with the time-domain representation is that it allows the steady-state effects of frequency-dependent circuits (filters) to be examined more readily. Consider for example, passing a signal having the frequency spectrum of Fig. 1-7(b) through a bandpass filter having the frequency response illustrated in Fig. 1-7(c). Since the bandpass filter will permit only the $2f_1$ frequency to be passed to the output, the output signal will be a sinusoid of frequency $2f_1$, as shown in Fig. 1-7(d) and (e).

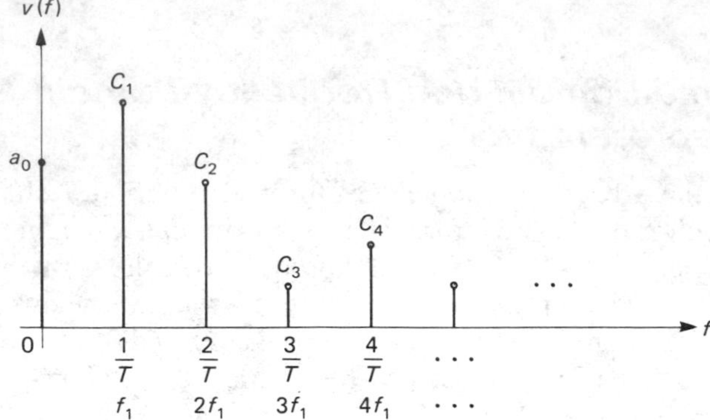

Figure 1-6 Frequency-domain representation.

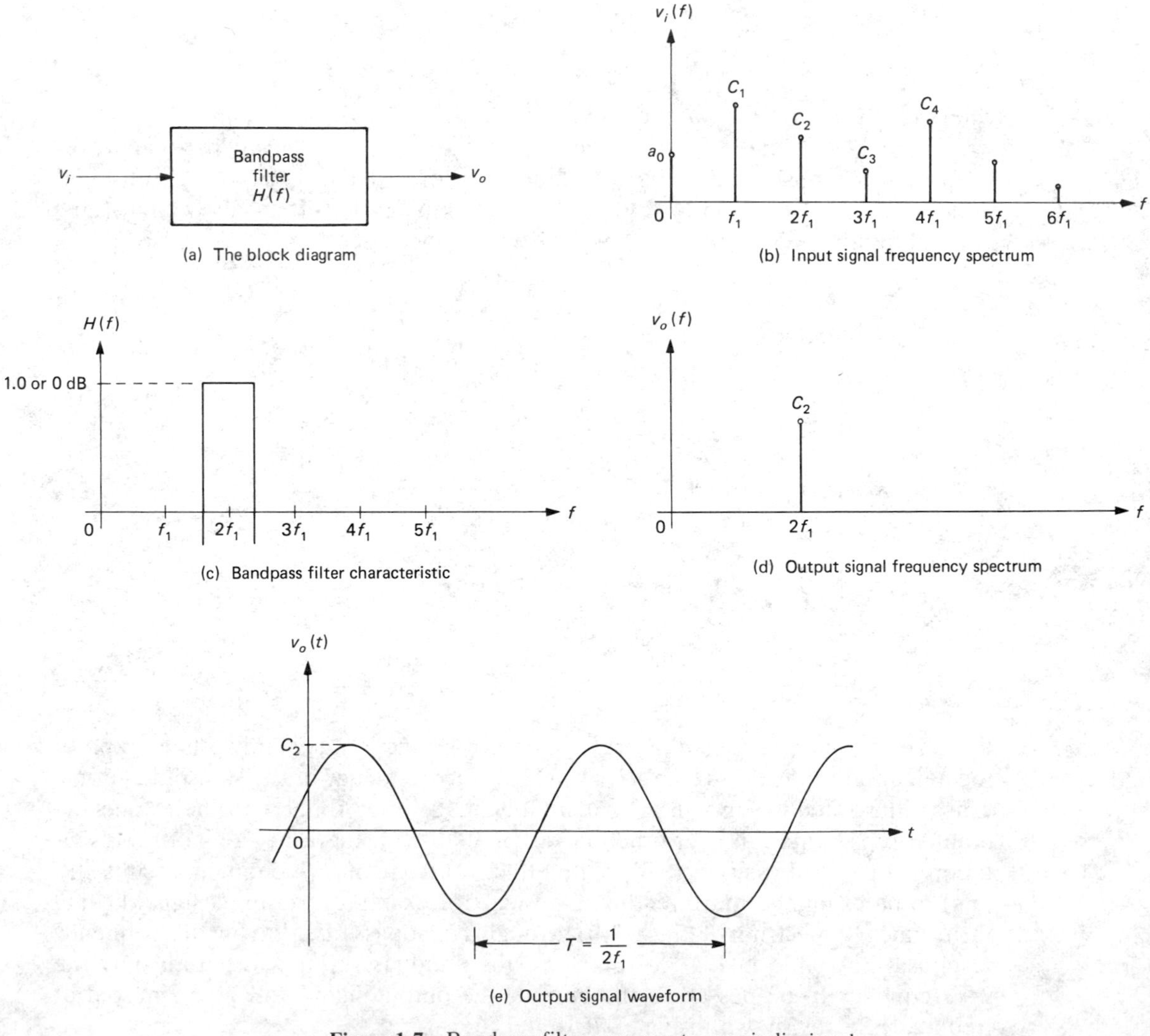

(a) The block diagram

(b) Input signal frequency spectrum

(c) Bandpass filter characteristic

(d) Output signal frequency spectrum

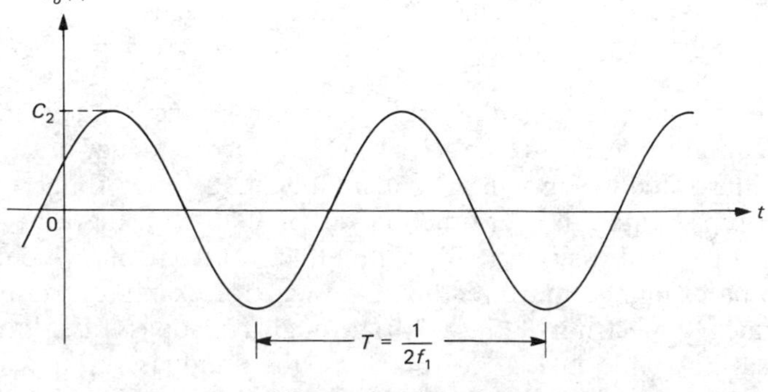

(e) Output signal waveform

Figure 1-7 Bandpass filter response to a periodic signal.

The test instrument used to obtain a time-domain representation of a signal in circuit is the *oscilloscope*. The corresponding instrument used to obtain a frequency-domain representation of a signal is called a *spectrum analyzer*. It operates on the basic principle that the magnitudes of the various harmonic components may be determined by using an adjustable center-frequency bandpass filter to pick off and measure one harmonic component amplitude at a time.

1-6.2 Basic Waveform Manipulations

Addition

An oscilloscope monitors a voltage waveform in a circuit as shown in Fig. 1-8. In this time-domain representation the waveform appears to be quite complicated. If, however, the same point is monitored with a properly adjusted spectrum anaylzer, the display shown in Fig. 1-9 would appear. From this frequency-domain representation it is obvious that the waveform is simply the sum of two sinusoids of frequencies 1 kHz and 2 kHz. The original waveform is, in fact, the summation of the two sinusoids shown in Fig. 1-10.

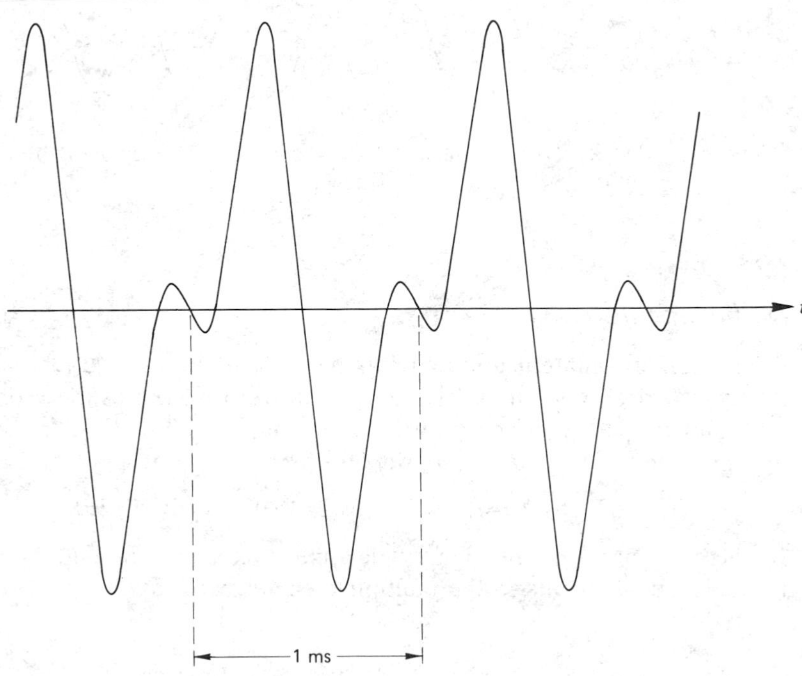

Figure 1-8 Observed waveform on an oscilloscope.

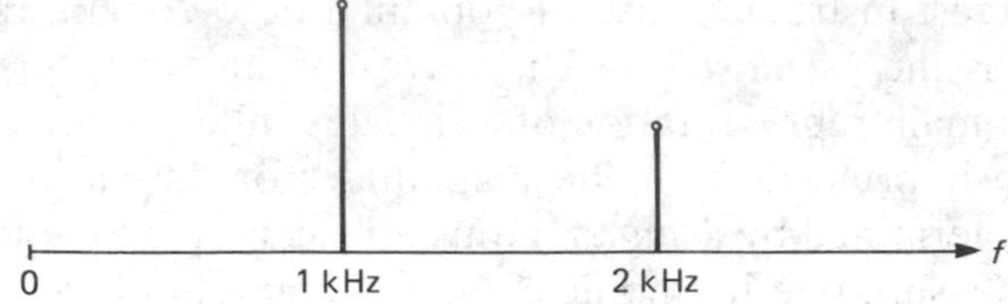

Figure 1-9 Observed frequency spectrum on a spectrum analyzer.

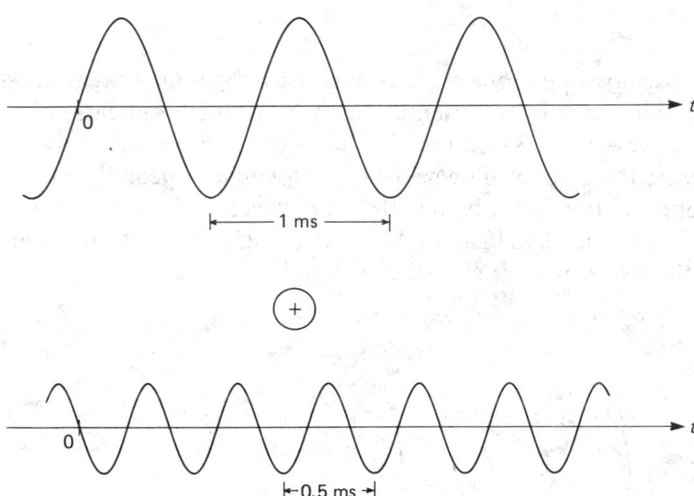

Figure 1-10 Sinusoids contained in the waveform of Figure 1-8.

Multiplication

A number of communication circuits perform multiplicative-type action between two waveforms to obtain a desired signal waveform. Let us therefore examine the time and frequency effects of waveform multiplication. First let us consider the multiplication of two sinusoidal waveforms:

$$V(t) = [B\cos(2\pi f_2 t)] \times [A\cos(2\pi f_1 t)] = AB\cos(2\pi f_2 t)\cos(2\pi f_1 t) \quad (1\text{-}4)$$

In the time domain, the signal will appear as shown in Fig. 1-11. To find the frequency components in the multiplied signal, we may make use of the trigonometric identity:

$$\cos\alpha\cos\beta = \tfrac{1}{2}\cos(\alpha - \beta) + \tfrac{1}{2}\cos(\alpha + \beta) \quad (1\text{-}5)$$

Thus equation (1-4) can be expressed as

$$V(t) = \frac{AB}{2}\cos[2\pi(f_2 - f_1)t] + \frac{AB}{2}\cos[2\pi(f_2 + f_1)t] \quad (1\text{-}6)$$

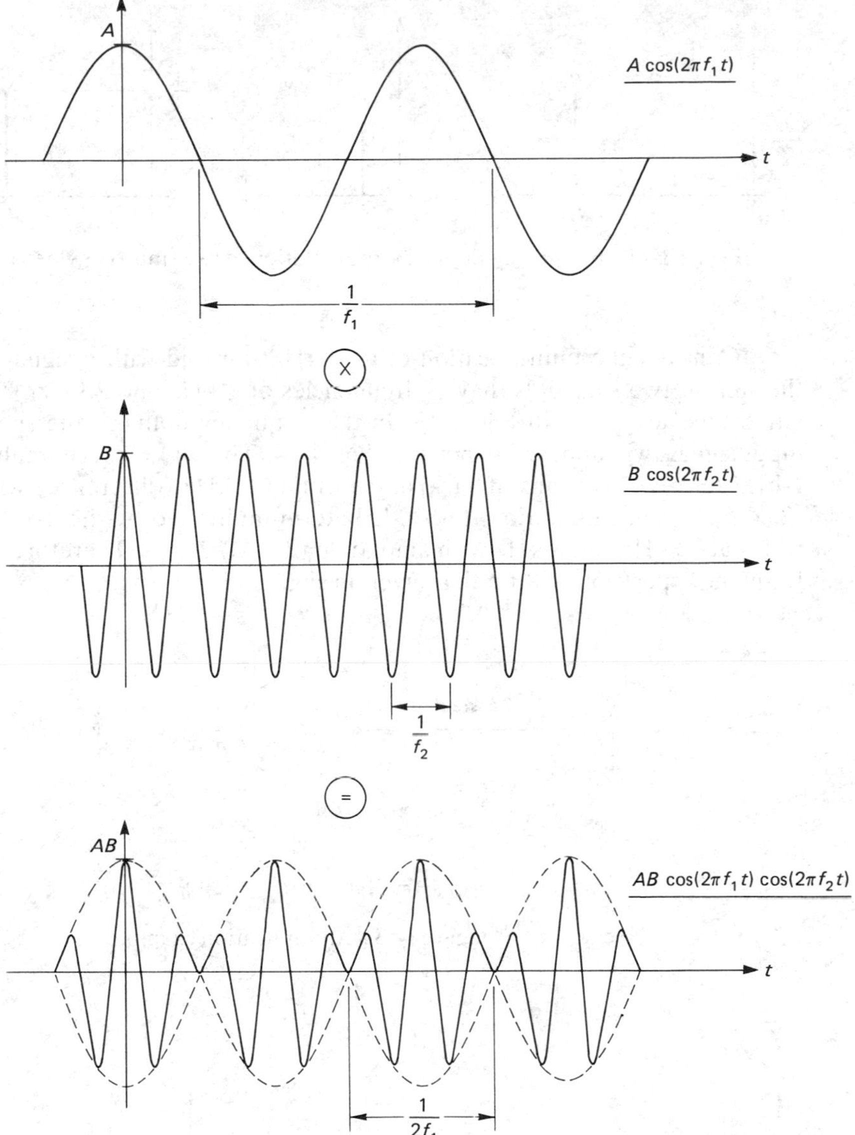

Figure 1-11 Time-domain representation of two sinusoids when multiplied.

In this form we may readily see that the frequency-domain representation is simply the sum of two sinusoids, as shown in Fig. 1-12. Note that two sinusoids when multiplied together generate two new sinusoids at the "sum" $(f_2 + f_1)$ and "difference" $(f_2 - f_1)$ frequencies. The amplitude of the two new sinusoids is the same and equal to the logarithmic mean $(AB/2)$ of the original components. This result may be extended to cover the multiplication of any two periodic waveforms by performing term-by-term multiplication of the harmonic components of each waveform.

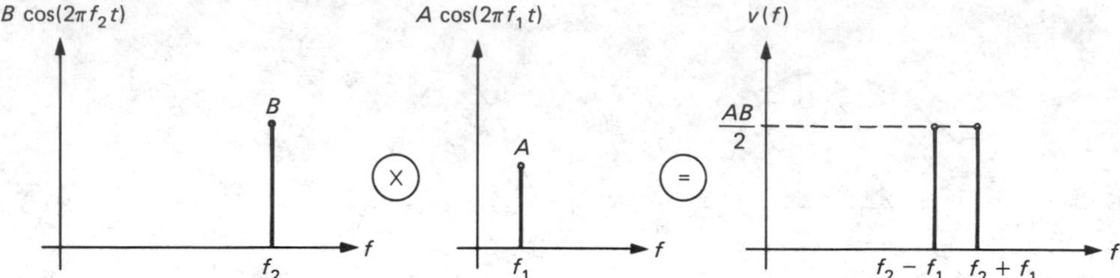

Figure 1-12 Frequency-domain representation of two sinusoids when multiplied.

Consider the multiplication of a 10-kHz sinusoid with a signal consisting of the sum of two sinusoids, having frequencies of 1 kHz and 2 kHz. The respective amplitudes are given in Fig. 1-13. In the frequency domain, the spectrum of the input signals will appear as shown in Fig. 1-14. The 10-kHz term multiplied by the 1-kHz term yields terms at 11 kHz (sum) and 9 kHz (difference) with amplitude $(2 \times 6)/2 = 6$. In addition, the 10-kHz term multiplied by the 2-kHz term yields terms at 12 kHz and 8 kHz with amplitude $(2 \times 3)/2 = 3$. Therefore, the resultant frequency spectrum will be that given in Fig. 1-15.

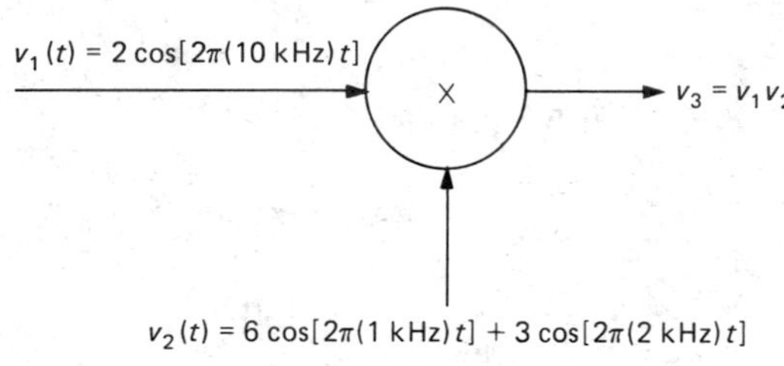

Figure 1-13 Multiplication example.

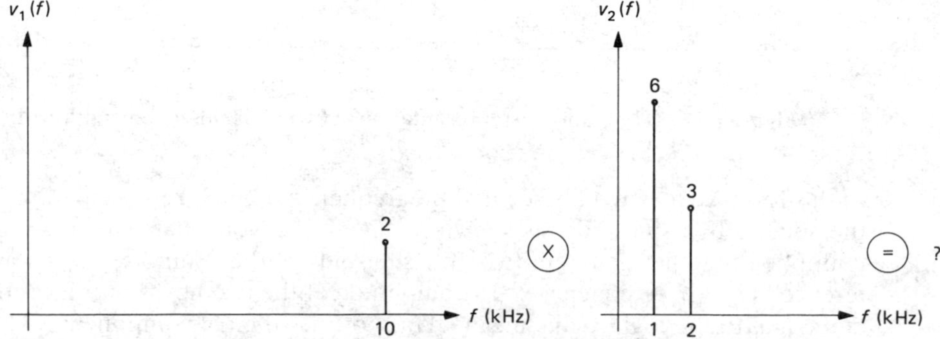

Figure 1-14 Frequency spectrum of the input signals.

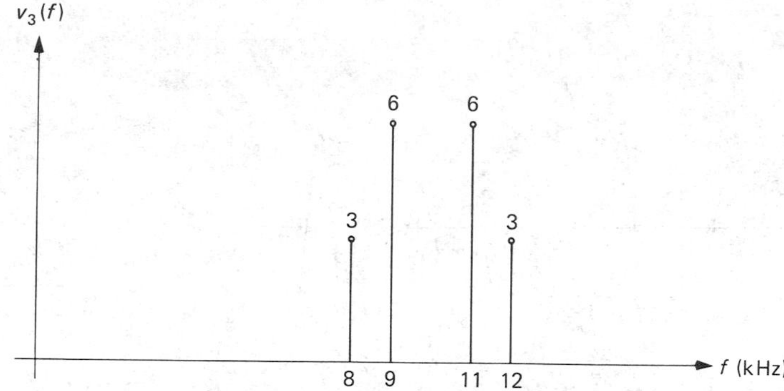

Figure 1-15 Output frequency spectrum.

1-7 APPLICATIONS OF THE WAVEFORM MANIPULATION TECHNIQUES

The examples selected to demonstrate the use of the manipulation techniques fall in two basic areas. The first examples examine the spectrum of an AM waveform and how the carrier frequency of an AM waveform may be heterodyned down to an intermediate frequency. These examples are included to explain some of the basic operations that occur in a superheterodyne receiver. The second set of examples deals with the switching (or sampling) function and the applications examined are oriented more toward digital communication.

1-7.1 Amplitude Modulation

AM waveforms and the various types of circuitry required to generate and receive AM signals are examined in detail in Chapter 6. We thus restrict our present examination to the simple block diagram shown in Fig. 1-16. Note that for simplicity the information signal to be impressed on the amplitude of the carrier wave is assumed to be a single audio tone at frequency f_a. When viewed on an oscilloscope (in the time domain), the resulting AM waveform $v_3(t)$ would appear as shown in Fig. 1-17.

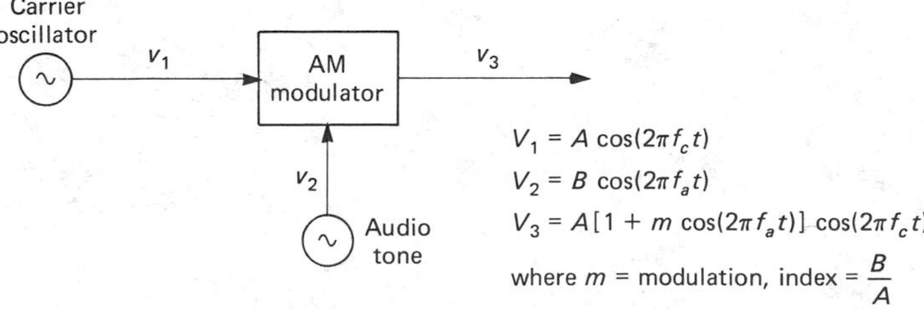

Figure 1-16.

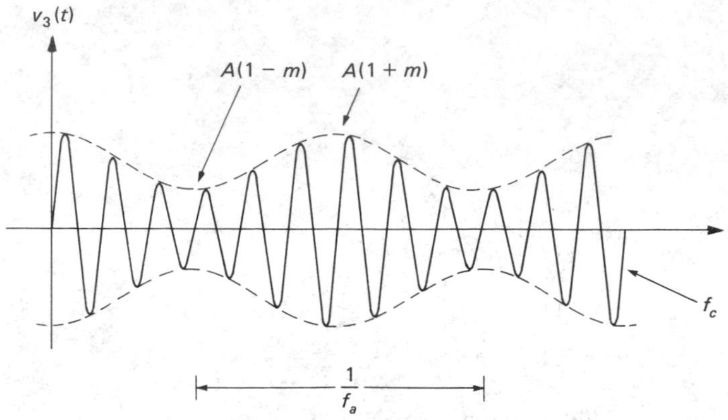

Figure 1-17 AM waveform in the time domain.

The frequency-domain representation may be obtained by multiplying the components of $v_3(t)$ (Fig. 1-18). The zero-frequency (dc) term multiplied by the f_c term yields a term at $f_c + 0 = f_c$ and a term at $f_c - 0 = f_c$, both with amplitude $(A \times 1)/2 = A/2$, the net result being a single component at f_c with amplitude A. The f_a term multiplied by the f_c term yields a term at $f_c + f_a$ and a term at $f_c - f_a$, both with amplitude $(mA \times 1)/2 = mA/2$, as shown in Fig. 1-19. The extra frequency components appearing on either side of the carrier are referred to as the *sidebands*.

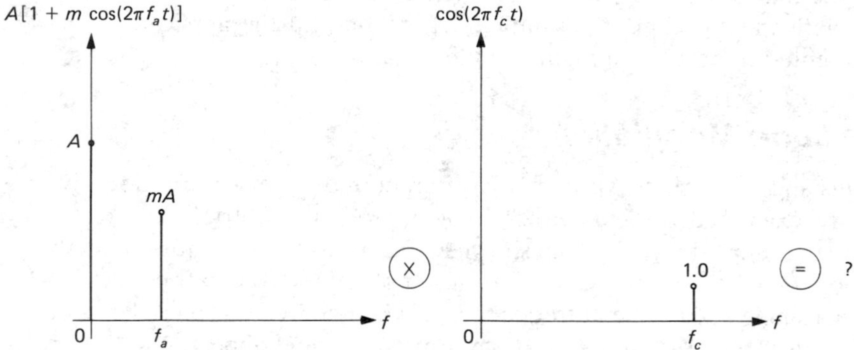

Figure 1-18 Determining the frequency component in an AM waveform.

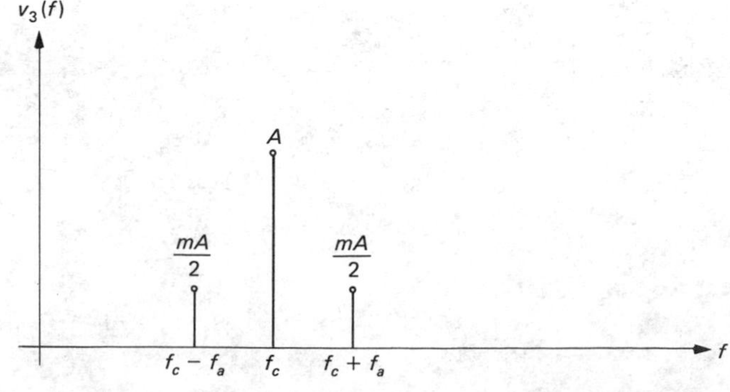

Figure 1-19 AM waveform in the frequency domain.

1-7.2 Heterodyne Conversion

Consider what would happen if the AM waveform of the preceding example was multiplied by a further sinusoidal waveform, as shown in Fig. 1-20. The $(f_c - f_a)$ term when multiplied by the f_{LO} term yields components at $[f_{LO} + (f_c - f_a)]$ and $[f_{LO} - (f_c - f_a)]$ with amplitudes $[(mA/2) \times 2]/2 = mA/2$. Similarly, there will be terms at $(f_{LO} + f_c)$ and $(f_{LO} - f_c)$ with amplitudes A, and finally, terms at $[f_{LO} + (f_c + f_a)]$ and $[f_{LO} - (f_c + f_a)]$ with amplitudes mA/2. The resulting spectrum is shown in Fig. 1-21. Examining Fig. 1-21, one may recognize two AM waveforms added together.

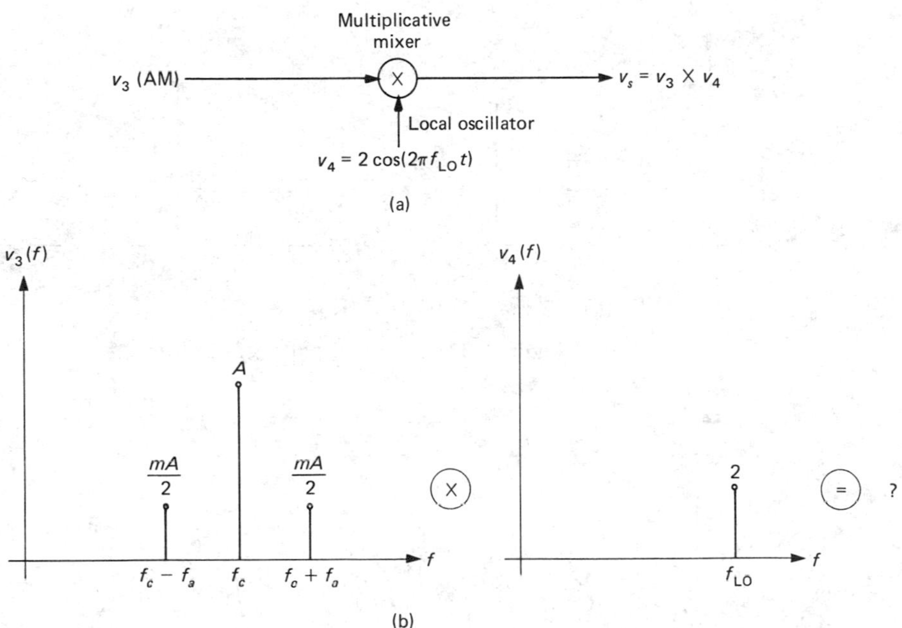

Figure 1-20 Effect of an ideal multiplicative mixer on an AM waveform.

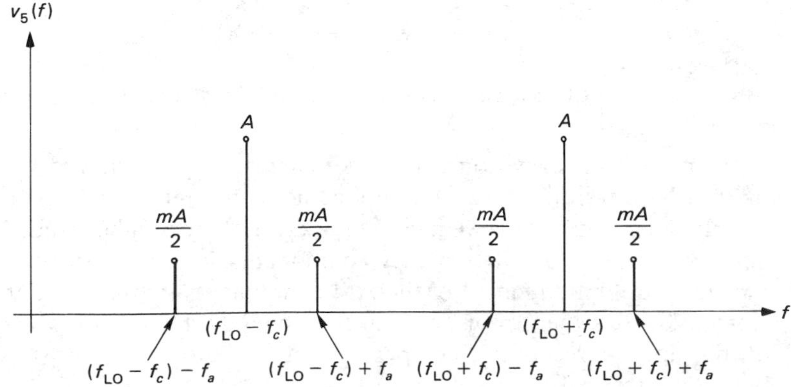

Figure 1-21 Spectrum of the AM waveform after mixing with the local oscillator.

Let us now consider what would occur if V_5 were passed through a bandpass filter which passed the three lower-frequency components only as shown in Fig. 1-22. Comparing this with the results from the last example of AM waves, it follows that the time-domain representation is as shown in Fig. 1-23. The resulting waveform is essentially the same as the original AM waveform except that the amplitude-modulating signal is now impressed on a carrier with frequency $f_{LO} - f_c$ rather than a carrier with frequency f_c.

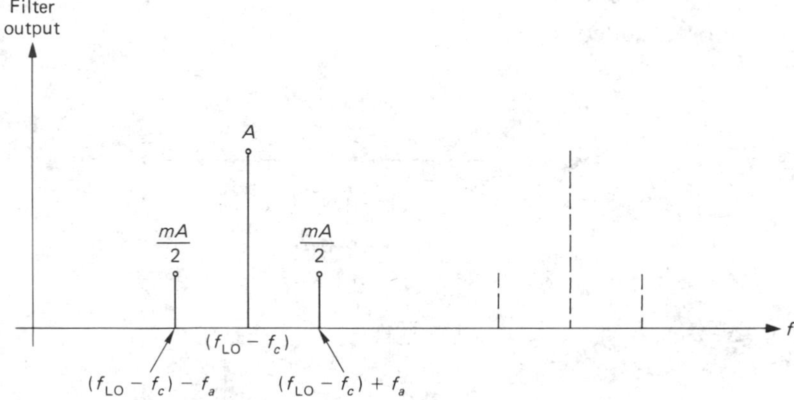

Figure 1-22 Mixer output after filtering.

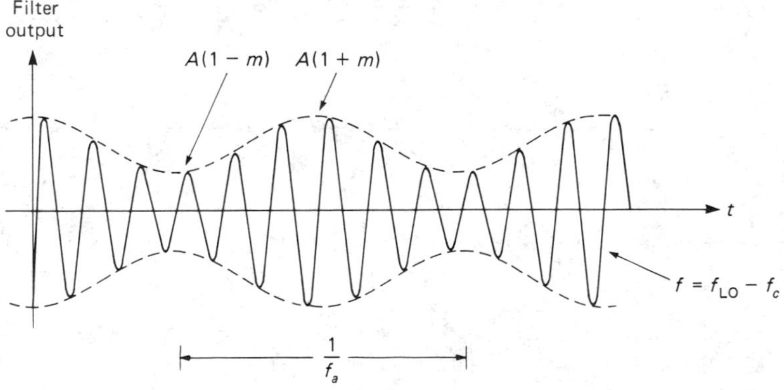

Figure 1-23 Time-domain representation of the filtered waveform.

In the frequency domain the net change may be viewed simply as a shifting or translation of all the AM components to the left (down in frequency) by the amount f_{LO}. Similarly, if we had filtered off the three higher-frequency components and rejected the lower-frequency components, the net result would be a shift to the right (up in frequency) of the AM components by the amount f_{LO}. This ability to translate a signal up or down in frequency (heterodyne) is the operation basis of the superhet receiver. The operation of a superhet receiver is examined in more detail in Section 1-8.

1-7.3 Switching Function

The analysis of a number of communication waveforms may most easily be undertaken if we first are familiar with a waveform referred to as the *switching* (or *gate*) *function*. The switching function $S(t)$ shown in Fig. 1-24 is simply a regular train of pulses of amplitude A. To obtain a frequency-domain representation of the switching function we must determine the Fourier expansion for $S(t)$. It can be shown that the Fourier expansion is

$$S(t) = \underbrace{\frac{At_p}{T}}_{a_0} + \sum_{n=1}^{\infty} \underbrace{\frac{2At_p}{T} \operatorname{sinc}\left(\frac{nt_p}{T}\right)}_{C_n} \cos \frac{2\pi nt}{T} \tag{1-7}$$

where the "sinc" function is defined as

$$\operatorname{sinc}(x) = \frac{\sin \pi x}{\pi x}$$

and the constants a_0 and C_n are the magnitude terms for the various harmonic components of $S(t)$. The sinc function has the general form shown in Fig. 1-25.

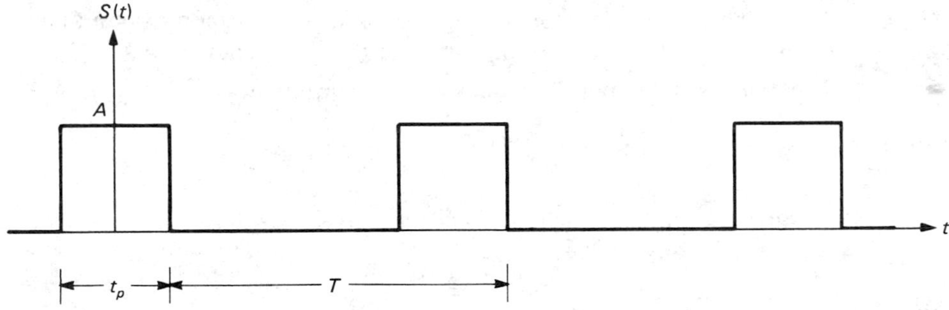

Figure 1-24 Time-domain representation of the switching function.

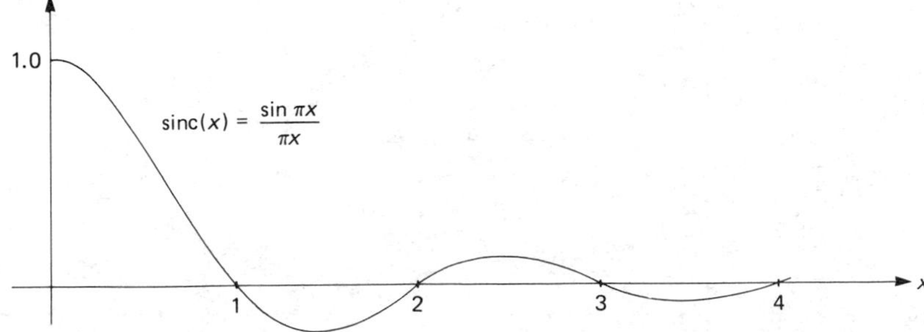

Figure 1-25 The sinc(x) function.

Since the amplitude constants (C_n) of each harmonic contains a "sinc" term, the amplitude of harmonics will lie within an envelope with the "sinc" shape. Thus in general the frequency-domain representation of the switching function has the form shown in Fig. 1-26.

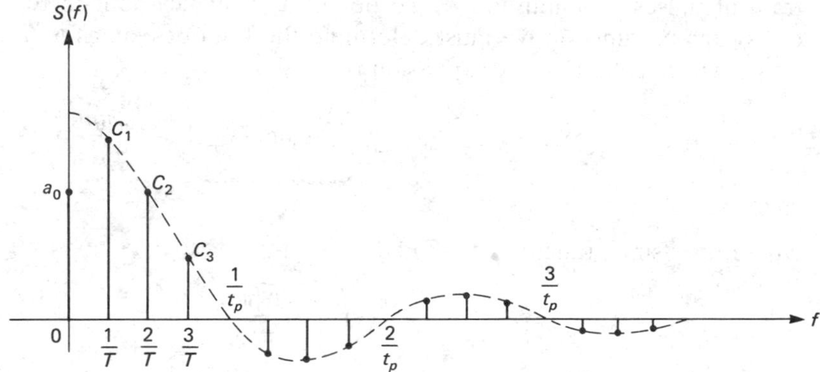

Figure 1-26 Frequency-domain representation of the switching function.

Note the following features of the switching function:

1. The basic pulse frequency is $1/T$ and thus the harmonic components occur at integer (0, 1, 2, 3, . . .) multiples of this frequency.
2. The discrete amplitude values of each harmonic,

$$C_n = \frac{2At_p}{T} \operatorname{sinc}\left(\frac{nt_p}{T}\right) \qquad \text{for } n = 1, 2, 3, 4, \ldots \tag{1-8}$$

will lie within an envelope defined by the continuous function:

$$E(f) = \frac{2At_p}{T} \operatorname{sinc}(ft_p) \tag{1-9}$$

which has zero crossing at $1/t_p$, $2/t_p$, $3/t_p$, and so on, as can be seen in Fig. 1-27.

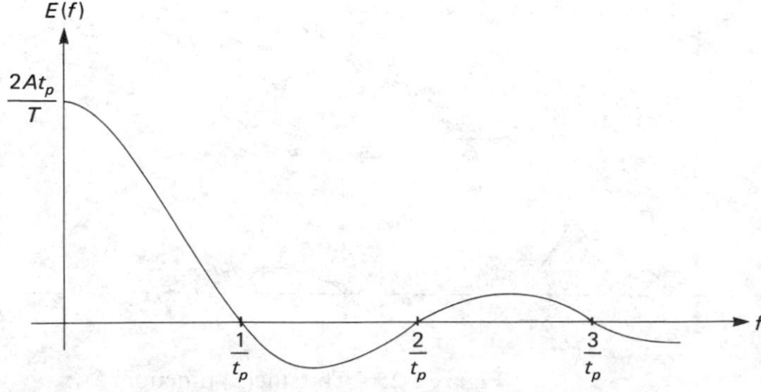

Figure 1-27 Envelope of the switching function.

3. The plot of some amplitude values as negative numbers simply means that the sinusoids for these components have an additional 180° phase shift, that is,

$$-C_5 \cos\frac{2\pi 5t}{T} = C_5 \cos\left(\frac{2\pi 5t}{T} + \pi\right) \qquad (1\text{-}10)$$

A spectrum analyzer, since it determines the amplitude of the harmonic components only, will display all the components as positive, as shown in Fig. 1-28.

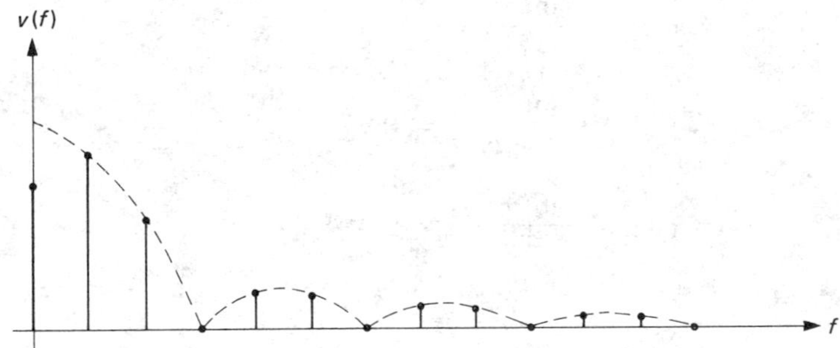

Figure 1-28 Spectrum analyzer display for a pulse waveform.

EXAMPLE 1-1: Switching Function

Consider the pulse train shown in Fig. 1-29. Let us determine the frequency-domain representation by first sketching the amplitudes and determining the frequencies; then we will calculate the amplitude of the first few components.

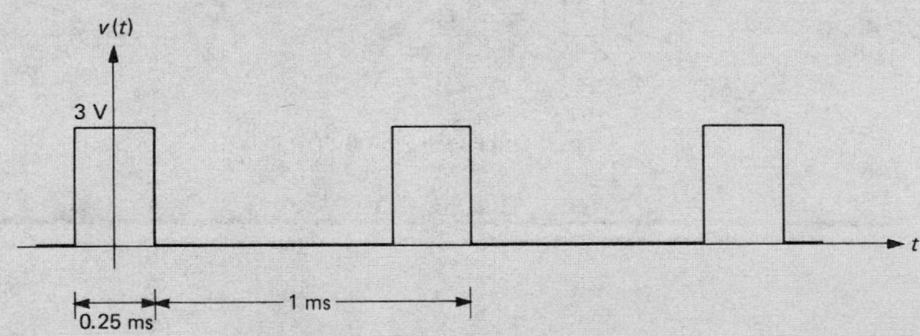

Figure 1-29 Pulse train in the time domain.

Solution The basic period (T) is 1 ms, therefore we should expect components at 1 kHz ($1/T$), 2 kHz ($2/T$), 3 kHz ($3/T$), and so on, as well as a dc (zero-frequency) component. The amplitude of these components must lie within a sinc envelope with zero crossings at 4 kHz ($1/t_p$), 8 kHz ($2/t_p$), 12 kHz ($3/t_p$), and so on. The resulting spectrum is sketched in Fig. 1-30. The component amplitudes may be calculated as follows:

$$a_o = \frac{At_p}{T} = \frac{(3\text{ V})(0.25\text{ ms})}{1\text{ ms}} = 0.75\text{ V}$$

$$C_n = \frac{2At_p}{T}\,\text{sinc}\left(\frac{nt_p}{T}\right) = (1.5\text{ V})\,\text{sinc}\left(\frac{n}{4}\right) = (1.5\text{ V})\,\frac{\sin(\pi n/4)}{\pi n/4}$$

$$C_1 = (1.5\text{ V})\,\frac{\sin(\pi/4)}{\pi/4} = 1.35\text{ V}$$

$$C_2 = (1.5\text{ V})\,\frac{\sin(2\pi/4)}{2\pi/4} = 0.95\text{ V} \qquad (1\text{-}11)$$

$$C_3 = (1.5\text{ V})\,\frac{\sin(3\pi/4)}{3\pi/4} = 0.450\text{ V}$$

$$C_4 = (1.5\text{ V})\,\frac{\sin\pi}{\pi} = 0.0\text{ V}$$

$$C_5 = (1.5\text{ V})\,\frac{\sin(5\pi/4)}{5\pi/4} = -0.27\,0\text{ V}$$

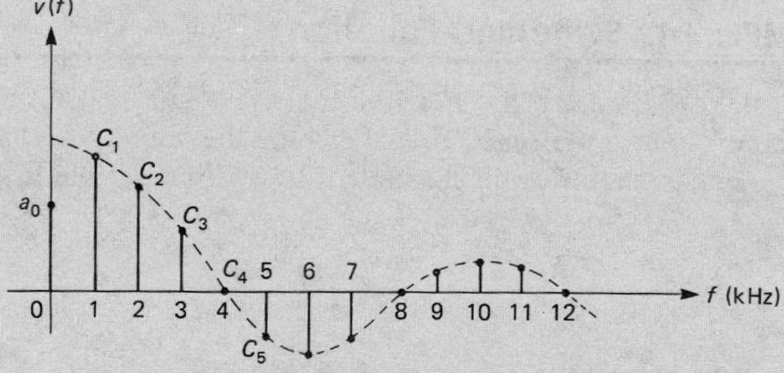

Figure 1-30 Sketch of $v(f)$.

1-7.4 Amplifier Bandwidth

Consider what occurs when we pass the pulse waveform of the previous example through an amplifier. From Fourier analysis we may view the waveform as being made up of a number of harmonically related sinusoids each of which has a particular amplitude and phase relation to all the other sinusoids.

Each of the harmonic components of the input pulse waveform will be affected by the amplifier. Assuming that amplitude and phase Bode plots of the amplifier transfer function are available, the amplitude and phase of each harmonic component leaving the amplifier can be predicted. A time-domain representation of the output waveform can be obtained by adding together the harmonic components leaving the amplifier. Note, however, that if some of the components experience different gains and phase shifts from the others, then when all the components are added back together to form the output, they will not have the same shape as the original input pulse waveform. The output would thus be a distorted replica of the input.

Any real amplifier will begin to roll off in gain beyond some point in frequency (the break frequency). Those harmonic components of a pulse waveform that are beyond this frequency will receive less gain (and more phase shift) and the output waveform will thus be distorted. This distortion may, however, be minimized if the amplifier has sufficient bandwidth to pass all the lower-frequency harmonics that have "significant" amplitudes. The loss of the higher-frequency components will have little effect if their amplitudes are small compared to the harmonic components that are passed.

A fairly coarse "rule of thumb" for determining the "significant" amplitude harmonics of a pulse waveform is that if all the harmonic components below the first zero crossing in the sinc function are passed, the output will be a reasonable representation of a pulse waveform. Thus an amplifier should have a bandwidth of at least 0 to $1/t_p$ in frequency if it is to pass a pulse waveform. As a specific example, let us consider a binary-coded message with the general format shown in Fig. 1-31. The most rapidly changing signal we would have to transmit would be a 101010101 . . . alternating character string. If the channel has sufficient bandwidth to carry this *worst-case* signal, it would be able to carry any other pattern of 1's and 0's sent at the same rate. We may view the worst-case signal as simply a regular pulse train with $t_p = 1\ \mu s$ and $T = 2\ \mu s$. Thus the amplifier/line should have a bandwidth of at least zero to 1 MHz ($1/t_p$) if it is to carry the signal with acceptably low distortion.

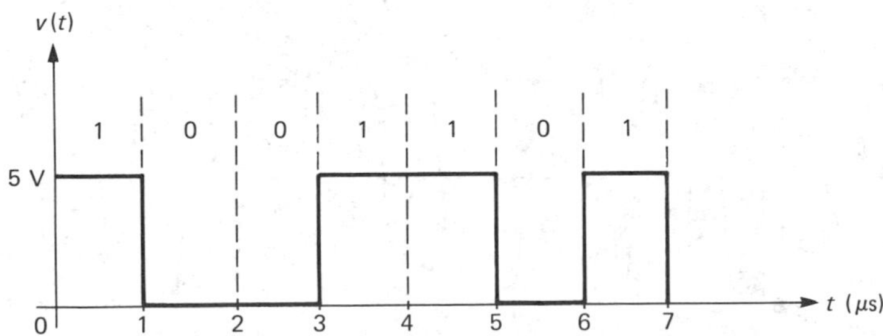

Figure 1-31 Binary-coded message.

1-7.5 Tone Bursts

Let us consider the transmission of the tone burst waveform illustrated in Fig. 1-32. In order to establish what sort of bandwidth is required on our channel to carry this waveform, we must determine the frequency-domain representation for the waveform. We may note that the later waveform could be generated by multiplying a switching function of period 20 ms. with a 1-kHz sinusoidal as illustrated in Fig. 1-33.

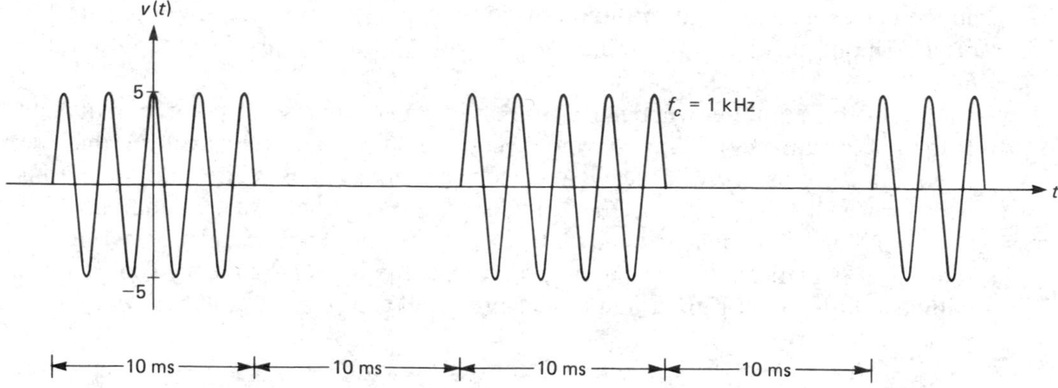

Figure 1-32 Tone burst waveform.

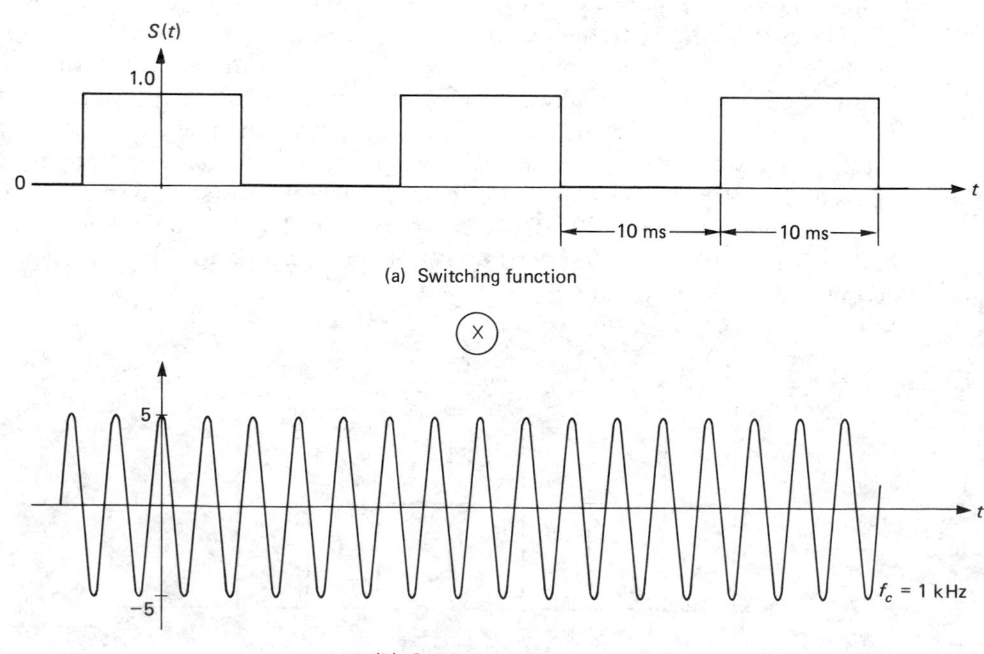

Figure 1-33 Derivation of a tone burst signal.

The frequency-domain representations of the waveforms of Fig. 1-33 and the resultant signal are shown in Fig. 1-34. The resultant spectrum is obtained by taking the sum and difference frequency of parts (a) and (b). From this representation we see that the waveform has multiple frequency components. However, using our first zero crossing of the sinc function criteria, the channel should have the bandwidth to pass frequencies between 900 and 1100 Hz as a minimum.

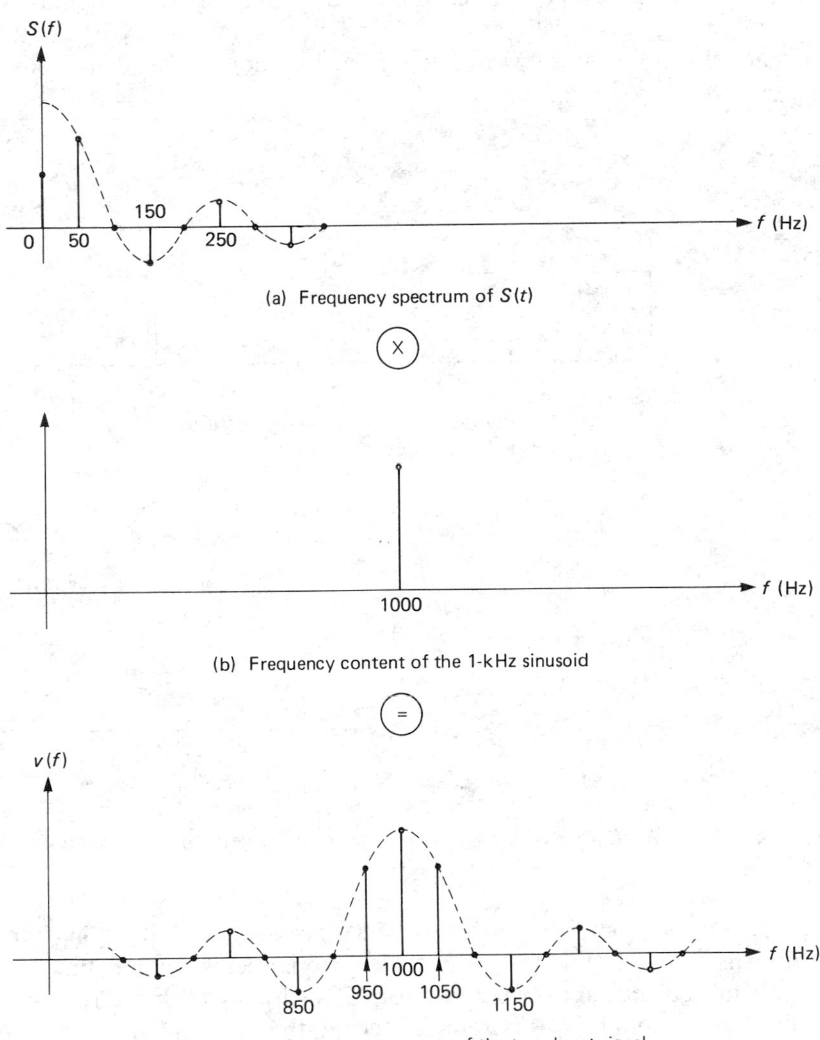

Figure 1-34 Derivation of the frequency spectrum of a tone burst signal.

1-7.6 Frequency-Shift-Keying Modems

Let us consider the problem of connecting a portable terminal unit to a remote computer via a standard voice-grade telephone line. Both ends of the system will be transmitting information in some type of serial binary code, similar to that shown in Fig. 1-35. A standard telephone line is a bandlimited channel that will pass frequencies between 300 and 3400 Hz. Direct transmission of the binary code on the channel cannot be accomplished reliably, as some signals, such as a long string of 1's (essentially dc), will not be passed correctly by the channel, as the channel will not pass low-frequency (such as dc) signals. This is illustrated in Fig. 1-36.

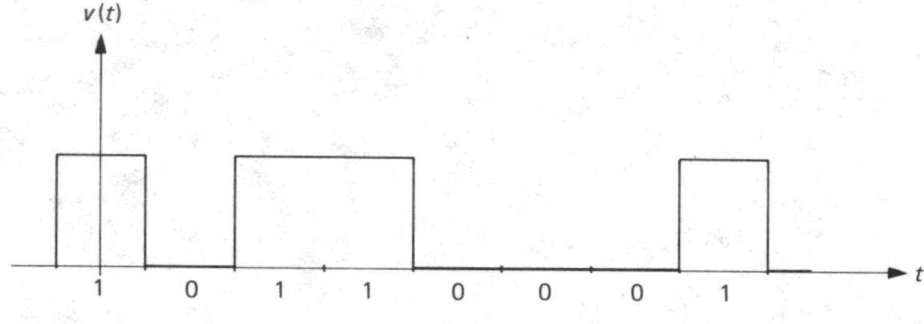

Figure 1-35 Serial binary code.

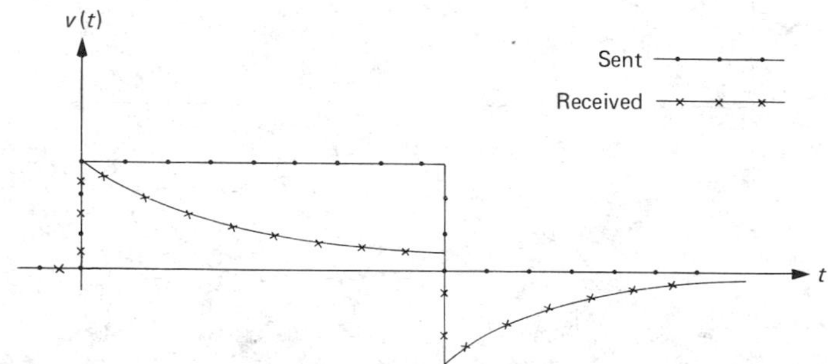

Figure 1-36 Effect of low-frequency filtering on a signal.

To overcome this difficulty (FSK) frequency-shift-keying modems may be interfaced to the line (see Fig. 1-37). The modems convert the 1's and 0's of the transmitted code into two audio tones which may be more readily carried on the line. For example, a 0 is represented by frequency f_2 and a 1 is represented by frequency f_1 as shown in Fig. 1-38. At the receiver modem, frequency-selective filters or phase-locked loops may be used to determine which of the two frequencies were sent and thus output a 1 or a 0 to the end device. The FSK waveform of Fig. 1-38 is presented in the time domain, and again the frequency-domain representation can provide more information as to the limitations of transmitting this signal on a bandlimited channel.

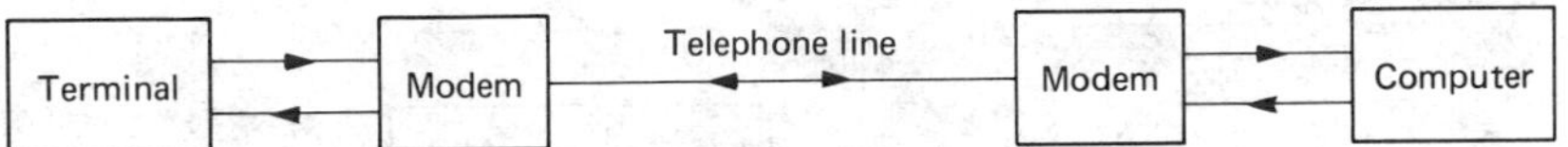

Figure 1-37 Modem arrangement.

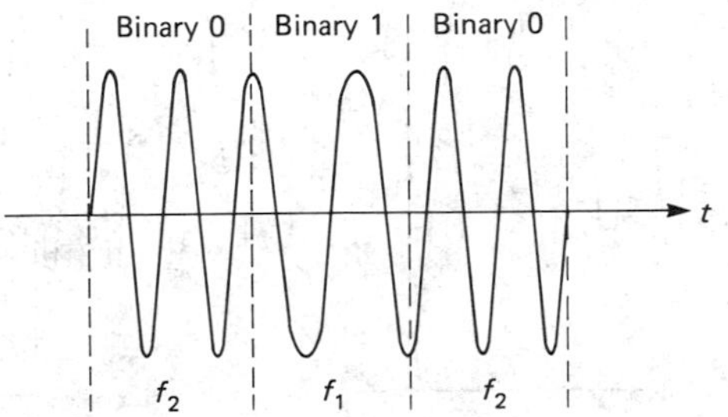

Figure 1-38 Frequency shift keying.

To this end consider that the waveform $v(t)$ consists of alternating frequencies f_1 and f_2 as shown in Fig. 1-39. This waveform could be considered to be made up of the sum of two multiplied waveforms as outlined in Fig. 1-40. If the frequency-domain representations of the waveforms of Fig. 1-40 are multiplied out and added together, the resulting spectrum contains not only components at f_1 and f_2 but also multiple sideband components on either side of f_1 and f_2, as shown in Fig. 1-41. Note that the first sidebands of the two carrier frequencies f_1 and f_2 are separated from the carrier frequencies by the frequency of the switching function (f_d).

The rate of data transmission (bits/second) is equal to twice the frequency of the switching function. If the data transmission rate is increased, the sidebands move farther away from f_1 and f_2. If excessive transmission rates are attempted, the sidebands of, say, f_1 will start to fall into the f_2 frequency band. The FSK receiver, unable to differentiate between the sideband and valid f_2 signal, will begin giving erroneous results and data transmission will fail. This effect of interfering frequency components is called *aliasing*.

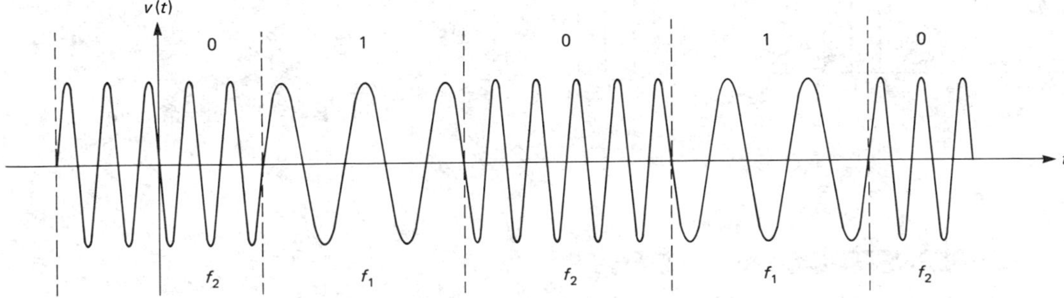

Figure 1-39 FSK signal.

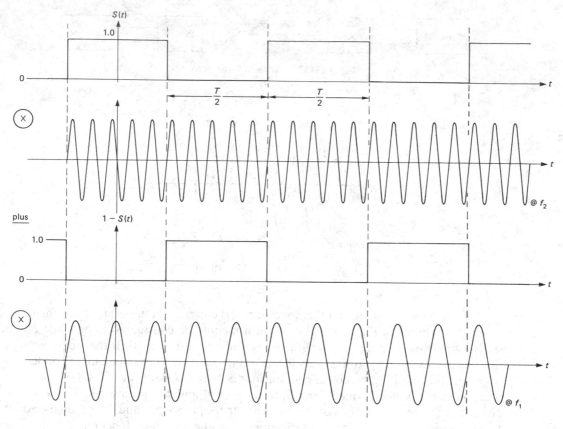

Figure 1-40 Equivalent FSK signal.

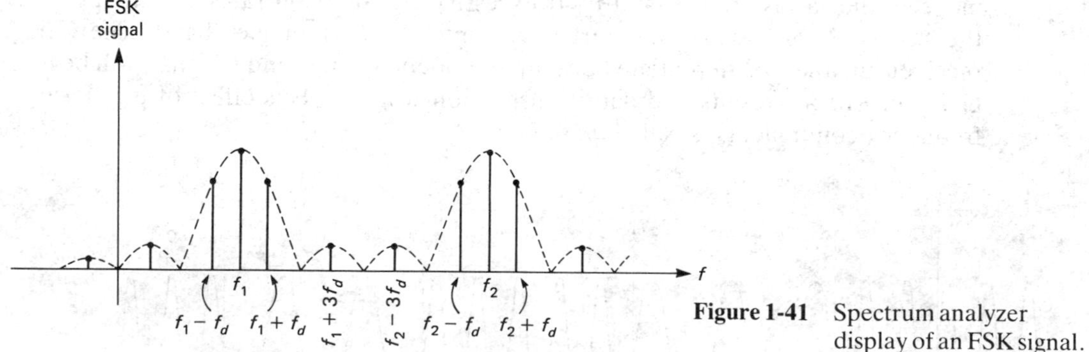

Figure 1-41 Spectrum analyzer display of an FSK signal.

1-8 SUPERHETERODYNE RECEIVER

A communication receiver must select and amplify only the relatively narrow band of frequencies associated with the particular channel of interest. Typically very sharp filtering characteristics are required to reject channels immediately adjacent to the desired channel, and very large gains, typically in excess of 100 dB, are required before the incoming signal can be "detected." The detector circuit is the one that recovers the information signal back off the high-frequency carrier. If the information signal is a voice channel, an additional audio amplifier stage will be used to deliver the voice signal to a speaker.

Some of the earliest radio receivers built attempted to amplify and then detect the band of frequencies associated with the desired channel. This was what was referred to as the *tuned radio-frequency* (TRF) *receiver*. The TRF receiver, however, had two major drawbacks. The foremost drawback has to do with its ability to filter out the desired channel selectively while rejecting channels immediately adjacent in frequency. It is a relatively straightforward task to design a bandpass filter that passes a fixed band of frequencies and then rolls off very rapidly outside the band. If, however, the filter is simultaneously required to be adjustable over a wide range of frequencies, the design requirements become quite phenomenal. Some of the earlier radio receivers built on the TRF principle were, as a result, mechanical marvels in which the tuning knob had to adjust a large number of variable capacitors simultaneously.

The second drawback of the TRF receiver is the tendency for its amplifier circuits to become unstable and burst into oscillations. The physical layout of the amplifier circuits is quite critical. Because of the very high gain required, only a very minute amount of stray feedback is required before the tuned amplifiers burst into oscillations (feedback oscillators are discussed in Chapter 4).

The *superheterodyne receiver* avoids these two drawbacks and as a result is almost universally used in analog receiver applications. At the heart of these communication receivers is the heterodyne principle. The term *heterodyne* indicates a "mixing" of two or more frequencies in a nonlinear circuit. The nonlinear aspect of the mixed circuit is used to generate a multiplicative action between the signals injected into a mixer circuit. As we have seen earlier, multiplying two sinusoids (f_1, f_2) together results in output components at the sum ($f_1 + f_2$) and difference ($f_1 - f_2$) frequencies.

The block diagram of a typical superhet receiver is shown in Fig. 1-42. Let us consider how frequency-domain signals are processed by the superhet receiver. For simplicity we shall consider the carrier frequency of the incoming signal only. The information impressed on the carrier will result in sideband components around the carrier, but all of these should be within the bandwidth allocated to the channel

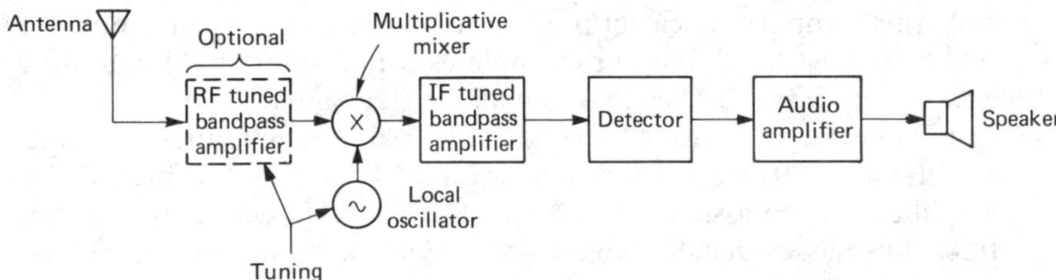

Figure 1-42 Single-conversion superhetarodyne receiver.

and therefore pass through the receiver along with the carrier, provided that an adequate bandwidth is passed by the filters of the receiver.

A large number of signal frequencies (channels) will be delivered by the receiver antenna to the input of the superhet receiver. These radio-frequency (RF) signals are amplified by the tuned RF amplifier stages and some degree of initial selective filtering is performed by this stage. This initial filtering is typically done with a simple *LC* tuned circuit (see Chapter 2) where a variable capacitor is adjusted by the radio tuning knob. The tuning at this point is kept mechanically simple and adjustable over a large range. The filtering is, as a result, not terribly selective, and typically several adjacent channels will pass through the amplifier along with the desired channel bandwidth of frequencies.

The incoming frequencies (f_s) are then applied to a mixer stage along with a local oscillator signal (f_o). The mixer output ideally contains sum ($f_s + f_o$) and difference ($f_s - f_o$) frequency outputs. Any real mixer circuit will also typically contain signals at frequencies f_s, f_o, $2f_s$, $2f_o$, $2f_s - f_o$, $2f_o - f_s$, and so on. The additional components are generated because the mixer circuit, rather than being an ideal multiplier, is typically just a nonlinear device, and the multiplicative action occurs as a result of the nonlinear characteristic. A properly designed mixer stage will optimize the generation of sum and difference frequency components and minimize the other frequency components.

The signal leaving the mixer is thus quite complex containing (at a minimum) the sum and difference frequencies of f_s and f_o. Since several channels (different f_s values) passed the RF stage, quite a large number of frequency components will be present. The signal leaving the mixer then passes through the IF amplifier. The intermediate-frequency (IF) amplifier is a tuned amplifier. The center frequency of the filtering is fixed (does not adjust with tuning), and as a result a very selective bandpass filter with sharp rejection outside the bandpass can be produced. A number of "standard" intermediate frequencies are used in most receivers. The most common intermediate frequency in use for AM standard broadcast band is, for example, 455 kHz. The standardization of the IF to 455 kHz allows the filter components to be mass produced such that excellent characteristic filters can be purchased quite cheaply.

Of the multiple frequency components leaving the mixer, only the frequency component at the intermediate frequency will be amplified and passed farther on into the receiver. It is at this point that a single channel is selected by the receiver. If you examine Fig. 1-42, you may note that the tuning is indicated not only to adjust the RF stage filtering, but also, simultaneously, to adjust the frequency at which the local oscillator operates. To examine how the local oscillator frequency performs a tuning effect, let us assume some specific signal frequencies from the AM broadcast band. For our example we will assume that the desired station has a carrier frequency of 800 kHz and all the sidebands of the carrier are within a 10-kHz bandwidth (795 to 805 kHz). We will also assume that two other stations at 820 kHz and 780 kHz pass through the tuned RF stages. The intermediate-frequency amplifier will be assumed to be tuned to 455 kHz with a 10-kHz bandpass. The filter thus passes signals between 450 and 460 kHz but severely rejects any signals outside this frequency range. The various frequencies and filter shapes are sketched on Fig. 1-43.

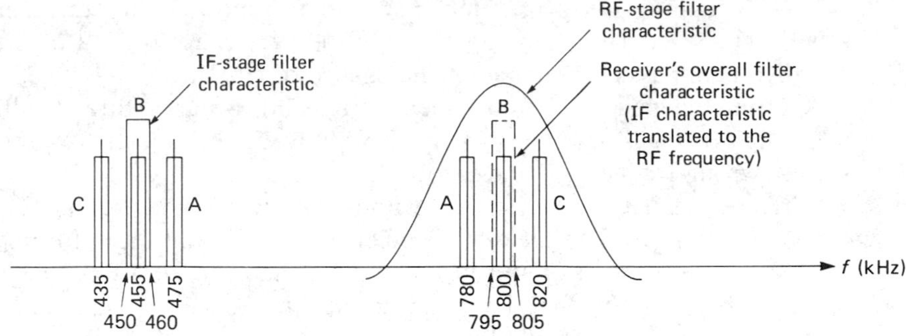

Figure 1-43 Signal and filter characteristics in a superheterodyne receiver.

If the tuning is set correctly when the radio tuning knob is set to 800 kHz, the RF stage will be tuned to pass 800 kHz and the local oscillator will be set to operate at 1255 kHz (f_s + IF). Of the multiple-frequency-component signals exiting the mixer, we are interested mostly in the difference frequency components ($f_o - f_s$). The desired station is translated (see Section 1-7.2) from 800 kHz to 1255 − 800 = 455 kHz. The undesired stations will be translated to 1255 − 820 = 435 kHz and 1255 − 780 = 475 kHz. The desired station (now with a carrier frequency of 455 kHz rather than 800 kHz) will be passed and amplified by the IF stage, whereas the undesired stations at 435 kHz and 475 kHz are outside the passband of the IF stages and will be rejected.

One may note that the band of frequencies to be received by the radio is established by the fixed IF filters rather than the more broadly tuned RF stage. The effective bandpass characteristic of the receiver reflected back to the RF band is shown in Fig. 1-43. Tuning with this receiver is relatively simple. If, for example, the tuning knob is turned up to 820, the RF tuned circuit is shifted to center on 820 kHz and the local oscillator circuit is simultaneously set up to 1275 kHz from the previous value of 1255 kHz. The result, as shown in Fig. 1-44, is that the difference frequency components out of the mixer are all 20 kHz higher and the new station now falls in the fixed bandwidth of the IF amplifier stages.

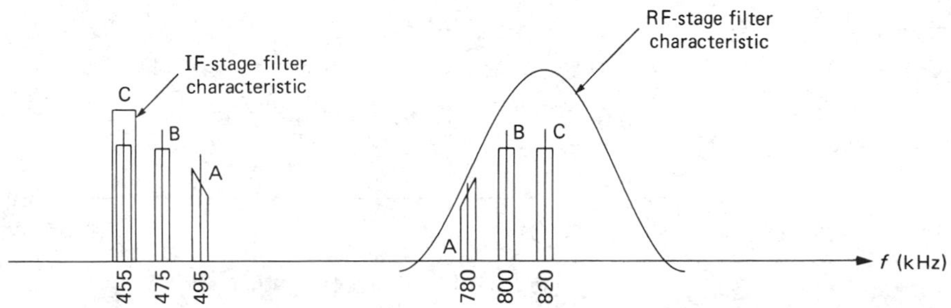

Figure 1-44 Signal spectrum after retuning.

The superhet receiver is much simpler to tune than an equivalent TRF receiver. The selectivity is excellent due to the very sharp roll-off of the receiver's effective bandpass characteristic. The chance of circuit instability can be reduced by splitting the gain required between the RF and IF stages. Stray feedback signals from the IF stages appearing at the RF input will be rejected by the tuned RF stages (wrong frequency), and feedback oscillations are avoided.

The superhet receiver is not without its own unique faults. The principal problem occurs as a result of the mixer circuit shifting the information on the original carrier signal (f_s) to a new carrier at the difference frequency. The local oscillator frequency (f_o) must be selected such that this difference frequency carrier is at the intermediate frequency:

$$|f_s - f_o| = \text{IF} \tag{1-12}$$

There are, however, two possible selections for the local oscillator frequency. High-side injection may be used. This is the case where the local oscillator frequency (f_o) is set higher than the carrier signal (f_s) by a frequency offset equal to the IF. Alternatively, low-side injection may be used. In this case the local oscillator frequency (f_o) is selected to be less than the carrier frequency (f_s) by an offset equal to the IF. In either case the *absolute value* of the difference frequency out of the mixer will equal the IF value.

Consider the case of a 1000-kHz signal being received by an AM superhet radio which has an IF of 455 kHz and is operated using high-side injection. These frequencies are indicated on Fig. 1-45. The local oscillator must operate at 1455 kHz in order to convert the signal at 1000 kHz to one at 455 kHz. Note that there is a second frequency indicated on Fig. 1-45 at 1910 kHz. This second frequency is called the image frequency (f_i) and is also offset from the local oscillator frequency by 455 kHz (the IF). *If* a signal at the image frequency is being picked up, and if the signal reaches the mixer input, a severe problem will develop. The desired channel will be converted to the IF by high-side injection, but the signal at the image frequency will also be converted to the IF, by low-side injection. The result is two channels passing through the IF-stages. The resulting audio signal will be the composite of the two information signals, neither of which will be intelligible due to the presence of the other.

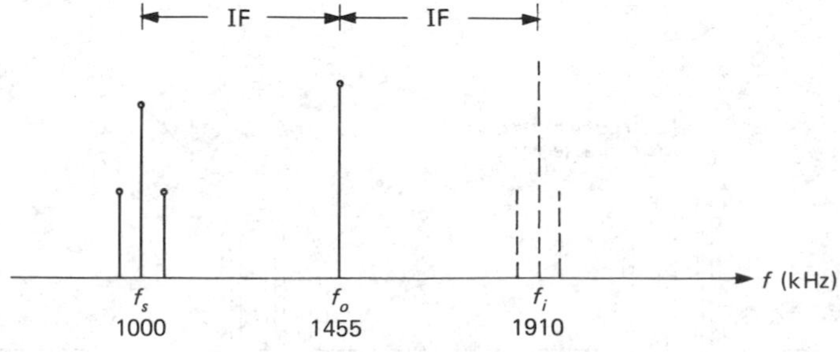

Figure 1-45 Image frequencies.

A second problem that occurs, due to the same effect, is called *double spotting*. This occurs when the same radio station is picked up at two different settings of the tuning dial. Consider what occurs if the station is initially correctly received by high-side injection. If the dial is then turned lower (f_o drops), there will be a point where the local oscillator is lower than the station frequency by the IF. At this point the original station is now at the image frequency of the current dial setting. Assuming that there is no station broadcasting at this new frequency, only the former station will be heard.

Both these problems can be eliminated if a tuned RF stage is used in the receiver rather than letting the incoming signal go directly into the mixer stage. The tuned RF stage does not have to filter well enough to receive only one channel since the IF stage will do this. It should, however, be able to reject a signal at the image frequency of the desired signal. Note that the image frequency is separated from the desired frequency by two times the IF.

1-9 FULL-FUNCTION SUPERHETERODYNE RECEIVER

Figure 1-46 is a block diagram of a superhet receiver operating in the citizen's band (CB) range of frequencies. The receiver is a superhet; however, a number of additional features have been included. The first major difference is that this is a double-conversion superhet. By this we mean that the incoming RF signal is first converted down to a 10.7-MHz carrier (the first IF) and then that signal is in turn down-converted to a 455-kHz carrier (the second IF). Double (or even triple) conversion is required in superhets operating at higher RF frequencies. The reason for this may be determined by considering the AM broadcast band example shown in Fig. 1-45. In that case the tuned RF stage had to pass the desired signal at 1000 kHz while rejecting any image signal present 2 × IF away at 1910 kHz. If this same single-conversion receiver was to be used with an RF signal at 27 MHz, the RF stage would have to pass a signal at 27 MHz but reject any signal 2 × IF away at 27.9 MHz. The spread between the signal and its image is in this second case only a small percentage of the operating frequency and separating two signals so close together is difficult. A better alternative is to increase the value of the IF (to 10.7 MHz in this case) and thereby increase the frequency spread (2 × IF) between the desired signal and its image. Again the use of essentially "standard" frequencies for IF values allows mass production of the filter components, and good-quality filters at relatively inexpensive cost are available. For the system shown in Fig. 1-46 the use of standard components would typically involve selecting the first IF stages to operate at 10.7 MHz with a 200-kHz bandwidth (a widely available "standard" filter) and the second IF to operate at 455 kHz with a 10-kHz bandwidth.

There are two other features denoted in Fig. 1-46, AGC and squelch. The AGC (automatic gain control) circuit monitors the detected audio signal, and by use of a long-time constant filter circuit, derives a dc potential that is proportional to the strength of the detected signal. This dc voltage is used to adjust the gain of

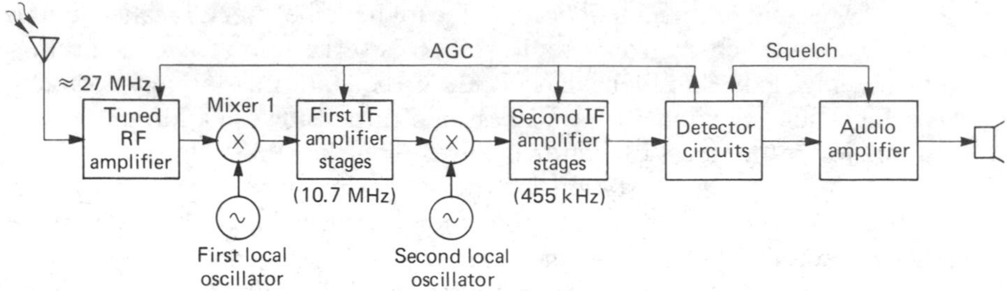

Figure 1-46 Block diagram of a typical CB receiver.

the various RF and IF amplifiers. High-level detected signals reduce the applied gain, and low-level detected signals increase the applied gain. The net result is that, within reasonable limits, the signal level delivered to the detector circuitry is essentially constant over a very wide range of RF receiver signal levels (typically, a range of 30 to 40 dB). Such a circuit is particularly essential to any mobile receiver. A car radio, for example, will experience huge fluctuations in received RF signal strength as it moves around obstacles such as hills or tall buildings. If these fluctuations were not removed by the AGC circuitry, the audio signal leaving the car speaker would also undergo huge fluctuations in volume. Most vehicle operators would take exception to a radio operating in this manner.

The final feature to be noted in the receiver shown in Fig. 1-46 is squelch. Squelch circuits are normally not found on broadcast band receivers because stations in these bands are normally transmitting continuously. Squelch circuits are more typically found in transceivers such as CBs where the remote transmitter is not transmitting all the time. When the remote transmitter stops sending, the received signal strength drops. The AGC circuitry responds by pushing the receiver gain up to try and maintain the signal level. With no signal present and gains set to maximum, the signal out of the speaker is a loud swoosh of static (noise). This output is unpleasant to listen to and normally the squelch circuitry is adjusted to reject it. The squelch circuits compare the received signal strength (via the AGC signal) to a level set by the radio operator via a squelch control potentiometer. If the incoming RF signal is strong enough to "break squelch" (is larger than the squelch threshold setting), the audio amplifier following the detector is turned on and the audio signal reaches the speaker. If, however, the RF signal detected is weak (or just noise), the squelch circuitry shuts off the audio amplifier. The speaker output is thus quiet until the next transmission is sent.

Problems

1-1. For a signal frequency of 600 Hz, the wavelenth would be:
(1) 5000 km
(2) 300 km
(3) 500 km
(4) 5 km

1-2. Identify the graphs in Fig. P1-2 as to whether they represent FDM or TDM.

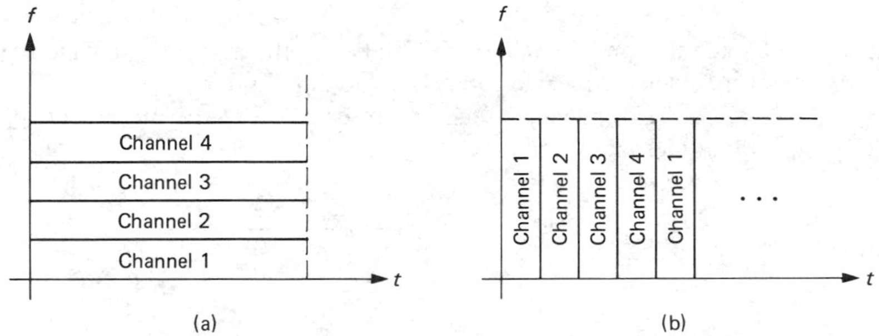

Figure P1-2.

1-3. A cable TV distribution system is an example of what form of multiplexing?

1-4. Adding two sine-wave frequencies together produces:
(1) Sum and difference frequencies only
(2) Sum and difference frequencies plus the original frequencies
(3) Original frequencies only
(4) Sum and difference frequencies centered about a carrier frequency

1-5. Multiplying two sine-wave frequencies together produces:
(1) Sum and difference frequencies only
(2) Sum and difference frequencies plus the original frequencies
(3) Original frequencies only
(4) Sum and difference frequencies centered about a carrier frequency

1-6. Peform the following (graphical) multiplication and complete Fig. P1-6 as to the resultant frequencies and their respective amplitudes.

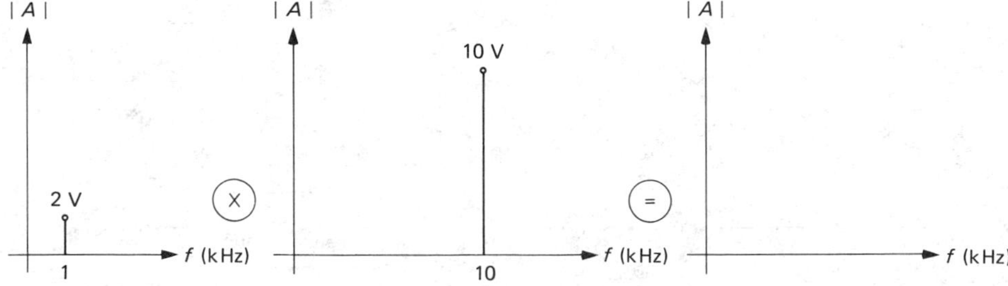

Figure P1-6.

1-7. A square-law mixer is modeled as shown in Fig. P1-7. For A = cosine wave, amplitude 10, frequency 630 kHz and B = cosine wave, amplitude 3, frequency 1085 kHz, sketch $v_o(f)$ and label all frequencies and amplitudes.

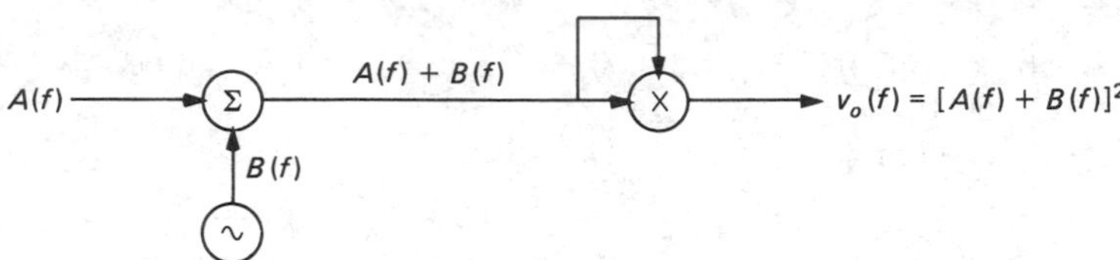

Figure P1-7.

1-8. **(a)** Sketch the time-domain representation of the input V_1 shown in Fig. P1-8. Denote all relevant frequencies, periods, and amplitudes.

(b) Sketch the frequency-domain representation of the output signal V_o. Denote all amplitudes and frequencies.

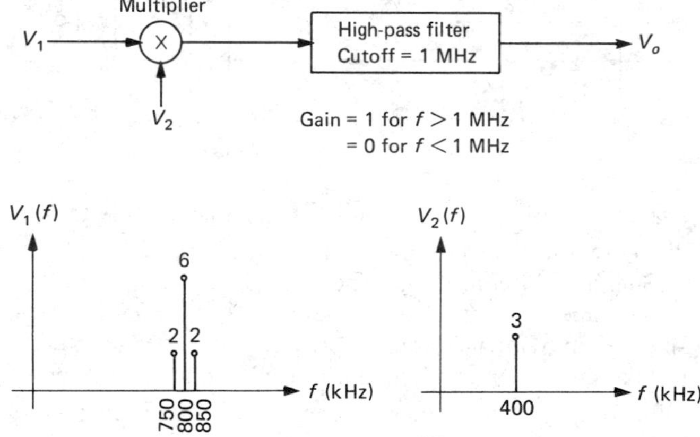

Figure P1-8.

1-9. Sketch the output frequency spectrum of the circuit shown in Fig. P1-9. All oscillators are set for 1-V peak output.

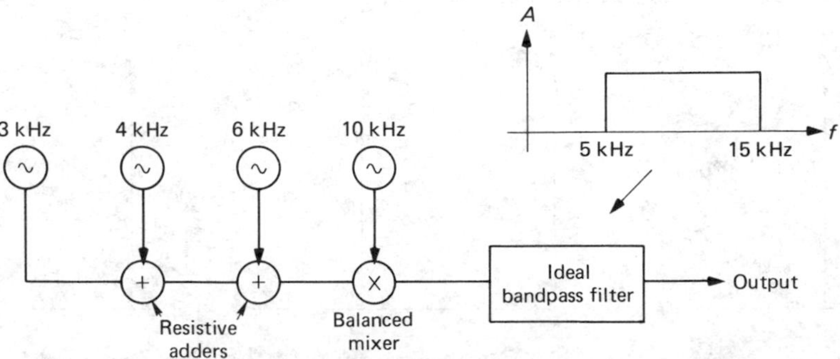

Figure P1-9.

1-10. The frequency spectrum of the switching function shown in Fig. P1-10:
(1) Consists of harmonics of 4 kHz with odd harmonics missing
(2) Consists of harmonics of 4 kHz with no components missing
(3) Consists of harmonics of 1 kHz with no components missing
(4) Consists of harmonics of 1 kHz with odd harmonics missing
(5) Consists of harmonics of 1 kHz with harmonics such as 4 kHz, 8 kHz, and so on, missing

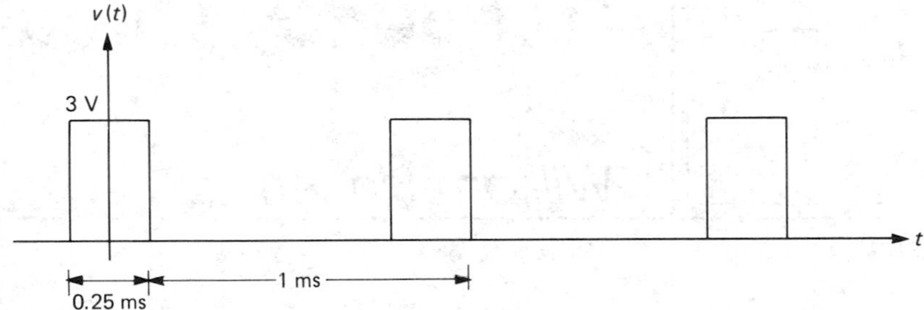

Figure P1-10.

1-11. The pulse train shown in Fig. P1-11 is to be sent through a communication system.
(a) Calculate the amplitude of the fundamental frequency.
(b) Sketch a frequency-domain representation for this waveform. Show all components up to the tenth harmonic and label the frequency of each. Show the amplitude envelope but do not calculate the actual individual amplitudes.
(c) What bandwidth must the communication system possess if it is to pass a reasonable representation of this waveform?

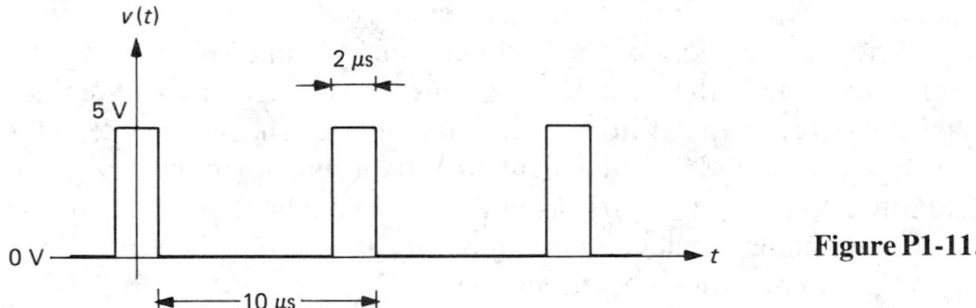

Figure P1-11.

1-12. A single-conversion superhet receiver has a 455-kHz IF stage. The local oscillator frequency has been selected to be larger than the incoming frequency. If the radio is tuned to pick up 1100 kHz and has no tuned RF stage, another station operating at what frequency can cause interference?
(1) 1555 kHz
(2) 2010 kHz
(3) 645 kHz
(4) 190 kHz

1-13. Explain two methods for reducing image frequency interference in superheterodyne receivers.

1-14. A receiver has a 10.7-MHz IF section. A carrier frequency of 98 MHz is received when the local oscillator is adjusted to 87.3 MHz. At what frequency may image interference occur?

Passive Radio-Frequency Circuits

William Sinnema

2-1 INTRODUCTION

In this chapter we consider practical passive RF circuits which contain only resistance, inductance, and capacitance employed to select or reject bands of RF frequencies. Since most of these circuits include one or more resonant circuits, we consider initially the frequency response of such circuits and define such quantities as quality factor, bandwidth, and resonant frequency as they relate to this response.

We will find it convenient to model an impedance ($Z = R + jX$) or an admittance ($Y = G + jB$) as either a two-element series circuit equivalent or as a two-element parallel circuit equivalent (Fig. 2-1). Given an impedance $Z = R_s + jX$, the corresponding admittance is

$$Y = G_p + jB_p = \frac{1}{Z} = \frac{1}{R_s + jX_s} = \frac{R_s - jX_s}{R_s^2 + X_s^2} \tag{2-1a}$$

Thus,

$$G_p = \frac{R_s}{R_s^2 + X_s^2} \quad \text{and} \quad B_p = \frac{-X_s}{R_s^2 + X_s^2} \tag{2-1b}$$

Given an admittance (i.e., $Y = G + jB$), on the other hand, results in a corresponding impedance of

$$Z = R_s + jX_s = \frac{1}{Y} = \frac{1}{G_p + jB_p} = \frac{G_p - jB_p}{G_p^2 + B_p^2} \tag{2-2a}$$

Thus

$$R_s = \frac{G_p}{G_p^2 + B_p^2} \quad \text{and} \quad X_s = \frac{-B_p}{G_p^2 + B_p^2} \tag{2-2b}$$

The characterization of a device in terms of its impedance suggests that we can represent it by an equivalent circuit consisting of a resistance and reactance in series, while the characterization of the same device by its admittance implies an equivalent circuit consisting of a conductance and susceptance in parallel.

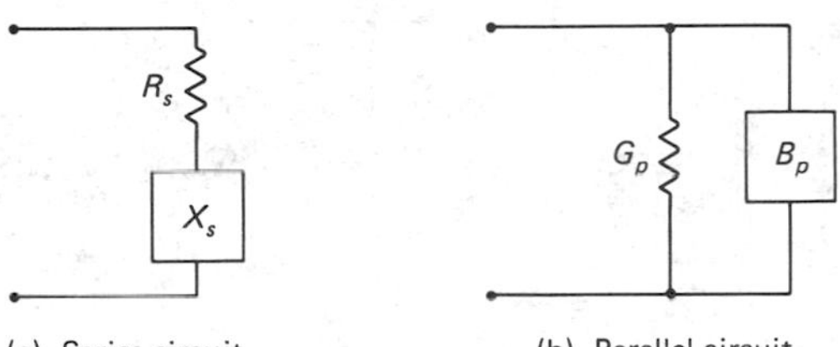

Figure 2-1 Two-element circuit representations.

2-2 QUALITY AND DISSIPATION FACTORS

Practical inductors have an inductance rated in henrys, and because of coil windings, also have a finite resistance. Because current tends to flow near the surface of conductors at high frequencies due to the phenomenon known as the *skin effect*, the resistance increases as the square root of the frequency. This resistance can be reduced by making the coils from heavy-gauge silver-plated wiring or tubing. This does have the disadvantage of increasing the coil size. Inductors store energy in the magnetic field surrounding the coil. The ratio of this stored energy to the energy that is lost in the coil is known as the *quality factor Q*.

Practical capacitors have a capacitance rating in farads, and because of leakage through the dielectric, also have a large resistance in parallel with the capacitance. Capacitors store energy in the electric field residing between the capacitor plates. As before, the ratio of the stored energy to the energy that is lost in the capacitor is given by Q. For capacitors, it is more common to use the inverse term: *dissipation factor D*. Modern capacitors experience very little leakage and thus have very large Q's.

It can be shown that for series or parallel circuits, the quality and dissipation factors are given by

$$Q = \frac{1}{D} = \frac{|X_s|}{R_s} = \frac{|B_p|}{G_p} \tag{2-3}$$

where the variables are as illustrated in Fig. 2-1.

For an indicative circuit, the quality factor simplifies to

$$Q = \frac{\omega L_s}{R_s} = \frac{R_p}{\omega L_p} \tag{2-4}$$

where L_s and R_s are the series representation and R_p and L_p are the parallel representation (Fig. 2-2). Similarly, for a capacitive circuit, the quality factor is given by

$$Q = \frac{1}{\omega R_s C_s} = \omega R_p C_p \tag{2-5}$$

where R_s and C_s are the series representation and R_p and C_p the parallel representation.

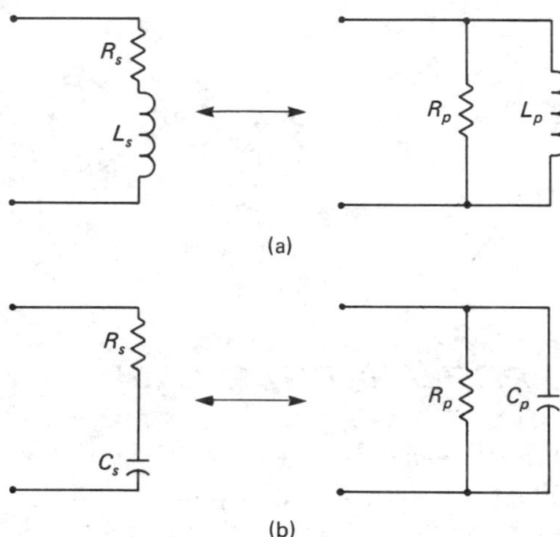

Figure 2-2 Series and parallel representations of an inductor (a) and a capacitor (b).

2-3 SERIES-TO-PARALLEL/PARALLEL-TO-SERIES CONVERSIONS

When converting a series circuit to a parallel circuit, or vice versa, it is convenient to express equations (2-2) in terms of the circuit Q, as the Q of either of the equivalent circuits must be identical if the circuits are to behave identically. Substituting equation (2-3) into expression (2-2), we obtain

$$R_s = \frac{G_p}{G_p^2(1 + B_p^2/G_p^2)} = \frac{1}{G_p(1 + Q^2)}$$

Letting $R_p = 1/G_p$, the parallel-to-series resistance converion is given by

$$R_s = \frac{R_p}{1 + Q^2} \tag{2-6}$$

Similarly,

$$X_s = \frac{B_p}{B_p^2\,[(G_p^2/B_p^2) + 1]} = \frac{1}{B_p(1/Q^2 + 1)}$$

or

$$X_s = \frac{X_p}{1 + 1/Q^2} \tag{2-7}$$

Equations (2-6) and (2-7) can also be rewritten to obtain the series-to-parallel conversions:

$$R_p = R_s(1 + Q^2) \tag{2-8}$$

$$X_p = X_s(1 + 1/Q^2) \tag{2-9}$$

One should note that if Q is rather large, X_p can be assumed to be equal to X_s. In most of our conversion examples, we shall use equation (2-8) along with the Q equations.

EXAMPLE 2-1

Convert the series circuit of Fig. 2-3 to its parallel equivalent.

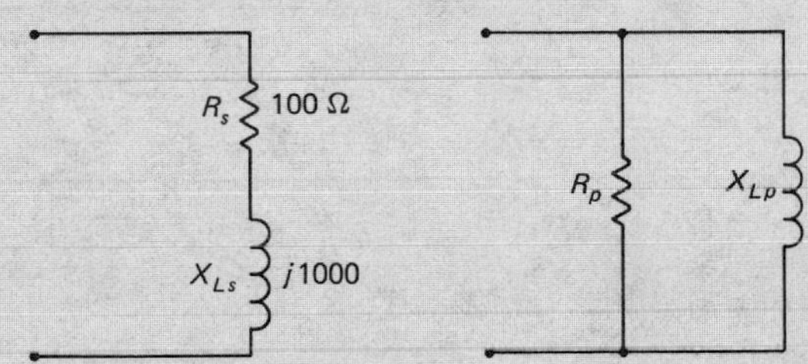

Figure 2-3 Circuit for Example 2.1.

Solution

$$Q_s = \frac{X_{Ls}}{R_s} = \frac{1000}{100} = 10$$

From equation (2-8),

$$R_p = 100(1 + 10^2) = 10.1 \text{ k}\Omega$$

Since both circuits must have the same Q,

$$Q = \frac{R_p}{X_{Lp}}$$

Thus

$$X_{Lp} = \frac{R_p}{Q} = \frac{10{,}100}{10} = 1010 \ \Omega$$

2-4 SERIES RESONANT CIRCUITS

Consider the series circuit of Fig. 2-4(a), where R represents the sum of the coil resistance and the physical resistance placed in the circuit. The loop current in this circuit is given by

$$I = \frac{V}{Z} = \frac{V \angle 0^\circ}{R + j(X_L - X_c)}$$

$$= \frac{V}{\sqrt{R^2 + (X_L - X_c)^2}} \Bigg/ \underline{\phi = -\tan^{-1}\frac{X_L - X_c}{R}} \qquad (2\text{-}10)$$

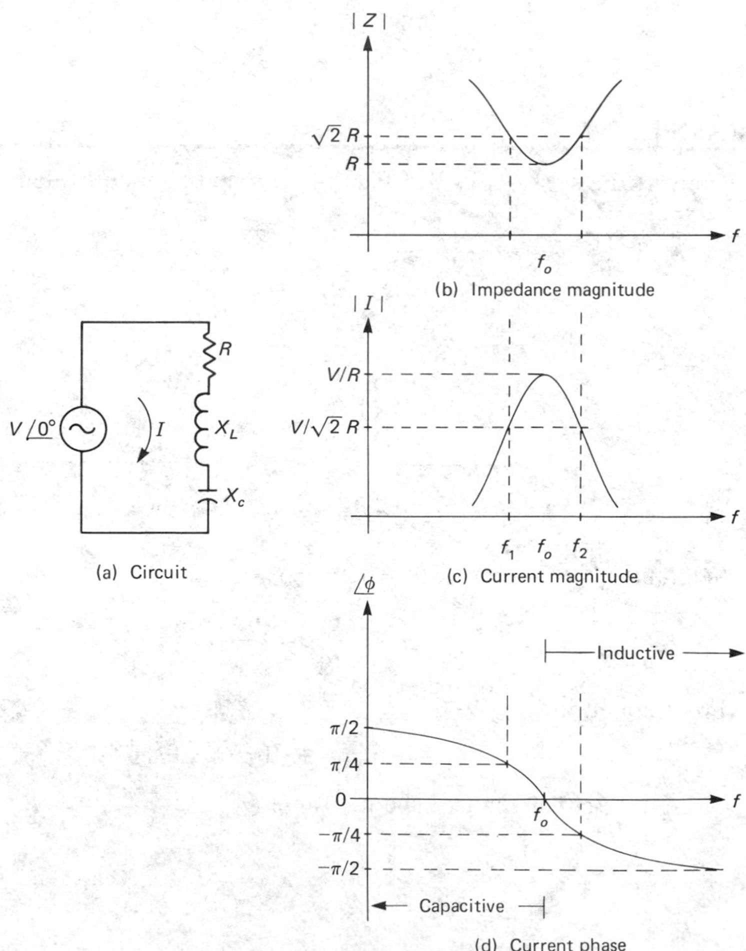

Figure 2-4 Response of a series resonant circuit.

Assuming R to be independent of frequency, the current will be at a maximum and be in phase with the applied voltage when

$$X_L = X_c$$

or

$$2\pi f_o L = \frac{1}{2\pi f_o C}$$

or when

$$f_o = \frac{1}{2\pi\sqrt{LC}} \tag{2-11}$$

This is called the *series resonant frequency* of the circuit. The current will be reduced from the maximum value (V/R) as the frequency varies from the resonant position. In terms of power, at resonance the power absorbed by the resistance will be at its maximum (V^2/R). When the current is reduced to $1/\sqrt{2}$ of its resonant value (i.e., $V/\sqrt{2}\,R$), the power absorbed by the resistance will be halved as $I^2R = V^2/2R$. This is called the *half-power point* or in decibels, the *3-dB point*. The 3-dB point thus occurs when the denominator of equation (2-10) becomes equal to

$$\sqrt{2}\,R = \sqrt{2R^2} = \sqrt{R^2 + (X_L - X_c)^2}$$

or when

$$|X_L - X_c| = R \tag{2-12}$$

At the half-power point the net reactance is thus equal to the resistance and the phase angle is $\pm\pi/4$. That is, when X_L exceeds X_c and $X_L - X_c = R$, the current phase angle is $-\pi/4$; and when X_c exceeds X_L, the phase angle is at $\pi/4$ at the half-power point. At very low frequencies the current phase angle approaches $\pi/2$ as the capacitive reactance becomes predominant, whereas at very high frequencies the current phase angle approaches $-\pi/2$ as the inductive reactance predominates. This is illustrated in Fig. 2-4(c).

The bandwidth of a circuit or device is often defined as the frequency spread between the half-power of 3-dB points. In our case this would be $f_2 - f_1$, as observed from Fig. 2-4(b). At the upper half-power frequency point f_2, from equation (2-12),

$$2\pi f_2 L - \frac{1}{2\pi f_2 C} = R$$

To express R in terms of Q, divide both sides by $2\pi f_o L$, and noting that $R/2\pi f_o L = 1/Q$,

$$\frac{f_2}{f_o} - \frac{1}{(2\pi)^2 LCf_o f_2} = \frac{1}{Q}$$

Since $1/2\pi\sqrt{LC} = f_o$, this equation can be rewritten as

$$\frac{f_2}{f_o} - \frac{f_o}{f_2} = \frac{1}{Q}$$

or

$$f_2^2 - \frac{f_o}{Q} f_2 - f_o^2 = 0$$

Applying the quadratic equation to find the roots, we obtain

$$f_2 = \frac{f_o}{2Q} \pm \sqrt{\left(\frac{f_o}{2Q}\right)^2 + f_o^2} \tag{2-13}$$

Only the plus square-root radical can be used, as the frequency must remain positive.

Similarly, for the lower half-power frequency point f where $X_c > X_L$, we have, from equation (2-12),

$$\frac{1}{2\pi f_1 C} - 2\pi f_1 L = R$$

This can be expressed as

$$f_1^2 + \frac{f_o f_1}{Q} - f_o^2 = 0$$

resulting in roots of

$$f_1 = \frac{f_o}{2Q} \pm \sqrt{\left(\frac{f_o}{2Q}\right)^2 + f_o^2} \tag{2-14}$$

Only the positive square-root radical can be used, as the frequency must remain positive. The 3-dB bandwidth is thus

$$\begin{aligned} \mathrm{BW}_{3\mathrm{dB}} &= f_2 - f_1 \\ &= \frac{f_o}{\mathrm{Q}} \end{aligned} \tag{2-15}$$

This shows that a high Q results in a narrow bandwidth. Since

$$Q = \frac{\omega_0 L}{R} = \frac{1}{R}\sqrt{\frac{L}{C}} \tag{2-16}$$

Q can be increased by reducing the coil resistance or by increasing the L/C ratio. Figure 2-5 shows the effect of Q on the current in a series resonant circuit. The higher the Q, the better the circuit is able to discriminate the passband frequencies from the adjacent frequencies. The circuit is said to be more "selective" for the higher values of Q.

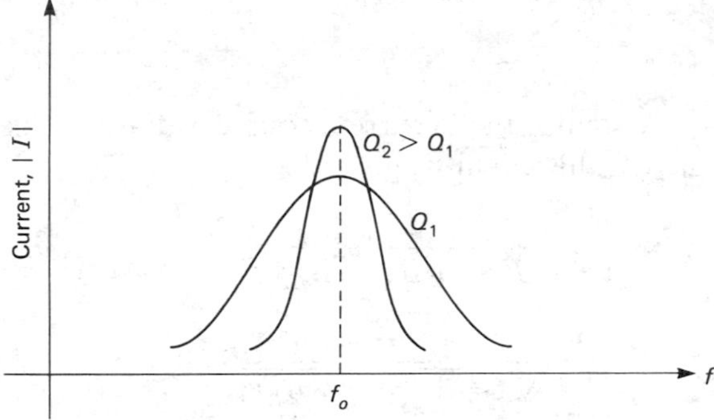

Figure 2-5 Effect of Q on selectivity.

Since the same current flows through each component, at series resonance the capacitor voltage, V_c, will bc cqual to

$$|V_c| = V\frac{X_c}{R}$$

As the ratio of X_c/R is equal to Q,

$$|V_c| = QV \tag{2-17}$$

The output voltage across the capacitor at series resonance is thus equal to Q times the input voltage.

2-5 PARALLEL RESONANCE (ANTIRESONANCE)

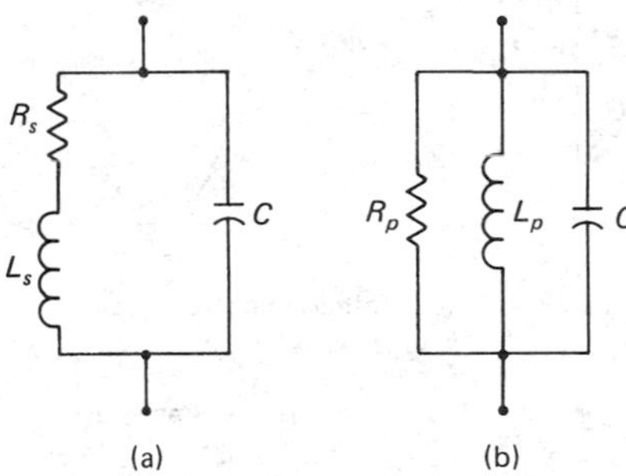

Figure 2-6 (a) Parallel tuned circuit and (b) equivalent parallel tuned circuit.

A parallel tuned LC circuit appears electrically as shown in Fig. 2-6(a), with the assumption that the capacitance has negligible leakage. We can represent the tuned circuit with a more convenient but yet equivalent form by using the series-to-parallel conversion expression given by equations (2-8) and (2-9). If we assume a reasonably high Q, say greater than 10, the parallel expressions for R_p and X_p are approximated by

$$R_p = R_s Q^2 \tag{2-18}$$

$$X_p = X_s \tag{2-19}$$

or

$$L_p = L_s = L$$

where $Q = X_s/R_s = R_p/X_p$. Assuming that a current source is drawing this network [Fig. 2-7(a)], the voltage across this circuit is given by

$$V = \frac{I}{Y} = \frac{1}{G_p + j(B_c - B_L)}$$

$$= \frac{I}{\sqrt{G_p^2 + (B_c - B_L)^2}} \qquad (2\text{-}20)$$

where $G_p = 1/R_p$, $B_c = 1/X_c$, and $B_L = 1/X_L$.

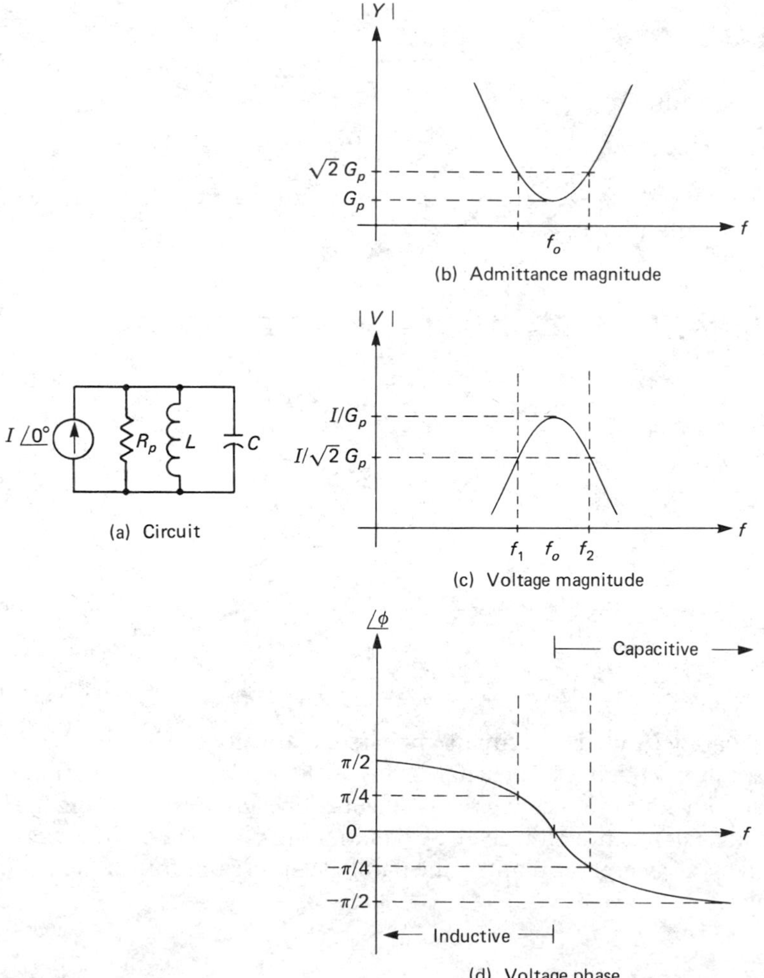

Figure 2-7 Response of a parallel resonant circuit.

Equation (2-21) is of exactly the same form as (2-10), resulting in curves similar to those shown in Fig. 2-4, except that Y replaces Z and V replaces I. These curves are shown in Fig. 2-7. As for the series circuit, the 3-dB bandwidth for the parallel circuit is given by

$$\mathrm{BW}_{3\mathrm{dB}} = \frac{f_o}{Q} \tag{2-21}$$

where

$$f_o = \frac{1}{2\pi\sqrt{LC}}$$

$$Q = \frac{X_L}{R_s} = \frac{R_p}{X_p} = \frac{1}{R_s}\sqrt{\frac{L}{C}} = R_p\sqrt{\frac{C}{L}} \quad \text{at } f = f_o \tag{2-22}$$

At parallel resonance, the admittance is a minimum and equal to G_p. The impedance is a maximum and equal to $1/G_p = R_p$. This maximum resistance value, often called the *dynamic resistance*, is equal to

$$\begin{aligned} R_p &= Q^2 R_s \\ &= \frac{2\pi f_o L}{R_s} Q R_s = 2\pi f_o L Q \\ &= \frac{Q}{2\pi f_o C} \\ &= \frac{L}{R_s C} \end{aligned} \tag{2-23}$$

If the circuit Q is not large, the circuit appears purely resistive at the resonant frequency, given by

$$f_o = \frac{1}{2\pi}\sqrt{\frac{1}{L_s C} - \frac{R_s}{L_s^2}} \tag{2-24}$$

This shows that for a parallel circuit, the resonant frequency depends on the circuit resistance. The parallel dynamic resistance is then given by $R_s\,(1 + Q^2)$. The dynamic resistance does not quite coincide with the maximum impedance modulus obtainable.

EXAMPLE 2-2

Design a parallel resonant bandpass filter having a bandwidth of 4 kHz and centered at 100 kHz (Fig. 2-8). The dynamic resistance is to be 10 kΩ at the center frequency.

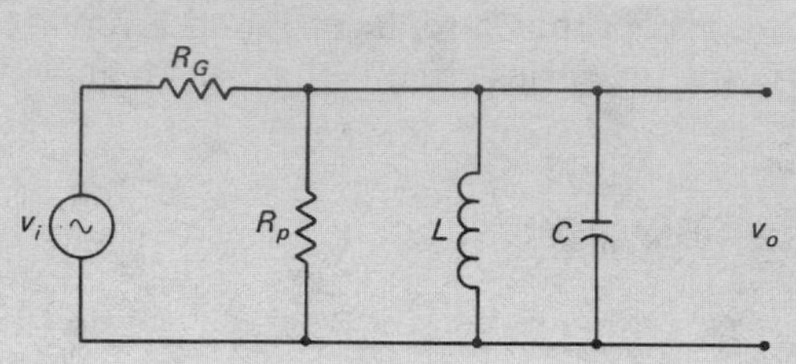

Figure 2-8 Circuit for Example 2.2.

Solution At the center frequency, the dynamic resistance must be 10 kΩ. Thus R_p = 10 kΩ. From equation (2-22), Q is given by

$$Q = \frac{f_o}{\text{BW}} = \frac{100 \text{ kHz}}{4 \text{ kHz}} = 25$$

Equation (2-23) can be used to find L and C. Thus

$$L = \frac{R_p}{2\pi f_o Q} = \frac{10 \text{ k}\Omega}{2\pi \times 100 \text{ kHz} \times 25} = 637 \ \mu\text{H}$$

$$C = \frac{Q}{2\pi f_o R_p} = \frac{25}{2\pi \times 100 \text{ kHz} \times 10 \text{ k}\Omega} = 4 \text{ nF}$$

Although the value of Q can be achieved with standard components, higher Q values require the special crystal, ceramic, or mechanical filter.

2-6 FILTER CHARACTERIZATION

A filter is a device that introduces a relatively small loss over a range of some frequencies called the *passband* and relatively larger loss at other frequencies called the *stopband.* The region between the passband and stopband has somewhat flexible boundaries and is called the *transition region.* The general purpose of filters used in receivers and transmitters is to pass a given band of frequencies and to attenuate everything outside the desired passband. If the source and load resistances are unequal, it is possible to combine the process of filtering and matching into one network. Such networks are commonly used to connect a transmitter power amplifier to an antenna.

Filter characteristics are typically plotted as shown in Fig. 2-9 for a low-pass filter. In Fig. 2-9(a) the magnitude of the output voltage V_o is plotted against frequency f. V_m is the voltage that would be developed across the load if there were no loss (insertion loss) in the passband of the filter. It is quite common to plot relative attenuation versus frequency as shown in Fig. 2-9(b), where relative attenuation is defined as the ratio of the peak output voltage V_p to the voltage output V_o at the frequency being considered.

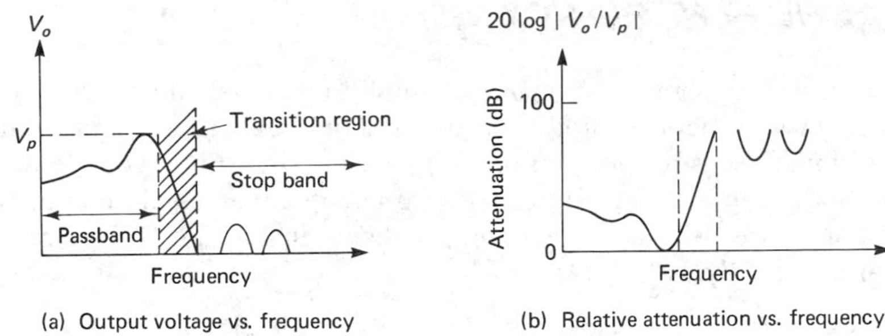

Figure 2-9 Lowpass filter output voltage and relative attenuation characteristics.

Filters are placed into four different classes: low pass (LP), high pass (HP), bandpass (BP), and bandstp (BS), with each class representing a different amplitude response. The frequency-domain amplitude characteristic for each class with the associated symbols are shown in Fig. 2-10. The ideal phase response should be linear over the passband in each case, as explained further in the next section.

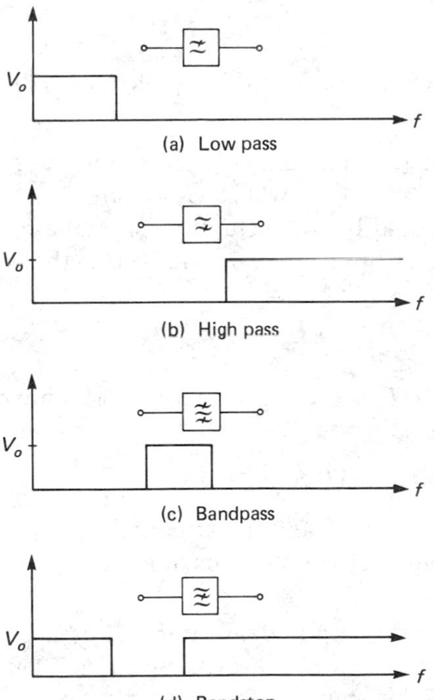

Figure 2-10 Ideal filter amplitude response complete with symbols.

Although we will not pursue transient behavior of filters, one should be aware that when a pulse hits a filter, the response will tend to ring. That is, a damped sinusoidal distortion will be superimposed on the edges of the pulse. For this reason, when sweeping an IF strip or a narrow-band filter, the sweep speed should be not so rapid as to produce ringing.

2-7 IDEAL FILTER FREQUENCY RESPONSE

An ideal filter should attenuate (or amplify) all passband frequencies by the same factor, while the undesired frequencies should experience infinite attenuation. In addition, the desired portions of the input signal may undergo a fixed delay, whereas the undesired portions can experience any delay. This all assures that the shape of the desired signal is preserved as shown in Fig. 2-11. This can be expressed mathematically as

$$v_o(t) = Kv_i(t - T_p) \tag{2-25}$$

where K represents the level change of the input voltage v_i and T_p is the delay.

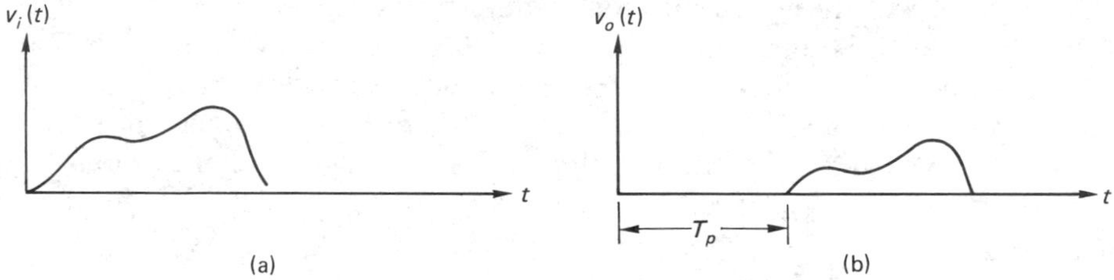

Figure 2-11 Input signal (a) and output signal (b) of an ideal filter.

The constant time delay T_p in the time domain is expressed in the frequency domain as a linear phase shift of $\beta(f) = -\omega T_p$, where $\omega = 2\pi f$. That is, for an ideal filter, the amplitude response should be a constant and the phase response should be a linear function of frequency. Thus for frequencies within the passband,

$$\frac{V_o(f)}{V_i(f)} = K\underline{/-\omega T_p} \tag{2-26}$$

where K is the amplitude factor and $-\omega T_p$ is the phase factor. It is quite common to label the phase function $\beta(f)$. Thus

$$\beta(f) = -\omega T_p \tag{2-27}$$

This is illustrated in Fig. 2-12. Outside the passband the attenuation should rapidly approach infinity so as to reject the frequency components not in the desired signal band.

To illustrate the effects of an ideal and nonideal filter, let us apply an approximation of the square wave of Fig. 2-13 to each of these filters. The mathematical expression (Fourier series) for this square wave is given by

$$v_i(t) = \frac{4}{\pi}\left(\cos 2\pi ft - \frac{1}{3}\cos 3\cdot 2\pi ft + \frac{1}{5}\cos 5\cdot 2\pi ft - \cdots + \frac{1}{n}\sin\frac{n\pi}{2}\cos n\cdot 2n\mathrm{ft}\right) \tag{2-28}$$

where f represents the fundamental frequency component of $1/T = 1/10^{-3} = 1$ kHz [see W. Sinnema, *Digital, Analog and Data Communications* (Reston Publishing Company, Inc., Reston, Va., 1982), App. A].

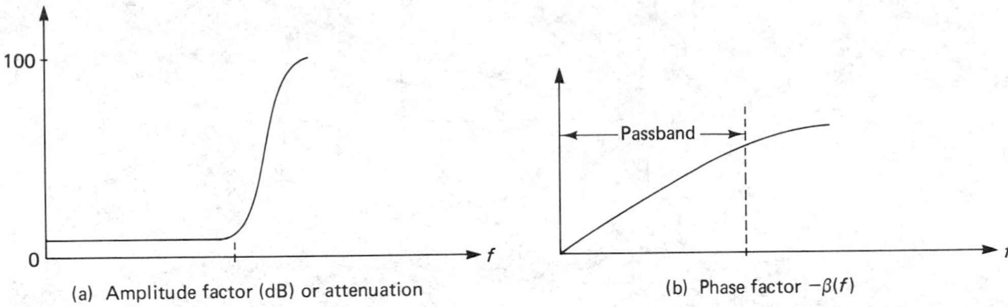

Figure 2-12 Amplitude (a) and phase (b) characteristics of a filter.

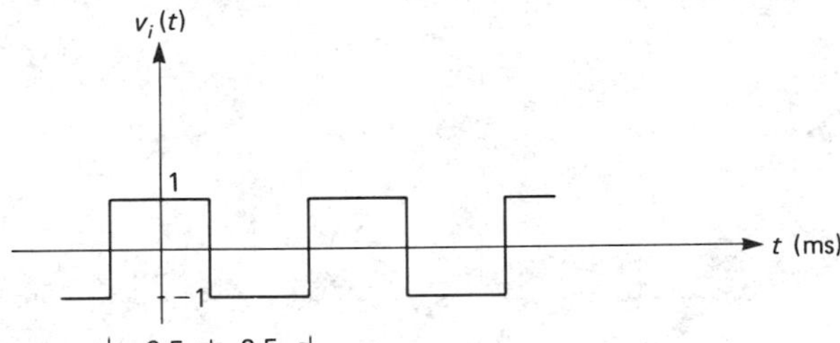

Figure 2-13 Square wave.

Although a large number of terms are required to approximate closely the true square wave, we will assume that the first six terms are adequate. This results in the input signal plotted in Fig. 2-14(a). If this signal is applied to an ideal filter with a constant amplitude factor of $K = 0.5$ and a linear phase shift of 0.2^r/kHz [i.e., $\beta(1\text{ kHz}) = 0.2^r$, $\beta(3\text{ kHz}) = 0.6^r$, etc.], the output signal will be represented by

$$v_o(t)_{\text{ideal}} = \frac{2}{\pi}\left[\cos(2\pi ft - 0.2) - \frac{1}{3}\cos(3\cdot 2\pi ft - 0.6) + \cdots - \frac{1}{11}\cos(11\cdot 2\pi ft - 2.2)\right] \tag{2-29}$$

The result is plotted in Fig. 2-14(b). The output waveform is merely the attenuated version of the input waveform with no distortion.

If the approximated square-wave signal [six terms of expression (2-28)] were applied to a nonideal filter having a constant amplitude factor of $K = 0.5$ and a constant phase shift of 0.2^r at all the harmonics, the output signal will be represented by

$$v_o(t)_{\text{nonideal}} = \frac{2}{\pi}\left[\cos(2\pi ft - 0.2) - \frac{1}{3}\cos(3\cdot 2\pi ft - 0.2) - \cdots - \frac{1}{11}\cos(11\cdot 2\pi ft - 0.2)\right] \tag{2-30}$$

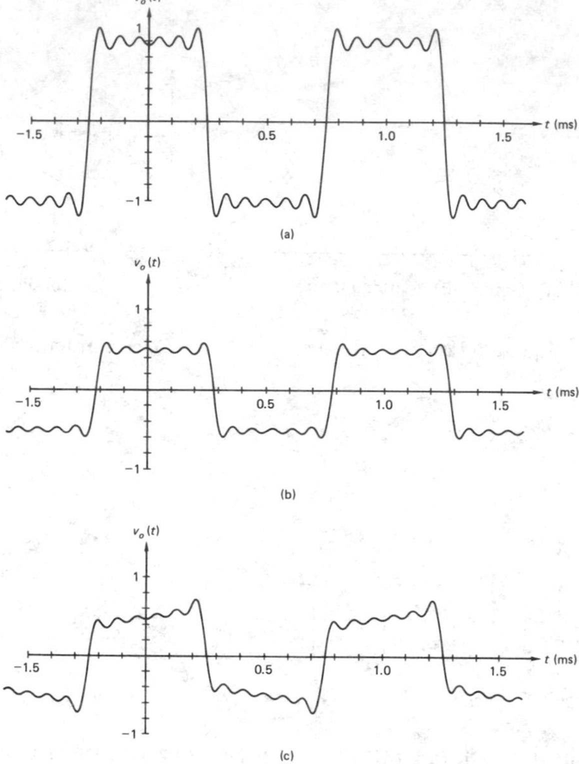

Figure 2-14 Signal response of ideal (b) and nonideal filters (c); (a) input signal to filters; (b) output signal for ideal filter; (c) output signal for nonideal filter.

The result plotted in Fig. 2-14(c) clearly indicates phase distortion.

In certain applications, such as when considering conditioned or unconditioned telephone channels, it is common to deal with time delays rather than the phase function. Corresponding to equation (2-27), the phase delay is given by

$$T_p(f) = \frac{\beta(f)}{\omega} \tag{2-31}$$

Another time delay parameter is the group or envelope delay T_g, given by

$$T_g(f) = \frac{-d\beta(f)}{d\omega} \tag{2-32}$$

[For a derivation, refer to W. Sinnema, *Digital, Analog and Data Communications* (Reston Publishing Company, Inc., Reston, Va., 1982), App. C].

For the ideal filter with a linear phase response, the phase and group delays are identical. In general, however, the phase response is nonlinear, and the various frequency components will experience different delays, resulting in a distorted signal. Figure 2-15 shows the graphical differences between the phase and group delays for a nonlinear system.

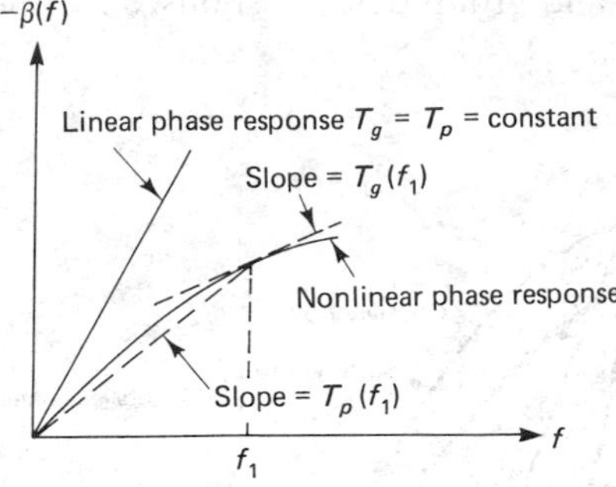

Figure 2-15 Relationship of phase and group delay to the phase shift.

The difference between phase delay and group delay can be illustrated with a narrow-band AM signal. The envelope (or intelligence) of the AM signal is delayed by a time equal to the group delay, while the carrier is delayed by a time equal to the phase delay. For communications it is important that the group delay over its bandwidth be fairly constant so as not to distort the information.

2-8 FILTER ARCHITECTURES

Filter synthesis and design is a complete field in itself and entire books are devoted to the subject. The intent here is to familiarize the reader with the nomenclature of modern filters and to describe the main electrical characteristics for the most common designations. All the characteristics considered are given in low-pass form, but they may be extended to high-pass, bandpass, and bandstop forms.

Bessel filter. The Bessel filter has a very linear phase response and a fairly gentle roll-off in the transition band. It also suffers in passband flatness.

Butterworth filter. The Butterworth filter has a maximally flat passband with moderately sharp roll-off of the skirt that monotonically decreases with frequency. It has a slightly nonlinear phase response.

Chebyshev filter. The Chebyshev filter rolls off faster than the Butterworth filter but at the expense of significant passband ripple. The phase response is highly nonlinear in the transition region.

For a given order of filter that is equal to the number of reactive elements that contribute to the amplitude/phase characteristics of the filter, there is a direct trade-off between the passband ripple and the rate of roll-off. For a faster roll-off (increased stopband attenuation), the passband ripple is increased. The number of maxima and minima in the passband is equal to the order of the filter.

Elliptic or Cauer filter. The elliptic filter has a very fast roll-off but exhibits ripple in both the passband and stopband. Specifying an allowable equal ripple in the passband and required equal ripple in the stopband, the maximum rate of roll-off between the passband and stopband is obtained by the elliptic filter for a given number of filter elements.

Figure 2-16 shows typical amplitude and group delay responses for the filters discussed.

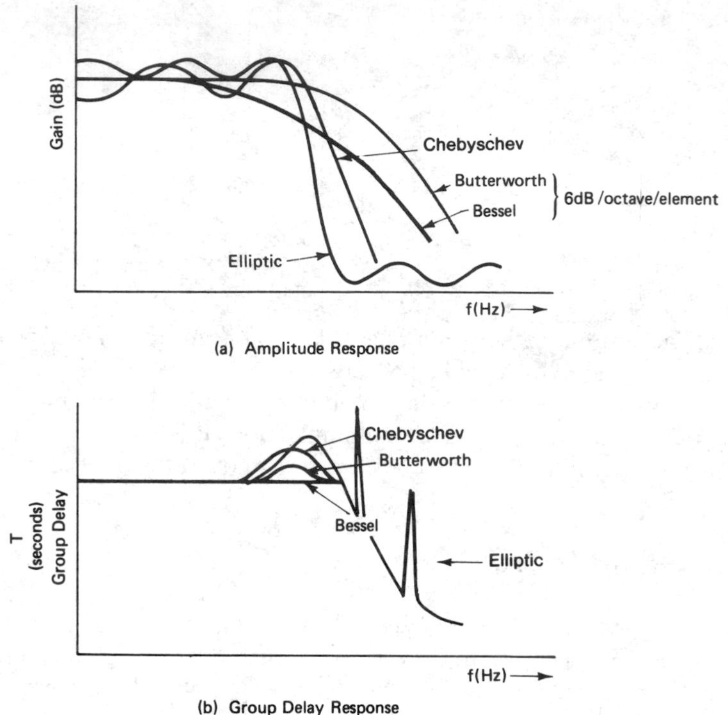

Figure 2-16 Filter characteristics: (a) amplitude response; (b) group delay response.

The order of a filter or the number of "poles" of a filter is equal to the number of reactive elements in the filter. The term *pole* or *zero* comes from pole–zero technique in network analysis and synthesis. In general, as the number of poles is increased, the steeper the roll-off and the higher the stopband attenuation. The asymptotic slope of the filter skirt in the transition region is usually expressed in dB/octave of frequency change. (An octave is a 2:1 ratio of frequencies.) Filters roll off at 6 dB/octave for each pole in the network, as shown in Fig. 2-17.

A summary of analog filter types is given in Table 2-1.

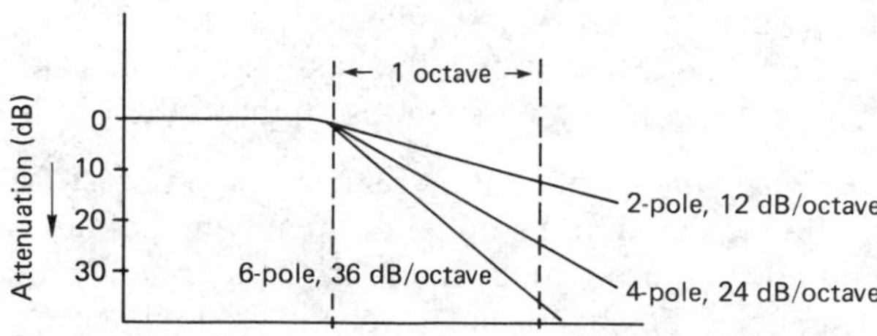

Figure 2-17 Effect of number of poles on roll-off rate.

TABLE 2-1 Summary of Analog Filter Types

	Type of Filter			
Characteristic	Bessel	Butterworth	Chebyshev	Elliptic or Cauer
Amplitude response	Monotonic decrease in amplitude in passband and stopband; 6 dB/octave/pole roll-off well outside the passband	Maximally flat amplitude response with nearly zero slope over the passband; 6 dB/octave/pole roll-off in stopband	Equal ripple amplitude in passband; higher rate of roll-off in the transition region than in the Bessel or Butterworth filters; 6 dB/octave/pole roll-off well within the stopband	Equal ripple amplitude in passband and stopband; highest rate of roll-off in the transition region
Phase or time response	Maximally flat time delay or nearly linear phase delay	Nonlinear phase delay	Ripple in phase delay	Poor phase and time delay characteristics
Transient response	Zero overshoot	High overshoot	High overshoot	Highest overshoot
Applications	Matched filter to receive pulse trains in a noisy environment	Used where phase distortion of the signal is not important (e.g., signal rejection, filtering a fixed local oscillator)	Used when high selectivity and ultimate rejection is required	Not used where signal distortion is a strong consideration

2-9 BANDPASS FILTERS: DEFINITIONS AND TYPES

The purpose of many filters used in receivers and transmitters is to pass a given band of frequencies and to attenuate everything outside the desired passband. The main types of bandpass filters presently in use are:

1. *LC* (tuned transformer) filters
2. Ceramic filters
3. Crystal filters
4. Mechanical filters
5. Surface acoustic wave filters
6. Helical resonators

Each of these six types has different characteristics; thus each type will be more or less suitable for particular applications. The tuned transformer type is still the most common, probably followed by the ceramic filter. Much effort is being put into incorporating some form of ceramic filter with silicon integrated circuits.

Whatever the type of filter, certain measurable characteristics can be used to compare specific filters. The characteristics defined refer to Fig. 2-18.

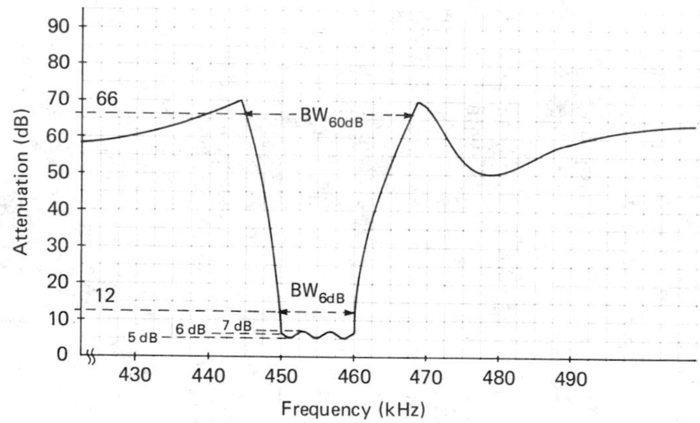

Figure 2-18 Typical bandpass filter response.

Bandwidth Bandwidth is measured when the attenuation is 6 dB greater than the insertion loss. It is the number of hertz between the upper and lower 6 dB attenuation points. Remember that 6 dB down corresponds to the point where the output voltage is 0.5 of the output voltage in the passband. In Fig. 2-18 the BW = 10 kHz.

Skirt Selectivity Skirt selectivity is the ratio of frequency change to attenuation change. It refers to the slope of the sides of the attenuation curve. Observe that in Fig. 2-18 the skirt selectivity is different on the low-frequency end compared to the high-frequency end:

$$\text{skirt selectivity} = \frac{(450 - 445)\ \text{kHz}}{(66 - 12)\ \text{dB}} = \frac{5\ \text{kHz}}{54\ \text{dB}}$$

$$= 0.093\ \text{kHz/dB at low end}$$

$$\text{skirt selectivity} = \frac{(467.5 - 460)\ \text{kHz}}{(66 - 12)\ \text{dB}} = \frac{7.5\ \text{kHz}}{54\ \text{dB}}$$

$$= 0.139\ \text{kHz/dB at high end}$$

Insertion Loss Insertion loss is the average amount of decibel loss in signal at the output of the filter compared to the input measured in the passband of the filter. For Fig. 2-18 the insertion loss is 6 dB.

Shape Factor The shape factor is the ratio of the bandwidth measured at an attenuation 60 dB greater than the insertion loss to the bandwidth measured at 6 dB greater than the insertion loss:

$$\text{shape factor} = \frac{(467.5 - 445)\ \text{kHz}}{(460 - 450)\ \text{kHz}} = \frac{22.5\ \text{kHz}}{10\ \text{kHz}} = 2.25{:}1$$

The steeper the attenuation curve of the filter is, the closer the shape will be to 1.

Ripple Ripple is the maximum deviation of attenuation from the insertion loss measured in the passband of the filter. In Fig. 2-18 the ripple is plus and minus 1 dB. That is, 6 dB − 5 dB and 7 dB − 6 dB.

Stopband Rejection Stopband rejection is the attenuation measured outside the passband. It is 50 − 6 = 44 dB minimum and 70 − 6 = 64 dB maximum for Fig. 2-18.

Stability Stability refers to the amount of change in passband frequencies due to time and temperature changes. As an example, a crystal filter might undergo changes in passband frequencies of ± 5 parts per million per year for a year or two. Also, the passband frequencies might vary ±10 parts per million per degree C between 0 and 50°C.

Output Termination Most filters will meet specifications only when driven by and terminated by specific impedances. As an example, one type of ceramic filter specifies a 2-kΩ termination on the input and output.

2-9.1 LC Filters

LC filters can be used from about 1 kHz to 100 MHz but tend to have Q values of less than 100. At frequencies of less than 1 kHz the inductive values tend to become excessive. Active filters employing RC components and an operational amplifier provide reasonable filtering over the audio-frequency range down to fractions of a hertz. Above 100 MHz, parasitic capacitance and inductance cause

LC components to react strangely. At these higher frequencies, distributed elements such as transmission lines and waveguides are used.

2-9.2 Ceramic Filters

Ceramic filters utilize the piezoelectric effect: When an electric field is applied to such material, the material is physically deformed. By reversing the process, mechanical movement is converted into electrical energy. Typically, the filter consists of a thin disk of zirconate-titanate vibrating in a radial mode whereby the disk expands and contracts radially. Its diameter determines the resonant frequency.

The equivalent circuit of a single disk and its associated impedance versus frequency curve is shown in Fig. 2-19. When several piezoelectric disks are placed in a ladder network [Fig. 2-20(a)], a desired bandpass characteristic can be obtained. In this case the low-impedance frequency point of the series resonator (f_s) is made equal to the high-impedance point of the shunt resonator (f_p), both tuned to the desired passband frequency f_o. The points of maximum signal attenuation will occur at f_p of the series resonator and f_s of the shunt resonator. An attenuation curve such as that illustrated in Fig. 2-20(b) is the result.

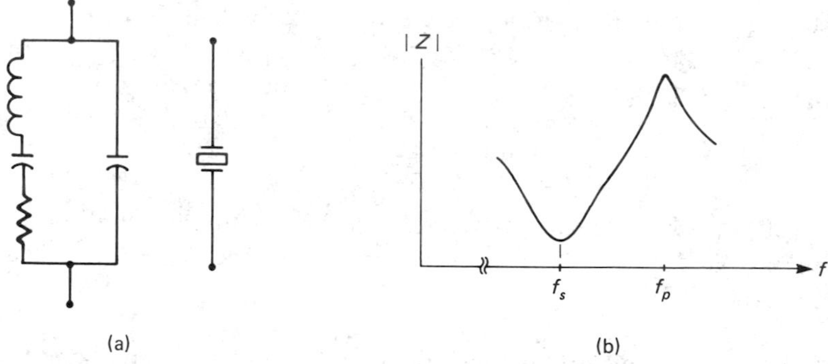

Figure 2-19 Equivalent circuit (a) and impedance curve (b) of a ceramic resonator.

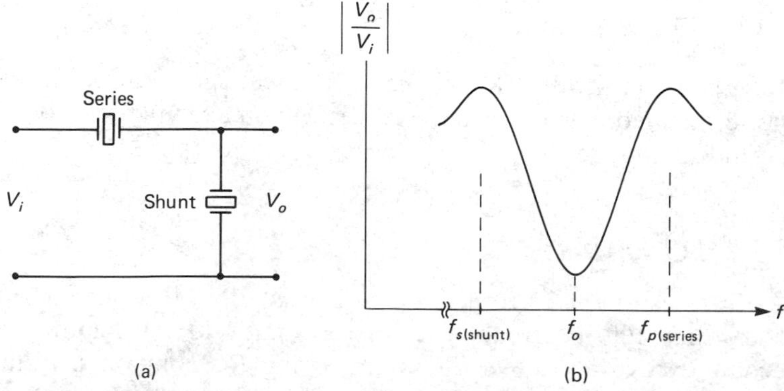

Figure 2-20 Two-resonator circuit (a) with response curve (b).

A typical four-disk ceramic filter used in SSB, AM, and FM transceivers operating at 455 kHz has a shape factor of 3, a 6-dB bandwidth range of 4 to 12 kHz, and a stopband rejection of 60 dB. To achieve this stopband rejection, tuned *LC* circuits are combined with the ceramic filter to take care of the poor attenuation in the wings of the ceramic filter. Thus the *LC* filter provides excellent attenuation in the rejection band, while the ceramic filter provides an excellent shape factor near the passband region.

The ceramic filter's center frequency and bandwidth vary as the operating temperature changes. Its stability is 0.2% of the center frequency from −40 to 85°C. Thus a 455-kHz ceramic filter may change 1 kHz over the temperature range −40 to +85°C. The frequency shift may become excessive for many applications if the temperature exceeds 85°C.

Ceramic filters are generally restricted to two frequency slots: 8 to 50 kHz and 300 to 800 kHz. Broadband ceramic filters are available at 4.5 and 10.7 MHz.

2-9.3 Crystal Filters

The quartz crystals discussed in Section 4-4 can also be used as a filter element. They have excellent temperature stability and have bandwidths of less than a few tenths of 1% of the operation frequency. Values of Q in the range 10,000 to 100,000 are quite common. An equivalent circuit of a 10-MHz quartz crystal is shown in Fig. 2-21.

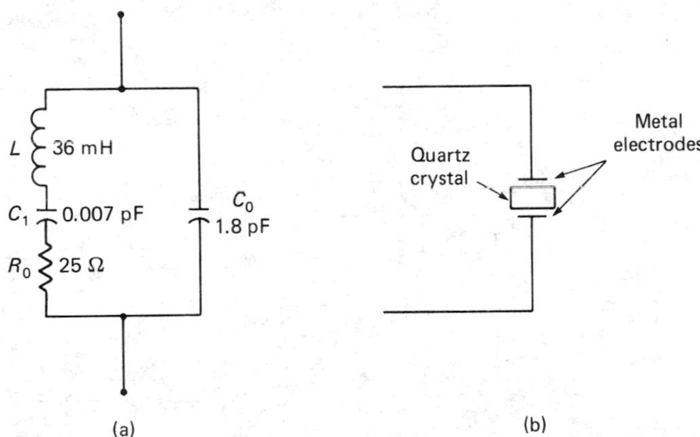

Figure 2-21 Equivalent circuit of a 10-MHz quartz crystal (a) and crystal symbol (b).

Associating the crystal with a transformer (or inductor) as is done in half-lattice networks (Figs. 2-22 and 2-23) causes a spreading between the series and parallel resonant frequencies for each crystal, and provides an additional parallel resonant frequency at a lower frequency. For the half-lattice network the series resonant frequency for crystal X_1 is made equal to the lowest parallel resonant frequency for X_2, while the series resonant frequency for X_2 is made equal to the highest parallel resonant frequency for X_1, as shown in Fig. 2-24. In this way one crystal or the other is near series resonance in the passband, while just outside the passband either one crystal or the other is parallel resonant. There is a slight dip at the center frequency of the response curve which can be minimized by proper selection of R_0. The BW at the 3-dB points is approximately 1.5 times the crystal-frequency spacing.

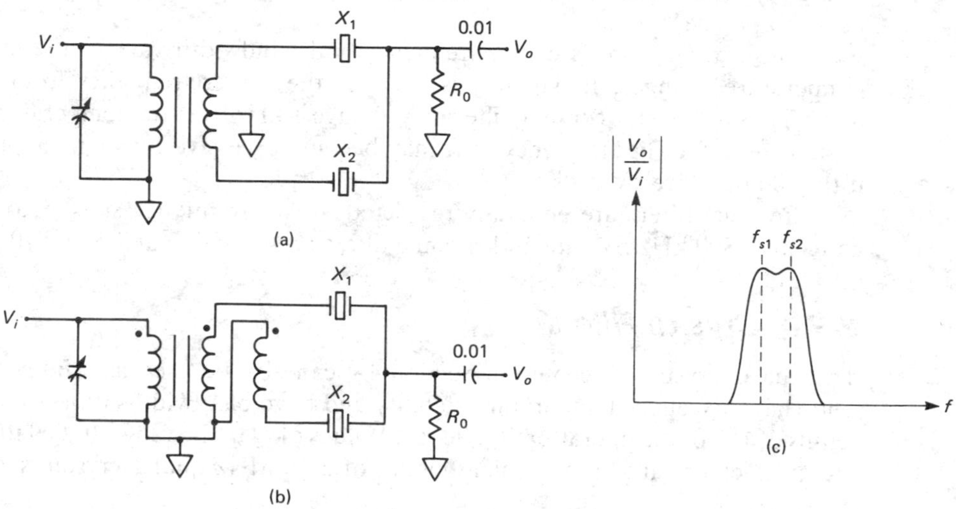

Figure 2-22 Half-lattice crystal filter schematic (a) and its trifilar equivalent (b) with response curve (c).

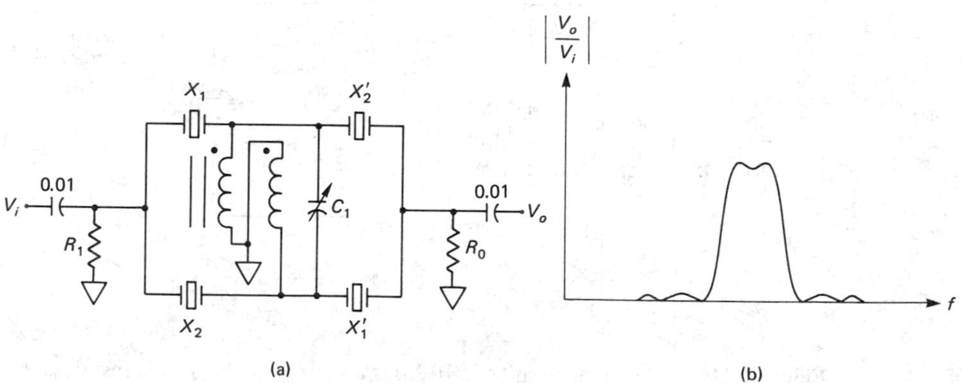

Figure 2-23 Cascaded half-lattice filter (a) with response curve (b): $X_1 = X_1'$, $X_2 = X_1'$.

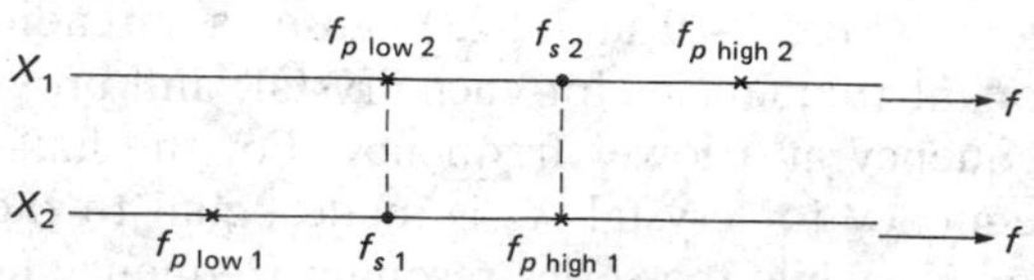

Figure 2-24 Crystal resonant frequencies for a half-lattice bandpass filter.

Skirt selectivity can be improved by cascading two half-lattice filters as shown in Fig. 2-24. C_1 is adjusted for a symmetrical response and R_1 and R_0 are selected to provide a minimum amount of ripple in the passband.

The transformers of Figs. 2-23 and 2-24 are wound trifilar or bifilar on ferrite cores having the appropriate frequency characteristic and are tuned to the center passband frequency. Crystal filters tend to be more expensive than ceramic filters because of the effort involved in grinding the crystals to the proper frequencies.

2-9.4 Mechanical Filters

Mechanical filters are extremely stable and are obtainable with shape factors as low as 1.2, making them excellent devices where sharp selectivity is required. They utilize the magnetostrictive effect whereby a magnetic field causes contraction and expansion of materials such as nickel and ferrite. The mechanical movement of such material is transformed into electrical energy by the inverse magnetostrictive effect.

The construction of a filter assembly is shown in Fig. 2-25. The signal current causes the nickel driving wire to expand and contract, resulting in the vibration of the first resonant disk. This mechanical vibration can be coupled through several disk resonators by using coupling wires, which act as springs welded to the peripheries of the disks. As these circular disks or plates can vibrate in an infinite number of modes, it is necessary to select a mode of operation that is well separated from all others. Each individual disk can have a Q of up to 10,000. At the output the mechanical vibrations are again converted to an electrical signal. Because of the magnetostrictive transducers the insertion loss can be quite high, reaching on the order of 25 dB.

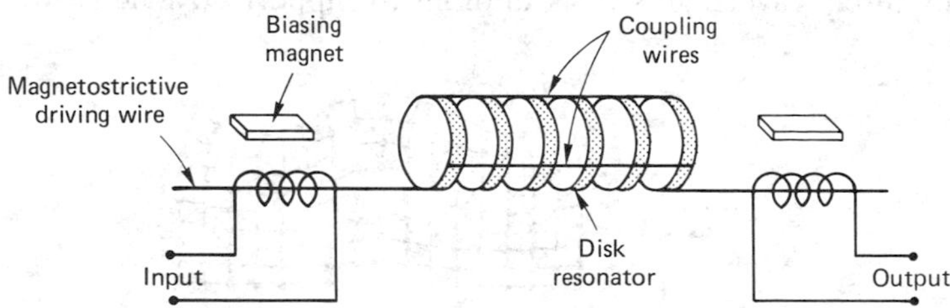

Figure 2-25 Six-disk mechanical resonator.

Figure 2-26 shows a typical spurious response curve of a mechanical filter as a result of the different modes that can be set up. In this example, the maximum spurious signal is attenuated by about 30 dB when compared to the passband signal. The input signal must be kept below the overload level (typically, around 10 to 15 V), where saturation begins to set in. The time delay in the passband varies from about 0.5 to 1 ms. Because these filters tend to be rather expensive, they are used only when sharp selectivity and extreme temperature stability is required.

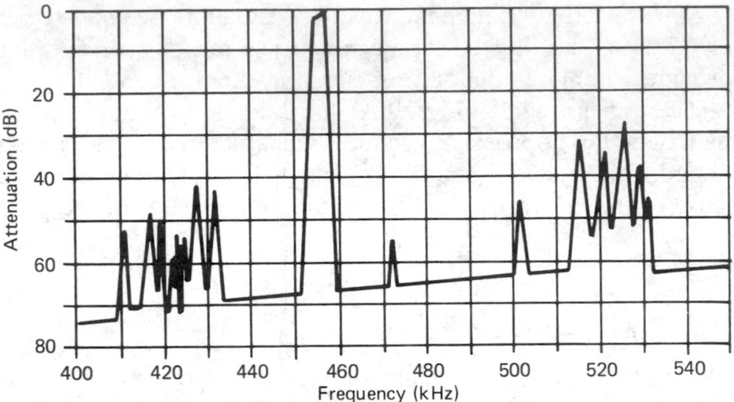

Figure 2-26 Spurious response of a mechanical filter.

2-9.5 Surface-Acoustic-Wave Filters

Surface waves can exist on solid materials in much the same manner as the familiar surface waves on liquid media. The surface acoustic wave (SAW) on a solid arises from the elastic nature of the solid, while the gravitational pull on a liquid causes its surface waves. These waves, often called *Rayleigh waves* after their discoverer, travel at speeds of 1 to 4 km/s. This results in a wavelength of about 30 μm at a frequency of 100 MHz ($\lambda = v/f$). This is on the order of 10^5 times smaller than for its electromagnetic counterpart. Because of these short wavelengths, reasonably sized SAW devices can be designed down to frequencies of around 5 MHz. The disturbance of the wave (Fig. 2-27) becomes negligible within about three wavelengths from its surface. Thus the substrate on which the SAW travels needs to be only three wavelengths thick in order to support a Rayleigh wave.

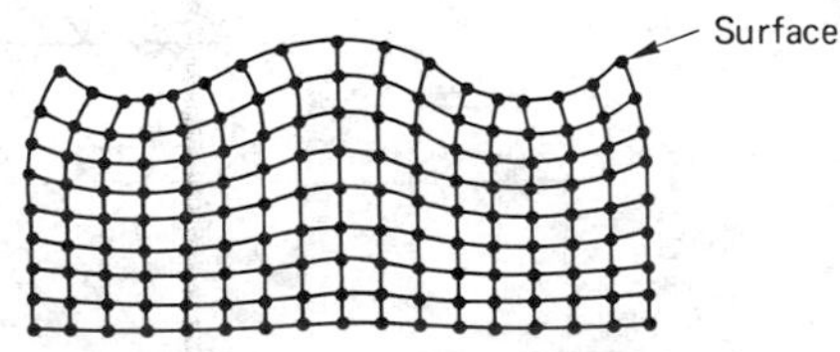

Figure 2-27 Acoustic surface wave displacement.

Surface acoustic waves can be generated and detected using a metallized interdigital transducer structure on a piezoelectric substrate, as shown in Fig. 2-28. The metal comb, usually made of aluminum, is several wavelengths long and capable of launching a place surface wave in the substrate. The piezoelectric substrate undergoes mechanical deformation in response to the applied electric field. After the SAW travels some distance through the substrate, the mechanical (acoustic) deformation is converted into an electrical signal. The electrodes can be made by standard photolithographic techniques.

To understand the operation of a SAW filter, it is best to consider its operation in the time domain. The SAW filter is a transversal filter consisting of interleaved metal electrode fingers on a piezoelectric substate, as in Fig. 2-28(a). In the equivalent circuit of Fig. 2-28(b), D is the delay caused by the spacing between the fingers and A is the weighting or magnitude of the delayed signal, determined by the finger overlap. If each delay is equal and the weighting of each finger is uniform, the frequency response will approximate the sinc function. By varying the length, width and spacing of the fingers, different filter responses are obtained.

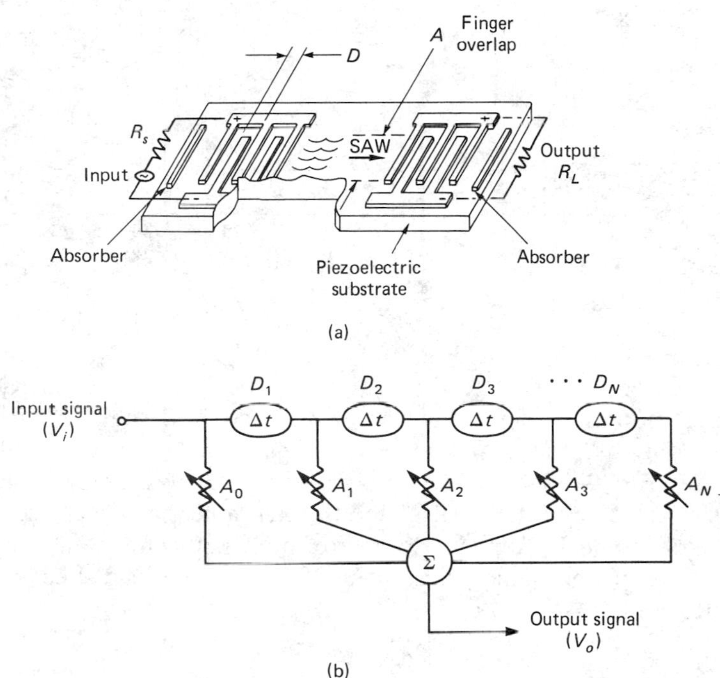

Figure 2-28 Layout of SAW device (a) with corresponding transversal filter equivalent (b).

The center passband frequency of the filter is given by $f_o = v/2D$, where $2D$ is the periodicity of the comb and v is the surface wave velocity. Common substrates used are lithium niobate (LiNbO3), lithium tantalate ($LiTaO_3$), bismuth germanium oxide ($Bi_{12}GeO_{20}$), and ST quartz. Quartz has excellent temperature stability, while lithium niobate exhibits excellent electric field to acoustic coupling efficiency.

SAW filters can readily be produced in printed-circuit-board-compatible packages. They are highly reproducible and do not require individual tuning to obtain the desired response. SAW filters typically have insertion losses of around 20 dB, introduce a capacitance of about 15 pF, and have shape factors as low as 1.15. They tend to become physically large below 10 MHz, and because of the submicron resolution required to fabricate the comb structure at high frequencies are limited to about 2 GHz. They are commonly used in TV receivers as IF filters (Fig. 2-29) or channel filters and are readily available at the common 70-, 400-, and 600-MHz center frequencies.

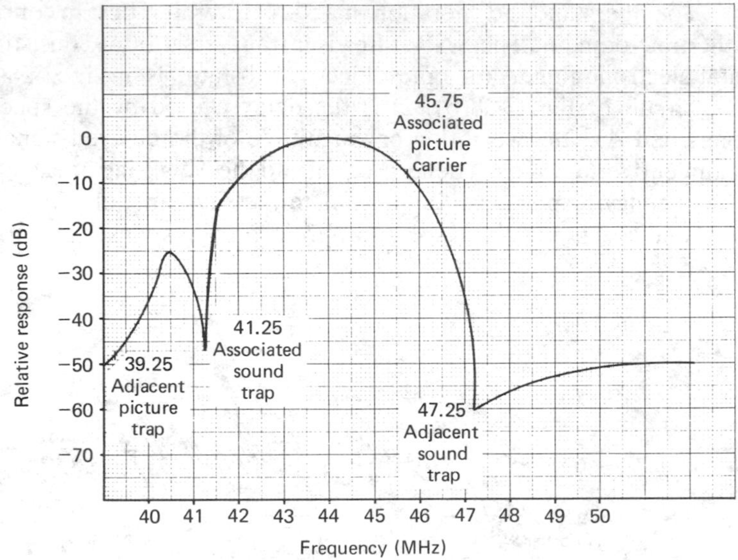

Figure 2-29 SAW TV IF filter response.

Direct electrical feed-through is a particular problem with SAW devices. When mounting a SAW device double-sided PCB should be used, with one side used only as ground. The package is grounded flush to the PCB and the case soldered to the board. The output and input leads are separated by creating an earth barrier by a series of plated-through holes, as illustrated in Fig. 2-30. Amplifiers should be kept in close proximity to the filter and any input feed lead prevented from running close to the SAW output lead. Leads should not be run beneath the SAW device.

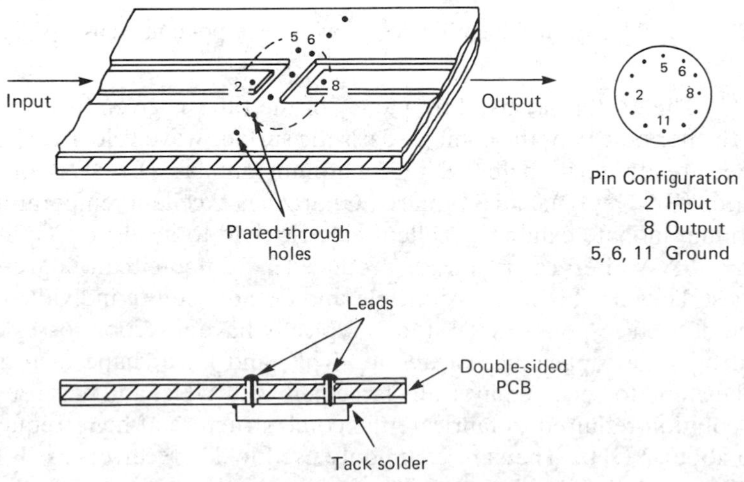

Figure 2-30 Mounting of SAW filter.

2-9.6 Helical Resonators

To maintain high-Q resonant circuits in the VHF band, quarter-wavelength coaxial cavities can be used. Since a quarter-wavelength at 50 MHz comes to 1.5 m, the cavities in this frequency band are quite large, taking up substantial space. A better choice is the helical resonator shown in Fig. 2-31, as it is quite a bit smaller. The helical resonator is a shielded resonant section of helically wound transmission line, with the shield being usually cylindrical or rectangular in shape. One end of the helical winding is soldered to the shield and the other end is left open-circuited.

Figure 2-31 A helical resonator with top and side plates removed.

At 144 MHz, a typical coil consists of seven turns of No. 14 gauge wire having a coil diameter of 17 mm. The rectangular shield is approximately 3 by 3 by 3.5 cm in size and the coil length stretches to about 7 mm from the far ends of the shield. Signals are coupled into and out of the helical resonator by direct taps on the coil, typically about one-eighth to one-quarter turn from the soldered end for 50-Ω loads. Coupling can also be performed with inductive loops located near the soldered end of the coil, the plane of the loop being perpendicular to the helix axis. Helical resonators are conveniently coupled together by providing an open window in the adjacent shield, as illustrated in Fig. 2-32.

Frequency tuning can be performed by locating an air-variable capacitor between the open end of the helix and the shield end plate. Compression or expansion of the helix windings can also cause a small change in the resonant frequency. The Q of a helical resonator is lower than that of a comparable coaxial cavity, being on the order of 700 for the 144-MHz helical resonator described earlier.

Figure 2-32 Coupled helical resonator assembly.

2-10 WAVETRAPS

A high-Q bandstop filter with a narrow rejection notch is known as a wavetrap, or simply just a trap. Occasionally, a very narrow band of interfering frequencies must be removed from a region close to or within the signal band. An example of the use of a wavetrap is the suppression of the second harmonic of a signal emitted from a general radio service band transmitter (27 MHz), as the harmonic would otherwise fall into channel 2 of the lower VHF television band (54 MHz). Wavetraps are also used commonly in the IF sections of television receivers to reduce interference due to the sound and picture carriers of adjacent channels. A wavetrap can be a shunt series resonant circuit or a series–parallel resonant circuit, or a combination thereof, as shown in Fig. 2-33.

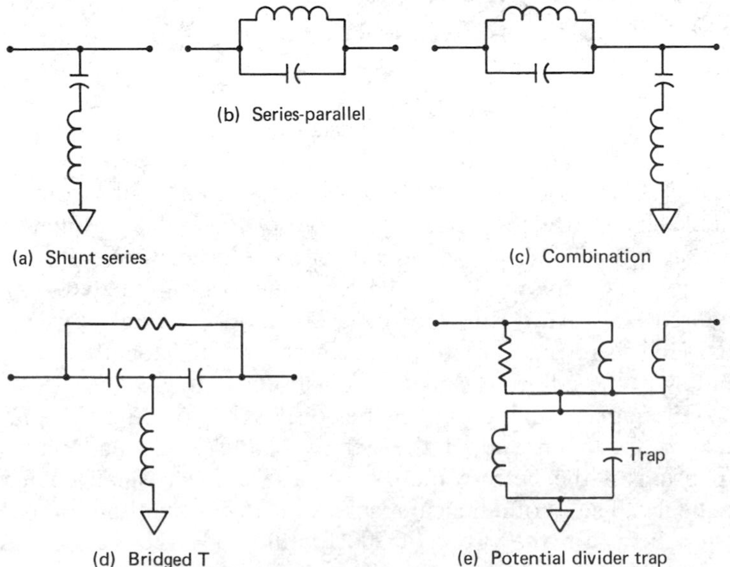

Figure 2-33 Wavetrap circuits.

For large attenuations, the bridged-T circuit of Fig. 2-33(d) effectively short-circuits the signals of the frequency to which the network is tuned. A delta–wye transformation can be used to analyze this circuit. In the circuit of Fig. 2-33(e), at the trap frequency a larger portion of the signal appears across the trap circuit than across the transformer primary. Off resonance, the major portion of the applied signal appears across the transformer primary because the trap will have a very small impedance.

2-11 RADIO-FREQUENCY TRANSFORMERS

Transformers are frequently used to couple RF circuits. The coupling occurs because of the common flux linking the primary and secondary coils, as illustrated in Fig. 2-34. In this circuit, the primary current produces a flux ϕ_1 which is directly proportional to the number of turns of the primary winding N_p and the current value I_p. Since the inductance of a winding varies as the square of the number of turns, the flux produced is proportional to the square root of the primary inductance L_p.

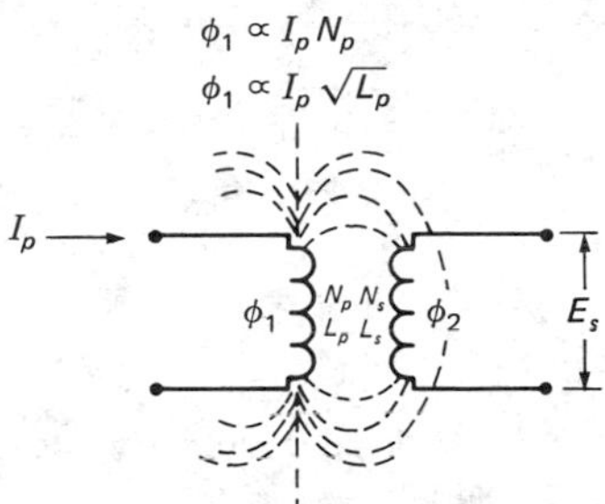

Figure 2-34 Flux paths of air-cored coupled coils.

The number of flux lines common to both the primary and secondary windings is defined in terms of a coupling coefficient k. If all the flux lines produced by the primary link to the secondary, as is approximately the case if an iron or ferrite core is used, the coefficient of coupling is 1. In general,

$$k = \frac{\text{common flux linkage between primary and secondary}}{\text{total flux produced by the primary}}$$

The common flux linkages ϕ_{12} can thus be expressed as

$$\phi_{12} = k\phi_1$$

As the voltage induced across the secondary is proportional to the time rate of change of ϕ_{12} as well as the number of turns of the secondary winding, for a sinusoidal signal

$$E_s \propto \omega k \phi_1 N_s = \omega k I_p \sqrt{L_p}\,\sqrt{L_s}$$

or

$$E_s \propto \omega k \sqrt{L_p L_s}\, I_p$$

where L_s is the inductance of the secondary winding. The constant $k\sqrt{L_sL_p}$ is called the *mutual inductance M*. Thus

$$M = k\sqrt{L_sL_p} \tag{2-33}$$

Because the induced voltage is proportional to the time rate of change of the flux, for a sinusoidal signal a 90° phase shift occurs, resulting in an induced voltage of

$$\begin{aligned} E_s &= \pm j\omega M I_p \\ &= \pm jX_m I_p \end{aligned} \tag{2-34}$$

2-11.1 Measurement of L_p, L_s, M, and k

The various inductance values and the coupling coefficient can be measured by employing the setup shown in Fig. 2-35. A high-Q primary and secondary are assumed. To measure L_p, the secondary should be opened so that zero second current flows, thereby eliminating any induced voltage or reflected impedance from the secondary circuit to the primary circuit. The appropriate signal should be applied to the primary, including dc biasing if required to represent actual circuit conditions.

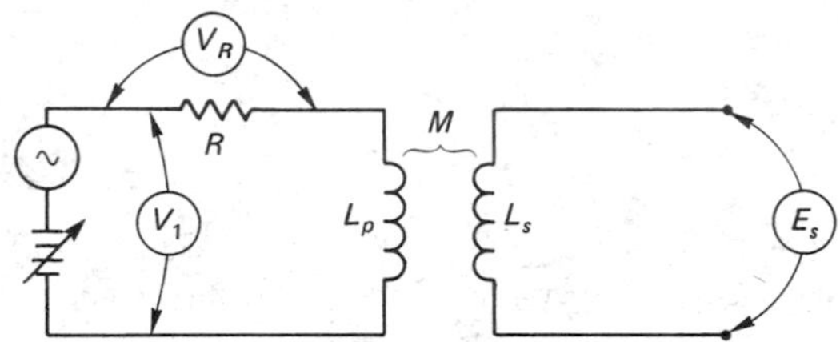

Figure 2-35 Hookup for measurements of inductance values.

Then the applied ac voltage V_1, the ac resistor voltage V_R, and the open-circuited secondary voltage E_s should be measured. The phase angle θ between V_1 and V_R is computed as

$$\theta = \cos^{-1}\frac{V_R}{V_1} \tag{2-35}$$

As Fig. 2-36 illustrates, this allows us to determine X_{Lp} and therefore L_p.

$$X_{Lp} = R\tan\theta \tag{2-36}$$

and

$$L_p = \frac{X_{Lp}}{2\pi f} \tag{2-37}$$

Since the secondary voltage is given by $E_s = X_m I_p$,

$$X_m = \frac{E_s}{I_p} = \frac{E_s}{V_R/R}$$

and

$$M = \frac{X_m}{2\pi f} \tag{2-38}$$

Reverse the transformer winding positions of Fig. 2-35 and determine L_s in like fashion. The coupling coefficient is calculated from the expression [see equation (2-33)]

$$k = \frac{M}{\sqrt{L_s L_p}} \tag{2-39}$$

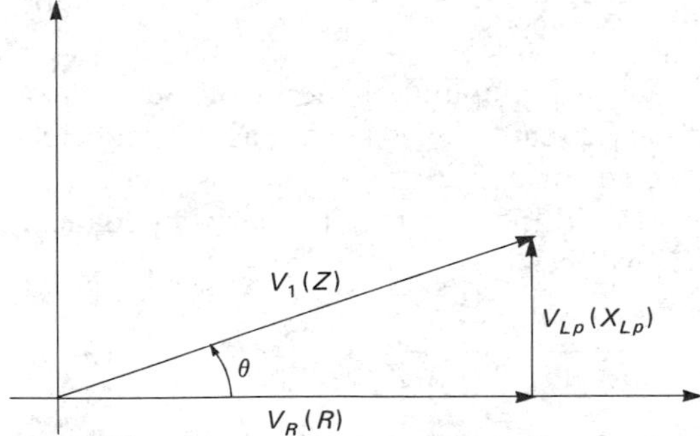

Figure 2-36 Voltage and impedance phase relationships.

2-11.2 Reflected Impedance

With a load Z_L connected across the secondary terminals, the secondary current I_s will be (Fig. 2-37)

$$I_s = \frac{E_s}{Z_s + Z_L}$$

where E_s is the induced voltage in the secondary winding and $Z_s = R_s + jX_{Ls}$, the resistance and inductive reactance of the secondary coil.

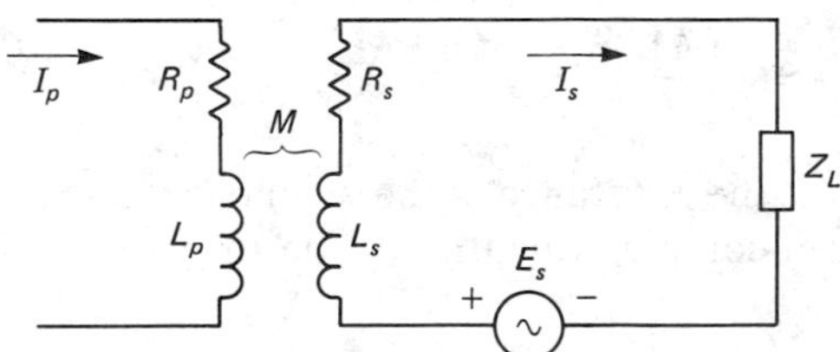

Figure 2-37 Transformer circuit showing induced secondary voltage.

Substituting equation (2-34) for E_s, we obtain

$$I_s = \frac{\pm j\omega M I_p}{Z_s + Z_L}$$

The secondary current causes a secondary flux, which will link with the primary inducing a voltage $E_p = +j\omega MI_s$ in the primary. Replacing I_s by the previous expression, we obtain for the induced voltage

$$E_p = \frac{\pm j\omega M(\pm j\omega MI_p)}{Z_s + Z_L}$$

$$= \frac{\pm(\omega M)^2 I_p}{Z_s + Z_L}$$

This equation is of the form $E_p = ZI_p$, indicating that an impedance

$$\frac{E_p}{I_p} = Z_{Rp} = \frac{(\omega M)^2}{Z_s + Z_L} \tag{2-40}$$

is reflected into the primary circuit. Because of coupling, some of the secondary impedance is reflected back into the primary circuit; this is known as the reflected impedance Z_{Rp}. The positive sign is selected for expression (2-40), as a negative resistance cannot be obtained in a passive circuit. Including the primary impedance, $R_p + jX_{Lp}$, the total input impedance Z'_p will be

$$Z'_p = R_p + jX_{Lp} + \frac{(\omega M)^2}{R_s + jX_{Ls} + Z_L} \tag{2-41}$$

Let us consider briefly the character of the reflected impedance for various load conditions.

Open Secondary Circuit If the secondary circuit is open, the denominator of expression (2-40) goes to infinity and the reflected impedance is zero. Thus if a transformer feeds a high-impedance circuit such as a common-source FET, the reflected impedance is negligible.

Resistive Secondary Circuit If the total secondary impedance $Z_s + Z_L$ is resistive, as is the case when the secondary is tuned (series or parallel), the reflected impedance is purely resistive.

Reactive Secondary Circuit When the impedance of the secondary circuit is capacitive, the term $Z_s + Z_L$ is $-jX_c$ and the reflected impedance becomes

$$E_{Rp} = \frac{(\omega M)^2}{-jX_c} = j\,\frac{(\omega M)^2}{X_c}$$

indicating an inductive reactance. In practice, the resistance of the secondary cannot be ignored and the reflected impedance is a complex quantity of the form

$$R + jX_L$$

If the secondary impedance is inductive, the reflected impedance will be capacitive in nature.

EXAMPLE 2-3

The parameters of an air-coupled transformer at 2 MHz are determined to be $L_p = 100\ \mu H$, $L_s = 10\ \mu H$, $k = 0.05$, $Q_p = 100$, and $Q_s = 40$. Determine:

(a) The mutual inductance

(b) The reflected and primary impedance when:

(1) Z_L is an open circuit

(2) Z_L is a short circuit

(3) $Z_L = -j\,250\ \Omega$ (capacitive reactance)

Solution

(a) $M = k\sqrt{L_s L_p} = 0.05\sqrt{(10 \times 10^{-6})(100 \times 10^{-6})} = 1.58\ \mu H$

(b) $X_{Ls} = \omega L_s = 2\pi(2 \times 10^6)(10 \times 10^{-6}) = 126\ \Omega$

$$R_s = \frac{X_{Ls}}{Q_s} = \frac{126}{40} = 3.2\ \Omega$$

$$\omega M = 2\pi(2 \times 10^6)(1.58 \times 10^{-6}) = 19.9\ \Omega$$

$$X_{Lp} = \omega L_p = 2\pi(2 \times 10^6)(100 \times 10^{-6}) = 1257\ \Omega$$

$$R_p = \frac{X_{Lp}}{Q_p} = \frac{1257}{100} = 12.6\ \Omega$$

(1) $Z_L = \infty$

$$Z_{Rp} = \frac{(\omega M)^2}{R_s + jX_{Ls} + Z_L} = 0$$

$$Z'_p = R_p + jX_{Lp} + Z_{Rp} = 12.6 + j1257\ \Omega$$

(2) $Z_L = 0$

$$Z_{Rp} = \frac{(19.9)^2}{3.2 + j126} \simeq \frac{396}{j126} = -j3.1\ \Omega$$

$$Z'_p = 12.6 + j1257 - j3.1 = 12.6 = j1254\ \Omega$$

Note that the secondary inductive reactance reflects as a capacitive reactance in the primary.

(3) $Z_L = -j\,250\ \Omega$

$$Z_{Rp} = \frac{(19.9)^2}{3.2 + j126 - j\,250} = \frac{396}{3.2 - j124} \simeq \frac{396}{-j124} = j3.2\ \Omega$$

$$Z'_p = 12.6 + j1257 + j3.2 = 12.6 + j1260\ \Omega$$

If the secondary circuit is made resonant (i.e., $Z_L = -jZ_{Ls}$), only a resistive component of $(\omega M)^2/R_s$ is reflected into the primary.

To examine more closely the response curve of a typical transformer circuit, consider the doubly tuned parallel circuit of Fig. 2-38. Models of this circuit are given in Figs. 2-39 and 2-40, with the circuit values and various expressions given in Table 2-2.

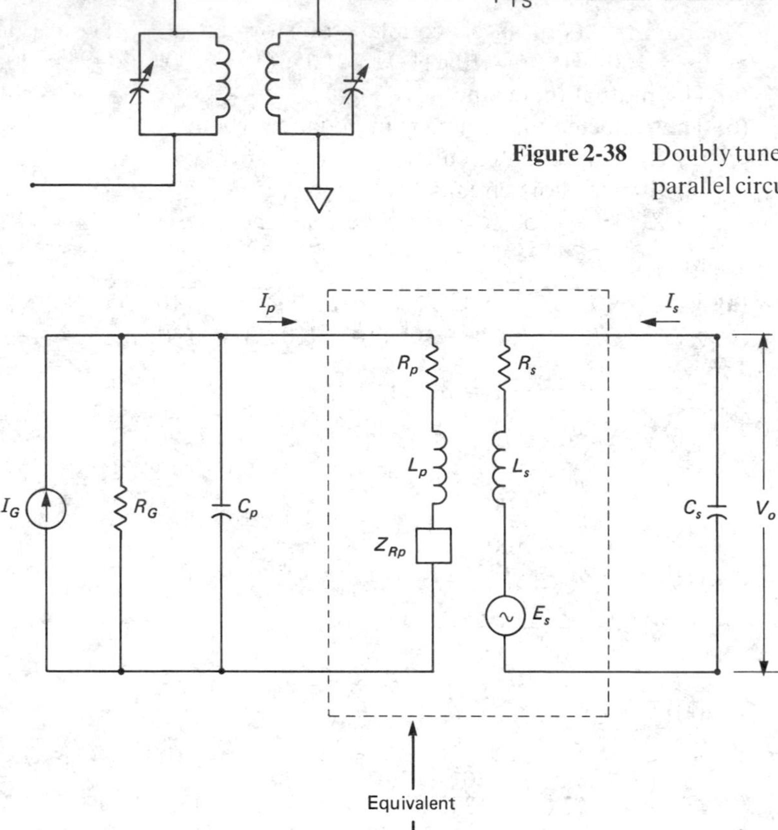

Figure 2-38 Doubly tuned parallel circuits.

Figure 2-39 Double-tuned parallel transformer model.

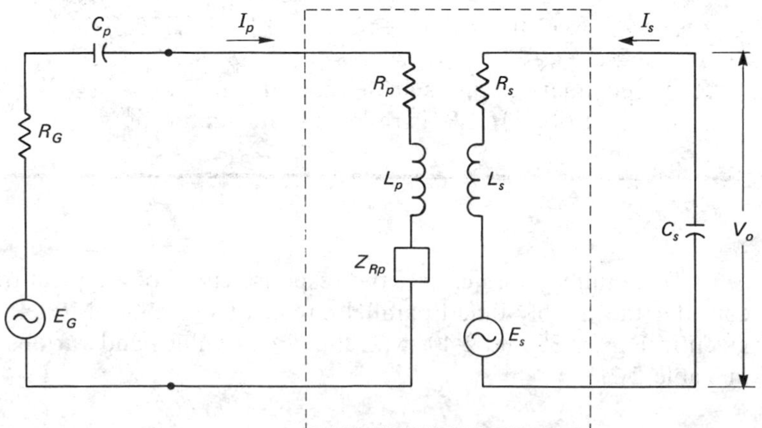

Figure 2-40 Double-tuned series transformer model.

TABLE 2-2 Doubly-Tuned Series Transformer Circuit Values and Circuit Relations

Circuit Values

$E_G = 1\text{ V}$	$R_s = 4\ \Omega$
$R_G = 0\ \Omega$	$L_s = 200\ \mu\text{H}$
$R_p = 4\ \Omega$	$C_s = 1.25\text{ nF}$
$L_p = 200\ \mu\text{H}$	
$C_p = 1.25\text{ nF}$	

Circuit Equations

$$Z_{Rp} = \frac{(\omega M)^2}{R_s + j\omega L_s + 1/j\omega C_s} \qquad M = k\sqrt{L_p L_s}$$

$$V_o = V_c = I_s X_{cs} = Q_s E_s \qquad A_v = \frac{V_o}{E_G}$$

$$I_p = \frac{E_G}{R_G + R_p + j\omega L_p + 1/j\omega C_p + Z_{Rp}} \qquad I_s = \frac{E_s}{R_s + j\omega L_s + 1/j\omega C_s} \qquad \omega_r = \frac{1}{\sqrt{L_s C_s}} = \frac{1}{\sqrt{L_p C_p}}$$

$$E_s = j\omega M I_p$$

EXAMPLE 2-4

For the circuit values given in Table 2-2 and for $k = 0.001$, find at the resonant frequency, the mutual inductance, and

(a) The self-, reflected, and total primary impedances
(b) The primary current
(c) The secondary impedance
(d) The secondary open-circuited voltage
(e) The secondary current and output voltage

Solution At resonance,

$$\omega_r = \frac{1}{\sqrt{L_p C_p}} = \frac{1}{\sqrt{L_s C_s}}$$

$$= \frac{1}{\sqrt{200 \times 10^{-6} \times 1.25 \times 10^{-9}}} = 2 \times 10^6 \text{ rad/s}$$

$$M = k\sqrt{L_p L_s} = 0.001 \times 200 \times 10^{-6} = 0.2 \times 10^{-6}$$

(a) *Primary circuit*

$$\text{Self-impedance} = R_g + R_p + j\omega L_p + \frac{1}{j\omega C_p}$$

$$= R_p \text{ (at resonance the reactances cancel)}$$

$$= 4\angle 0^\circ\ \Omega$$

$$\text{Reflected impedance} = Z_{Rp} = \frac{(\omega M)^2}{R_s}$$

$$= \frac{(2 \times 10^6 \times 0.2 \times 10^{-6})^2}{4} = 0.04 \angle 0^\circ$$

$$\text{Total input impedance} = 4.04 \angle 0^\circ \ \Omega$$

(b) *Primary current*

$$I_p = \frac{E_g}{\text{input } Z} = \frac{1}{4.04} = 0.247 \angle 0^\circ \text{ V}$$

(c) *Secondary impedance*

$$R_s + j\omega L_s + \frac{1}{j\omega C_s}$$

$$= R_s$$

$$= 4 \angle 0^\circ \ \Omega$$

(d) *Secondary open-circuited voltage*

$$E_s = j\omega M I_p$$

$$= 2 \times 10^6 \times 0.2 \times 10^{-6} \times 0.247 \angle 90^\circ$$

$$= 0.099 \angle 90^\circ \text{ V}$$

(e) *Secondary current*

$$I_s = \frac{E_s}{\text{sec. } Z} = \frac{0.099 \angle 90^\circ}{4} = 0.0247 \angle 90^\circ$$

$$\text{Output voltage} = QE_s = 100 \times 0.099 = 9.9 \angle 0^\circ \text{ V}$$

$$= I_s X_{Cs} = 0.0247 \angle 90^\circ \frac{1}{2 \times 10^6 \times 1.25 \times 10^{-9} \angle 90^\circ}$$

$$= 9.9 \angle 0^\circ \text{ V}$$

These results are tabulated in Table 2-3, as are those for several other coupling coefficients. The effects of coupling on the response for the double tuned RF transformer are plotted in Fig. 2-41. These were plotted by computer because of the many calculations involved.

TABLE 2-3 Summary of Circuit Characteristics

	Primary Circuit Characteristics				Secondary Circuit Characteristics				
	Self-Impedance	Reflected Impedance	Input Impedance	Current	Self-Impedance	Open Circuit Voltage	Current	Output Voltage	Voltage Transfer Ratio
	$R_G + R_p + j\omega L_p + \frac{1}{j\omega C_p}$	Z_{rp}	$R_G + R_p + j\omega L_p + \frac{1}{j\omega C_p} + Z_{rp}$	I_p	$R_s + j\omega L_s + \frac{1}{j\omega C_p}$	E_s	I_s	V_o	V_o/V_i
Undercoupled $k = 0.001$ $M = 0.2 \times 10^{-6}$	$4 \angle 0°$	$0.04 \angle 0°$	$4.04 \angle 0°$	$0.247 \angle 0°$	$4 \angle 0°$	$0.1 \angle 90°$	$24.7 \angle 90°$	$9.9 \angle 0°$	19.9 dB
Critical coupling $k = 0.01$ $M = 2 \times 10^{-6}$	$4 \angle 0°$	$4 \angle 0°$	$8 \angle 0°$	$0.125 \angle 0°$	$4 \angle 0°$	$0.5 \angle 90°$	$0.125 \angle 90°$	$50 \angle 0°$	33.98 dB
Optimum coupling $k = 0.0163$ $M = 0.0096$	$4 \angle 0°$	$10.6 \angle 0°$	$14.6 \angle 0°$	$0.068 \angle 0°$	$4 \angle 0°$	$1.63 \angle 90°$	$0.111 \angle 90°$	$44.6 \angle 0°$	33 dB
Overcoupling $k = 0.1$ $M = 20 \times 10^{-6}$	$4 \angle 0°$	$400 \angle 0°$	$404 \angle 0°$	$2.47\text{ mA} \angle 0°$	$4 \angle 0°$	$10 \angle 90°$	$0.025 \angle 90°$	$9.9 \angle 0°$	19.9 dB

At resonance $\omega_r = \frac{1}{\sqrt{L_s C_s}} = \frac{1}{\sqrt{L_p C_p}} = \frac{1}{\sqrt{(200\ \mu\text{H})\ (1.25\ \text{nF})}} = 2\ \text{Mrad/s}$

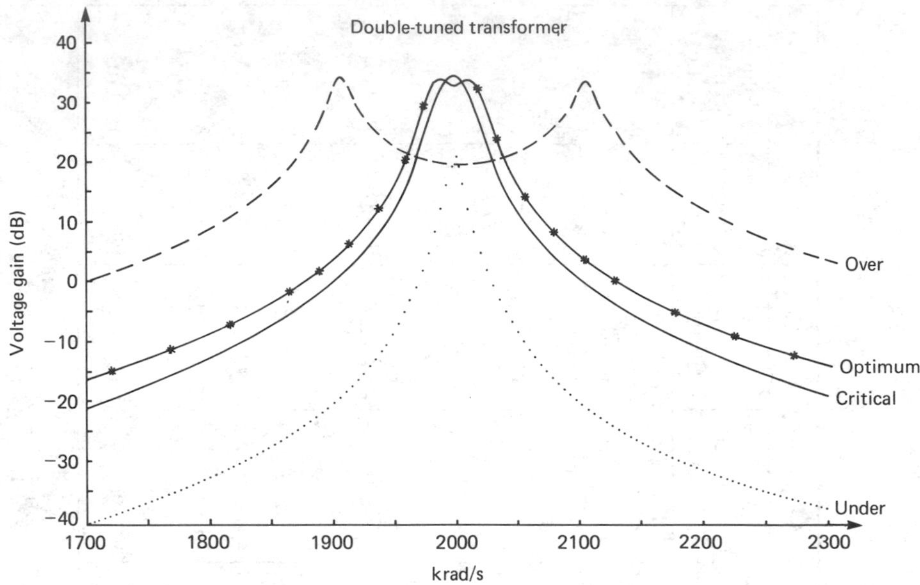

Figure 2-41 Response curves for a doubly-tuned inductive-coupled RF transformer for various coupling coefficients.

For the undercoupled condition, where the coupling coefficient is sufficiently low that the reflected resistance is negligible, as the resonant condition is approached, the primary current increases. Although ωM is low because of the loose coupling, the output voltage V_o rises because of the almost constant Q values of the primary and secondary circuits. Off resonance, the output voltage rapidly drops off.

As the coupling is increased, the reflected impedance increases, causing the Q of the primary to decrease. This causes a broadening of the response curve, as shown. The increased coupling also raises the output voltage. As the coupling is increased still further, the reflected resistance into the primary reaches the resistance of the primary circuit. This results in an optimum output voltage, as the increase in ωM is just offset by a decrease in Q. This condition is known as *critical coupling*. Critical coupling occurs when

$$R_{Rp} = \frac{(\omega M)^2}{R_s}$$

or

$$M = \frac{\sqrt{R_{Rp}R_s}}{\omega} \tag{2-42}$$

Substituting M into equation (2-39) yields

$$k = k_c = \frac{M}{\sqrt{L_p L_s}} = \sqrt{\frac{R_{Rp} R_s}{\omega L_p \omega L_s}}$$

$$= \frac{1}{\sqrt{Q_p Q_s}} \quad (2\text{-}43)$$

where $Q_p = \omega L_p / R_{Rp}$ and $Q_s = \omega L_s / R_s$.

If the primary and the secondary circuits have the same Q values as in Example 2-4:

$$k_c = \frac{1}{Q} \quad (2\text{-}44)$$

$$Q = \frac{\omega L_s}{R_s} = \frac{2 \times 10^6 \times 200 \times 10^{-6}}{4} = 100$$

and

$$k_c = \frac{1}{100} = 0.01$$

Beyond critical coupling, the primary Q at resonance drops more than the increase in ωM, so that the resonant peak is lower than at the critical coupling point. This is termed *overcoupling*. At frequencies below resonance, the secondary appears capacitive (with respect to the induced voltage E_s, the secondary is a series resonant circuit), reflecting an inductive reactance into the primary. This added inductance causes the primary to resonate at a lower frequency, causing a hump in the response curve below the previous maximum points.

Similarly, above resonance, the secondary circuit appears inductive, reflecting a capacitive reactance into the primary. As this capacitance is in series with the primary circuit, the net series capacitance is reduced, causing the primary circuit to resonate at a slightly higher frequency. Figure 2-41 clearly shows the double-humped shape response curve for the overcoupled condition. In practice, a more-flat-topped response is desirable. This is called the optimum response and occurs at around $1.5k_c$.

The doubly tuned transformer has a much faster roll-off at the skirts than if the transformer was singly tuned. Furthermore, singly tuned transformers do not display the overcoupled response discussed previously. Singly tuned transformers are often employed with transistors, as the lower output resistance displayed by these devices tends to load down the primary transformer circuit. To prevent this, the transistor should tap into the coil as shown in Fig. 2-19(a). Figure 2-42 shows a typical singly tuned stage along with the response curve.

Although inductive coupling is undoubtedly the most common method of electrically interconnecting tuned circuits, capacitive coupling is also used at times. In addition, series rather than parallel resonance is used occasionally, especially when low-impedance circuits are encountered. Several variations of coupled tuned circuits are illustrated in Fig. 2-43.

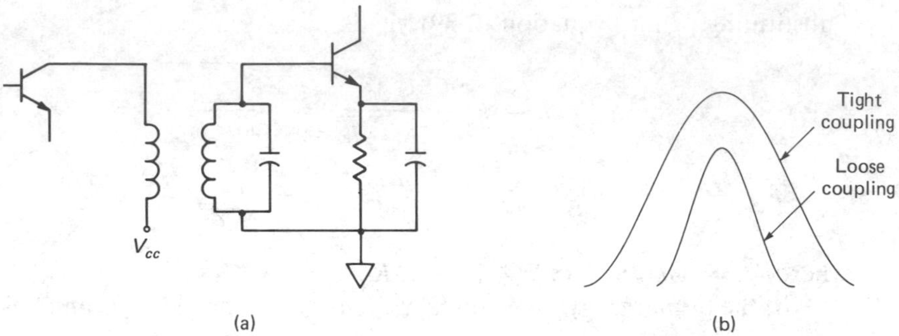

Figure 2-42 Single-tuned inductive coupled circuit (a) and response curve (b).

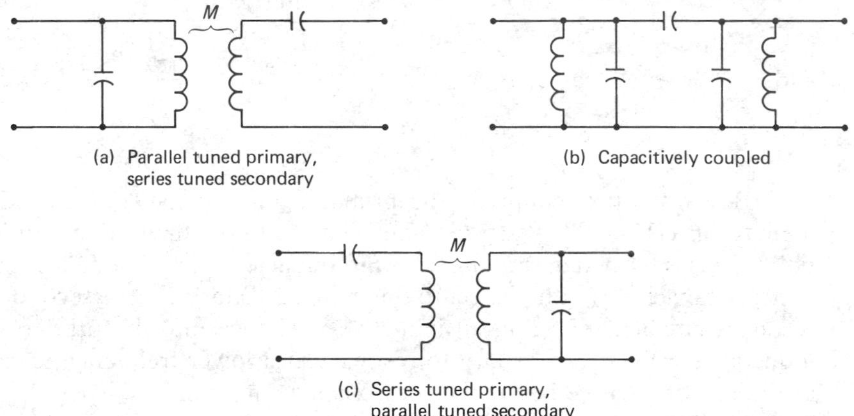

Figure 2-43 Some alternative configurations of coupled tuned circuits.

2-11.3 Stagger Tuning

If a much broader bandwidth than the 10 kHz nominally obtainable by the previous tuned circuits is desired, as in the case of a television 6-MHz intermediate-frequency amplifier, stagger tuning is used. Stagger tuning is the technique of cascading a number of high-Q tuned stages, each tuned at a slightly different resonant frequency. As shown in Fig. 2-44, the resulting overall response is rather flat in the passband, with the skirts having a rather sharp roll-off.

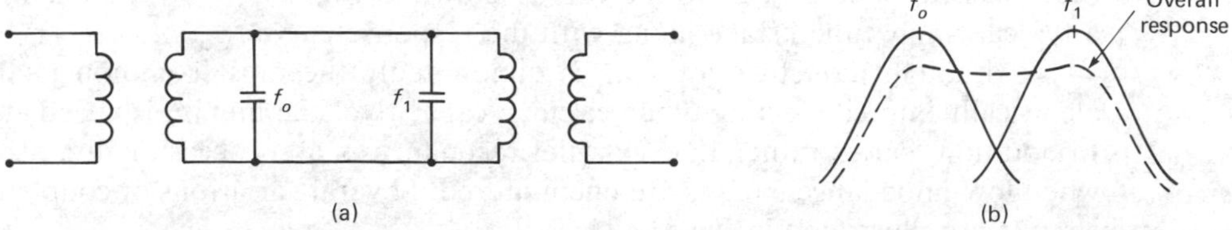

Figure 2-44 Stagged tuned circuit (a) with response (b).

2-12 DUPLEXERS

To allow for the provision of simultaneous transmission and reception of radio signals, the transmitter can operate over a narrow band of frequencies that is different from the receiver operating frequency band. Because transmitters always transmit some power on frequencies above and below the assigned channel, there is a strong likelihood that this "transmitter broadband noise radiation" appears in the receiver, masking the receiver's desired signal and thereby reducing the receiver's sensitivity. It is important that the transmitter noise appearing in the receiver frequency slot be eliminated prior to reaching the receiver.

In addition, even if the receiver has excellent selectivity characteristics, a strong signal from a nearby transmitter may still get through the front-end stages of a receiver causing a degradation in the receiver's performance. This phenomenon is called *receiver desensitization* and may occur even though the transmitter frequency is several megahertz from the receiver frequency.

To reduce both of these forms of interferences, the electrical isolation between the receiver and transmitter must be increased. The higher the transmitter output power and the closer the transmitter and receiver frequencies, the greater the isolation that must be provided. One method of assuring this isolation is to employ separate transmitter and receiver antennas and to physically separate them vertically or horizontally. A vertical separation of 10 ft between vertical dipoles at 150 MHz provides about 35 dB of isolation, whereas a horizontal separation of 50 ft is necessary for the same isolation. This technique can be quite costly.

A second and more common method of reducing Tx–Rx interference is by the use of duplexers. The duplexer requires only a single antenna for receiving and transmitting and provides the necessary isolation with the employment of high-Q helical or cavity resonators. An example of such a system is the mobile repeater illustrated in Fig. 2-45. The system used band-reject cavities or notch filters which present a short circuit to the loop coupler when at resonance and an open circuit when off-resonance. As will be discussed in Section 13-5, impedances along a transmission line repeat themselves at an integral number of half-wavelength intervals, and a quarter-wavelength line results in an inverse normalized impedance transformation [i.e., $Z_{in}/Z_o = 1/(Z_L Z_o)$]. Thus at a distance of $\lambda/2$, λ, $\frac{3}{2}\lambda$, and so on, from a short on a transmission line, the input impedance is also 0 Ω. At a distance $\lambda/4$ from a short circuit, the impedance is infinite or an open circuit. At a distance of $\lambda/4$ from an open circuit, the input impedance is 0 Ω or a short.

Consider the duplexer operation when transmitting at 153.00 MHz. The Tx filters are off-resonance and therefore present an open circuit to the loop couplers. At filter junctions A and B a half-wavelength distant, the impedances are again infinite, and therefore the Tx filters have no effect on the transmitted 153.00-MHz signal. The Rx filters, on the other hand, are resonant at 153.00 MHz, presenting a short to the loop couplers. This is reflected as a short circuit to the D and E junctions, preventing any of the Tx signal to appear at the receiver. Furthermore, the short circuit at point D (or E) appears as an open circuit at point C a quarter-wavelength away (or $\frac{3}{4}\lambda$ for point E). Thus the receiving arm at the antenna coupling junction, point C, appears as an open and the entire Tx signal is directed toward

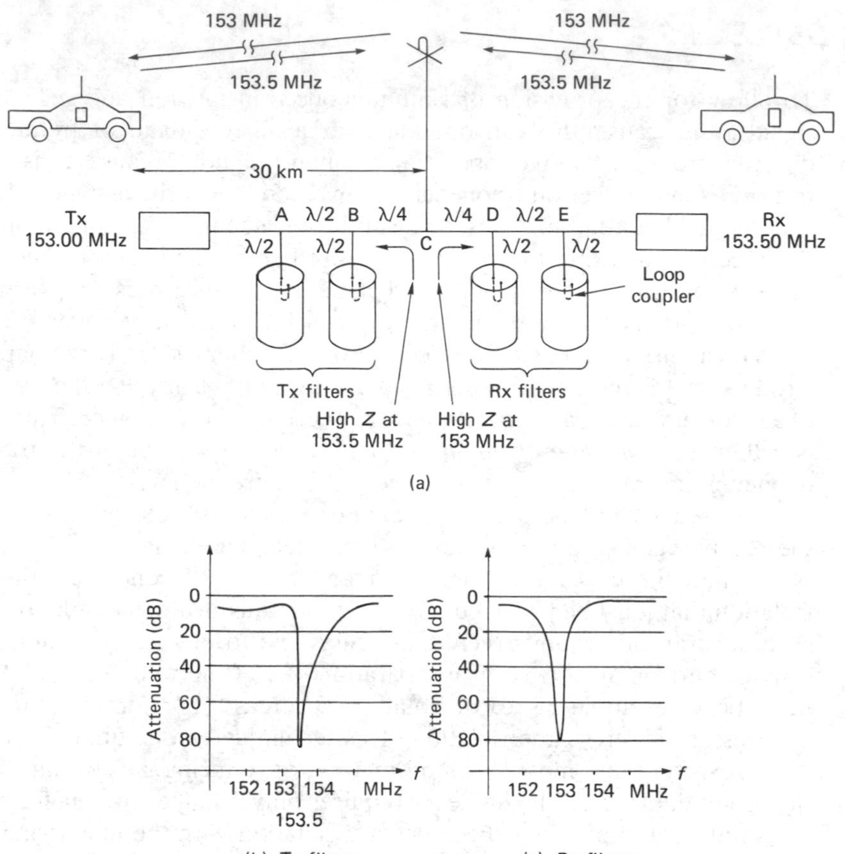

Figure 2-45 Mobile repeater station employing a duplexer (a) having transmitter (b) and receiver (c) band rejection characteristics as shown.

the antenna. The band-reject filters in the receiver section notch out the transmitter energy, thereby preventing receiver desensitization.

Any transmitter noise energy that appears at and near the receive frequency of 153.5 MHz is rejected by the Tx band-reject filters, because the Tx filters present a short circuit to junctions A and B at their 153.5-MHz resonant frequencies. The net result is a good match between the Tx and antenna at the Tx frequency, with an insertion loss of typically less than 1 dB. Similar action occurs when receiving—only then all of the signal from the antenna at 153.5 MHz is directed to the receiver and not to the transmitter. The detailed explanation for this will not be given here but would be based on the same analysis methods as those used for the transmitter frequency.

Problems

2-1. Convert the series circuits of Fig. P2-1 to their parallel equivalents.

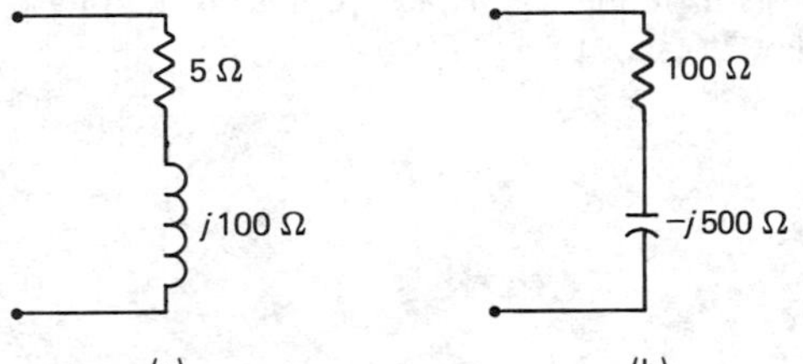

Figure P2-1.

2-2. Convert the parallel circuits of Fig. P2-2 to their series equivalents.

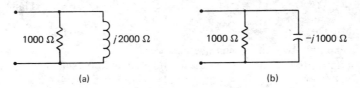

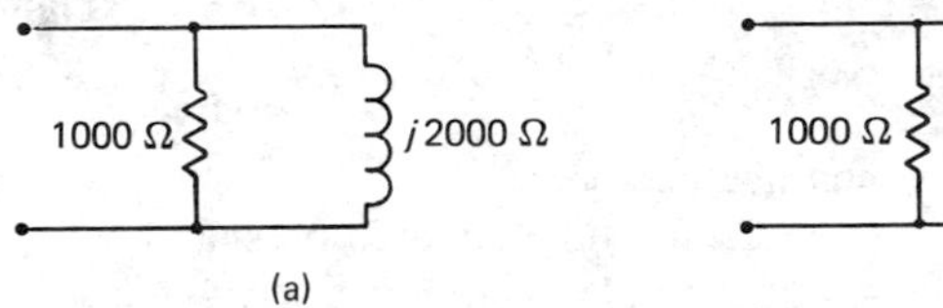

Figure P2-2.

2-3. For the series *LRC* circuit of Fig. P2-3, find:
- **(a)** Resonant frequency
- **(b)** Series impedance at resonance
- **(c)** Circuit Q
- **(d)** Circuit bandwidth
- **(e)** Voltage across R, L, and C at resonance

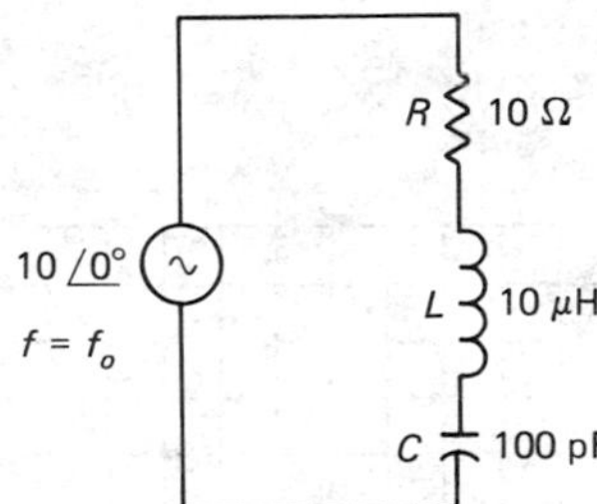

Figure P2-3.

2-4. For the parallel *LRC* circuit of Fig. P2-4, find:
- **(a)** Parallel resonant frequency
- **(b)** Parallel impedance at resonance
- **(c)** Circuit Q
- **(d)** Circuit bandwidth
- **(e)** Current through R_p and L at resonance

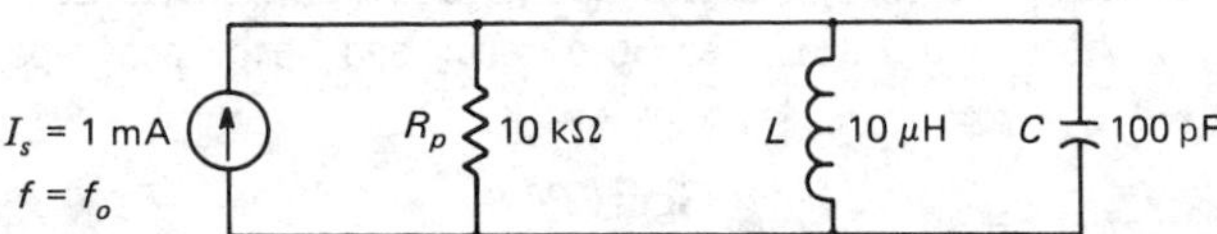

Figure P2-4.

2-5. If the inductor in the parallel circuit of Problem 2-4 has a resistance of 10 Ω, find:
(a) Resonant frequency at which the circuit is purely resistive
(b) Parallel impedance at resonance

2-6. Sketch the symbol used for the **(a)** low-pass, **(b)** bandpass, and **(c)** high-pass filter.

2-7. A delayed sinusoid can be expressed as $v = A \sin(\omega t - \beta)$. If all sinusoidal signals are to experience the same time delay as represented in Fig. P2-7, how should the phase shift, β, vary with frequency? *Hint:* Represent the phase delay in terms of an equivalent time delay T_p (i.e., $\beta = \omega T_p$).

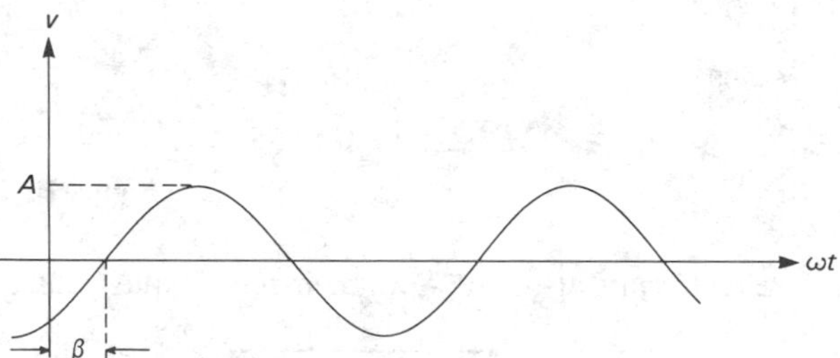

Figure P2-7.

2-8. Match the filter type to the characteristic.

(1) Butterworth	__ flat time delay in passband
(2) Chebyshev	__ ripple in stopband
(3) Bessel	__ flat amplitude response in passband
(4) Elliptic	__ ripple in passband only

2-9. From the curve in Fig. P2-9 determine:
(a) Insertion loss
(b) Ripple (I)
(c) Shape factor
(d) Maximum stopband rejection
(e) Minimum stopband rejection
(f) High-frequency end selectivity

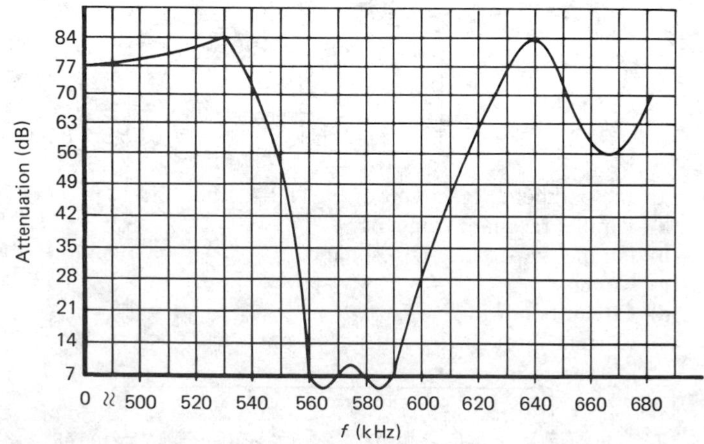

Figure P2-9.

2-10. The graph shown in Fig. P2-10 was obtained from a practical filter. Determine:
- **(a)** BW_{6dB}
- **(b)** Q
- **(c)** Shape factor
- **(d)** Insertion loss
- **(e)** Stopband rejection

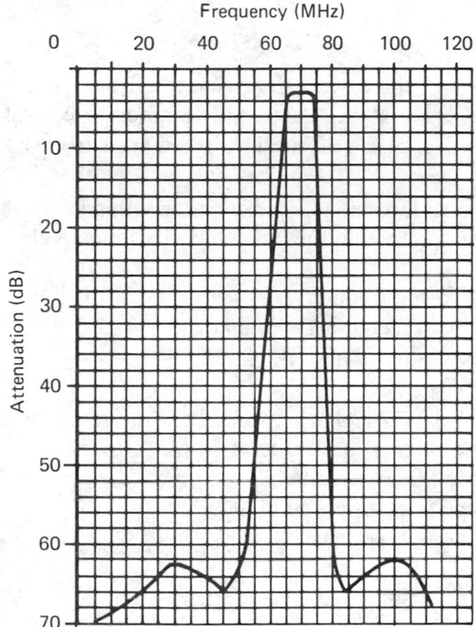

Figure P2-10.

2-11. The BW_{-6dB} of a filter is 10 kHz. The BW_{-60dB} for the same filter is 22 kHz. The upper and lower skirts are symmetrical about the center frequency. For this filter, calculate:
- **(a)** Shape factor
- **(b)** Upper and lower skirt selectivity

2-12. What is the purpose of wavetraps in communication circuits? Sketch a simple wavetrap circuit.

2-13. For the doubly tuned paralleled inductively coupled circuit of Fig. P2-13:
- **(a)** Find the resonant frequencies of the primary and secondary circuits.
- **(b)** Determine if the circuit is critically coupled, undercoupled, overcoupled, or optimally coupled.
- **(c)** Find V_o at resonance when $E_g = 10$ V.
- **(d)** Find $|V_o|$ at resonance if the inductor spacing is reduced to obtain critical coupling.

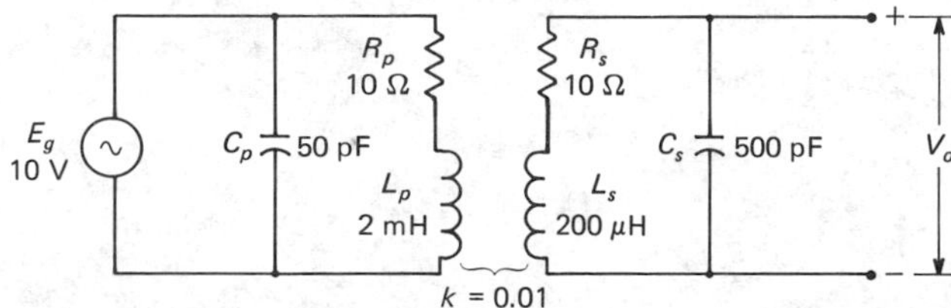

Figure P2-13.

2-14. What is the purpose of stagger tuning in some IF amplifiers?

2-15. Explain why the receiver is directly coupled to the antenna through the duplexer of Fig. 2-45.

2-16. For an ideal bandpass filter the shape factor should be:
(1) 0
(2) 1
(3) 2
(4) 100
(5) 36-24-36

2-17. Match the filter type with the correct filter characteristics.

(1) *LC* filter	___ extremely temperature stable
(2) Ceramic filter	___ used at frequencies from 1 kHz to 100 MHz
(3) Mechanical filter	___ utilizes the piezoelectric effect

2-18. In order that a filter introduces no phase distortion within its passband, the filter must have linear phase response within the passband. This means that phase angle within the passband must:
(1) Increase with increasing frequency
(2) Remain constant with increasing frequency
(3) Decrease with increasing frequency
(4) Phase out to nothing

2-19. Two frequencies, $f_1 = 1$ kHz and $f_2 = 3$ kHz, are in the passband of an *ideal* filter. In passing through the filter, f_1 undergoes a 12-μs delay.
(a) What time delay is experienced by f_2?
(b) What phase change is experienced by f_1?
(c) What phase change is experienced by f_2?

Matching Circuits

William Sinnema

3-1 IMPEDANCE MATCHING

When connecting a transmitter to an antenna or a driver to a power amplifier, it is important to ensure that as little loss as possible occurs within the interconnecting network and that the maximum amount of power is transferred. As slight system nonlinearities result in the production of harmonics, it is also important that these be attenuated by the network. In brief, when interconnecting power stages or connecting power stages to a load, the connecting network must provide (1) maximum power transfer, (2) minimum insertion loss, and (3) filtering of the harmonic frequencies.

To achieve each of these ends, low-loss matching circuits are employed. The low loss is obtained by using only reactive components in the matching network, and filtering is obtained by ensuring that the network is of a low-pass type with a sufficiently high Q. It is common practice to use an unloaded Q of about 10 for the matching circuit with load. When combined with the source or driving impedance, the unloaded Q is effectively halved and is known as the loaded Q.

The most commonly used matching circuit forms are shown in Fig. 3-1. Tuned transformers have already been considered in Section 2-11, and will not be discussed further here. The down-transformation circuit transforms the load R_L into a smaller value as seen by the input terminals. The up-transformation circuit [Fig. 3-1(c)] transforms the load R_L into a larger value.

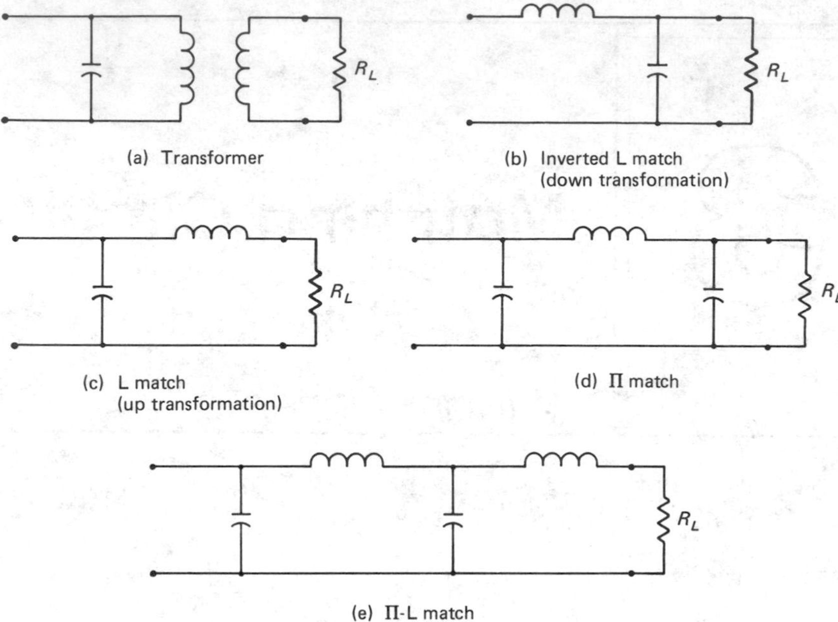

Figure 3-1 Low-pass impedance-matching networks.

In our analysis of these matching networks, we rely heavily on the expressions derived in Sections 2-2 and 2-3: the quality-factor expressions and the series-to-parallel conversions. In particular, we will use the relations (2-3), (2-4), and (2-8), which are rewritten here:

$$Q = \frac{X_s}{R_s}$$

X_s

R_s

$$Q = \frac{R_p}{X_p}$$

X_p R_p

$$R_p = (Q^2 + 1)R_s$$

To aid us in comparing the responses of these matching low-pass filter networks, we will in all cases, excepting for the inverted L circuit, match a 50-Ω load to a 3000-Ω driving source impedance at a frequency of 1 MHz. This is illustrated in Fig. 3-2.

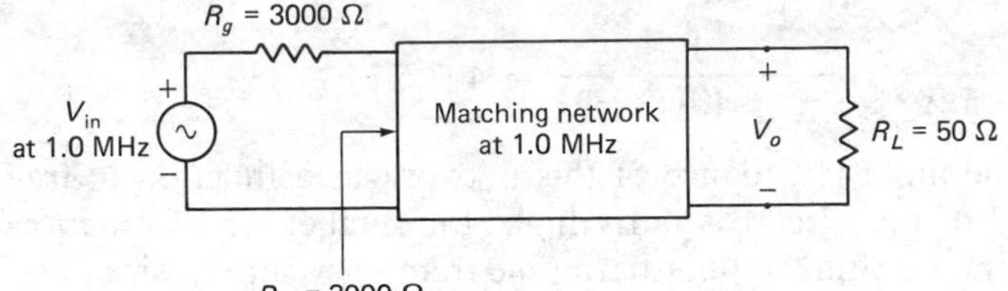

Figure 3-2 Matching conditions used in examples.

3-1.1 Inverted L Match

The inverted L match of Fig. 3-1(b) provides a down transformation, as will be shown when the parallel circuit of Fig. 3-3(a) is transformed into its series equivalent. Consider the design of an inverted L match for transforming a load of $R_L = 3000\ \Omega$ to an input resistance of $R_{in} = 50\ \Omega$ at 1 MHz.

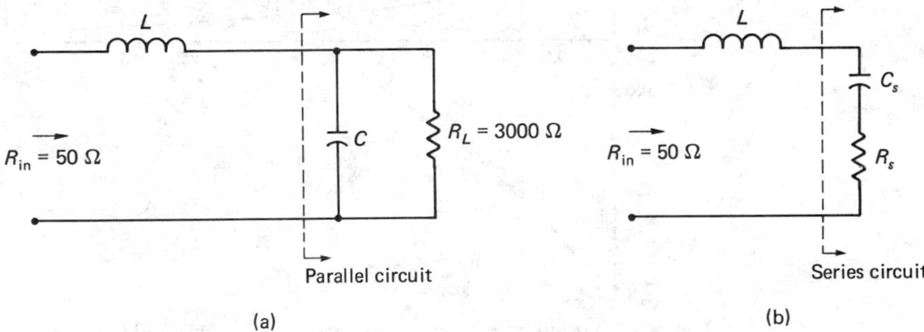

Figure 3-3 (a) Inverted L nctwork and (b) series equivalent.

Using the parallel-to-series conversion expression, we have

$$R_s = \frac{R_L}{Q^2 + 1} = 50 = \frac{3000}{Q^2 + 1}$$

$$Q = 7.7 \text{ for this network}$$

This means that

$$X_L = X_{cs} = QR_s$$

$$= 7.7 \times 50 = 384\ \Omega$$

Thus

$$L = \frac{X_L}{2\pi f} = \frac{384}{2\pi \times 10^6} = 61\ \mu\text{H}$$

As the parallel circuit must have the same Q as the series circuit,

$$X_c = \frac{R_L}{Q} = \frac{3000}{7.7} = 391\ \Omega$$

Therefore,

$$C = \frac{1}{2\pi f X_c} = \frac{1}{2\pi \times 10^6 \times 391} = 407.5 \text{ pF}$$

Figure 3-4 shows the input impedance of this network as a function of frequency. Although the Q of this circuit is fairly high, for smaller transformation ratios, R_p/R_s, the Q suffers, resulting in poor harmonic frequency suppression. For this reason, the Π and Π–L circuits are used more commonly.

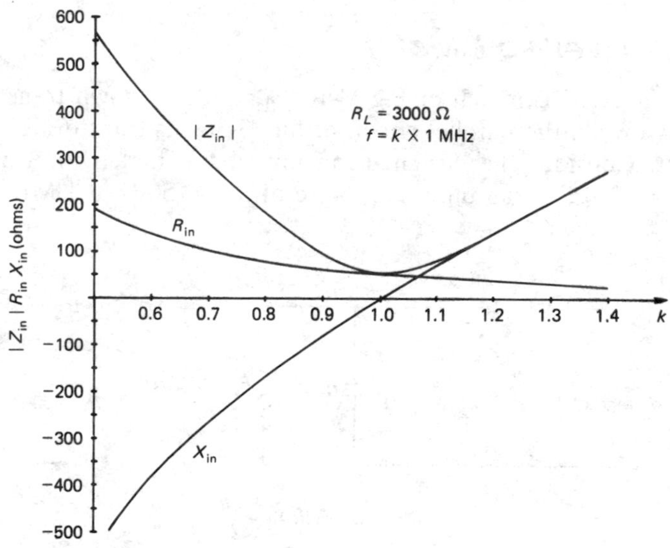

Figure 3-4 Input impedance characteristics of an inverted L network.

3-1.2 L Match

Consider the design of the L match that transforms a load of $R_L = 50\ \Omega$ to an input resistance of $R_{in} = 3000\ \Omega$ at 1 MHz. Figure 3-5 illustrates the procedure, which is similar to that of the preceding example. Using the series-to-parallel conversion expression,

$$R_p = (Q^2 + 1)R_L$$

$$3000 = (Q^2 + 1)50$$

$$Q = 7.7$$

This means that

$$X_{Lp} = X_c = \frac{R_p}{Q}$$

$$= \frac{3000}{7.7} = 391\ \Omega$$

Thus

$$C = \frac{1}{2\pi f X_c} = \frac{1}{2\pi \times 10^6 \times 391}$$

$$= 407.5 \text{ pF}$$

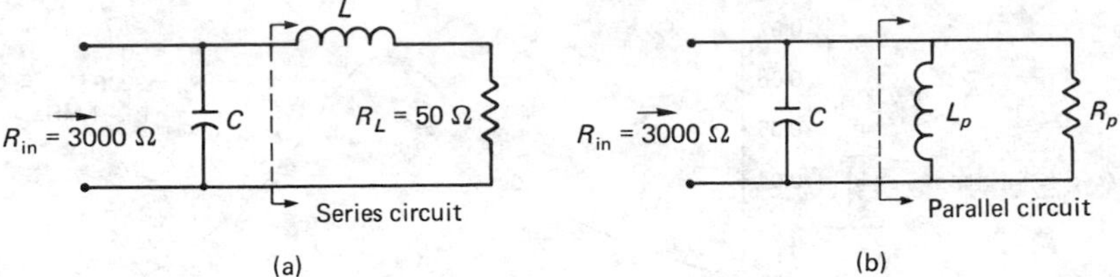

Figure 3-5 (a) L networks and (b) series equivalent.

Since the series circuit must have the same Q as the parallel circuit,

$$X_L = QR_L = 7.7 \times 50 = 384\ \Omega$$

$$L = \frac{X_L}{2\pi f} = \frac{384}{2\pi \times 10^6} = 61.1\ \mu\text{H}$$

Figure 3-6 shows the input impedance of this network as a function of frequency. Note that at the resonant point, the input impedance is real and equal to 3000 Ω.

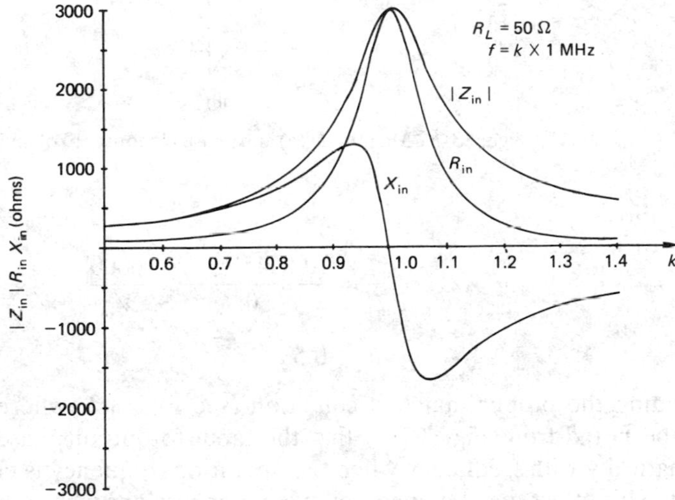

Figure 3-6 Input impedance characteristics of an L match.

The magnitude and phase response of the L-match network is given in Fig. 3-7. At the resonance point, the power delivered by a 3000-Ω source is given by

$$P_{in} = \frac{V_{in}^2}{6000}$$

since the input impedance of the L match is 3000 Ω. The power delivered to the 50-Ω load from Fig. 3-7(a) is

$$P_{out} = \frac{V_{out}^2}{R_L} = \frac{(0.0645V_{in})^2}{50}$$

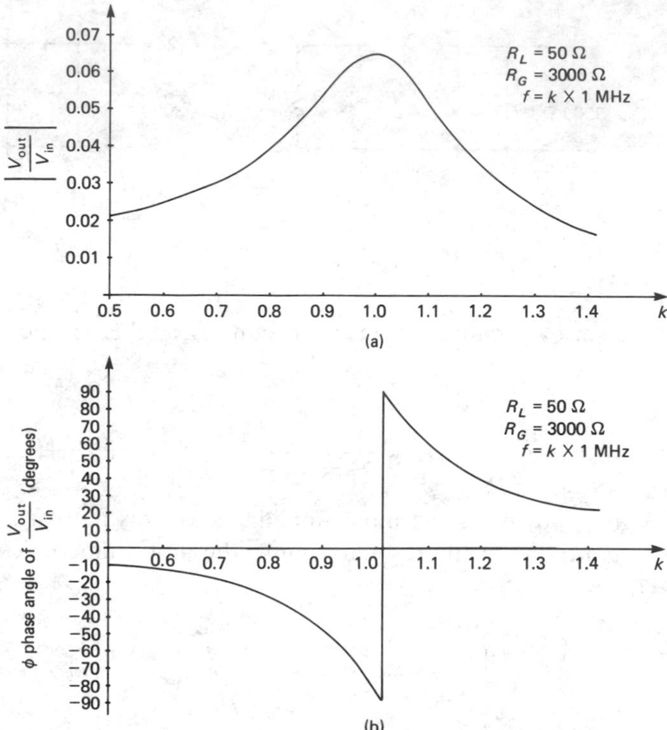

Figure 3-7 Magnitude (a) and phase response (b) of an L match.

The ratio

$$\frac{P_{\text{out}}}{P_{\text{in}}} = \frac{(0.0645V_{\text{in}})^2}{50} \times \frac{6000}{V_{\text{in}}^2}$$

$$= 0.5$$

indicating the proper matched condition (see Fig. 3-2, where V_{in} is defined). It may be noted from Fig. 3-7(b) that the input-to-output phase shift changes very dramatically with frequency when the operating frequency is close to the resonant point. This effect may be unacceptable in applications where the load being inpedance transformed by the L section is phase sensitive. An example would be a case where the load to be matched is an antenna array (see Chapter 15 for a description of how signal phase affects the radiation pattern of an array of antennas).

Although the L section offers a significant improvement over the simple series of parallel resonant sections in allowing impedance transformation, one significant limitation exists. With the transformation ratio signal to $(1 + Q^2)$, low transformation ratios automatically imply low Q. Specifying one immediately sets the other! For example, matching a 100-Ω load to a 50-Ω source requires a Q of 1. Such a low Q value is intolerable in communications applications where harmonic filtering is required. One solution to this dilemma involves using the cascade connection of an L section and an inverted L section to form a Π section. One section realizes high network Q and the other is used to set the required overall transformation ratio. In our analysis the first section will be designed to have a sufficiently large Q to filter off undesired harmonic frequencies. The second section will then be used to shift the impedance such that the overall desired impedance transformation is obtained.

3-1.3 Π Match

To design the Π match of Fig. 3-1(d) we break the network up into two sections as shown in Fig. 3-8: an up-transformation section to obtain the required Q, and a down-transformation section to satisfy the required resistive transformation. The R_m or R_{midpoint} is a fictitious resistance value which is used as a crutch in the derivation of the L and C values. We will consider the design of a Π network that provides an impedance match of $R_L = 50\ \Omega$ to a source resistance of 3000 Ω while maintaining an unloaded Q of 10 or a loaded Q of 5.

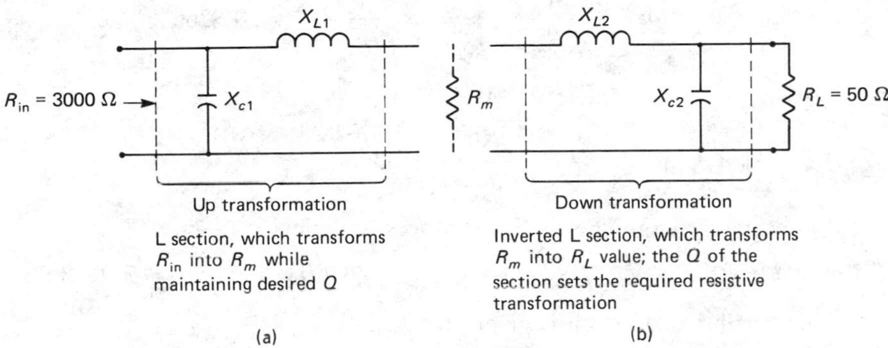

Figure 3-8 Formation on the Π matching network.

Input Section (L Section)

Q = 10
X_{L1}
X_{c1}
R_m
R_{in} = 3000 Ω
Series circuit
X_{c1}
X_{Lp}
R_p
R_{in} = 3000 Ω
Parallel circuit

$$\textit{Step 1. } X_{c1} = \frac{R_p}{Q} = \frac{3000}{10} = 300\ \Omega$$

$$\textit{Step 2. } R_m = \frac{R_p}{1 + Q^2} = \frac{3000}{1 + 10^2} = 29.703\ \Omega$$

$$X_{L1} = QR_m = 10 \times 29.7 = 297.03\ \Omega$$

Output Section (Inverted L Section) Having a Quality Factor Q_2

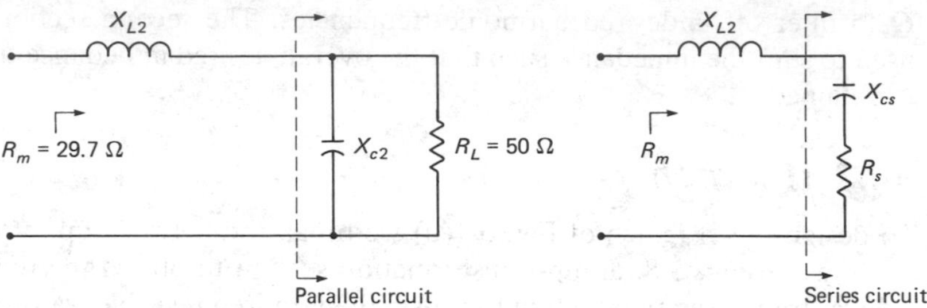

Step 3. $R_s = \dfrac{R_L}{1 + Q_2^2} = R_m$

$$Q_2 = \sqrt{\frac{R_L}{R_m} - 1} = \sqrt{\frac{50}{29.7} - 1} = 0.8266$$

$$X_{L2} = Q_2 R_s = (0.8266)(29.7) = 24.55\ \Omega$$

Step 4. $X_{c2} = \dfrac{R_L}{Q_2} = \dfrac{50}{0.8266} = 60.49\ \Omega$

Step 5. Combine the inductances in series:

$$X_L = X_{L1} + X_{L2} = 297.03 + 24.55 = 321.58\ \Omega$$

The resultant Π match is given in Fig. 3-9.

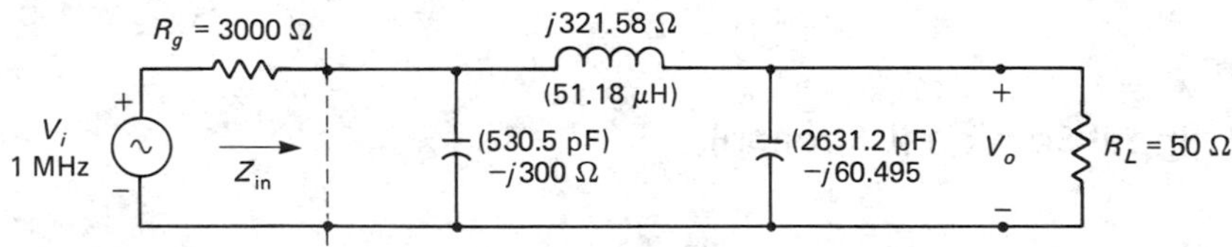

Figure 3-9 Π matching network.

Figure 3-10 shows the input impedance characteristics of the Π matching network of Fig. 3-9 as a function of frequency, and Fig. 3-11 shows the transfer response. All of these curves are plotted at frequencies near the operating frequency. To observe how the low-pass filter characteristics of the filter suppress harmonics, the voltage transfer response in decibels is plotted over a much larger frequency range. This is illustrated in Fig. 3-12.

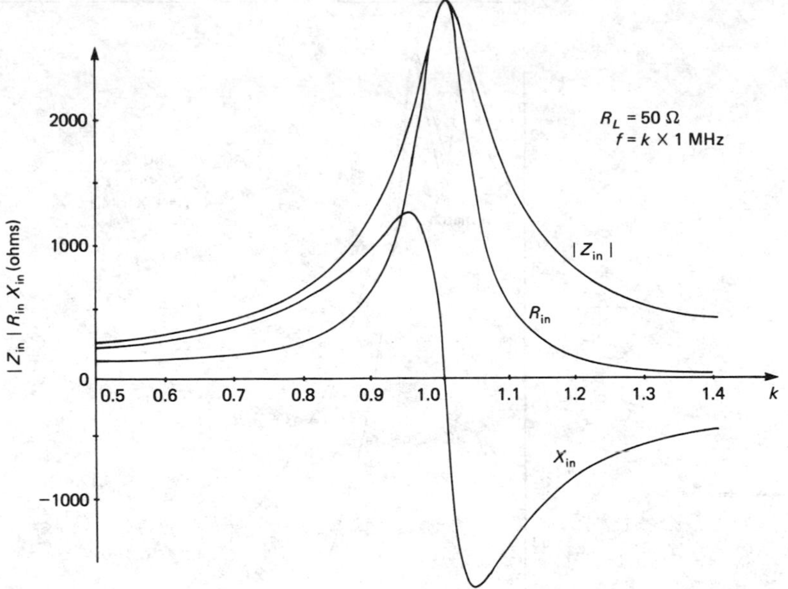

Figure 3-10 Input impedance characteristics of the Π match of Figure 3-9.

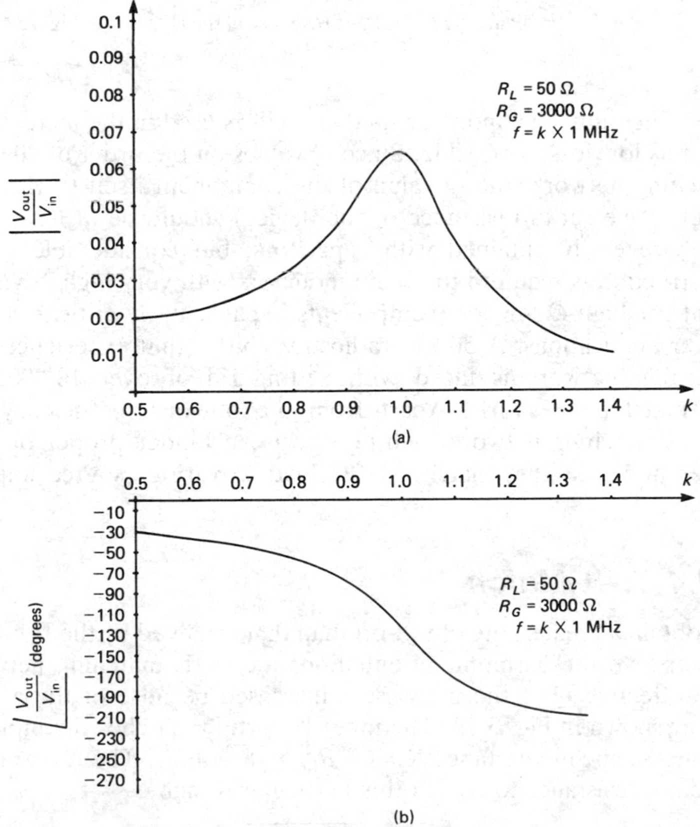

Figure 3-11 Magnitude (a) and phase response (b) of the Π match.

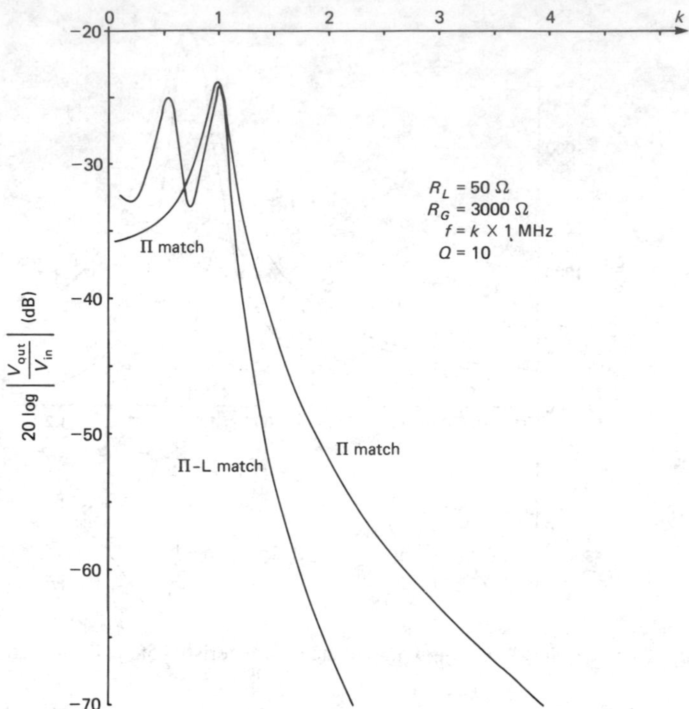

Figure 3-12 Voltage response of the Π match of Figure 3-9 and Π–L match of Figure 3-15.

The individual inductors and capacitors used in the matching network should be of as low loss as possible. Since Q values on the order of 10 are required in the matching network, the Q value of the component itself (series reactance divided by effective series resistance of the device) should be at least 100. Q values this high are easily obtained with capacitors, but considerable care in design and construction is required to obtain inductors with very high Q values. The requirement for high-Q low-loss components is particularly critical in high-power transmitter applications. A 50-kW radio transmitter that experiences a 1 dB loss in its matching network is faced with getting rid of over 10,000 W of heat from the matching network. Apart from the obvious inefficiency, difficulties such as the matching network melting down can hinder proper operation. For some other networks that match a 50-Ω load to various source impedances, refer to Table 3-1.

3-1.4 Π–*L* Match

To obtain a greater rate of roll-off than that achieved by the Π network and thereby to improve the harmonic attenuation, the Π–L matching network is employed. The effective Q of the network is increased because of the additional reactance L_2, as shown in Fig. 3-13. The input Π portion matches the input resistance to the geometric mean resistance $R = \sqrt{R_{in} \times R_L}$ ohms. The X_{L2} transforms this intermediate resistance to R_L. In this particular design,

$$R = \sqrt{3000 \times 50} = 387.398\ \Omega$$

TABLE 3-1 Π Networks That Match a 50-Ω Load to Various Source Impedances[a]

Q	R_{in}	X_1	X_2	X_3	X_4
		Π-Network Values (Ω)			
8	3000	−375.00	382.5542	−173.2051	
8	2750	−343.75	356.5016	−117.2604	
8	2500	−312.50	328.7586	−91.2871	
8	2250	−281.25	300.0000	−75.0000	
8	2000	−250.00	270.4791	−63.2456	
10	3000	−300.00	321.5834	−60.4858	
10	2750	−275.00	297.1778	−54.6729	
10	2500	−250.00	272.5235	−49.5074	
10	2250	−225.00	247.6236	−44.8211	
10	2000	−200.00	222.4734	−40.4888	
		Π-L Network Values (Ω)			
8	3000	−375.00	494.7104	−72.8549	129.8650
8	2750	−343.75	456.3520	−69.7073	126.6511
8	2500	−312.50	417.7783	−66.3802	123.1977
8	2250	−281.25	378.9629	−62.8462	119.4592
8	2000	−250.00	339.8733	−59.0706	115.3750
10	3000	−300.00	400.0911	−63.8365	129.8650
10	2750	−275.00	368.9983	−60.9351	126.6511
10	2500	−250.00	337.7391	−57.8778	123.1977
10	2250	−225.00	306.2931	−54.6418	119.4592
10	2000	−200.00	274.6345	−51.1982	115.3750

[a]Matching networks studied (all match to 50 Ω).
Source: R. W. Johnson, Response of Π–Π L and tandem quarter wave-line matching networks, *Ham Radio*, February 1982, p. 13.

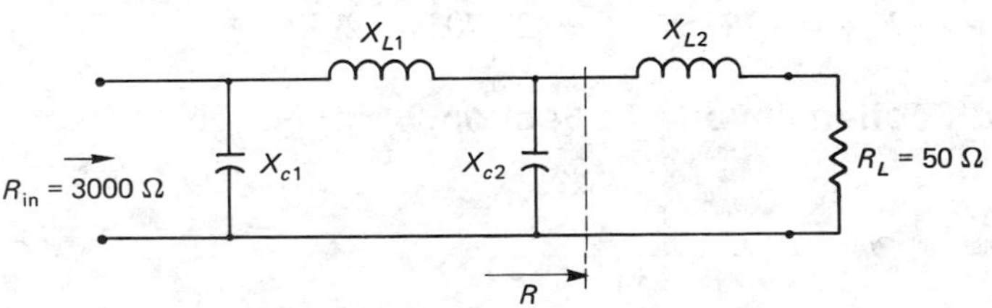

Figure 3-13 Π-L matching network.

Similar to that of the Π match, the network can be broken up into three sections, an up transformation to obtain the desired Q of 10, a down transformation to match the geometric mean resistance R to the midpoint resistance R_m of the Π network, and an up-transformation section to satisfy the resistive transformation from the load to the geometric mean resistance. This is illustrated in Fig. 3-14.

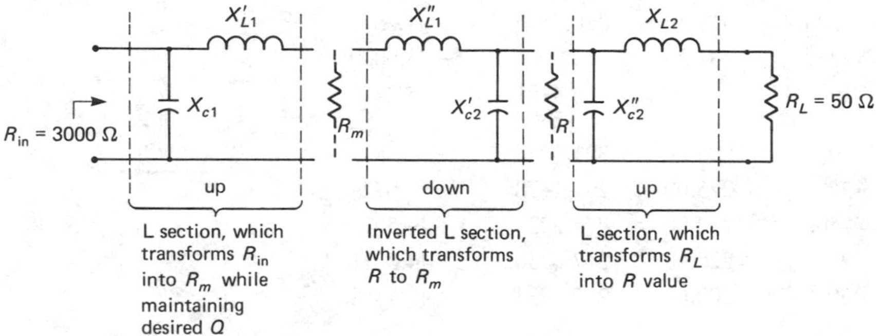

Figure 3-14 Formation of the Π-L matching network.

First Section (L Section)

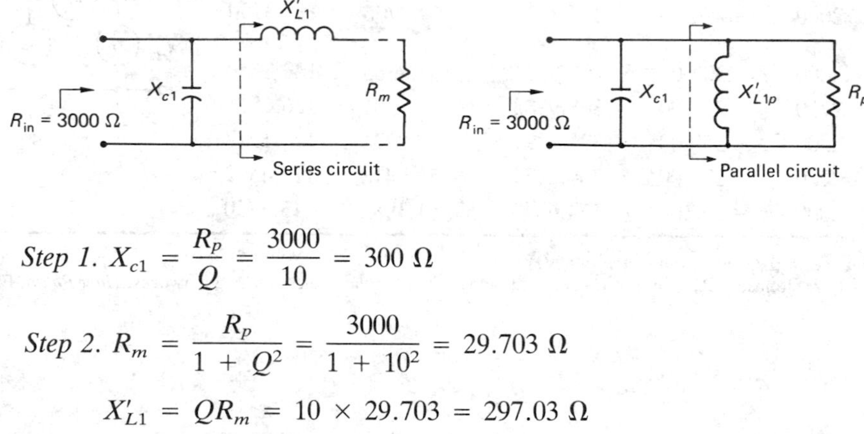

$$\textit{Step 1.}\ X_{c1} = \frac{R_p}{Q} = \frac{3000}{10} = 300\ \Omega$$

$$\textit{Step 2.}\ R_m = \frac{R_p}{1 + Q^2} = \frac{3000}{1 + 10^2} = 29.703\ \Omega$$

$$X'_{L1} = QR_m = 10 \times 29.703 = 297.03\ \Omega$$

Second Section (Inverted L Section)

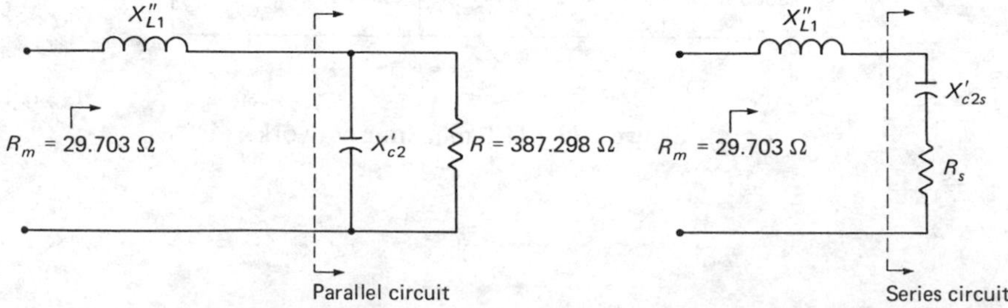

$$\textit{Step 3. } R_s = \frac{R}{Q_2^2 + 1}$$

$$Q_2 = \sqrt{\frac{R}{R_s} - 1} = \sqrt{\frac{387.298}{29.703} - 1} = 3.4697$$

$$X''_{L1} = X'_{c2s} = Q_2 R_s = 3.4697 \times 29.703$$

$$= 103.06\ \Omega$$

Thus

$$X_{L1} = X'_{L1} + X''_{L1} = 297.03 + 103.06$$

$$= 400.09\ \Omega$$

$$\textit{Step 4. } X'_{c2} = \frac{R}{Q_2} = \frac{387.298}{3.4697} = 111.62\ \Omega$$

Third or Output Section (L Section)

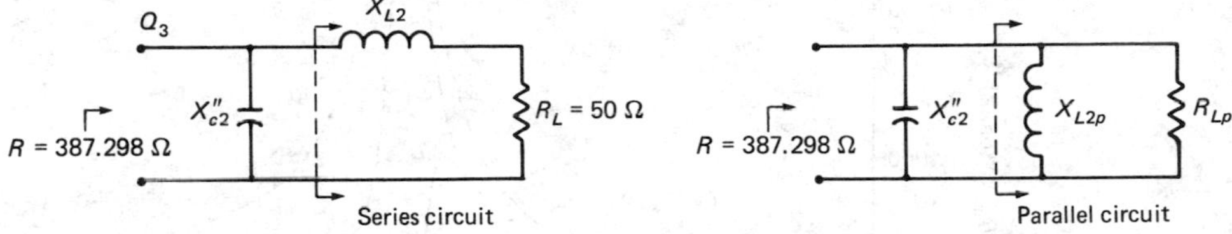

$$\textit{Step 5. } Q_3^2 + 1 = \frac{R_{Lp}}{R_L} = \frac{387.298}{50}$$

$$Q_3 = 2.597$$

$$X''_{c2} = \frac{R_{Lp}}{Q} = \frac{387.298}{2.597} = 149.116\ \Omega$$

Thus

$$X_{c2} = X'_{c2} \parallel X''_{c2} = \frac{111.62 \times 149.116}{111.62 + 149.116}$$

$$= 63.836\ \Omega$$

$$\textit{Step 6. } X_{L2} = Q_3 R_L$$

$$= 2.597 \times 50 = 129.865\ \Omega$$

The resultant Π–L match is given in Fig. 3-15. For several other Π–L networks that match a 50-Ω load to several source impedances, refer to Table 3-1.

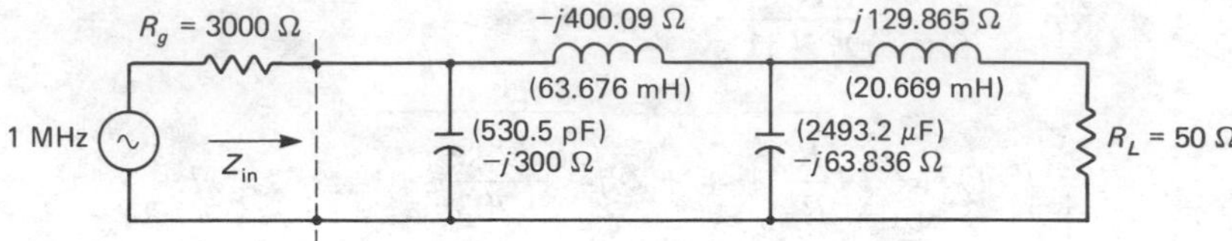

Figure 3-15 Π-L matching network.

Figure 3-16 shows the input impedance characteristics of the Π–L matching network of Fig. 3-15 as a function of frequency around the operating point. When comparing this to the Π match frequency characteristics (Fig. 3-10), we observe that the maximum impedance peaks do not occur at the resonant resistance frequency point for either case. The effective Q of the Π–L network is higher for the same design Q in the Π–L than for the Π, as can be seen from the sharper resonant peak. Harmonic suppression is also improved by the Π–L match, as illustrated in Fig. 3-12.

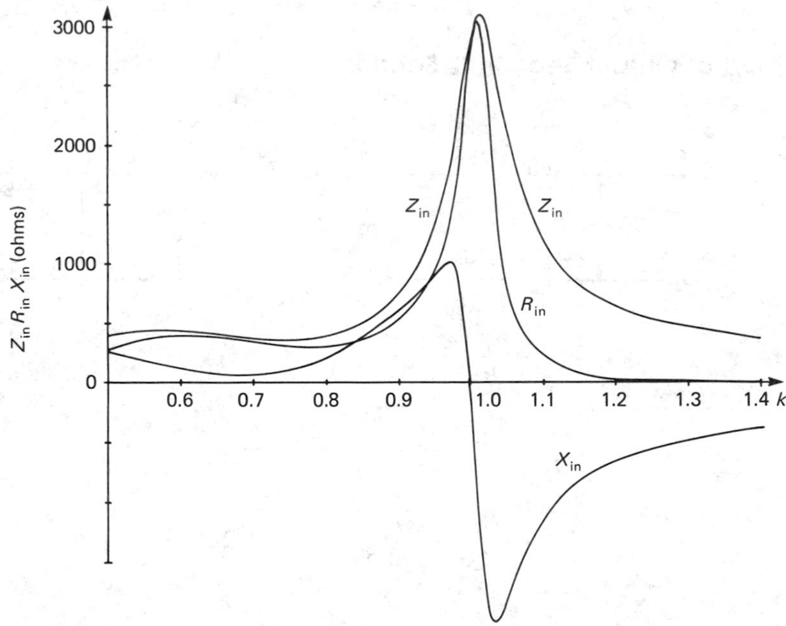

Figure 3-16 Input impedance characteristics of the Π–L match of Figure 3-15.

3-2 TAPPED CIRCUITS

To provide an impedance match between one electronic circuit and another so that an improved transfer of power occurs, or to prevent the loading down of a tuned circuit which results in a reduced Q, a tapped inductive or capacitive circuit is often

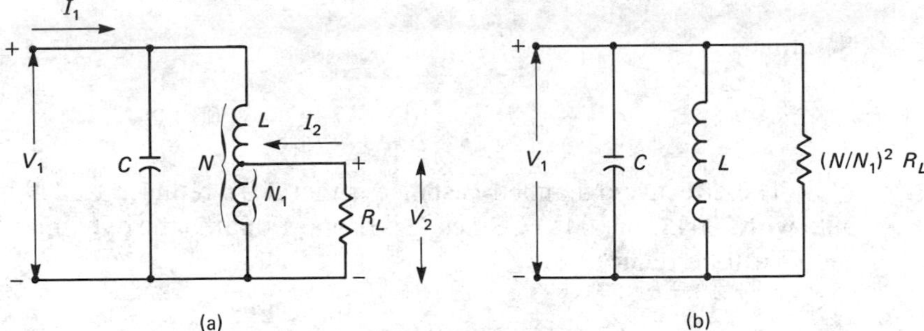

Figure 3-17 Tapped coil as a transformer (a) with the load referred to the primary (b).

used. The tapped coil of Fig. 3-17(a) is similar to an autotransformer in which each winding is linked to the same magnetic flux. The voltage division achieved by the tapped inductor is directly proportional to the number of turns. Thus

$$V_2 = \frac{N_1}{N} V_1 \tag{3-1a}$$

or

$$V_1 = \frac{N}{N_1} V_2 \tag{3-1b}$$

If the coil is assumed to be loss-less, the input and output power must be identical. Thus

$$V_1 I_1 = V_2 I_2 \quad \text{and} \quad I_2 = \frac{V_1}{V_2} I_1 = \frac{N}{N_1} I_1 \tag{3-2a}$$

or

$$I_1 = \frac{N_1}{N} I_2 \tag{3-2b}$$

For a load resistance $R_L = V_2/I_2$, the equivalent resistance reflected back into the input will be

$$\frac{V_1}{I_1} = \frac{N}{N_1} V_2 \times \frac{N}{N_1 I_2} = \left(\frac{N}{N_1}\right)^2 R_L \tag{3-3}$$

This increased resistance reflected across the tuned circuit reduces the loading effect of R_L on the Q of the circuit.

EXAMPLE 3-1

An unloaded tuned circuit has a Q_u of 100 and a dynamic resistance of 200 kΩ. If a 2.0-kΩ load is tapped into the coil at the $N/N_1 = 10$ turns ratio point, find the new or loaded circuit Q.

Solution

$$\left(\frac{N}{N_1}\right)^2 R_L = 10^2 \times 2\ \text{k}\Omega = 200\ \text{k}\Omega$$

The effective dynamic resistance is therefore reduced to 200 kΩ in parallel with 200 kΩ or 100 kΩ. Since Q is proportional to the dynamic resistance, the Q will be reduced to

$$Q_{\text{new}} = 100 \times \frac{100\ \text{k}\Omega}{200\ \text{k}\Omega} = 50$$

This Q is still significantly higher than the Q would have been if the 2 kΩ were placed directly across the tank circuit. The Q in the later case would have been

$$Q = 100 \times \frac{2\ \text{k}\Omega}{200\ \text{k}\Omega} = 1$$

EXAMPLE 3-2

A tuned transformer having a primary winding inductance of 50 μH and an unloaded Q_u of 50 is attached to two loads, as illustrated in Fig. 3-18. Obtain **(a)** the resonant frequency, **(b)** the loaded Q, and **(c)** the 3-dB bandwidth. Assume that Q is sufficiently large enough so that equation (2-19) holds.

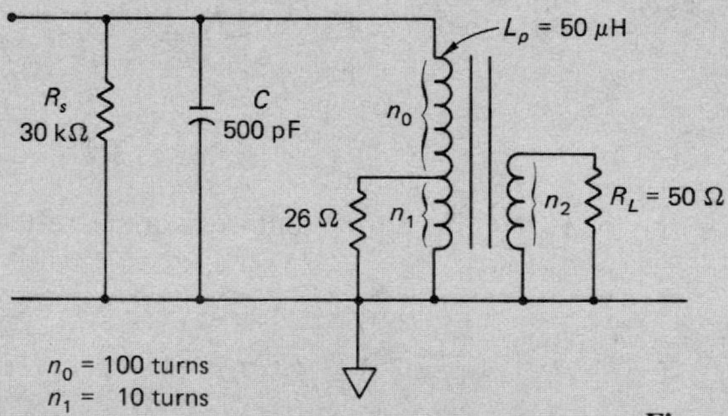

Figure 3-18 Circuit for Example 3-2.

Solution

(a) $f_o = \dfrac{1}{2\pi\sqrt{LC}} = \dfrac{1}{2\pi\sqrt{50 \times 10^6 \times 500 \times 10^{-12}}} = 1\ \text{MHz}$

(b) The dynamic resistance of the primary coil is

$$R_{\text{coil}} = Q_u X_L = 50(2\pi \times 10^6 \times 50 \times 10^{-6}) = 15.7\ \text{k}\Omega$$

The 26-Ω load reflected up to the top of the tank is

$$R_1 = 26\left(\frac{n_0 + n_1}{n_1}\right)^2 = 26\left(\frac{110}{10}\right)^2 = 3.15\ \text{k}\Omega$$

The 50-Ω load reflected up to the top of the tank is

$$R_2 = 50\left(\frac{n_0 + n_1}{n_2}\right)^2 = 50\left(\frac{110}{5}\right)^2 = 24.2\ \text{k}\Omega$$

The net loading across the tank then is

$$R = R_{\text{coil}} \parallel R_1 \parallel R_2 \parallel R_s$$

$$G = \frac{1}{R} = \frac{1}{15.7\ \text{k}\Omega} + \frac{1}{3.15\ \text{k}\Omega} + \frac{1}{24.2\ \text{k}\Omega} + \frac{1}{30\ \text{k}\Omega} = 4.55 \times 10^{-4}\ \text{S}$$

$$R = 2.2\ \text{k}\Omega$$

Therefore,

$$Q = \frac{R}{X_L} = \frac{2.2 \times 10^3}{2\pi \times 10^6 \times 50 \times 10^{-6}} = 7$$

(c) $\text{BW} = \dfrac{f_o}{Q} = \dfrac{10^6}{7} = 142\ \text{kHz}$

When the loading effect of a source on a tuned circuit must be reduced, the source is tapped into the coil as illustrated in Fig. 3-19(a). As outlined previously, the apparent resistance placed across the tuned circuit wil be stepped up from the generator resistance value by a factor of $(N/N_1)^2$. The source voltage will be stepped up by the factor N/N_1, as shown in Fig. 3-19(b).

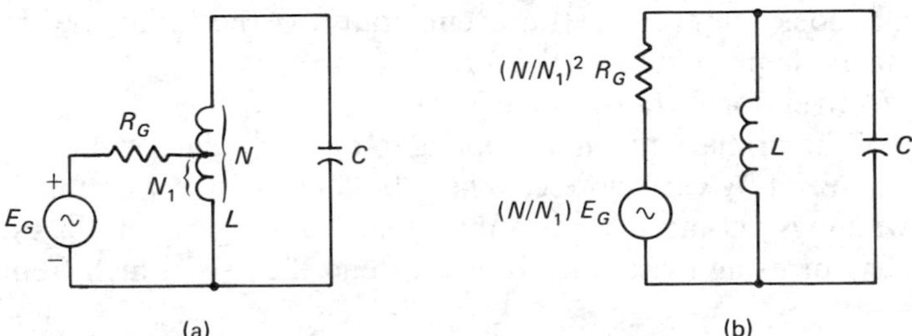

Figure 3-19 Low-impedance source tapped into a coil (a) with the source referred to the secondary (b).

The capacitor tap illustrated in Fig. 3-20(a) can also be used to couple a local resistance to a tank circuit to reduce the loading effect of R_L. For $R_L >> \omega C_2$, the capacitor tap referred to the primary appears as shown in Fig. 3-20(b). The price paid for the reduced loading on the tuned circuit is a reduced output voltage of

$$V_2 = V_1 \frac{C_1}{C_1 + C_2} \tag{3-4}$$

If $R_L/\omega C_2$ is not greater than about 5, the approximate expressions shown in Fig. 3-20 are not valid and more detailed circuit analysis is required.

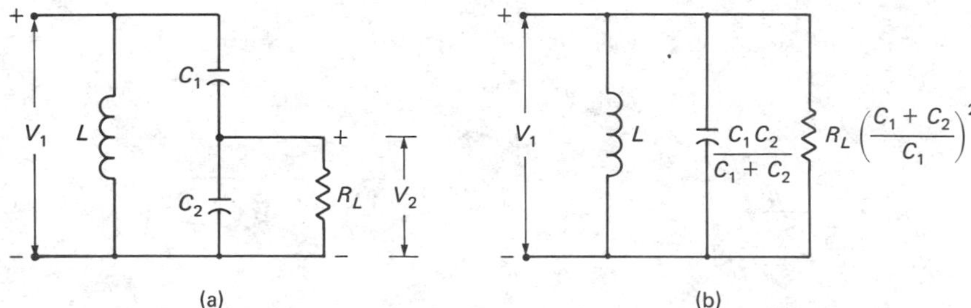

Figure 3-20 Capacitor tap (a) with the load referred to the input (b).

Capacitive coupling is not commonly used in oscillators and medium- to high-power amplifiers, as it readily couples to the output the high-order distortion products that are present in such applications. The auto transformer couples power from the primary to the secondary via a high-permeability core and care must be taken to ensure that the core does not go into the saturation region but that the flux density remains within the linear portion of the $B-H$ curve.

To keep the core losses to a minimum, high-permeability and low-loss materials with narrow $B-H$ curves are used, as the area inside the $B-H$ curve represents the power lost within the core. Ferrites composed of iron, nickel, and zinc are most commonly used for the HF/VHF frequency bands. Core losses, and a portion of the winding losses, heat the core, and precautions must be taken so that the core does not reach the Curie temperature of the ferrite, permanently destroying the magnetic properties of the ferrite.

Conventional transformers suffer from poor high-frequency response due to some of the primary flux not coupling to the secondary windings. Flux leakage can be minimized by very close coupling, both capacitively and magnetically between the windings. Using a copper tube that fits snuggly within the core holes as a primary, or using twisted leads, can extend the useful high-frequency limit.

3-3 TRANSMISSION-LINE TRANSFORMERS

To overcome the saturation and leakage problems encountered with the conventional power RF transformer, the transmission-line transformer has been designed. Transmission-line transformers are very broadbanded and can operate up to several hundred megahertz. As we will see, they can only realize impedance transformation ratios of $(n + 1)^2$, where n is an integer. Their low-frequency performance, however, is poorer than the conventional transformer.

As illustrated in Fig. 3-21(a), the conventional RF transformer transfers power from the primary to the secondary winding via magnetic coupling through the ferrite core. That is, a magnetic field is established by the primary current, and this resultant magnetic flux couples to the secondary winding, setting up an induced voltage in the secondary winding.

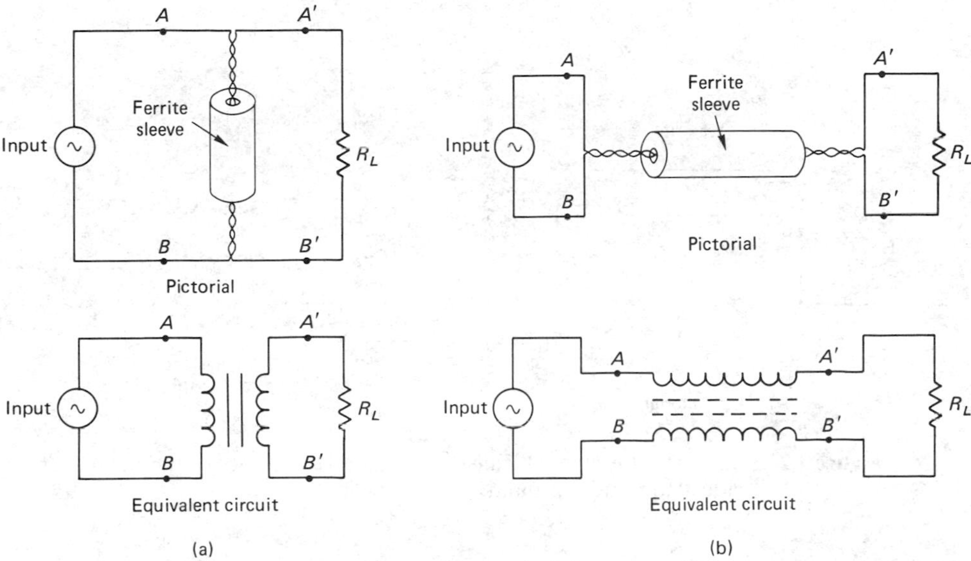

Figure 3-21 Conventional (a) and transmission line (b) transformers.

The transmission-line transformer consists of a transmission line passing through a ferrite core as illustrated in Fig. 3-21(b). Since equal and opposite currents flow in each conductor, the net magnetizing ampere-turns in the core are zero and no magnetic field exists. Thus saturation is not a problem and only a small-cross-section ferrite core is required.

Furthermore, the two conductors form a balanced transmission line, and the coupling from input to output is the result of transmission-line action, eliminating the problem of leakage flux lines. Power is transferred from the input to the output through the transmission line.

Any unbalanced currents or common-mode currents are suppressed by the core, as these establish a net magnetic field and therefore result in a large inductive reactance. Various forms of transmission-line transformers using toroids, sleeves, and twisted or coaxial transmission lines are depicted in Fig. 3-22.

The line inductance and capacitance which determine the characteristic impedance of the transmission line depend on the wire size, spacing, and the dielectric used in the case of the coaxial line. If twisted-pair wire is used, the characteristic impedance $Z_0 = \sqrt{L/C}$ decreases as the number of turns per unit length increases (Fig. 3-23).

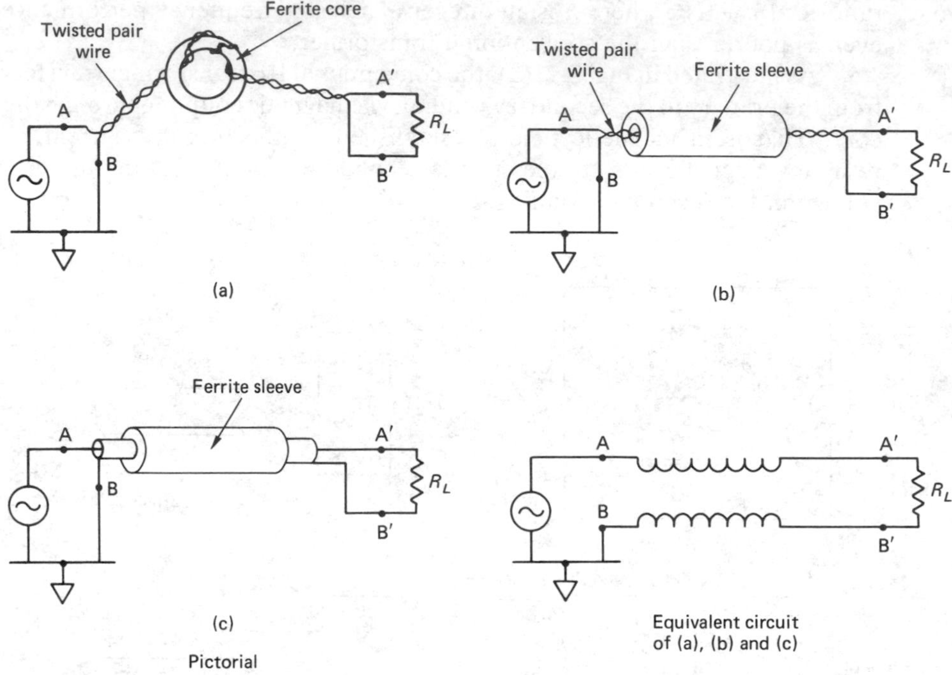

Figure 3-22 Transmission-line 1:1 transformers, used to balance the impedance from each side of the load to ground.

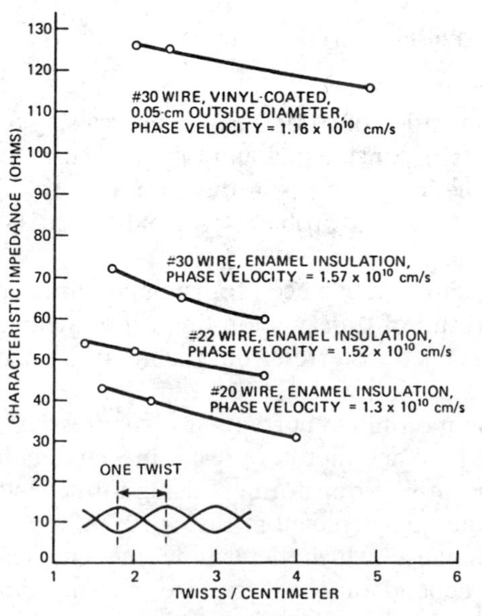

Figure 3-23 Characteristic impedance of a twisted pair transmission line as a function of the number of twists per centimeter. (From H. L. Krauss and C. W. Allen, "Designing toroidal transformers to optimize wideband performance." Reprinted with permission from *Electronics* magazine, copyright © August 16, 1973, VNU Business Publications Inc.)

The concept of a transmission line's characteristic impedance is examined in more detail in Chapter 13. Consider, however, what occurs as energy propagates (at the speed of light) along a transmission line from the source toward the eventual load. There will be a potential developed between the two insulated conductors of the line and current flows along the length of the conductors. The ratio of the potential across the line to the current down the line is called the *characteristic impedance* of the line. Note that the ratio has units of ohms since it is volts divided by amperes. The characteristic impedance is *not*, however, a real resistance. It is simply a ratio of two quantities. As will be described in Chapter 13, it is normally desirable for the eventual load terminating the transmission line to have the same impedance as the characteristic impedance (a matched load).

Transmission-line transformers are usually designed so that the line length is less than a tenth of a wavelength at the maximum intended frequency. The shorter the physical line length, the greater the higher cutoff frequency of the transformer. For more detailed analysis of the transmission-line transformer, refer to Section 6.3 of the book by Jack Smith, *Modern Communication Circuits* (New York: McGraw-Hill Book Company, 1986).

The length of a transmission line in wavelengths is given by

$$\frac{l}{\lambda} = \frac{l}{2\pi/\beta} = lf\sqrt{LC} \tag{3-5}$$

where l is the physical length of the line, L the inductance per unit length of the line, C the capacitance per unit length of the line, f the frequency of operation, and β the phase shift (rad/m) that the current or voltage experiences as it moves down the line.

If the line is matched with its characteristic impedance and is electrically short and loss-less, the input and output voltages are identical as are the input and output currents. Thus for the equivalent circuit of Fig. 3-22, redrawn in Fig. 3-24,

$$I_1 = I_2 = I \tag{3-6}$$

$$V_1 = V_2 = V \tag{3-7}$$

and

$$Z_{\text{in}} = \frac{V_1}{I_1} = \frac{V_2}{I_2} = R_L \tag{3-8}$$

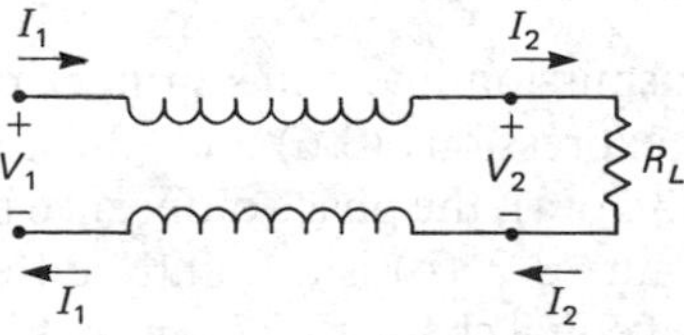

Figure 3-24 A one-to-one impedance transformer.

If the line length becomes an appreciable portion of a wavelength in length, equations (3-6) and (3-7) no longer hold. This limits the high-frequency response of the transmission-line transformer.

As the current flows in opposite directions in the two winding conductors, no magnetic field exists outside the conductors and the core or ferrite sleeve has no effect on the internal magnetic field or the characteristic impedance of the line. That is, the ferrite sleeve has no effect on the differential currents. The ferrite sleeve, however, has a strong suppression effect on common mode currents (Fig. 3-25).

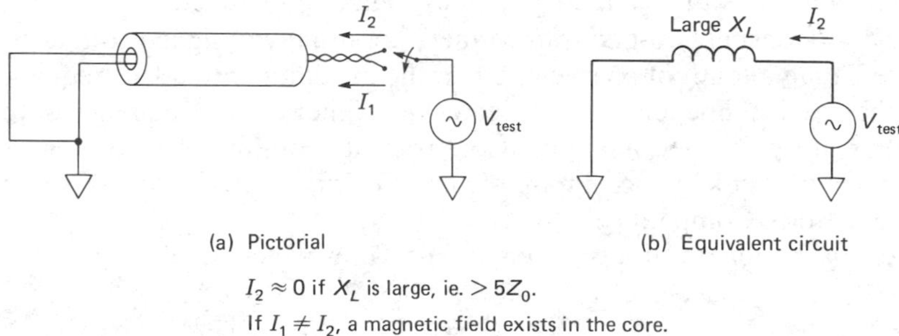

Figure 3-25 Test for common mode current.

Applying a signal to either of the output leads causes a magnetic field to exist outside the conductor. The high permeability ferrite results in an increased inductance L, preventing current from flowing because of the large inductive reactance. Because both output leads have equal (high) impedance to ground, the output is "balanced" and floats above ground. The result is a 1:1 balun (*bal*anced–*un*balanced).

The ferrite sleeve determines the lower cutoff frequency of the balun as the inductive reactance decreases with frequency. The impedance presented to common-mode currents should be greater than five times the load impedance. The energy is coupled from the input to the output through the dielectric medium of the transmission line and not through the core or sleeve.

Step-Up Transformer

The numerous transmission-line transformers can be analyzed by assuming the ideal conditions of expressions (3-6) and (3-7). Consider the step-up transformer of Fig. 3-26. In Fig. 3-26(a), the phase difference between points A and B is assumed to be negligible. Figure 3-26(b) is a slightly different form of Fig. 3-26(a), in that a transmission line of equal characteristic impedance and length as the ferrite loaded line connects A to B. Point B will experience the same phase delay in both cases. The ferrite assures that only a differential current will exist. These transmission-line transformers, called *equal-delay transmission-line transformers*, allow for physical separation of the transformer input and output connections.

If the line is ideal,

$$I_{RL} = 2I \tag{3-9}$$

$$V_{RL} = V \tag{3-10}$$

and

$$R_L = \frac{V_{RL}}{I_{RL}} = \frac{V}{2I} \tag{3-11}$$

$$V_i = 2V \tag{3-12}$$

$$I_i = I \tag{3-13}$$

The input impedance thus is

$$Z_i = \frac{V_i}{I_i} = \frac{2V}{I} \tag{3-14}$$

Substituting $2R_L$ for V/I from equation (3-11), we obtain

$$Z_1 = 4R_L \tag{3-15}$$

The line Z_0 should be equal to the geometric mean of R_L and Z_1 for maximum bandwidth. Thus $Z_0 = \sqrt{R_L(4R_L)} = 2R_L$.

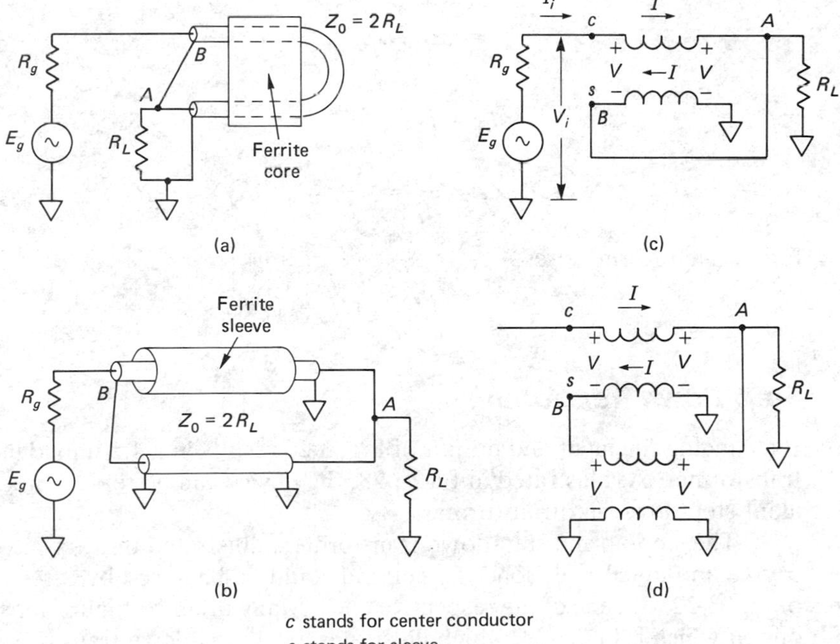

Figure 3-26 Pictorial diagrams (a) and (b) and the schematics (c) and (d) of a 4:1 impedance step-up transformer.

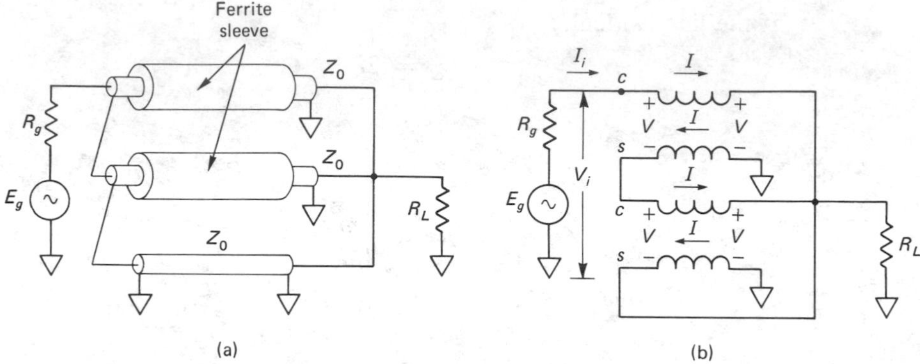

Figure 3-27 Pictorial diagram (a) and schematic (b) of a 9:1 impedance step-up transformer.

By using multiple cores, larger impedance ratios can be obtained. For the two-core impedance step-up transformer of Fig. 3-27,

$$I_{RL} = 3I \tag{3-16}$$

$$V_{RL} = V \tag{3-17}$$

and

$$R_L = \frac{V_{RL}}{I_{RL}} = \frac{V}{3I} \tag{3-18}$$

$$V_i = 3V \tag{3-19}$$

$$I_i = I \tag{3-20}$$

$$Z_i = \frac{V_i}{I_i} = \frac{3V}{I} = 3(3R_L) = 9R_L \tag{3-21}$$

For N such ferrite cores,

$$Z_i = (N + 1)^2 R_L \tag{3-22}$$

Step-Down Transformer

Exchanging the input and output of Fig. 3-26 results in a 4:1 impedance step-down transformer. As illustrated in Fig. 3-28, the device can be represented by an equivalent step-down autotransformer.

The step-up and step-down transformers illustrated thus far have unbalanced source and unbalanced load. Either end could be balanced by cascading the balun of Fig. 3-22 with one of the respective ends. Many other configurations are possible, one of which is Fig. 3-29, which illustrates a 4:1 step-down transmission line transformer.

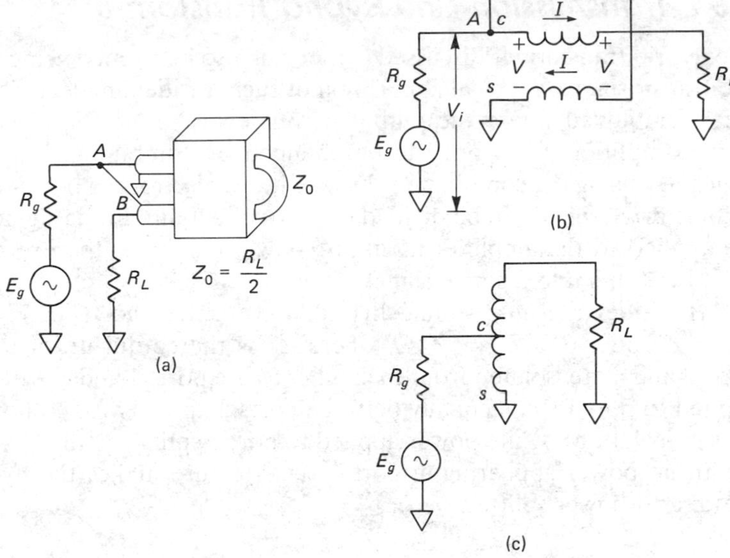

Figure 3-28 Pictorial diagram (a), schematic (b) and equivalent circuit (c) of a 4:1 step-down transformer, ie $Z_i = R_L/4$.

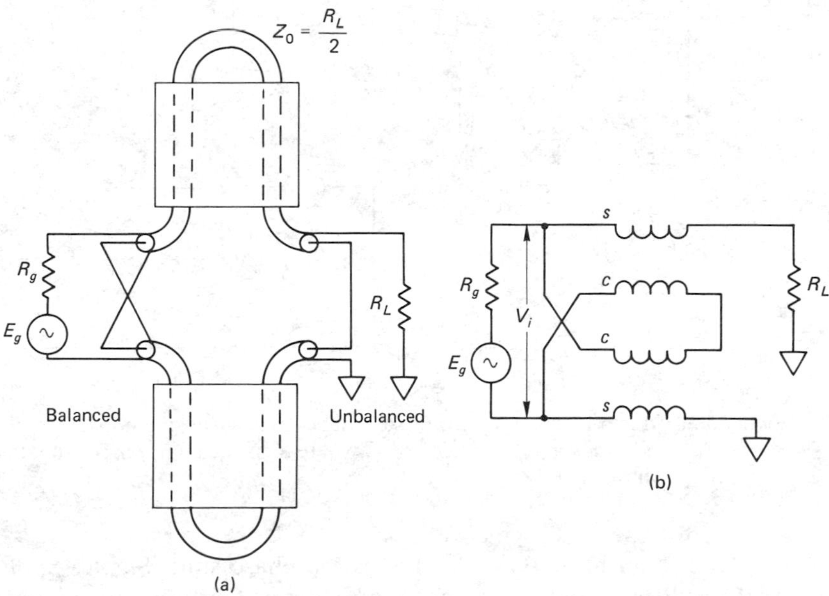

Figure 3-29 Pictorial diagram (a) and schematic (b) of a 4:1 step-down balanced input, unbalanced output transformer, ie. $Z_i = R_L/4$.

3-3.1 Transmission-Line Hybrid Transformer

The hybrid transformer discussed earlier can also be formed using the transmission-line transformer. A typical application of such a wideband transformer is the combiner, employed where the output power levels cannot be achieved by a single power amplifier stage, but can be obtained by combining the powers from two stages. By using the combiner in reverse, it can also form a power splitter, whereby a single drive signal can be divided into two equal-amplitude signals whose outputs are applied to the amplifier inputs. A power combiner than recombines the amplified outputs into a single signal.

For the transmission-line hybrid transformer shown in Fig. 3-30, if $Z_1 = Z_2 = Z$ and $Z_3 = Z_4 = Z^*/2$, where Z^* is the conjugate of the impedance Z, ports 3 and 4 are isolated from each other as are ports 1 and 2. Furthermore, power applied to port 1 splits equally between ports 3 and 4, with no power appearing at port 2 (isolation), while power applied to port 3 splits equally between ports 1 and 2 with no power appearing at port 4. This feature makes the hybrid transformer useful as a power splitter.

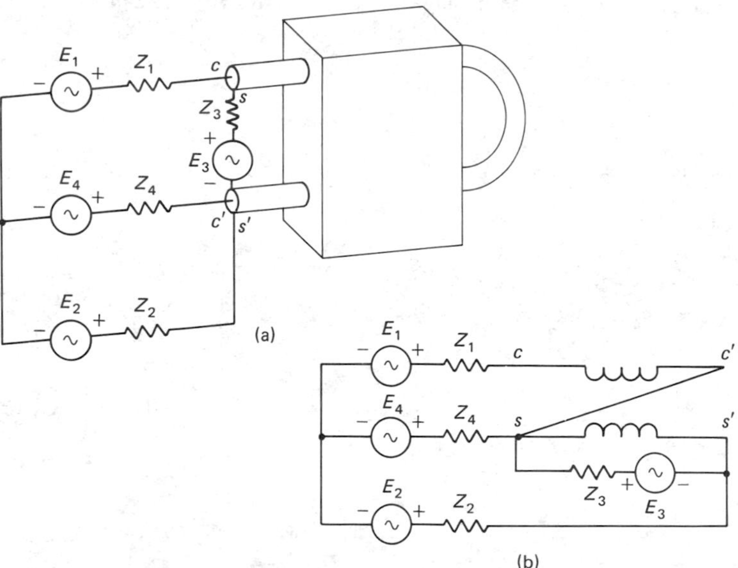

Figure 3-30 Pictorial diagram (a) and schematic (b) of a transmission line hybrid transformer. $Z_1 = Z_2 = Z$, $Z_3 = Z_4 = Z^*/2$ where Z^* is the complex conjugate of the impedance Z.

One other property of interest is the phase shift. Signals applied to ports 1 and 2 result in an output at port 3, which is proportional to the difference of the input signals, whereas at port 4 the output is proportional to the sum of the input signals. This feature makes the hybrid transformer useful as a power combiner.

Problems

3-1. **(a)** For the matching network shown in Fig. P3-1, determine L and C.
(b) What is one obvious disadvantage of the network?
(c) Indicate a possible solution to part (b).

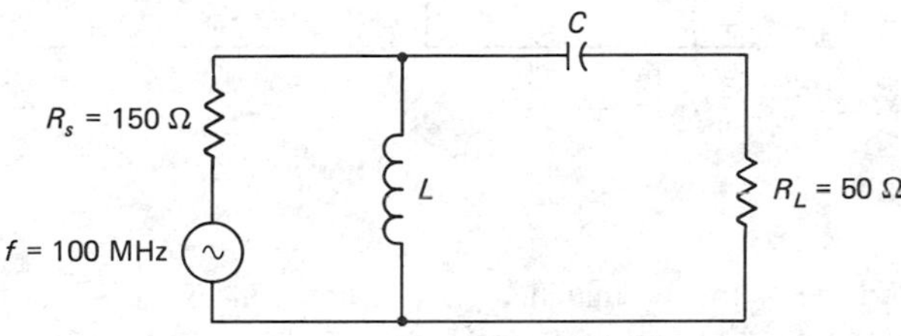

Figure P3-1.

3-2. Determine L and C of the resonant circuit (Fig. P3-2) so that it operates between a source resistance of 150 Ω and a load resistance of 1000 Ω. The loaded Q must be 20 at a resonant frequency of 50 MHz. Assume loss-less components and no impedance matching.

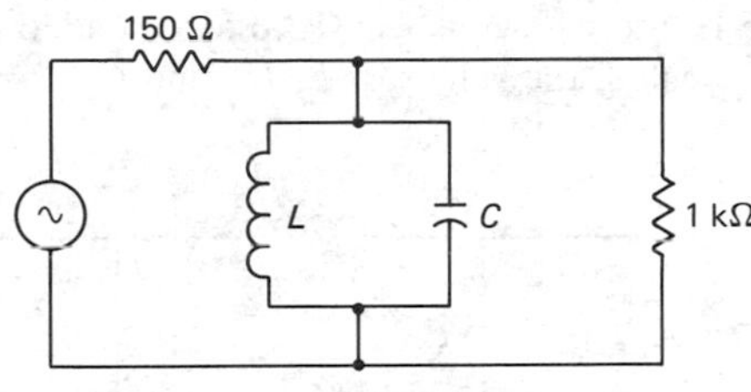

Figure P3-2.

3-3. For the matched network shown in Fig. P3-3:
(1) $R_s > R_L$
(2) $R_s < R_L$
(3) $R_s = R_L$

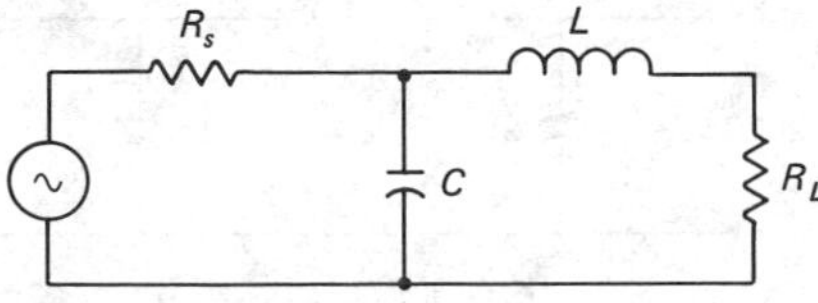

Figure P3-3.

3-4. For the matching networks shown in Fig. P3-4, determine:
(a) If the transformation is up or down
(b) The input resistance
(c) The unknown reactance

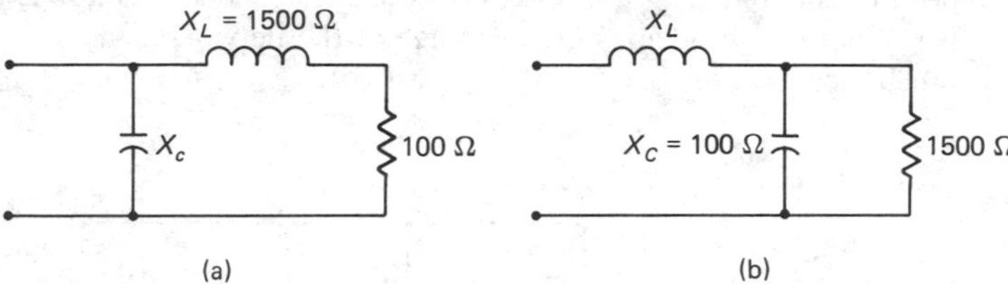

Figure P3-4.

3-5. Find the Q for the matched circuit shown in Fig. P3-5. Is this Q acceptable for most communications applications?

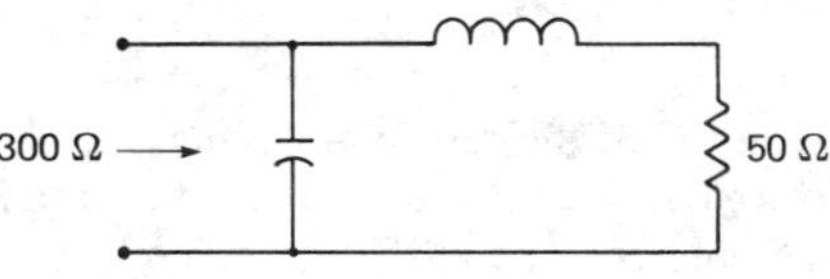

Figure P3-5.

3-6. Design the Π filter shown in Fig. P3-6 for a loaded Q of 5 and resonant frequency of 27 MHz. Calculate the values of L, C_1, and C_2.

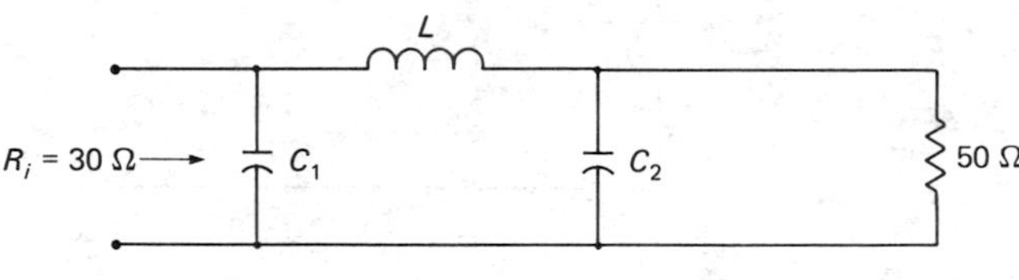

Figure P3-6.

3-7. Design the Π filter shown in Fig. P3-7 for a loaded Q of 5 and resonant frequency of 5 MHz. Determine X_L, L, X_{c1}, C_1, X_{c2}, and C_2.

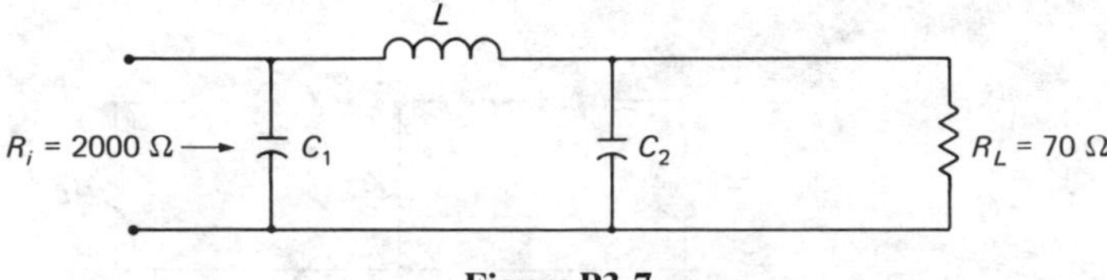

Figure P3-7.

3-8. Find the impedance looking back into the network from the output terminals of Fig. 3-15. What conclusions can be drawn with regard to power transfer and to matching a 3000-Ω load to a 50-Ω generator?

3-9. Design a Π–L match having an unloaded Q of 10 that matches a 50-Ω load to a 2000-Ω generator.

3-10. For the circuit of Fig. P3-10, find the resonant frequency, the input impedance at resonance, and the loaded Q if the unloaded Q of the coil is 100.

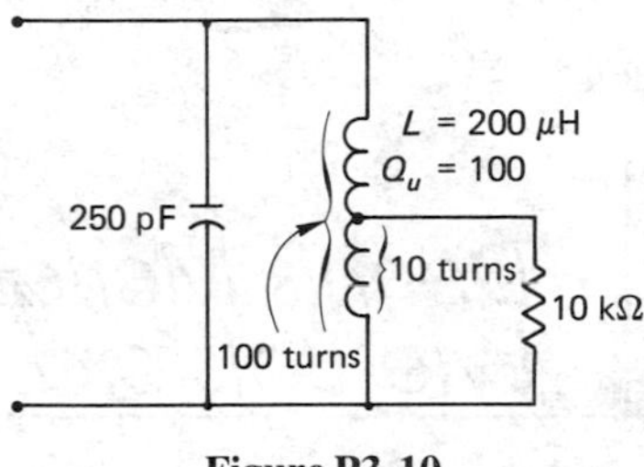

Figure P3-10.

3-11. Determine the loaded Q, R_{in}, and L for Fig. P3-11 if $C_1 = 200$ pF, $C_2 = 1000$ pF, and $R_L = 200\ \Omega$. The circuit is resonant and operating at $f = 1$ MHz. (*Hint:* Convert R_L, C_2 to a series equivalent.)

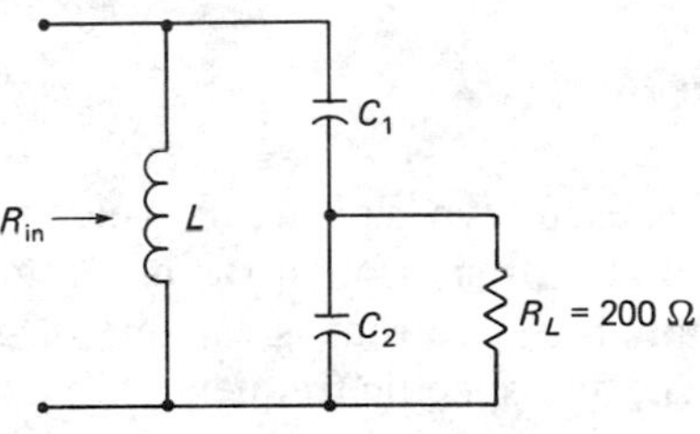

Figure P3-11.

3-12. What factors determine the low- and high-frequency limits of a transmission-line transformer?

3-13. Prove that the transmission-line transformer configuration of Fig. 3-28 is a 4:1 impedance step-down transformer.

3-14. Prove that the transmission-line transformer of Fig. 3-29 is a 4:1 step-down transformer.

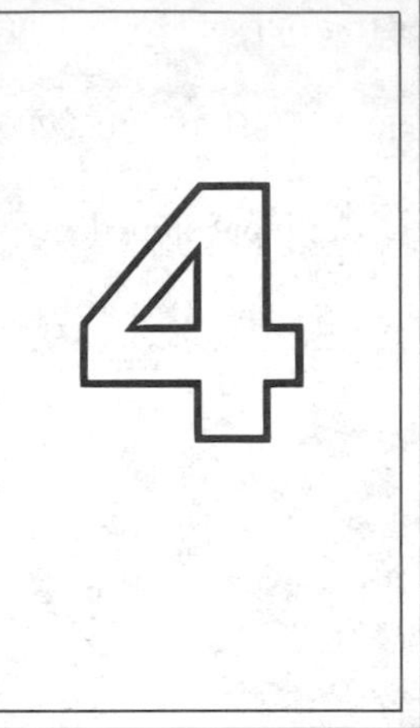

Oscillators

Dennis Morland
Lyle Botsford

4-1 INTRODUCTION

One of the most useful types of circuit in electronic communications is the oscillator, for without sources of sinusoidal signals, communication could not occur. An *oscillator* is simply a device that is capable of generating and supplying alternating electrical energy at a specific frequency. In doing this it converts energy from a dc source to an alternating signal. Oscillators come in a great variety of configurations. In this chapter we deal with the most common types.

In general, oscillators can be classified into three broad classifications: (1) multivibrator types (nonsinusoidal), (2) negative resistance types, and (3) harmonic or sinusoidal types. Here we deal mainly with harmonic oscillators, although multivibrator types based on CMOS and timer integrated circuits are also covered. Negative-resistance types are specialized high-frequency designs and are not treated here. Because of the importance of phase-locked loops (PLLs) in new designs, they are examined in this chapter as well.

4-2 GENERAL FEEDBACK SYSTEM

To begin the treatment of harmonic oscillators, let us consider a generalized feedback system (Fig. 4-1) and then develop a model for oscillators. In Fig. 4-1, x_i is the input to the amplifier, which is the sum of the stimulus, x_s, and the feedback,

x_f: $x_i = x_s + x_f$. The gain of the system is defined as the output, x_o, divided by the input, x_s. That is, $A = x_o/x_s$. B is the feedback ratio, $B = x_f/x_o$. K is the amplifier gain.

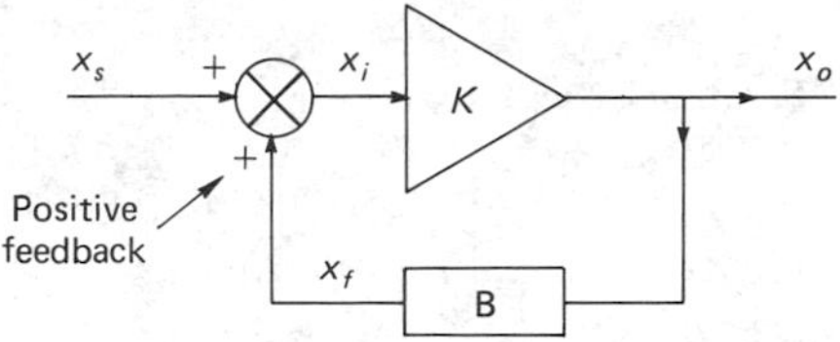

Figure 4-1 Generalized feedback system.

$$x_i = x_s + x_f = \frac{x_o}{A} + Bx_o = \frac{x_o}{A} + BKx_i$$

$$\frac{x_o}{A} = x_i - BKx_i = x_i(1 - BK) = \frac{x_o(1 - BK)}{K}$$

Therefore,

$$A = \frac{K}{(1 - BK)} \tag{4-1}$$

Note that as the product BK approaches 1, $(1 - BK)$ approaches zero and the system gain (A) approaches infinity. This implies that if the condition $BK = 1$ is met, a sustained output can occur without an input. In practice, an initial stimulus is given by noise or other impulse and then self-sustaining oscillations quickly build up, as shown in Fig. 4-2. Since an oscillator begins this operation as a linear small-signal amplifier, its small-signal parameters may be used in determining starting conditions.

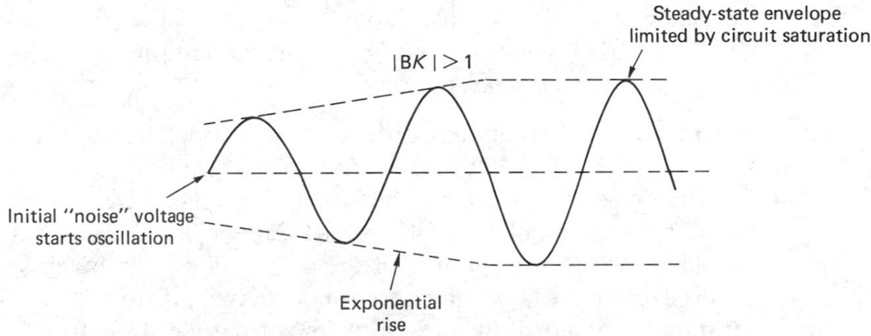

Figure 4-2 Typical oscillator startup waveform (with $BK > 1$).

Classically, the oscillator configuration becomes that shown in Fig. 4-3. Note: Assuming that the amplifier inverts and the summing network does not provide phase shift, the feedback network must provide the additional 180° of phase shift. This is a common situation for practical oscillators.

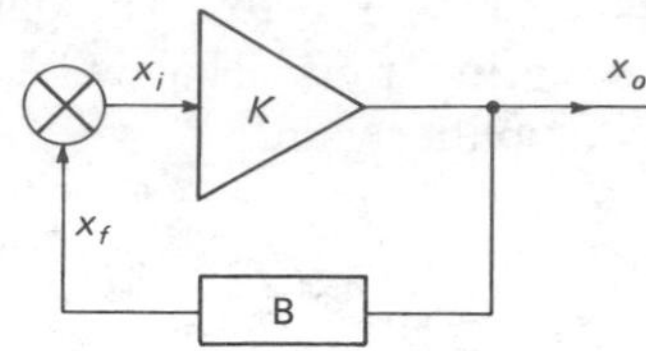

Figure 4-3 Feedback oscillator configuration.

From the closed-loop gain equation $A = K/(1 - BK)$, we may summarize three possible cases:

1. $|BK| < 1$ no sustained oscillations at the output.
2. $|BK| = 1$ sustained, linear oscillations.
3. $|BK| > 1$ increased oscillations without limit: this is the ideal situation, but in practice the magnitude of oscillations are limited by (a) the available power supply voltage, (b) circuit resistive losses, (c) device nonlinearity near cutoff, and (d) device nonlinearity near saturation.

4-2.1 Conditions Required for Oscillation

From feedback theory as examined above, we must have $BK = 1$. But remember that the ratios B and K are complex numbers. More explicitly, to achieve oscillation it is necessary that:

1. The loop phase shift (angle of BK) must equal zero or some multiple of 360° (positive feedback).
2. The magnitude of the loop gain, $|BK|$, must be greater than or equal to one for sustained oscillation. It might be noted that if $BK = 1$, then theoretically, the circuit will oscillate, but in this case circuit operation becomes highly dependent on device characteristics, temperature, aging, and loading circuit, in which case it may stop oscillating. In practice, $BK > 1$ is used to ensure reliable operation under practical operating conditions. These two conditions are known as the *Barkhausen criteria.*

The frequency of oscillation with reactive components in the feedback network is therefore the frequency where the loop phase shift is exactly zero. Note that if Barkhausen's criteria are satisfied at more than one frequency, it is possible to have simultaneous oscillation at several frequencies. Usually, care is taken to avoid this situation. The magnitude of the output is limited by circuit saturation. If the gain is excessively large, the output may be very distorted. To better understand harmonic oscillator principles, a review of parallel *LC* "tank" circuit action is in order.

4-2.2 Tank Circuits

Consider the parallel *LC* circuit shown in Fig. 4-4.

1. With the switch in position 1, the capacitor charges to supply voltage E, and energy is stored in the electric field $W = (Cv^2)/2$.

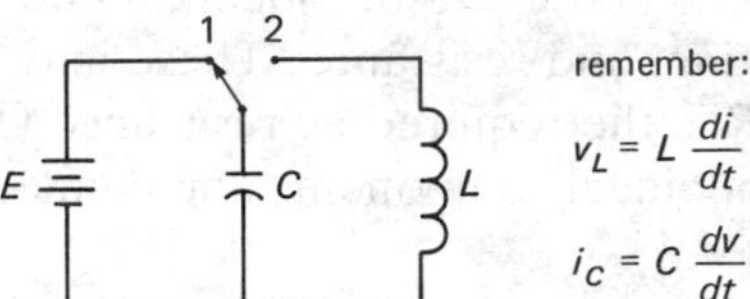

Figure 4-4 Switched *LC* circuit.

2. The switch is then moved to position 2, where it remains.
   a. The capacitor (discharging voltage source) attempts to discharge through the inductor, and hence current flow is established through the inductor.
   b. At the instant the capacitor voltage is zero, maximum current is flowing and hence the energy that was stored in the capacitor's electric field is now stored in the inductor's magnetic field by conservation of energy (assuming ideal components) $W = (Li^2)/2$.
   c. The inductor (discharging current source) opposes a change in current flow, maintaining a depleting but constant current direction (collapsing magnetic field) until the current through the inductor is zero, at which time the energy from the magnetic field has once again been transferred to the electric field. A diagrammatic representation would be as shown in Fig. 4-5.

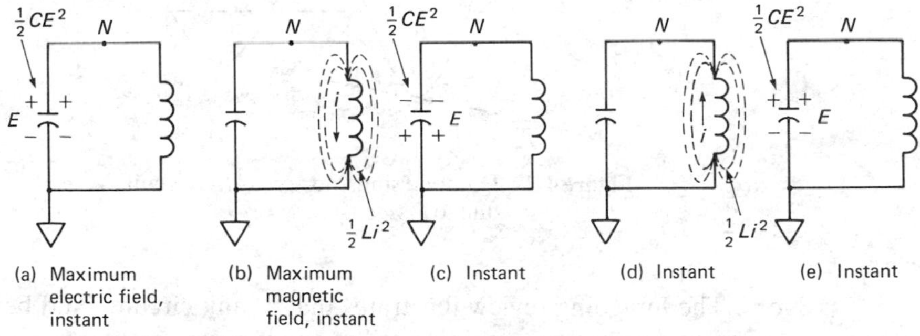

Figure 4-5 Changing states in a *LC* circuit.

The voltage at *N* with respect to ground (across the tank circuit) would appear as in Fig. 4-6. This action, called *oscillatory* action, repeats with a periodic exchange of energy between *L* and *C* or the tank circuit's natural or resonant frequency, $f = 1/(2\sqrt{LC})$, and the voltage waveform generated across *LC* is sinusoidal in nature.

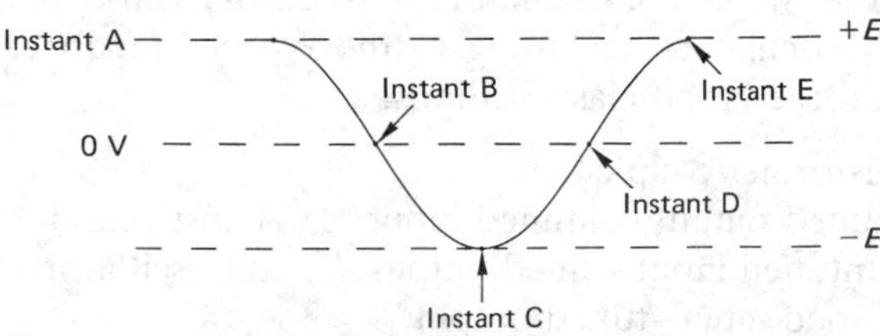

Figure 4-6 Waveform across the tank circuit.

For nonideal or practical cases, L and C are not ideal and do have losses in the form of radiation, dielectric heating, and resistance. These may all be lumped into an equivalent I^2R loss, where R is the required ac resistance (R_{ac}) necessary to account for all the losses in both nonideal components and connecting circuitry. As a result, the circuit exhibits a Q:

$$Q = \frac{1}{R_{ac}}\sqrt{\frac{L}{C}} \tag{4-2}$$

and a damped sinusoid results in a frequency

$$f = \frac{1}{2\pi\sqrt{LC}}\sqrt{\frac{Q^2}{1 + Q^2}} \tag{4-3}$$

as shown in Fig. 4-7.

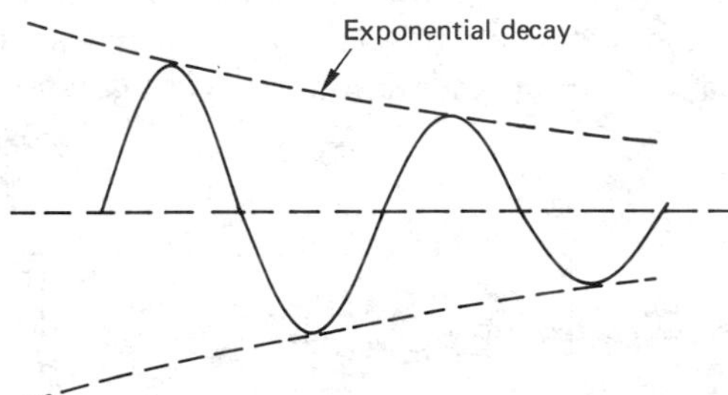

Figure 4-7 Damped sinusoid across LC circuit, due to losses.

Note: The foregoing review illustrates that a tank circuit could be utilized to generate continuous wave sinusoidal oscillations if some method of compensating for inherent energy losses is provided. In actual fact, this is accomplished by utilizing an active device with regenerative or positive feedback.

4-3 CLASSIFICATION OF HARMONIC OSCILLATORS

Although all harmonic oscillators must satisfy Barkhausens's criteria, it is useful to classify oscillators according to the type of feedback that is employed. This results in three broad classifications:

1. Transformer coupled
   a. Tuned output–untuned input — Armstrong or "tickler"
   b. Untuned input–tuned output — coil oscillators
   c. Tuned input–tuned output
   d. Blocking oscillator logically "fits" in this classification but behaves as a multivibrator type so will not be considered at this time

2. Reactance–coupled *LC* configurations
   a. Hartley
   b. Colpitts
   c. Ultra-audion
   d. Clapp
   e. Crystal oscillators (Miller and Pierce)
3. Reactance-coupled *RC* configurations
   a. Phase-shift oscillator
   b. Wien bridge oscillator
   c. Quadrature oscillator

We now consider these oscillator classifications in greater detail.

4-3.1 Transformer-Coupled Types

As the name implies, the B network consists of a transformer that couples the output to the input. Since transformers are capable of introducing 180° or 0° phase shift, careful attention must be given when connecting transformers into the circuit. A capacitor is frequently placed in shunt (tuned) with the input or output to "tune" the amplifier and hence set its frequency of oscillation. In all cases Barkhausen's criterion $BK = 1$ must be satisfied by proper phasing of the transformer and proper turns ratio selection. General diagrams for transformer-coupled oscillators would be as shown in Fig. 4-8. Note: the amplifiers must be properly biased gain devices.

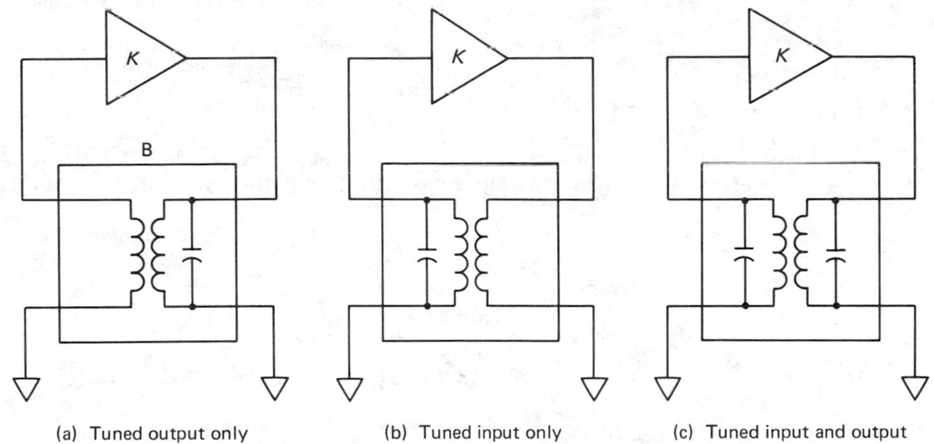

Figure 4-8 Transformer-coupled oscillators.

Consider the following practical circuits. Figure 4-9 shows a FET used in the configuration of Fig. 4-8(a). Figure 4-10 shows a similar configuration using an NPN transitor. If single-base-resistor biasing is used, the Q point shifts and the device oscillates at its "preferred" operating point. Figure 4-11 includes typical component values. Note: The tank circuit in the base circuit may be loaded by the low input impedance of the transistor. A low-tank-circuit Q results in a larger frequency drift and a more distorted output waveform with many harmonic frequencies.

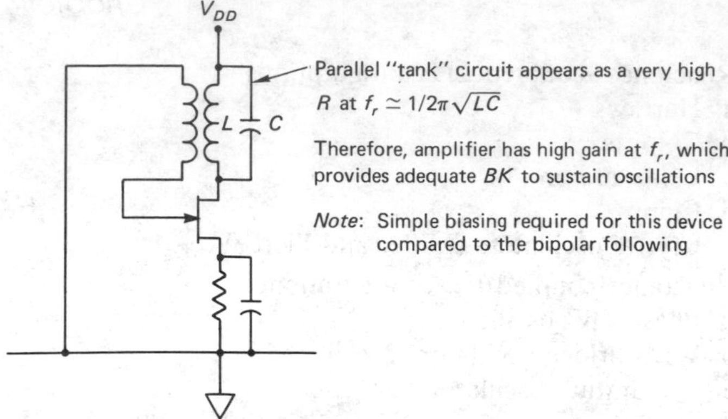

Figure 4-9 FET tuned output, untuned input.

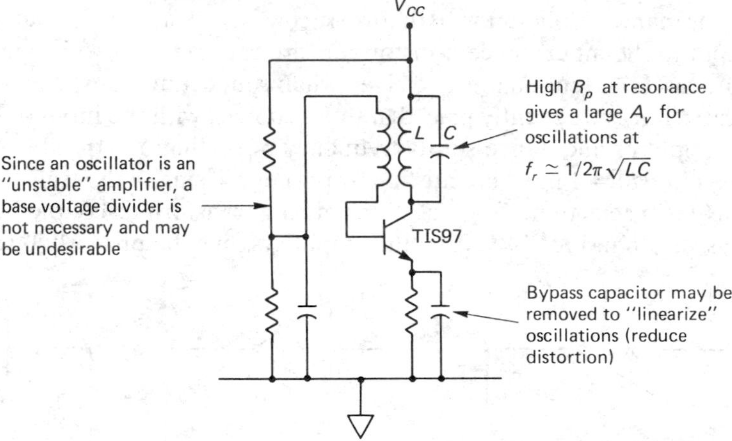

Figure 4-10 Tuned output, transformer oscillator with NPN transistor.

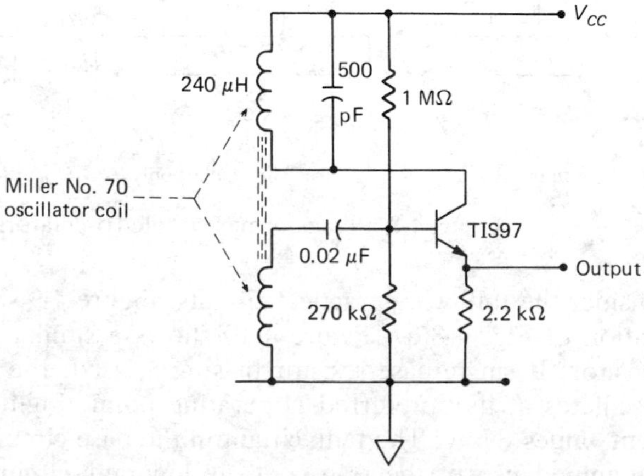

Figure 4-11 Typical tuned-output oscillator (460 kHz).

4-3.2 Signal Bias

The bias requirement of an oscillator changes as it moves from startup to steady-state operation. This requirement can be met by using signal bias. *Signal bias*, also called *gate-leak bias* for FETs, *grid-leak bias* for vacuum tubes, and *base-leak bias* for transistors, is actually nothing more than signal clamping. The FET in Fig. 4-12 does not seem to have any bias, but in fact the rectifying action of the device does change the bias.

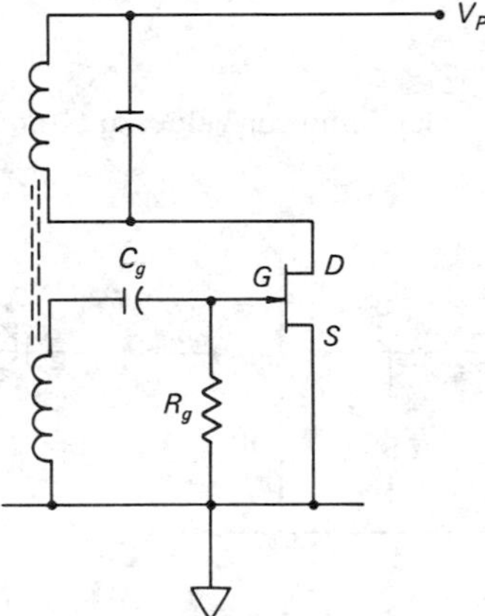

Figure 4-12 FET oscillator with gate-leak bias.

We will examine the circuit of Fig. 4-13 more closely. The transfer characteristic of this device is shown in Fig. 4-14. The circuit as shown has no bias (i.e., $V_{gs} = 0$, $I_d = I_{dss}$). Consider first the positive half-cycle of signal, shown in Fig. 4-15. For the positive half-cycle the gate-to-source junction is forward biased and the C_g charges up to V_p, the forward voltage drop of the diode. The gate-to-source diode is reverse biased after peak voltage has occurred, and hence the capacitor discharges through R_g and the generator resistance as shown in Fig. 4-16. Note that the voltage across R_g from this discharge reverse-biases the gate-to-source junction and the amplifier is now biased at approximately $V_{gs} = V_p$. The voltage

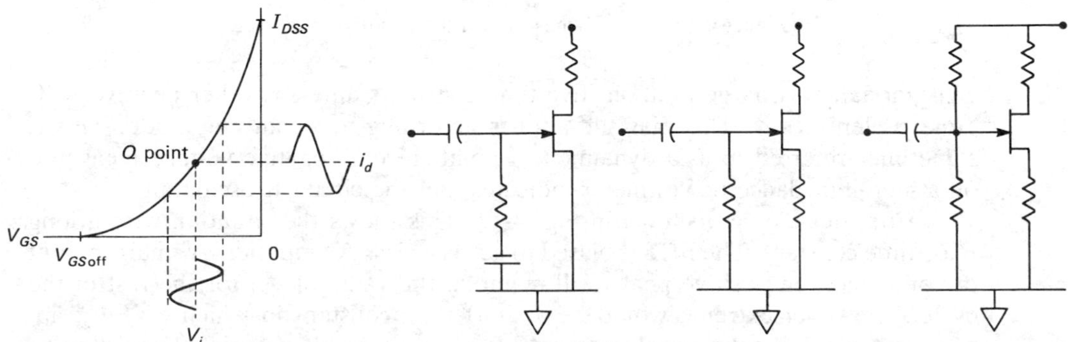

Figure 4-13 FET amplifier with gate-leak bias.

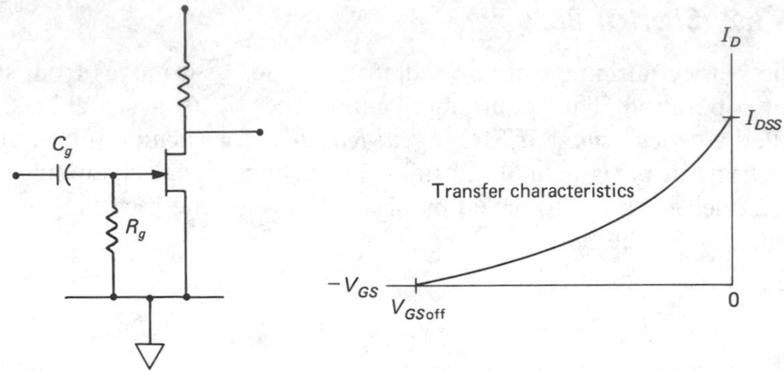

Figure 4-14 Class A, B, or C amplifier, depending on values of C_g and R_g.

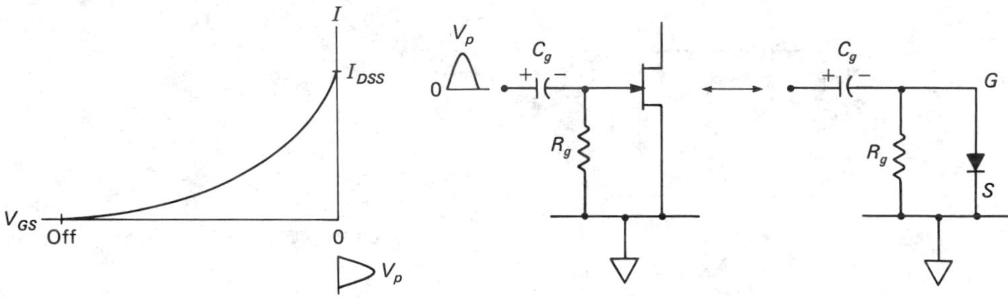

Figure 4-15 Positive half-cycle applied to gate of n-channel FET.

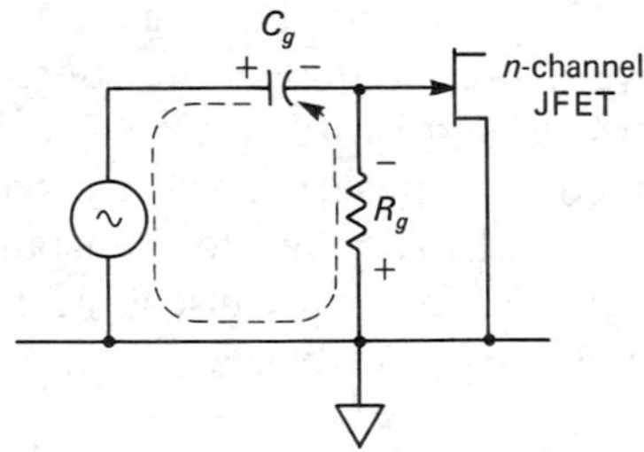

Figure 4-16 Discharge path for the negative half-cycle.

will gradually discharge in about five time constants unless another positive half-cycle replenishes it. This bias self-adjusts according to signal amplitude. This is sometimes referred to as a dynamic test point since a negative voltage measured on a high-impedance dc voltmeter indicates that the circuit is oscillating.

Consider the case shown in Fig. 4-17. This shows the situation with a long R_gC_g time constant. The FET is biased to act as a class A amplifier. A small amount of conduction on positive peaks will maintain the value of V_{gs} for the rest of the cycle. The other extreme would be a short time constant in which case V_{gs} approximately follows the signal and a class B amplifier results or a very large positive half-cycle could easily bias the device into class C operation.

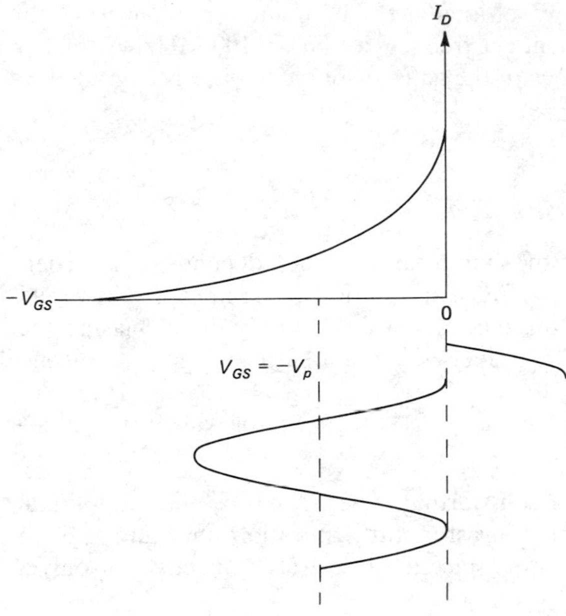

Figure 4-17 Bias situation with a long R_gC_g time constant.

4-3.3 Merits of Signal Bias

Recall the equations (from solid-state theory)

$$I_D = I_{DSS}\left(1 - \frac{|V_{GS}|}{|V_{po}|}\right)^2 \tag{4-4}$$

$$g_m = \frac{2I_{DSS}}{V_{po}}\left(1 - \frac{|V_{GS}|}{|V_{po}|}\right) \tag{4-5}$$

where V_{po} is the pinch-off voltage. These equations imply that the transconductance (g_m) of the FET and the voltage gain of the circuit will approach a maximum as the gate-to-source voltage approaches zero. This bias technique may therefore be used to achieve maximum gain from an amplifier (with allowable distortion) by letting it automatically set its bias level according to the magnitude of the incoming signal. It also means that maximum gain is realized on the first positive half-cycle. This is desirable from an oscillator standpoint because it implies that when the oscillator starts it will exhibit maximum voltage gain, ensuring $BK >> 1$, hence ensuring that the circuit will start oscillating. Once oscillations are in progress, bias is developed (according to how strong the device is oscillating or magnitude of swing) which reduces the gain, thereby moving BK toward $BK = 1$, which is a more linear mode of operation. Many oscillator configurations will utilize both signal bias and some minimum standard biasing. The device is then operating at the Q point, which is the sum of the two types.

A note of caution: An un-bypassed R_s should be incorporated in most FET circuits to limit the magnitude of the gate charging current to levels below that listed in specification sheets. Transistors can develop signal bias as well under certain cases, and weirdly shaped waveforms at the base and emitter will be observed. This technique is effective only at frequencies below 100 MHz. At higher frequencies the rectification efficiency of the gate-to-source voltage is degraded by junction capacitance.

4-3.4 Design Considerations

For transformer-coupled oscillators the amount of feedback depends on the coefficient of coupling between the coils. In most applications, commercial coils would be used. This means that design considerations would simply entail biasing the device at the manufacturer's specified levels to obtain sufficient gain to sustain oscillation.

Depending on the amount of feedback (B), the oscillator waveform can assume different shapes.

Condition 1: ratio just sufficient for self-starting. The peak oscillation amplitude is just large enough to drive the transistor into saturation and cutoff. Such oscillation is a quasi-class A oscillator, since the transistor is inactive for only a very small portion of the cycle.

Condition 2: heavier feedback in which evident distortion of the oscillator waveform is observed. The frequency-determining element will tend to filter out some of the distortion. The device is then operating essentially as class B.

Condition 3: extreme values of feedback; the device is inactive for several cycles of oscillation. This results in a damped sinusoid at the operating frequency (impulse excitation). The device is operating as a class C oscillator, which is used when it is desirable to obtain the highest possible power output.

Note: If the feedback ratio is set too critically, it may result in the oscillator failing to operate under conditions of (1) low supply voltage, and (2) falloff of device gain due to aging, temperature drift, and so on.

4-3.5 Reactance-Coupled LC Configurations

The general configuration for LC-coupled oscillators is shown in Fig. 4-18, where K indicates an active inverter, whether transistor, FET, vacuum tube, IC, and so on; and X_1, X_2, and X_3 are reactive components, either L or C. Recall that for a common-emitter (CE) amplifier the voltage gain depends on the load (Fig. 4-19). Now consider such a CE amplifier used with the circuit of Fig. 4-18. Assuming no input loading,

$$Z_L = X_2 \| (X_1 + X_3)$$

$$B_v = \frac{X_1}{X_1 + X_3} \quad \text{feedback voltage ratio} \tag{4-6}$$

Recall that for sustained oscillations it is required that the circuit have positive feedback (total loop phase shift equal to zero degrees) and adequate gain so that

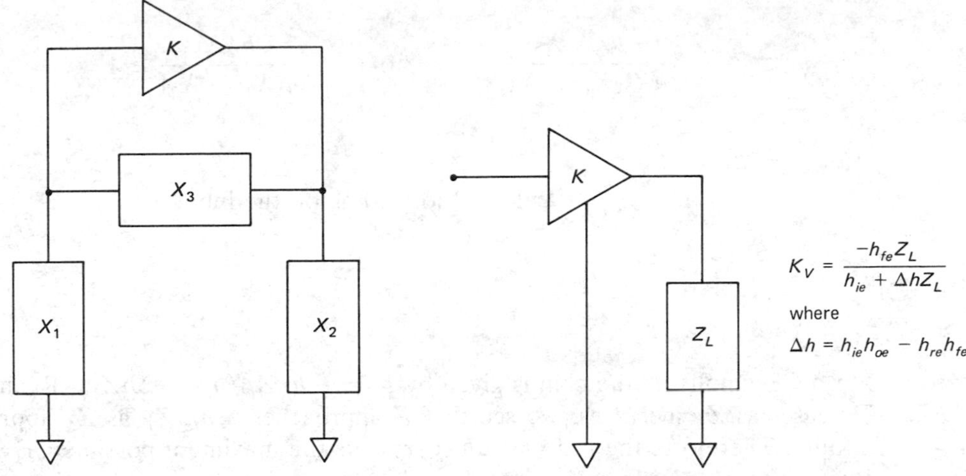

Figure 4-18.

Figure 4-19 Common-emitter amplifier gain.

the product of the gain and feedback terms will be a positive real number greater than or equal to 1. Hence to satisfy the Barkhausen criterion $B_v K_v \geq 1$,

$$\frac{X_1}{X_1 + X_3} \frac{-h_{fe}Z_L}{h_{ie} + \Delta h Z_L} \geq 1$$

$$\frac{X_1}{X_1 + X_3} \frac{-h_{fe} \dfrac{X_2(X_1 + X_3)}{X_1 + X_2 + X_3}}{h_{ie} + \Delta h \dfrac{X_2(X_1 + X_3)}{X_1 + X_2 + X_3}} \geq 1 \qquad (4\text{-}7)$$

$$\frac{X_1}{X_1 + X_3} \frac{-h_{fe}X_2(X_1 + X_3)}{h_{ie}(X_1 + X_2 + X_3) + \Delta h X_2(X_1 + X_3)} \geq 1 + j0$$

$$\frac{\overbrace{-h_{fe}X_1X_2}^{\text{real number}}}{\underbrace{h_{ie}(X_1 + X_2 + X_3)}_{\text{imaginary number}} + \underbrace{\Delta h X_2(X_1 + X_3)}_{\text{real number}}} \geq 1 + j0$$

For this to equal a real number,

$$h_{ie}(X_1 + X_2 + X_3) = 0$$

$$X_1 + X_2 + X_3 = 0$$

so

$$X_1 + X_3 = -X_2$$

and

$$\frac{-h_{fe}X_1X_2}{\Delta h X_2(X_1 + X_3)} \geq 1 \qquad \text{or} \qquad \frac{-h_{fe}X_1}{\Delta h(X_1 + X_3)} \geq 1$$

or

$$\frac{h_{fe}X_1}{\Delta h X_2} \geq 1 \qquad \text{(no loading on the input)}$$

$$\frac{h_{fe}}{\Delta h} \geq \frac{X_2}{X_1} \tag{4-8}$$

For a CE amplifier, the gain is given by $K = -h_{fe}Z_L/(h_{ie} + \Delta h Z_L)$. Realizing K increases as Z_L increases, we see that K approaches $-h_{fe}/Z_L$ as Z_L approaches infinity. Therefore, the ratio $h_{fe}/\Delta h$ represents the maximum possible CE voltage gain. Since $h_{fe}/\Delta h$ is a real positive number, X_2/X_1 is also a real positive number, so X_1 and X_2 are *identical reactances types* (both C or L). Since $(X_1 + X_2 + X_3) = 0$, X_3 is the *opposite reactance type* from X_1 and X_2.

Recall the maximum possible voltage gains for devices, $K_{v(\max)}$:

$$\text{transistor} = \frac{h_{fc}}{\Delta h} \qquad \text{triode} = u$$

$$\text{pentode} = g_m r_p \qquad \text{FET} = g_m r_d$$

It is now possible to extrapolate for the other types of active devices the gain criteria necessary for oscillations.

$$\text{transistor } \frac{h_{fc}}{\Delta h} \geq \frac{X_2}{X_1} \tag{4-9}$$

$$\text{pentode } g_m r_p \geq \frac{X_2}{X_1} \tag{4-10}$$

$$\text{triode } u \geq \frac{X_2}{X_1} \tag{4-11}$$

$$\text{FET } g_m r_d \geq \frac{X_2}{X_1} \tag{4-12}$$

There are only two possible combinations:

1. The case in which X_1 and X_2 are capacitive and X_3 is inductive defines a *Colpitts oscillator*.
2. The case in which X_1 and X_2 are inductive and X_3 is capacitive defines a *Hartley oscillator*.

The ac circuits would therefore be as shown in Figs. 4-20 and 4-21. At this point it is appropriate to recall the Π-type network (Fig. 4-22) and the T-type network (Fig. 4-23) from impedance-matching network theory. The Π type is used to match R_g to a smaller R_L at one frequency (a resonant matching network). The T type is used to match R_g to a larger R_L, also at one frequency. From the figures it is obvious that the Hartley and Colpitts oscillator configurations use, in effect, a Π type of impedance-matching network to connect between the amplifier's output impedance and the amplifier's input impedance at one frequency (resonance).

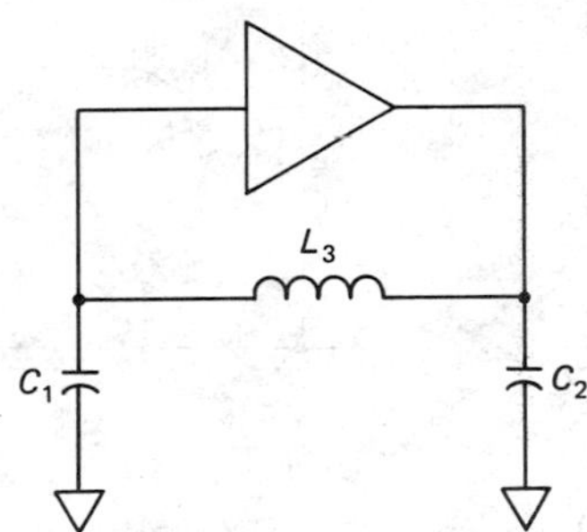

Figure 4-20 Colpitts oscillator.

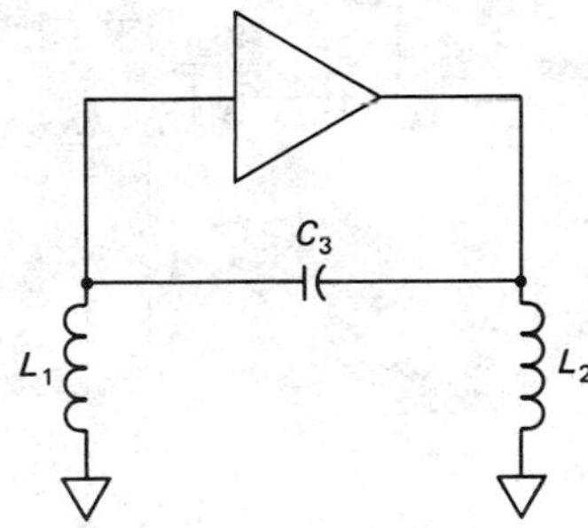

Figure 4-21 Hartley oscillator.

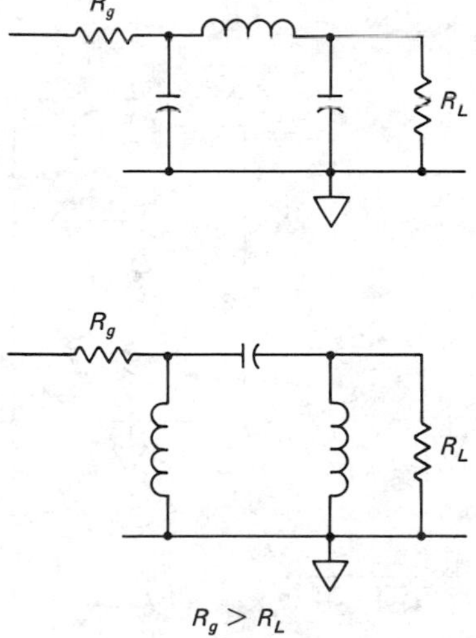

Figure 4-22 Π network.

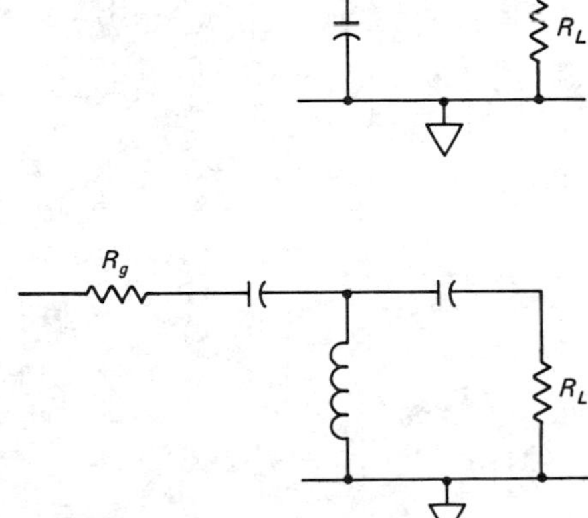

Figure 4-23 T network.

4-3.6 Practical Circuits

Figure 4-24 shows a Hartley oscillator using a NPN transistor, and Fig. 4-25 depicts a Colpitts oscillator. Other oscillators, using FETs, are shown in Figs. 4-26 and 4-27.

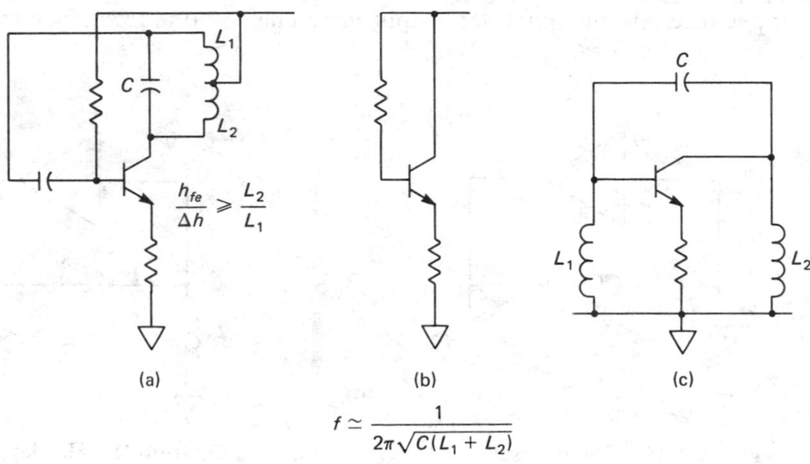

Figure 4-24 (a) Hartley oscillator; (b) dc circuit; (c) ac circuit.

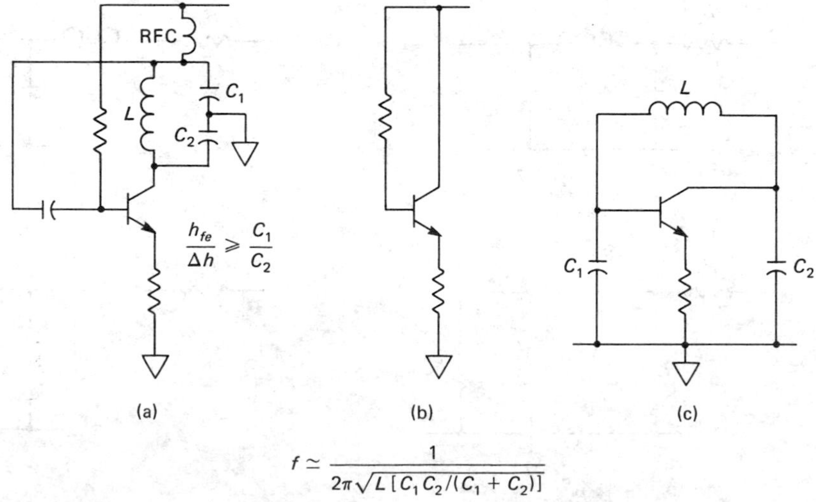

Figure 4-25 (a) Colpitts oscillator; (b) dc circuit; (c) ac circuit.

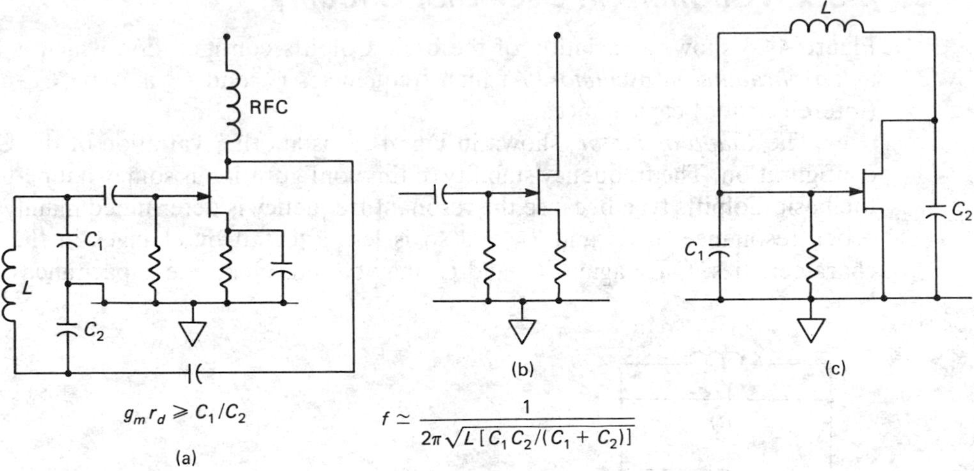

Figure 4-26 (a) FET Colpitts oscillator; (b) dc circuit; (c) ac circuit.

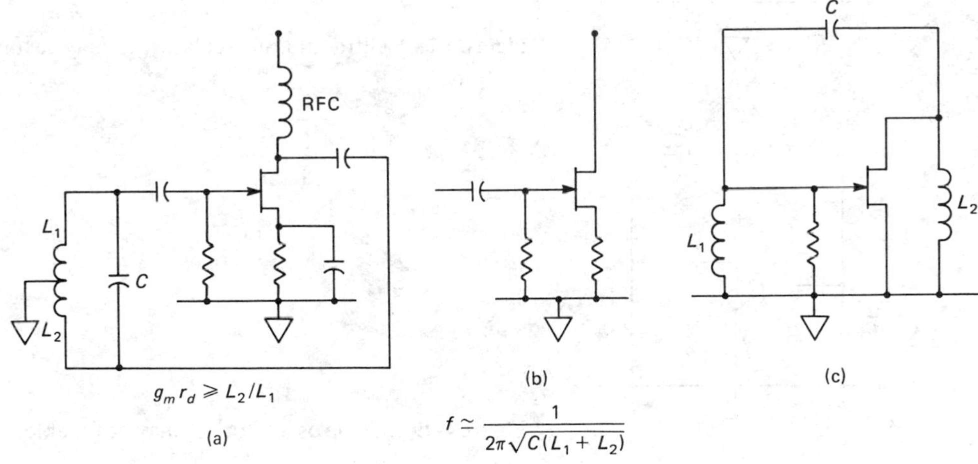

Figure 4-27 (a) FET Hartley oscillator; (b) dc circuit; (c) ac circuit.

4-3.7 Variations in Oscillator Circuitry

Figure 4-28 shows a variation of the basic Colpitts configuration which is known as an *ultra-audion oscillator*. At high frequencies C_1 and C_2 may be C_{be} and C_{ce} (interelectrode) capacitances.

The *Clapp oscillator*, shown in Fig. 4-29, is another variation of the Colpitts configuration. The frequency stability of this configuration is somewhat better than the basic Colpitts type because the resonant frequency is determined mainly by the series resonance of C_3 and L_3 and so is less affected by changes in the device characteristics. Once again, C_1 and C_2 may be interelectrode capacitances.

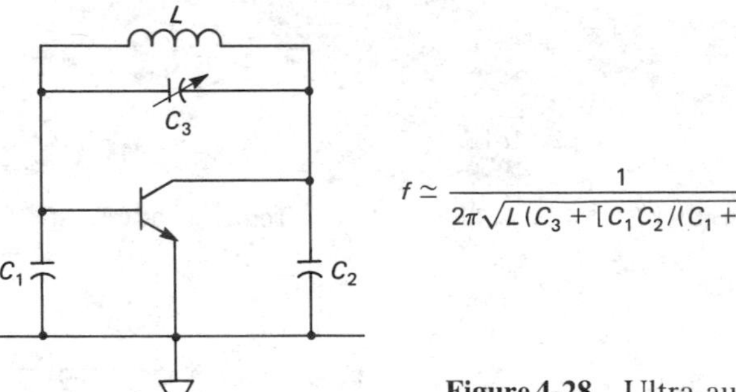

Figure 4-28 Ultra-audion oscillator. C_3 may be tunable.

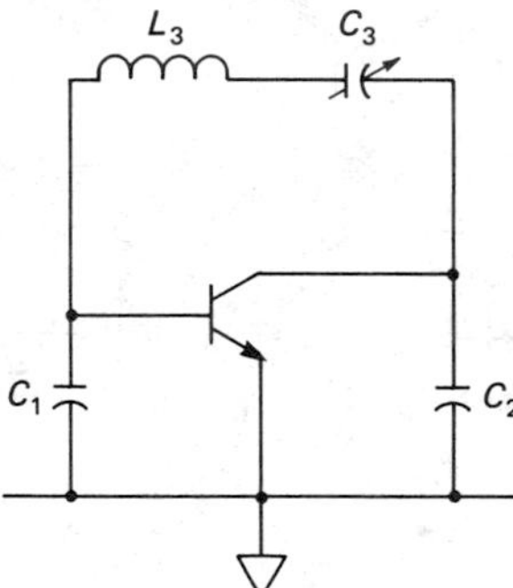

Figure 4-29 Clapp oscillator. C_3 may be tunable.

4-4 CRYSTAL OSCILLATORS

4-4.1 Introduction

Before discussing crystal oscillators it is necessary to note some characteristics of crystals. Among a number of crystalline substances, quartz, Rochelle salt, barium titanate, and tourmaline most exhibit piezoelectric properties. If a slab of one of these crystalline substances is properly cut, a mechanical stress will produce an emf across the slab. Conversely, an applied emf will produce a mechanical stress. Rochelle salt produces the greatest piezoelectric reaction and is widely used in audio components such as microphones and phonograph pickups. Quartz has the best properties for radio-frequency (RF) work. It has good mechanical strength,

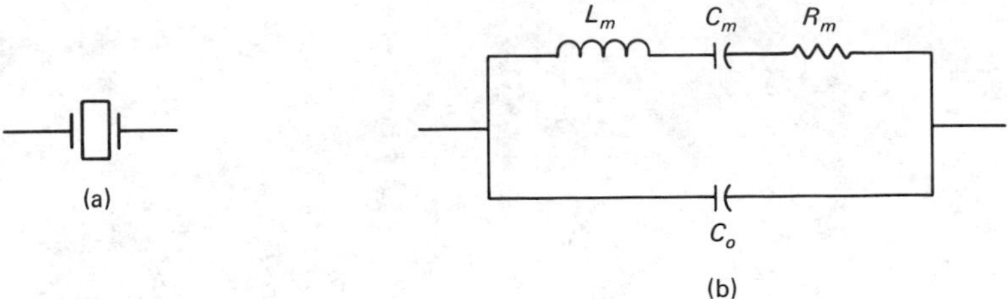

Figure 4-30 (a) Quartz crystal; (b) equivalent circuit.

a high Q value, a low-temperature drift, and a high degree of electrical stability. A quartz crystal is shown schematically in Fig. 4-30(a).

When properly cut and ground, a quartz crystal acts like a parallel resonant circuit of high Q and can be used as a tank circuit in many applications. A quartz crystal also has a series resonance, which is frequently used in crystal filters. An equivalent-circuit model of a quartz crystal is depicted in Fig. 4-30(b). Electrical connections are provided by depositing a metallic plating on the opposite faces of the crystal. The crystal must be mounted in a special holder, and since it is quite brittle, it must be handled carefully.

A typical 428-kHz crystal is less than 1.0 × 1.0 × 0.1 cm in size and its equivalent values are $C_0 = 5.8$ pF, $C_m = 0.042$ pF, $L_m = 3.3$ H, and $Q = 23{,}000$. The resonant frequency is inversely proportional to its size and thickness. Crystals are commercially available from 2 kHz to about 35 MHz on fundamental and 150 MHz on overtones.

There are many ways of cutting a quartz plate from a crystal. Quartz has a hexagonal crystalline structure with three major axes: the Z axis (optical axis), the Y axis (crystal face or mechanical axis), and the X axis (corners or electrical axis). The orientation of the crystal cutting with respect to these axes determines the electrical properties of the crystal, which include: (1) suitability for different frequency ranges, (2) suitability for filters, (3) suitability for oscillators, and (4) different or zero-temperature coefficients.

The AT-cut crystal has the least variation with temperature and is commonly used in oscillator circuits. Temperature stability can be further improved with compensation circuits. Figure 4-31(a) shows the orientation of some crystal cuts, and Fig. 4-31(b) depicts typical frequency variation with temperature. For example, if a 10-MHz crystal deviates by ±0.002%, this represents a (0.0020/100) × (10 MHz) = 200 Hz or 20 ppm (parts per million) frequency error.

A plot of crystal reactance versus frequency is given in Fig. 4-32. There is a series resonance, due to L_m and C_m, and at a slightly higher frequency there is a parallel resonance which involves C_0 as well. The series resonance frequency is

$$f_s = \frac{1}{2\pi\sqrt{L_m C_m}} \tag{4-13}$$

The parallel resonance frequency is

$$f_p = \frac{1}{2\pi\sqrt{L_m\,[C_0C_m/(C_0 + C_m)]}} \tag{4-14}$$

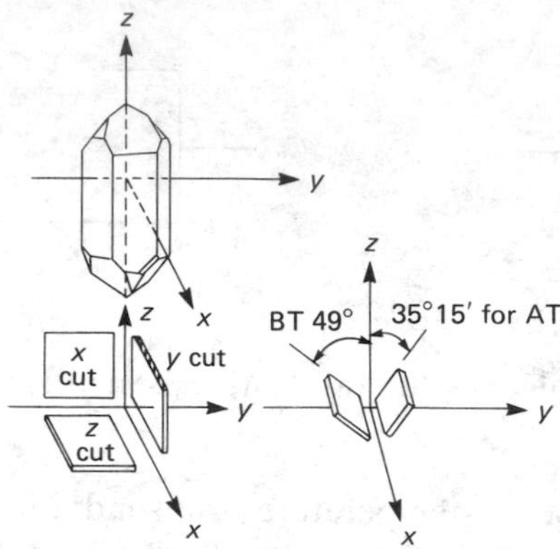

Figure 4-31 (a) Crystal cuts.

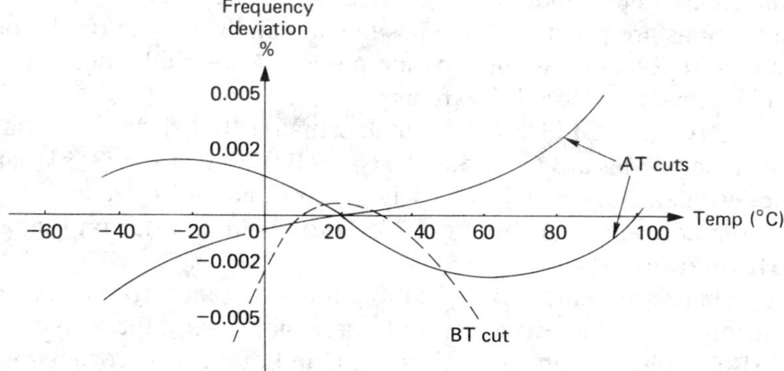

Figure 4-31 (b) Typical frequency/temperature curves.

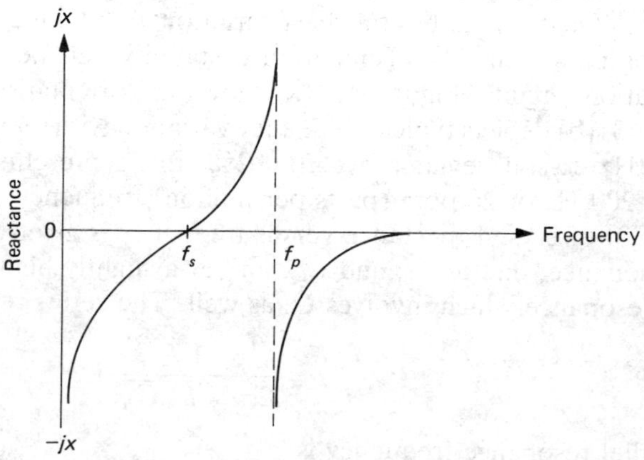

Figure 4-32 Reactance versus frequency for a quartz crystal.

The reactance plot of Fig. 4-32 applies to the fundamental resonance mode of the crystal. But crystals can oscillate in more than one way, so a crystal will actually have a number of f_s and f_p resonant frequencies that are near multiples of the fundamentals. These harmonic or overtone frequencies can be used to extend the frequency range of crystal oscillators. The *LC* resonant circuit in the crystal oscillator makes sure that the oscillator is operating on the correct overtone. Some crystal oscillators' configurations, defined with reference to the general configuration of Fig. 4-18, are listed in Table 4-1.

TABLE 4-1 Crystal Oscillator Configurations

Oscillator Name	X_1	X_2	X_3
Miller	Inductive crystal	Inductive tuned tank	C_3
Pierce	C_1	C_2	Inductive crystal

4-4.2 Miller Oscillator

The Miller oscillator uses the basic Hartley configuration with a crystal taking the place of one of the inductors, as in Fig. 4-33(a). Since thc crystal has to act as an inductance, the operating frequency will be a bit less than the parallel resonant frequency of the crystal. The tuned circuit on the output, instead of being just an inductor, also ensures operation at the right overtone (or fundamental). The feedback capacitor can often be eliminated since the inherent junction capacitance is usually large enough, as in the FET oscillator of Fig. 4-33(b).

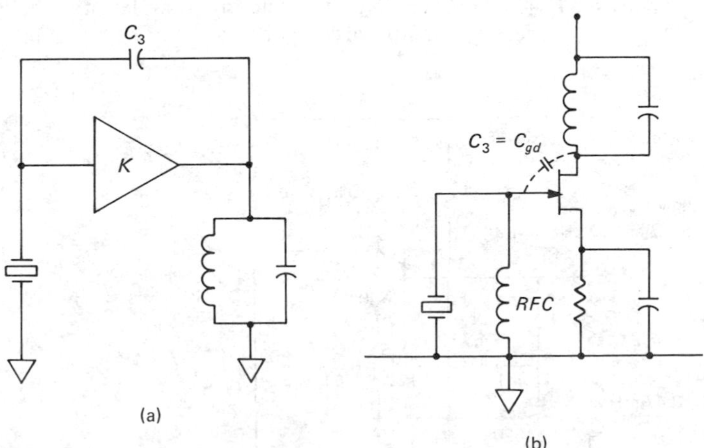

Figure 4-33 (a) Miller crystal oscillator and (b) FET example.

4-4.3 Pierce Crystal Oscillator

This oscillator employs the basic Colpitts configuration with a crystal in place of the inductor, as shown in Fig. 4-34(a). Operating frequency will be a bit less than the normal parallel resonant frequency. As Fig. 4-34(b) illustrates, sometimes the inherent junction capacitances can serve for C_1 (C_{be}) and C_2 (C_{ce}). Many variations are possible, of course. For example, Fig. 4-35 shows a modified Pierce oscillator which has an inductor and resistor added to the feedback path. The addition of these components creates a tank circuit with a very low Q, which provides nearly 180° of phase shift for the feedback signal. The operating frequency in this case will be closer to the series resonant frequency of the crystal. The tank circuit will help to prevent operation at an unwanted (overtone) frequency. With the values shown, the circuit of Fig. 4-35 produced a 270-mV_{pp} waveform at the base and a very clean 6-V_{pp} waveform at the junction of the crystal and the inductor, with a 15-V supply.

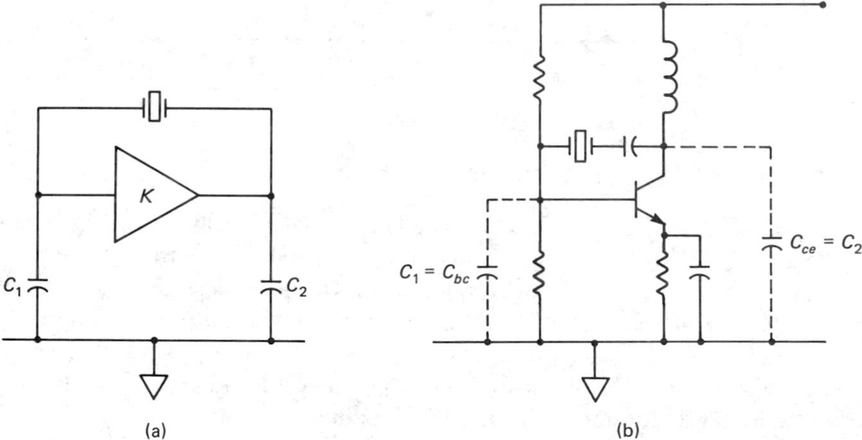

Figure 4-34 (a) Pierce configuration and (b) example with junction capacitance serving for normal components.

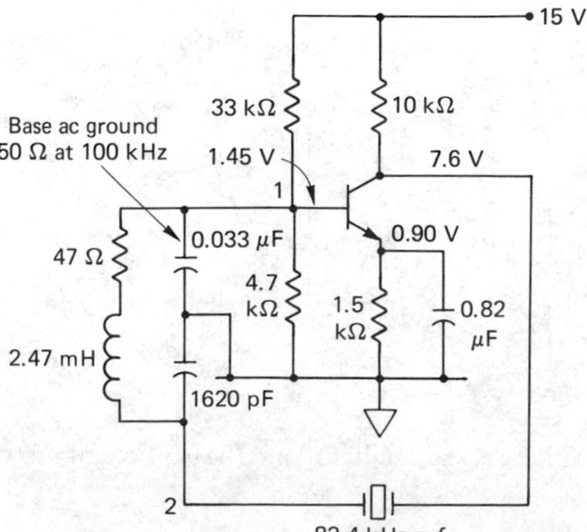

Figure 4-35 Modified Pierce crystal oscillator.

4-4.4 Operational Amplifier Crystal Oscillator

Operational amplifiers, or op amps, can be used in place of other active devices in the various types of oscillator circuits that have been examined. A simple example of this is given in Fig. 4-36, which shows a crystal (with optional trimmer capacitor) providing the feedback path to the positive (in phase) input of the device. Operation would occur at the series resonant frequency of the crystal. The resistors would be chosen as appropriate to set the necessary gain. Optional resonant circuits could be added for overtone control.

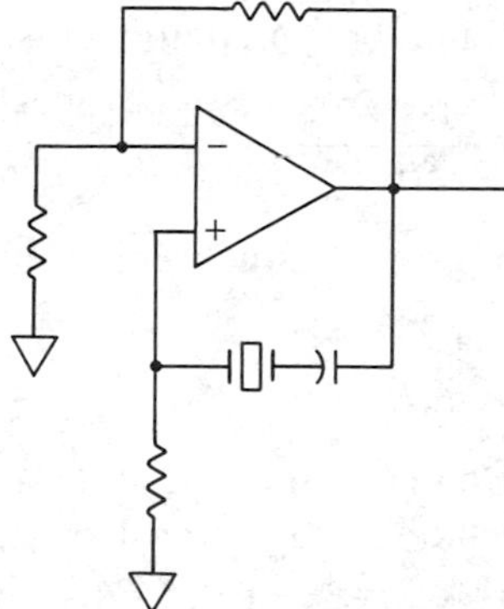

Figure 4-36 Op-amp crystal oscillator.

4-4.5 Reactance-Coupled RC Configurations

For low frequencies (audio or a few hertz to several hundred kilohertz) the values of the inductors and capacitors for Colpitts and Hartley, or for that matter, all *LC* oscillators, become impractical in size and cost. Therefore, simple *RC* (phase shift) circuit networks are commonly used.

4-5 PHASE-SHIFT OSCILLATOR

Consider the simple *RC* network of Fig. 4-37. Recall that a simple *RC* network (single-section high-pass filter) may introduce up to 90° phase shift. This would imply that two sections would be sufficient to provide 180° phase shift, in theory, and hence if applied between input and output of an inverting amplifier, Barkhausen's criteria would be satisfied and oscillations would result. In practice, three

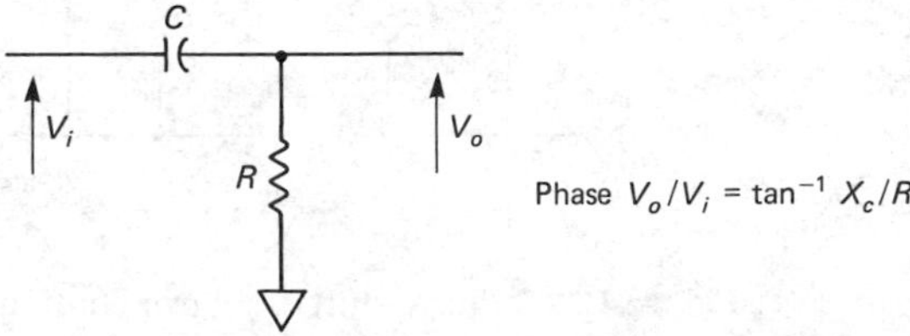

Figure 4-37 *RC* voltage divider provides phase shift.

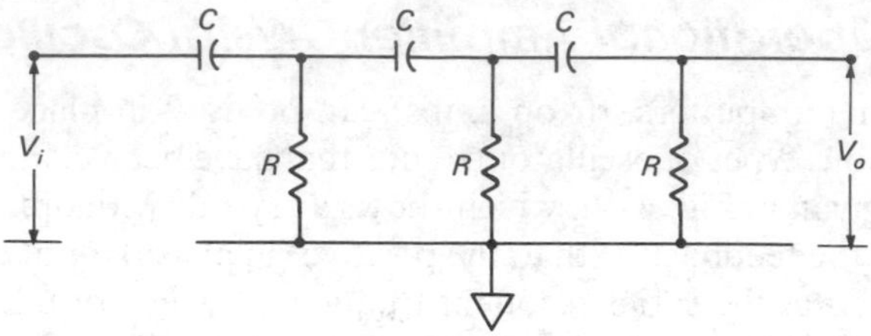

Figure 4-38 Three-section phase-shift network.

or more sections are used. A three-section phase-shift network, as in Fig. 4-38, works well and requires a relatively small number of components.

$$\frac{v_o}{v_i} = \frac{1}{1 - 5\,\alpha^2 - j(6\,\alpha - \alpha^3)} \tag{4-15}$$

where $\alpha = 1/\omega RC$. The phase shift of the transfer function is 180° for $\alpha^2 = 6$, or

$$f = \frac{1}{2\pi\sqrt{6}\,RC} \tag{4-16}$$

At this frequency the attenuation is 1/29 (so gain must be at least 29).

To satisfy the gain criteria it may be shown that a transistor must have a beta that satisfies the relation

$$h_{fc} \geq 23 + \frac{29R}{R_c} + \frac{4R_c}{R} \tag{4-17}$$

from which optimization of R_c/R at 2.7 will result in a transistor requiring a h_{fe} of at least 44.5 to satisfy oscillator criteria. A typical example is shown in Fig. 4-39. Such a circuit, when connected, would exhibit severe distortion due to the fact that the TIS97 had an $h_{fe} >> 44.5$. A gain adjust (unbypassed R_e) would have to be incorporated so that the gain could be changed and adjusted for a linear sinewave output. Overdriving the transistor also causes a change in operating frequency.

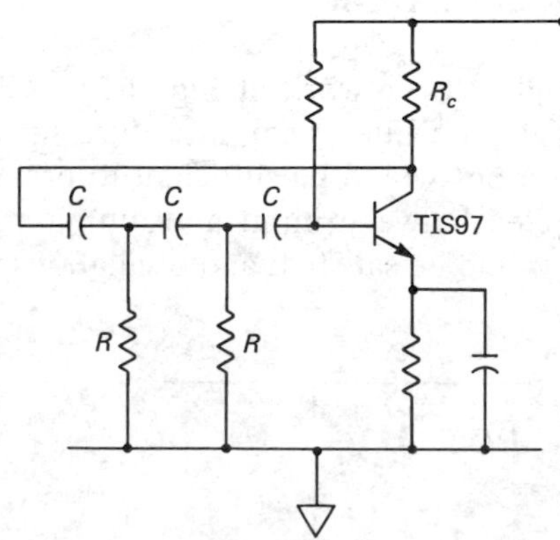

Figure 4-39 Phase-shift oscillator configuration.

4-5.1 Op-Amp Phase-Shift Oscillator

As shown in Fig. 4-40, an op amp can be used to provide the necessary gain and is a very simple oscillator to design. One of the gain-setting resistors can be made adjustable to ensure the correct value. Another variation of this type is shown by Fig. 4-41.

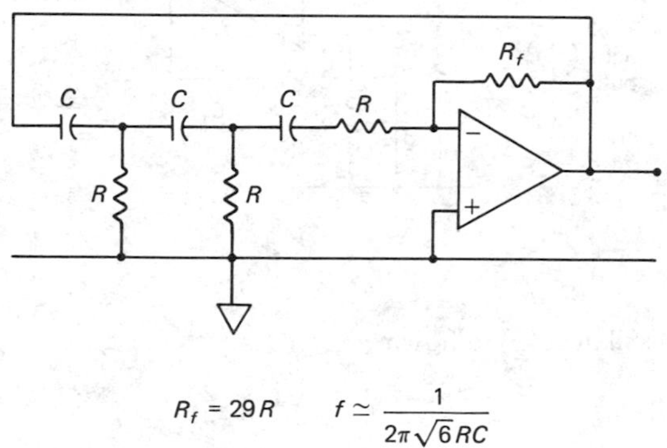

Figure 4-40 Op-amp phase-shift oscillator.

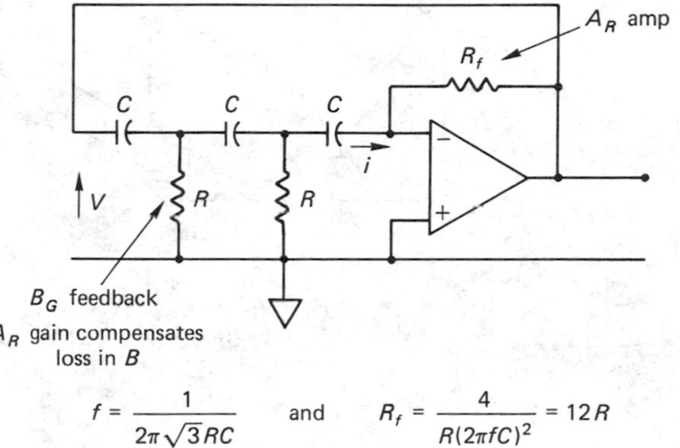

Figure 4-41 Op-amp phase-shift oscillator.

The R_f resistor is critical between linear oscillations and no oscillation at all. To avoid this problem, the feedback resistance can be modified as shown in Fig. 4-42. The setting on the 10-kΩ resistor is not critical and can easily be adjusted for a sinusoidal or distorted output waveform. The output amplitude will vary with this setting as well. A sample circuit was tested with $R = 5.6$ kΩ, $R_f = 68$ kΩ, and $C = 0.01$ μF, implying that $f = 1.5$ kHz. The measured oscillator frequency was 1.35 kHz.

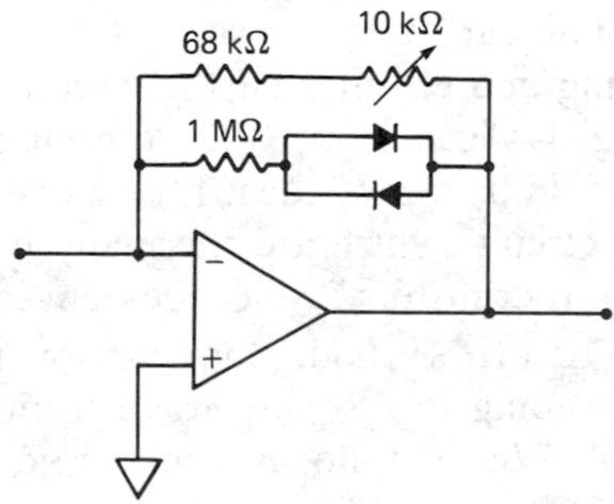

Figure 4-42 Modified gain control.

4-6 WIEN BRIDGE OSCILLATOR

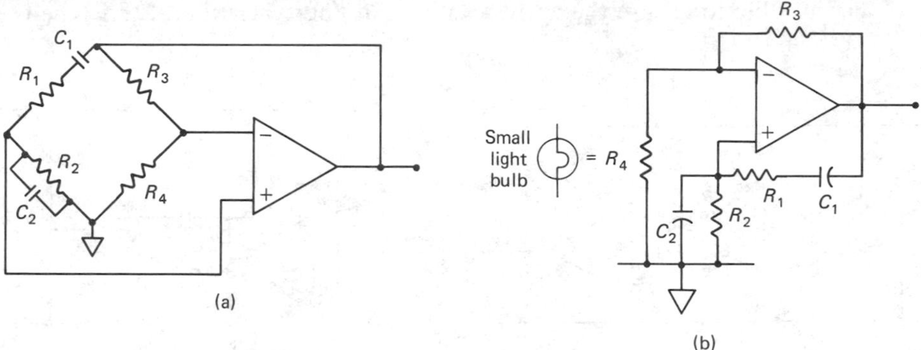

Figure 4-43 (a) Wien bridge oscillator; (b) redrawn.

Another popular oscillator that provides a clean output is the Wien bridge type shown in Fig. 4-43(a). When the bridge is balanced, both inputs of the op amp receive the same signal. If the op amp has a high gain, and it usually will, the circuit will oscillate at a frequency very close to that which would achieve balance, that is,

$$f = \frac{1}{2\pi\sqrt{R_1R_2C_1C_2}} \tag{4-18}$$

Or if $R_1 = R_2 = R$ and $C_1 = C_2 = C$, then

$$f = \frac{1}{2\pi RC} \tag{4-19}$$

and minimum voltage gain $K = 3$ for sinusoidal output (i.e., $R_3 = 2R_4$). In practice, R_4 is normally a nonlinear resistance such as an incandescent lamp bulb, to give a high K upon starting and ensure that the circuit will oscillate. Alternatively, the configuration of Fig. 4-44 can be used. The output should be a large fixed resistance, so a buffer stage would be a good idea. Lower distortion may be obtained by using amplitude-limiting circuits which are thermally limited. The limiting elements of such circuits can be thermistors or incandescent lamps.

Without the 10 kΩ + diode combination the circuit will oscillate, but to prevent the output going to a square wave from a sine wave, the setting of the 1 kΩ is very critical. With the diodes, there is no problem with the adjustment of the 1-kΩ resistor, as it serves mainly as an amplitude adjustment.

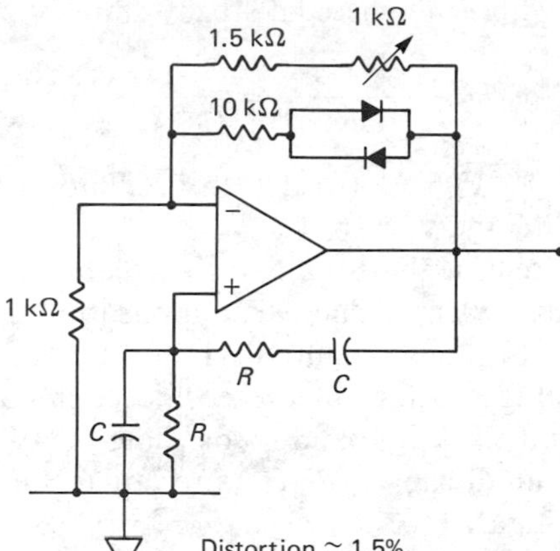

Figure 4-44 Alternative configuration for Wien bridge oscillator.

4-7 QUADRATURE OSCILLATOR

A quadrature oscillator is a circuit that provides two outputs which are 90° out of phase with each other (sine-wave, co-sine-wave output). The analysis starts with considering the circuit of Fig. 4-45. The following equations develop the voltage transfer ratio from input to output.

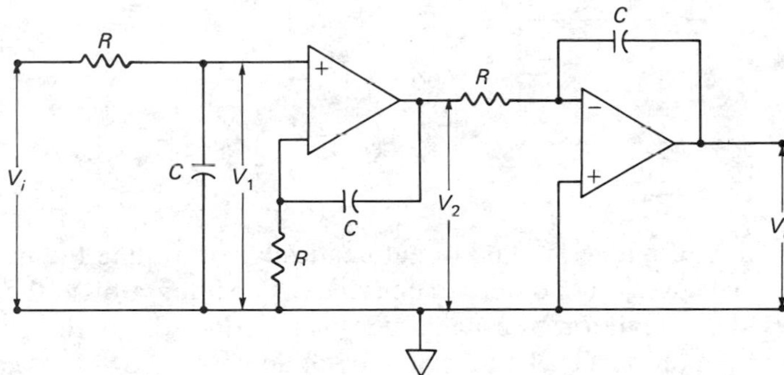

Figure 4-45 Quadrature oscillator development.

$$\frac{V_1}{V_i} = \frac{1/sC}{R + 1/sC} = \frac{1}{1 + s\tau} \tag{4-20}$$

$$\frac{V_2}{V_1} = \frac{R + 1/sC}{R} = \frac{(\text{SRC} + 1)/sC}{R} = \frac{1 + s\tau}{s\tau} \tag{4-21}$$

$$\frac{V_o}{V_2} = \frac{1/sC}{R} = -\frac{1}{s\tau} \tag{4-22}$$

$$\frac{V_o}{V_i} = \frac{1}{1 + s\tau}\,\frac{1 + s\tau}{s\tau}\,\frac{-1}{s\tau} = \frac{-1}{(s\tau)^2} \tag{4-23}$$

Assuming that s can be replaced with $j\omega$ (sinusoidal steady state)

$$\frac{V_o}{V_i} = \frac{-1}{(j\omega)^2(RC)^2} = \frac{-1}{-1(\omega RC)^2} = \frac{1}{(\omega RC)^2} \tag{4-24}$$

If V_o is connected to V_i, that is, $V_o/V_i = 1$ (to get an oscillator), the resonant angular frequency will be $\omega = 1/RC$ rad/s.

Another advantage of this circuit is that stabilizing oscillation without introducing excessive distortion is relatively easy. The usual application for this circuit is in fixed frequency operation in the range from 10 Hz to 10 kHz. Figure 4-46 shows a variation in the gain-setting resistors of the cosine circuit which makes operation more reliable. Distortion with this kind of oscillator is often less than 1%. In theory, all resistors (R) and all capacitors (C) are equal, but typically R'' is lower to ensure that oscillations start.

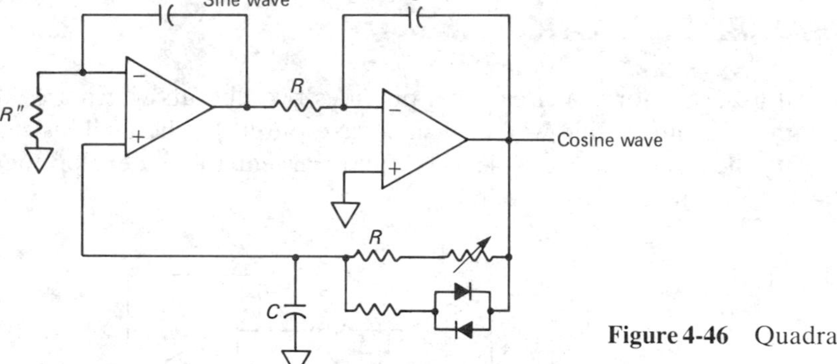

Figure 4-46 Quadrature oscillator.

An alternative method of automatically controlling the gain is shown in Fig. 4-47, which uses zener diodes to limit the cosine integrator. If the zeners are directly across the integrator, the amplitude will be limited at the zener voltage level. Adding the output-level adjustment potentiometer allows the output voltage swing to have peak values above the zener voltage.

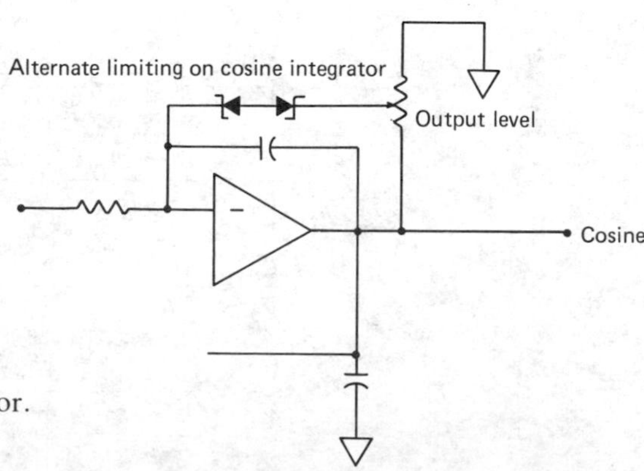

Figure 4-47 Zener limiting on integrator.

4-8 CMOS OSCILLATORS

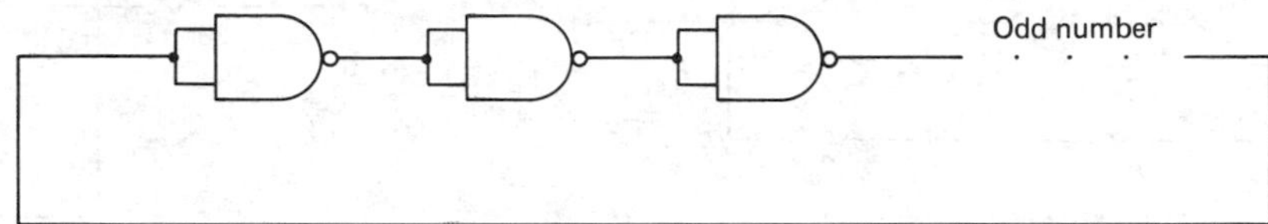

Figure 4-48 CMOS oscillator using propagation delay (square wave).

Oscillators used to generate timing signals for digital applications are often made using ordinary logic gates. Although all forms of logic devices can be used for this purpose, only CMOS devices will be considered here. A very simple form of this kind of oscillator is shown in Fig. 4-48, which has an odd number of inverting gates connected in a ring. First assume that at some moment the input to the first gate has gone low. After the propagation delay time of the gate, its output will go high. Proceeding through two more gates, the output of the last gate will also go high, becoming the input of the first gate. The process repeats, producing a square-wave output. The frequency of oscillation for this configuration will be

$$f = \frac{1}{2nt_p} \tag{4-25}$$

where t_p is the propagation delay per gate and there are n gates. Since the frequency is inversely proportional to the propagation time, the frequency will not be stable if the propagation time varies. With CMOS logic devices, however, the propagation time increases significantly as the supply voltage (V_{DD}) is decreased or as the ambient temperature increases. It also changes with loading or circuit stray capacitance. For a supply voltage of 15 V and a circuit capacitance of 12 pF, the propagation delay may be about 10 ns. For a three-stage oscillator this results in a frequency of

$$f = \frac{1}{2 \times 3 \times 10 \text{ ns}} = 16.7 \text{ MHz}$$

If the supply voltage is dropped to 5 V, the propagation delay might increase to 50 ns. To improve the stability of the oscillator, it is necessary to make the operation independent of the active device and dependent on external passive elements. Figure 4-49 shows an oscillator that is nearly independent of supply voltage variations. This is accomplished by putting a resistance in series with the input to the inverter. If $R_1 = R_2$, the oscillator will operate theoretically at a frequency of about

$$f \simeq \frac{0.48}{RC} \tag{4-26}$$

Because of the input diode protection circuit that is usually found in CMOS devices to provide protection from static discharge (as in Fig. 4-50), the voltage V_1 will be

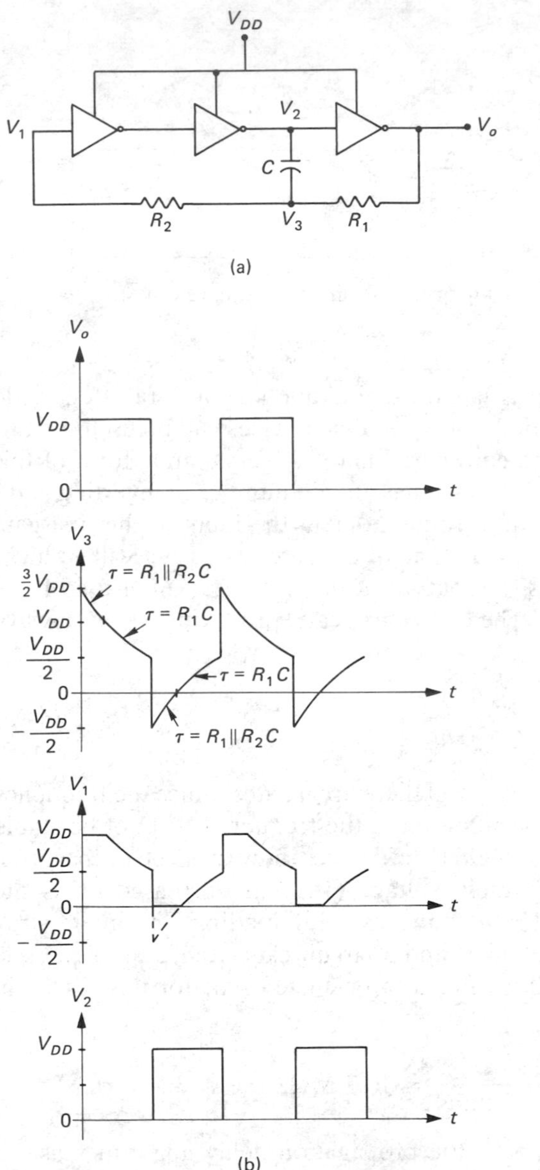

Figure 4-49 (a) Supply independent CMOS oscillator and (b) waveforms.

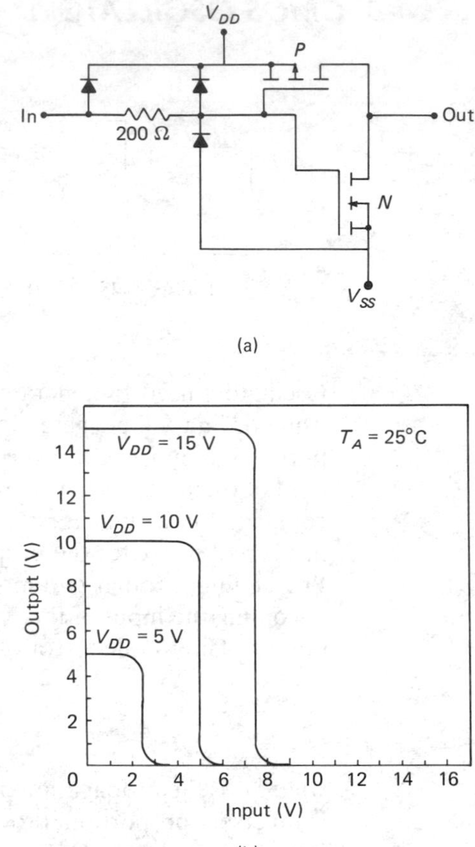

Figure 4-50 (a) CMOS input protection circuit and (b) typical transfer characteristics.

clamped to V_{DD} when V_3 exceeds V_{DD} and to zero volts when V_3 is more negative than ground. Current flows through R_2 during these clamping periods.

When V_1 reaches the threshold (about 50% of V_{DD}), the first inverter begins to switch to the opposite state. The second inverter also starts to change state, which is reflected as change in V_3, since the voltage across a capacitor cannot change instantly. This change in V_3 reinforces the initial switching stimulus. Such positive feedback causes the switching action to be almost instantaneous.

For improved frequency stability, quartz crystals or ceramic resonators can be used as the feedback element in CMOS oscillators. Figure 4-51 illustrates a typical CMOS crystal oscillator that can be used as the timing reference for a digital clock (dividing 32.768 kHz by 2^{15} gives a 1-Hz signal). This circuit is identical to the Pierce oscillator, where the crystal operates as the inductive feedback element. The two capacitors form the other legs of the Pierce oscillator configuration. The 10-MΩ resistor biases the input at about $V_{DD}/2$. This resistor value is not critical, but it should be high enough to prevent loading of the feedback network, yet low enough in comparison to the amplifier input resistance so as to have a small dc voltage drop. The value for the resistor on the output (R) is also not critical and can even be eliminated. But it does provide an improved frequency stability and tuning adjustment.

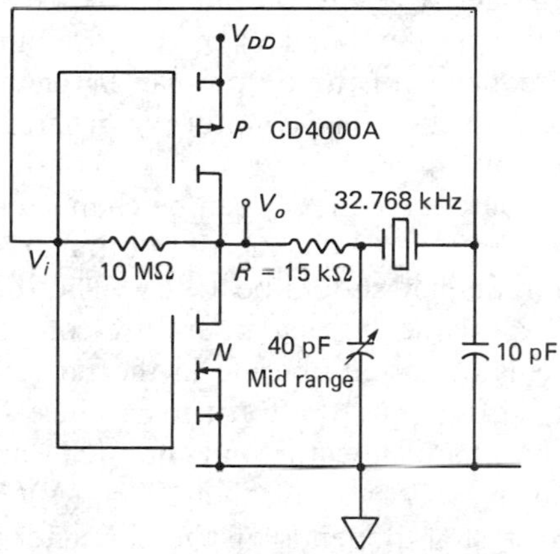

Figure 4-51 CMOS crystal oscillator (Pierce).

4-9 PHASE-LOCKED LOOPS

Phase-locked loops (PLLs) are systems that involve voltage-controlled oscillators along with phase detectors, amplifiers, filters, and programmable counters. They have a number of applications, such as FM demodulators, frequency multiplication, and frequency synthesis. In this section we examine the use of PLLs to synthesize frequencies for use as local oscillators in receivers and transmitters. Typical PLL synthesizer circuits will be analyzed.

PLL synthesizers are often used in modern equipment in place of more conventional oscillator circuits because they provide a more precise method of controlling the local oscillator frequency and because they are easy to adapt to digital (computer) control and digital displays. Some disadvantages also exist. PLL synthesizers are usually more complicated and expensive than conventional circuits, and unless careful design techniques are used, the signals produced can have a higher-phase noise component than those of conventional oscillators.

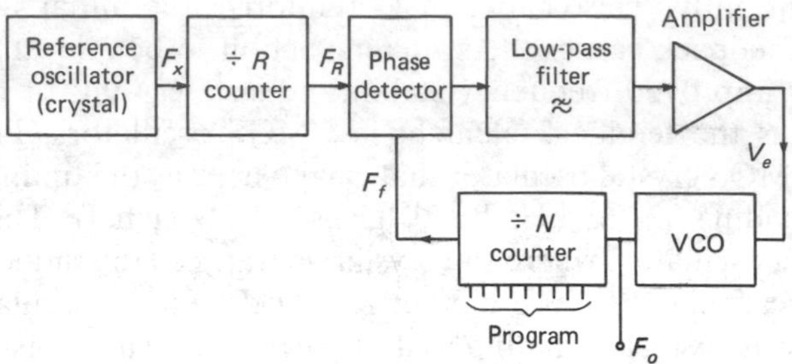

Figure 4-52 Basic phase-locked-loop frequency synthesizer.

A basic design for a phase-locked-loop frequency synthesizer appears in Fig. 4-52. The object of the design is to produce a stable output frequency (F_o) from the voltage-controlled oscillator. The frequency of this output will depend on the reference frequency (F_R) and on the division number (N) of the programmable counter. In fact, $F_o = N \times F_R$. The frequency of the output can be changed by changing the number by which the counter divides. The counter is a programmable digital device that can be loaded with a binary number from a keyboard, digital memory, or a controlling computer. The output frequency can be changed only by increments of F_R. That is, if we want to be able to change the output frequency by increments of 1 kHz, the reference frequency should be 1 kHz. The reference frequency is derived from a highly stable source such as a temperature-compensated crystal oscillator. Since such crystal oscillators often operate in the range 1 to 10 MHz, a separate "divide by R" circuit is often placed after it to produce F_R. The feedback control loop, consisting of the phase detector and the VCO and programmable counter, keep $F_R = F_f$, and since $F_f = F_o/N$, then $F_o = (N)(F_R)$.

Since the operation of this control system depends on the characteristics of its components, their operation should be understood. The phase detector is a component that produces a dc output voltage proportional to the phase difference between the two input sine waves, which must be of the same frequency. This seemingly difficult task can actually be performed by a number of simple circuits, including a balanced modulator (multiplier), an AND gate, or an exclusive-OR gate. To illustrate this concept, consider the case where a multiplier is fed with two sine waves, $v_1(t) = A \cos(2\pi ft + \theta)$ and $v_2(t) = B \cos(2\pi ft)$, which have the same frequency but differ in amplitude and also have a relative phase difference [Fig. 4-53(a)]. The output of the multiplier will be the product of the two waves, that is,

$$\text{output} = [v_1(t)][v_2(t)] = AB[\cos(2\pi ft + \theta)][\cos(2\pi ft)]$$

$$= \frac{AB}{2}[\cos(2\pi ft + \theta - 2\pi ft) + \cos(2\pi ft + \theta + 2\pi ft)]$$

$$= \frac{AB}{2}[\cos\theta + \cos(4\pi ft + \theta)] \tag{4-27}$$

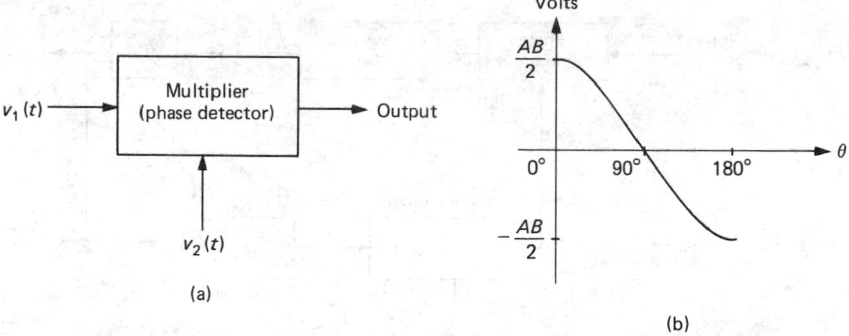

Figure 4-53 (a) Phase detector and (b) voltage/phase response.

The second term in the final relation for the output spectrum is a high-frequency term of twice the frequency of the input signals. It is removed by the low-pass filter, leaving only the first term, which is a dc component whose value depends on θ. If the phase difference θ is 90°, this dc voltage will be $\cos 90° = 0$. If the phase varies to either side of 90°, the output voltage will vary accordingly.

The VCO is usually designed and adjusted so that its free-running frequency, which it produces when the input error voltage (V_e) is zero, is equal to the required frequency (NF_R). Under these conditions, the phase difference will be 90°. Any tendency of the VCO to drift above or below this frequency would cause a phase shift that would create a feedback error voltage which would correct the frequency error. The phase detector is usually designed so that the input sine waves are clipped into pseudo-square waves before being applied to the multiplier. This results in the voltage/phase curve of the detector being closer to the ideal (a straight line) [Fig. 4-53(b)].

If the division number N is changed, a phase shift and error voltage will occur that will cause the VCO output frequency to shift by whatever amount is necessary to make F_o/N equal to F_R, for the new value of N. The controllable range of the VCO and the linearity of the phase detector will limit the possible values that can be used for N if the system is to remain "in lock." The maximum and minimum VCO frequencies define the *lock range* and relate to maximum and minimum values for N. Within this allowed range, any desired multiple of the reference frequency can be obtained by choosing the appropriate value for N. In a frequency synthesizer, a programmable counter is used for the divide-by-N circuit. In this case, the output frequency can be chosen by changing the binary number that is fed to the counter's programming pins. One problem with this approach is that the programmable counters that are available in the standard logic families (TTL or CMOS) are limited to below 50 MHz (roughly). Two methods are commonly used to extend the basic design to the VHF, UHF, or even microwave regions. The first method is to use fixed or dual-modulus prescalers (see Figs. 4-54 and 4-55), which are made using fast logic types such as ECL or GaAs. The second method is to use a mixer and reference oscillator, as in Fig. 4-56.

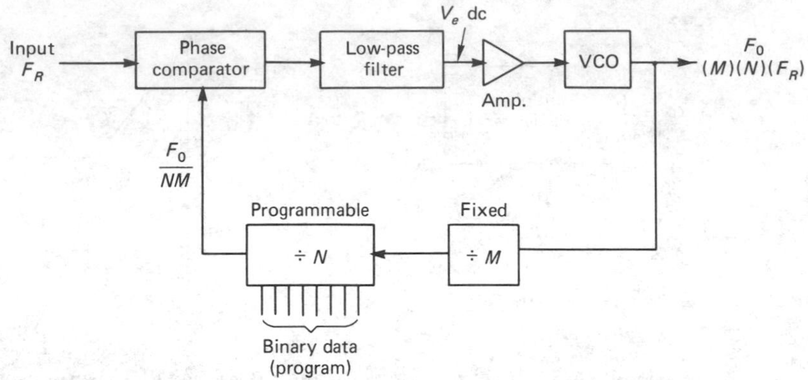

Figure 4-54 Frequency synthesizer with fixed prescaler.

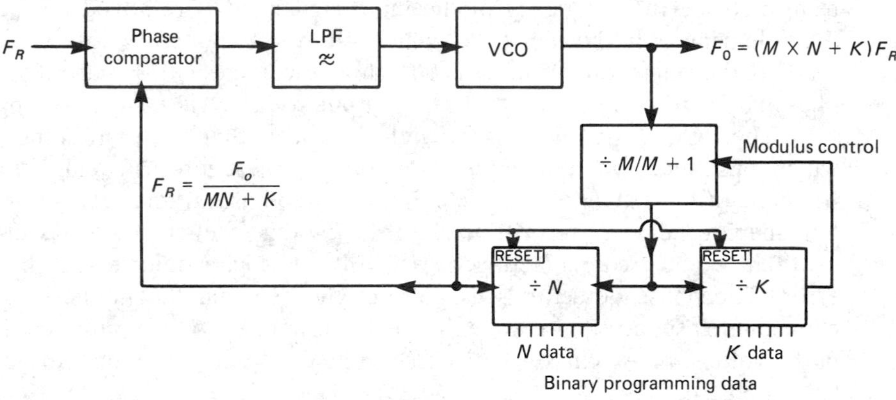

Figure 4-55 Frequency synthesizer using a dual-modulus prescaler.

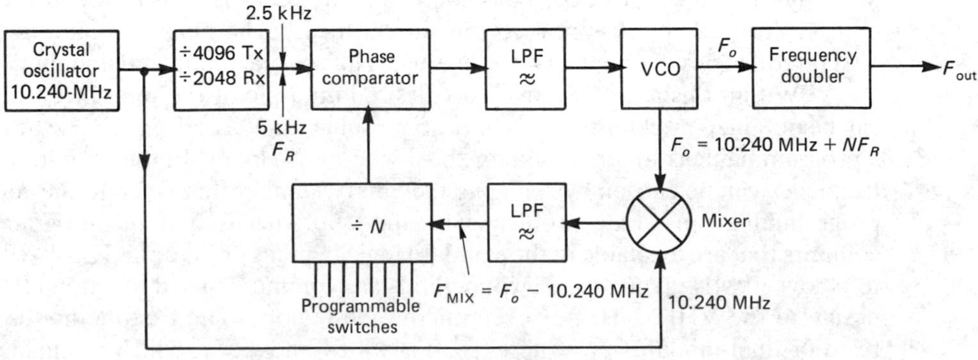

Figure 4-56.

Another relevant term with regard to a PLL is *capture range*. If the system is not initially in lock, it should be able to adjust itself until it achieves lock. Generally, the capture range is smaller than the lock range. Since this section is aimed at analyzing the operation of typical PLLs, not designing them, an analysis of the parameters that determine lock and capture ranges will not be undertaken.

4-9.1 Fixed-Modulus-Prescaler Frequency Synthesizer

Fixed-modulus prescalers are available which operate at frequencies into the gigahertz range. When inserted between the VCO and the programmable counter, their job is to reduce the output frequency (F_o) to a range (F_o/M) the programmable counter can handle. The output frequency with this configuration is given by $F_o = NMF_R$. The value for M ranges typically from 4 to 256. One drawback of this arrangement is that the minimum increment by which the output frequency can change is now MF_R, not F_R. That is, if N is increased by 1, the output frequency will increase by $MF_R = (N + 1)MF_R - NMF_R$. If the reference frequency is decreased (by $1/M$) to compensate for this decreased resolution, the loop will take longer to lock up and the delay will also cause the noise (phase jitter) to increase.

4-9.2 Dual-Modulus-Prescaler Frequency Synthesizer

The system shown in Fig. 4-55 is a bit more elaborate than the fixed modulus type but avoids some of its problems. At the start of the count cycle, both of the programmable counters are loaded with their division numbers from their binary data inputs, and the dual-modulus prescaler is set to divide by $M + 1$. The output from the prescaler decreases both programmable counters in parallel. When the K counter reaches zero [after $(M + 1)K$ cycles of F_o], its output sets the dual-modulus prescaler to divide by M instead of by $M + 1$. The N counter continues to count down to zero (it is presumed that N is greater than K). When the N counter reaches zero, the N and K counters are reset to their respective count numbers and the prescaler is reset to $M + 1$. The N counter therefore reaches zero after $K(M + 1) + (N - K)M$ cycles of F_o. As a result,

$$F_R = \frac{F_o}{K(M + 1) + (N - K)M}$$

$$F_o = F_R[K(M + 1) + (N - K)M] = F_R(KM + K + MN - KM) \tag{4-28}$$

$$F_o = F_R(MN + K)$$

Since N and K are selectable, any desired integer can be obtained for $(MN + K)$. Therefore, the output frequency can change in minimum increments of F_R so a major disadvantage of the fixed-modulus prescaler is eliminated.

4-10 TIMER-BASED OSCILLATORS

Another class of oscillators makes use of timer or multivibrator integrated circuits. Such devices generally operate by allowing a capacitor to charge up through a resistor. When a preset threshold voltage is reached, a switching operation occurs which can reset the device, thereby producing a periodic output whose frequency is determined by the R and C timing components. Typical devices are the 74LS221 dual monostable multivibrator, NE555/ICM7555 timer, and 74LS628 voltage-controlled oscillator. The reader is directed to standard data books for device details. To illustrate the concept, Fig. 4-57 shows an ICM7555 (Signetics) general-purpose CMOS timer configured for astable (oscillator) operation. When power is applied to the circuit, the capacitor C is in the discharged state and the trigger input is at a low voltage. This triggers the timer so that the internal discharge transistor (n-channel FET) is turned off, allowing the capacitor C to charge up through R_A and R_B. The output is high at this point. When the voltage on the capacitor reaches the threshold level $\frac{2}{3}V_{CC}$ (as determined by the internal voltage divider), the output goes low and the internal discharge transistor is turned on. This allows the capacitor to discharge through R_B. When the voltage on this capacitor drops to $\frac{1}{3}V_{CC}$, the trigger comparator switches, causing the flip-flop to turn the internal transistor off. The output goes high and another cycle begins. The frequency (f) of the (nonsymmetrical) output is determined by R_A, R_B, and C but is relatively independent of V_{CC}. If $R_A < R_B$, then

$$f = \frac{1.38}{(R_A + 2R_B)C} \tag{4-29}$$

The duty cycle of the output is given approximately by

$$D = \frac{R_A + R_B}{R_A + 2R_B} \tag{4-30}$$

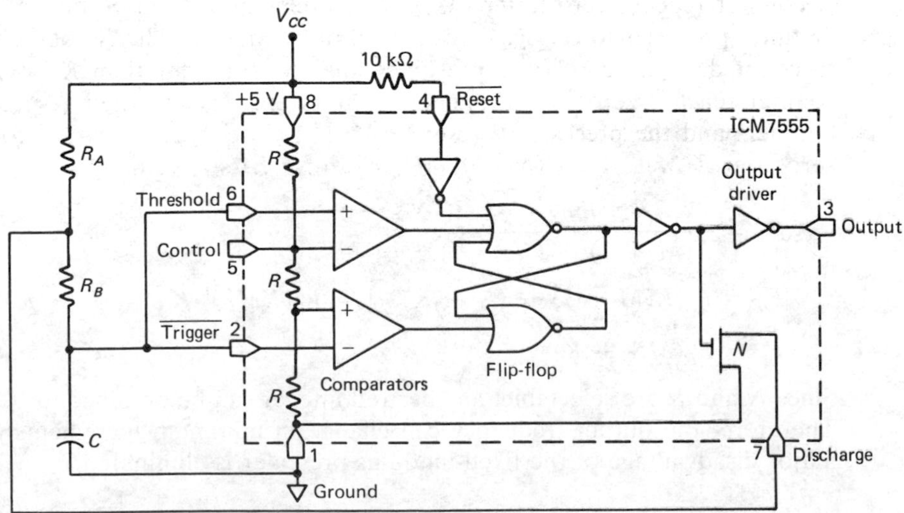

Figure 4-57 Astable oscillator using ICM7555 CMOS timer.

Operation over a frequency range of 1 Hz to 500 kHz is possible with proper selection of the timing components, but a minimum value of 3000 Ω is recommended for R_B to ensure reliable start of oscillations. For example, choosing $R_A = 4700\ \Omega$, $R_B = 12{,}000\ \Omega$ with $C = 0.001\ \mu F$ gives an oscillation frequency of 48.08 kHz and a duty cycle of about 58%.

Problems

4-1. With reference to the FET Colpitts oscillator of Fig. 4-26, choose the appropriate value for L if $C_1 = 39$ pF, $C_2 = 120$ pF, and an oscillation frequency of 5.00 MHz is desired. Assume that the FET is properly biased.

4-2. If the series resistance term in the equivalent-circuit model of a quartz crystal is increased, the Q of the crystal will:
(1) Go up
(2) Go down
(3) Stay the same
(4) Change in an unpredictable manner

4-3. If a small capacitor is added in parallel with a quartz crystal, the parallel resonant frequency will:
(1) Go up
(2) Go down
(3) Stay the same
(4) Change in an unpredictable manner

4-4. Crystal oscillators are often used instead of LC oscillators in communication transceivers because:
(1) They are much simpler to build
(2) They have a wider tuning range
(3) They produce the purest sine waves
(4) They have better frequency stability

4-5. A quartz crystal is used in an overtone mode when:
(1) A higher operating frequency is desired
(2) A more muscial tone is desired
(3) A lower operating frequency is desired
(4) Simultaneous operation at both the series resonant and parallel resonant frequencies is required

4-6. With reference to Fig. 4-49, calculate the expected oscillation frequency if $R_1 = R_2 = 6800\ \Omega$ and $C = 470$ pF.

4-7. What determines the frequency stability of a phase-locked-loop (PLL) frequency synthesizer?
(1) The magnitude of the division number used in the programmable dividers
(2) The stability of the voltage-controlled oscillator
(3) The bandwidth of the low-pass filter
(4) The stability of the reference oscillator

4-8. What would be the output frequency of the PLL synthesizer in Fig. P4-8?

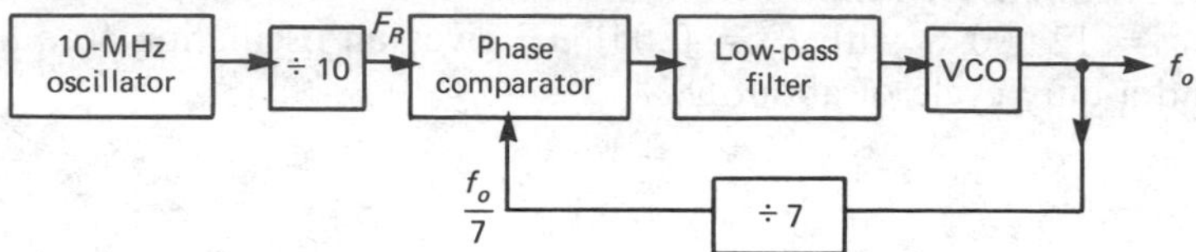

Figure P4-8 PLL synthesizer.

4-9. The frequency synthesizer in Fig. P4-9 is used to produce a signal frequency in the range 26.020 to 26.800 MHz inclusive, with channel separation of 20 kHz.
(a) If $f_1 = 15$ MHz, calculate K, N_{min}, and N_{max}.
(b) If N is limited to two decimal digits (i.e., 99), calculate the minimum value that can be used for f_1 if the output is to be in the range specified.

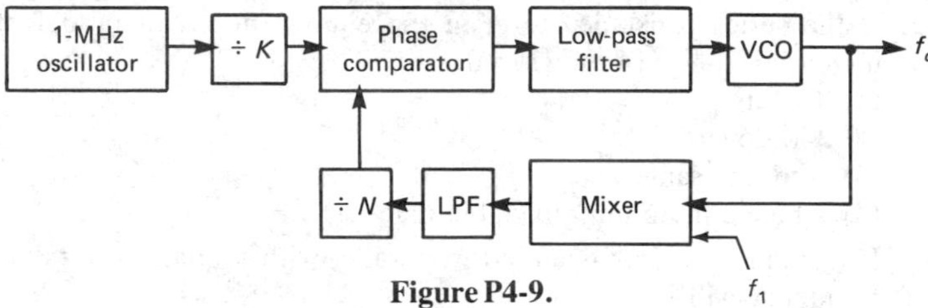

Figure P4-9.

4-10. For the PLL frequency synthesizer of Fig. P4-10, what should the dividing counter (N) be set to for an output frequency of 2.1 MHz?

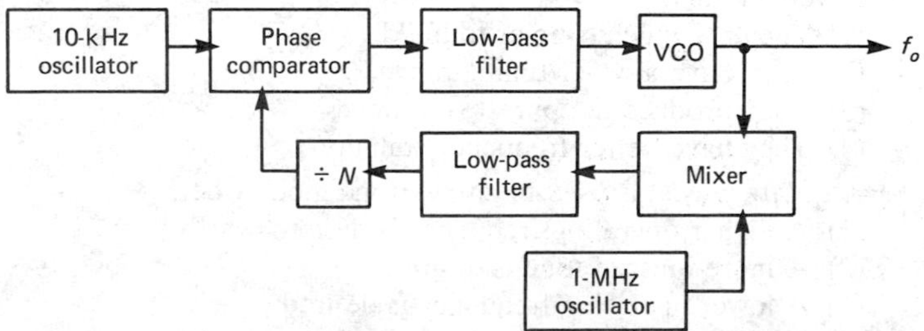

Figure P4-10.

4-11. Consider the circuit of Fig. P4-11. Assume that the low-pass filters have cutoff frequencies appropriate to pass the lower sideband components of signals fed to them. If the crystal oscillator produces a frequency of 10 MHz, what would the output frequency (f_o) be for $N = 2$? Assume that $f_o > f_2$.

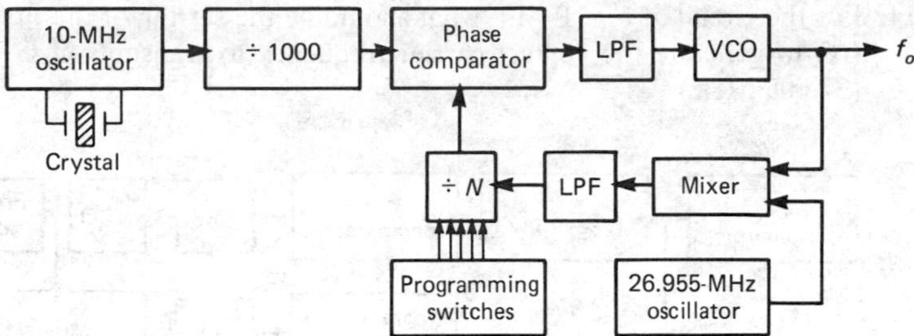

Figure P4-11.

4-12. Assume that the programmable counter in Fig. P4-12 can take on values in the range 81 to 125 inclusive. For the output frequency shown (26.86 MHz), calculate:
(a) VCO frequency, assuming that $f_{vco} > 30.82$ MHz
(b) Mixer 2 output frequency, under the assumption above
(c) The required value for the programmable divider (N)

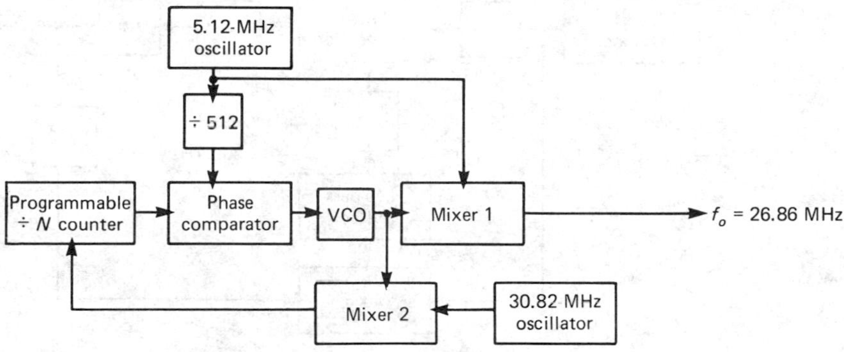

Figure P4-12.

4-13. The PLL synthesizer of Fig. P4-13 is used in a 40-channel GRS (citizen's band) transceiver.
(a) Derive the general expression for f_o in terms of f_1, N, and K if $f_o/2 > f_1$.
(b) If $K = 4096$ in transmit mode, what value is required for N if the output frequency (f_o) is to be 26.965 MHz (channel 1)?
(c) For a channel spacing of 10 kHz, what value of N is required for channel 2?

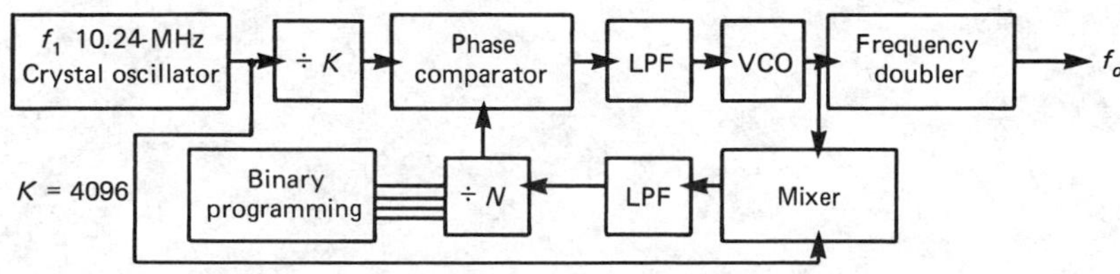

Figure P4-13.

4-14. For the circuit of Fig. P4-14, what should be the setting of the programmable counter (N) to generate the correct carrier frequency to transmit on GRS (CB) channel 14 (27.005 MHz)?

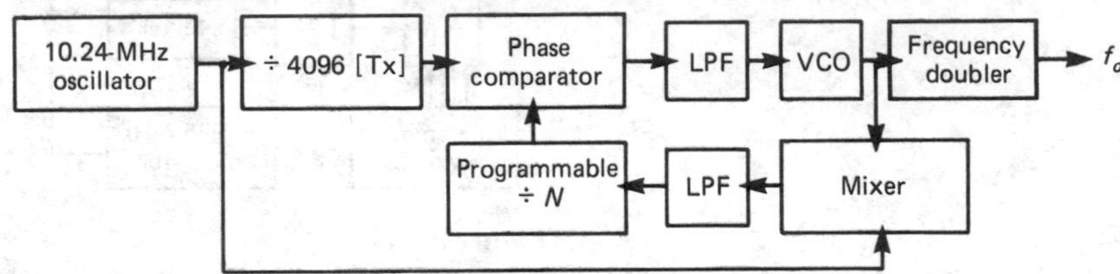

Figure P4-14.

4-15. Determine f_o in Fig. P4-15 assuming that the high-pass filter passes the sum frequency component only.

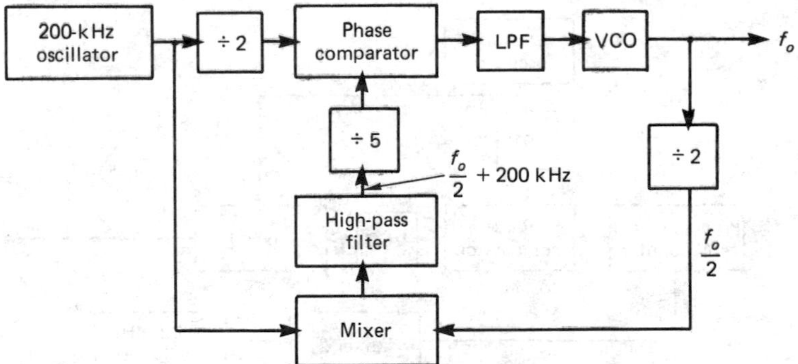

Figure P4-15.

4-16. Referring to Fig. 4-56, which shows an oscillator based on a CMOS timer, calculate the required value for R_B if R_A is 1800 Ω, C is 470 pF, and an oscillation frequency of 360 kHz is desired.

Active Radio-Frequency Circuits

Alex Krieger

5-1 INTRODUCTION

The purpose of this chapter is to present some basic concepts and ideas about the fundamental active radio-frequency (RF) circuits most often encountered in typical RF receivers and transmitters: namely, amplifiers and mixers. Although their purposes and operation are basically simple, particular applications in the RF domain often impose certain constraints and requirements on them which may lead to design configurations which are not as straightforward and easy to analyze and understand as they are intended to be. Although the circuit configurations discussed in this chapter will handle frequencies up to some 300 MHz, the concepts and concerns apply to any frequency range, and the typical RF is usually taken to cover only up to about 1 to 2 GHz.

It is the intention in this chapter to introduce the reader to some of these peculiar applications, issues, constraints, and requirements, and the various less-than-obvious circuit configurations they may lead to, and their rationale. The overall objective is for the reader to be aware of the various issues involved in a given application so as to better appreciate and understand any related receiver or transmitter schematic diagram and its operation. Thus the emphasis is not on the design of these circuits, the detailed math behind them, and the issues involved, but rather on a qualitative feeling for what is going on and for the practical considerations that may have led the original designer to choose one circuit configuration over another. For more detailed information on RF design, the reader is referred to other excellent and well-respected references, some of which are listed at the end of this chapter.

5-2 COMMUNICATIONS AND SIGNAL QUALITY

In any communication system, the primary concern is the quality of the message delivered. Thus we want the end user to receive the "cleanest" possible copy of the original message sent, with as little corruption, distortion, and alteration, whether intentional or not, as possible. This may at first seem rather obvious. However, we must realize and accept the simple fact of life that anything (electronic device, computer, paper, voice, transducer, human being, etc.) involved in transferring this message (voice, computer data, television picture, phone facsimile, atmospheric temperature measured at a remote weather station, etc.), from its original source to the very end user, inherently either picks up random noise and adds it to the message it is entrusted with, before passing it on, or else is simply not capable of reproducing it exactly, and therefore distorts it a little bit.

This message signal quality is conventionally measured by the *signal-to-noise (S/N) ratio*, which is defined as

$$\text{SNR} = \frac{\text{signal power level}}{\text{noise (and distortion) power level}}$$

which we will obviously want to be as high as possible. For typical voice communication, it is usually accepted that the voice is still intelligible with an *S/N* value of some 12 to 15 dB. For television, however, a moderate-quality picture requires an *S/N* of at least about 40 dB. For high-quality music, one may demand as high as 65 dB (vinyl records) to 85 dB (digital compact disk), or more. Thus a principal concern with all the circuits considered in this chapter is their performance with respect to how accurately they reproduce the information signal presented to them, or how badly they distort or alter it from its original shape. This concept of the signal quality and how to respect it is the main element to keep in mind when dealing with the analysis, operation, selection, and design of any electronic circuit to be entrusted with processing a message.

5-2.1 Some Basic Concerns with RF Circuits

As opposed to lower audio and baseband frequency circuits, the basic challenge with RF circuits has to do with the inherently high frequencies required to convey the information by carrier modulation. Thus the approach to, the underlying assumptions, and the design philosophy of such circuits are quite different from those in the case of lower frequencies. Some issues apply to both, such as noise, nonlinear distortion and saturation, and power supply filtering. However, in the context of communication by radio signals, inherent characteristics must constantly be kept in mind if they are to be dealt with effectively, and if the circuit configurations, and how they are intended to process these signals, are to be appreciated. These factors include the following:

1. The high frequencies of the signals to be processed, which means that stray capacitances and inductances of simple wiring, even within a transistor, begin

to interfere with the desired circuit operation, and must be accounted for properly lest the operation (e.g., an amplifier) become unstable and oscillate, or produce other undesirable results.

2. The different kinds of carrier modulation that can be used to convey information, each requiring a different amplifier structure.
3. The simultaneous sharing of the radio-frequency spectrum with other users. This requires that the circuits be tuned to reject undesired signals, lest there be interference and failure to communicate clearly.
4. The great attenuation of the signal power levels at long distances. This requires considerable receiver gains, leading to potential instability and oscillation, which is usually further encouraged by the stray elements mentioned above.
5. The possible need for the receiver to discern a very weak signal in the presence of another strong one nearby in the frequency spectrum, which requires that the receiver have a wide enough dynamic range to properly amplify and separate the weak signal without overdriving the amplifier into saturation by the strong signal.
6. When transmitting, the signals are being generated, and their levels are very strong. There is therefore no worry about losing information due to distortion, since it is available in its original form. However, we must ensure that we are sending a recognizable copy of it. Furthermore, even if this aspect is dealt with properly, we are also concerned about preventing the generation of extra interference. Thus, when transmitting a signal, the issues are obviously quite different than when receiving it.
7. The relative signal power level determines the nature of the amplifier. When small, up to 1 V peak or so, the primary concern is for linearity. However, when larger, in addition to linearity, we are also concerned with the cost of processing the signal, in terms of energy (i.e., efficiency). This is especially so with transmitters.
8. At high frequencies, the signal wavelengths are relatively small and times of propagation are relatively large. Therefore, impedances must be well matched. Otherwise, the signal energy "bounces" back from its intended destination and is lost, either reducing the system efficiency or degrading the signal quality.

Again, one way or another, the prime concern is the information signal quality and how it is affected (i.e., corrupted) by passing through a given circuit, and how can this corruption be minimized.

5-3 RF AMPLIFIERS

Assuming that the reader is reasonably familiar with basic electronics, diodes, transistors, and amplifiers, let us now discuss some specific characteristics we need be more concerned with in communications.

5-3.1 General Operating Characteristics

Consider a simple one-transistor amplifier such as in Fig. 5-1. The output collector circuit is tuned, so it will amplify only a narrow band of frequencies, those at which we wish to transmit or receive a signal. For practical purposes, if we are not concerned with the internal construction details, we may represent it by its equivalent symbol. Looking at it as a black box, we are interested in applying an input signal and obtaining at the output an identical copy, although larger, in voltage, current, or power level.

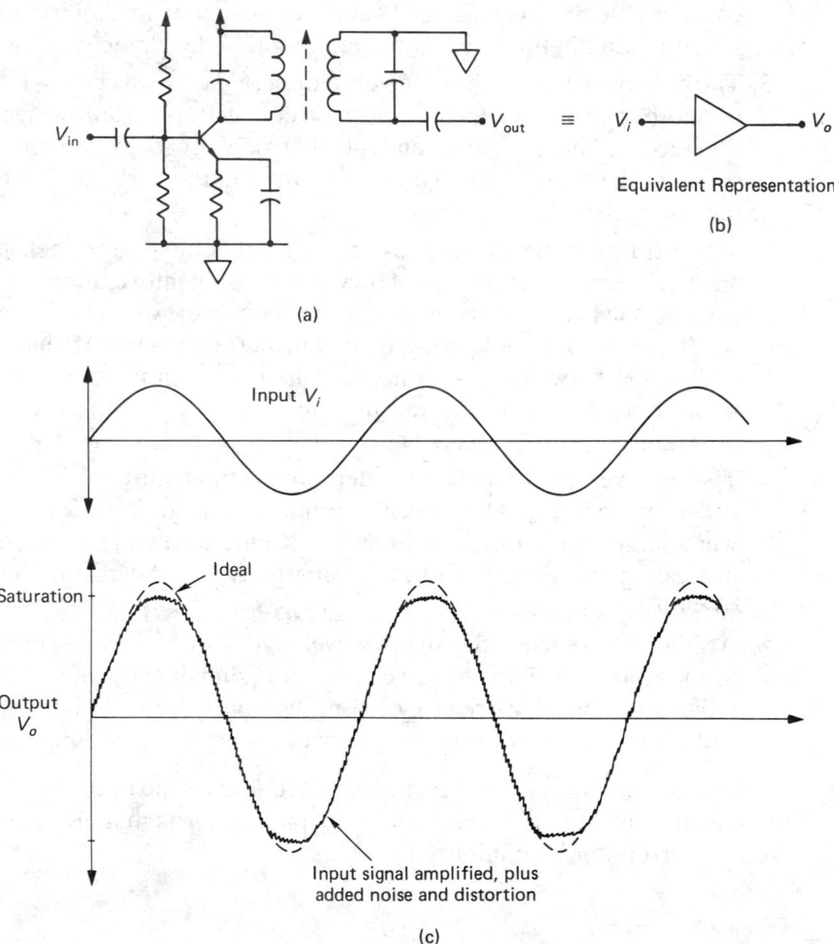

Figure 5-1 Basic transistor amplifier operation — amplify input, plus noise, plus distortion.

Note that the input signal may be of any shape. However, it is conventionally accepted that any complicated signal may be considered as being made up of many sine waves. Thus the sine wave is considered as the "fundamental" signal. By the principle of linear superposition, to the extent to which it is valid, what applies to two separate signals taken individually also applies to them together. Therefore, we can analyze what happens to a simple sine wave, and then can generalize the results to more complicated situations. Note that this principle of linear superposition requires a linear circuit, with no distortion. Although this is usually desired and can be assumed, it is *not* always the case, as we shall quickly see. When this principle is violated, we get into trouble, so we wish to be able to anticipate when it works and when it does not, and what can be done about it. Thus this factor is of principal concern in the following discussion.

Consider now the resulting output of the amplifier. Although ideally, we would like it to be an exact duplicate of the input, we invariably find that it is different. The differences may be small and of no concern, or very significant and render the signal useless. These variations are due to two factors. The first is noise, caused by the random movement of electrons and picked up from external radiations, or "added" to the signal from within the amplifier itself. Noise causes the signal to not be as well "focused," not as "sharp," as observed on a scope. Noise is always present, and is unavoidable. Its effects are most damaging to weak signals. Therefore, it is preferable that the signals be as large as possible if we wish to minimize this kind of corruption.

The second factor is nonlinear distortion, which results when too much is demanded of the amplifier, that is, when the output signal is limited in amplitude due to the fixed power supply. This manifests itself in the "clipping" of the signal at the extremes, due to amplifier saturation, as shown in Fig. 5-1. Therefore, the input signal should be kept small enough to avoid this. What exactly is "small" is relative and depends on the particular amplifier configuration. For example, for a class A amplifier as in Fig. 5-1, the signal level should be restricted to about 60 mV or so, peak to peak.

This general amplifier behavior may be summarized by a transfer function, as shown in Fig. 5-2. Thus, ideally, the input–output relationship should be a perfectly straight line, extending infinitely in both directions. In reality, its range is limited, as indicated by the saturation, and in addition, there is an uncertainty about its one-to-one relationship, due to the randomness of the noise. The extent of this randomness can be expressed by the relative "smearing" of this line, typically in microvolts rms (root mean square). For example, a typical receiver front-end RF preamplifier "adds on" about 200 nV or so of noise for a typical receiver equivalent noise bandwidth of 10 kHz.

The practical consequences of these effects are important, since they effectively corrupt the information signal with which they are entrusted. Consider again a slightly different situation, as in Fig. 5-3. A large signal is applied to the tuned amplifier of Fig. 5-1. If too large, the original message contained in the signal carrier envelope is again distorted. As discussed in Chapter 6, the frequency spectrum of the input AM signal consists of the original carrier plus two sidebands, determined by the modulation frequency, as shown in Fig. 5-4. However, when distorted it results in extra "splatter," or spurious undesired sidebands, as in Fig. 5-4(c). Note that these are evenly spaced at f_m, the original modulating frequency. If this were a transmitter output, this would cause interference to adjacent channels and would be unacceptable, as should be quickly confirmed by a visit from the appropriate authorities.

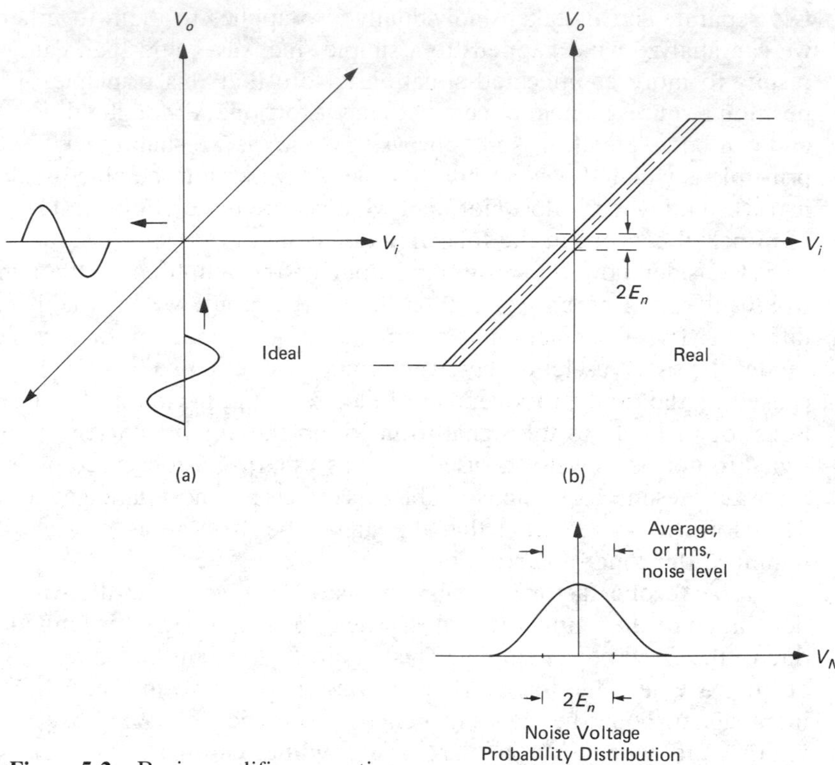

Figure 5-2 Basic amplifier operation description characteristics.

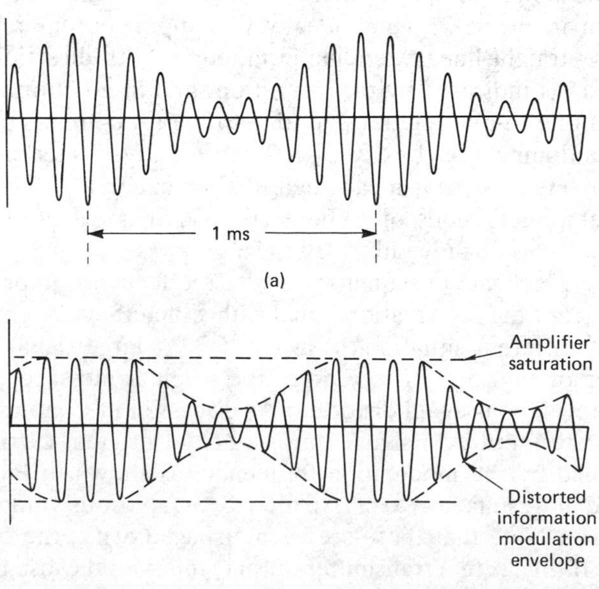

Figure 5-3 A.M. signal message envelope distortion due to amplifier saturation in the time domain.

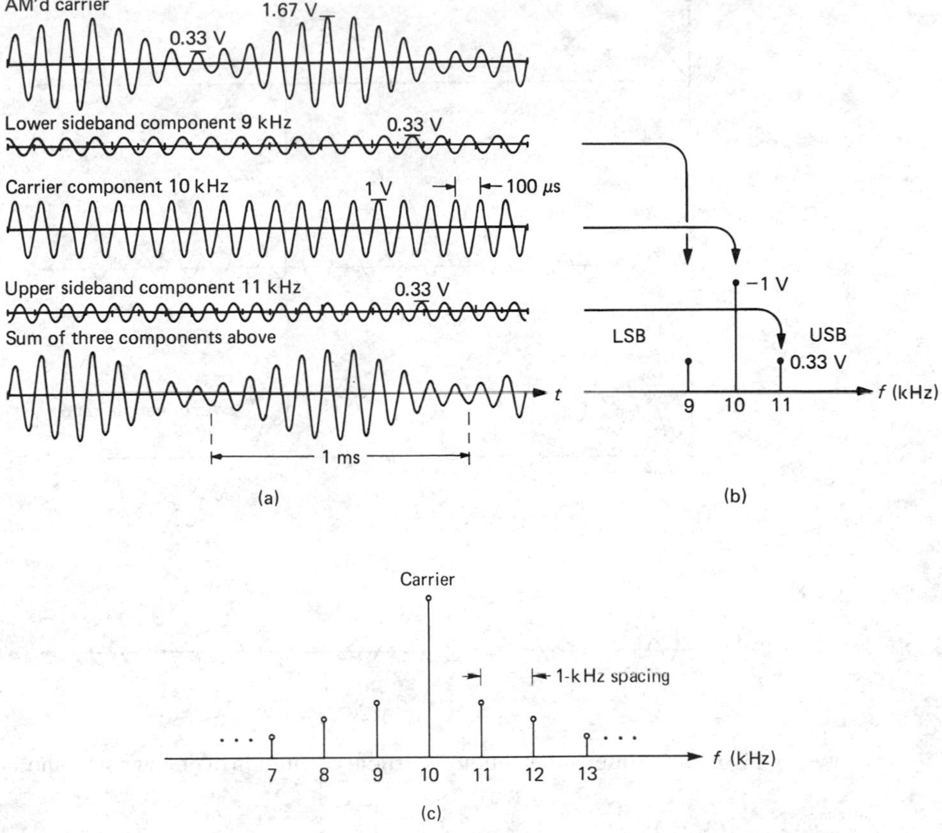

Figure 5-4 Time domain (scope) (a) and frequency spectrum (b) representations of sine-wave modulated A. M. carrier—carrier at 10 kHz, modulating frequency at 1 kHz.

Consider now the case where two signals are received simultaneously at a receiver antenna, as shown in Fig. 5-5. Such will inevitably be the case, considering the thousands of radio transmissions throughout the world at any given time. In this case we might be interested in receiving the weaker signal. This may be easy enough if the two signals are at very different frequencies. In such a case, a simple tuning filter can be used to tune in only the desired signal, and filter out the undesired one. However, if their frequencies are close enough, such as with two adjacent channels, as seen in Fig. 5-6, this filtering may not be as simple, and is usually relegated to the intermediate-frequency (IF) amplifier, in the typical superheterodyne receiver. So, prior to this filtering, or separation process, the weak signal must be amplified, but then the strong signal will also be amplified. Again, if the large signal is strong enough, the desired weaker signal may be altered as shown in Fig. 5-5(d). By the time the two signals can be separated, the desired signal has been distorted. Note that it has been altered, or "modulated," according to the other one. For this reason, the resulting distortion is commonly referred to

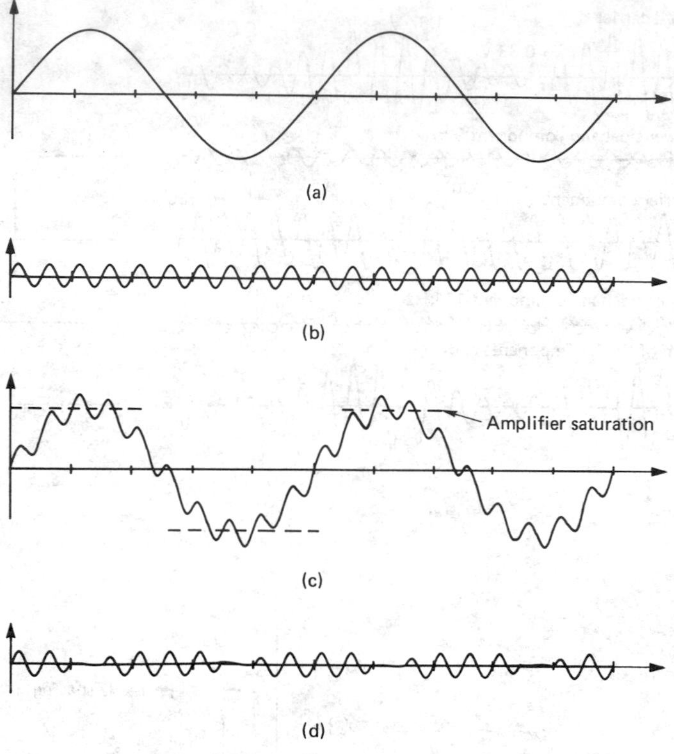

Figure 5-5 Inter-modulation distortion: corruption of one signal by another.

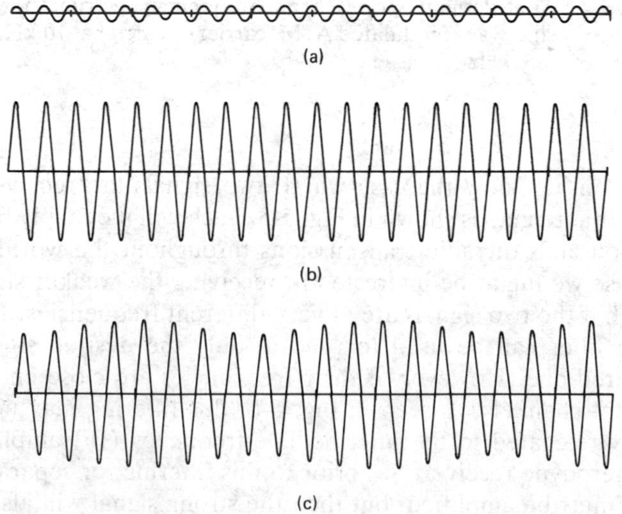

Figure 5-6 A weak signal in presence of a strong signal, at receiver input amplifier, at two different, but close frequencies, both within receiver's front-end pre-selector bandwidth.

as *intermodulation distortion* (IMD), since the two signals "intermodulate" each other. If strong enough, the offending signal may effectively completely "desensitize" the receiver to the desired signal, and thus "block" it out, although this offending signal is itself filtered out before making it to the speaker, and is never heard.

As in the case of the distorted AM signal discussed earlier, such IMD distortion results in new undesired frequency components being generated. Given two signals at frequencies f_1 and f_2, these new distortion-products appear at new frequencies, evenly spaced at every $(f_1 - f_2)$ Hz, according to the following relationship:

Third-order IMD products at $2 f_1 \pm f_2$ and $2 f_2 \pm f_1$
Fifth-order IMD products at $3 f_1 \pm 2 f_2$ and $3 f_2 \pm 2 f_1$
Seventh-order IMD products at $4 f_1 \pm 3 f_2$ and $4 f_2 \pm 3 f_1$
etc.

These odd-order products are usually of greatest concern since half of them end up within the filter bandwidth, in the immediate vicinity of the original signals, and so cannot be filtered out. On the other hand, there are also even-order products, but their resulting frequencies fall outside the filter's bandwidth, and so can easily be filtered out, and are therefore not cause for worry. This is discussed further in Section 5-3.5.

For example, refer to Fig. 5-7. This represents a typical receiver block diagram. As indicated, both the desired and undesired signals are accepted by the broad input filter. Again, "broad" is relative to the filter's center frequency. At best, one can hope for a filter Q of maybe 50. For a typical AM receiver tuned to 1 MHz, this results in a bandwidth of about 20 kHz, which spans more than one (desired) station. This receiver input preselector filter's bandwidth typically cannot be made any narrower. Thus if the larger signal overdrives the amplifier into saturation, distorting the desired signal, the overall result is intermodulation between the two signals, *and* also *any two* other frequency components present at the receiver input within that filter bandwidth. Thus many new components appear. The most troublesome are the sidebands immediately adjacent to the original desired signal component, for these cannot be removed at all by any process. The stronger signal, however, is far enough away (at a different channel frequency), that it will be possible to filter it out in a subsequent receiver stage (the IF amplifier).

The typical amplifier transfer function presented earlier is commonly presented in terms of these odd-order IMD products, as in Fig. 5-8. This basically summarizes the important characteristics that we are interested in from the point of view of communications signals. Note that the nature of these distortion products are such that their level increases faster than the desired output product, as the input signal level is increased. At the point of saturation, they are effectively strong enough to render the desired signal completely useless, as has already been established. Thus the input signal is limited as to how large it can be and still be useful. In this particular case, in Fig. 5-8, "too large" amounts to about -25 dBm, or 3 μW, which is equivalent to some 12 mV rms into 50 Ω. Note that, usually, the third-order products are the most predominant and thus the only ones presented in such a display, even though the higher products are indeed also present.

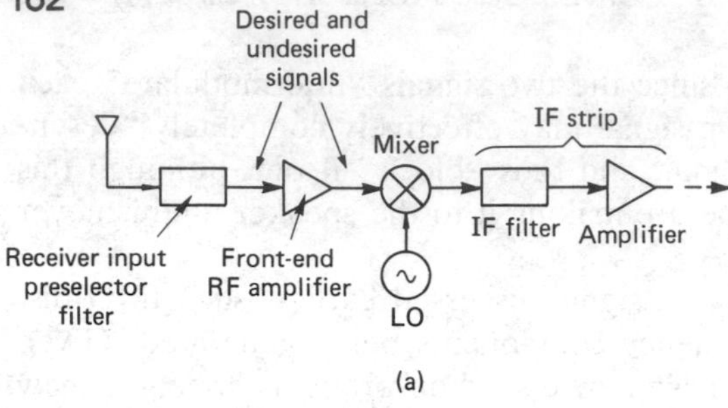

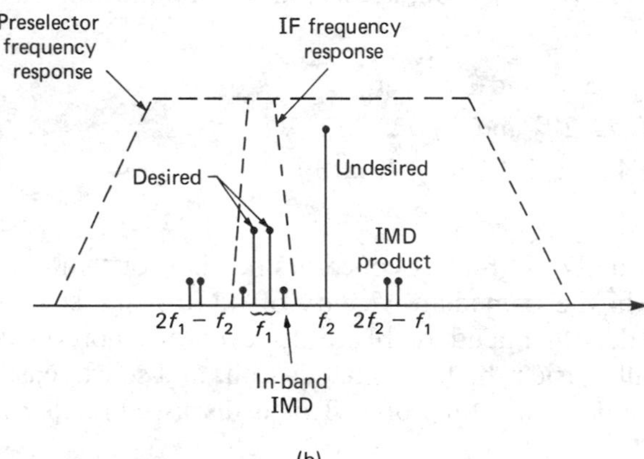

Figure 5-7 Significance of amplifier distortion in communication receiver.

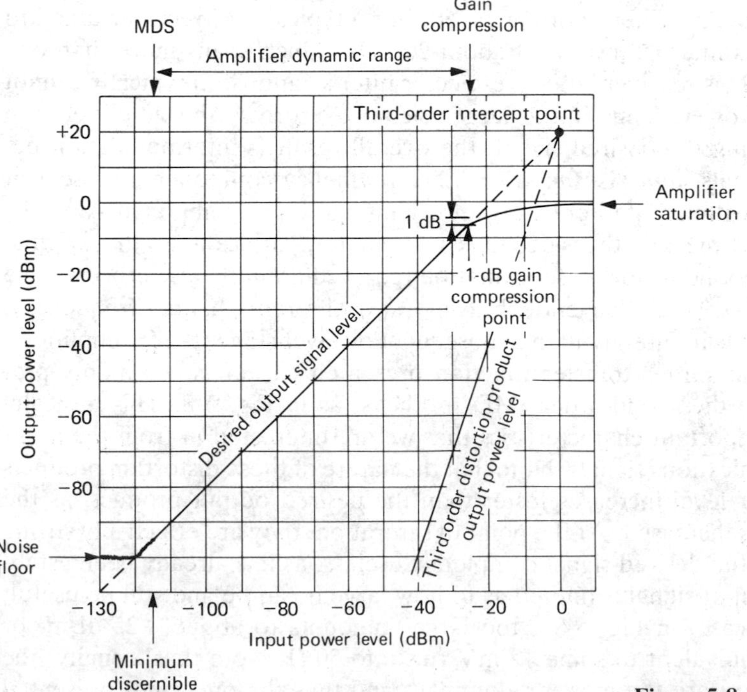

Figure 5-8 Amplifier distortion characteristics.

Consider, for example, that two signals applied together, as picked up from the antenna, of equal power but at two different frequencies, say at 10.001 and 10.002 MHz, add up to a total input power level of −40 dBm (0.1 μW, thus 0.05 μW each). Then the total output signal power would be at −20 dBm. This represents a gain of 20 dB. The most significant distortion products would appear at a level of −100 dBm (it is purely coincidental in this example that they are at the same level as the noise floor). Thus the net output signal to noise and distortion ratio would be (−20 dBm) − (−100 dBm) −3 dB = 77 dB. (recall that the noise and distortion powers add together in watts, and here they have the same level). The frequency spectrum would appear as in Fig. 5-9. Other higher-order distortion products would normally also appear, although at lower levels, for example at 10.002, 10.003, etc. and at 10.000, 9.999, etc., spaced every 1 kHz (10.002 to 10.001 MHz).

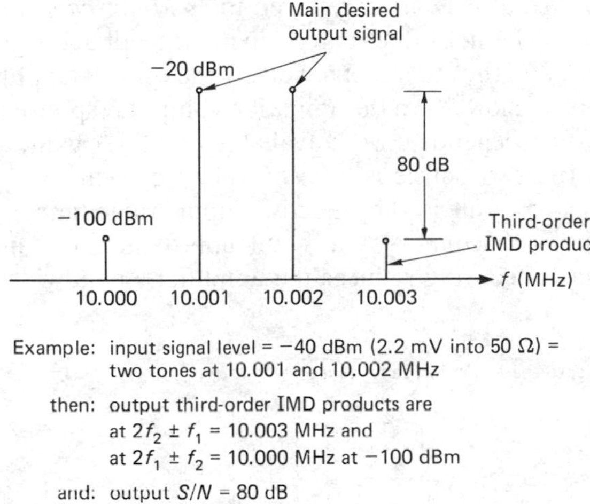

Example: input signal level = −40 dBm (2.2 mV into 50 Ω) = two tones at 10.001 and 10.002 MHz

then: output third-order IMD products are at $2f_2 \pm f_1$ = 10.003 MHz and at $2f_1 \pm f_2$ = 10.000 MHz at −100 dBm

and: output S/N = 80 dB

Figure 5-9 Amplifier output frequency spectrum and power levels: example.

At the other extreme of Fig. 5-8, if the input signal is reduced to zero, there is still an output signal present: noise. This is commonly referred to as the noise floor. Thus if the output signal is to be intelligible, the input must be strong enough to overcome the amplifier's own internally generated noise. For the output signal to be intelligible, it should typically be at least a few decibels stronger than the equivalent input noise level. How many decibels depends on the application. A voice may be understood with an S/N value of, say, 10 dB. With slow digital data, the level may be as low as 3 dB. This minimum required level is referred to as the *minimum discernible signal* (MDS) level.

Thus the amplifier can properly handle any signal level in between the MDS and that of saturation. This range is referred to as the *dynamic range*. For a good receiver it can be in excess of 100 dB. Referring to the earlier case of IMD, two signals can be effectively separated by the receiver (i.e., the undesirable filtered out), provided that their power level ratio does not exceed the receiver's dynamic range. Note that if we are only willing to accept a signal-to-noise and distortion ratio of 40 dB or greater (e.g., out of the speaker), the minimum signal level should be −80 dBm, and at most about −25 dBm. Thus, for this purpose, the amplifier's effective dynamic range would be only 55 dB.

5-3.2 Feedback Capacitance and Neutralization

As indicated earlier, because of the relatively high frequencies of RF signals, they tend to radiate easily from where they are carried (wires, components, transistors), to undesirable destinations, by whatever appropriate path exists (e.g., by air), whether or not such a path has been intentionally provided in the design. Such a path inherently and unavoidably exists inside any transistor between any two leads, and also at the semiconductor junctions. These stray paths are characterized by the junction capacitances, typically 1 to 5 pF or so, and they behave as indicated in Fig. 5-10(a). This is often referred to as *Miller capacitance*, which results in the *Miller effect.* The net effect is to provide a feedback signal from the output back to the input, and thus to affect adversely the overall amplifier's frequency response, for example, to slow down the amplifier's output response and reduce its bandwidth. This net effect depends on the available amplifier gain, and translates mathematically into the two fictitious but equivalent capacitances shown in Fig. 5-11. The most dramatic result is the effective input capacitance, which is larger than the actual Miller capacitance by a factor approximately equal to the amplifier gain. This is what effectively reduces the amplifier's bandwidth.

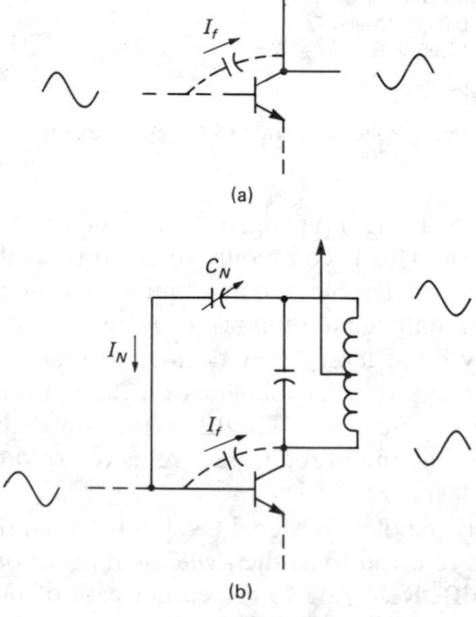

Figure 5-10 Transistor amplifier neutralization.

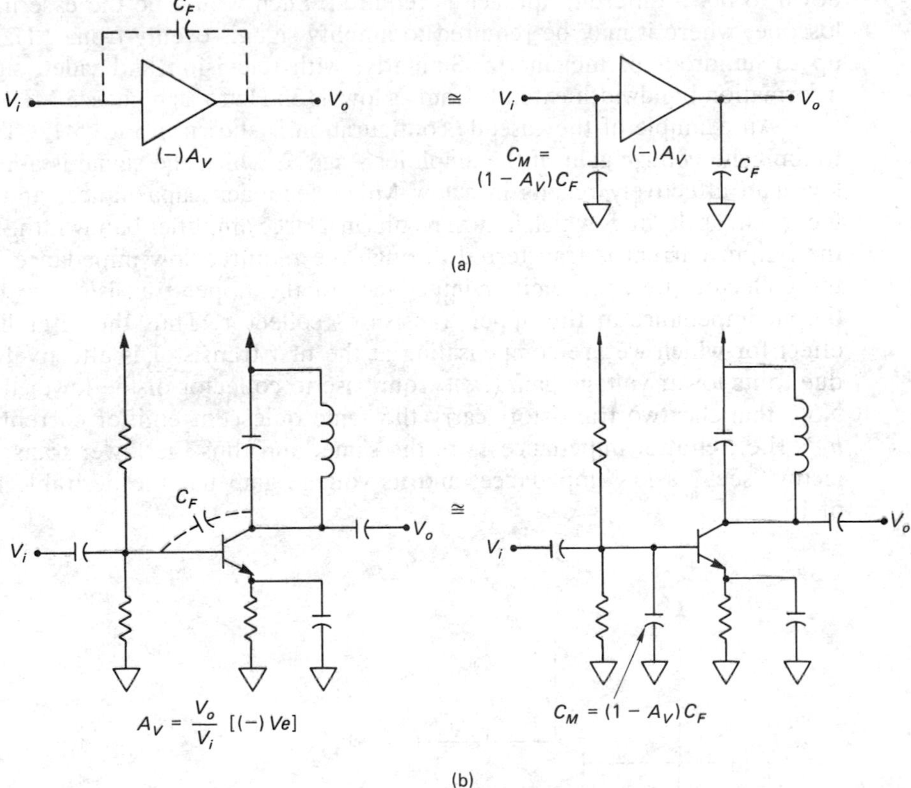

Figure 5-11 Effect of Miller feedback capacitance.

To compensate for this feedback path, a common approach is to intentionally provide another path, also derived from the output, but with the opposite signal polarity, using a center-tapped transformer, as shown in Fig. 5-10(b). This is referred to as *neutralization* of the undesirable feedback path. The exact amount of neutralization required can be adjusted with the trimmer capacitor C_n. Note that the neutralization path effectively provides positive feedback, so the adjustment must be made carefully. Too much positive feedback will result in unwanted oscillations, and although the receiver may be well tuned at the time of servicing, these oscillations may occur only when it is taken outside to a different temperature. Note also that the output is tuned, so that this approach is limited in bandwidth, although the center frequency can be extended.

Cascode Amplifier Configuration

As indicated above, although the neutralization approach allows us to operate the amplifier at higher frequencies, it works only on a tuned output over a restricted bandwidth. However, if we wish to operate over a wider bandwidth, extending down to dc, a different aproach is required. Such would be the case in an oscilloscope, where it may be required to amplify signals evenly from 0 Hz (i.e., dc) up to hundreds of megahertz. Similarly, with television and video signals, the information bandwidth extends from as low as 30 Hz to as high as 5 MHz or more.

An example of the cascode configuration is shown in Fig. 5-12. The idea is to limit the *voltage* gain of the amplifier stage to which the signal is applied. This low gain effectively results in a low Miller feedback capacitance, and thus low feedback, as desired, which in turn results in a large amplifier bandwidth. However, the output current is transferred through the required low impedance "seen" at the collector (i.e., the emitter impedance of the upper transistor) to the actual output impedance at the upper transistor's collector. Thus the offending Miller effect for which we are compensating at the first transistor is effectively reduced due to its lower voltage gain (i.e., from base to collector of the lower transistor). Note that the two transistors carry the same quiescent emitter current, so their h_{ib}'s (i.e., emitter impedances) are the same, and thus the lower transistor's collector "sees" a low impedance, and its voltage gain has the desirable low value of 1.

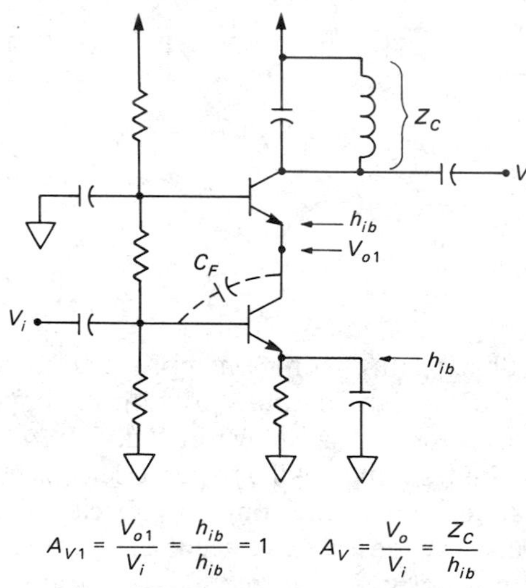

Figure 5-12 Reducing Miller effect by cascade configuration.

5-3.3 Amplifier Gain Control

As indicated earlier, the signals received at the antenna may vary in strength over a very wide range. If the overall receiver gain remained fixed, as the receiver was tuned to different stations, some would be loud, and others would not be heard at all. Consider the inconvenience of having to adjust the receiver volume while tuning across the dial, as some stations come in strong or weak, while driving a car. With respect to television signals, which carry frame synchronization in addition to picture information, the receiver may not be able to lock onto a weak signal and therefore would not reproduce any picture at all.

Thus there is a very obvious need for some sort of automatic amplifier gain control (AGC). This is absolutely necessary with AM signals, as too much gain applied to an already strong signal will drive the receiver amplifiers into saturation, whereas weak signals will pick up too much noise. With FM signals and receivers, there is no such need, since the information is not conveyed in the carrier amplitude, and thus there is no worry about amplifier saturation.

The basic idea in such electronic amplifier gain control consists of varying the biasing on a transistor. Recall that for a common-emitter (or source, for a JFET) transistor amplifier, the voltage gain depends on the ratio of output impedance to emitter (source) impedance. If this emitter impedance consists only of h_{ib} this parameter can be controlled by the amount of emitter current (or gate voltage on a JFET). This emitter current can easily be controlled by the base biasing voltage, which for example, can be derived from a signal-strength detection circuit, such as an AM detector, in order to control the amplifier gain, as in Fig. 5-13(f).

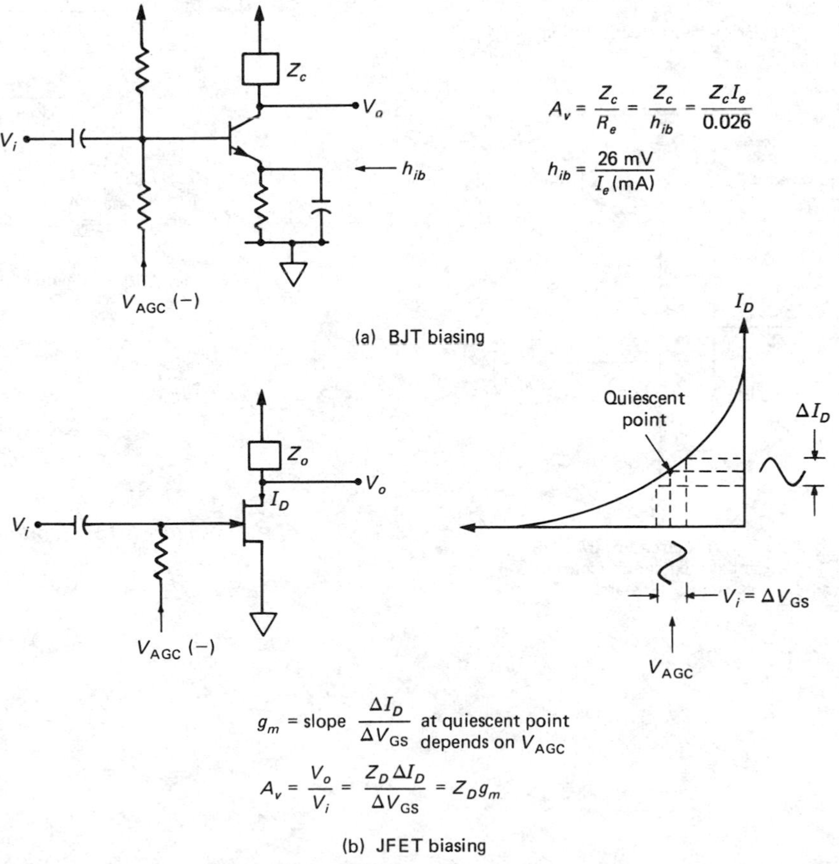

Figure 5-13 Methods of amplifier gain control.

(continued)

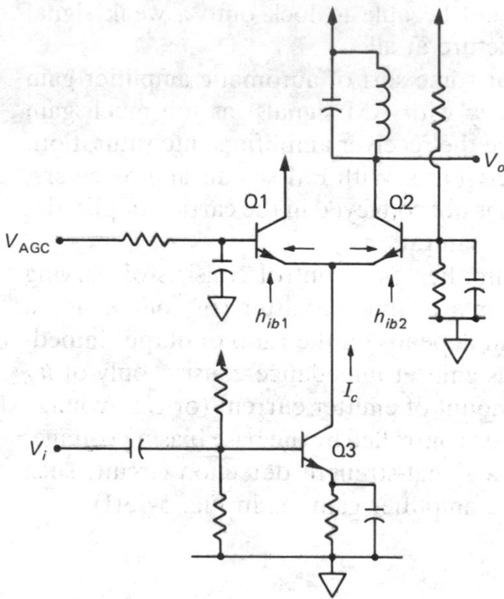

(c) Variable current splitting

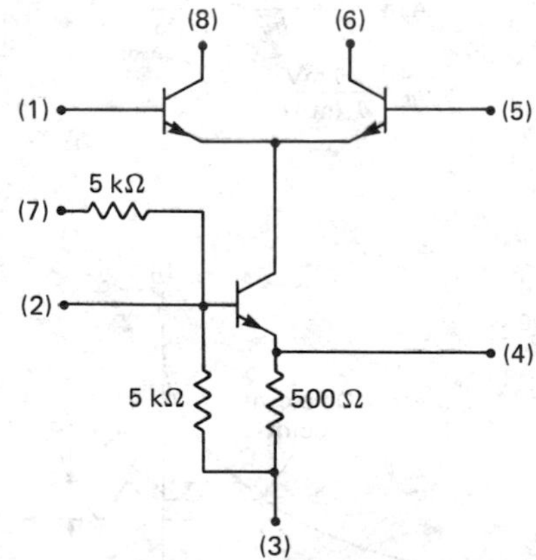

(d) Internal circuit of CA3028 integrated circuit (courtesy of RCA solid state)

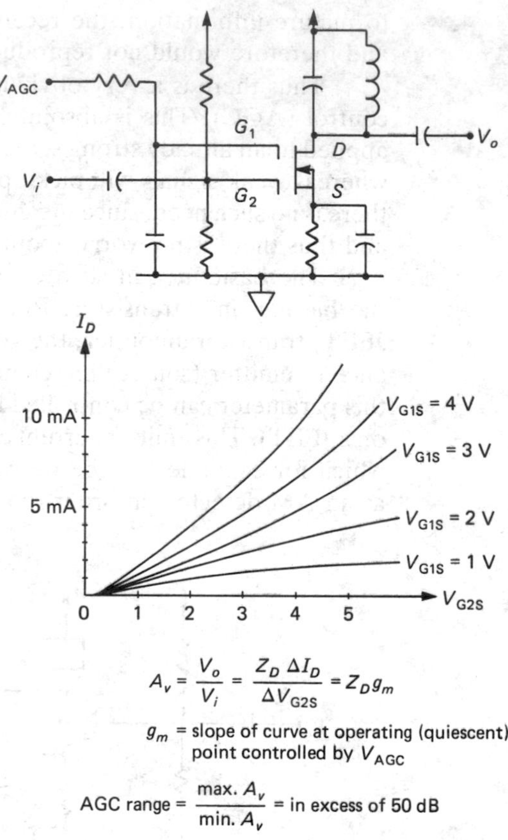

$$A_v = \frac{V_o}{V_i} = \frac{Z_D \, \Delta I_D}{\Delta V_{G2S}} = Z_D g_m$$

g_m = slope of curve at operating (quiescent) point controlled by V_{AGC}

$$\text{AGC range} = \frac{\text{max. } A_v}{\text{min. } A_v} = \text{in excess of 50 dB}$$

(e) With dual-gate MOSFET

Figure 5-13 *(continued).*

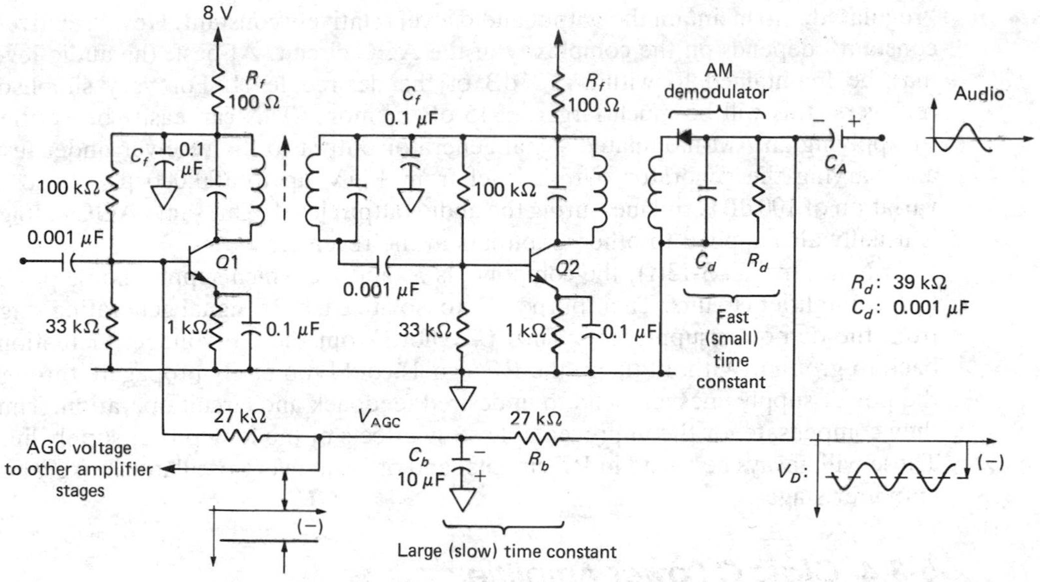

(f) IF amplifier with AGC

Figure 5-13 *(continued).*

Such transistor amplifier gain control is shown in Fig. 5-13. In circuit (a), a BJT base bias voltage is controlled from the AGC line, as in (f). In (b), the bias voltage applied to a JFET gate determines the slope of the device transconductance (output current versus input voltage) function. The steeper the slope, the higher the gain. In circuit (c), the collector signal current from Q3 splits into the emitters of Q1 and Q2. The bias voltage applied to the base of Q1 determines the amount of current drawn away by Q1 from Q2, and thus from the output, thus controlling the output signal strength. In effect, due to the uneven biasing of Q1 and Q2, the collector current from Q3 "sees" two different impedances, h_{ib1} and h_{ib2}, and thus splits accordingly. Such a transistor configuration is available on a chip, such as the CA3028 differential amplifier from RCA shown in (d). In (e) a dual-gate MOSFET is used to effectively control, with one gate, the device's transfer function from the other gate to the output. This effectively results in controlling its gain by varying the input–output function slope, similarly to the JFET case in (b).

An example of AGC is shown in Fig. 5-13(f). Such a setup would be typical of a simple IF amplifier. After the IF signal has been amplified sufficiently, it is rectified and then applied to an audio low-pass filter, C_d and R_d (see V_d in the diagram). The average voltage at this point is negative, due to the diode's orientation, and this average dc voltage is proportional to the IF signal level. This average dc signal is recovered by a smoothing low-pass circuit, R_bC_b, with a relatively large time constant. Thus the resulting dc voltage tracks only the slowly varying fluctuations in signal intensity. This negative-going AGC voltage is applied to the base of Q1, thus reducing its voltage gain. With no input signal, Q1's gain is at its highest, and is lowered as an input signal is applied. Thus the overall gain is "regulated," to maintain the output audio level relatively constant. How "relatively constant" depends on the complexity of the AGC circuit. At best, the audio level may be maintained to within ±3 dB of the desired level. For very simplistic receivers, this will be much larger, ±15 dB or more. This can easily be verified by applying an AM-modulated signal generator output to the receiver under test, and varying the generator output level from 1 μV up to 100,000 μV or so (a variation of 100 dB), and measuring the audio output level. This same AGC voltage is usually also applied to other amplifiers in the receiver.

Note, in Fig. 5-13(f), the components R_f and C_f, which supply the dc power to the amplifier circuits. Their purpose is to isolate each RF signal generating stage from the dc power supply line. Thus C_f "shorts" out the RF voltage fluctuations back to ground. Without them, the RF signals could too easily propagate through the power supply lines and lead to undesired feedback and circuit operation. Thus they compensate for the unpreventable stray effects of the long power supply line. These will always be found in RF circuits, and often at every oscillator, mixer, and amplifier stage.

5-3.4 Class C Power Amplifiers

When transmitting, power is taken from a source, typically dc (e.g., a battery), or rectified from an ac supply, and then converted to a high-frequency oscillation for launching from an antenna. For this signal eventually to get to the destination receiver with enough strength to be picked up, it must be sent with a lot of power. This may range from a few milliwatts for very short distances of a few hundred feet, to hundreds of kW when thousands of miles are involved. With such high power levels, an important concern is that of *power conversion efficiency*. This is conventionally defined as

$$\begin{matrix}\text{efficiency}\\ \text{(of power}\\ \text{conversion)}\end{matrix} = \frac{\text{ac output power transmitted}}{\text{dc power input}} \qquad \text{percent}$$

In this respect, the class C amplifier is the most efficient in its mode of operation. Recall that a class A amplifier can theoretically produce up to almost 50%, and a class B up to about 78%, whereas the class C configuration can in practice achieve upward of 80% (and theoretically upward of 95%). For that reason a class C amplifier is used whenever possible for precisely such a conversion of dc power to ac oscillations. The basic simple class C amplifier configuration is shown in Fig. 5-14.

Recall that a class C amplifier is biased to conduct for less than 180° of an entire cycle of oscillation. The conduction angle may typically vary from 30° up to almost 180°. See, for example, Fig. 5-15(c) and (d). Thus, in effect, the applied sine wave is grossly distorted (i.e., clipped). This may at first seem contrary to the basic assumption of properly reproducing the applied input signal, as emphasized earlier. However, what must be realized is that the sole purpose of such an amplifier is to efficiently *generate* high-power ac oscillations, not to *amplify* (i.e., duplicate) an input signal to a higher power level. Thus, in this respect, the original concerns of signal quality are still respected but are handled differently.

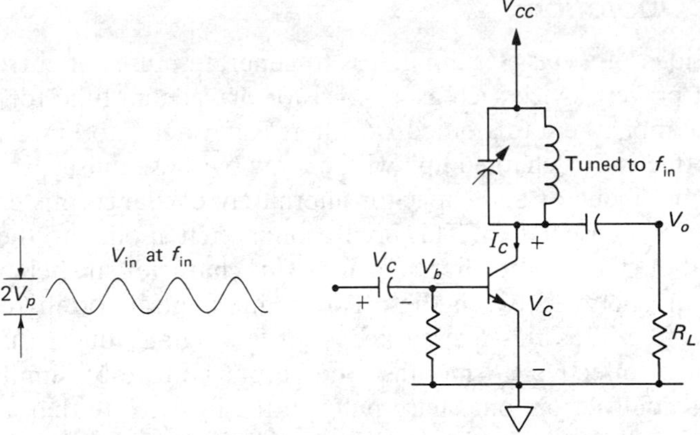

Figure 5-14 Basic class *C* amplifier.

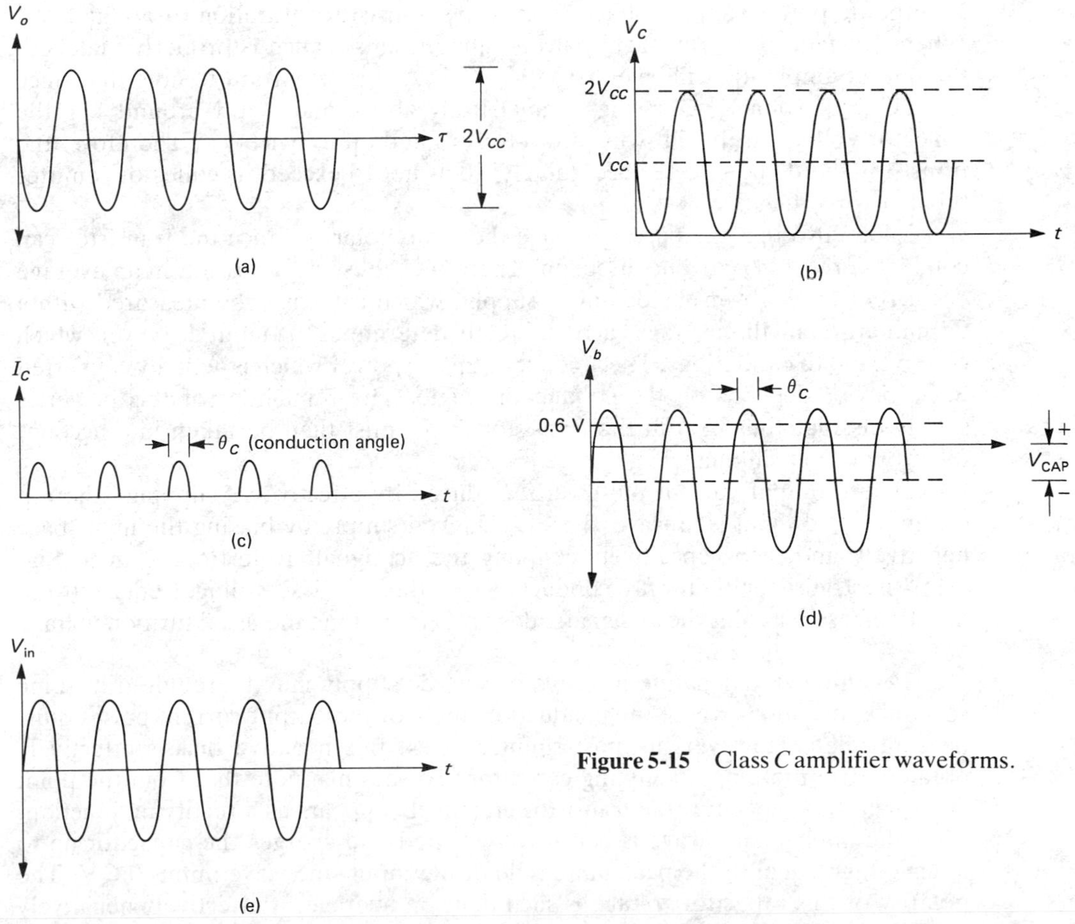

Figure 5-15 Class *C* amplifier waveforms.

Class C Operation

The basic idea of a class C amplifier is to generate pulses of current, produced by some high power device such as a transistor or vacuum tube for very high power levels, and apply them to a tuned *LC* filter. Refer again to Fig. 5-15. The *LC* tank circuit is effectively "charged up" with energy by these pulses, so, by their natural behavior, the inductor and capacitor alternately exchange this energy, producing a nice sinusoidal oscillation. This oscillation is then applied to the antenna, or the next amplifier stage, for further boosting. This characteristic behavior of sinusoidal exchange of energy between these two components is often referred to as the "flywheel effect," as this energy keeps on "swinging" due to its "momentum." This behavior effectively "smooths" the output voltage by simply swinging naturally, sinusoidally, from one sharp pulse to the next. Note that a good smoothing effect requires the use of a high Q *LC* tank circuit. As a result, the class C amplifier will only operate within a narrow frequency band around the tank resonant frequency.

The key point of interest to us is that the energy derived from the dc power source in the form of pulses is converted very efficiently to the desired output oscillations. The peak amplitude of the output oscillation can range from 0 V to V_{cc}, the dc power supply voltage (minus the transistor saturation of about 1 V). When designed properly, such that the antenna impedance is properly matched, the output amplitude will be equal to $V_{cc} - V_{\text{sat}}$. It is important to note that since the *LC* tank voltage will swing symmetrically above and below around V_{cc}, the collector voltage itself will swing down to zero and up to twice V_{cc}. Therefore, the transistor should be selected accordingly, so as not to exceed its collector–emitter breakdown voltage.

Note also that the pulses all have the same polarity, since the transistor can conduct current in only one direction. Therefore, these pulses result in an average dc current drawn from the dc power supply, which can easily be measured with a dc ammeter, and this average current directly determines the input dc power, which is converted to oscillations. The amount of this dc power which is actually converted to ac power depends on the conduction angle. The remainder of this power is simply dissipated as heat in the transistor. This must then be taken into account for proper heat sinking.

These output current pulses are produced by effectively "clipping" the top off an applied input sine wave. This is achieved simply by biasing the input base negatively and then capacitively coupling the ac signal. Referring again to Fig. 5-15, the transistor then only conducts when the net base voltage, equal to the negative base bias plus the ac signal, adds up to more than the usual turn-on voltage of 0.6 V.

As implied, a separate negative-biasing dc supply may be required in some specific applications where the conduction angle of the output current pulses must be controlled. However, in most simpler cases, this negative bias is effectively obtained from the input coupling capacitor. To see this, note that from the point of view of this capacitor, the transistor effectively appears as a rectifying junction. Thus the applied sine wave is half-wave rectified and charges the capacitor up to a dc voltage equal to the peak amplitude of the input sine wave minus 0.6 V. The polarity of this capacitor voltage is such that, on average, it effectively negatively biases the base. It is then necessary to provide the extra base resistor to partially discharge the capacitor on the opposite polarity of the cycle. This allows the capacitor to couple the top peaks to the transistor base and turn it on. Otherwise, the capacitor would charge up fully and no more base current pulses would be delivered to be boosted to the output tank circuit. This base-biasing resistor is typically found to be under 100 Ω, sometimes even under 10 Ω in higher-power amplifiers.

Output Frequency Spectrum and Filtering

Note that the output sine wave resulting from this flywheel effect is in fact not so perfect. During every cycle, between current pulses, energy is drawn out of the *LC* tank circuit to the antenna, such that the amplitude of the oscillations decays, and so the sine wave is not so "clean." From the point of view of the frequency spectrum, the current pulse train consists of a fundamental sine-wave component, and many more higher-frequency harmonics, as shown in Fig. 5-16. Thus the parallel *LC* tank presents a high impedance to the fundamental component but a much

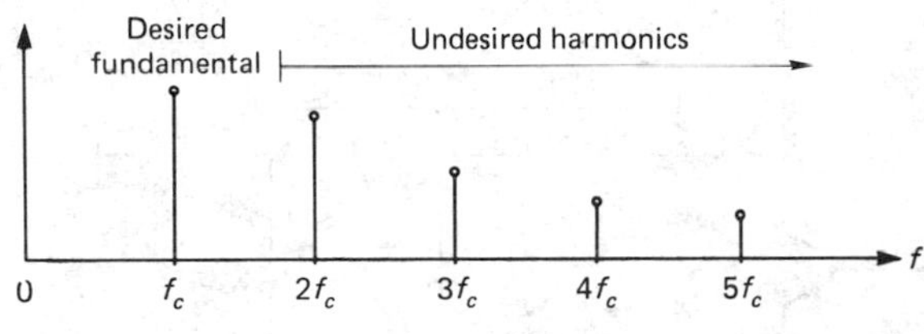

(a) Frequency spectrum of current pulses of Fig. 5-15 (c)

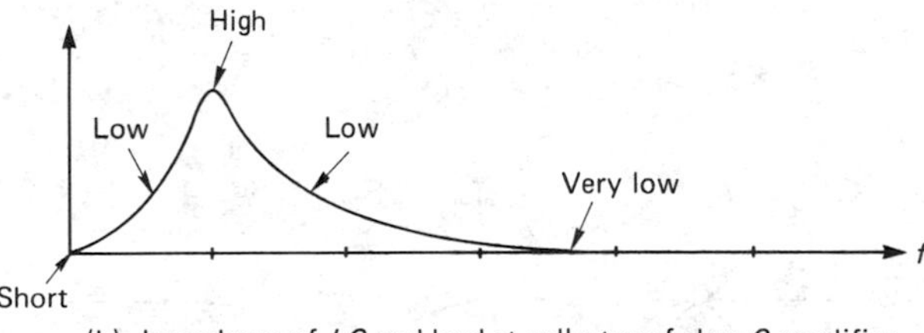

(b) Impedance of *LC* and load at collector of class *C* amplifier

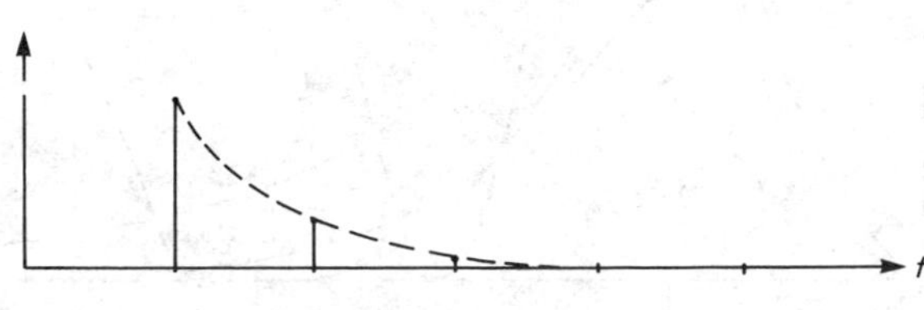

(c) Frequency spectrum of near-perfect sine wave at load

Figure 5-16 Frequency domain representations of current pulses, collector load impedance, and output voltage for a class *C* amplifier.

lower impedance at the harmonic frequencies, due to the capacitor's lower reactance at the higher frequencies. Thus these harmonics are "shorted out" (not quite) by the capacitor and do not contribute as much to the net output voltage. In effect, they are greatly attenuated. However, this *LC* tank typically has a low *Q* value of only about 5 or so, and does not attenuate these harmonics enough to prevent interference. Therefore, an extra harmonic rejection low-pass filter is provided prior to sending the final signal to the antenna, as shown in Fig. 5-17. In addition, extra sections may be included to further emphasize the fundamental, such as the series *LC* combination tuned to f_c, or to short out or block specific harmonics, such as series and/or parallel *LC* combinations tuned to $2f_c$ and $3f_c$, respectively. Note their notching effect in the overall frequency response compared to low-pass filtering alone. Usually, the regulations set by the Department of Communications (or the FCC in the United States) will specify that the harmonics be attenuated at least 80 dB below the carrier component. This typically amounts to spurious emissions of only a few hundred microwatts.

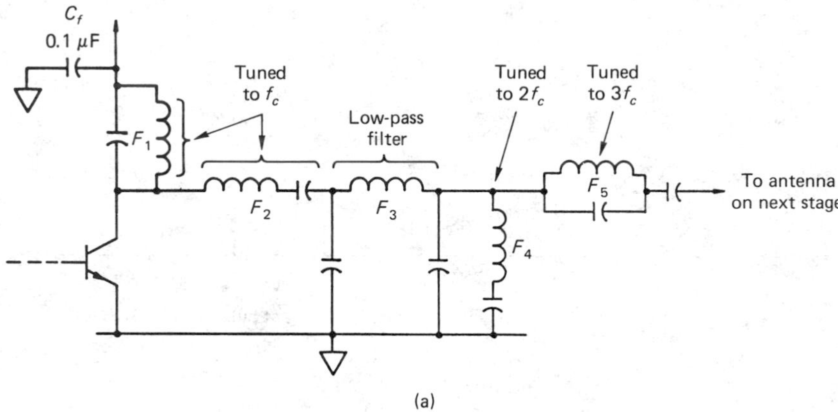

(a)

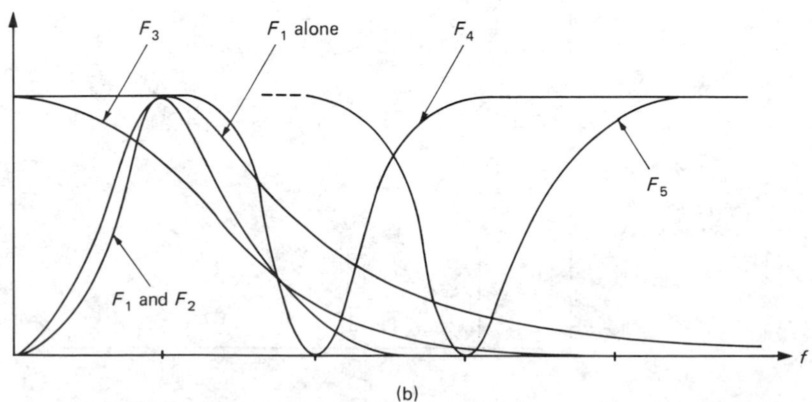

(b)

Figure 5-17 Class *C* output harmonic filtering.

Tuning a Class C Amplifier

As indicated previously, the class C amplifier is a tuned amplifier. Therefore, a typical circuit may include some fine-tuning components in order to maximize the power transferred into and out of an amplifier stage. Such is the function of trimmer capacitors C_i and C_o in Fig. 5-18. For very simple one- or two-stage amplifiers, these may be tuned merely by monitoring the final output power to the antenna. However, for best tuning it is necessary to monitor the ac power transfer from each stage to the next. This is accomplished by "sampling" the average dc supply current to each stage, using an emitter resistor, and "smoothing" the resulting pulses with a filter capacitor, as shown in Fig. 5-18. This in effect is simply a rectifier and a filter, as found in dc power supplies. However, because of the relative power and current levels at the transistor input and output, we find that the rectification process by the base–emitter junction is effectively "power-assisted" by the collector current to charge up the emitter capacitor at the very high frequency.

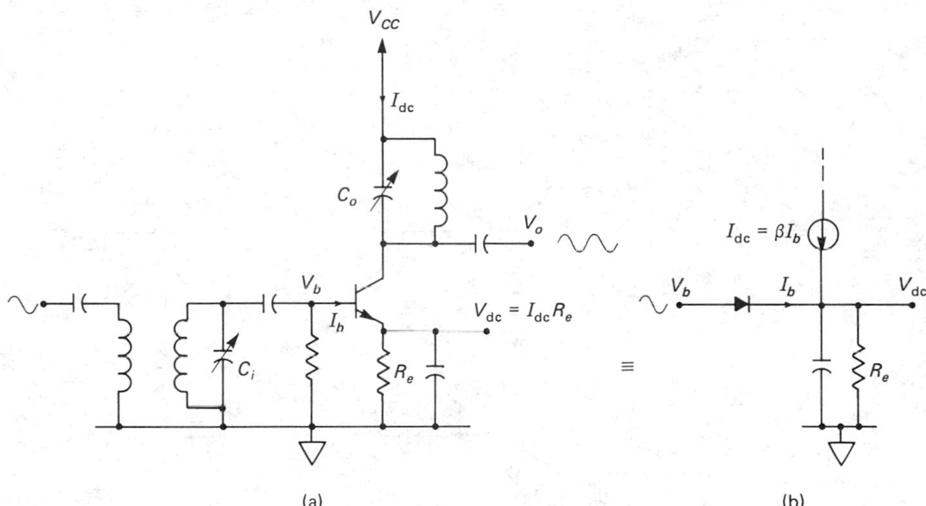

Figure 5-18 Tuning a class *C* amplifier.

A dc voltage, which is easily measured, is now available at the emitter for monitoring its performance and efficiency. To obtain maximum power from it, it will be necessary to tune C_i for maximum dc voltage at the emitter, and then tune C_o for a minimum. In the second adjustment of the output LC tank, the optimum occurs when maximum ac power is delivered for the minimum dc input power, as indicated by the dip in the emitter voltage reading.

Output Power Adjustment

In some applications, such as mobile transceivers, it may be desirable to make possible easy adjustment of the output power level, depending on the anticipated distances to be covered without wasting excessive energy. In cellular radio applications, for example, the land radio station computer remotely commands the mobile radio to adjust the power level as necessary. This is to maintain a strong enough signal level to avoid possible interruption of the communication, but at the same time to keep it low enough to prevent interference to adjacent cells.

As indicated earlier, the amplitude of the output oscillations is determined primarily by the power supply voltage, V_{cc}. Therefore, simply by providing a variable dc regulator, the output power level can easily be adjusted as desired, as shown in Fig. 5-19. This is accomplished by the potentiometer and emitter follower Q2. For best overall dc (i.e., battery) power efficiency, and for simplest design considerations, the output power may be adjusted by controlling the power drive to, and thus also the conduction angle of, the last stage, as suggested in the diagram.

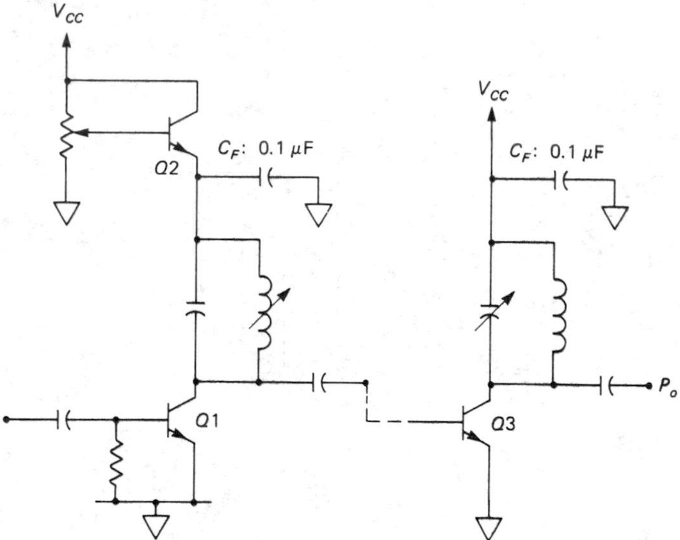

Figure 5-19 Adjusting output power level from a class *C* amplifier.

In some applications it may be desirable to provide automatic level control (ALC) of the output power, either to protect the final amplifier stage transistor from destruction due to improper antenna impedance matching, or to prevent overmodulation and interference to adjacent channels. In such a case, the output level may be monitored by rectification and filtering, just as in the AM receiver demodulator discussed earlier, and the resulting dc voltage may be amplified and used to control the output power level as indicated above.

Frequency Multiplication

Recall that the principal feature of the class C amplifier is the output *LC* tank, which is "kicked" and made to "swing" from pulse to pulse. Just like a child on a swing, this *LC* tank gets enough energy in each pulse that it does not really need an extra "push" with every cycle. It has enough momentum to keep on oscillating through more than one cycle between pulses (Fig. 5-20). This suggests that we can tune the output *LC* circuit to a multiple of the input driving frequency and so achieve frequency multiplication. For example, drive the input at 12 MHz and tune the output to 36 MHz. However, just as the swinging child eventually runs out of energy and does need a good push, the *LC* circuit loses its energy to the antenna. Thus, in practice, such frequency multiplication can efficiently be accomplished up to three times only. For higher multiplication factors, the power conversion efficiency drops off too quickly to make it worthwhile, and thus requires additional amplification. Thus, if it is desired to start with a crystal oscillator at, say 12.1 MHz, and transmit at 145.2 MHz, we will require an overall multiplying factor of 12. This must then be accomplished by cascading, say, a class C frequency tripler, followed by two frequency doublers.

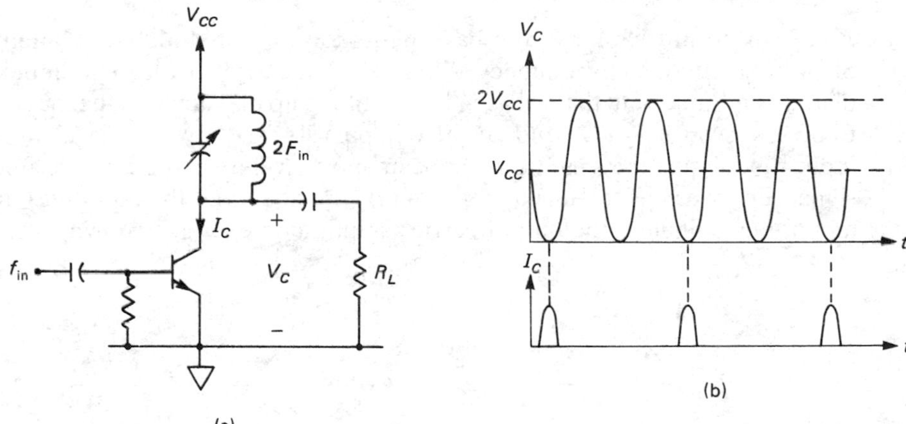

Figure 5-20 Frequency multiplication using a class *C* amplifier.

In cases such as the last example of times 12 multiplication, it is common first to multiply the frequency as required, with all the multiplier stages operating at low power levels, and then finally, to provide two or three stages of power amplification at the same final output frequency, thus maximizing the overall power efficiency. However, if it is desired to produce a higher power level with the multiplication process, then a push-pull doubler, as shown in Fig. 5-21, may be preferable. In this case, the input center-tapped transformer feeds alternate polarities of each input cycle to the two transistors. Each one conducts alternately, but with their collector currents tied together they drive the output *LC* tank at twice the input frequency.

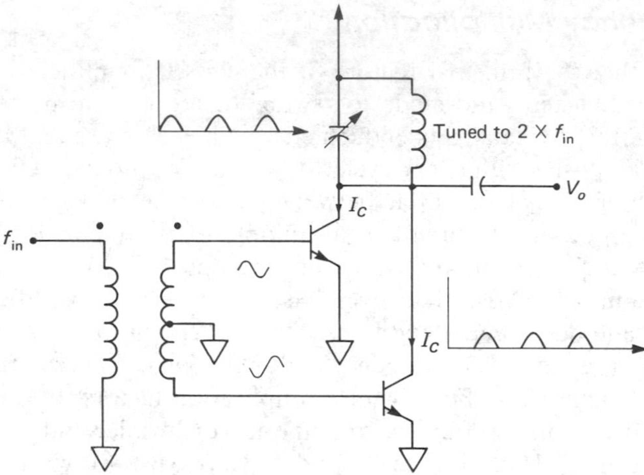

Figure 5-21 Class *C* frequency doubler.

Amplitude Modulation

As discussed in Chapter 6, the basic definition of amplitude modulation consists of starting with a high-frequency oscillation, the carrier, with some initial unmodulated amplitude, and then varying its amplitude up and down as the message signal to be sent swings positive and negative in polarity, as shown in Fig. 5-22. Thus the input modulating message signal appears in (a), whereas the final desired output signal sent over the antenna appears in (c). Provided that the input message signal amplitude is limited, the output carrier amplitude never goes down to zero.

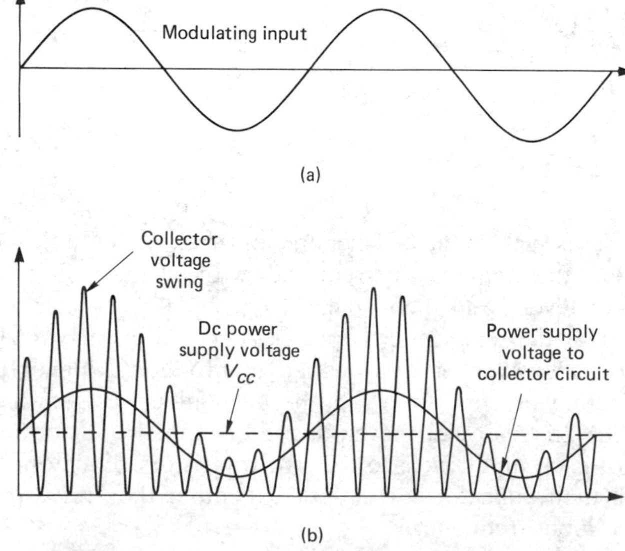

Figure 5-22 Generating A.M. using a class *C* amplifier.

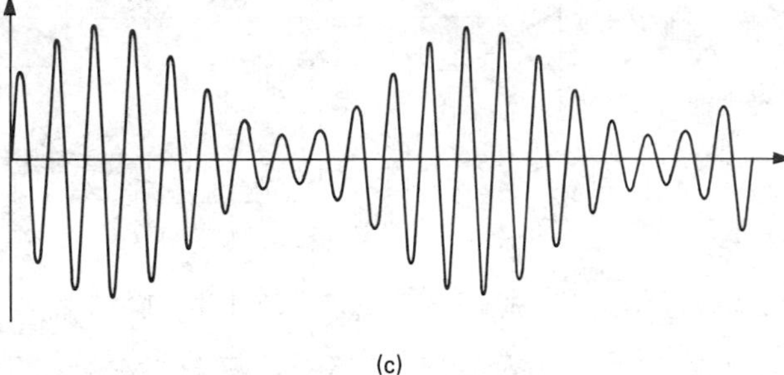

(c)

Figure 5-22 *(continued).*

Considering the previous discussion on how to adjust the output power level of a class C amplifier, it is then not difficult to imagine how to modulate, or control, its amplitude in accordance with an input audio signal. The idea consists simply of amplifying the input audio signal and then applying it to the amplifier's dc power supply, thus effectively modulating the power supply. This is shown in Fig. 5-23. Consider first that no input audio signal is applied. Then the transformer secondary simply appears as a dc short, which results in a constant-amplitude, unmodulated output carrier. However, as an input signal is applied to the transformer primary, it induces the same voltage at the secondary, which adds to, or subtracts from, the power supply voltage. The resulting voltage supplies the power to the final class C power amplifier, thus modulating the output carrier amplitude, as desired originally. Recall that the collector voltage swings up to twice its own supply voltage, not the dc power supply voltage. Therefore, if the audio transformer output can itself be driven by the audio amplifier down to zero, and up to twice V_{cc} (recall transformer-coupled class A amplifiers), the collector voltage will itself swing up to four times V_{cc}. Again, this must be taken into account when selecting a transistor, lest it be destroyed.

Because the modulation process is applied to the final stage before going to the antenna, the final modulation process must be carefully controlled. Should the input audio signal be too strong, for example, shouting into the microphone, this could result in overmodulation, whereby the carrier amplitude envelope would be clipped. This would result in additional undesirable, spurious sidebands, or "splatter," extending to a wider bandwidth than legally allocated, causing severe interference to the adjacent channels. For this reason all such AM amplifiers include some form of amplitude control, usually referred to as automatic level control. The idea is very similar to AGC. In this case, the transformer output voltage amplitude is monitored and rectified by diode D1 and filter capacitor C_1. The resulting dc signal is then used to control the audio amplifier gain.

As indicated earlier, a simple sinusoidally modulated AM signal consists of a carrier and two sideband components. The carrier component power remains constant regardless of the amount of modulation, and results only from the V_{cc} supply voltage times the *average* dc current I_c indicated in the diagram. However, the additional power for the sidebands must be supplied by the audio amplifier through the transformer. Therefore, the audio amplifier must be designed to deliver

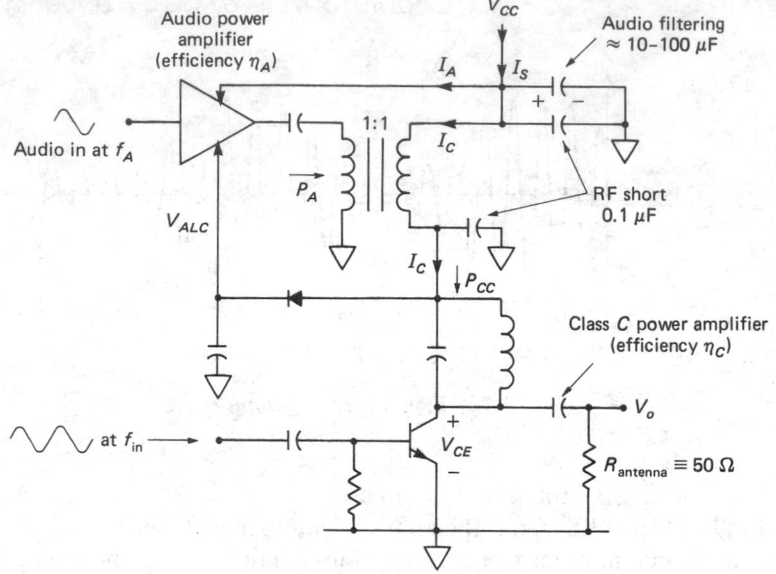

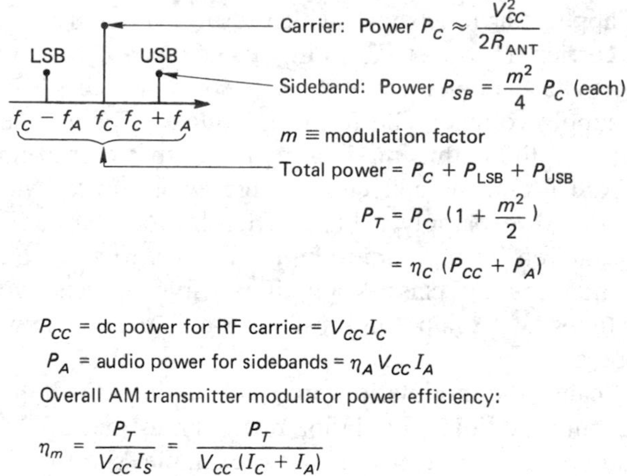

Frequency Spectrum and Power Distribution for Sinusoidal Audio Modulation
LSB
USB
$f_C - f_A$ f_C $f_C + f_A$
Carrier: Power $P_C \approx \dfrac{V_{CC}^2}{2R_{ANT}}$
Sideband: Power $P_{SB} = \dfrac{m^2}{4} P_C$ (each)
$m \equiv$ modulation factor
Total power $= P_C + P_{LSB} + P_{USB}$
$P_T = P_C \left(1 + \dfrac{m^2}{2}\right)$
$= \eta_C (P_{CC} + P_A)$
P_{CC} = dc power for RF carrier $= V_{CC} I_C$
P_A = audio power for sidebands $= \eta_A V_{CC} I_A$
Overall AM transmitter modulator power efficiency:
$\eta_m = \dfrac{P_T}{V_{CC} I_S} = \dfrac{P_T}{V_{CC}(I_C + I_A)}$

Figure 5-23 Generation of A.M. by collector modulation of class *C* amplifier.

that sideband power, plus the power lost due to its own inefficiency *and* that lost due to the RF amplifier's inefficiency. Also, in determining the *total* (battery) power efficiency, one must take into account the audio amplifier current drain, I_a, its conversion efficiency, the total power supplied to the RF amplifier, and its efficiency.

For example, consider a simple CB transceiver powered from a battery of 13.5 V (with the car engine running), with a final class C amplifier efficiency of 75% and an audio power amplifier with an efficiency of 60%. When transmitting with no voice, a RF power meter indicates 8 W. Thus we can expect the supply current to the final class C stage, I_c, to be 0.79 A (*not* the total battery current, since the battery also supplies some minimal amount to the audio amplifier and the rest of the circuit). Furthermore, with a 1-kHz signal applied to the microphone input and the level adjusted for a measured modulation index of 0.8 (by observing the transmitter output signal on a scope), the RF power meter should confirm this with a RMS reading of 10.56 W, and the measured battery current should be at least 1.21 A (0.79 A dc to the class C amplifier, and 0.42 A dc to the audio amplifier plus a bit). Thus the audio amplifier consumes 5.67 W from the battery and supplies 3.41 W through the audio transformer to the sidebands (before conversion loss in the class C amplifier). The RF amplifier receives a total of 14.1 W and converts it to 8 W in the carrier and 2.56 W in the sidebands. The battery supplies a total of 16.33 W (plus a bit), with an overall efficiency of 65%.

Wrong Use of a Class C Amplifier

As has been emphasized earlier, the primary purpose of the class C amplifier is to generate ac oscillation and boost the amplifier to higher power levels; it is not to reproduce an input signal faithfully. However, it may be interesting to consider what would happen if we did attempt to use it to amplify an arbitrary input signal.

Consider, for example, that we have generated an AM signal, but with a low-power class C modulator, and now wish to boost it to a higher level using a class C amplifier. The situation then appears as shown in Fig. 5-24. The signal is capacitively coupled to the base, and the input capacitor charges up to the peak of

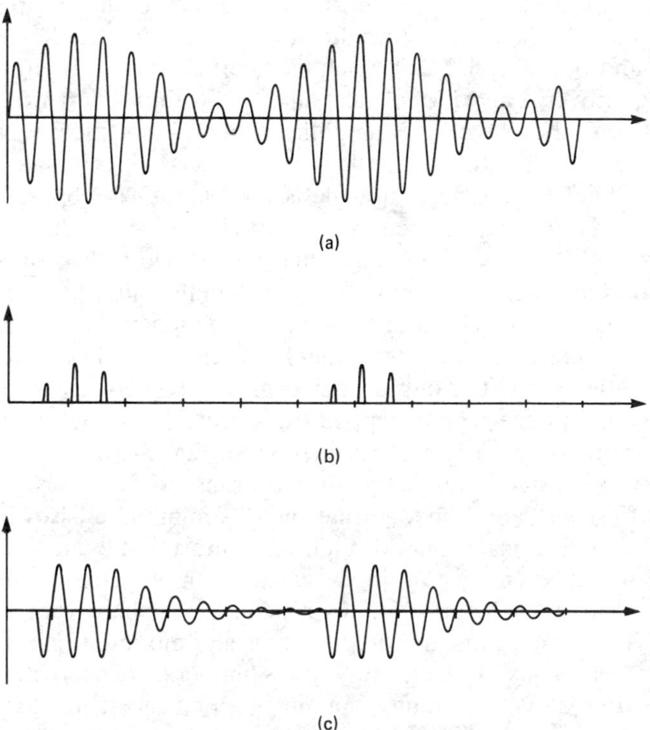

Figure 5-24 Wrong use of class C amplifier.

the AM signal envelope. However, when the signal amplitude drops, because of the modulation, the transistor drive stops. The result would be bursts of collector current pulses at the higher signal amplitudes only. The output *LC* tank is therefore momentarily driven, and responds with oscillations. But when the drive stops, its energy decays. The net result is a very distorted output. The original information is badly corrupted; it will be received, but not intelligibly. In addition, just as with overmodulation, this will cause very severe interference to adjacent channels.

5-3.5 Class A-B Linear Amplifiers

In order to handle the kind of signal discussed in the preceding section, we have to find a different solution. The most straightforward answer is to use a class A amplifier. However, as pointed out earlier, it is not the most efficient, and when transmitting, power conversion efficiency is an important factor. A commonly adopted solution is the linear class A-B amplifier. It is called linear simply because it has a linear transfer function, as opposed to the class C amplifier, which has been shown to be very nonlinear and not at all suited to the amplification of such signals. In addition, unlike the class C amplifier, which must be tuned to operate properly, this amplifier usually operates over a rather broad range of frequencies and can typically handle a frequency range of up to five octaves, such as 1 to 30 MHz.

The need for such amplifiers typically arises in single-sideband (SSB) transmitters, where the nature of the signal is such that it can be generated only at very low power levels, and then must be boosted to the desired power level. This contrasts with the class C amplifier which can generate the power directly at high levels and therefore be that much more efficient. But then it is restricted to generating only AM and FM signals, and only within a narrow frequency band.

Recall that the class B amplifier is biased just slightly on, such that if a sine wave were applied to the input, it would be effectively half-wave rectified at the output. Thus the conduction angle is 180°. Class A-B operation simply translates to a slightly greater conduction angle of about 185°, reducing nonlinearity due to crossover distortion. The idea is then to use two transistor amplifiers in a push-pull configuration, such that each one handles alternate polarities of the input signal (Fig. 5-25). The input transformer secondary is center-tapped, just as with the class C push-pull frequency doubler discussed earlier, and alternately feeds the two transistors. At the output, however, the two collector currents feed opposite ends of yet another center-tapped transformer, together recombining the divided components of the original signal (see also Fig. 5-26).

An important point is that the two transistors are biased on, and so are never both off at the same time, so that there is minimal crossover distortion, as there would be with a class C biased amplifier. Unfortunately, it does mean that dc power is consumed, even with no input signal to amplify. However, this is required to ensure minimum distortion. This biasing is supplied from a matched diode. To improve the matching, the diode is typically mounted next to the transistors, on the same heat sink. This ensures that its temperature characteristics will track those of the transistors, and thus maintain proper bias. If not for that, the transistors could overheat due to thermal runaway and be permanently destroyed.

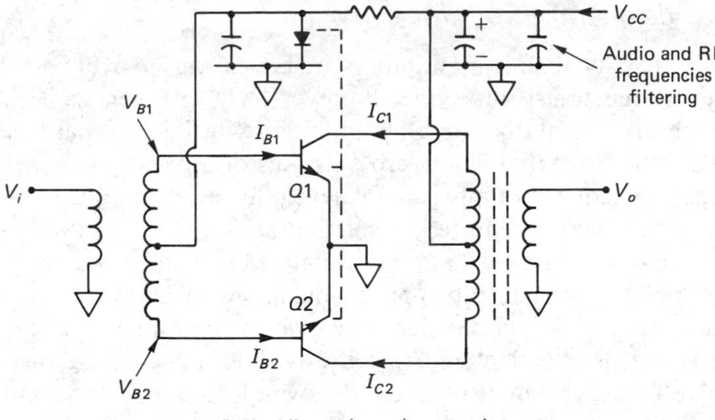

Figure 5-25 Basic class *AB* linear amplifier.

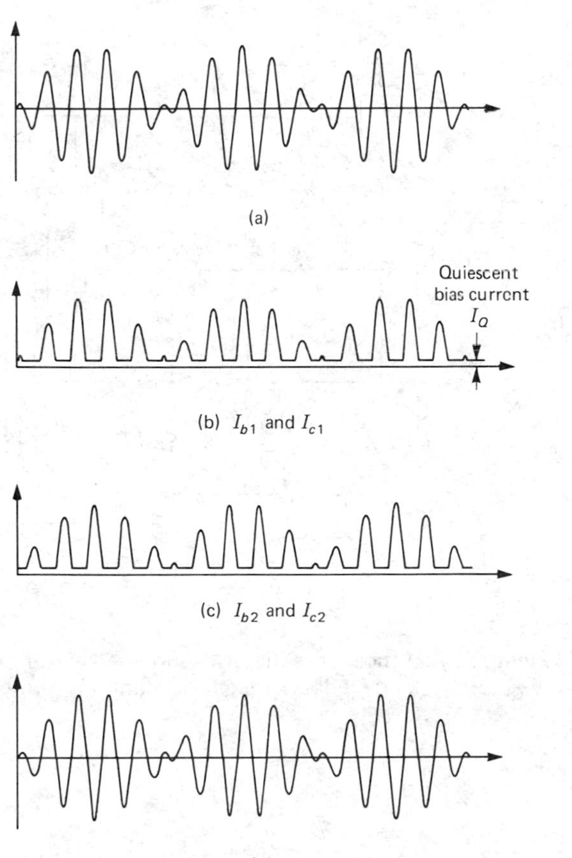

Figure 5-26 Linear class *AB* push-pull amplifier operation.

Improving Amplifier Linearity

As indicated above, linear amplifiers are typically used over wide frequency ranges. However, the transistors typically never have an even gain over such a range. Furthermore, when operating an amplifier with large signal levels, nonlinear distortion, other than that due to crossover distortion, becomes inevitable, due to the device's semiconductor physics (at levels lower than saturation). In such cases, negative feedback is required to compensate for these factors, as shown in Fig. 5-27. Recall that a primary feature of negative feedback is that the feedback elements (more specifically, their ratio) predominantly determine the overall response. Thus any deviation of the output signal relative to what it should be, as dictated by the input signal, is effectively corrected for, thus reducing the distortion. This same negative feedback will also flatten the overall frequency response. As indicated in Fig. 5-28, the frequency response of the feedback compensation is basically inverse to that of the transistors alone. Together, then, they work to even out the gain over a much wider frequency range. The principle is very similar to that of op amps with negative feedback.

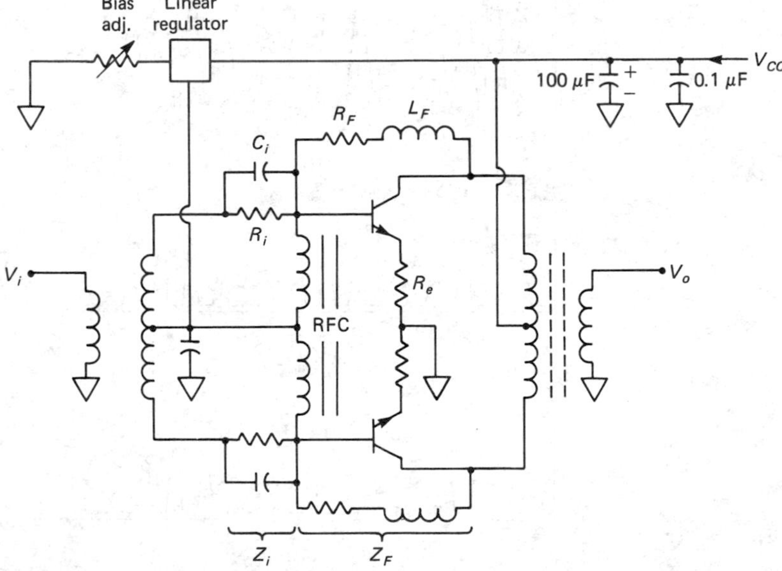

Figure 5-27 Linear amplifier with improved biasing for better linearity, and feedback for gain leveling and better linearity.

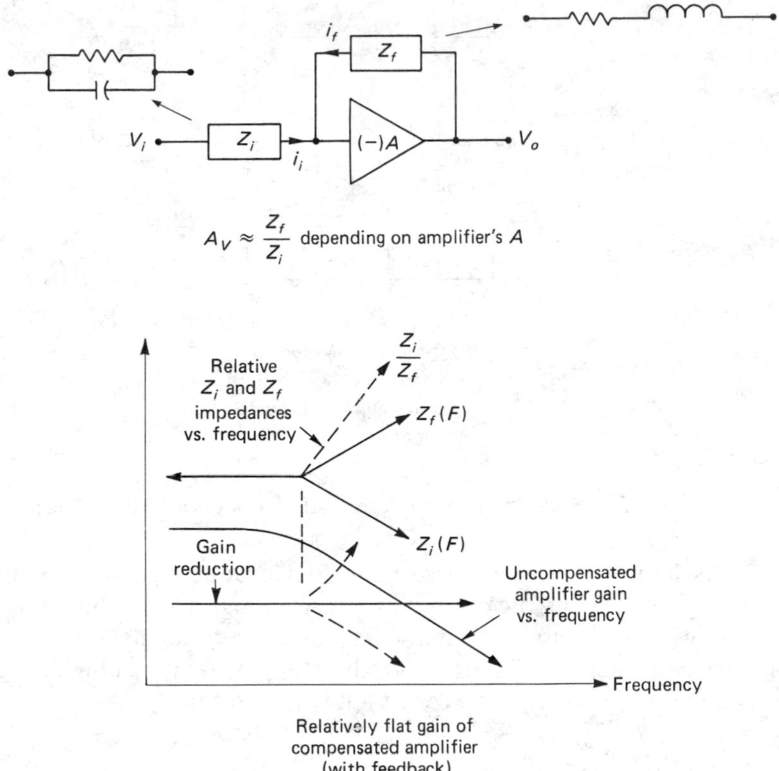

Figure 5-28 Improving amplifier performance by using negative feedback.

Amplifier Output Distortion and Two-Tone Test

After all that has been said and done the linear amplifier input–output transfer function is still quite far from being perfectly linear. At very large output signal levels, the feedback network operates well, but will not entirely suppress distortion products. To determine this aspect of the amplifier's overall performance, a two-tone test is performed. This consists basically of applying two audio tones to the audio input, say 1 and 2 kHz, if testing an entire transmitter. If testing only the RF power amplifier, two closely separated RF signals (separated by 1 kHz) can be combined and fed directly to its input. The two signals are applied at the same power level, and increased together. Eventually, the extra IMD products appear, as shown in Fig. 5-29. These products appear above and below the original desired components and are spaced by the same amount as the original two (here 1 kHz). Their maximum level allowed will depend on the legal requirements and criteria, but typically, at maximum output power, the third-order products should be at least 30 dB below the main components.

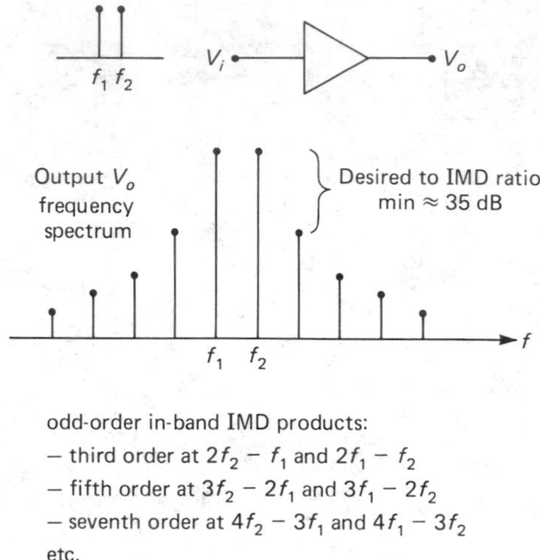

odd-order in-band IMD products:
– third order at $2f_2 - f_1$ and $2f_1 - f_2$
– fifth order at $3f_2 - 2f_1$ and $3f_1 - 2f_2$
– seventh order at $4f_2 - 3f_1$ and $4f_1 - 3f_2$
etc.

Figure 5-29 Linear amplifier IMD products for a two-tone input.

As indicated earlier, these odd-order products cannot be eliminated, since no filter can be constructed to be narrow enough to cut them out. Furthermore, just as with the class C amplifier, even-order products will also be generated. However, as indicated in Fig. 5-30, these appear at much higher frequencies, and so can easily be taken care of with a low-pass filter.

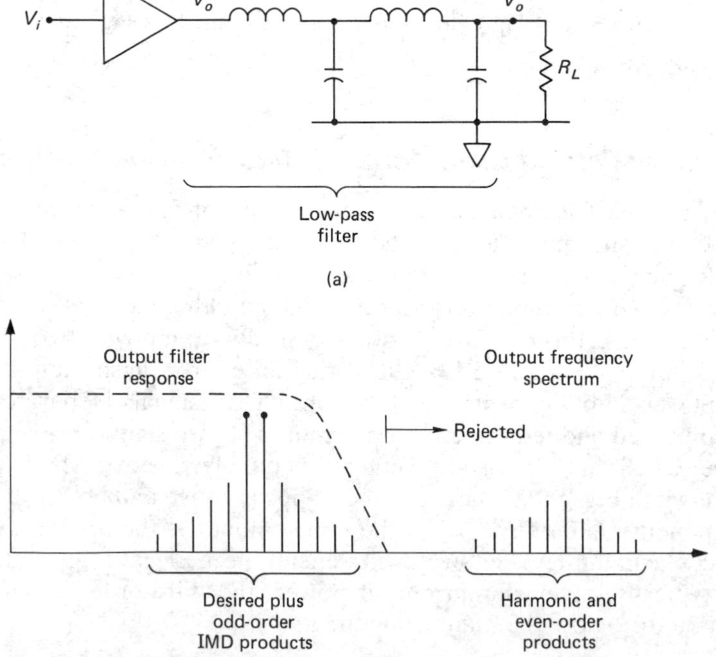

Figure 5-30 Amplifier output distortion suppression.

5-4 MIXERS

The mixing operation is just about as fundamental as amplification in all radio communication systems. The basic idea consists of translating an input signal from one range of frequencies to another. A simple example is any ordinary AM radio, where the input station the receiver is tuned to, say, 1280 kHz, is translated, or converted, to an intermediate frequency (IF) of 455 kHz within the receiver. (Refer to Figs. 5-31 and 5-7.) Without going into details, the reason for such conversion is that the receiver design is greatly simplified, and the resulting operation is much more reliable. A simple explanation is that the receiver selectivity (i.e., effective bandwidth) can be made much narrower at a lower frequency. This process would then allow us to filter out the strong interfering signal indicated in Fig. 5-7. Any input frequency the receiver is tuned to is converted to the same IF, such that the main receiver gain is provided by one unchanging, fixed-tuned IF amplifier. Similarly, a cable television converter is tuned to a desired input station frequency, and translates it to one output frequency, regardless of the input station tuned to. Thus the television receiver only needs to be tuned to that output frequency once and does not have to be retuned every time we wish to change channels.

Thus, as can be seen, the mixing idea is to take a desired input signal, at whatever frequency it exists, and translate it to another frequency, at which it is more convenient to perform the main signal processing on it, such as filtering with a fixed crystal or mechanical filter, or amplification, or demodulation, or scrambling, and so on. For example, for various reasons, such as spectrum availability and demand, legal, technological, cost, practical considerations, component availability, and so on, cellular radio has been assigned to the UHF frequencies. However, signal processing, as outlined above, is much more convenient to perform at lower frequencies, say below 10 MHz. Thus, upon reception of the signal from the antenna, one of the first things done to it is to translate it down to the selected IF.

We discuss next the basic mixing operation, first from an ideal point of view, and then some practical circuit configurations that realize this function.

5-4.1 Frequency Mixing and Translation

To appreciate the mixing operation, a necessary review of some elementary trigonometry is in order. Realizing that all signals are considered to consist of sine waves, recall that the only known mathematical operation that effectively "speaks" of our desired objective of frequency translation is represented by the trigonometric identity

$$\cos A \times \cos B = \tfrac{1}{2}[\cos(A + B) + \cos(A - B)]$$

Similarly,

$$\sin A \times \sin B = \tfrac{1}{2}[\cos(A - B) - \cos(A + B)]$$

For example, if we let $A = 2\pi f_1 t$ and $B = 2\pi f_2 t$, where f_1 and f_2 are, say, 10 kHz and 1.5 kHz, the indicated product of two such sine waves results is the sum of two new ones, at 8.5 and 11.5 kHz. This is shown in Fig. 5-31(a), where the mixer is represented by a multiplier which operates on two applied sinusoidal input voltages. The resulting output voltage consists of the higher-frequency sine wave, whose amplitude is "modulated," or "controlled," by the lower-frequency sine wave. This higher-frequency sine wave, commonly referred to as the *carrier*, will effectively carry the lower-frequency message signal over the air from the transmitter to the receiver antennas. Now, this same output result can be obtained by *adding* two sine waves at the sum and difference of the original frequencies. Therefore, we can say that this amplitude-modulated sine wave effectively consists of these two newly produced components. This can be verified by applying this output signal to a narrow bandpass filter tuned to one of these two frequencies. The filter then responds to only one of these components and rejects the other, as shown in Fig. 5-31(b).

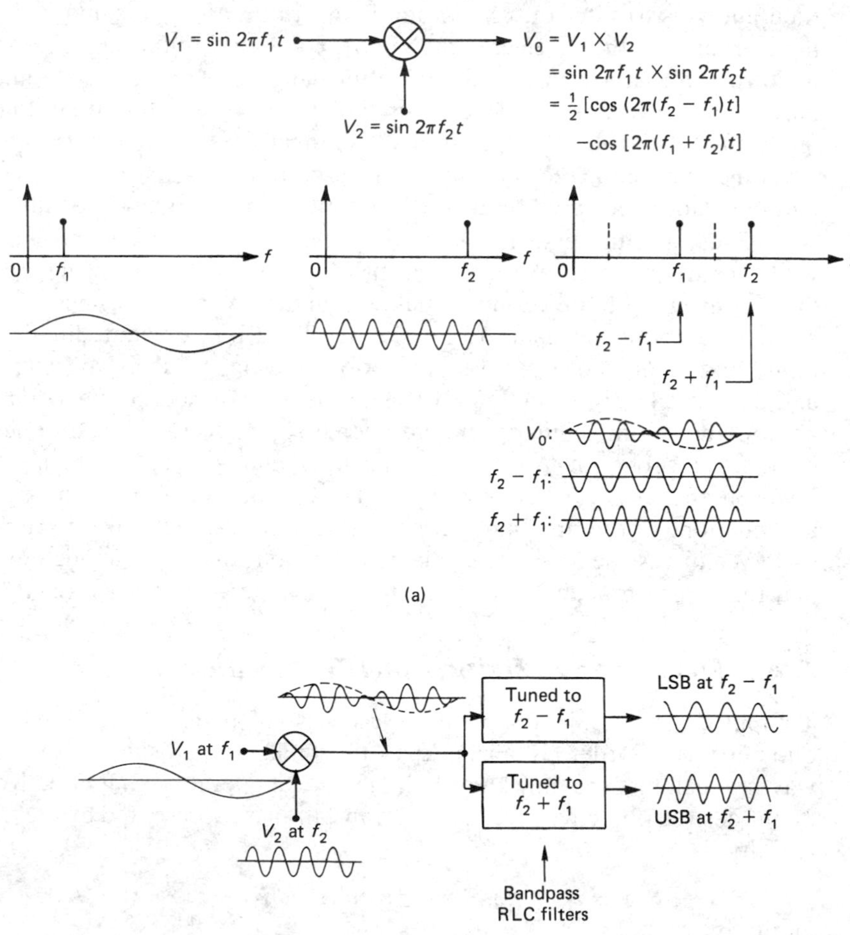

Figure 5-31.

This concept is identical to the beating of two guitar strings slightly out of tune with each other. The two strings vibrate at different frequencies, generating two separate components, say at 999 and 1001 Hz, respectively (the difference and sum of the frequencies to be indicated next). Note that these two components can be separately produced, heard, and distinguished from each other. When played together, however, their sum, the total sound heard, is that of a single tone, at 1000 Hz, whose loudness is modulated at a lower frequency, 1 Hz (thus the two added components are at 1000 + 1 Hz and at 1000 − 1 Hz). This, in effect, describes the reverse order of this trigonometric equation indicated above.

Ideal Mixer Operation and Practical Realization

The next question is how to produce such a multiplying operation electronically, given that such oscillations are available from oscillators, or in groups as a message, such as an audio voice signal (recall that any signal is regarded as consisting of many sine-wave components). Electronic multipliers are available, such as Exar's XR2208 or Motorola's MC1495, based on a transistor configuration referred to as a Gilbert cell, to be discussed further shortly. However, the principal limitation of such devices when used as true multipliers is the very limited dynamic range they can afford, typically about 40 dB or so. This is not sufficient to handle signals that can span as much as 80 dB or even more.

A solution commonly adopted is to produce a slight approximation of the desired output signal. Rather than attempting to multiply two sine waves together, we will modulate a higher-frequency square wave with a lower-frequency sine wave (or other lower-frequency message signal), as suggested in Fig. 5-32. Then we can simply apply the resulting output to a bandpass filter and so retain only the desired output frequency components, effectively "round off" the sharp corners, and end up with the originally desired output signal. As indicated in the figure, such multiplication effectively amounts to multiplication of the lower-frequency signal and the higher-frequency square wave. Referring to Fig. 5-33(c), recall that a square wave consists of an infinite amount of harmonics, at odd multiple harmonics of the fundamental frequency, and it is these higher-frequency components that produce the sharp square corners of the square wave. Thus the effective multiplication process produces sum and difference frequencies between the lower-frequency input signal and each of the square-wave harmonics, as shown in the frequency spectrum. The resonant bandpass filter thus rejects these higher-frequency components.

In practice, such an approach can effectively handle a much greater dynamic range, and is limited only by the resulting switching noise. Because of the convenience of this approach, considerable effort is directed at developing devices to handle even larger dynamic ranges (see reference 13, for example).

Referring to the intermediate result produced by modulating the square wave in Fig. 5-32(d), one can see that it can be produced "mechanically" by a double-pole, double-throw switch, as shown in Fig. 5-33(a), where the output signal is the same as the input, but where the polarity alternates at the rate of the high-frequency carrier square wave. This switching concept thus leads to the actual circuit configurations, discussed next.

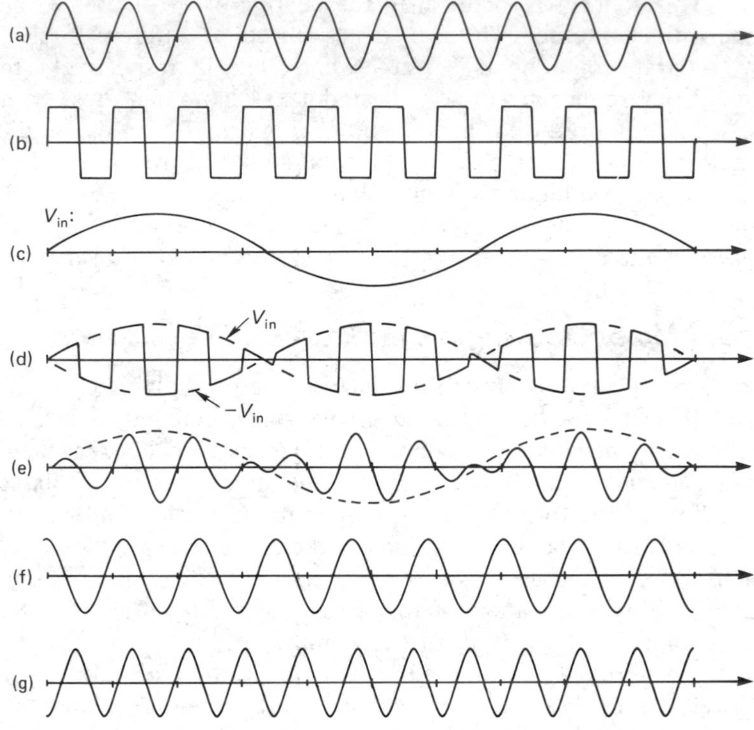

Figure 5-32 Practical mixer realization by operating components in switching mode (saturation/cutoff : ON/OFF).
(a) strong sinusoidal L.O.
(b) switching effects: saturation/cutoff operation due to strong L.O.
(c) modulating signal V_{in}
(d) modulated L.O., switched modulating signal
(e) filtered modulated L.O.
(f)(g) lower (difference), and upper (sum) frequency sidebands.

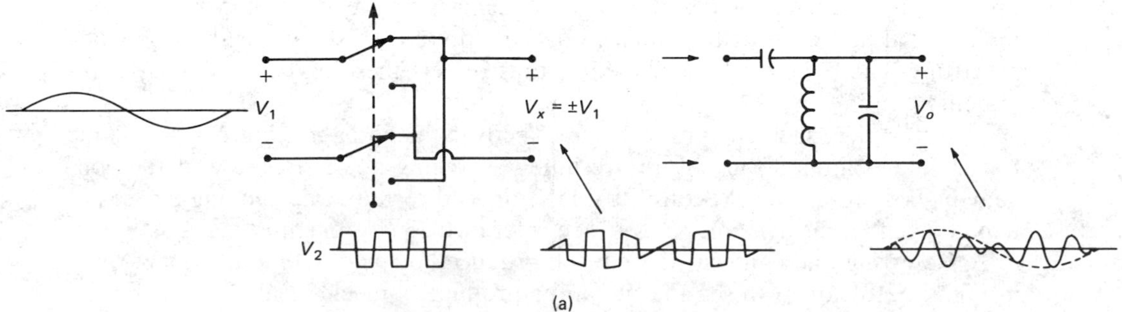

Figure 5-33 Balanced mixer operation: (a) practical realization: switching and filtering *(continued)*.

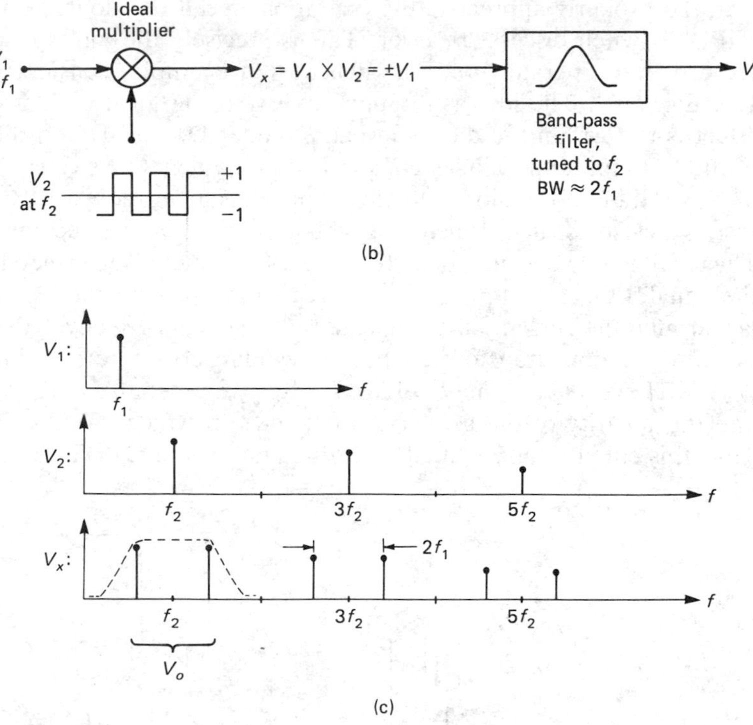

Figure 5-33 Balanced mixer operation: (b) mathematical equivalent; (c) frequency spectra.

5-4.2 Double Balanced Mixers

Referring to the trigonometric identity discussed previously, the practical realizations, and the output frequency spectrum, note that the two original frequency components themselves do not appear at the output; only their sum and difference frequency components appear. This is the ideal objective. In such a case, both of the applied input signals are canceled, or "balanced" out of the output—hence the name of the following mixer configurations. In practice, this perfect balance depends very much on proper symmetrical component matching, such as the diodes, transistors, resistors, and transformers used, and is only approximated. Typically, however, the levels of the original mixer input frequency components can be reduced by more than 70 dB.

Diode Ring Mixer

The typical diode ring mixer configuration is shown in Fig. 5-34. There are three ports, one for the input carrier, or local oscillator injection; a second one for the input modulating signal; and a third to draw the output from. From a structural point of view, any port can serve any one of these three functions.

To properly appreciate the operation, recall the double-pole, double-throw (DPDT) switch discussed earlier. This is precisely the mode in which the diodes are driven to operate (i.e., as switches). For example, consider the setup in case 1 of Fig. 5-34. The idea is to apply a very strong local oscillator to port 2. The effect is to turn on the diodes in pairs, diodes D1 and D3 on the positive swings of the LO input, and diodes D2 and D4 on the negative swings. When turned on, or forward biased, a diode effectively presents a very low ac dynamic impedance, just as a closed switch. When reverse biased, however, it becomes an open switch. The result is as shown in (b) and (c). The transformers help in electrically decoupling the signal, LO, and output ports. The reader may verify that, by Kirchhoff's voltage law around the current paths indicated, the currents effectively balance each other out in the output transformer (port 3) windings, therefore nulling the LO out to port 3. However, the input voltage is effectively "connected" to the output port, and the polarity of the connection depends on which diode switches are closed. Thus this circuit realizes the desired function of the DPDT switch.

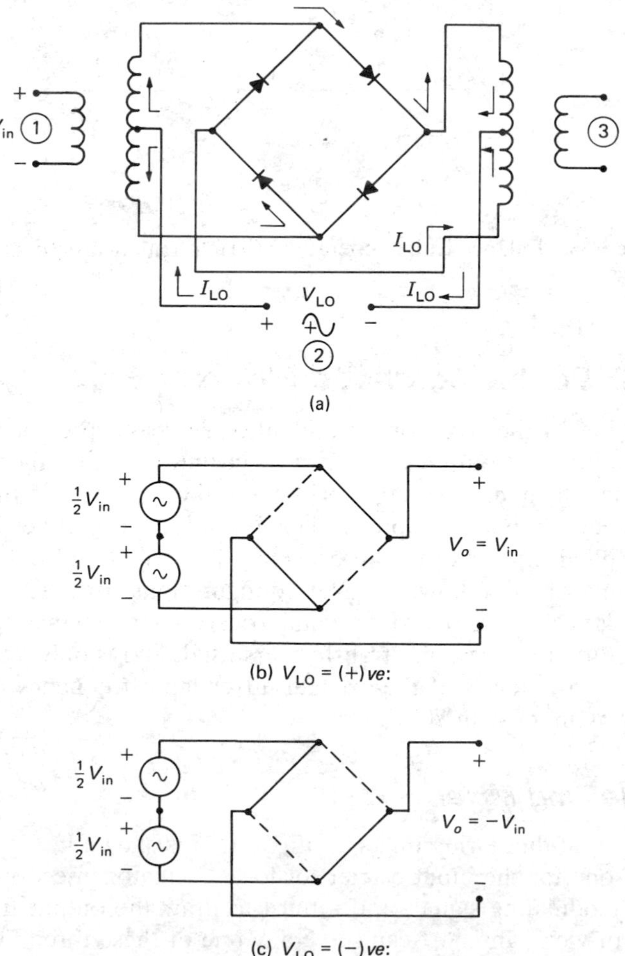

Figure 5-34 Diode ring balanced mixer.

Consider next the arrangement shown in Fig. 5-35. Here the LO feeds into port 3, whereas the audio modulating signal is coupled to port 1 and the output is taken from port 2. In this case the diode pairs that are switched together are different, (i.e., diodes D2 and D3, and D1 and D4). With the input transformer secondary center tap grounded, we again have two sources of V_{in}, of opposite polarity. However, the diodes connect only one to the output, thus alternating V_{in}'s polarity at the output. The reader should again verify that with proper balancing of the components, the currents cancel out and none of the LO itself appears either at the output or at port 1.

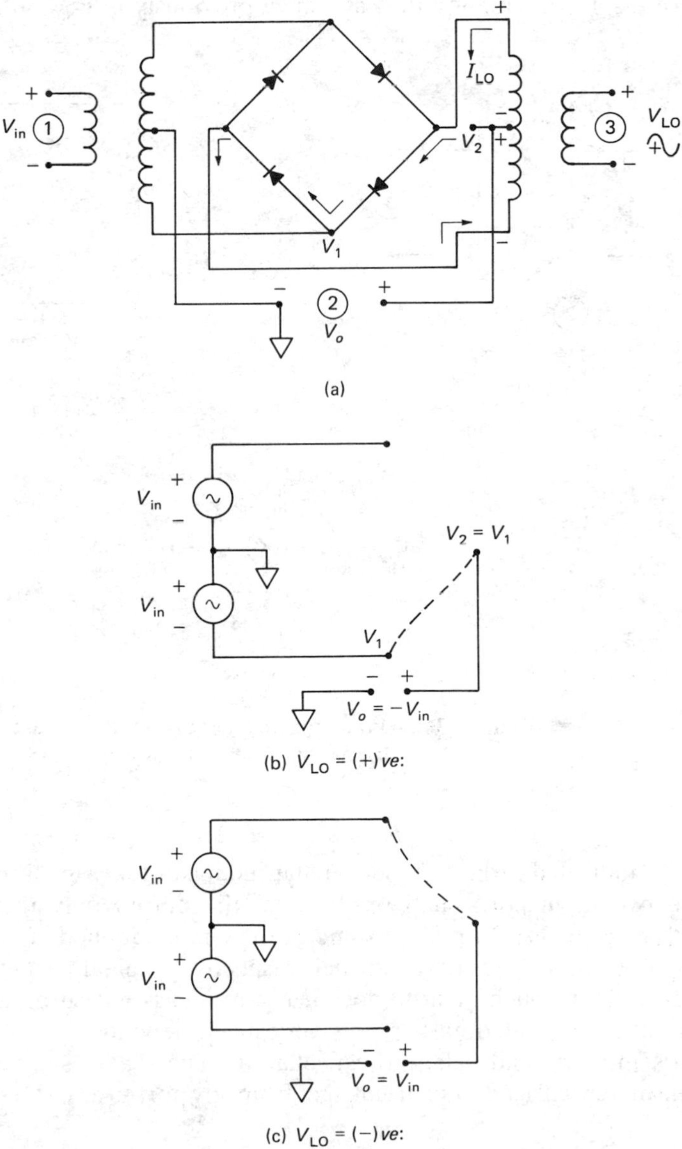

Figure 5-35 Diode ring balanced mixer.

These two configurations are summarized and compared in their effective operation in Fig. 5-36. Note that the first circuit realizes a DPDT switch, whereas the second arrangement realizes a SPDT switch, driven by the LO. For all practical purposes, either one amounts to the same multiplying function indicated in (c). Note that proper operation requires that the diodes operate as switches (i.e., fully on or fully off). This requires that the LO signal be quite strong, while the input signal level should be less than the LO by at least 6 dB or more. If not, then the diodes turn on only partially, and we end up with the very IMD products we originally sought to avoid with this circuit. We find that a mixer's performance, or linearity, can be expressed in terms of its output distortion products, in the same kind of display as an amplifier, as shown previously in Fig. 5-8.

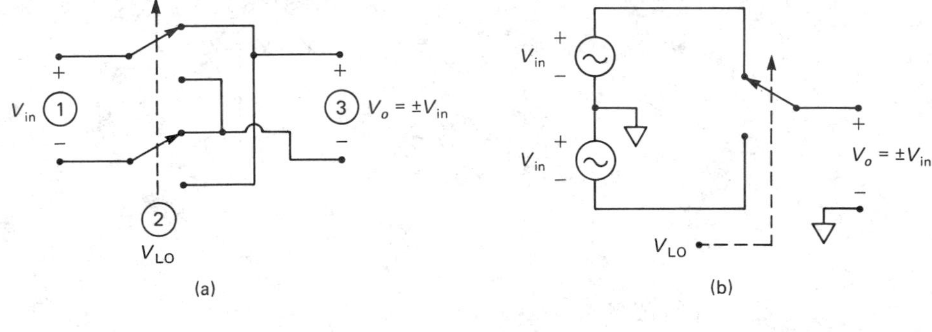

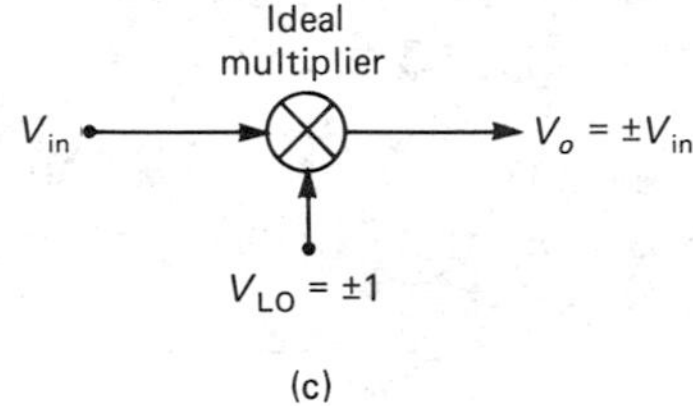

Figure 5-36 Diode ring mixer equivalents: (a) case 1; (b) case 2; (c) general equivalent.

As indicated earlier, the ideal balanced mixer takes in all the applied energies at the two input ports, and produces at the output only the desired sum and difference products. In reality, some of the energy applied at each port will "leak through" to both other ports, at their respective original frequencies, as shown in Fig. 5-37. How much or little does leak through is measured by the port-to-port isolation. As indicated earlier, this very much depends on component matching, signal symmetry, and balanced cancellation. The reader should easily verify that any mismatch will lead to such leakage, from any port to any other port, as suggested in Fig. 5-37.

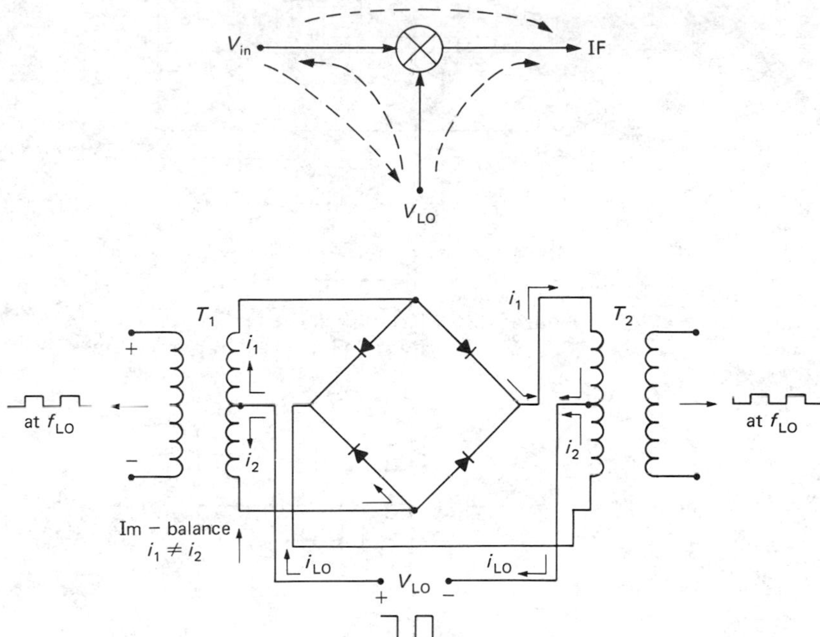

Leakage from LO port to both other ports due to imperfect cancellation of i_1 and i_2 resulting from imbalanced T_1 secondary winding

Figure 5-37 Balanced mixer—port-to-port isolation and leakage.

IC Mixer

Although diode ring balanced mixers are preferred for their better noise performance, they cannot provide gain. In addition, they can be relatively expensive. A popular and inexpensive alternative consists of the Gilbert cell arrangement, such as the MC1496, shown in Fig. 5-38. An equivalent configuration is shown in Fig. 5-39(a).

For practical purposes, the circuit is made to behave as in Fig. 5-39(b). Thus we have an input differential amplifier, with a differential output signal current. These two differential output currents are effectively switched by the LO. Again, the LO must be strong enough to turn the top transistors cleanly on and off, that is, at least 600 mV_{pp}. The resulting output current I_o consists of the dc quiescent bias current I_{dc} and an ac signal. This ac signal is the input audio signal, alternating in polarity according to the LO. The output *LC* filter may be tuned to any desired frequency component, whether it be the sum frequency, the difference, or with a bandwidth wide enough to include both if the input is intended to modulate the LO (as opposed to synthesizing a new frequency).

A similar device is Signetics' NE602 balanced mixer and oscillator, described in Fig. 5-40. It uses the same Gilbert cell configuration, shown in (c). A test circuit is suggested in (d).

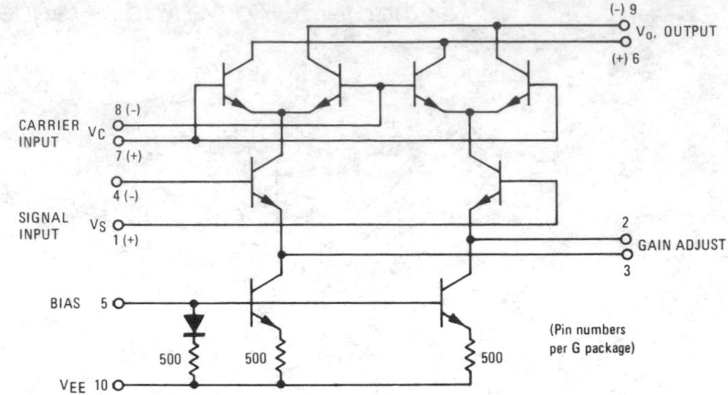

Figure 5-38 MC1496 balanced mixer. (Copyright Motorola Inc. Used by permission.)

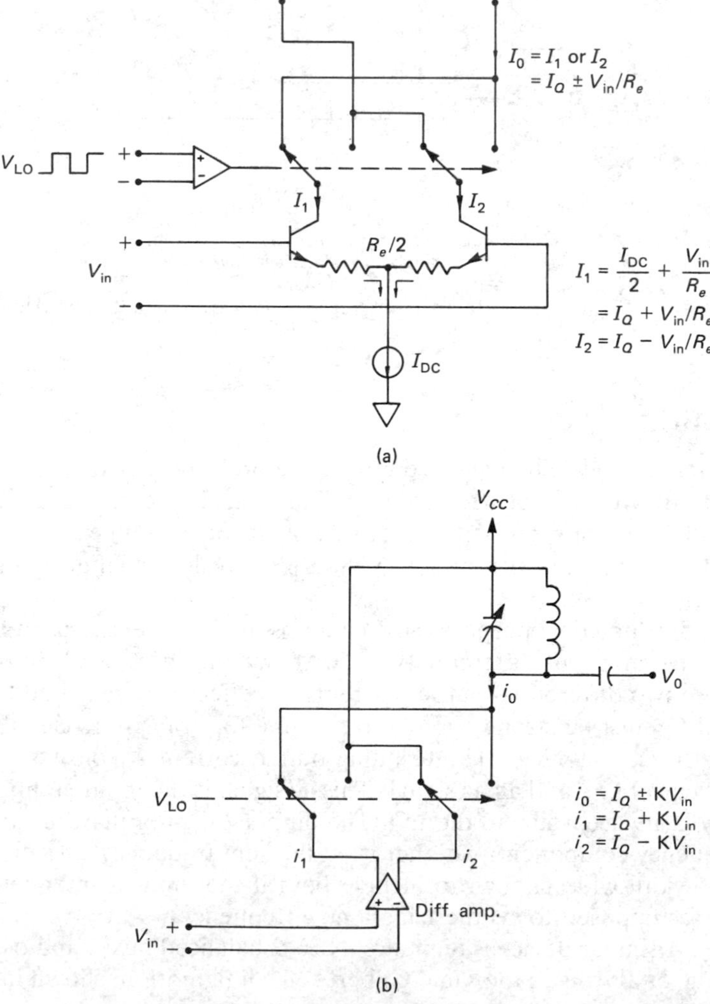

Figure 5-39 Operating equivalent of 1496 balanced modulator.

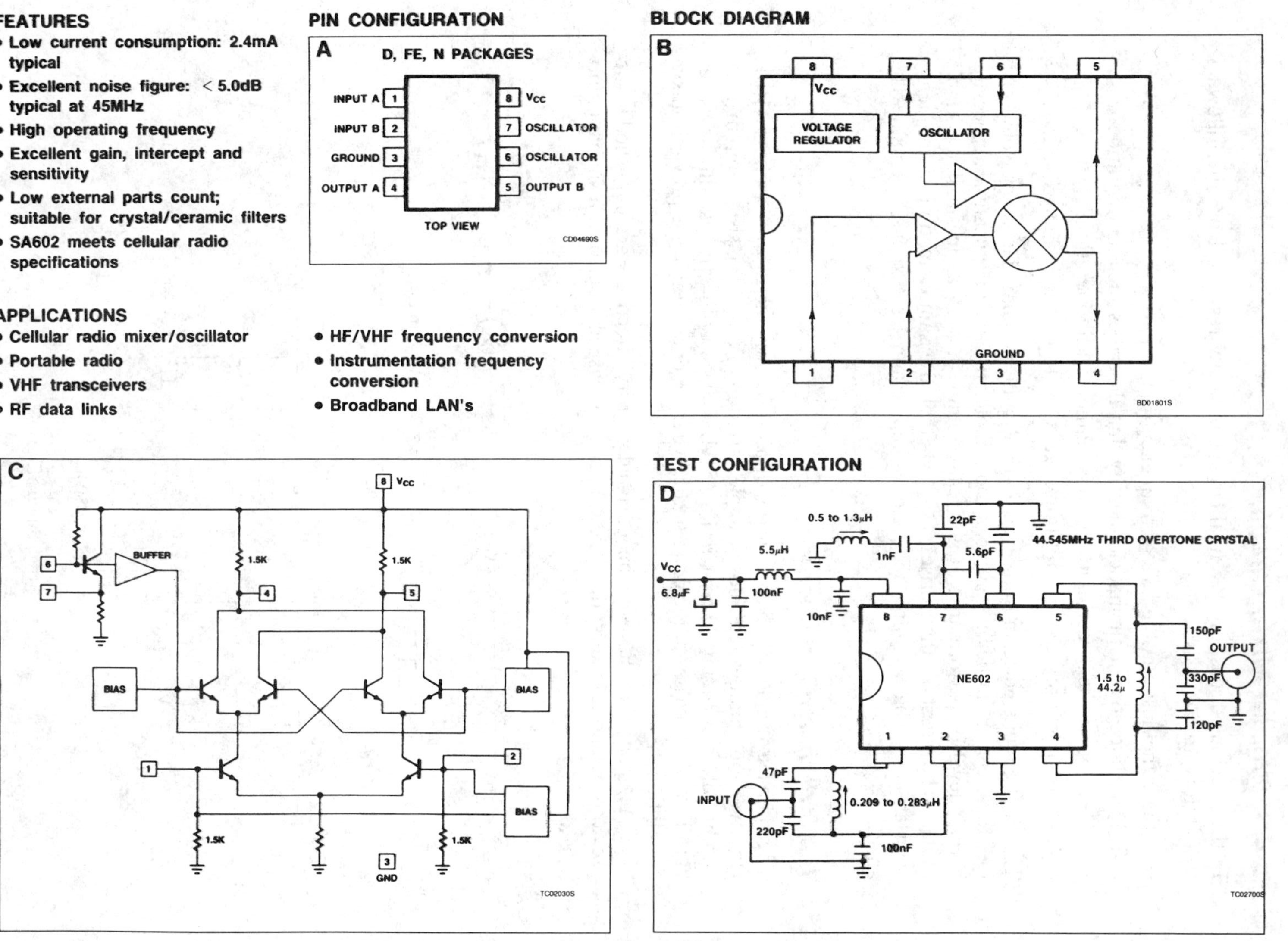

Figure 5-40 (a) and (b) Signetics' NE602 double balanced mixer and oscillator; (c) and (d) Signetics' NE602 equivalent circuit and test application circuit for cellular radio. (Courtesy of Signetics Corp.)

5-4.3 Simpler Mixer Circuits

In the previous discussion, the final circuits basically resulted from the following straightforward analysis:

1. State the desired objective of frequency translation.
2. Determine the mathematical identity that describes such an operation.
3. Develop an ideal realization for this mathematical operation.
4. Determine a practical circuit to realize it, or to realize it approximately.

The final circuits perform quite well and are quite popular. However, some may find these circuits rather expensive or too complex for some applications, especially if their level of performance is not really required, such as in smaller portable radios. The question then arises as to whether the same function of frequency translation can be accomplished with a simpler and less expensive circuit. The answer is yes.

The Switching Operation

The simplest frequency mixer consists of a diode only, as shown in Fig. 5-41(b). However, one more often finds a simple variation of a one-transistor amplifier. A number of such circuits appear in Fig. 5-44. The end objective is still the same, but the analysis begins at the point of the switching operation in the previous mixer description.

As opposed to the switching of signal polarity effectively accomplished in the previous balanced mixers, the following circuits basically perform a simple on/off

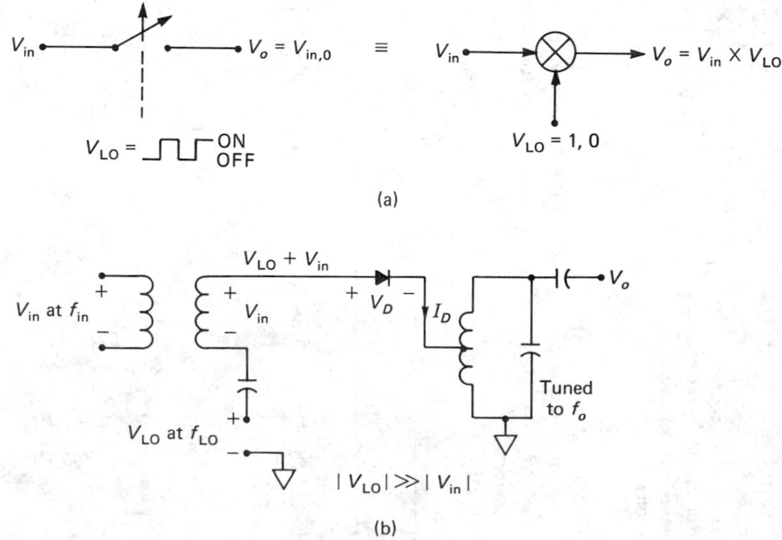

Figure 5-41.

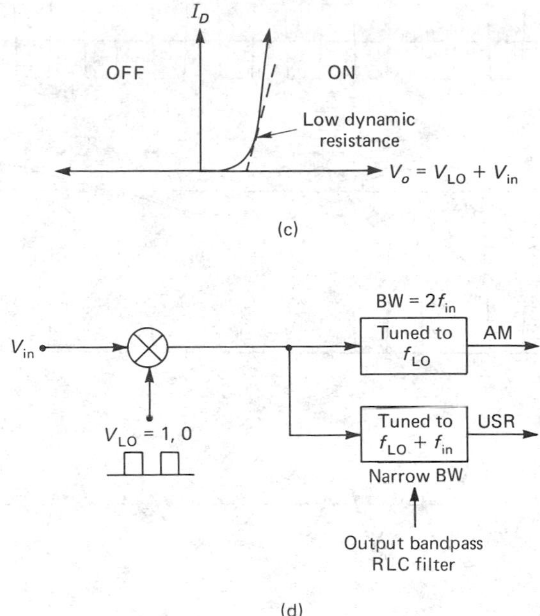

Figure 5-41 *(continued).*

switching, as shown in Figs. 5-41(a) and (d) and 5-42. Note, however, that the biasing requirements of practical circuits end up adding a dc offset to the actual input signal. The end effect does not interfere with the objective, but must nevertheless be taken into account. Thus the carrier signal very rapidly switches the input modulating signal (along with its dc offset) on and off to the output. This can be represented by an equivalent multiplication operation, whereby we are multiplying together the input signal by a square pulse train alternating between 1 and 0 V.

To confirm that the desired objective is accomplished, we must again analyze this operation from the frequency spectrum point of view. Refer to Figs. 5-42 and 5-43. Recall that the square wave consists of a dc component, at 0 Hz, and a number of harmonics, at odd multiples of the pulse-train frequency. Thus the multiplier output consists of products of the input signal and each one of these harmonics, and each produces sum and difference frequencies. All that need be done at this point is to filter out the undesired frequency components with a bandpass filter, centered at the desired component, and with an appropriate bandwidth.

At this point it is important to note that *both* of the signals applied to the mixer, that is, the input signal, including the effective dc offset, *and* the carrier, appear at the switch output. This is quite different from double balanced mixers, where each of these two signals was effectively balanced out. This factor must be taken into account, and in certain applications may very well disqualify the simpler solution in favor of a balanced mixer.

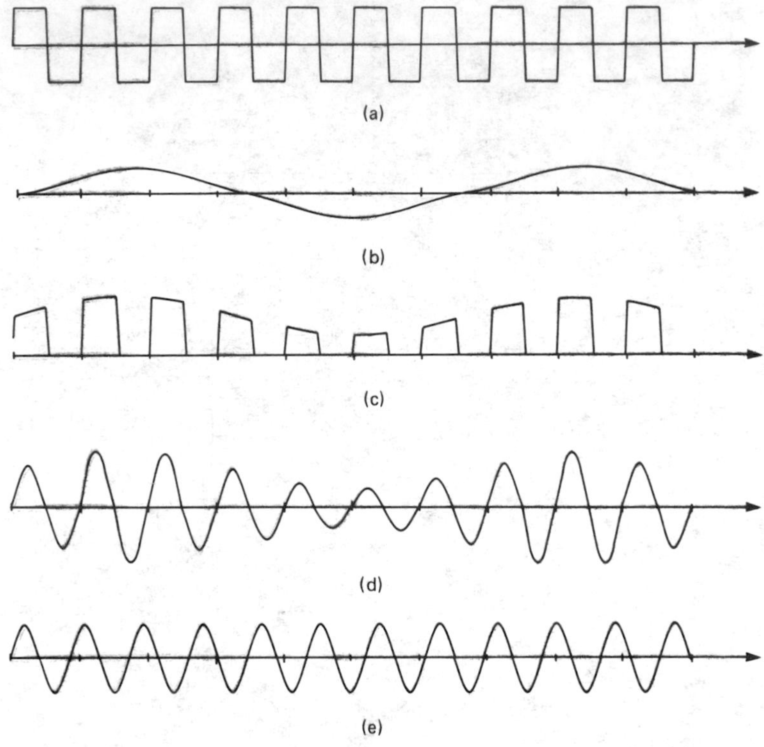

Figure 5-42 Simple diode (switching) mixer waveforms, for Figure 5-41(b).

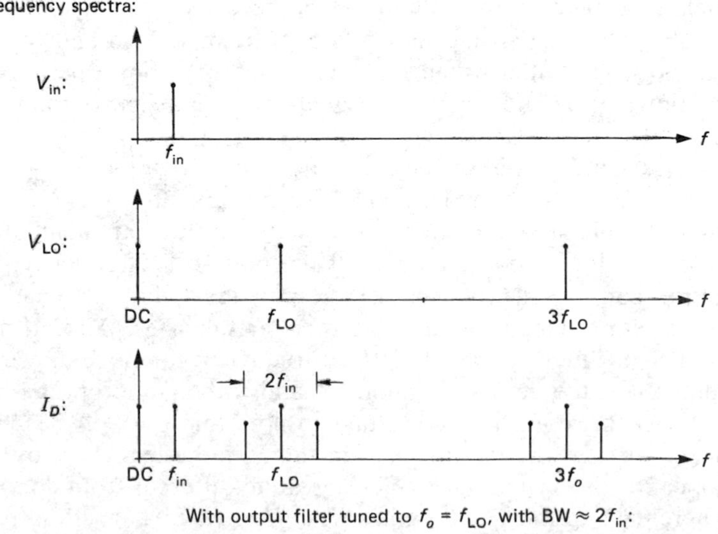

Figure 5-43 Frequency spectra at various points in simple diode mixer.

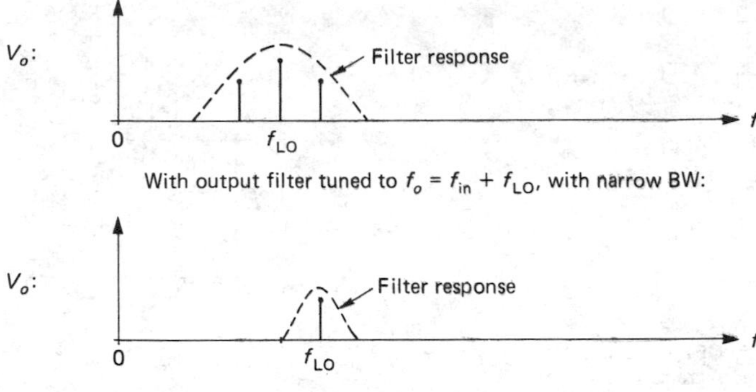

Figure 5-43 *(continued).*

Simple Diode Switching

Just as with the diode ring mixer, we can again use a diode as a switch. Again, this is accomplished by driving it on and off, by alternately biasing in its forward region of low dynamic resistance, and then in the reverse region, at the rate of the carrier frequency. Simply by adding the input voltage, the desired intermediate output is produced, and simple filtering results in the desired frequency translated output signal (see Figs. 5-41 to 5-43).

Transistor Amplifier Switching

A more common realization of such a switching mixing operation is an ordinary transistor amplifier. While the one-diode mixer will inevitably attenuate the signal, since it provides no gain, the transistor mixer will also amplify it. Various common alternatives are presented in Fig. 5-44. In circuit (a), the carrier is used at the emitter to turn the bias on and off, and so also the transistor. The actual input signal is applied at the base and thus does not interact with the applied local oscillator. In circuit (b), however, both the signal and the local oscillator are applied at the base. To reduce their interaction, the local oscillator is made strong, but its coupling capacitor is made small, for example, 50 pF. This configuration presents an advantage over the preceding one: the local oscillator drive need not be as powerful, since it feeds into the higher base impedance, as opposed to the much lower emitter impedance. The same configurations apply to JFET devices. These may be preferred for their lower noise figure and more linear operation, although they are typically somewhat more expensive than BJT devices.

To reduce the mutual interaction between the local oscillator and the input signal further, the dual-gate MOSFET is preferred. In circuit (c) (Fig. 5-44), each signal feeds into its own gate. The two gates are internally physically separated and extremely small, thus presenting a very low capacitance of only a few picofarads. On the other hand, this device is more expensive.

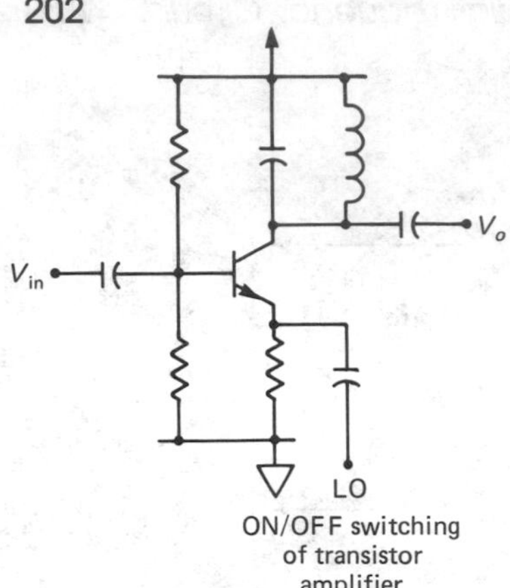

(a) Emitter-driven BJT mixer

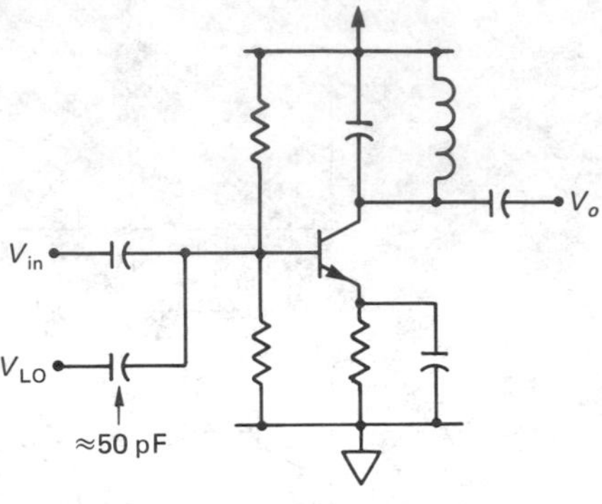

(b) Base-driven BJT mixer

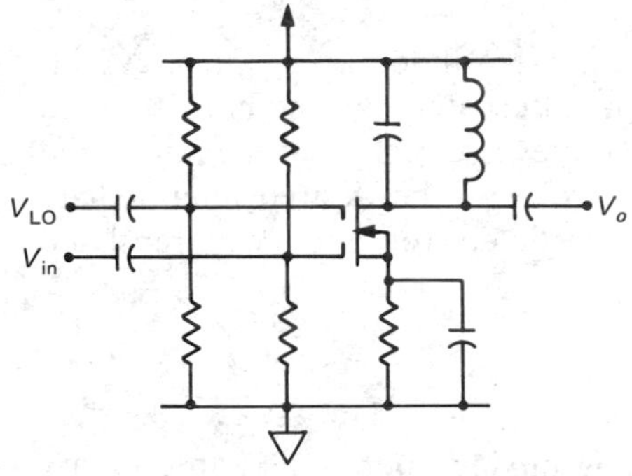

(c) Dual-gate MOSFET mixer

Figure 5-44 Simple transistor mixing.

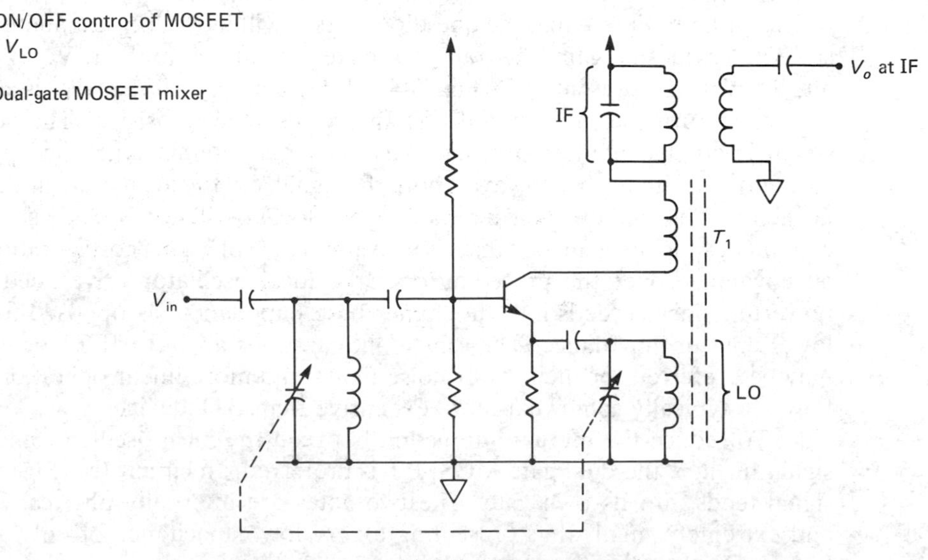

(d) Self-excited autodyne mixer

If we really wish to reduce costs, both the local oscillator and mixer/amplifier functions may be combined in one single transistor, as shown in circuit (d). The idea is to start with an oscillator circuit, with appropriate tuned positive feedback, usually by a transformer. This inevitably results in strong oscillations, as normally desired, driving the transistor nearly to the on and off extremes. Next, the input signal is applied directly to the input, base of a BJT or gate of a JFET. The transistor then amplifies it and effectively switches it on and off according to its own oscillations. At the output, collector or drain, a tuned *RLC* filter circuit must also be provided to retain and pass on the desired frequency components. If properly designed, the oscillator tuned circuit and output selection filter do not interact, so that the transistor can indeed oscillate, amplify, *and* perform the desired mixing operation.

Single Balanced Mixers

Again, the factors of cost, performance requirements, component availability, personal experience, and preferences for different circuit configurations determine the circuit selected. As indicated earlier, the double balanced mixer ideally completely balances out both the input signal *and* the local oscillator signal. However, an important concern is about the relative strength of resulting output signal and the strength of the frequency components. If this signal is too strong, this may result in undesired saturation and intermodulation products. Now, if the input signal is small, as is usually the case, then only the local oscillator must be balanced out to prevent such output saturation. Thus we need only a single balanced mixer circuit, which then results in a simpler and less expensive configuration.

Typical circuit configurations appear in Fig. 5-45. The basic idea is the same for each: a differential amplifier is switched on and off by the LO. It should be quite apparent that the LO currents that result from the lower switching transistor are effectively balanced out of the output signal by the center-tapped output transformer. To ensure the best attenuation of the LO at the output, a trim pot should be provided to balance out the two transistor currents, as in circuit (a).

The same principle applies to circuit (b), where the LO and the input voltages are simply added together to the JFET gates. Again, note here that the LO signal does not appear out of the input signal's port. In circuit (c), a pair of dual-gate MOSFETs are used as switches.

Nonlinear and Square-Law Mixing

As emphasized so far, all mixer circuits were purposely driven by the LO to operate as switches. This is the only way to operate the double balanced mixers. However, with the single and unbalanced circuits, we must now also consider what occurs if the LO signal is not that strong. The result and analysis depends on the devices used.

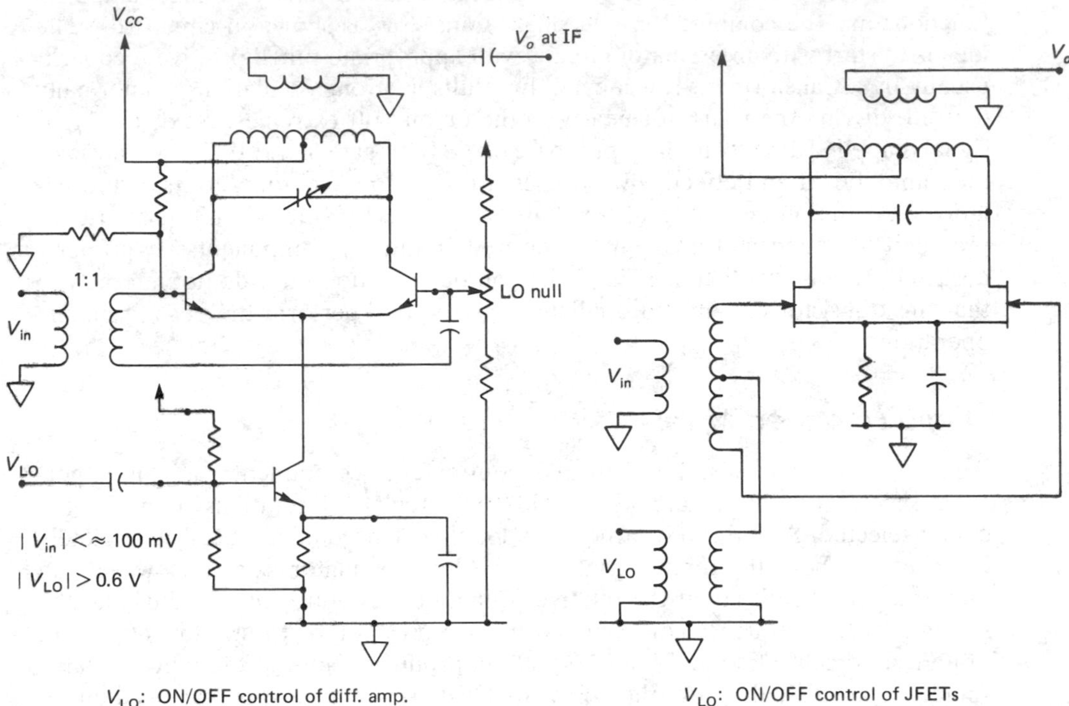

V_{LO}: ON/OFF control of diff. amp.

LO currents cancel out at balanced output transformer.

(a)

V_{LO}: ON/OFF control of JFETs

(b)

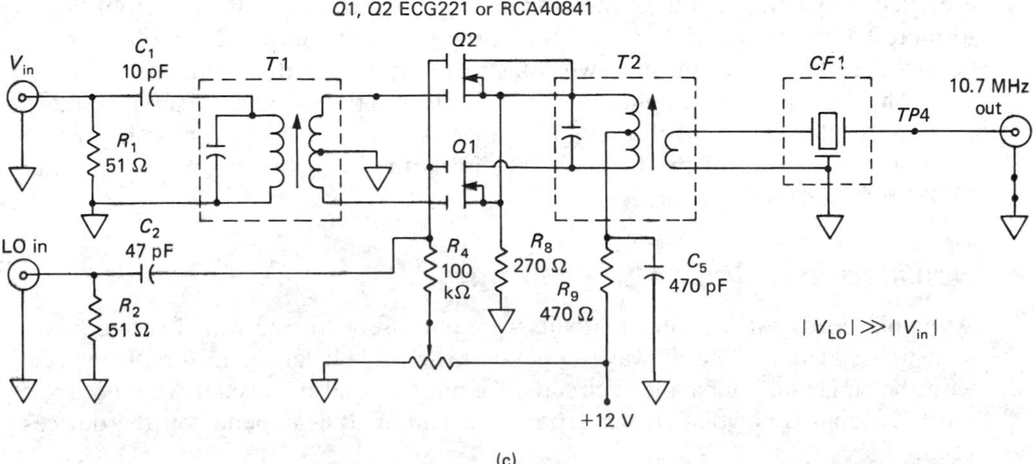

(c)

Figure 5-45 Single balanced mixers.

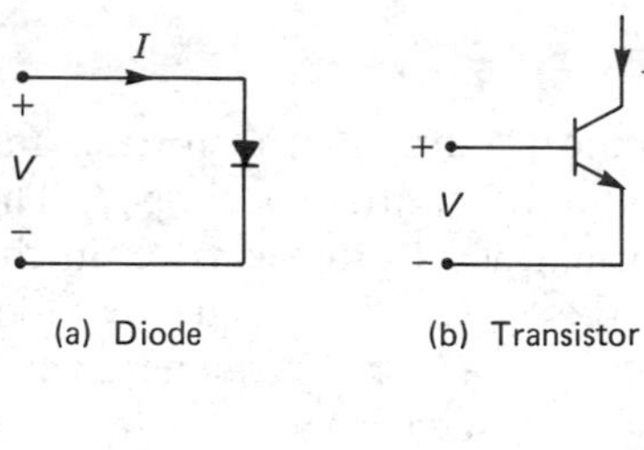

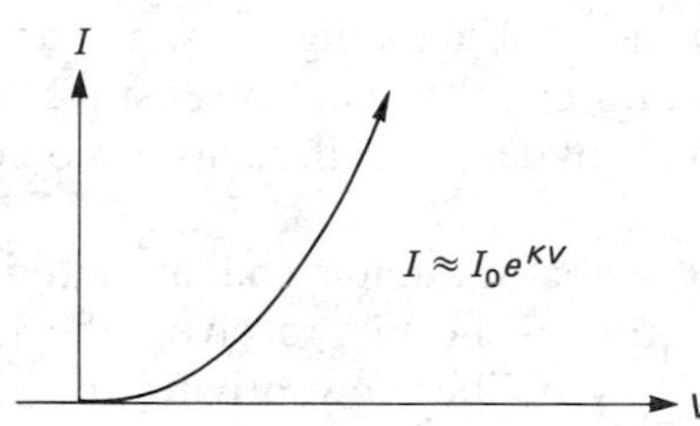

$$I = I_0 e^{KV} = I_0(1 + av + bv^2 + cv^3 + dv^4 + \cdots)$$

– Spectral analysis, if v consists of two sine waves V_1 and V_2 at two different frequencies f_1 and f_2, $|V_1| \approx |V_2|$

$$V = V_1 + V_2 = \cos(2\pi f_1 t) + \cos(2\pi f_2 t)$$

then: – V consists of components at f_1 and f_2

– $V^2 = \frac{1}{2}(\cos[2\pi(f_1 - f_2)t] + \cos[2\pi(f_1 + f_2)t])$ consists of components at $f_1 \pm f_2$

– V^3 consists of components at $2f_1 \pm f_2$ and $2f_2 \pm f_1$

– V^5 consists mainly of components at $3f_1 \pm 2f_2$ and $3f_2 \pm 2f_1$

– V^7 consists mainly of components at $4f_1 \pm 3f_2$ and $4f_2 \pm 3f_1$

Figure 5-46 Diode/transistor transfer function.

When using diodes and BJTs, recall that their transfer function is basically exponential, as shown in Fig. 5-46. The basic consequence of interest to us concerns the frequency components that result. All imaginable order intermodulation products result; this is the nonlinear mode of operation, whereby the original signal is purposely distorted. This distortion is necessary to produce the sum and difference frequencies. However, their relative strength depends on the level of the LO. Note also that the LO signal is always a sine wave. However, when strong enough, it effectively simply switches the devices on and off. When smaller, this is no longer the case.

When moderately strong (100 to 200 mV_{pp}), the resulting output consists predominantly of the original frequency components, and only their sum and difference frequencies. This is the square-law mode of operation, whereby the higher-order products are basically of negligible intensity. The desired output component is then easily retained by an *LC* filter. When stronger, without going into the switching mode (200 to 400 mV_{pp}), the undesirable odd-order products appear. The difficulty with this approach is the necessity to control the LO signal level properly, which may prove rather inconvenient, which is why it is simply preferable to overdrive the devices into the switching mode.

It is inevitable that violent switching of the devices will produce significant switching pulses, transients, and noise. In sensitive cases, this is undesirable. For this reason the more gradual nonlinear mixing by intermodulation approach is preferable. However, as indicated, it is difficult to keep the undesirable odd-order products low. In this respect, the JFET and MOSFET are very much appreciated, because of their simple square-law transfer function. The reader should mathematically verify that indeed, only the original components and their sum and difference products result at the output, regardless of the signal levels. As long as the total gate–source voltage does not exceed the pinch-off voltage, the behavior remains according to the square-law function. This in effect makes the FET mixer design easier to realize and is not as critical—hence its popularity.

5-5 LIMITING AND FM IF AMPLIFIERS

Thus far, emphasis has been placed on the importance of amplifier distortion and signal corruption, as is easily observed when the signal amplitude is clipped. The obvious reason for this emphasis is that the information is often carried in the amplitude of the signal, for example as in AM or the SSB. With frequency modulation, however, the information signal modulates the carrier's frequency, while the carrier amplitude remains constant. Therefore, in such a case, we are not worried about overdriving an amplifier into saturation and clipping the signal amplitude. Furthermore, the typical final FM detection circuitry is also sensitive to amplitude variations, as in an AM receiver, which is not desirable here, as it should respond only to frequency shifts.

Thus, whereas in an AM receiver the output detector audio signal amplitude is kept relatively constant by automatic gain control, in the FM receiver, the IF signal amplitude is controlled, or rather limited, by deliberately driving the IF amplifier into saturation. The IF amplifier is thus designed to provide a very large gain, such that a receiver input signal of as little as 5 μV or less will saturate the final IF amplifier stage prior to feeding the resulting constant-amplitude signal to the FM detector circuit. Any weaker signal at the receiver input will result in a correspondingly weaker signal at the detector and a degraded audio output signal quality. Refer to Fig. 5-47. Depending on the type of FM detector (by phase shift such as the ratio detector, or quadrature detector, or PLL), it may be necessary to provide a bit of further filtering with a resonant *LC* circuit before applying the clipped IF signal to the detector.

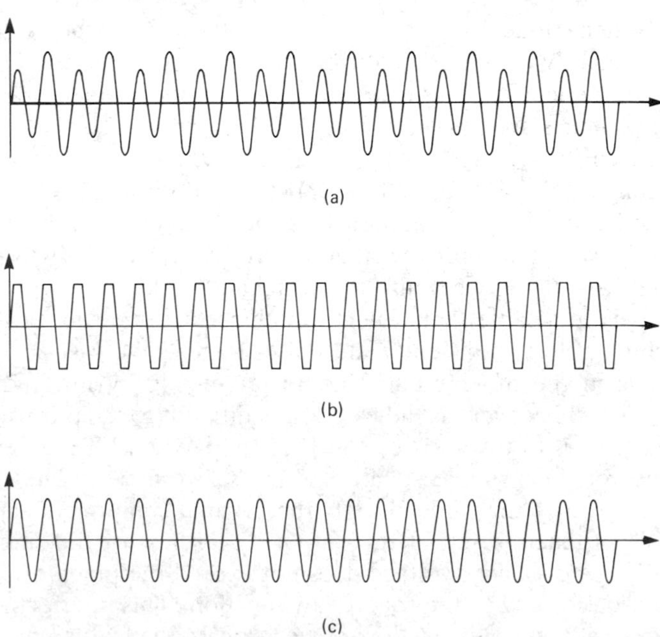

Figure 5-47 Simulated FM signal plus noise (a), out of limiter after saturation (b), and bandpass filtered (c).

In addition to improving the FM detector's performance in the presence of signal amplitude fluctuations, any noise and spikes accompanying the signal are also effectively "clipped off," thus reducing their contribution to the corruption of the output audio signal. Refer again to the simulated "noisy" input (a), "clipped" (b), and "cleaned up" (c) signals in Fig. 5-47. Note, however, that the noise effect is not entirely removed, since the noise also contributes to phase noise in the carrier, as observed in the signal zero crossings, which is what the FM detector effectively demodulates.

A number of such FM IF amplifier/limiters are available in integrated-circuit (IC) packages, such as Signetics' NE604, shown in Fig. 5-48. A typical FM IF

FEATURES

- Low-power consumption: 2.3mA typical
- Logarithmic Received Signal Strength Indicator (RSSI) with a dynamic range in excess of 90dB
- Separate data output
- Audio output with muting
- Low external count; suitable for crystal/ceramic filters
- Excellent sensitivity: 1.5μV across input pins (0.27μV into 50Ω matching network) for 12dB SINAD (Signal to Noise and Distortion ratio) at 455kHz
- SA604 meets cellular radio specifications

PIN CONFIGURATION

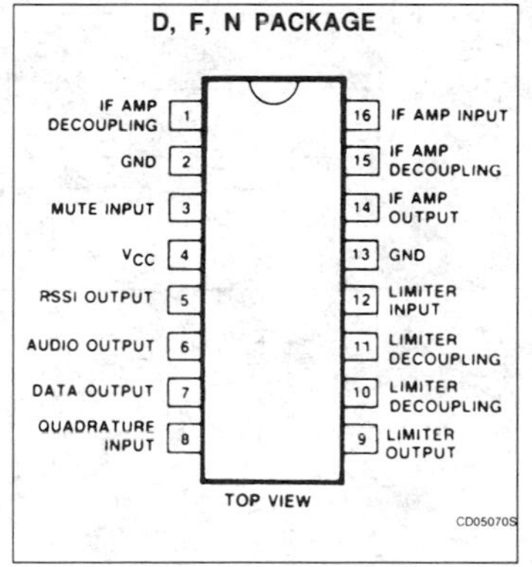

APPLICATIONS

- Cellular Radio FM IF
- Communications receivers
- Intermediate frequency amplification and detection up to 15MHz
- RF level meter
- Spectrum analyzer
- Instrumentation

BLOCK DIAGRAM

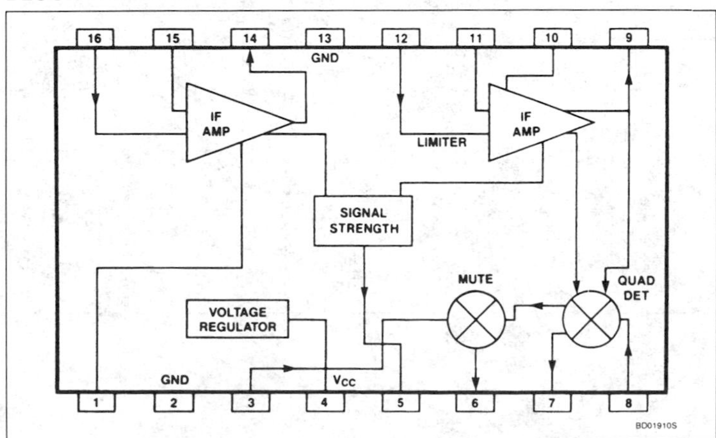

Figure 5-48 Signetics' NE604. (Courtesy of Signetics Corp.)

amplifier/limiter/detector application is shown in Fig. 5-49, using RCA's CA3090 FM stereo decoder, which follows directly after their CA3089 IF amplifier/limiter. Further integrating more functions directly inside the IC is Motorola's MC3362 complete dual-conversion FM receiver, which has an input signal sensitivity of 0.7 μV. Its principal features and a suggested test circuit are shown in Fig. 5-50.

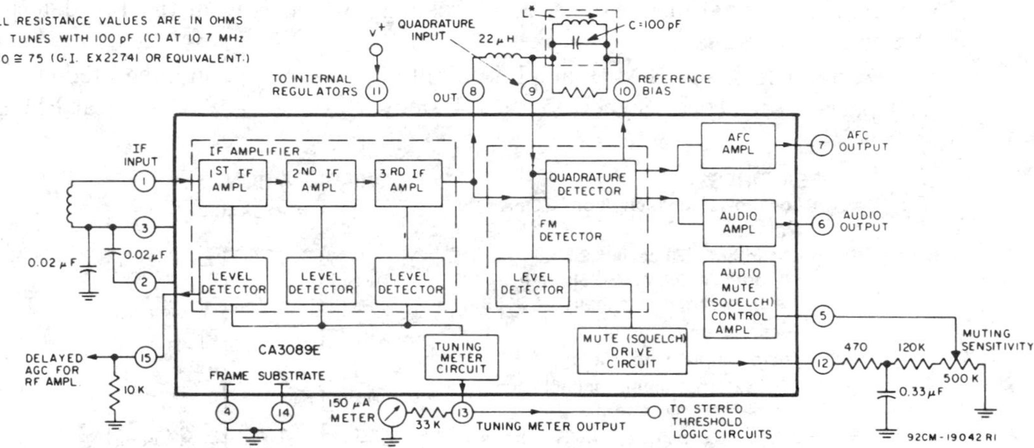

Figure 5-49 (a) Block diagram and pin-out of CA3089 FM IF amplifier/limiter/detector. (Courtesy of GE Solid State.)

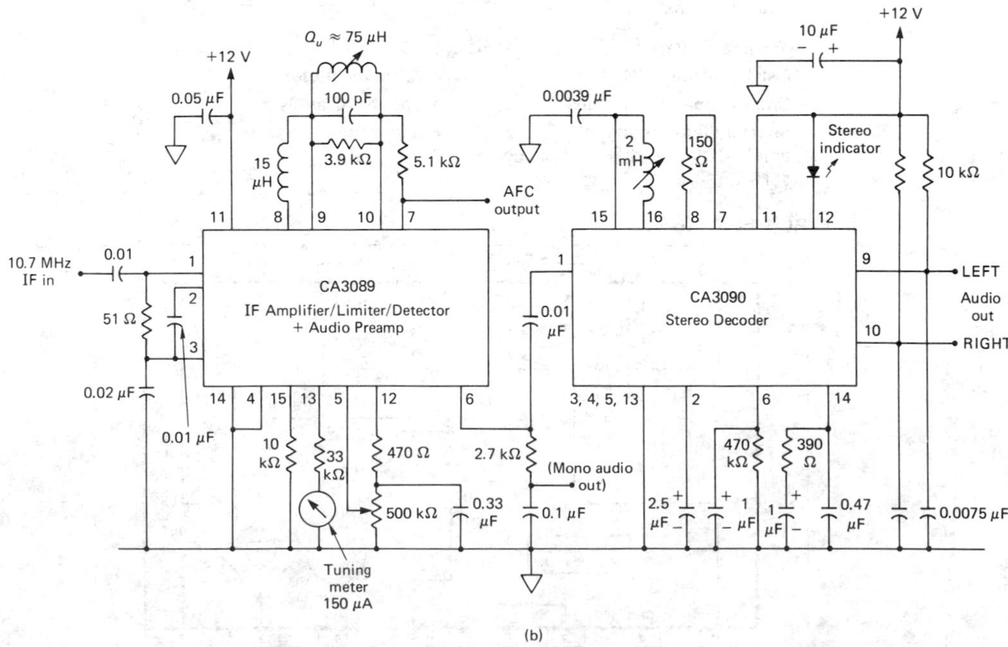

Figure 5-49 (b) FM 10.7 MHz IF amplifier, limiter, detector, and stereo decoder.

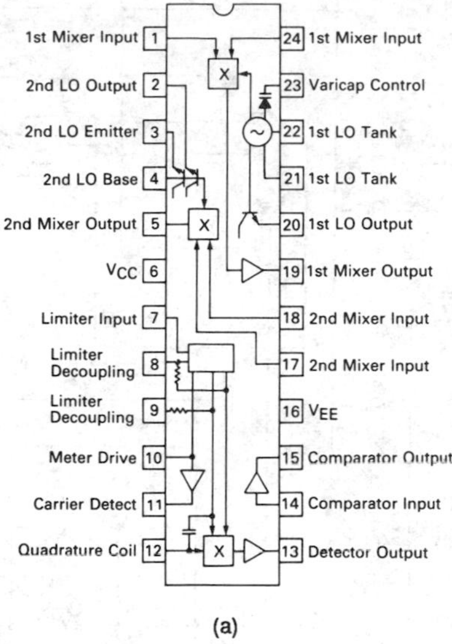

(a)

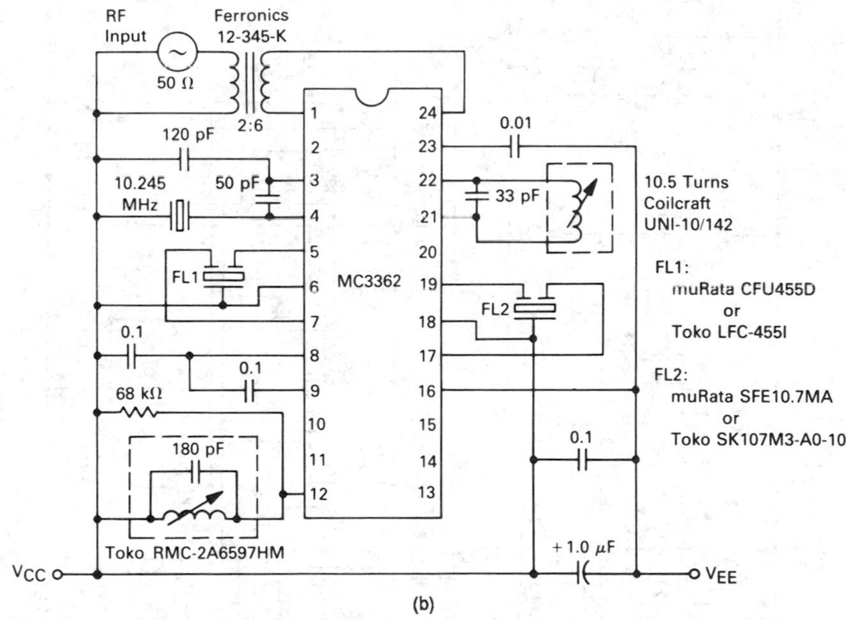

(b)

Figure 5-50 (a) Pin-out and block diagram of Motorola MC3362 dual conversion FM receiver IC. (Copyright of Motorola, Inc. Used by permission.)
(b) Test circuit for MC3362

(continued).

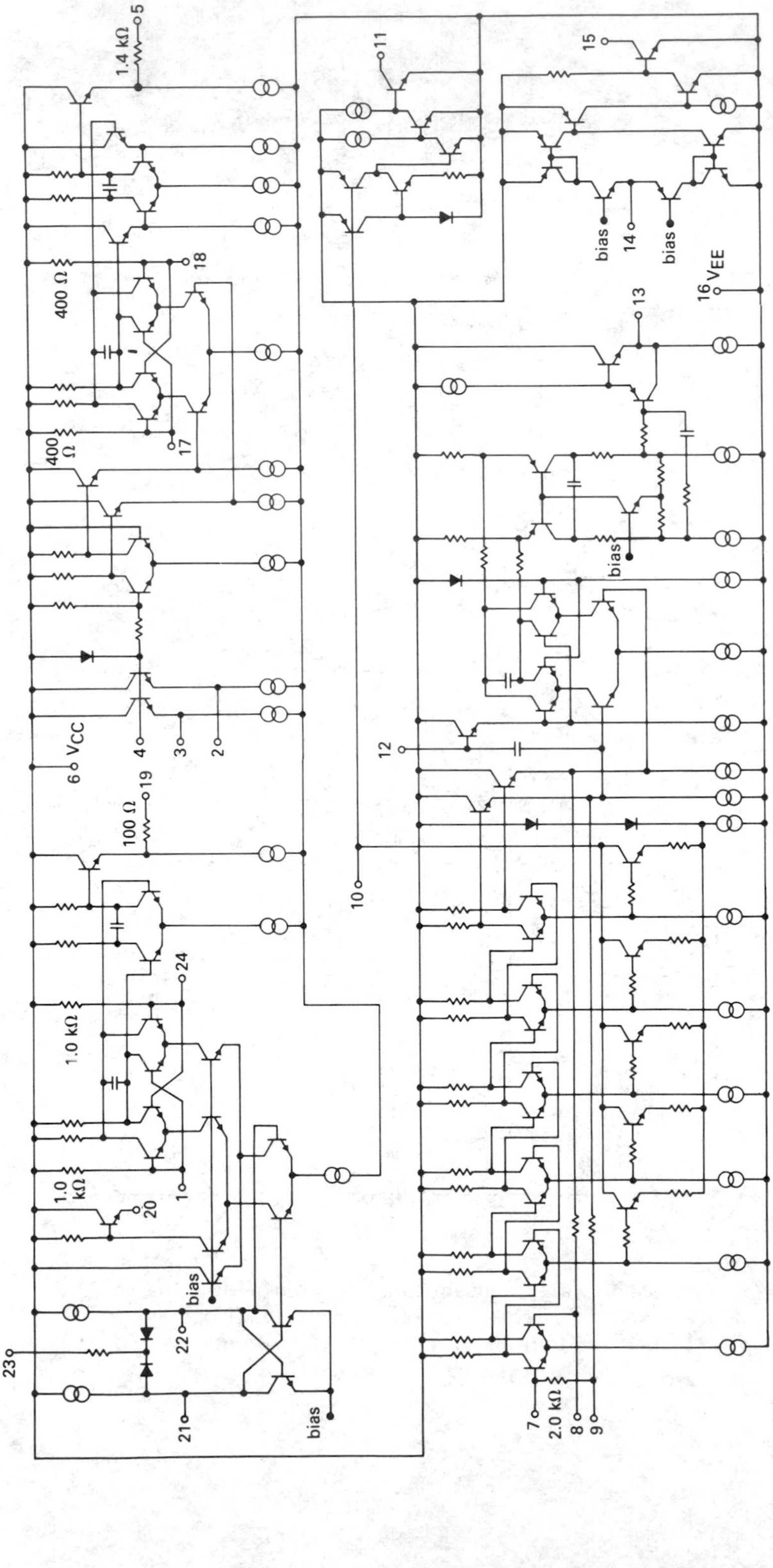

Figure 5-50 (c) Internal circuit schematic of MC3362 *(continued)*.

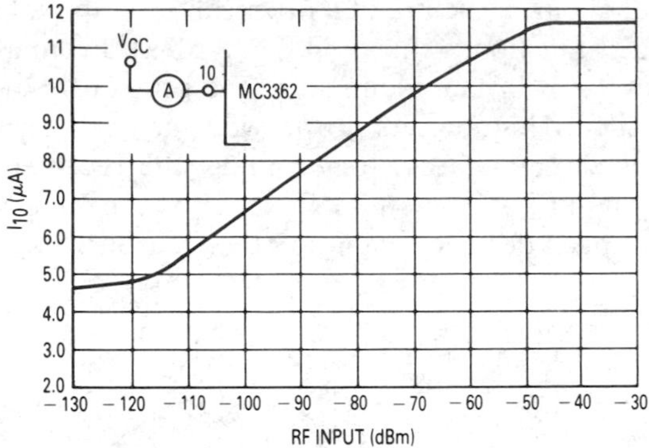

Figure 5-50 (d) RSSI output DC current on pin 10 versus input signal level.

5-6 LOGARITHMIC AMPLIFIERS AND SIGNAL STRENGTH INDICATION

One of the most difficult aspects to deal with in communication systems in general is the wide dynamic range of the signals to be processed simultaneously. As discussed earlier, we may be interested in one weak signal at the receiver input, but may also have to deal with another very strong one at another frequency too close nearby to be able to filter it out at the front end. We must then amplify them together, and do some further processing before we can get rid of the stronger one.

From an instrumentation perspective, then, such as with a spectrum analyzer, we wish to be able to measure these two signals and their relative strengths, and to compare them on the same scale. When two such signals are in a power ratio of, say, 80 dB (voltage ratio of 10,000) or more, this may not at all be resolved by the naked eye on an ordinary scope face or on a meter. Typically, one may visually resolve to maybe 2 or 3% of the vertical scale of a scope face, and just a bit better on a meter face. Similarly, with a communications receiver, one may wish to have a visual indication on a meter of the relative signal strength, which may be as weak as 1 μV or as strong as 100,000 μV or more. This represents a range of 100 dB that the end user expects to be displayed on one single meter scale. It is then more convenient to measure such signals on a logarithmic scale. Thus one amplitude unit, such as one vertical division on a scope face, represents a constant ratio, typically 10 dB. On the same scope face, eight divisions high, one can simultaneously observe two such signals at a power ratio of 80 dB.

To obtain such a vertical mode of signal display, a logarithmic amplifier is required. The principal characteristic of a logarithmic input–output relationship is that the amplifier differential gain (i.e., the slope of the input–output transfer function) is inversely proportional to the input signal level (or approximately so with any real circuit). Thus a weak 10-μV signal may be amplified by a factor of 100,000 to 1 V, whereas a 10-mV signal would only be amplified by a factor of 400, to 4 V. Note that on a semilog graph (with a logarithmic input horizontal axis), this logarithmic transfer function appears as a straight line.

The functional structure of a practical log amp realization which accomplishes such a transfer function is shown in Fig. 5-51(a). From a structural *and* operational point of view, it is in fact identical to the previously discussed FM IF amplifiers. Thus most such IC amplifiers do provide such a logarithmic signal-strength-indicating output dc signal. See, for example, the received signal strength indication (RSSI) output pin 5 on Signetics' NE604 or on pin 10 of Motorola's MC3362. These devices have been designed, among other reasons, for cellular radio applications,

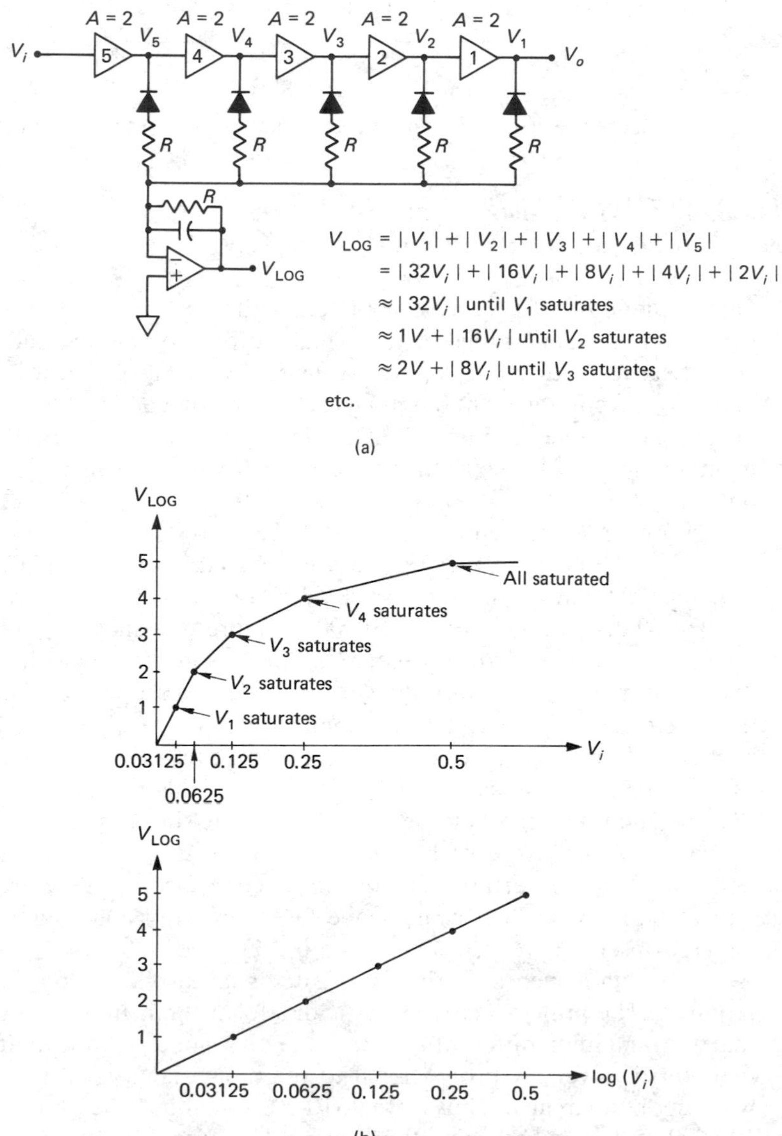

Figure 5-51 (a) Logarithmic amplifier and input–output relationship; (b) logarithmic input–output transfer function of circuit (a).

where it is necessary somehow to measure the approximate distance of the mobile radio-telephone from the land-based telephone system antenna. This is accomplished by measuring the strength of the signal received from the mobile. According to this signal-strength indication, the land-based computer may send a command to the mobile radio to increase or decrease its transmitted output power level. Similarly, RCA's CA3089 FM IF amplifier/detector device also provides such an output dc current signal on pin 13, for the purpose of driving a signal level indicating meter movement.

Circuit Operation

The operation of such a logarithmic amplifier is quite straightforward. As shown in Fig. 5-51, a number of equal-gain amplifiers are cascaded to provide a large overall ac gain. For very weak input signals, the amplifier behaves linearly, and the output is not saturated. However, as the input signal level increases, the last stage eventualy saturates and the output V_o ac signal amplitude remains constant, as desired for proper FM detection. Now, as the input signal level is increased further, the other stages will also saturate, one after the other, at proportionally higher input levels (every time the input doubles, the next previous stage saturates). If the ac output of each stage is rectified, and all resulting dc signals are summed together, the resulting dc output signal $V_{\log}$ will *increase* at a progressively *slower rate*, as the net gain of the unsaturated amplifiers reduces, by a factor of 2, every time another amplifier saturates. Thus we have the desired (approximate) logarithmic input–output relationship. Although each amplifier is expected to saturate only at discrete input signal levels, such that we should expect the corners and straight-line segments indicated in the transfer function of Fig. 5-51(b), in practice this occurs in a more progressive manner, as the diodes gradually turn on, such that the final net result is more gradual and closer to the shape desired.

Referring to the internal circuit diagram of the MC3362 in Fig. 5-50(c), one can see the cascaded amplifiers/limiters at the bottom left, driven from pins 7 and 8. The transistors at the emitters of the last five differential stages effectively perform the magnitude-level detection, implemented by the diodes in Fig. 5-51(a), and their outputs are summed at pin 10, as the output op amp does in Fig. 5-51. The resulting logarithmic function of the signal strength indication current output is shown in Fig. 5-50(d). Note that the dynamic range handled by this output spans roughly 70 dB.

Problems

5-1. Refer to the AM class C amplifier/modulator circuit in Fig. 5-23. Determine the battery supply current I_s, given that (1) the unmodulated carrier output power is measured at 20 W, (2) the modulation factor is increased to 70%, (3) the audio amplifier's efficiency is 55%, and (4) the class C amplifier's efficiency is 65%.

5-2. Referring again to the class C AM modulator, if the battery voltage can get as high as 14 V, what should be the minimum transistor collector–emitter breakdown voltage rating?

5-3. We wish to design a cable TV converter. The signal from the cable consists of many radio and television channels, spanning a frequency spectrum from about 50 to 250 MHz. The television channels are 6 MHz wide, whereas the FM radio channels are 200 kHz wide. In the converter, the input signal is applied to a 250-MHz low-pass filter, to reject anything above that for proper mixer operation. Next, this signal is mixed with a voltage-controlled oscillator output, with its frequency tunable from 600 to 800 MHz. This frequency is tuned according to the desired input channel. The mixer output feeds into a 10-MHz-wide bandpass filter tuned to 550 MHz, and then into a second mixer, driven by a fixed-frequency oscillator. Finally, this mixer output feeds into another bandpass filter, tuned to 54 MHz, and with a 6-MHz bandwidth. The filter output then feeds to the output connector to the television set.

(a) Sketch the block diagram of such a converter.

(b) Determine the required frequency of the second oscillator, such that when the VCO is tuned to 600 MHz, an input signal at 50 MHz will be translated to the output.

(c) To what frequency must the VCO be tuned to convert an input signal at 155 MHz to the output?

5-4. Show that a single balanced mixer, such as the CA3028, does require a trim pot to adjust the base bias on one of the differential stage transistors in order to cancel out the local oscillator at the output. Assume that the two transistors are not well matched. (*Note:* It should prove useful to work with an appropriate circuit diagram and some example biasing current values.)

5-5. A normal receiver, with no input signal, usually produces hiss out of the speaker. This is because the AGC sets the overall gain to its highest, and thus amplifies the internally generated noise. This is a good sign. Consider the setup shown in Fig. P5-5, where a receiver's sensitivity is boosted by a simple externally connected preamplifier. Both are tuned to 4.8 MHz. There seems to be a problem. With no signal applied to the preamp, nothing is heard out of the speaker, although some hiss is heard when the preamp is disconnected. Furthermore, when a signal generator is connected, AM modulated with a 1-kHz tone, tuned to 4.8 MHz, then a very distorted 1 kHz is heard and observed on the scope. As the generator frequency is swept up, nothing is heard out of the speaker until 9.6 MHz. Then again, the same distorted sound is heard. Determine the problem, and a solution.

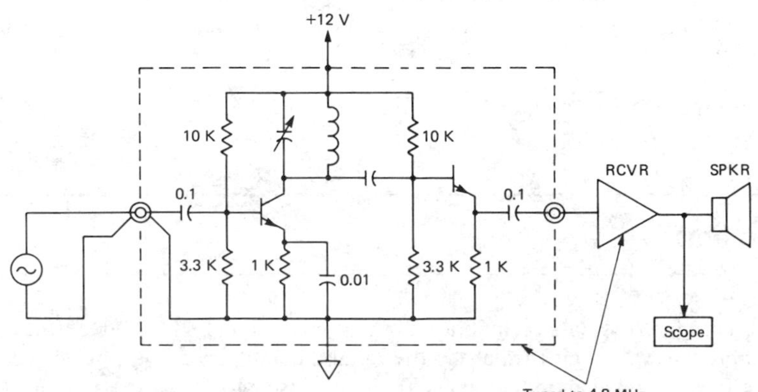

Figure P5-5.

5-6. Refer to the transmitter output circuit shown in Fig. P5-6.
(a) Identify the class of operation of amplifier Q1. Explain.
(b) Determine the function of D1, C_5, and pot R_A.
(c) What would you expect if the wiper on R_A was moved down toward ground? How would V_1, V_2, V_3, and V_4 respond?
(d) Determine the function of L_1–L_2–C_3–C_4.
(e) What could you modify, or add, to improve the rejection of the second harmonic, at 54 MHz, at the antenna?

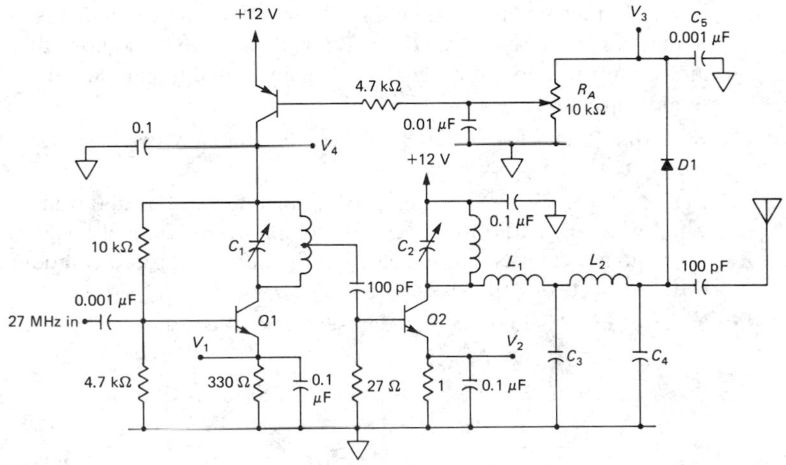

Figure P5-6.

5-7. Referring to Figs. 5-34 and 5-35:
(a) Determine the third way that the local oscillator, audio message (or second frequency source), and output can be connected to the diode ring mixer, and determine the equivalent "switched" configuration. Note that this would be a more appropriate way of feeding in the audio message for such mixing, such as in a SSB transmitter.
(b) Verify that indeed, ideally, the local oscillator, and audio signal currents balance out, respectively, at all the other ports.

5-8. Refer to Fig. 5-8. If two signals are applied simultaneously to this amplifier, at 11 and 12 MHz, at power levels of −80 and −40 dBm, respectively:
(a) What frequency components would you expect to observe at the output?
(b) If the 11-MHz signal level were increased up to −15 dBm, what frequency components would you then expect to observe at the output?

5-9. Referring to Fig. 5-5, if signal A were at 5 kHz and signal B at 100 kHz, what general features would you expect in the frequency spectrum of the signal in (d)?

5-10. Refer again to Fig. 5-8.
(a) If we apply an input signal and desire a minimum output signal-to-noise ratio of 40 dB (if this were a TV receiver input preamp, for moderate picture quality), what is the minimum signal level that can be applied?
(b) In this case, what is the largest interfering signal that can be applied without overloading the amplifier? What, then, is the effective amplifier dynamic range?

5-11. Refer to Fig. 5-11.

(a) If the transistor collector-to-base capacitance is 5 pF and the voltage gain is 50 (actually -50, since the common-emitter configuration inverts the output voltage), determine the equivalent input Miller capacitance. (*Note:* It will not be "negative.")

(b) If the two base biasing resistors present together (in parallel) a net resistance of 5 kΩ, determine this amplifier's 3-dB bandwidth.

(c) Determine the bandwidth if this amplifier is modified to a cascode configuration, as in Fig. 5-12, with the same capacitance and resistance parameters as above.

5-12. Referring to the AGC circuit in Fig. 5-13(f), determine how the AGC operation would change if a diode were placed across R_b with its "arrow" pointing to the right. (Consider the circuit's response if a signal is suddenly applied, maintained for a while, and then suddenly removed.)

5-13. Referring to Fig. 5-17, sketch the overall frequency response of all the filter sections together, F_1 to F_5.

5-14. Refer to Fig. 5-29. A SSB transmitter, tuned to 72 MHz and transmitting on the USB, is fed with two audio tones, at 1.0 and 2.5 kHz, of equal level. Determine the first 10 frequency components expected in the output (desired components, and first four IMD products on either side).

5-15. Refer to Fig. P5-15, which is representative of a double-conversion superheterodyne receiver.

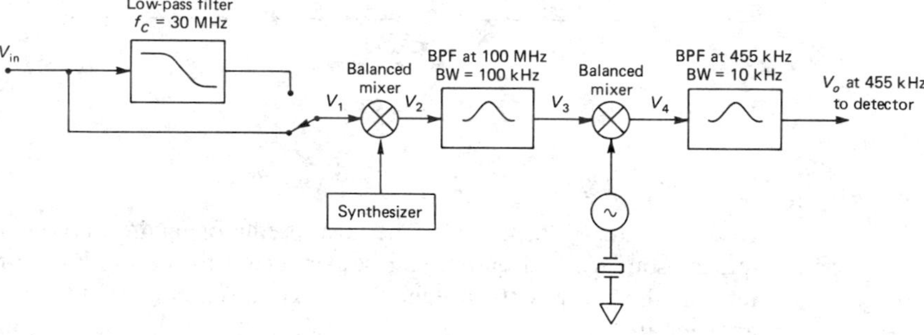

Figure P5-15.

(a) Determine the frequency f_2 of the crystal oscillator required to produce the desired output at 455 kHz. (There are two possibilities.)

(b) We are interested in the radio-frequency range, at the antenna input, from 0 to 30 MHz. What must be the tuning range of the local oscillator synthesizer? (Again, there are two possibilities.)

(c) Assuming that the synthesizer is designed to generate the lower range determined above, what must be its output frequency if we wish to tune in to a radio signal at 4.3 MHz?

(d) In part (c), with the input preselection filter switch set to the lower position, as we listen to this signal at 4.3 MHz, there is another signal, called the image, which can also be received and heard, interfering with the desired one at 4.3 MHz. Determine this image frequency.

(e) Consider now that this switch is set to the upper position and that the same two signals are still picked up by the antenna. In addition, two more signals are picked up, at 4.29 and 4.52 MHz. Determine the frequency spectrum diagrams for the voltages at V_{in}, V_1, V_2, V_3, V_4, and V_o. Use the lower-frequency f_2 determined in part (a). Identify which stage rejects which signal present at the input V_{in}, as the desired signal progresses to the output V_o.

5-16. Refer to the typical MC1496 balanced mixed application, as suggested by Motorola, in Fig. P5-16. Consider, in addition, that the output at pin 6 is coupled through a 0.01-μF capacitor to a 100-kΩ load resistor.

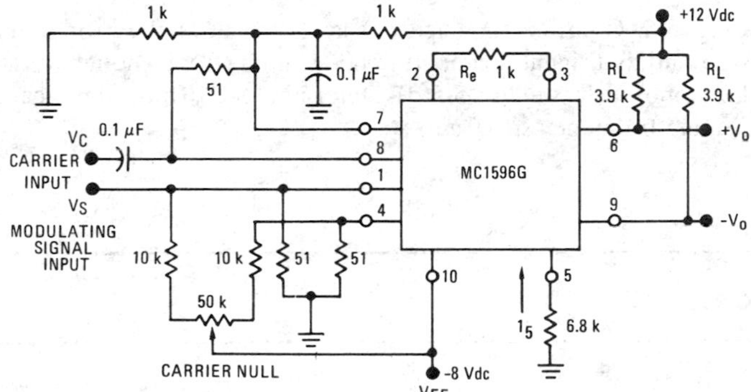

Figure P5-16

(a) For the power supply values indicated, determine the expected quiescent currents at the output pins 6 and 9.

(b) Now consider that the "carrier null" trim pot is misadjusted such that the quiescent current at pin 6 is 100 μA more than at pin 9. If a 2-MHz sinusoidal signal is applied at the carrier input, with an amplitude of 1 V_{pp}, sketch the output voltage AC component, showing the phase relationship to the input carrier signal. (Consider the modulating signal input to be exactly zero.) Also, determine and sketch the frequency spectrum of this output voltage up to 20 MHz, indicating the components' amplitudes.

(c) Now assume that this trim pot is properly readjusted such as to balance the lower transistors' (the differential amplifier) bias currents. If a modulating signal is applied at a frequency of 5 kHz with an amplitude such that the output current at pin 6 swings 200 μA_{pp}, sketch the voltage pin 9 and the output voltage across the load resistor. Also determine the output frequency components up to 20 MHz.

(d) If the same modulating signal as above is applied, while the 50-kΩ biasing trim pot is adjusted as in part (b), how will the output current at pin 9 and the output spectrum at the load resistor be different from those in part (c)?

5-17. Consider an AM signal such as in Fig. 5-22(c), produced out of an amplifier stage inside a receiver, at a carrier (or IF) frequency of 500 kHz and with an amplitude envelope swinging at a rate of 5 kHz, from a low of 0.5 mV up to a peak of 3 mV. This signal is applied to a very high voltage gain (say 100,000) IF amplifier whose output swing saturates at a peak swing of 1 V_{pp}.

(a) Sketch the output voltage of this IF amplifier, indicating the phase relationship with the applied input AM signal. (Ignore any internal phase delay from the input to the output.)

(b) If the original AM signal is applied to the balanced mixer in Fig. P5-16 (properly biased) at pin 1, through a 0.01-μF coupling capacitor and the IF amplifier output above is applied to pin 8, then from the output sketch determined in part (a), determine and sketch qualitatively the output voltage signal (i.e., do not be concerned with the exact amplitudes, only with the shape).

(c) Will the output envelope frequency be at 5 kHz, 10 kHz, 15 kHz, 500 kHz, 1 MHz, or other? Explain. Will the output IF frequency still be 500 kHz, 1 MHz, 2 MHz, 3 MHz, or other? Explain.

(d) If the mixer output is applied to an *RC* low-pass filter with a 16-kHz corner frequency, determine the filter's output frequency components.

5-18. Refer to Fig. 5-5. Consider that signal *B* is at 1 kHz, applied to an audio amplifier being tested, and that signal *A* consists of 60-Hz hum strong enough to cause distortion of the 1-kHz tone, such as in Fig. 5-5(d). Determine the 10 most significant components of the resulting frequency spectrum.

References

1. H. L. Krauss, C. W. Bostian, and F. H. Raab, *Solid State Radio Engineering.* New York: John Wiley & Sons, Inc., 1980.
2. D. DeMaw, *Practical R.F. Design Manual.* Englewood Cliffs, N.J.: Prentice Hall, Inc., 1982.
3. *Reference Data for Radio Engineers*, 6th ed. Indianapolis, Ind.: Howard W. Sams & Company, Inc., 1985.
4. Mark J. Wilson, ed., *The 1987 A.R.R.L. Handbook for the Radio Amateur.* American Radio Relay League, 1987.
5. C. Boswick, *R.F. Circuit Design.* Indianapolis, Ind.: Howard W. Sams & Company, Inc., 1977.
6. W. Hayward and D. DeMaw, *Solid State Design for the Radio Amateur.* Newington, Conn.: American Radio Relay League, 1977.
7. W. H. Hayward, *Introduction to Radio Frequency Design.* Englewood Cliffs, N.J.: Prentice Hall, Inc., 1982.
8. U. L. Rohde and T. T. N. Bucher, *Communications Receivers, Principles and Design.* New York: McGraw Hill Book Company, 1988.
9. J. K. Hardy, *High Frequency Circuit Design.* Reston, Va.: Reston Publishing Co., Inc., 1979.
10. Martin Giles, ed., *Audio/Radio Handbook.* Santa Clara, Calif.: National Semiconductor, 1980.
11. *Linear Integrated Circuits*, Technical Series IC-42. Somerville, N.J.: RCA, 1970.
12. *R.F. Power Transistor Manual*, Technical Series RFM-430. Somerville, N.J.: RCA, 1971.
13. Edwin S. Oxner, Commutation Mixer Achieves High Dynamic Range. *R. F. Design Magazine*, February 1986.

Amplitude Modulation

William Sinnema

6-1 GENERATION OF AM

As noted in Section 1-3, the amplitude of the carrier is made proportional to the instantaneous amplitude of the modulating voltage in AM transmission. To understand how such a waveform is typically obtained, consider the collector modulated class C AM modulator of Fig. 6-1. The information signal is introduced into the collector potential, causing it to vary accordingly. This technique is known as *high-level modulation*, as the modulation occurs at the collector, where the modulator must supply all the information power. If the information signal is injected at the transistor base, the process is termed *low-level modulation*, as the modulating signal also experiences some gain. In class C operation the bias is adjusted so that current flows for less than 180° of the input voltage cycle. The carrier causes the transistor to act as a switch, oscillating back and forth between the conducting and nonconducting states at the carrier frequency rate.

The common-emitter transistor is a class C amplifier whose base is driven by an RF oscillator at the carrier frequency of $f_c = \omega_c/2\pi$. Its collector voltage is effectively the sum of the supply voltage, V_{cc}, and the modulating voltage, v_m. The RFC acts as a low impedance to the modulating signal but prevents the radio frequency from entering the modulator or power supply circuitry. The RF bypass capacitor further helps to shunt any RF signal to ground.

With no modulation, the collector voltage is just a sinusoid operating between 0 and $2V_{cc}$ at the carrier frequency. For a sinusoidal modulating signal, the effective collector voltage is that shown by the heavy line in Fig. 6-1(b) (i.e., $V_c = V_{cc} + v_m$). To assure that the collector voltage does not drop below zero volts, the maximum value v_m can take is V_{cc}. As described earlier when discussing class C amplifiers, when properly tuned the instantaneous collector voltage will be able to drop to nearly zero volts on the negative RF swing and thus to $2V_c$ on the positive swing. Since the maximum value of V_c is $2V_{cc}$, the peak RF voltage will reach $4V_{cc}$. A designer must assure that the transistor used will be able to withstand this high voltage. For a slowly varying collector voltage of $V_c = V_{cc} + v_m$, the instantaneous collector voltage will be of the form

$$v_c = V_c + V_c \cos \omega_c t$$

The steady or dc component of the resulting collector voltage is prevented from reaching the output by the coupling capacitor C_c. The output voltage waveform will thus appear as shown in Fig. 6-1(c). Because of the coupling factor of the RF

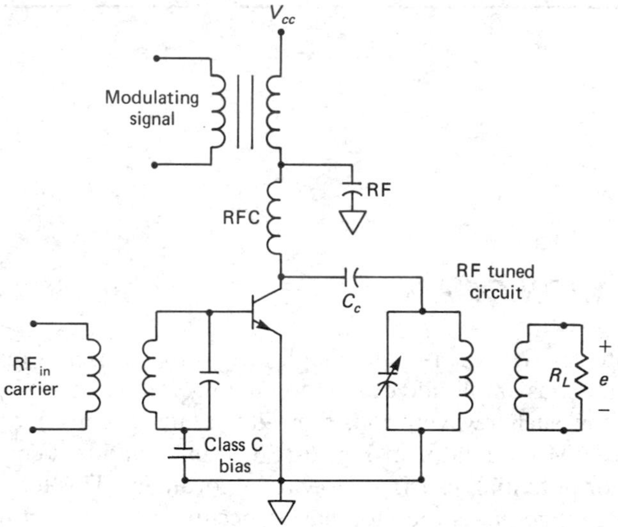

(a) Collector modulated transistor class C AM modulator

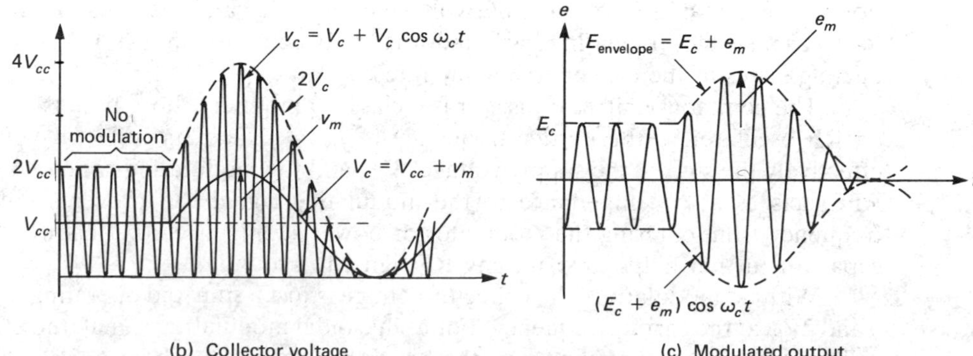

(b) Collector voltage

(c) Modulated output

Figure 6-1 Generation of an AM signal, illustrating circuit diagram and waveforms.

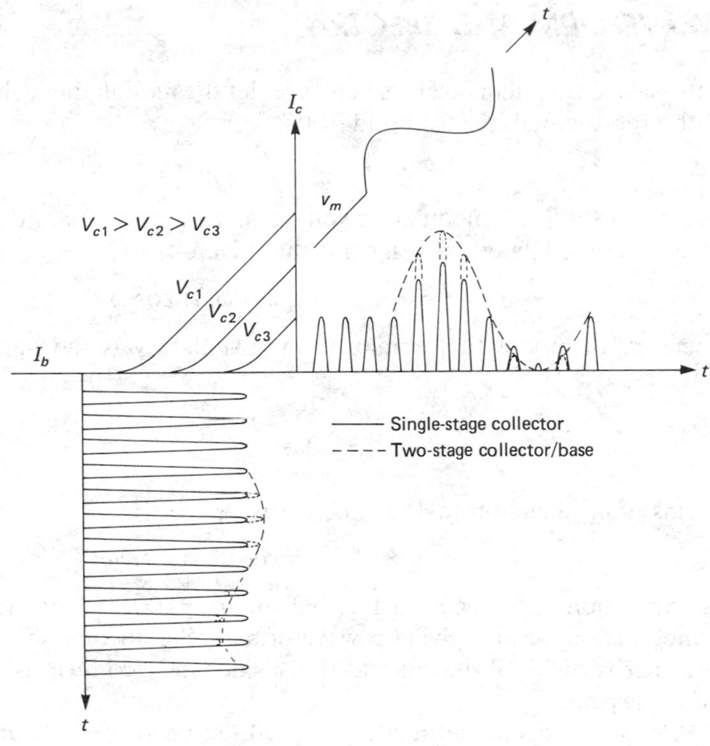

(d) Graphical analysis of collector–base modulation

Figure 6-1 *(continued).*

tuned circuit, the output ac voltage will not necessarily be the same as the ac collector voltage. Nevertheless, the peak carrier voltage at the output E_c, will be proportional to the collector supply voltage V_{cc}, and the swing of the envelope will be proportional to the modulating signal v_m (i.e., $E_c \propto V_{cc}$, $e_m \propto v_m$).

The envelope of the output am waveform of Fig. 6.1(c) can be expressed as

$$E_{\text{envelope}} = E_c + e_m \tag{6-1}$$

The corresponding AM signal voltage will thus be given by

$$e = (E_c + e_m) \cos \omega_c t \tag{6-2}$$

where $E_c + e_m$ represent the peak voltage of the sinusoidal signal.

Because of nonlinearities in the characteristic curve of the transistor [Fig. 6-1(d)], it is impossible to obtain a fully modulated output and still maintain linearity. To obtain a fully modulated output and still maintain linearity, an additional modulating signal must be introduced into the base circuit. This is usually accomplished by collector modulation of the preceding stage. This is illustrated in Fig. 6-1(d) by the dashed-line response. Note that with the collector modulation of the preceding stage, the base current of the modulator is also modulated, causing both a deepening and a strengthening of the modulated carrier at the output.

6-2 AM WAVEFORM AND SPECTRA

For the sake of mathematical convenience, let the modulating voltage in expression (6-2) be represented by a sinusoid of the form

$$e_m = E_m \cos \omega_m t \tag{6-3}$$

where E_m is the peak modulating voltage and $f_m = \omega_m/2\pi$ is the modulating frequency. The modulated AM signal is thus given by

$$e = (E_c + E_m \cos \omega_m t) \cos \omega_c t \tag{6-4}$$

A measure of modulation employed in AM theory is the modulation index m defined as

$$m = \frac{E_m}{E_c} \tag{6-5}$$

In terms of m, equation (6-4) can be expressed as

$$e = E_c(1 + m \cos \omega_m t) \cos \omega_c t \tag{6-6}$$

This expression is sketched in Fig. 6-2 for $m = 0$, 0.5, and 1. If m exceeds 1, overmodulation occurs, giving a waveform similar to that shown in Fig. 6-2(d). The signal is highly distorted and other sidebands, known as *sideband splatter*, begin to appear.

Using the trigonometric identity (1-5), expression (6-6) can be expanded to

$$e = E_c \cos \omega_c t + \frac{mE_c}{2} \cos(\omega_c - \omega_m)t + \frac{mE_c}{2} \cos(\omega_c + \omega_m)t \tag{6-7}$$

This suggests that three frequencies are present in the AM signal: the carrier wave of peak amplitude E_c and frequency $f_c = \omega_c/2\pi$; the lower sideband signal of peak amplitude $mE_c/2$ and frequency $f_c - f_m$, since $\omega_c - \omega_m = 2\pi(f_c - f_m)$; and the

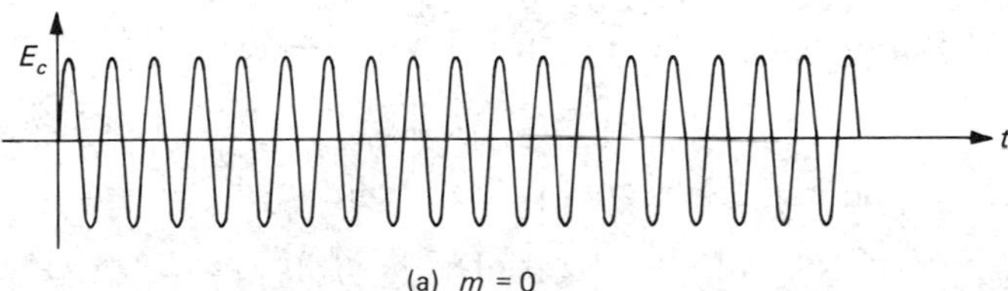

(a) $m = 0$

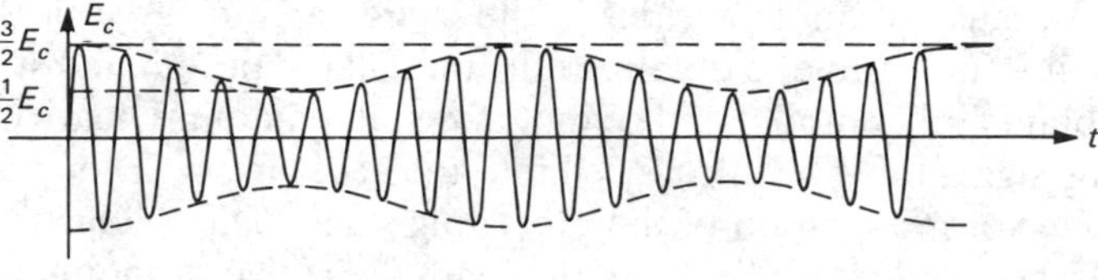

(b) $m = 0.5$

Figure 6-2 AM waveforms for several modulation indexes.

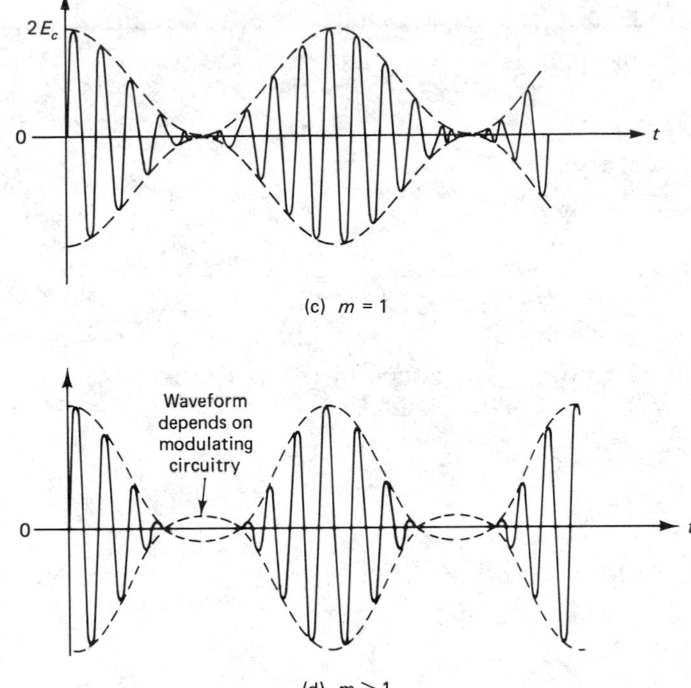

Figure 6-2 *(continued).*

upper sideband signal of peak $mE_c/2$ and frequency $f_c + f_m$. The voltage spectrum is sketched in Fig. 6-3(a). The power spectrum, which is proportional to the square of the voltage spectrum, is shown in Fig. 6-3(b), where the average carrier power, $P_c = (E_c/\sqrt{2})^2/R$. R represents the load resistance.

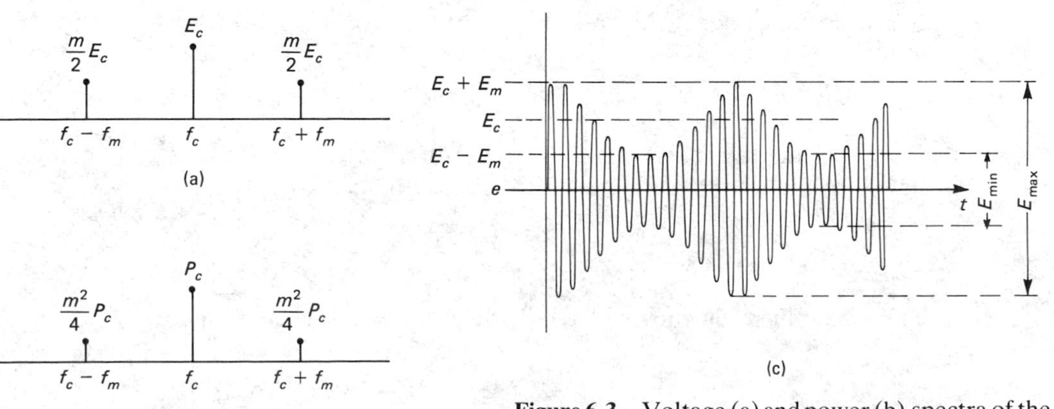

Figure 6-3 Voltage (a) and power (b) spectra of the sinusoidally modulated AM waveform (c).

EXAMPLE 6-1

Find the modulation index and voltage spectra for a 1-kV peak carrier wave of frequency 1.5 MHz modulated by an 800-V_{peak} 5-kHz sinusoidal signal. What is the average power contained in the carrier and the sidebands if the antenna load is 50 Ω?

Solution

$$m = \frac{E_m}{E_c} = \frac{800}{1000} = 0.8$$

The sideband frequencies are 1.5 ± 0.005 = 1.505 and 1.495 MHz. Since the peak voltage amplitude of the carrier frequency is 1000 V, the peak amplitude of each sideband is

$$\frac{m}{2} E_c = \frac{0.8}{2} \times 1000 = 400 \text{ V}$$

The average carrier power is

$$\frac{(1000/\sqrt{2})^2}{50} = 10 \text{ kW}$$

while the average power in each sideband is

$$\frac{m^2}{4} P_c = \frac{0.8^2}{4} \times 10{,}000 = 1.6 \text{ kW}$$

We can note from Fig. 6-3 that the carrier power remains fixed regardless of the modulation index. The carrier as such contains no information, causing rather low efficiencies. The large carrier does have the advantage of permitting the usage of inexpensive diode detectors in the receiver units. Either sideband can be used to obtain the information, but both can be added to increase the output information power.

The total average power in the modulated wave consists of the sum of the three components shown in Fig. 6-3(b).

$$P_T = P_c + \frac{m^2}{4} P_c + \frac{m^2}{4} P_c = P_c\left(1 + \frac{m^2}{2}\right) \tag{6-8}$$

If we consider the efficiency of the transmitter to be the ratio of the sideband power to the total power, the efficiency will be given by

$$\eta = \frac{(m^2/4)P_c + (m^2/4)P_c}{(1 + m^2/2)P_c} = \frac{m^2}{m^2 + 2} \tag{6-9}$$

The maximum efficiency will occur when $m = 1$, giving a value of

$$\eta = \frac{1}{1 + 2} = 0.33 \quad \text{or} \quad 33.3\%$$

6-3 MEASUREMENT OF MODULATION INDEX

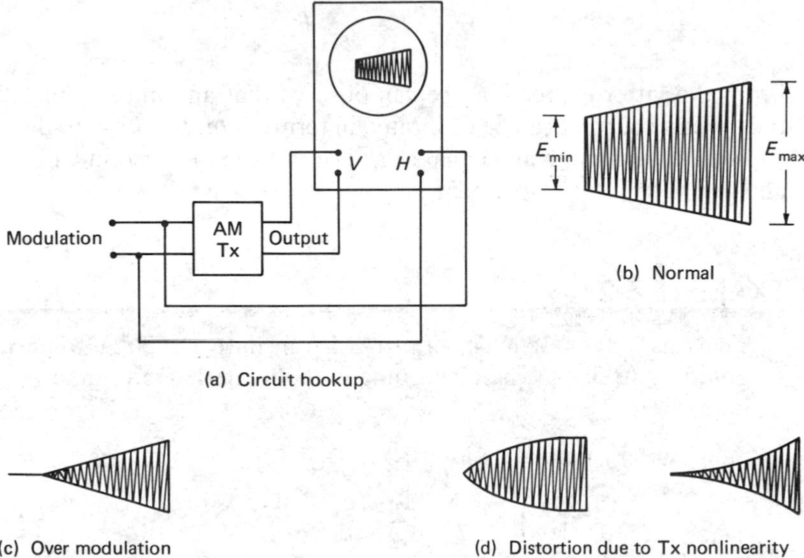

Figure 6-4 Trapezoidal displays for a sinusoidal modulating signal.

By observing the pattern illustrated in Fig. 6-3(c) on an oscilloscope, the modulation index can be calculated. An alternative but closely related method is that shown in Fig. 6-4(a). The modulating signal is applied directly to the horizontal plates of the oscilloscope, while the modulated signal is applied to the vertical deflection circuitry. The result is the trapezoidal pattern displayed in Fig. 6-4(b). The same notation is used here as in the preceding figure. From Fig. 6-3(c) we can note that

$$E_{\max} = 2(E_c + E_m) = 2(E_c + mE_c)$$

$$E_{\min} = 2(E_c - E_m) = 2(E_c - mE_c)$$

$$\frac{E_{\max}}{E_{\min}} = \frac{1 + m}{1 - m}$$

from which

$$m = \frac{E_{\max} - E_{\min}}{E_{\max} + E_{\min}} \tag{6-10}$$

A third alternative method of measuring m is to monitor the true effective or rms current, I, to the antenna. Let the total average power P_T be equal to

$$P_T = I^2R \tag{6-11}$$

If we let the unmodulated carrier rms current be I_c, equation (6-8) can be expressed as

$$P_T = I_c^2R\left(1 + \frac{m^2}{2}\right) \tag{6-12}$$

Equating the latter two expressions, we obtain for I

$$I = I_c\sqrt{1 + \frac{m^2}{2}} \tag{6-13}$$

From the latter expression, we can observe that an ammeter installed to monitor the antenna current can be calibrated in terms of m. For $m = 0$, the current reading will be at a minimum and equal to I_c. For $m = 1$, the current will be at a maximum and equal to $I_c\sqrt{1.5} = 1.225I_c$.

EXAMPLE 6-2

The rms antenna current of an AM transmitter is 30 A when unmodulated but rises to 36 A when sinusoidally modulated. Determine the modulation index.

Solution From expression (6-13):

$$\left(\frac{I}{I_c}\right)^2 = 1 + \frac{m^2}{2}$$

$$m = \sqrt{2\left[\left(\frac{I}{I_c}\right)^2 - 1\right]}$$

$$= \sqrt{2\left[\left(\frac{36}{30}\right)^2 - 1\right]} = 0.94 \quad \text{or} \quad 94\%$$

When several sinusoidal signals of different frequencies are added, the net rms voltage is the square root of the sum of the square of the component rms voltages. If a carrier is modulated simultaneously by the modulating voltages E_1, E_2, E_3, and so on, the total modulating voltage will be

$$E_T = \sqrt{E_1^2 + E_2^2 + E_3^2 + \cdots} \tag{6-14}$$

Divide through by E_c, we obtain

$$\frac{E_T}{E_c} = \sqrt{\frac{E_1^2}{E_c^2} + \frac{E_2^2}{E_c^2} + \frac{E_3^2}{E_c^2} + \cdots}$$

yielding a modulation index of

$$m_t = \sqrt{m_1^2 + m_2^2 + m_3^2 + \cdots} \tag{6-15}$$

where m_1, m_2, and so on, represent the modulation index for each sinusoidal component.

In practice, we cannot allow m_t to approach unity. At some time instants, the various signal components will tend to add up and full modulation will occur prior to m_t reaching 1. For complex waveforms, the modulation factor is of more importance. It is defined as the ratio of peak variation of the envelope from the amplitude of the unmodulated wave to the amplitude of the unmodulated wave, expressed as a percentage. This is illustrated in Fig. 6-5, where the modulation factor can be expressed as

$$\text{modulation factor} = \frac{E_c - E_{\text{min}}}{E_c} \times 100\% \tag{6-16}$$

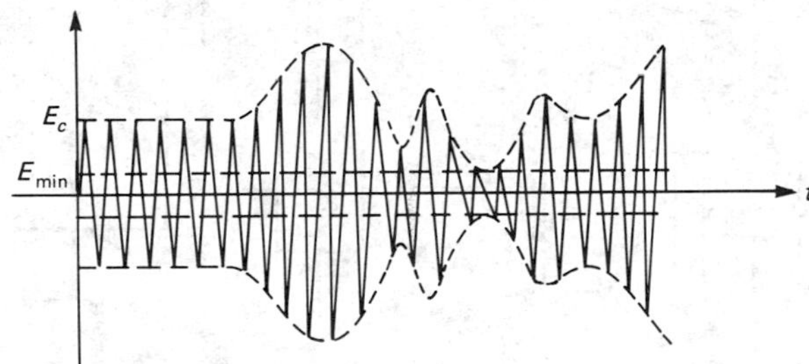

Figure 6-5 An AM waveform modulated by a complex signal.

In commercial AM it is desired to keep the modulation index or factor as close to 1 as possible, as this maximizes the power in the sidebands (information power) and thus the coverage area. During commercials, the sound level heard often undergoes a significant rise in volume.

In practice, the peak voltage is allowed to exceed $2E_c$, resulting in an effective modulation index greater than 1. The negative swing is never allowed to swing below the 0-V level, thereby preventing the generation of other sidebands and harmonics of the carrier frequency.

6-4 AM CITIZEN BAND TRANSCEIVER

An example of a low-power AM transmitter that incorporates many of the circuits discussed previously is the Midland 77-015 40-channel mobile citizen band (CB) transceiver. The transceiver block diagram, the phase-locked-loop block diagram, and the schematic are shown in Figs. 6-6 to 6-8. We will first concentrate on the transmitter section and then on the receiver section. CB transmitters operate in the general radio service band, 26.960 to 27.410 MHz. The channel center frequencies are listed in Table 6-1, together with the VCO frequencies for the transceiver under discussion.

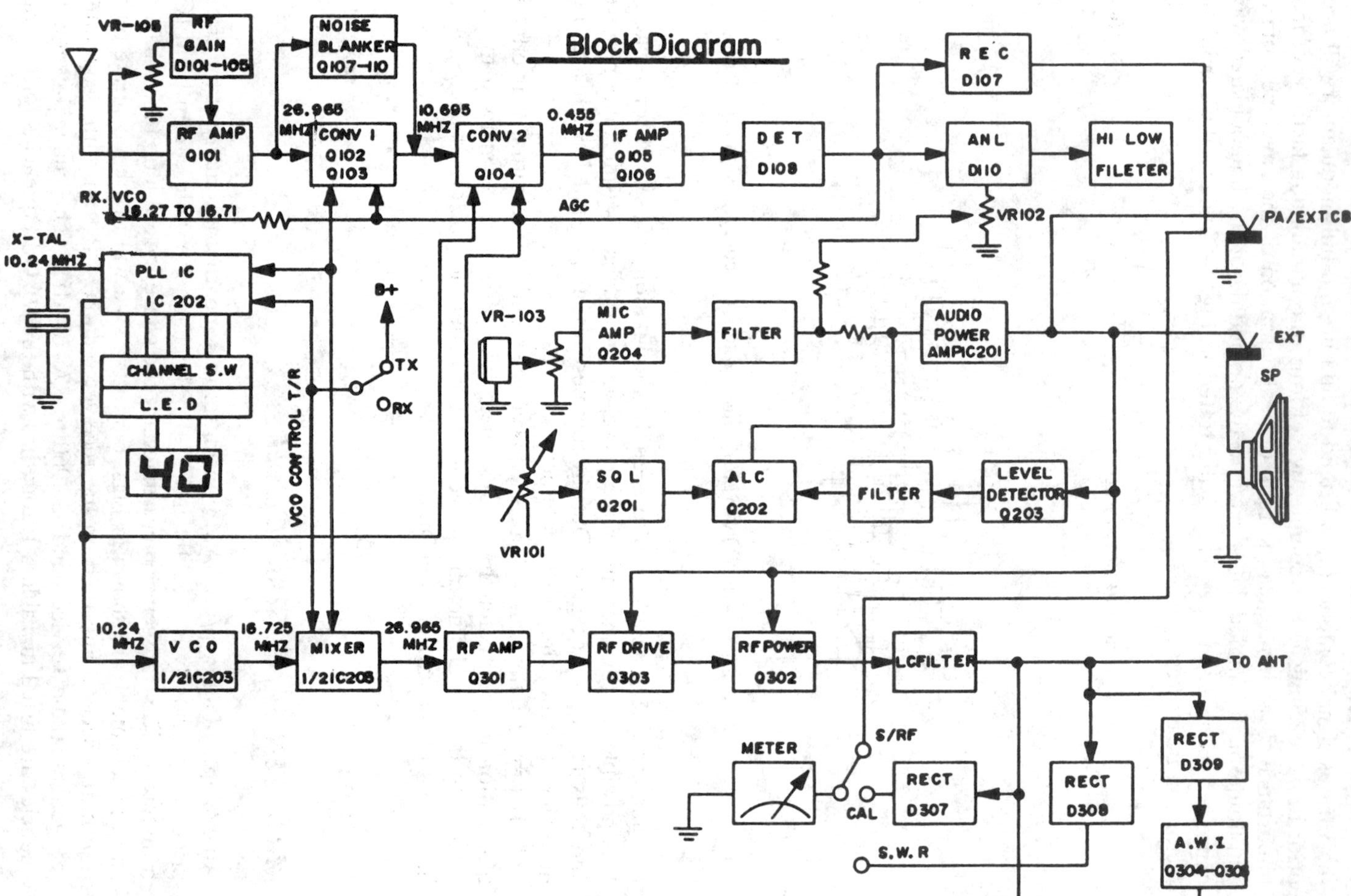

Figure 6-6 Block diagram of Midland 77-005 transceiver. (Courtesy of Midland International.)

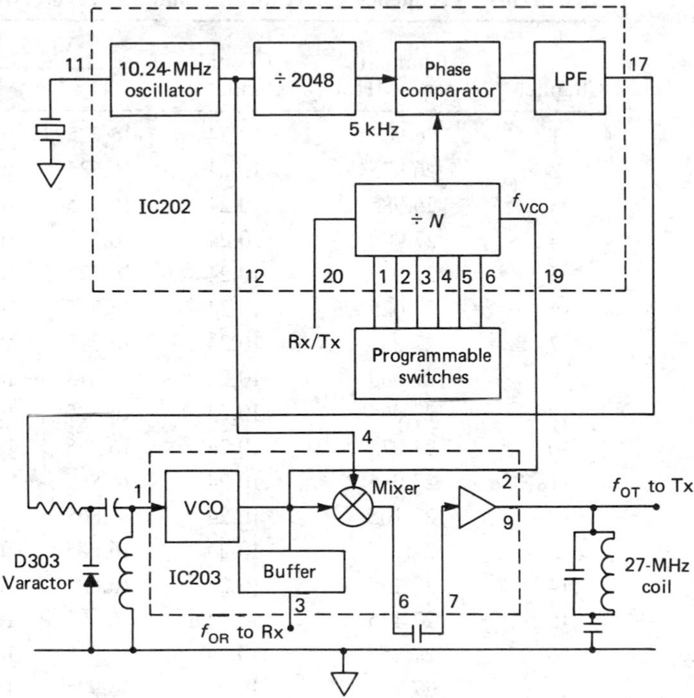

Figure 6-7 Block diagram of PLL circuit.

When discussing specific frequencies, we will assume operation on channel 1, 26.965 MHz. For the transmitter, this frequency is obtained when the VCO is at 16.725 MHz (determined by the channel selector switch) and mixed with the 10.24 MHz [i.e., (10.24 + 16.725) MHz = 26.965 MHz]. In the receive mode, the cathode of D301 is taken off chassis ground, causing the VCO to shift to 16.27 MHz. A forward bias is then also applied to IF amp Q106. When this VCO signal is mixed with a 26.965-MHz RF signal, a difference frequency of 10.695 MHz is obtained, a frequency that falls in the passband of the 10.7-MHz crystal filter. Mixing this signal with the 10.24-MHz crystal frequency, we obtain a difference or IF frequency of 455 kHz. Another RF signal frequency can be selected by changing the channel selector switch.

TABLE 6-1 Frequency Chart for Midland CB Transceivers

Channel	Channel Freq. (MHz)	Crystal Oscillator	VCO Tx	VCO Rx
1	26.965	10.24	16.725	16.27
2	26.975	10.24	16.735	10.28
3	26.985	10.24	16.745	16.29
4	27.005	10.24	16.765	16.31
5	27.015	10.24	16.775	16.32
6	27.025	10.24	16.785	16.33
7	27.035	10.24	16.795	16.34
8	27.055	10.24	16.815	16.36
9	27.065	10.24	16.825	16.37
10	27.075	10.24	16.835	16.38
11	27.085	10.24	16.845	16.39
12	27.105	10.24	16.865	16.41
13	27.115	10.24	16.875	16.42
14	27.125	10.24	16.885	16.43
15	27.135	10.24	16.895	16.44
16	27.155	10.24	16.915	16.46
17	27.165	10.24	16.925	16.47
18	27.175	10.24	16.935	16.48
19	27.185	10.24	16.945	16.49
20	27.205	10.24	16.965	16.51
21	27.215	10.24	16.975	16.52
22	27.225	10.24	16.985	16.53
23	27.255	10.24	17.015	16.56
24	27.235	10.24	16.995	16.54
25	27.245	10.24	17.005	16.55
26	27.265	10.24	17.025	16.57
27	27.275	10.24	17.035	16.58
28	27.285	10.24	17.045	16.59
29	27.295	10.24	17.055	16.60
30	27.305	10.24	17.065	16.61
31	27.315	10.24	17.075	16.62
32	27.325	10.24	17.085	16.63
33	27.335	10.24	17.095	16.64
34	27.345	10.24	17.105	16.65
35	27.355	10.24	17.115	16.66
36	27.365	10.24	17.125	16.67
37	27.375	10.24	17.135	16.68
38	27.385	10.24	17.145	16.69
39	27.395	10.24	17.155	16.70
40	27.405	10.24	17.165	16.71

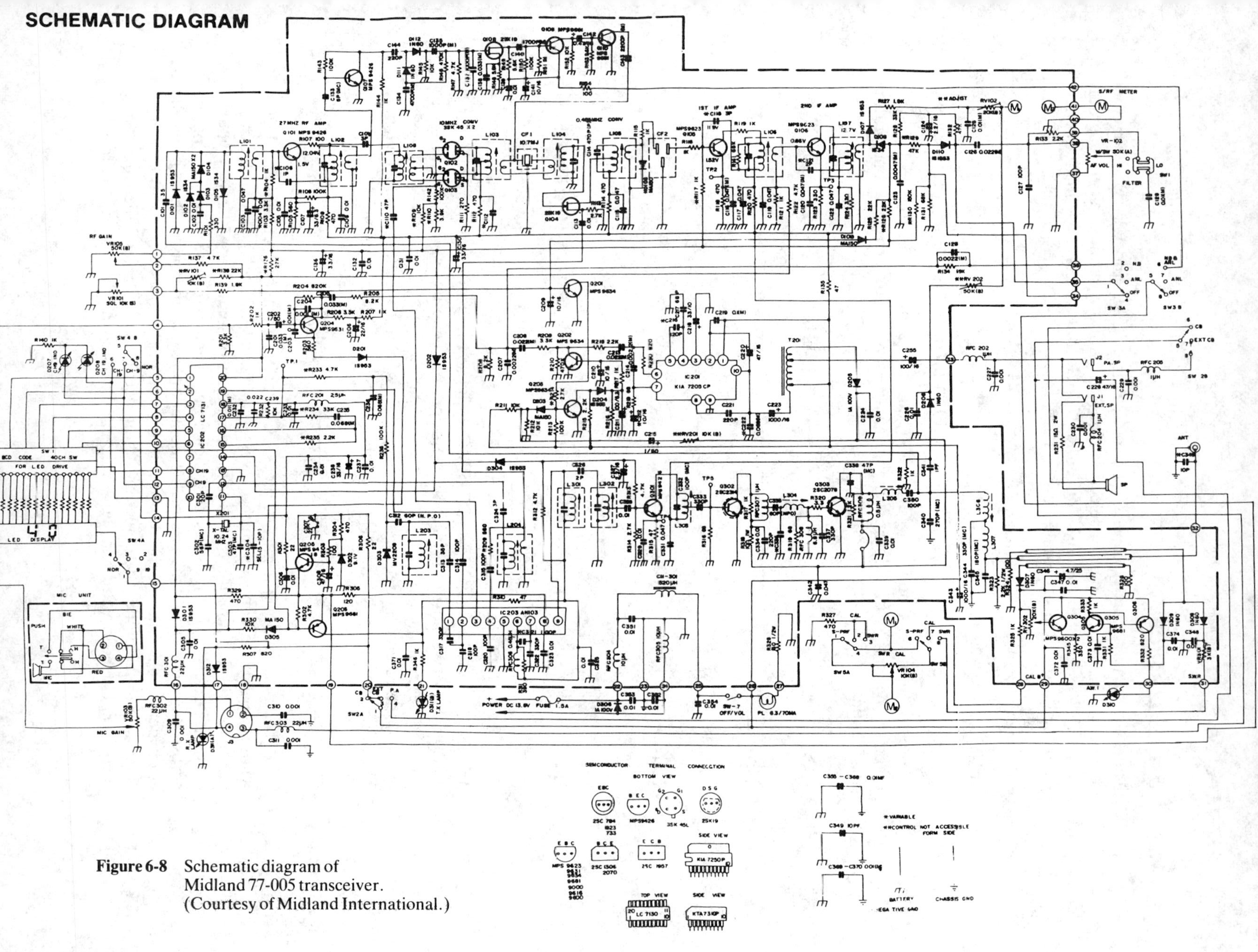

Figure 6-8 Schematic diagram of Midland 77-005 transceiver. (Courtesy of Midland International.)

6-4.1 PLL Circuit

The PLL circuitry chiefly consists of two IC chips, IC202 and IC203, a varactor diode, D303, and the channel selection circuitry. A more detailed block diagram of the PLL is given in Fig. 6-7. Since the 27-MHz coil selects the sum frequency from the balanced mixer located inside IC203,

$$f_{OT} = f_{VCO} + 10.24\text{ MHz}$$

or

$$f_{VCO} = f_{OT} - 10.24\text{ MHz}$$

where the terms f_{OT} and f_{VCO} are the frequencies indicated on the PLL block diagram. When the PLL is locked, the frequencies applied to the phase comparator are identical. Thus

$$\frac{f_{VCO}}{N} = \frac{10.24\text{ MHz}}{2048} = 5\text{ kHz}$$

Substituting in the previous expression for f_{VCO}, we obtain for N:

$$N = \frac{f_{VCO}}{5\text{ kHz}} = \frac{f_{OT} - 10.24\text{ MHz}}{5\text{ kHz}} \tag{6-17}$$

Thus for $f_{OT} = 26.965$ MHz (channel 1),

$$N = \frac{(26.965 - 10.24)\text{ MHz}}{5\text{ kHz}} = 3345$$

and $f_{VCO} = 3345 \times 5\text{ kHz} = 16.725$ MHz. For $f_{OT} = 26.975$ MHz (channel 2),

$$N = \frac{(26.975 - 10.24)\text{ MHz}}{5\text{ kHz}} = 3347$$

and $f_{VCO} = 3347 \times 5\text{ kHz} = 16.735$ MHz; and so on. The programmable switch and the Rx or Tx position select the appropriate divide-by-counter number N to give the desired channel.

In the receive mode, the mixer frequency is identical to the VCO frequency. Thus $f_{OR} = f_{VCO} = N \times 5$ kHz. For an f_{OR} of 16.27 MHz (channel 1),

$$N = \frac{16.27\text{ MHz}}{5\text{ kHz}} = 3254$$

The most important part of the VCO circuitry is the varactor diode, D303, whose capacitance depends on the reverse voltage applied to the cathode. In the vicinity of the PN junction of a diode, a depletion region exists which is devoid of current carriers and hence acts as a dielectric. When a reverse voltage is placed across the diode, the thickness of the depletion region increases, decreasing the equivalent capacitance of the diode as illustrated in Fig. 6-9. When a forward bias is applied to the diode, the depletion region becomes smaller and the capacitance increases. Under forward bias, however, significant current flows and the resulting

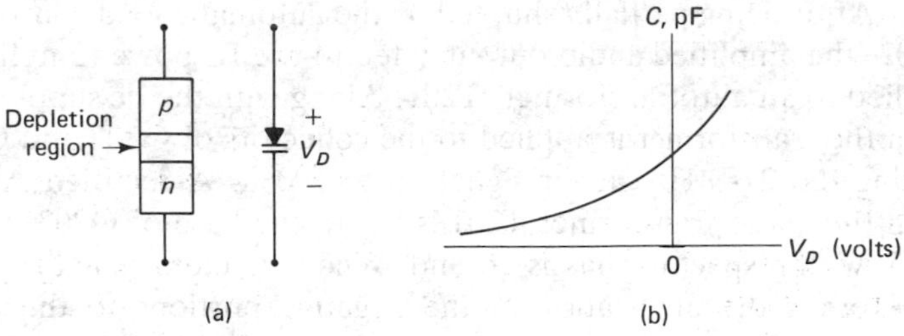

Figure 6-9 Varactor diode (a) and capacitance versus voltage characteristic (b).

forward resistance swamps any capacitance effect. The exact relationship between capacitance and reverse voltage depends on the variation of the impurity density with distance from the junction. In general

$$C = \frac{K}{(V_d + V)^n} \tag{6-18}$$

where V_d is the contact potential of the diode, V the applied reverse voltage, and K is a constant. n depends on the diffusion profile; $n = \frac{1}{2}$ for an "abrupt junction" and $n = \frac{1}{3}$ for a linear graded junction.

Figure 6-10 shows a typical circuit used with a varactor tuning diode. The voltage is applied through a large resister, R. Electronic tuning is commonly used in remote control and pushbutton tuning. It should be noted that besides the dc tuning, the RF voltage causes a somewhat impulsive current to flow, giving rise to various harmonics in the circuit.

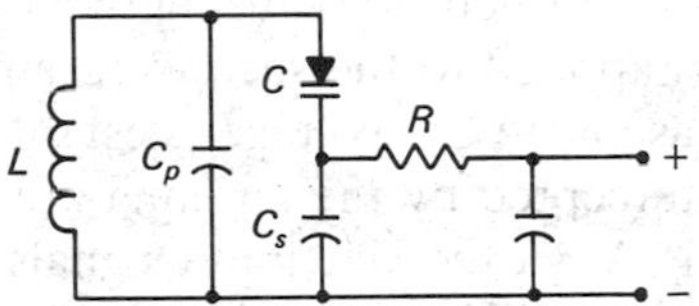

Figure 6-10 Varactor tuned circuit.

6-4.2 Transmitter Circuit

The transmit carrier frequency generated at pin 9 of IC203 is applied to the RF amplifier Q301 through the double-tuned circuits L301 and L302, then to the RF drive amplifier Q302 and the final RF power amplifier Q303. All these are tuned amplifiers. When the PLL is out of lock, the voltage on pin 14 of IC202 drops, disabling Q301 via D304 and thereby preventing the transmission of any illegal frequencies.

When the push-to-talk control switch is operated, the microphone is connected across pins 1 and 2 (chassis ground) of the microphone jack, thus allowing the microphone signal to be fed to the audio amplifier Q204 through the MIC gain control. Q204 is activated in the Tx mode as the emitter resistor R203 is forced to near ground via diode D201 and pins 4 and 2 of the microphone jack.

After being partially shunted by the automatic load control (ALC) transistor, Q202, the amplified audio output is fed to the IC power amplifier IC201 and then applied to an autotransformer, T201. Along with the dc supply voltage, the output from the transformer is applied to the collectors of Q302 and Q303, thereby modulating the 27-MHz carrier signal up to 100%. A rectified ALC voltage derived from the audio transformer T201 is fed to the base of Q203, which is normally at 0.7 V with respect to chassis ground. When an audio signal is present, diode D203 will permit the application of the negative portions to the base of Q203. The collector voltage will then experience voltage increases, which when rectified and filtered by D204 and C211 result in a more positive dc signal on the base of Q202. The collector-to-emitter impedance of Q202 is thus lowered, causing an increase in attenuation of the audio signal to the audio amplifier. The purpose of the ALC circuit is to prevent overmodulation. The modulated signal from the RF power amplifier is then passed through a low-pass Π-filter network, which attenuates the undesirable high-frequency signals. This network also serves as a match to the antenna for maximum power transfer. The magnetically/electrically coupled transmission lines monitor the incident and reflected signals, to assure proper match and to indicate the transmitted or received signal levels. When a mismatch occurs at the antenna, the reflected signal detected and rectified by D308 (also D309) is monitored by the meter to indicate the standing-wave ratio (SWR). If the reflected signal is large, indicating a badly mismatched or unconnected antenna, the rectified signal from D309 is amplified by Q305, turning on Q306 and thereby activating the automatic warning indicator (AWI) lamp.

6-4.3 Receiver Circuit

In the receive mode of operation, Q206 is turned off, causing the Tx RF amplifier Q301 to be turned off. Forward bias is applied to the second IF amplifier, Q106, in the receiver section and a proper bias and AGC voltage is established to Q101, Q102, Q103, and Q105. RF signals intercepted by the antenna are fed to the RF amplifier Q101 through C101 and L101. Any excessive input signals are limited by diodes D101 and D105. The amplified 27-MHz signal is mixed with the VCO frequency as selected by the channel switch. Since the first IF frequency centered at 10.695 MHz is sharply filtered by L103, L104, and ceramic filter CF1, the channel frequency selected is given by

$$f_{\text{RF}} = f_{\text{VCO}} + 10.695 \text{ MHz}$$

The first IF is then mixed with a second local oscillator of 10.24 MHz, giving a second IF of 455 kHz. This second ceramic filter, CF2, coupled with L105, determines the selectivity of the receiver. The 455-kHz IF signal is fed to two stages of amplification and detected by detector diode D108. D108 produces audio signal as well as a negative dc voltage for AGC. The negative voltage also produces some negative biasing to the cathode of the automatic noise limiting (ANL) diode D110. The biasing voltage has a time constant determined by R132 and C124. Therefore, any sharp negative-going pulses from D108 will back-bias D110 and will be clipped.

The detected audio signal is then fed to the audio amplifier IC201. The amplified audio finally is applied to the speaker.

Diode Detector

One of the simplest types of AM detectors is the diode detector circuit. If the diode is driven hard, that is, if either a large current or no current flows, it can be roughly approximated by the piecewise linear characteristic curve shown in Fig. 6-11(a). The diode either presents little resistance to the input signal when turned on or presents an open circuit when turned off. A better approximation is the one in Fig. 6-11(b), where the forward diode resistance is included. Since the frequencies generated are the same for either approximation, with the latter giving slightly lower-amplitude products, we will tend to use the simpler approximation (a).

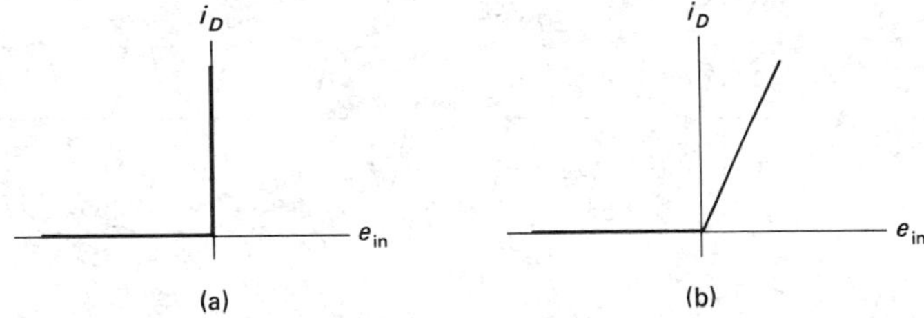

Figure 6-11 Approximated diode characteristic curves.

Let us consider applying the AM signal

$$e_{in} = E_c \cos \omega_c t + E \cos(\omega_c + \omega_m)t + E \cos(\omega_c - \omega_m)t \qquad (6\text{-}19)$$

to the switching diode, where E now represents the peak amplitudes of the side-bands of equation (6-7). This signal, represented in Fig. 6-12(a), crosses the time axis at a frequency of $f_c = \omega_c/2\pi$.

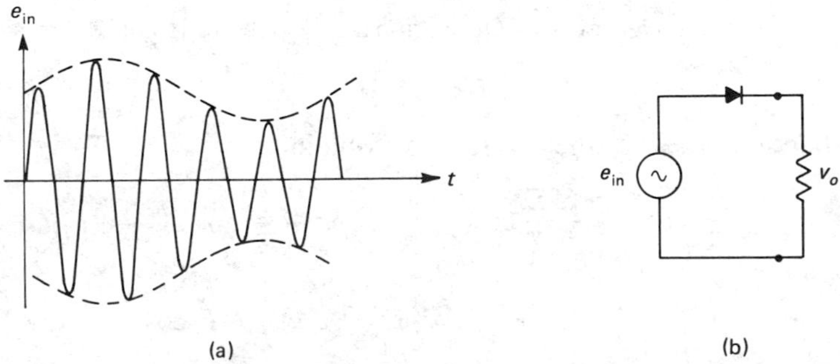

Figure 6-12 AM waveform (a) applied to a diode (b).

For a large carrier amplitude E_c, the diode will be driven into conduction for the positive excursions of the AM signal and into cutoff for the negative excursions. The rectification process that results is equivalent to multiplication of the modulated waveform signal by the switching function. $S(t)$ has a fundamental frequency equal to that of the carrier f_c, as illustrated in Fig. 6-13. The output signal will thus be

$$
\begin{aligned}
v_o &= e_{\text{in}} \times S(t) \\
&= [E_c \cos \omega_c t + E \cos(\omega_c + \omega_m)t + E \cos(\omega_c - \omega_m)t] \\
&\quad \times \left[\frac{1}{2} + \sum_{n=1}^{\infty} \frac{\sin(n\pi/2)}{n\pi/2} \cos n\omega_c t\right]
\end{aligned}
\tag{6-20}
$$

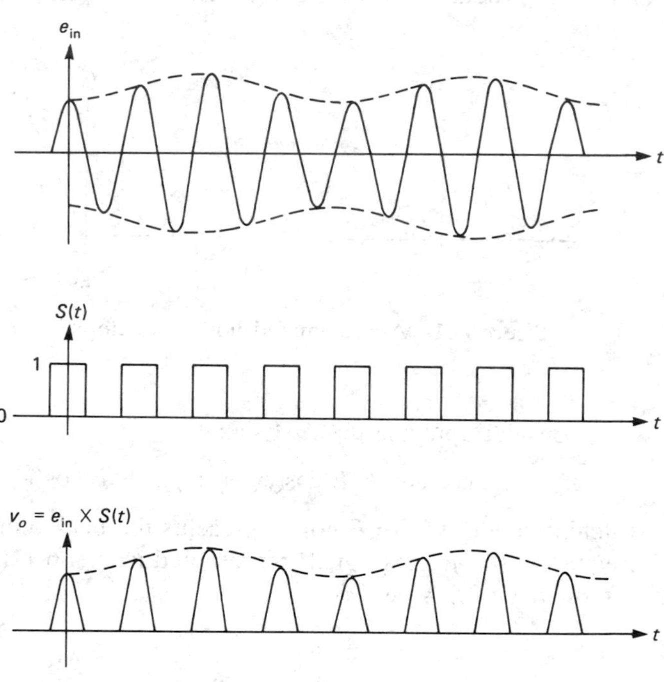

Figure 6-13 Derivation for the detected signal expression.

Expanding a few of these terms, we obtain

$$
\begin{aligned}
v_o = {} & \frac{E_c}{2} \cos \omega_c t + \frac{E}{2} \cos(\omega_c + \omega_m)t + \frac{E}{2} \cos(\omega_c - \omega_m)t \\
& + [E_c \cos \omega_c t + E \cos(\omega_c + \omega_m)t + E \cos(\omega_c - \omega_m)t] \left[\frac{2}{\pi} \cos \omega_c t \right. \\
& \left. + \sum_{n=2}^{\infty} \frac{\sin\,(n\pi/2)}{n\pi/2} \cos n\omega_c t\right]
\end{aligned}
$$

Employing identity (1-5) yields

$$v_o = \frac{E_c}{2}\cos\omega_c t + \frac{E}{2}\cos(\omega_c + \omega_m)t + \frac{E}{2}\cos(\omega_c - \omega_m)t$$
$$+ \frac{E_c}{\pi} + \frac{E_c}{\pi}\cos 2\omega_c t + \frac{E}{\pi}\cos\omega_m t + \frac{E}{\pi}\cos(2\omega_c + \omega_m)t$$
$$+ \frac{E}{\pi}\cos\omega_m t + \frac{E}{\pi}\cos(2\omega_c - \omega_m)t + R$$

where

$$R = [E_c\cos\omega_c t + E\cos(\omega_c + \omega_m)t + E_c\cos(\omega_c - \omega_m)t)]$$
$$\times\left[\sum_{n=2}^{\infty}\frac{\sin(n\pi/2)}{n\pi/2}\cos n\omega_c t\right] \tag{6-21}$$

The spectrum of this output signal is shown in Fig. 6-14.

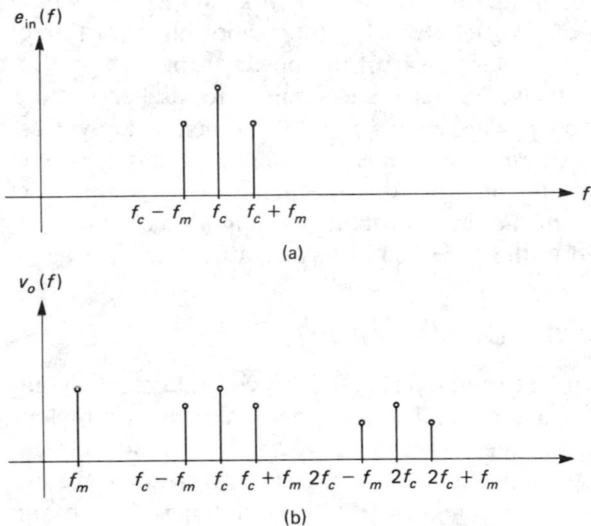

Figure 6-14 Frequency spectrum of AM signal applied to the diode detector (a) and of the output signal (b).

By applying the single low-pass (LP) filter shown in Fig. 6-15, the original information signal frequency f_m can be extracted. The signal from the LP filter will be the base-band signal with the dc component. The dc component's amplitude is a function of the carrier strength and can be used for the AGC circuit. That is, the original modulating or information signal from expression (6-21) is

$$\left(\frac{E}{\pi} + \frac{E}{\pi}\right)\cos\omega_m t = \frac{2E}{\pi}\cos\omega_m t$$

and the dc signal is E/π.

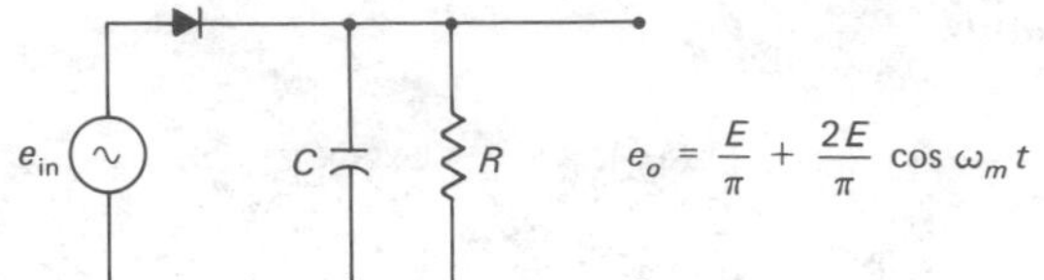

Figure 6-15 AM diode detector.

In the preceding discussion we analyzed the circuit on a frequency basis; we will now briefly explain its operation in the time domain. The waveshape of the signal applied to the diode detector is shown in Fig. 6-16(a). Since the diode can only conduct when forward biased, when the input voltage exceeds the output voltage the capacitor is charged and can follow rapid changes in input signal as the time constant is small (i.e., forward diode resistance × *C*). When the input voltage is less than the output voltage, the diode cannot conduct and the capacitor discharges with a time constant equal to *RC*.

If the time constant *RC* is too small, a large amount of ripple [Fig. 6-16(c)] occurs in the signal, indicating a sizable component of carrier frequency in the output. Considering the *RC* network as a low-pass filter, a small *RC* indicates a high cutoff frequency for the filter, allowing a portion of the AM signal [$f_c - f_m$, f_c, $f_c + f_m$ of Fig. 6-14(b)] to appear at the output. On the other hand, if *RC* is too large, several cycles are required to discharge the capacitor, as illustrated in Fig. 6-16(d). This results in a phenomenon known as *diagonal clipping*. On a frequency basis, this means that the high cutoff frequency ($f_c = 1/RC$) of the filter is too small, thus rejecting the higher audio frequencies. For these reasons, the selection of the time constant is a compromise. For our receiver, the diode D108 is placed in the circuit so as to give a negative average voltage.

Automatic Gain Control

AGC is used to maintain a relatively constant receiver output despite widely varying receive signal levels. The receive signals can vary by tens of decibels over seconds or minutes, due to changing weather and ionospheric conditions and are highly location dependent, much to the annoyance of mobile users. The level also changes abruptly when tuning to different stations, due to varying transmitter powers, terrain, and distances from the various transmitters.

In AM receivers, the amplifier gain is adjusted over a time span of 0.1 to 1 s to maintain a constant audio output. This is effected by controlling the gain of one or more stages in the receiver. Since the gain of a common-emitter transistor stage is directly proportional to the dc emitter current (i.e., $h_{ib} = 26 \text{ mV}/I_E$) and the transconductance gain (g_m) of a dual-gate FET is set by the dc bias on the second gate, by applying a negative dc feedback voltage to the appropriate stages, a constant audio output signal can be obtained. In the case of the Midland receiver, the large AGC filter capacitor C136 removes the audio component from the detector diode stage, leaving only a negative dc voltage, which is fed back to the RF amplifier (Q101) and the first converter (Q102/103) and IF amplifier (Q105) stages. As the received signal level tends to increase, the negative dc voltage feedback increases, causing a reduction in the gain of these stages, with the net result of an almost constant receive signal level.

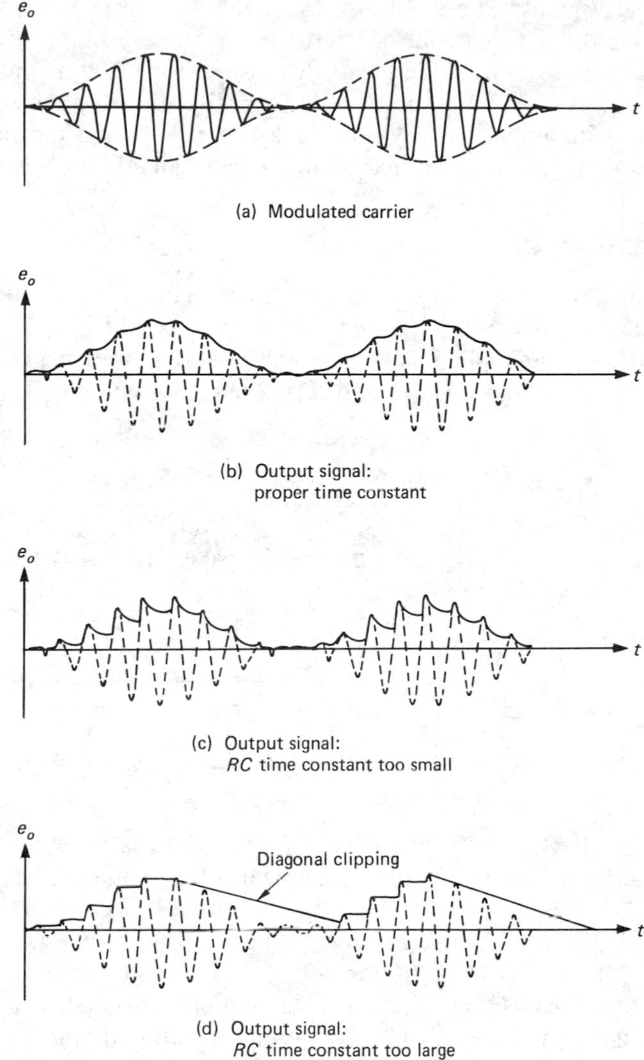

Figure 6-16 Diode envelope detector waveforms.

Squelch

The squelch circuit prevents a radio receiver from producing audio-frequency output when the signal drops below a predetermined value. This is to avoid a constant noisy background signal when listening for communications. The base of the squelch transistor Q201 is biased from the detector circuit D108. At low or no signal, the bucking negative signal applied to Q201 via D109 is reduced, turning it on. The bias to Q203 is decreased because of the low resistance presented by the collector of Q201 to one leg of the dc supply potential divider. This raises the collector voltage of Q203 and thus the bias to Q202. The audio signal is thus highly attenuated for weak signals. The cutoff level (squelch threshold level) can be controlled by adjusting the squelch control VR101.

Because of the large electrolytic capacitor C209 at the base of the squelch transistor, it does take a few hundred milliseconds for the squelch circuit to gain control. This is particularly noticeable when releasing the push-to-talk button; a noticeable pop will be audible if no receive signal is present.

Noise Blanker

To understand the operation of the noise blanker, let us first consider the full-wave doubler circuit hidden within it, as redrawn in Fig. 6-17. During the negative half-cycle of the applied ac voltage, diode D111 conducts, charging the capacitors to the peak voltage V_p as shown. D112 does not conduct during this time.

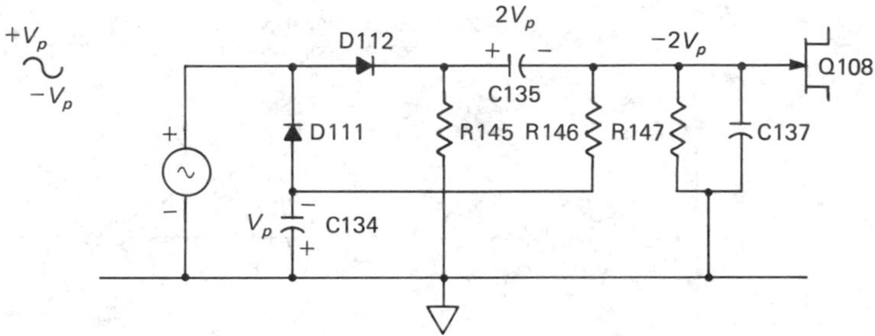

Figure 6-17 Full-wave voltage doubler.

During the positive half-cycle, D112 conducts. The amount of voltage delivered to the output is the sum of the peak signal voltage plus the voltage stored in C135. D111 is now nonconducting. If no load is connected to the output, the capacitor will remain charged and the output voltage will hold at $-2V_p$. The turn-on of Q108 is fast, but the turn-off time is controlled by the discharge of C137 (1000 pF) through R147 (4.4 kΩ). If impulse noise such as that caused by the engine ignition system is present, the brief positive output voltage will cause Q108 to conduct as well as the following two stages, Q109 and Q110. Q110 thus tends to load down the ceramic filter CF1, thereby preventing the ringing that would normally be caused by the noise pulse passing through L103. The time delay of the noise blanker circuit should be close to the time delay experienced by the signal passing through the L108 and L103 tuned circuits and mixer stage Q102 and Q103 for good impulse noise reduction.

Metering

A built-in meter with a function switch permits measurement of both the incoming signal and the transmitted signal; it is therefore commonly called an *S meter*. The received signal is rectified by D107 and filtered by R127/C126/RV102/RV202/C226 to provide a dc signal for the S meter. The calibration of the meter is arbitrary, but it can be used as a relative-strength meter. When transmitting, rectification and filtering are accomplished by D206/C226 and RV202.

Selectivity

The measure of the extent to which a receiver is capable of differentiating between the desired frequency band and all the other frequencies and disturbances is defined as *selectivity*. The selective circuits must be sharp enough to remove the interference from adjacent channels, yet broad enough to pass the desired channel with acceptable distortion. In the Midland receiver, the RF response is sharp enough to remove the image frequency but not sufficiently sharp to remove the adjacent channels. This is the second IF's function.

The receiver uses double conversion to obtain sufficient image rejection at the HF frequencies (refer to Section 1-8). Since the front-end RF response does not drop off fast enough, it is necessary to increase the IF frequency so that an increased frequency spacing to the image frequency is obtained. Since increasing the IF frequency causes a widening of its bandpass characteristics, a second lower IF frequency with a narrow bandpass is required.

The Midland unit employs a first IF of 10.695 MHz. For a 26.975-MHz CB signal frequency (channel 2), this allows sufficient selectivity to be obtained in the RF tuned stage to reject the image of the first IF at 5.585 MHz. However, with 10-kHz spacing between channels, many of the adjacent channels are passed to the first mixer. Figure 6-18 illustrates the RF and IF selectivities. The selectivity of the second IF ceramic filter, centered on 455 kHz and 6000 Hz wide at the 6-dB points, rejects the adjacent channels that appear at 445 and 465 kHz.

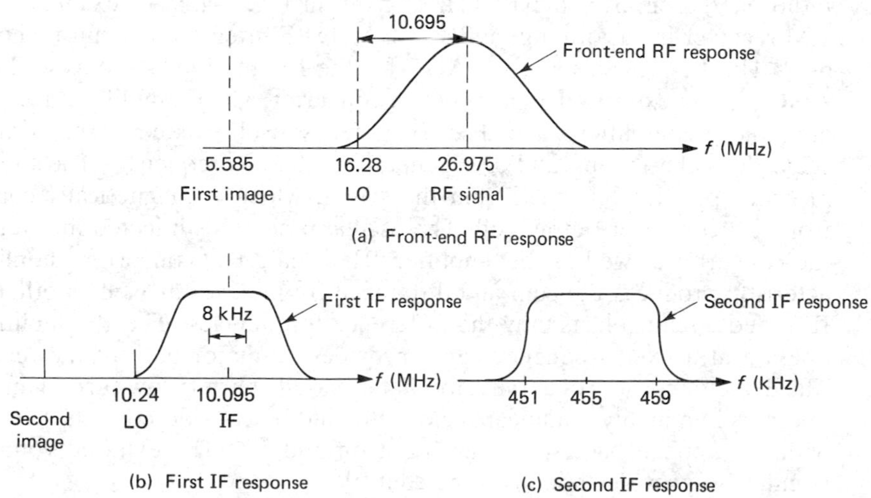

Figure 6-18 Midland double-conversion receiver response.

Sensitivity

A receiver's *sensitivity* refers to its ability to obtain at the output load an acceptable signal-to-noise ratio for a specified minimum input signal level. In measuring the sensitivity, a 30% modulated AM signal with a 1-kHz tone is applied to the RF input of the AM receiver, and reduced until the output signal plus noise and distortion-to-noise and distortion ratio is 10 dB. This particular form is called the 10-dB *SINAD test* (an acronym for "signal plus noise plus distortion to noise plus distortion ratio"). To remove the fundamental frequency, a 1-kHz band-reject filter is inserted prior to the audio load, leaving only the noise. The modulated RF input is adjusted until the SINAD ratio is 10 dB, giving the sensitivity of the receiver in microvolts. When making the SINAD measurement, the speaker can be replaced temporarily by a resistive load nominally equal to the speaker's impedance to provide a quiet atmosphere. Briefly, the procedure is as follows:

1. Apply a 30%, 1-kHz modulated AM signal to the RF input and note the signal plus noise level at the audio load. At this time the notch filter is bypassed.
2. With the notch filter inserted and tuned to 1 kHz, note the output noise level (i.e., harmonics and other nonlinear distortion products).
3. Adjust the generator's AM signal level until the ratio between steps 1 and 2 is 10 dB. The generator's output is the sensitivity of the receiver. For the Midland receiver, the sensitivity should be less than 0.7 μv.

6-5 AM BROADCAST SUPERHETERODYNE RECEIVER

The basic block diagram of the common AM receiver used in the standard AM radio band from 535 to 1605 kHz is given in Fig. 1-42. An example of a simple AM receiver incorporating the RCA CA 3088E integrated circuit is shown in Fig. 6-19. The IC provides internal AGC for the IF amplifier stage as well as delayed AGC for the external RF amplifier, a converter, an IF amplifier, a detector, and an audio preamplifier stage. The converter is a self-excited mixer in that a single stage does the mixing and also generates the LO frequency. The oscillator coil provides positive feedback from the signal, which is magnetically coupled back from the converter output coil. The oscillator signal is injected into terminal 2 of the converter as well as the amplified RF signal. The converter's nonlinear characteristic produces the sum and difference frequencies as well as others, and the IF tuned circuit selects only the difference frequencies. The IF amplifiers, which operate at a fixed frequency band, provide the major bulk of the receiver gain. The detector consists of a transistor that is biased at a very low current and therefore operates in a highly nonlinear region. It could be considered a "barely on" diode, with the applied base signal turning it on and off. The external volume control circuit also provides minimal tone control.

The external tuned circuits also provide the image rejection and selectivity, and are the limiting factors in reducing the physical size of the receiver. In the future ceramic filters may well be incorporated on the chip.

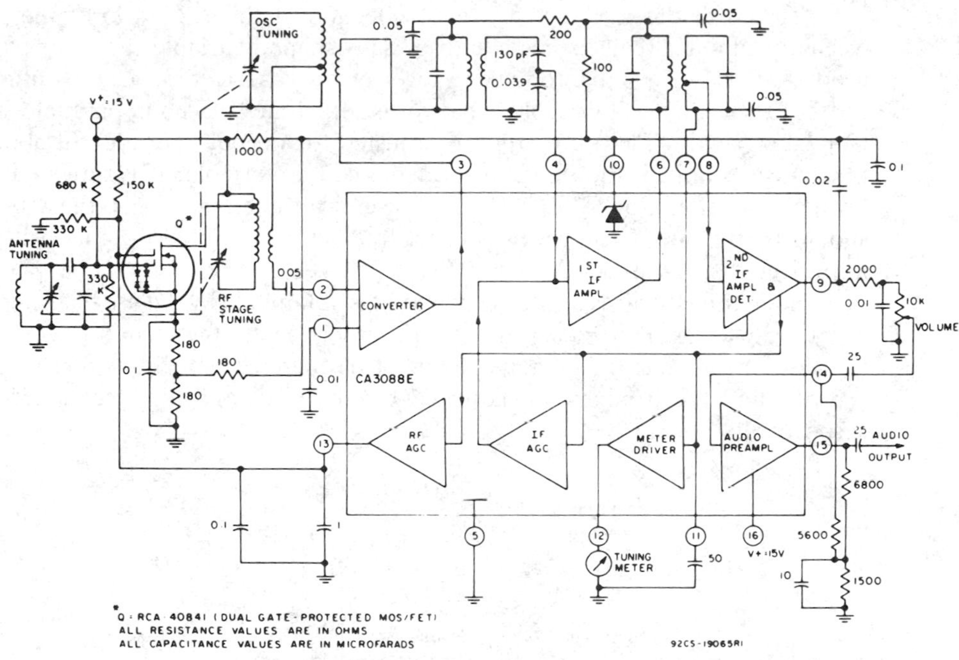

Figure 6-19 Typical AM broadcast receiver using the CA3088E. (Courtesy of GE Solid State.)

6-5.1 Tracking

To cover the entire AM radio band spectrum, the LO frequency must "track" the incoming RF signals such that their differences are a constant (the IF) frequency. This tracking is usually accomplished by ganging or mechanically linking the rotor assembly of the variable capacitors. Figure 6-20 shows a typical variable ganged capacitor.

Figure 6-20 Variable ganged capacitor.

To aid in the tracking, a careful selection of the L/C ratio of the RF and oscillator tuning circuits is made and specially shaped variable capacitor plates are used. As it is not possible to make a receiver track perfectly over the entire range of frequencies, a small variable capacitor is sometimes placed in parallel with each ganged capacitor [Fig. 6-21(a)]. It is adjusted for proper operation at about 1200 kHz. In addition, if the inductance of the coil is variable, it is adjusted to give proper operation around 650 kHz. Because of interaction effects, these two steps should be repeated. The result is exact tracking at these two points, with minor imperfections as shown in Fig. 6-21(b).

Sometimes a third exact tracking point is obtainable by the use of an adjustable padder capacitor. This capacitor, which is in series with the main tuning capacitor, is adjusted for exact tracking at the lowest frequency (plates fully meshed), while the coil and the trimmer are adjusted for the mid- and high-frequency points, respectively.

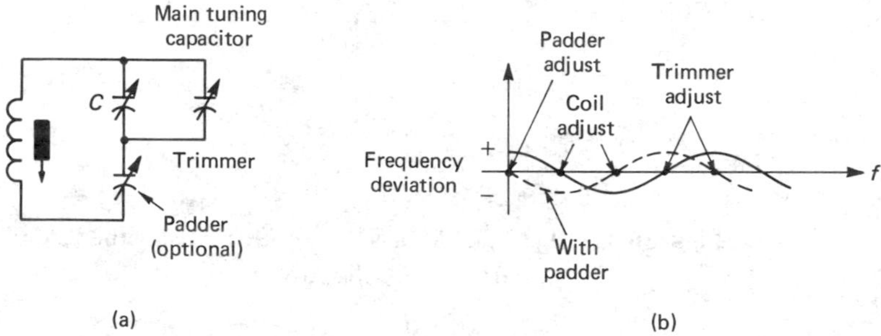

Figure 6-21 Tracking circuit showing adjustments.

6-5.2 Choice of Oscillator Frequencies

It is possible to cover the entire AM broadcast band with the use of a single variable LO capacitor if the LO frequency is greater than the RF frequency. For an IF of 455 kHz the LO must vary in frequency from 535 + 455 = 990 kHz to 1605 + 455 = 2060 kHz, giving a frequency ratio of

$$\frac{f_{LO(max)}}{f_{LO(min)}} = \frac{2060}{990} = 2.08$$

Since the oscillator frequency is inversely proportional to the square root of the oscillator capacitance or the oscillator capacitance is inversely proportional to the square of the oscillator frequency, the frequency ratio requires a capacitance ratio of $C_{LO(max)}/C_{LO(min)} = 2.08^2 = 4.33$. This is quite easily obtainable in practice.

The corresponding RF stage has a frequency and capacitance ratio of

$$\frac{f_{RF(max)}}{f_{RF(min)}} = \frac{1605}{535} = 3$$

$$\frac{C_{\mathrm{RF(max)}}}{C_{\mathrm{RF(min)}}} = 3^2 = 9$$

If the local oscillator frequency is lower than the RF frequency by 455 kHz, the LO must vary in frequency from 535 − 455 = 80 kHz to 1605 − 455 = 1150 kHz, resulting in a frequency ratio of

$$\frac{f_{\mathrm{LO(max)}}}{f_{\mathrm{LO(min)}}} = \frac{1150}{80} = 14.4$$

and thus a capacitance ratio of

$$\frac{C_{\mathrm{LO(max)}}}{C_{\mathrm{LO(min)}}} = 14.4^2 = 207$$

This is an impractical range for a single capacitor.

For such large capacitance ratios, other capacitors or coils must be switched into the circuit. This is common practice when several bands of frequencies are to be detected.

6-5.3 Double Spotting

For the higher frequencies, it is possible to detect the same station at two different settings of the tuning dial if the image frequency rejection is poor. To understand this phenomenon, called *double spotting*, consider the two RF or dial settings shown in Fig. 6-22. At the proper dial setting the RF stage is tuned at the signal frequency f_s. When mixed with the LO frequency, it gives the appropriate IF. When the dial or RF stage is tuned to twice the IF frequency below the signal frequency, the image or signal frequency can mix with the LO to give a suitable IF for detection. To reduce the problem of double spotting, the sharpness (selectivity) of the RF amplifier must be improved.

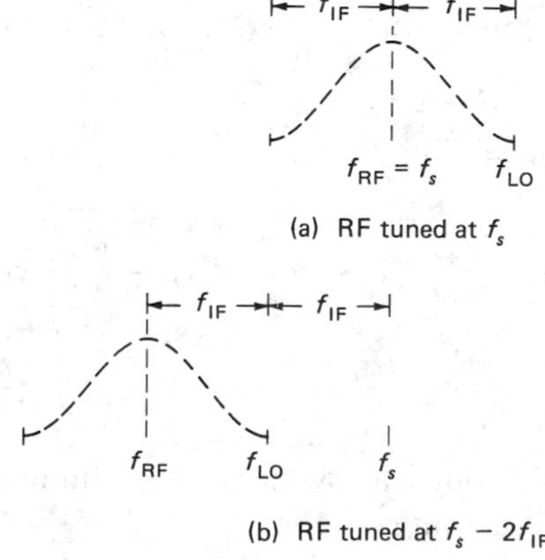

Figure 6-22 Illustration of double spotting.

6-6 SYNCHRONOUS AM DETECTION

When many signals are present at the input to a PLL, by setting f_o of the PLL to the desired frequency, the PLL will lock onto the band of frequencies $f_o \pm \frac{1}{2} f_{\text{lock}}$ and block off other signals. If the loop bandwidth (BW) is set narrow, the S/N value of the VCO output is much better than the input S/N. This principle can be used in AM synchronous demodulation (coherent detection), as shown in Fig. 6-23.

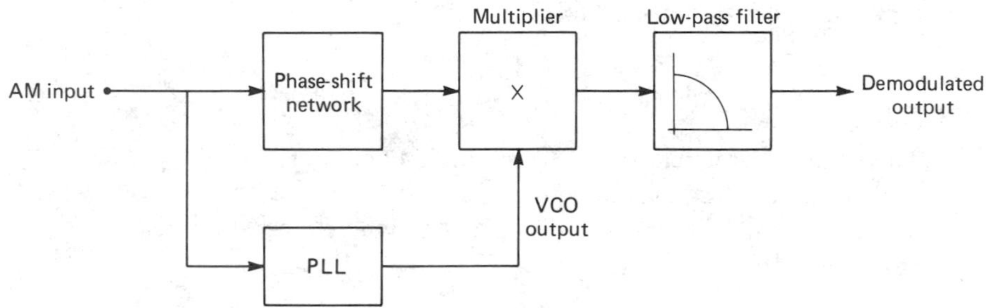

Figure 6-23 Coherent amplitude-modulation detection using PLL.

The 90° phase shift is necessary because the VCO output of the PLL has a 90° phase shift compared with the input. The phase shifter is usually an RC network. The VCO output of the PLL tracks the AM carrier but has no modulation. The multiplier will give the sum and difference frequency where the difference frequency is the demodulated signal. Further filtering (LP) will clean up the signal. This can be shown mathematically by multiplying the AM signal [equation (6-6)] by the carrier signal $\cos \omega_c t$, that is,

$$e_{\text{multiplier output}} = E_c(1 + m \cos \omega_m t) \cos \omega_c t \times \cos \omega_c t$$

Expanding this expression and using the identity

$$\cos a \cos b = \tfrac{1}{2} \cos(a + b) + \tfrac{1}{2} \cos(a - b) \tag{6-22}$$

we obtain

$$e_{\text{mixer output}} = \frac{E_c}{2}(1 + \cos 2\omega_c t) + \frac{mE_c}{2} \cos \omega_m t(1 + \cos 2\omega_c t)$$

$$= \frac{E_c}{2}(1 + \cos 2\omega_c t) + \frac{mE_c}{2} \cos \omega_m t$$

$$+ \frac{mE_c}{4} \cos(2\omega_c t + \omega_m)t + \frac{mE_c}{4} \cos(2\omega_c - \omega_m)t \tag{6-23}$$

Passing this signal through a low-pass filter eliminates all the frequencies except the dc and the modulating component:

$$\text{demodulated output} = \frac{E_c}{2} + \frac{mE_c}{2} \cos \omega_m t \tag{6-24}$$

The dc component is also usually blocked with a blocking capacitor, leaving only the original modulating signal at the output.

Coherent detection of AM has a higher degree of noise immunity than that for a conventional peak detector method. Another advantage is that the different signals are detected by tuning the VCO free-running frequency of the PLL, which is simply done by a potentiometer in the monolithic PLLs available.

6-7 AM STEREO

For many years, FM stations have been broadcasting stereo. As a consequence, stereo has become a preferred listening source and AM broadcasters are attempting to tap into this market by providing stereo AM service. Furthermore, since conventional AM broadcast signals operate at much lower carrier frequencies, they do not suffer the limited geographic coverage area of the FM service due to line-of-sight transmission. In addition, the multipath effects experienced in FM—the simultaneous arrival of a second time-delayed signal at the receiving antenna caused by reflections from tall structures, causing serious distortion of the received signal, is not a problem with AM stereo. The traveling automobile listener will obtain acceptable reception as long as the receiver is in an area of sufficient signal strength and has good AGC.

Although several AM stereo systems have been proposed and are in use (i.e., Magnavox AM/PM, Harris, Motorola C-QUAM, Belar AM-FM, Kahn Hazeltine), countries such as Canada, Australia, and many South American countries have standardized on the Motorola C-QUAM system. Our discussion is limited to this system.

As is the case when introducing an additional service such as AM stereo, when there are already millions of receivers in place, the new service must be compatible with existing radios. The Motorola C-QUAM stereo is compatible with the envelope detectors in existing AM radios. In stereo, two microphones are spatially separated, providing the L (left), and R (right) signals. The L and R audio components are matrixed to form $(L + R)$ and $(L - R)$ signals. Under monaural conditions, only $L + R$ is used to amplitude modulate the carrier.

To comprehend the generation and detection of AM stereo, the functions of a balanced modulator and a balanced demodulator must be understood. We shall not discuss the circuitry that performs these functions at this time, as this is done in Chapter 7, but we shall consider the mathematical concepts. This is really an expansion on the earlier synchronous detector section.

6-7.1 Balanced Modulator and Balanced Demodulator

The balanced modulator and the balanced demodulator or product detector perform fundamentally the same operation. They both act as multiplying circuits. Thus if two input signals, $\cos(\omega_1 t + \theta)$ and $\cos \omega_2 t$ are applied to one of the circuits, the output is equal to

$$[\cos(\omega_1 t + \theta)](\cos \omega_2 t) = \tfrac{1}{2}\cos[(\omega_1 + \omega_2)t + \theta] + \tfrac{1}{2}\cos[(\omega_1 - \omega_2)t + \theta] \quad (6\text{-}25)$$

applying identity (1-5). The resultant indicates that a sum and difference frequency is produced and that the phase angle remains unchanged. The latter is critical in understanding stereo AM operation.

6-7.2 Quadrature AM Stereo

Although pure quadrature AM–AM could be a legitimate method of generating AM stereo, it cannot be detected by the envelope detector found in most AM radios. It could, however, be detected by a synchronous detector. Let us consider the pure quadrature AM–AM system first, as it will enable us to understand the Motorola C-QUAM system, which is a modification of the quadrature system to enable the detection of the stereo AM signal by an envelope detector. In the latter case the stereo component is lost.

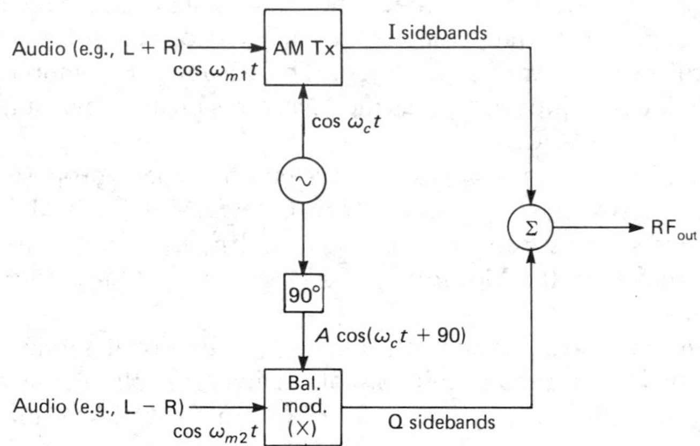

Figure 6-24 Quadrature AM transmitter.

Consider the quadrature transmitter shown in Fig. 6-24. As expressed earlier, the output from the AM transmitter section will consist of three components, i.e., for $E_c = 1$:

$$\cos \omega_c t + \frac{m}{2}\cos(\omega_c - \omega_{m_1})t + \frac{m}{2}\cos(\omega_c + \omega_{m_1})t \quad (6\text{-}26)$$

The sidebands are in phase with the carrier or *I sidebands*.

At the balanced modulator section, the carrier phase is shifted by 90°. The resulting output signal is given by

$$A\cos(\omega_c t + 90)(\cos \omega_{m2} t)$$
$$= \frac{A}{2}\cos[(\omega_c + \omega_{m2})t + 90] + \frac{A}{2}\cos[(\omega_c - \omega_{m2})t + 90] \quad (6\text{-}27)$$

The sidebands generated by the balanced modulator are 90° out of phase with the AM sidebands and are in "quadrature." These become our *Q sidebands*.

The $m/2$ and $A/2$ coefficients depend on the actual circuitry and amplitude of the audio signals, and as they are of little significance to our present discussion, will be assumed equal to 1 for convenience. Thus the total output signal will be

$$\begin{aligned} RF_{\text{Out}} &= \cos \omega_c t + \cos(\omega_c + \omega_{m1})t + \cos(\omega_c - \omega_{m1})t \\ &\quad + \cos[(\omega_c + \omega_{m2})t + 90] + \cos[(\omega_c - \omega_{m2})t + 90] \end{aligned} \tag{6-28}$$

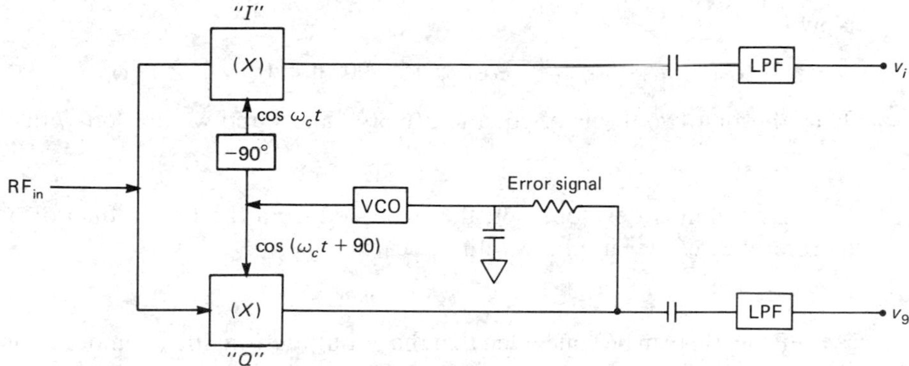

Figure 6-25 Quadrature AM detector.

To recover the audio signals separately at the receiver, a system of product detectors is used as shown in Fig. 6-25. At the output of the Q product detector, we obtain

$$\begin{aligned} \cos(\omega_c t + 90) \times RF_{\text{Out}} &= [\cos(\omega_c t + 90)](\cos \omega_c t + \cos(\omega_c + \omega_{m1})t \\ &\quad + \cos(\omega_c - \omega_{m1})t + \cos[(\omega_c + \omega_{m2})t + 90] \\ &\quad + \cos[(\omega_c - \omega_{m2})t + 90)] \\ &= \tfrac{1}{2}\cos(2\omega_c t + 90) + \tfrac{1}{2}\cos 90 \\ &\quad + \tfrac{1}{2}\cos(2\omega_c t + \omega_{m1}t + 90) \\ &\quad + \tfrac{1}{2}\cos(\omega_{m1} t - 90) + \tfrac{1}{2}\cos(2\omega_c t - \omega_{m1}t \\ &\quad + 90) + \tfrac{1}{2}\cos(\omega_{m1}t + 90) + \tfrac{1}{2}\cos(2\omega_c t \\ &\quad + \omega_{m2}t + 180) + \tfrac{1}{2}\cos \omega_{m2}t \\ &\quad + \tfrac{1}{2}\cos(2\omega_c t - \omega_{m2}t + 180) \\ &\quad + \tfrac{1}{2}\cos \omega_{m2}t \end{aligned} \tag{6-29}$$

The only dc output term present is the second one (i.e., $\frac{1}{2}\cos 90 = 0$). When the frequency and phase of the voltage-controlled oscillator (VCO) is locked in as assumed, the dc error signal is zero. If the VCO is not on the carrier frequency, $\cos[(\omega_c + \Delta\omega)t + 90]$, the second term of expression (6-29) would become $\frac{1}{2}\cos(\Delta\omega t + 90)$, giving an ac error signal, causing the frequency of the VCO to zero in on the input carrier frequency. Similarly, if the VCO does not have the proper phase [i.e., $\cos(\omega_c t + 90 + \Delta\theta)$], the second term becomes $\frac{1}{2}\cos(90 + \Delta\theta)$, giving a dc error signal, forcing the phase of the VCO to zero.

After passing through the LP filter, the only terms remaining are

$$v_q = \tfrac{1}{2}\cos(\omega_{m1}t - 90) + \tfrac{1}{2}\cos(\omega_{m1}t + 90) + \tfrac{1}{2}\cos\omega_{m2}t + \tfrac{1}{2}\cos\omega_{m2}t \qquad (6\text{-}30)$$

Now

$$\cos(\omega_{m1}t - 90) = -\cos(\omega_{m1}t - 90 + 180) = -\cos(\omega_{m1}t + 90)$$

Thus the first two terms of equation (6-30) cancel and we are left with

$$v_q = \cos\omega_{m2}t \qquad (6\text{-}31)$$

By a similar development, the signal fed from the I demodulator, which has its reference carrier in phase with the carrier, is

$$v_i = \cos\omega_{m1}t \qquad (6\text{-}32)$$

From this can be concluded that the ac output from the I demodulator provides the original $\cos\omega_{m1}t$ audio from the AM transmitter, and the Q demodulator provides the original $\cos\omega_{m2}t$ audio from the balanced modulator transmitter.

A normal envelope detector ignores any phase information in the RF signal of equation (6-28) and monitors only the rms total of the modulated signal. The simple vector addition results in a very distorted signal.

6-7.3 Motorola C-QUAM AM Stereo Encoder

The Motorola C-QUAM system takes the pure quadrature signal as just described and extracts the phase modulation components of the quadrature signal; the latter, in turn, phase modulates the RF carrier of the broadcast transmitter. At the same time it sends the L + R audio to the audio input of the transmitter as usual.

The block diagram of the C-QUAM encoder is given in Fig. 6-26. If we represent the carrier by the expression

$$\text{carrier} = A_c\cos\omega_c t$$

the output from the I and Q modulators can be expressed as

$$\text{I} = M_s[L(t) + R(t)]A_c\cos\omega_c t$$

$$\text{Q} = M_d[L(t) - R(t)]A_c\cos(\omega_c t + 90)$$

where M_s is the index of modulation for sum information $L + R$, and M_d is the index of modulation for difference information $L - R$. L and R have been shown to have the independent variable t to emphasize that these are time dependent.

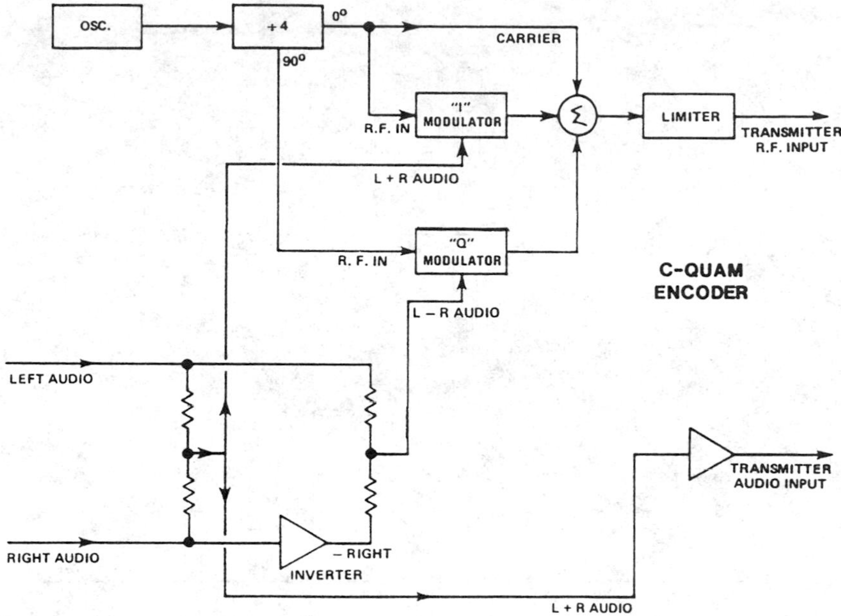

Figure 6-26 Block diagram of the Motorola C-QUAM encoder. (Copyright of Motorola, Inc. Used by permission.)

As proven in Appendix A, after amplitude limiting the sum of these three inputs, the following phase-modulated carrier is obtained:

$$\text{Tx, RF}_{\text{input}} = A_c \cos\left(\omega_c t + \tan^{-1}\frac{M_d[L(t) - R(t)]}{1 + M_s[L(t) + R(t)]}\right) \tag{6-33}$$

The signal applied to the transmitter audio input is just the $L(t) + R(t)$ audio signal.

The C-QUAM modulated carrier can thus be expressed as

$$E_c = A_c(1 + M_s[L(t) + R(t)]) \cos\left(\omega_c t + \tan^{-1}\frac{M_d[L(t) - R(t)]}{1 + M_s[L(t) + R(t)]}\right) \tag{6-34}$$

For stereo indication at the receiver, a 25-Hz pilot tone is also applied to the Q modulator. This gives the resultant modulated carrier of

$$E_c = A_c(1 + M_s[L(t) + R(t)])$$
$$\times \cos\left(\omega_c t + \tan^{-1}\frac{M_d[L(t) - R(t)] + 0.05 \sin 50\,\pi t}{1 + M_s[L(t) + R(t)]}\right) \tag{6-35}$$

Figure 6-27 shows two sets of AM stereo spectra obtained at the transmitter antenna under the conditions in (a) that only the transmitter audio input is varied with a 500-Hz tone ($L + R$) and in (b) that only the transmitter RF input is varied with a 500-Hz tone ($L - R$). In Fig. 6-27(a) the normal AM spectrum is obtained, whereas in (b) the multiple sidebands of a phase-modulated signal are observed.

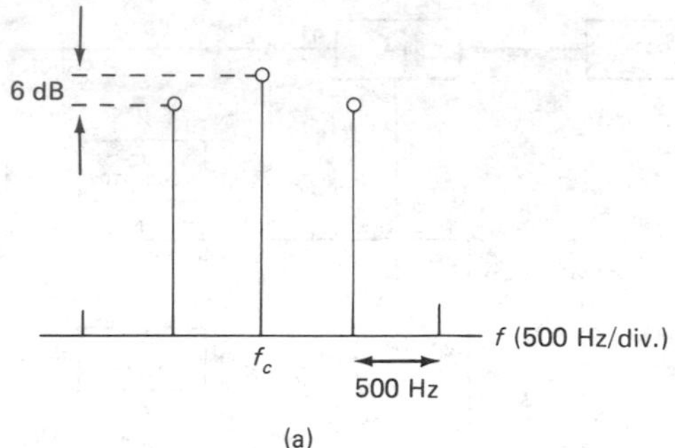

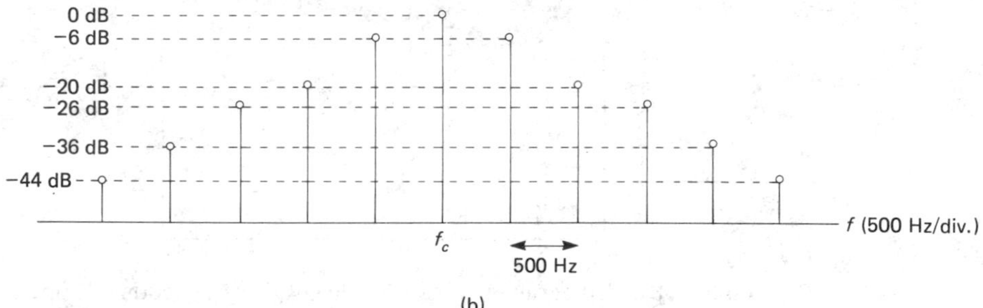

Figure 6-27 Frequency spectra of a stereo AM signal at the transmitting antenna input: (a) 500-Hz $L + R$ signal-to-transmitter audio input and only carrier-to-transmitter RF input; (b) 500-Hz $L + R$ signal plus carrier-to-transmitter RF input and no signal-to-transmitter audio input. Modulation index of 95%; vertical scale: 10 dB/div., horizontal scale: 500 Hz/div.

6-7.4 Incidental Phase Modulation

Incidental phase modulation (IPM) is the undesired phase modulation of an RF carrier. In the past, IPM has been of little concern to the broadcaster, as the common radio envelope detector is oblivious to moderate amounts of IPM. With the introduction of AM stereo, unintentional or IPM must be removed, as the phase information is now used to transmit program material. In the linear transmitter final amplifiers, care must be taken to neutralize all the stages so that the frequency does not slide around. Poor or aging low-voltage power supply filter capacitors can cause hum or IPM. Lead shielding should be used to prevent feedback of the radiation RF onto the audio cables.

6-7.5 Motorola C-QUAM Decoder

Figure 6-28 gives the block diagram of a C-QUAM decoder. Note that the demodulator contains a pure quadrature demodulator section, as shown previously in Fig. 6-25. The envelope detector section ignores the phase modulation and

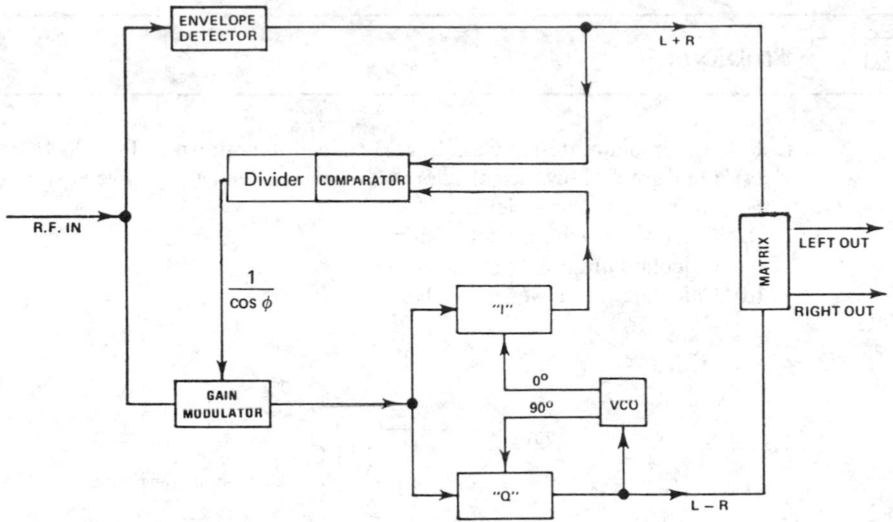

Figure 6-28 Motorola C-QUAM decoder. (Copyright of Motorola, Inc. Used by permission.)

extracts the $L + R$ signal as in a normal AM radio. Ignoring the 25-Hz pilot tone, the output of the Q detector is zero if a monaural signal is received. If a stereo signal is received, the envelope detector signal is divided by the I detected signal, giving an error signal to the gain modulator. The error signal AGC's the input level to the quadrature detector. This action makes the input signal to the I and Q demodulators look like pure quadrature signals and the Q audio output gives a perfect $L - R$ signal. The $L - R$ signal is combined with the $L + R$ signal in a matrix to give the right and left audio components [i.e., $(L + R) + (L - R) = 2L$, $(L + R) - (L - R) = 2R$].

Problems

6-1. For the optimum design class C, AM transmitter shown in Fig. P6-1:

(a) On a graph show the collector voltage waveform (v_c) for given conditions; indicate all frequencies and levels.

(b) Calculate m (the modulation index).

(c) Calculate the power in the carrier.

(d) Calculate the power in the USB.

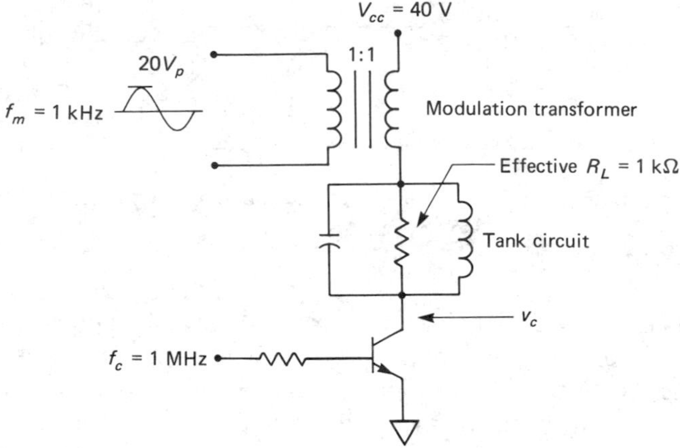

Figure P6-1.

6-2. A carrier wave $v_c = 50 \cos(6.28)10^6 t$ is amplitude-modulated by an audio voltage $v_m = 30 \cos(6.28)10^3 t$.

(a) Calculate m for the modulated wave.

(b) Determine the frequencies present and their respective amplitudes.

(c) Draw the power spectrum in decibels relative to the carrier if the modulated wave is loaded by a 50-Ω antenna. Label all frequencies.

6-3. If the carrier wave of Problem 6-2 is 80% modulated by a second tone of 10 kHz, determine the new frequencies and their power levels in decibels relative to the carrier. Also calculate the net modulation index when both tones are present simultaneously.

6-4. AM radio station CFRN broadcasts a total signal power of 50 kW at a carrier frequency of 1260 kHz. Assume a modulation index of $m = 1$. Show the formulas used.

(a) Determine the powers contained in the (1) carrier component of the AM signal, and (2) sideband components.

(b) What is the useful information efficiency of this signal?

(c) Determine the required bandwidth of a receiver used to recover the information in this signal completely if the signal information contained was 50 Hz to 5 kHz.

6-5. Sketch the frequency-domain [$v(f)$] representation of the AM waveform shown in Fig. P6-5. Indicate numerically the amplitude and frequency of all components.

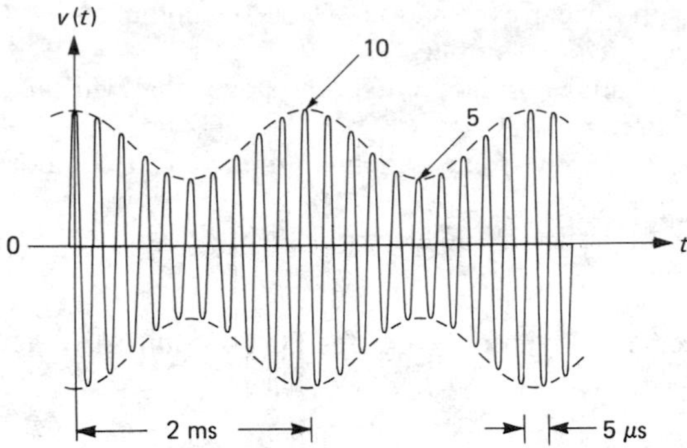

Figure P6-5.

6-6. A 1-MHz carrier voltage of peak value 10 V is amplitude modulated by a 1-kHz sinusoidal signal of peak value 5 V. The modulation index is:
(1) 0.15 (3) 5
(2) 0.5 (4) 15

6-7. If the modulating signal specified in Problem 6-6 is changed in frequency to 2 kHz, keeping the same amplitude, the modulation index will be:
(1) Unchanged (3) Halved
(2) Doubled (4) Multiplied by $\frac{1}{2}$

6-8. A carrier voltage is amplitude modulated by a complex audio signal which, over a short period of time, has a constant amplitude. If the modulation index measured during this time is m, the ratio in a load of

$$\frac{\text{total sideband power}}{\text{carrier power}}$$

during this time will be approximately:
(1) m (3) m^2
(2) $m/2$ (4) $m^2/2$

6-9. The converter circuit shown in Fig. P6-9 is used in the receiver section of the Midland 77-005 transceiver. For chan. no. 1 (26.965 MHz) the LO is at 16.27 MHz.

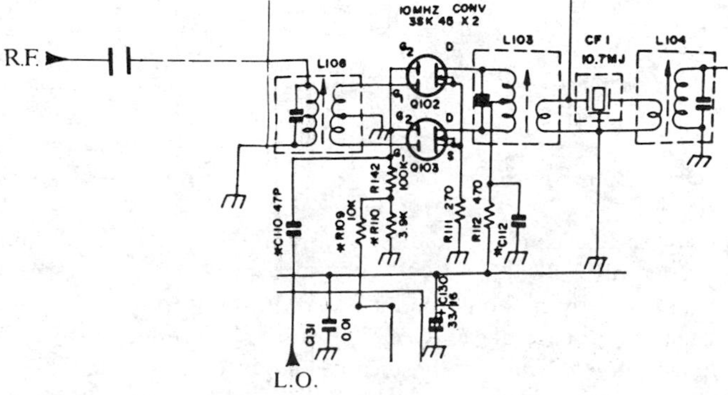

Figure P6-9 (Courtesy of Midland International.)

(a) Determine with a clear explanation of your technique, what frequencies will be present at the converter output.
(b) What is the advantage of this configuration over that which employs just a single dual-gate FET?
(c) Why are the L103/L104 tuned circuits employed in conjunction with the CF1 filter?
(d) Describe the difference in the electrical characteristics of the two grounds: ⏚ and ⏚.

6-10. Correctly connect the oscilloscope in Fig. P6-10 to the AM transmitter to give a trapezoidal display.

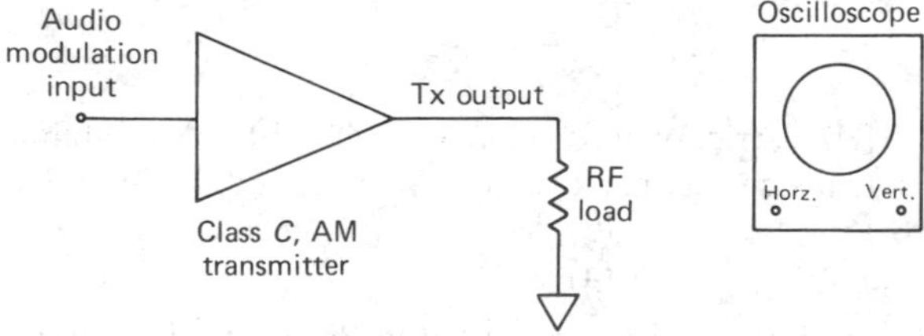

Figure P6-10.

6-11. At high frequencies a superhet receiver will use a double conversion system:
(1) To provide more separation between the desired signal frequency and the image frequency
(2) To allow the second IF stage to have a narrow bandwidth without requiring unreasonable large effective Q values in the circuits
(3) To allow electronic tuning to be used
(4) All of the above
(5) None of the above

6-12. Consider the test setup shown in Fig. P6-12.

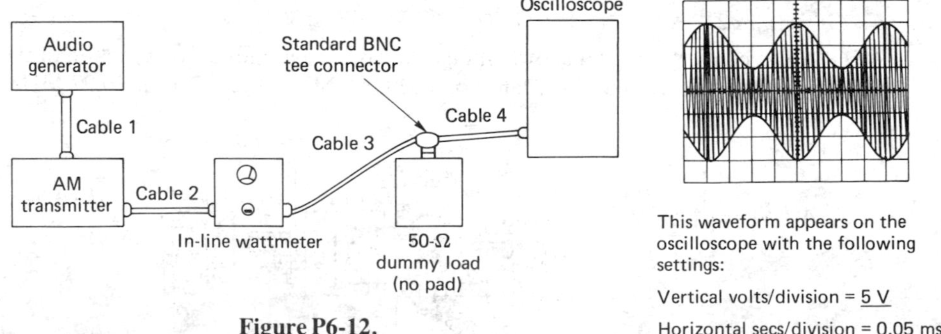

Figure P6-12.

(a) What are the percent modulation and modulating frequency indicated by this oscilloscope display?
(b) What forward power should be indicated on the wattmeter?
(c) Refer to the four coaxial cables of the test setup. If reflected power were indicated on the wattmeter, which of these cables might be responsible?
(d) If a spectrum analyzer were to be used in place of the oscilloscope, an attenuator should be placed in series with cable 4. Give two reasons why this attenuator is required.

(e) Assume that a spectrum analyzer was connected to the test circuit as described above. If the carrier was six divisions in amplitude in the linear display mode, how many divisions would each sideband be?

6-13. In a superhet receiver "tracking" refers to:
(1) The ability of the receiver to follow a signal if it shifts in frequency
(2) The ability of the receiver to adjust for the different signal strengths of the various stations which are to be received
(3) The ability of the receiver to block static output from the receiver when no input signal is present
(4) All of the above
(5) None of the above

6-14. Refer to the CB schematic diagram (Midland W 77-015) and match circuit components with their correct functions.

Q105	___	(1) Balanced mixer
Q302	___	(2) Demodulator
Q103, 102	___	(3) Auto. noise blanker
Q201	___	(4) Audio power amplifier
D108	___	(5) Squelch control
Q202, 203	___	(6) IF amplifier transistor
Q108	___	(7) RF power output transistor
D301	___	(8) Tx driver transistor
IC201	___	(9) Dc switching diode
IC202	___	(10) Overload protection diodes
IC203	___	(11) Mic gain control
T201	___	(12) Squelch transistor
VR101	___	(13) RF bypass capacitor
C106	___	(14) ALC
		(15) IF transformer
		(16) Low-level RF amplifier
		(17) PLL
		(18) Modulation transformer
		(19) Neutralization capacitor

6-15. **(a)** Label Fig. P6-15, the block diagram of a double-conversion superhet receiver.
(b) Indicate on the diagram which blocks would be associated with the following controls: (1) Channel frequency selector; (2) AF gain; (3) Receiver selectivity; (4) RF gain.

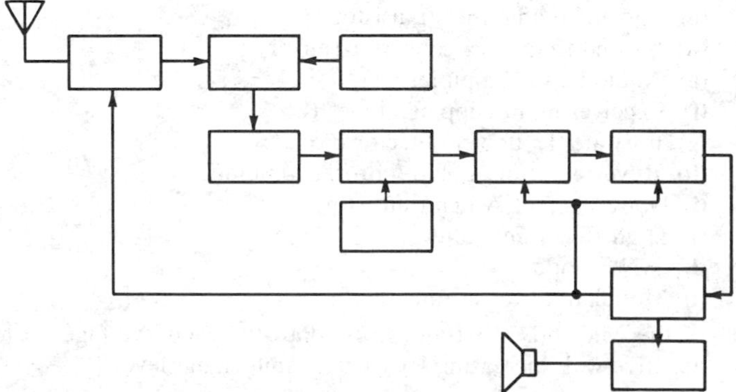

Figure P6-15.

6-16. Why is the local oscillator in an AM broadcast receiver selected to operate above the incoming carrier frequencies?

6-17. In MF-band receivers the local oscillator frequency is invariably above the RF signal frequency (rather than below it). The reason for this is:
(1) The percentage tuning range required of the oscillator is smaller than if its frequency were below the signal frequency.
(2) The mixing process would not work if the local oscillator frequency were below the signal frequency.
(3) It is easier to produce oscillations at a higher frequency.
(4) The frequency stability is better in an oscillator working at a higher frequency.

6-18. In a superhet receiver the local oscillator tuned circuit has a variable capacitor which is "ganged" to the variable capacitor(s) of the RF stage tuned circuits. The local oscillator tuned circuit generally also contains two preset capacitors. The purpose of these is:
(1) To increase the Q of the local oscillator tuned circuit
(2) To tune out stray inductance
(3) To make the local oscillator track correctly
(4) To provide the necessary feedback

6-19. An AM voltage has the representation in the frequency domain shown in Fig. P6-19.
(a) Find the value of the modulation factor, m.
(b) Sketch the waveform in the time domain, showing all voltages.
(c) Find the rms value of the given AM waveform.

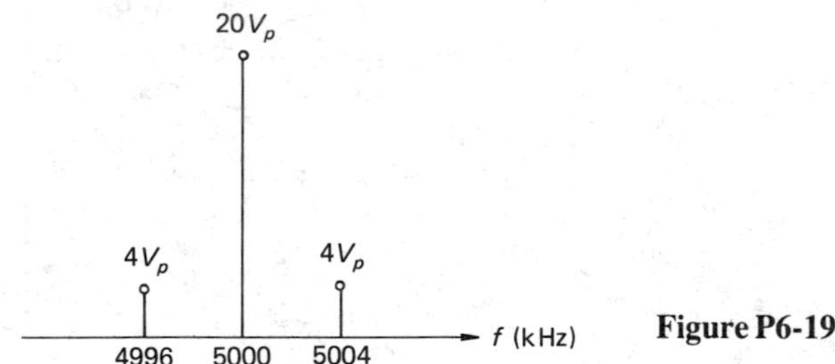

Figure P6-19.

6-20. From the Midland CB transceiver schematic, indicate the component by schematic number for the following circuit functions.
(a) RF power amplifier transistor
(b) AM detector diode
(c) Mic audio amplifier transistor
(d) Second frequency converter (mixer)
(e) Audio power amplifier
(f) Receiver input clipping diodes (2)
(g) Forward Tx power detector diode
(h) Reverse battery polarity protection diode
(i) Noise blanker switch transistor
(j) High IF ceramic filter
(k) ANL diode
(l) Modulation transformer

6-21. In the Midlands CB transceiver, does the AGC voltage go more positive or more negative with increasing Rx carrier input signal level?

6-22. Why is it not possible to have image frequency interference with a synchronous AM detector receiver?

6-23. An AM broadcast band receiver uses a 455-kHz IF and the local oscillator is injected higher than the incoming carrier frequency. A 1550-kHz carrier is received when the station selector is set to 1550 kHz.

(a) At what other station setting might the 1550-kHz carrier also be received?

(b) If this receiver picks up the same station at two different settings of the station selector, what is this referred to as?

6-24. Consider the test setup shown in Fig. P6-24(1). The waveform shown in Fig. P6-24(2) appears on the oscilloscope. The oscilloscope is set to 20 volts per division, 20 μs per division.

(a) What is the percent modulation indicated by this display?

(b) What modulating frequency is indicated by this display?

(c) What is the total power being delivered to the dummy load?

(d) What is the power contained in the upper sideband only?

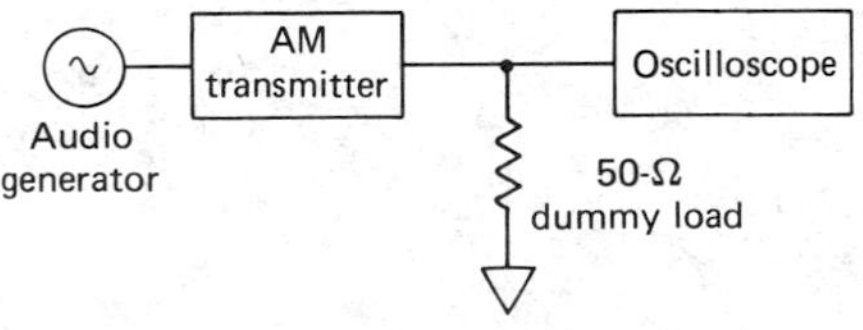

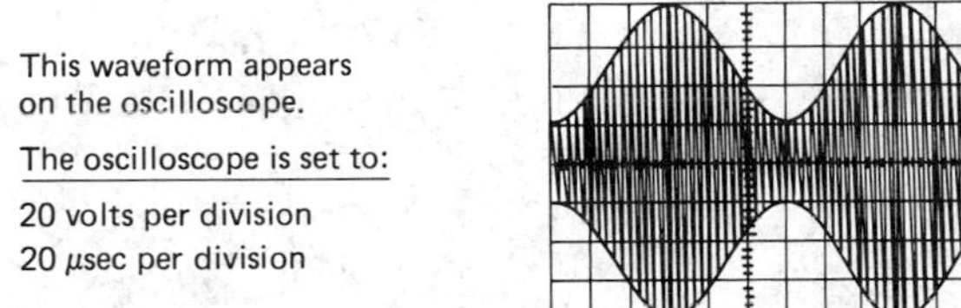

Figure P6-24.

6-25. Prove that the output signal from the I product detector of Fig. 6-25 after dc decoupling the low-pass filtering gives $v_i = \cos \omega_{m1} t$.

Single Sideband

William Sinnema

7-1 INTRODUCTION

We observed in Chapter 6 that in AM all the intelligence is contained in the sidebands and that the carrier frequency component, which is independent of the message, represents wasted power. The carrier can thus be suppressed or eliminated from the modulated wave, as it contains no intelligence. The result is double-sideband suppressed carrier (DSBSC), or double sideband for short (DSB). Although DSB overcomes the wasteful power of conventional AM, it is still wasteful in transmission bandwidth. As the two sidebands are images of each other, both having voltage components of $mE_c/2$, all the information can be conveyed by using only one of the sidebands. Suppressing one sideband reduces the transmission bandwidth and leads to single-sideband modulation (SSB). If a part of the other sideband is also transmitted, one obtains vestigial sideband. An expanded form of SSB is independent sideband (ISB). In such modulation systems, two independent sidebands are transmitted, each containing a different message.

Together with more efficient power and bandwidth utilization, SSB is less subject to the effects of selective fading when traversing through the ionosphere. In AM, the frequencies in one sideband can undergo a different phase shift as it refracts from the ionosphere and back to earth than the opposite sideband, causing partial or even complete cancellation of the two sidebands. This is because both sidebands contribute to the received signal. In SSB, only one sideband is present and this fading problem is minimized. Fading may still occur to some extent if the detected signal is the sum of refracted signals arriving over different paths.

In transmission, SSB has the disadvantage of requiring linear power amplifiers. Their power efficiency is much less than those of the class C amplifiers used in FM. Because of the frequency stability required of the local oscillators in the receiver, SSB receivers tend to be quite costly. This is one of the chief reasons why SSB has not replaced the standard AM radio transmission system.

7-2 DESIGNATION OF EMISSIONS

Many types of modulation schemes have evolved over the years to maximize the transmission reliability of the radio path or the information throughput. The World Administrative Radio Conference[1] has designated the various emissions according to their bandwidth and their classifications. The proper designation must be used when registering frequencies, reporting irregularities or infringements, or applying for a license with the ITU. The emissions are designated according to their bandwidth (shown in the first four characters) and their classification (the following three symbols). If deemed desirable, an additional two characters may be added in the future, making a total of 9.

7-2.1 Bandwidth

The bandwidth is expressed by one letter and three numerals. The letter occupies the position of the decimal point and indicates the bandwidth range. The three numerals permit bandwidth resolution to three significant figures within this range.

Bandwidth Range (Hz)	Symbol
0.001 to 999	H
1.00 k to 999 k	K
1.00 M to 999 M	M
1.00 G to 999 G	G

Examples are:

Bandwidth	Symbol
0.003 Hz	H003
0.2 Hz	H200
36.4 Hz	36H4
400 Hz	400H
3.2 kHz	3K20
15.5 kHz	15K5
200.4 kHz	200K
200.5 kHz	201K
200.6 kHz	201K
20 MHz	20M0
500 MHz	500M
5.925 GHz	5G93

1. ITU, "Final Acts of the World Administrative Radio Conference, Geneva, 1979," International Telecommunications Union, Geneva, Switzerland, 1980.

The three classifying symbols are:[2]

First symbol: type of modulation of the main carrier
Second symbol: nature of signal(s) modulating the main carrier
Third symbol: type of information transmitted

First Symbol

Emission of an unmodulated carrier	N
Emission in which the main carrier is amplitude modulated (including cases where subcarriers are angle modulated)	
Double sideband	A
Single sideband, full carrier	H
Single sideband, reduced carrier	R
Single sideband, suppressed carrier	J
Independent sidebands	B
Vestigial sideband	C
Emission in which the main carrier is angle modulated	
Frequency modulation	F
Phase modulation	G
Emission of pulses [if the carrier is directly modulated by a quantized signal (e.g., PCM), the emission should be designated by one of the above (A, N, R, J, B, C, F, or G)]	
Sequence of unmodulated pulses	P
A sequence of pulses	
Modulated in amplitude	K
Modulated in width/duration	L
Modulated in position/phase	M
Emission in which the carrier is angle modulated during the period of the pulse	Q
Emission that is a combination of the foregoing or is produced by other means	V
Cases not covered above, in which emissions consist of the main carrier modulated in a combination of two or more of the following modes: amplitude, angle, pulse	W
Cases not covered	X

Second Symbol

No modulating signal	O
A single channel containing quantized or digital information without the use of a modulating subcarrier (excludes TDM)	1

2. Extracted from William A. Luther, "Classification and Designation of Emissions," *IEEE Trans. Communications*, Vol. COM-29, No. 8, pp. 1127–1129, 1981.

A single channel containing quantized or digital information with the use of a modulating subcarrier (excludes TDM)	2
A single channel containing analog information	3
Two or more channels containing quantized or digital information	7
Two or more channels containing analog information	8
Composite system with one or more channels containing quantized or digital information, together with one or more channels containing analog information	9
Cases not covered	X

Third Symbol

No information transmitted	N
Telegraphy for aural reception	A
Telegraphy for automatic reception	B
Facsimile	C
Data transmission, telemetry, telecommand	D
Telephony (including sound broadcasting)	E
Television (video)	F
Combination of the above	W
Cases not covered	X

Let us now consider some examples using the foregoing designations.

Emission	Designation
Continuous wave (CW) with no modulation	NON
Continuous-wave telegraphy (Morse code) by AM; bandwidth of 100 Hz	100H A1A
ISB amplitude modulation, commercial-quality telephony with one channel per sideband; bandwidth of 6000 Hz	6K00 B8E
Single-sideband suppressed carrier, amplitude modulation sound broadcasting, bandwidth of 4966 Hz	5K00 J3E
Commerical radio-relay system, pulse modulated in position, voice circuits; bandwidth of 8 MHz	8M00 M7E
Vestigial sideband where some of the unwanted sideband is transmitted with full carrier AM, video plus FM sound, 6-MHz bandwidth	6M00 C8W

7-3 GENERATION OF SSB

In order to generate single sideband, the carrier is first eliminated. This is accomplished by the use of a balanced modulator. The basic principle of any balanced modulator is to mix the carrier with a message signal and produce only sidebands or the sum and difference frequencies at the output. In any balanced modulator circuit there is no output when there is no message signal.

The signal produced by a balanced modulator can be called *double-sideband suppressed carrier*. Such a signal is present, for instance, in the waveform making up left plus right (L + R) portion of the FM stereo composite signal. Once the carrier is suppressed, one of the two sidebands must be eliminated to produce SSB. This is accomplished either by filtering or by a phasing technique. The SSB signal is then heterodyned with a LO frequency for conversion to a higher-frequency band. Linear power amplifiers are employed to boost the power levels to a suitable level for launching by the antenna. A simple SSB Tx block diagram is shown in Fig. 7-1.

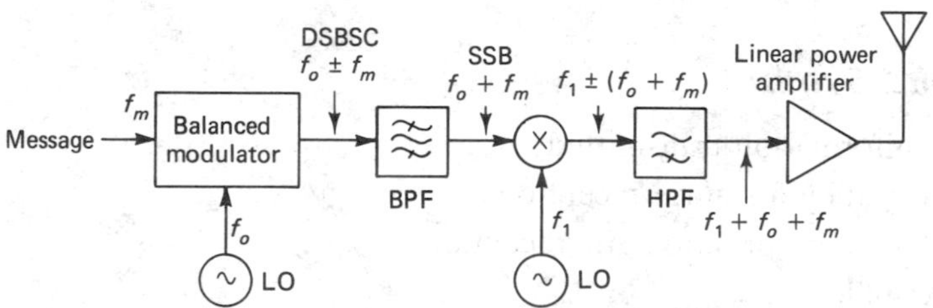

Figure 7-1 SSB transmitter.

7-3.1 Balanced Modulators

A very common type of balanced modulator that results in exceptional carrier suppression is the balanced ring modulator of Fig. 5-34. The high-amplitude carrier $V_{LO} = A \cos \omega_c t$ should be at least six to eight times the message voltage, V_{in}, for minimum distortion. The carrier alternately turns a pair of diodes off and on at a carrier frequency rate. The transformer removes any dc component in the message signal, thus preventing the application of a dc bias to the diodes.

The diodes should have low noise characteristics, closely matched forward and reverse resistance, good temperature stability, and fast switching times. To obtain high carrier rejection, small resistors of a few hundred ohms are often placed in series with each diode to obtain equal resistance in all legs when the diodes are forward biased. The hot-carrier diode is extremely fast and if formed on a common silicon chip with other diodes, can achieve extremely well matched characteristics and excellent temperature stability.

As described in some detail in Section 5-4.1, the balanced modulator generates double-sideband suppressed carrier (DSBSC) or sidebands of the message $m(t)$ around the odd harmonics of the carrier frequency. Figure 7-2(b) shows the resultant frequency spectrum.

Another balanced modulator is the IC Gilbert cell of Fig. 5-38. Such devices can also provide gain and have the major advantage of not requiring transformers. If the output is taken between pins 6 and 12, with resistors connected to these collectors and the power supply, the identical frequency response of Fig. 7-2(b) is obtained. The IC modulators tend to have poorer noise performance and carrier rejection than those of balanced ring modulators.

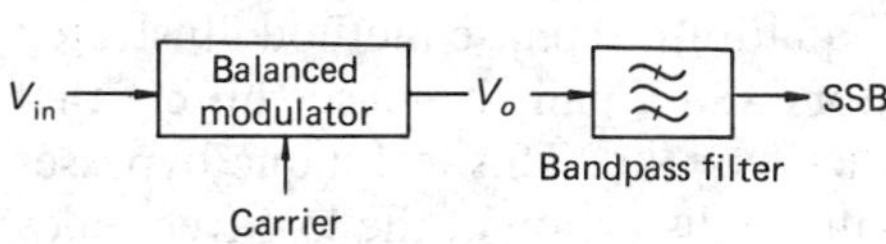

(a) Block diagram showing generation of SSB

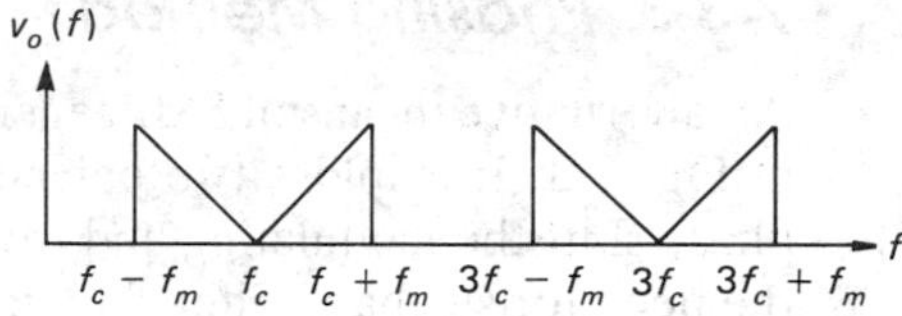

(b) Frequency spectra at the balanced modulator output

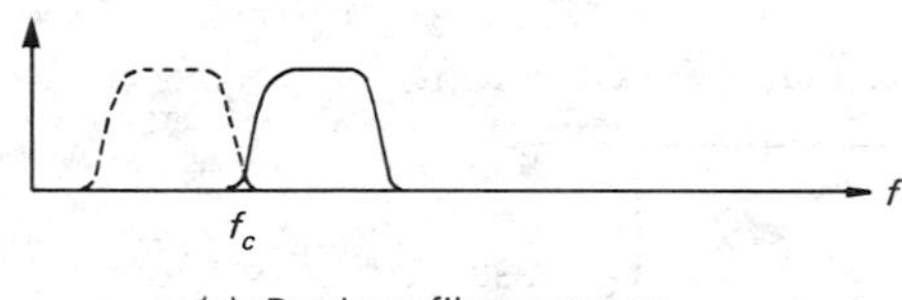

(c) Bandpass filter response

(d) SSB frequency spectrum (dashed line indicates transmission of the lower sideband; solid line indicates transmission of the upper sideband)

Figure 7-2 Generation of SSB employing the filter technique.

7-3.2 Filter Method

SSB is obtained by placing a bandpass filter having the characteristic shown in Fig. 7-2(c) after the balanced modulator. Figure 7-2(d) shows the resultant spectrum. This figure also illustrates the difficulties experienced when low frequencies are present, as any real filter does not have infinitely steep skirts. The lower-frequency components of the desired passband undergo greater attenuation than the mid- and high-frequency components if the filter is to reject the undesired passband. SSB has poor low-frequency response unless a portion of the undesired sideband is also present. For this reason, SSB is restricted primarily to voice communications, which have very little low frequency content. If facsimile or video is to be transmitted, both of which have significant low-frequency information, vestigial sideband is employed. In the later case, vestiges of the "other" sideband are also transmitted, but compensated by an equivalent attenuation of the lower frequencies in the main sideband.

The bandwidth of the SSB filter depends on the specific application. For voice communications it may range from 2 to 4 kHz, whereas for radio telegraphy (TTY) it may be only 100 Hz.

The type of bandpass filter used depends largely on the shape factor required. As noted in Section 2-9, mechanical filters exhibit the best shape factor, but at the expense of high cost. It is important that there be excellent isolation between the input and output so that sideband suppression will be determined by the shape factor rather than by signal leakage.

7-3.3 Phasing Method

An alternative means of SSB generation is the quadrature phase method illustrated in Fig. 7-3. It employs two balanced modulators and a pair of phase shifters that phase shift the modulating and carrier frequencies by 90°. This technique bypasses the need for sideband filters but generally results in distortion of the low-frequency modulating components because of the difficulty in designing a phase shifter that can produce a constant 90° phase shift over the entire audio-frequency range.

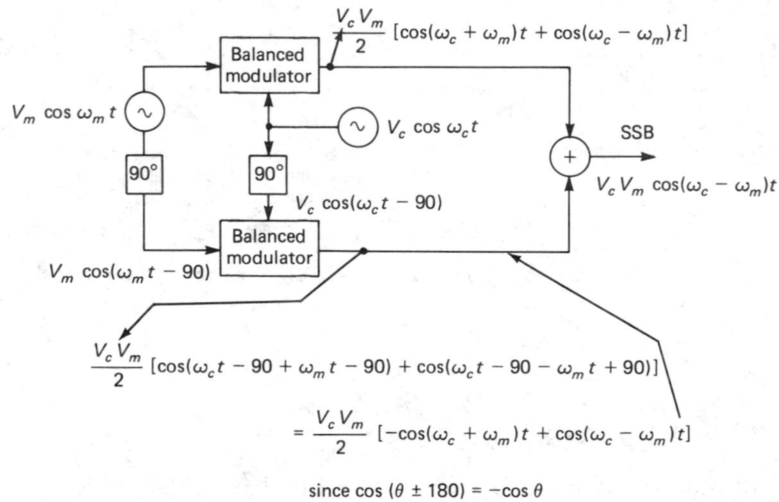

Figure 7-3 Phase-shift SSB modulator.

Figure 7-3 includes the expressions for the voltages at various points in the system. The balanced modulator gives an output voltage which is the multiplication of the inputs. The system as shown yields the lower sideband. Following a similar analysis, the upper sideband will be obtained if one of the phase shifters is moved to the opposite balanced modulator input. The major advantage of the phasing system is that the SSB signal can be generated directly at the operating frequency. No heterodyning is required.

Two popular networks that have been designed to produce two equal-amplitude signals in phase quadrature over the audio range 300 to 3000 Hz are the Norguard and Dome networks of Fig. 7-4. The outputs should observe at least 1-MΩ loads to maintain excellent carrier suppression. The amplitude and phase response of these networks, assuming ideal component values and input drive ratios, are also shown in Fig. 7-4. The shaded areas indicate the carrier suppression limit of 30 dB, and the audio phase-shift error should not encroach within this area. It should be noted that the carrier suppression is slightly less than 30 dB in the range 600 to 910 Hz for the Dome network.

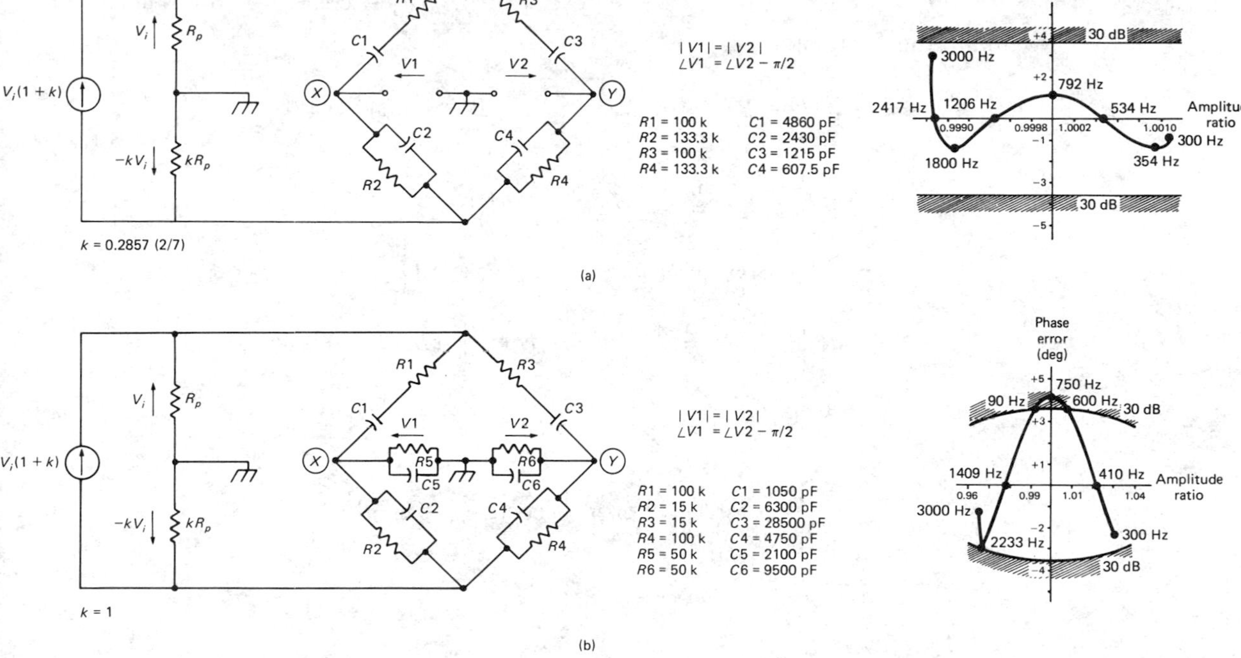

Figure 7-4 Norguard (a) and Dome networks (b), together with their error responses. (From A. I. Wade, Design criteria for SSB phase-shift networks. *Ham Radio*, June 1970.)

7-4 INTERMODULATION DISTORTION PRODUCTS

When a linear power amplifier is driven too hard, that is, when the applied signal is too large or the number of channels in an FDM system is excessive, the amplifier begins to operate in the nonlinear region. The initial effect of the nonlinearity shows up due to the square-law terms of the transfer function. Fortunately, the resultant frequency products fall outside the channel passband and therefore are readily removed. When the amplifier or repeater is driven such that the third-order terms must be included in the transfer function, products appear that fall into the passband. These are known as the third-order intermodulation products. These undesirable frequencies usually determine the maximum drive or the maximum output power of a stage.

To see how these intermodulation products are derived, consider a transfer function:

$$v_o = a_1 v_i + a_2 v_i^2 + a_3 v_i^3 \tag{7-1}$$

Let the input signal v_i consist of two components, that is,

$$v_i = A \cos \omega_a t + B \cos \omega_b t$$

Then, after employing the trigonometric identity $\cos a \cos b = \frac{1}{2}\cos(a+b) + \frac{1}{2}\cos(a-b)$ several times on the output signal,

$$v_o = a_1(A \cos \omega_a t + B \cos \omega_b t) + a_2(A \cos \omega_a t + B \cos \omega_a t)^2 + a_3(A \cos \omega_a t + B \cos \omega_b t)^3$$

we obtain the expression

$$\begin{aligned} v_o = {} & \frac{a_2}{2}(A^2 + B^2) + a_1(A \cos \omega_a t + B \cos \omega_b t) \\ & + a_2 \left\{ \frac{A^2}{2} \cos 2\omega_a t + \frac{B^2}{2} \cos 2\omega_b t \right. \\ & \left. + AB[\cos(\omega_a + \omega_b)t + \cos(\omega_a - \omega_b)t] \right\} \\ & + a_3 \left\{ \frac{3A}{4}(A^2 + 2B^2) \cos \omega_a t + \frac{3B}{4}(B^2 + 2A^2) \cos \omega_b t \right. \\ & + \frac{A^3}{4} \cos 3\omega_a t + \frac{B^3}{4} \cos 3\omega_b t \\ & + \frac{3A^2B}{4}[\cos(2\omega_a + \omega_b)t + \cos(2\omega_a - \omega_b)t] \\ & + \frac{3AB}{4}[\cos(2\omega_b + \omega_a)t \\ & \left. + \cos(2\omega_b - \omega_a)t] \right\} \end{aligned} \tag{7-2}$$

From the equations, the intermodulation frequencies are the $2f_a - f_b$ and $2f_b - f_a$ terms. These are very close to the desired channel frequencies f_a and f_b. These cannot be removed as they are within the passband; they are both in-band frequencies.

To illustrate this, consider the two-tone test outlined by the Canadian Department of Communications for use on mobile SSB radio telephone transmitters operating in the band from 1605 to 28,000 kHz: Radio Standards Specification RSS-123, Issue 1. Two tones of equal amplitudes and frequencies, 1 kHz and 1.6 kHz, when applied to the transmitter, are to result in equal-amplitude radio-frequency signals; and the highest-amplitude odd-order intermodulation product must be 26 dB below the level of either of the two test tones for the peak envelope power (PEP) up to 100 W. If we operate on channel 1, 26.900 MHz, we expect the USB passband to be from 26.900 to 26.905 MHz. The desired RF frequencies will be

$$f_a = f_c + 1 \text{ kHz} = 26.900 + 1 \text{ kHz} = 26{,}901 \text{ kHz}$$

$$f_b = f_c + 1.6 \text{ kHz} = 26.900 + 1.6 \text{ kHz} = 26{,}901.6 \text{ kHz}$$

The undesired "odd-order difference frequency intermodulation products" are

$$2f_a - f_b = 2f_c + 2 \text{ kHz} - (f_c + 1.6 \text{ kHz}) = f_c + 0.4 \text{ kHz} = 26{,}901.4 \text{ kHz}$$

$$2f_b - f_a = 2f_c + 3.2 \text{ kHz} - (f_c + 1 \text{ kHz}) = f_c + 2.2 \text{ kHz} = 26{,}902.0 \text{ kHz}$$

Both these products fall well within the passband.

When displayed on a spectrum analyzer, the IMD products appear similar to that shown in Fig. 7-5. The higher-order products are the result of even higher odd-order terms in the transfer function. In our example the decibel difference between the third-order IM product and test tone must be greater than or equal to 26 dB. The two-tone test is actually a test of the linearity of the SSB amplifiers.

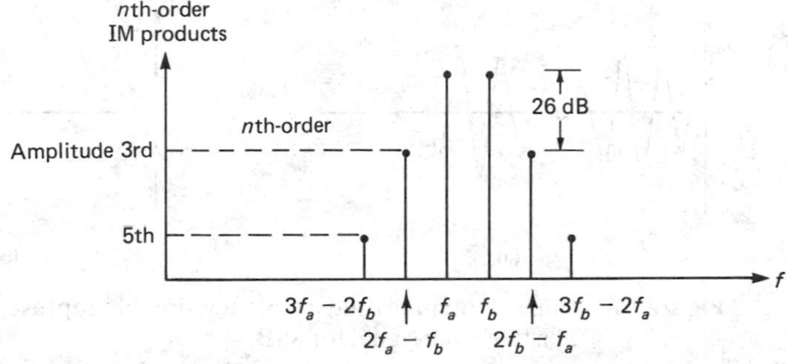

Figure 7-5 Output frequency spectrum showing desired IM products.

One should note from equation (7-2) that for equal-amplitude test frequencies, where $A = B$, the amplitudes of the desired frequencies vary as A and the intermodulation products as A^3. Thus if the two tones are increased by x dB, the third-order intermodulation products increase by $3x$ dB; that is, the ratio is

$$\frac{20 \log A^3}{20 \log A} = \frac{60 \log A}{20 \log A} = 3$$

This indicates that the third-order IM products increase at triple the rate of the desired test tone signals. When the difference between the IM products and test tone reduces to 26 dB (see Fig. 7-5), the maximum output power rating (PEP) of the unit is reached according to RSS-123 specification for transmitters producing less than 100 W. For higher output ratings, this difference must be even greater.

7-5 PEAK ENVELOPE AND AVERAGE POWER

Because the peak voltage swings introduce the maximum IM distortion products, a closely related term, the peak envelope power (PEP) rating of a transmitter is often given. PEP is a good specification for any transmitter, whether it operates as SSB, DSB, or AM. PEP indicates the maximum output capability of the transmitter (and thus the highest permissible voltage swing) before distortion becomes intolerable. PEP is the average power generated over one carrier cycle when the modulation envelope is at its maximum amplitude. PEP is normally larger than the average power (P_{avg}) transmitted. P_{avg} is determined by averaging the instantaneous power transmitted over one or more complete cycles of the modulation envelope.

A two-tone signal is usually employed to determine the PEP rating of a SSB transmitter. The Canadian Department of Communications recommends that two equal-amplitude tones of different (nonharmonic) frequencies be input to the SSB transmitter. Figure 7-6 shows the resulting SSB signal in both time and frequency domain.

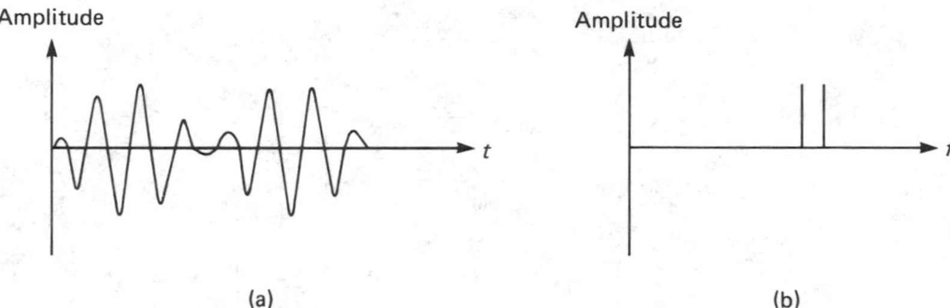

Figure 7-6 Time-domain (a) and frequency-domain representation (b) of a two-tone signal for SSB.

The null in the time-domain display is obtained when the two tones are exactly opposite in phase, and the maximum is obtained when the two tones are in phase. If E_{1m} and E_{2m} represent the peak voltages of the two tones, the peak envelope voltage (PEV) is given by $E_{1m} + E_{2m}$. This could also be called the peak instantaneous voltage.

The corresponding rms peak envelope voltages (not true rms) is given by

$$\text{PEV}_{\text{rms}} = \frac{\text{PEV}}{\sqrt{2}} = \frac{E_{1m} + E_{2m}}{\sqrt{2}} \tag{7-3}$$

This results in a peak envelope power, PEP, of

$$\text{PEP} = \frac{\text{PEV}^2_{\text{rms}}}{R} = \frac{[(E_{1m} + E_{2m})/\sqrt{2}]^2}{R} \tag{7-4}$$

where R is the load resistance.

For equal tones having rms values of E ($E = E_{1m}/\sqrt{2} = E_{2m}/\sqrt{2}$), respectively, the PEP will be

$$\text{PEP} = \frac{(E + E)^2}{R} = \frac{4E^2}{R} \tag{7-5}$$

The corresponding true or average power will be

$$P_{\text{avg}} = \frac{E^2_{1m}}{2R} + \frac{E^2_{2m}}{2R} \tag{7-6}$$

$$= \frac{E^2}{R} + \frac{E^2}{R} = \frac{2E^2}{R} \tag{7-7}$$

Thus $P_{\text{avg}}/\text{PEP} = \frac{1}{2}$ for two equal tones.

For N synchronized tones, the peak envelope power increases and the main peaks become narrower with steeper sides, as shown in Fig. 7-7. For N equal tones,

$$\frac{P_{\text{avg}}}{\text{PEP}} = \frac{1}{N} \tag{7-8}$$

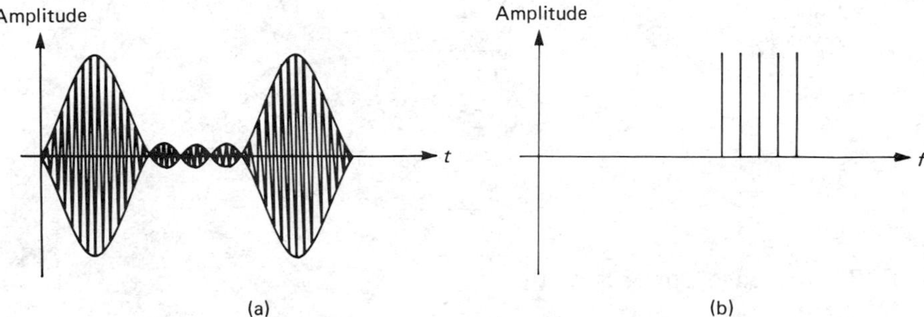

Figure 7-7 Time (a) and frequency (b) domain representations of a five-tone SSB signal.

To limit high peaks of the modulation envelope and thereby the IM products, automatic load control (ALC) or automatic drive control (ADC) is often employed in the final stages of a transmitter. ALC samples the output envelope and generates a bias when the envelope crest exceeds a set valuc. This bias is then routed to a previous stage to reduce the drive.

To further amplify the concept of PEP, let us consider an AM signal. A sinusoidally modulated AM signal consists of three frequency components.

$$e(t) = E_c \cos \omega_c t + \frac{mE_c}{2} \cos(\omega_c + \omega_m)t + \frac{mE_c}{2} \cos(\omega_c - \omega_m)t \tag{7-9}$$

where E_c is the peak carrier voltage, m is the modulation index, $\omega_c = 2\pi f_c$ (the carrier frequency), $\omega_c + \omega_m = 2\pi(f_c + f_m)$ (the USB frequency), and $\omega_c - \omega_m = 2\pi(f_c - f_m)$ (the LSB frequency). The peak envelope voltage is thus

$$\text{PEV} = E_c + \frac{mE_c}{2} + \frac{mE_c}{2} = E_c(1 + m) \tag{7-10}$$

giving a PEV_{rms} equivalent of (not true rms)

$$\text{PEV}_{\text{rms}} = \frac{E_c(1 + m)}{\sqrt{2}} \tag{7-11}$$

Thus the peak envelope power is

$$\text{PEP} = \frac{[E_c(1 + m)/\sqrt{2}]^2}{\text{R}} \tag{7-12}$$

The average power is directly related to the square of the total rms voltage. From general theory,

$$V_{\text{rms}}^2 = \sqrt{V_{1(\text{rms})}^2 + V_{2(\text{rms})}^2 + V_{3(\text{rms})}^2} \tag{7-13}$$

So from equation (7-10) the total rms voltage is

$$\begin{aligned} V_{\text{rms}}^2 &= \sqrt{\frac{(E_c)^2}{2} + \frac{1}{2}\left(\frac{mE_c}{2}\right)^2 + \frac{1}{2}\left(\frac{mE_c}{2}\right)^2} \\ &= \sqrt{\frac{E_c^2}{2} + \frac{m^2E_c^2}{4}} \\ &= \sqrt{\frac{2 + m^2}{4}E_c^2} \end{aligned} \tag{7-14}$$

The average power is written as

$$\begin{aligned} P_{\text{avg}} = P &= \frac{V_{\text{rms}}^2}{R} \\ &= \frac{E_c^2(2 + m^2)}{4R} \end{aligned} \tag{7-15}$$

EXAMPLE 7-1

Find the PEV_{rms}, PEP, and P_{avg} of the AM waveform shown in Fig. 7-8. Assume that $R = 50\ \Omega$.

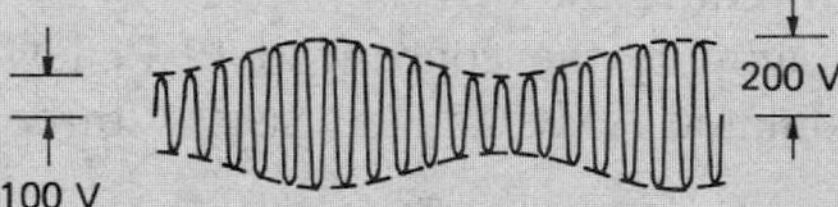

Figure 7-8.

Solution

$$E_c = 150\text{ V}$$

$$m = \frac{50}{150} = 0.333$$

$$\text{PEV}_{\text{rms}} = \frac{200}{\sqrt{2}}\text{ V}$$

$$\text{PEP} = \frac{(200/\sqrt{2})^2}{50} = 400\text{ W}$$

$$P_{\text{avg}} = \frac{150^2(2 + 0.33^2)}{4 \times 50} = 237.5\text{ W}$$

EXAMPLE 7-2

Find the PEV_{rms}, PEP, and P_{avg} of the three-tone signal (equal-amplitude tones) shown in Fig. 7-9. $R = 50\ \Omega$.

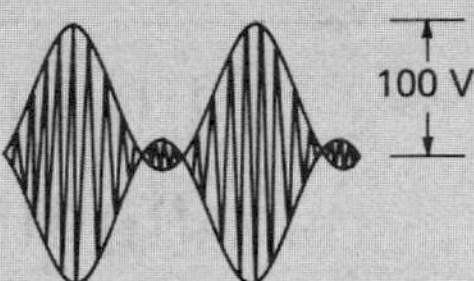

Figure 7-9.

Solution

$$PEV_{rms} = \frac{100}{\sqrt{2}}\ \text{V}$$

$$PEP = \frac{(100/\sqrt{2})^2}{50} = 100\ \text{W}$$

$$P_{avg} = \frac{PEP}{3} = 33.3\ \text{W}$$

EXAMPLE 7-3

Find the PEV_{rms}, PEP, and P_{avg} of the pulses continuous-wave (CW) signal shown in Fig. 7-10. $R = 50\ \Omega$.

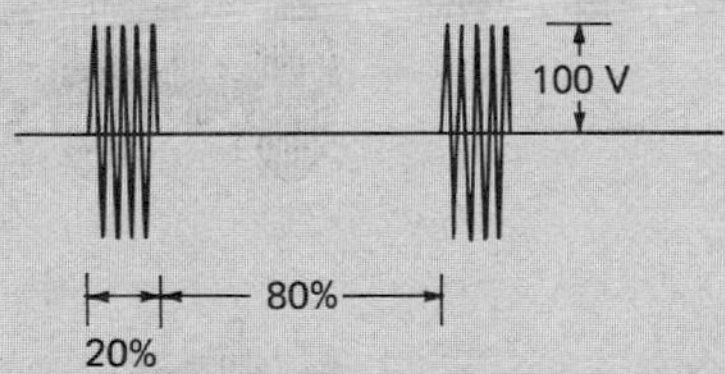

Figure 7-10.

Solution

$$PEV_{rms} = \frac{100}{\sqrt{2}}\ \text{V}$$

$$PEP = 100\ \text{W}$$

$$P_{avg} = \frac{20}{100} \times \frac{(100/\sqrt{2})^2}{50} = 20\ \text{W}$$

A waveform summary is shown in Table 7-1.

TABLE 7-1 Waveform Summary[a]

	Transmission Type and Scope Pattern	Frequency Spectrum (C: Carrier)
Table A CW	100 V	C
Table B AM 100% Mod.	200 V	C
Table C AM 73% Mod.	173 V	C
Table D SSB 1 tone	100 V	(C)
Table E SSB 2 tone	100 V	(C)
Table F SSB 3 tone	100 V	(C)
Table G SSB Voice	100 V	(C)
Table H Pulse	100 V 10% 90%	C

[a]Correlation of peak envelope power (PEP), carrier power (P_c), average heating power (P_{avg}), and percent modulati of AM signals of Table A, B, or C above.
$Z_0 = 50\ \Omega$. Carrier (or suppressed carrier) PEV was arbitrarily chosen at 100 V in all examples. $PEV_{rms} = PEV/\sqrt{2}$
Source: Bird Electronic Corporation, Bird Catalog GC-68, p. 57.

PEV_{rms} (arbitrary) (V)	PEP = PEV^2_{rms}/Z_0 (W)	Average (Heating) Power (W)	4311 in Peak Mode (W)	4311 in CW Mode (W)
$\frac{100}{\sqrt{2}}$	100	100	100	100
$\frac{200}{\sqrt{2}}$	400	150	400	100
$\frac{173}{\sqrt{2}}$	300	127	300	100
$\frac{100}{\sqrt{2}}$	100	100	100	100
$\frac{100}{\sqrt{2}}$	100	50	100	40.5
$\frac{100}{\sqrt{2}}$	100	33.3	100	—
$\frac{100}{\sqrt{2}}$	100	—	100	—
$\frac{100}{\sqrt{2}}$	100	10	100	—

7-6 DETECTION OF SSB

The SSB receiver is very similar to an AM superheterodyne receiver. Since the carrier is not present in a SSB signal, it must be reinserted at the receiver to generate the original modulating signal. As shown in the SSB receiver block diagram illustrated in Fig. 7-11, a carrier frequency is inserted into the "product detector" by the beat frequency oscillator (BFO). This carrier frequency could have been inserted at an earlier point in the receiver.

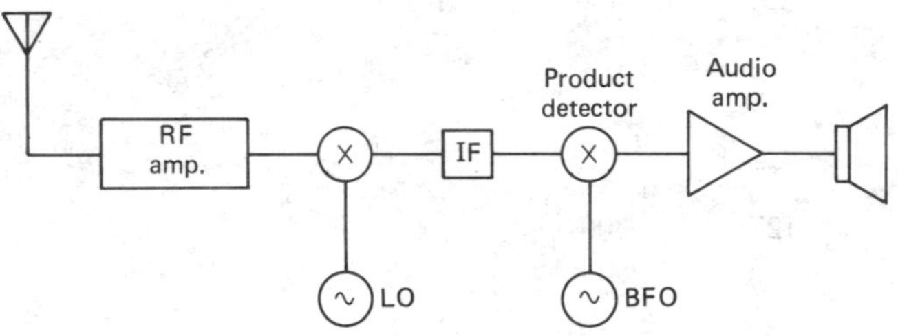

Figure 7-11 SSB receiver block diagram.

The product detector is another name for a mixer that is used to translate either the USB or LSB signal to baseband. The applied BFO frequency must be tightly controlled and be within about 50 Hz of the optional frequency to prevent the received voice signal from sounding like Donald Duck. This mixer could be any one of the mixers discussed in earlier chapters. If the receiver does not need to cover a complete band of frequencies, the BFO could well be a crystal oscillator. Even then, if the frequency of the transmitted signal is slightly in error or the LO drifts, a frequency adjustment control is usually available to the operator. This clarifier consists of a small variable capacitor in one of the oscillator circuits.

To illustrate the action of the product detector, consider a USB from 7.7985 to 7.8015 MHz as applied to a product detector, together with a 7.7985-MHz local oscillator signal. These numbers come from the SSB transceiver discussed in the next section. Ignoring all the frequency products except the sum and difference frequencies, we obtain at the output, frequency bands of 0 to 3 kHz and 15.597 to 15.6000 MHz. A simple *RC* low-pass filter readily rejects the sum frequencies, leaving only the desired baseband. In this case, the output signal remains erect, as shown in Fig. 7-12(a).

If the local oscillator is above the sideband signal, as shown in Fig. 7-12(b), inversion takes place. In this case the sideband was inverted by the transmitter and therefore needs to again be inverted in the receiver. A circuit that enables one to generate a DSB or SSB signal, and that can also act as a product detector, is shown in Fig. 7-13. If a modulating signal is applied to pin 7 and a carrier is applied to pin 6, a DSB signal can be obtained at pin 8. If the carrier frequency is adjusted to suit the ceramic filter passband, a SSB signal can be obtained at pin 14 or 16. The USB or LSB signal is passed by the filter.

If, on the other hand, a SSB signal is applied to pin 7 and the carrier or local oscillator signal is applied to pin 6, the sum and difference frequencies can be observed at pin 8. The baseband signal can be obtained at pin 10.

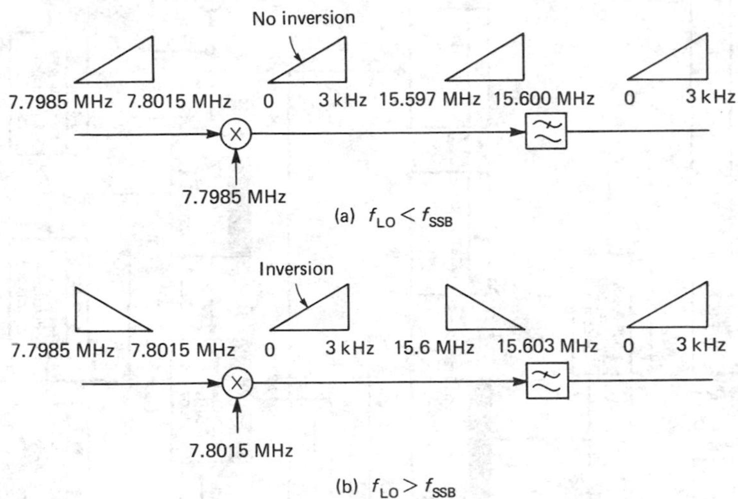

Figure 7-12 SSB detection circuit illustrating inverted and erect products.

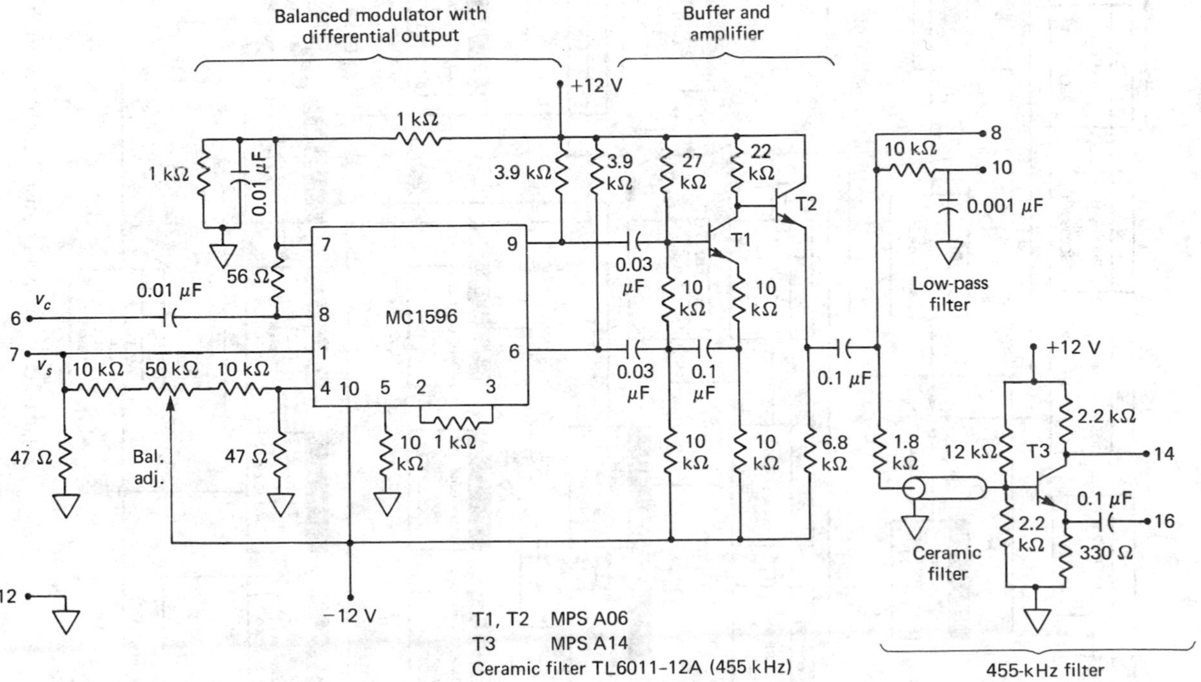

Figure 7-13 Experimental circuit that can act as a DSB or SSB modulator, or product detector.

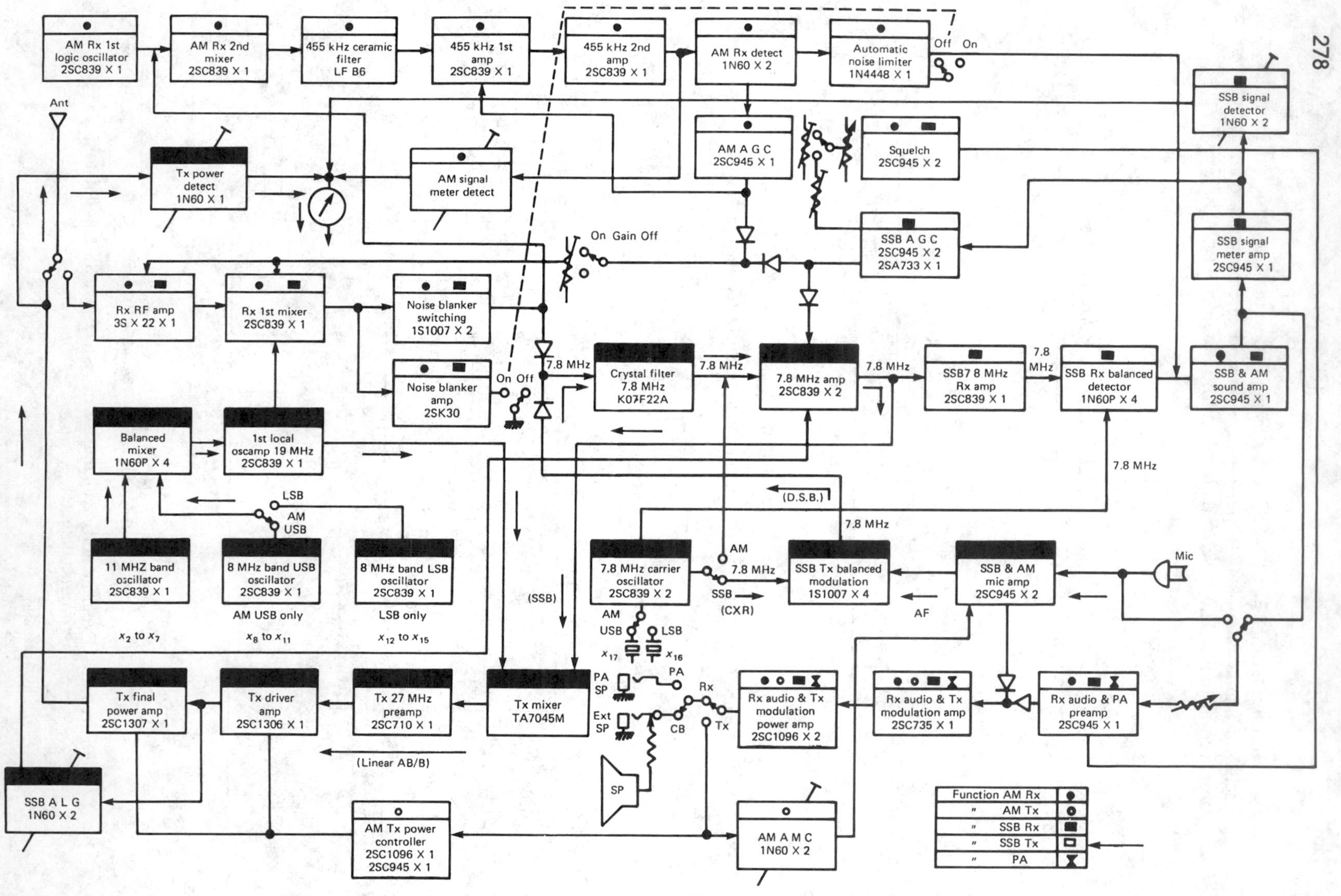

Figure 7-14 Spartan transmitter block diagram. (Courtesy of Resdel Industries.)

7-7 SPARTAN COURIER 25 CHANNEL CB TRANSCEIVER

The Spartan SSB transceiver is designed to operate in the 27-MHz Canadian GRS band under M.O.C. RSS-136. This transceiver can operate in three modes: AM (6K00A3E), USB, and LSB (3K00J3E). The transceiver operates from a companion 13.8-V power supply. The transmitter operates at the rated output power of 4 W in the AM mode and at 12 W PEP output in the SSB mode. The odd-order difference intermodulation is 25 dB or better down from the two-tone test level. All the spurious harmonics are rated at 60 dB down from the two-tone test signals on the carrier, depending on the transmitting mode. We will consider only the SSB mode in our discussions on the Spartan.

Transmitter Block Diagram

The first local oscillator frequency is generated by two sets of crystal oscillators, one for USB and another for LSB, as shown in Fig. 7-14. The first local oscillator frequency is synthesized by the combination of crystals X2 through X7, X8 through X11 for USB, and X2 through X7, and X12 through X15 for LSB. X17 is used for the carrier oscillator for USB, whereas X16 is used for generating LSB. The X16 or X17 frequency is mixed with the audio signal in a balanced modulator to obtain a DSB signal. The 7.8-MHz crystal permits only one of the sidebands to pass through unattenuated, depending on the selection of the X16, X17 crystals. The resultant SSB signal is further unconverted in frequency by mixing with the synthesized LO frequency. The 27-MHz SSB signal is then amplified and radiated into space.

Receiver Block Diagram

The receiver is a single conversion (7.8-MHz IF) receiver in the SSB mode. The local oscillator frequency for the IF is again generated by two sets of crystal oscillators, one for USB and another for LSB, as shown in Fig. 7-15. Demodulation is obtained by a balanced demodulator where a carrier is inserted at 7.7985 MHz for USB reception and 7.8015 MHz for LSB reception. Local oscillator frequencies for all channels are synthesized by the combination of X2 through X7 oscillators, with X8 through X11 oscillators for USB, and X2 through X7 oscillators with X12 through X15 oscillators for LSB.

The required selectivity is provided by the crystal filter. It is a crystal lattice having a bandwidth of 2.1 kHz at 6 dB and 5.5 kHz at 60 dB attenuation. The receiver also has a noise blanker, a squelch circuit, and an AGC circuit.

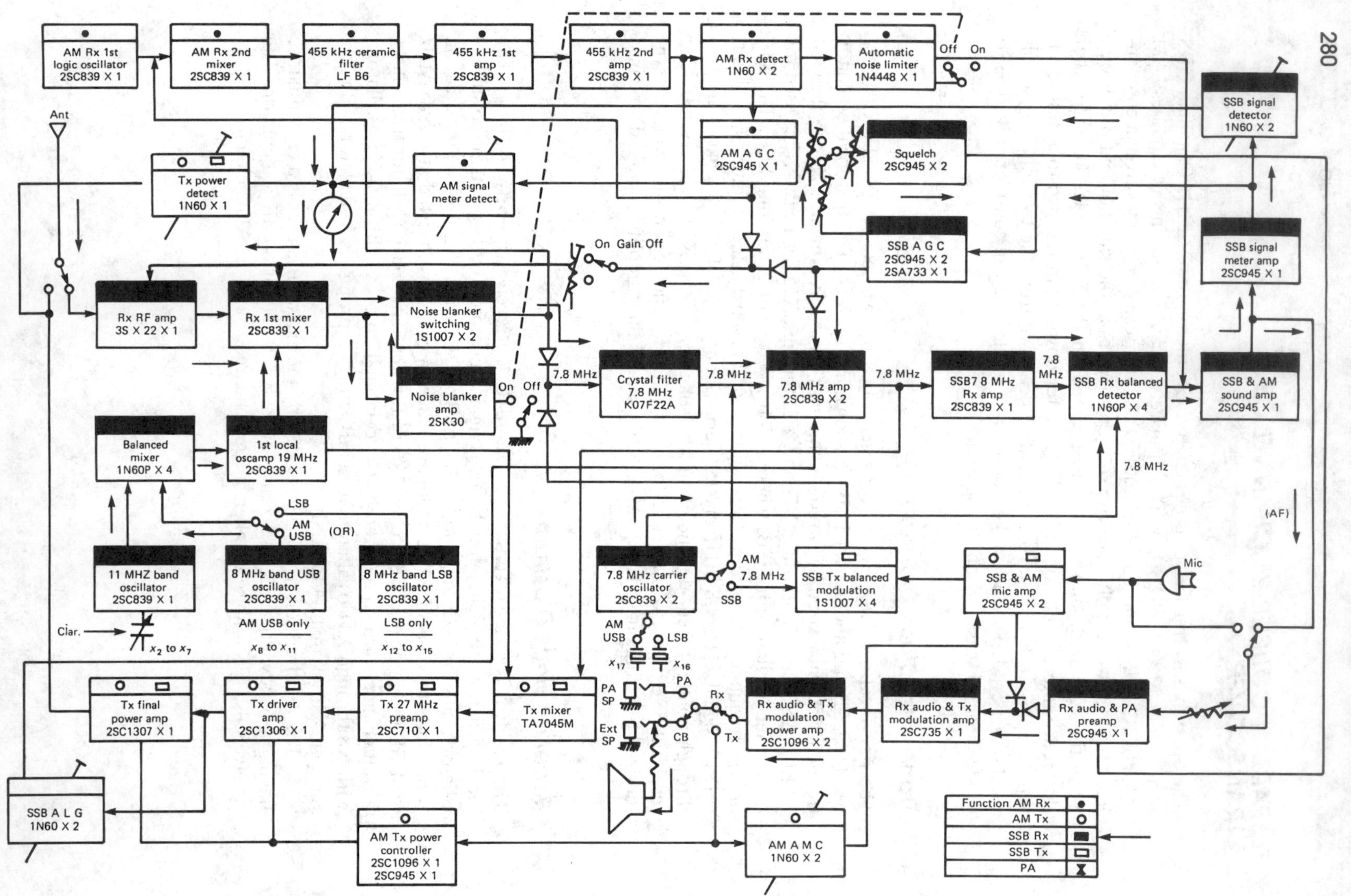

Figure 7-15 Spartan receiver block diagram. (Courtesy of Resdel Industries.)

Frequency Synthesis

To understand how the USB or LSB frequency bands are obtained, we will derive the various frequencies for channel 1 (26.965 MHz) using Tables 7-2 and 7-3, which specify the various crystal frequencies and their combinations. Table 7-2 lists the crystal frequencies and Table 7-3 indicates the crystal combinations for each channel.

Figure 7-16 shows the various oscillator frequencies and signal frequency bands at the various locations in the Spartan unit when using channel 1 in either the USB or LSB mode. Notice that when transmitting LSB, inversion takes place. This is rectified in the receiver by injecting a carrier frequency into the product detector that is greater than the incoming LSB signal frequency.

TABLE 7-2 Spartan Courier Crystal Frequencies

Crystal	Frequency (MHz)
X2	11.805000
X3	11.855000
X4	11.905000
X5	11.955000
X6	12.005000
X7	12.055000
X8	7.361500
X9	7.371500
X10	7.381500
X11	7.401500
X12	7.358500
X13	7.368500
X14	7.378500
X15	7.398500
X16	7.801500
X17	7.798500

For a more detailed description of the CB transceiver, consider the schematic diagram shown in Fig. 7-17. We will initially discuss the transmitter section located in the central and lower portions of the schematic and then the SSB receiver section, the central and upper portions of the schematic. We will not deal with the AM receiver located in the upper right portion of the circuit diagram or with the AM transmitter. The 13.8-V dc supplies power for either the transmitter or receiver, depending on the switch S3-4, S3-5 and S1-3 settings.

TABLE 7-3 Crystal Frequency Combinations for Spartan Courier Transceiver

Channel		Transmitter		Receiver		Channel		Transmitter		Receiver	
		USB	LSB	USB	LSB			USB	LSB	USB	LSB
1	×2	11.805	11.805	11.805	11.805	13	×5	11.955	11.955	11.955	11.955
	× 16 or × 17	7.7985	7.8015	7.7985	7.8015			7.7985	7.8015	7.7985	7.8015
	× 8 or × 12	7.3615	7.3585	7.3615	7.3585			7.3615	7.3585	7.3615	7.3585
2		11.805	11.805	11.805	11.805	14		11.955	11.955	11.955	11.955
		7.7985	7.8015	7.7985	7.8015			7.7985	7.8015	7.7985	7.8015
	× 9 or × 13	7.3715	7.3685	7.3715	7.3685			7.3715	7.3685	7.3715	7.3685
3		11.805	11.805	11.805	11.805	15		11.955	11.955	11.955	11.955
		7.7985	7.8015	7.7985	7.8015			7.7985	7.8015	7.7985	7.8015
	× 10 or × 14	7.3815	7.3785	7.3815	7.3785			7.3815	7.3785	7.3815	7.3785
4		11.805	11.805	11.805	11.805	16		11.955	11.955	11.955	11.955
		7.7985	7.8015	7.7985	7.8015			7.7985	7.8015	7.7985	7.8015
	× 11 or × 15	7.4015	7.3985	7.4015	7.3985			7.4015	7.3985	7.4015	7.3985
5	×3	11.855	11.855	11.855	11.855	17	×6	12.005	12.005	12.005	12.005
		7.7985	7.8015	7.7985	7.8015			7.7985	7.8015	7.7985	7.8015
		7.3615	7.3585	7.3615	7.3585			7.3615	7.3585	7.3615	7.3585
6		11.855	11.855	11.855	11.855	18		12.005	12.005	12.005	12.005
		7.7985	7.8015	7.7985	7.8015			7.7985	7.8015	7.7985	7.8015
		7.3715	7.3685	7.3715	7.3685			7.3715	7.3685	7.3715	7.3685
7		11.855	11.855	11.855	11.855	19		12.005	12.005	12.005	12.005
		7.7985	7.8015	7.7985	7.8015			7.7985	7.8015	7.7985	7.8015
		7.3815	7.3785	7.3815	7.3785			7.3815	7.3785	7.3815	7.3785
8		11.855	11.855	11.855	11.855	20		12.005	12.005	12.005	12.005
		7.7985	7.8015	7.7985	7.8015			7.7985	7.8015	7.7985	7.8015
		7.4015	7.3985	7.4015	7.3985			7.4015	7.3985	7.4015	7.3985
9	×4	11.905	11.905	11.905	11.905	21	×7	12.055	12.055	12.055	12.055
		7.7985	7.8015	7.7985	7.8015			7.7985	7.8015	7.7985	7.8015
		7.3615	7.3585	7.3615	7.3585			7.3615	7.3585	7.3615	7.3585
10		11.905	11.905	11.905	11.905	22		12.055	12.055	12.055	12.055
		7.7985	7.8015	7.7985	7.8015			7.7985	7.8015	7.7985	7.8015
		7.3715	7.3685	7.3715	7.3685			7.3715	7.3685	7.3715	7.3685
11		11.905	11.905	11.905	11.905	23		12.055	12.055	12.055	12.055
		7.7985	7.8015	7.7985	7.8015			7.7985	7.8015	7.7985	7.8015
		7.3815	7.3785	7.3815	7.3785			7.4015	7.3985	7.4015	7.3985
12		11.905	11.905	11.905	11.905						
		7.7985	7.8015	7.7985	7.8015						
		7.4015	7.3985	7.4015	7.3985						

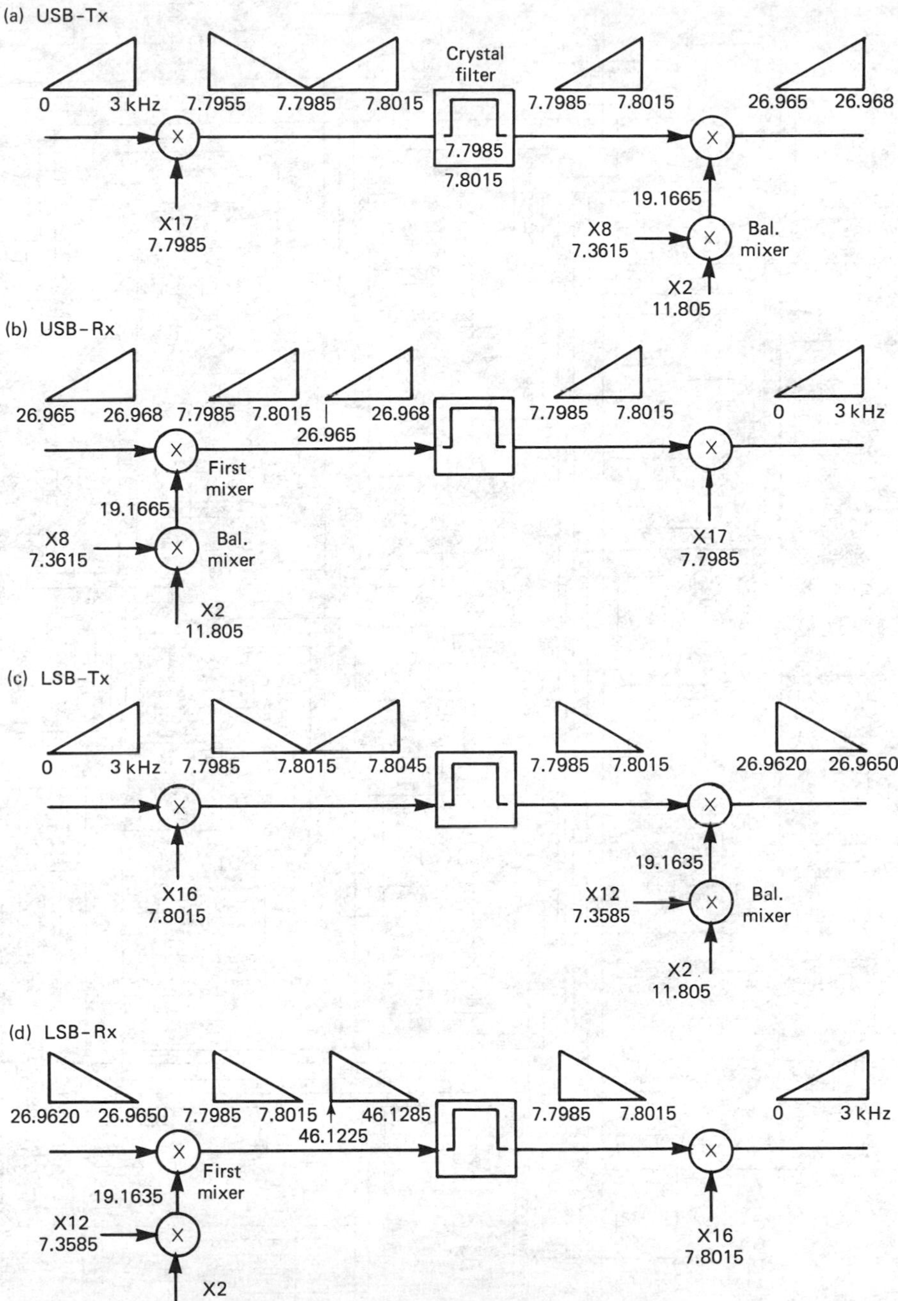

Figure 7-16 Oscillator and signal frequencies in the Spartan Courier when receiving or transmitting on channel 1, USB, or LSB.

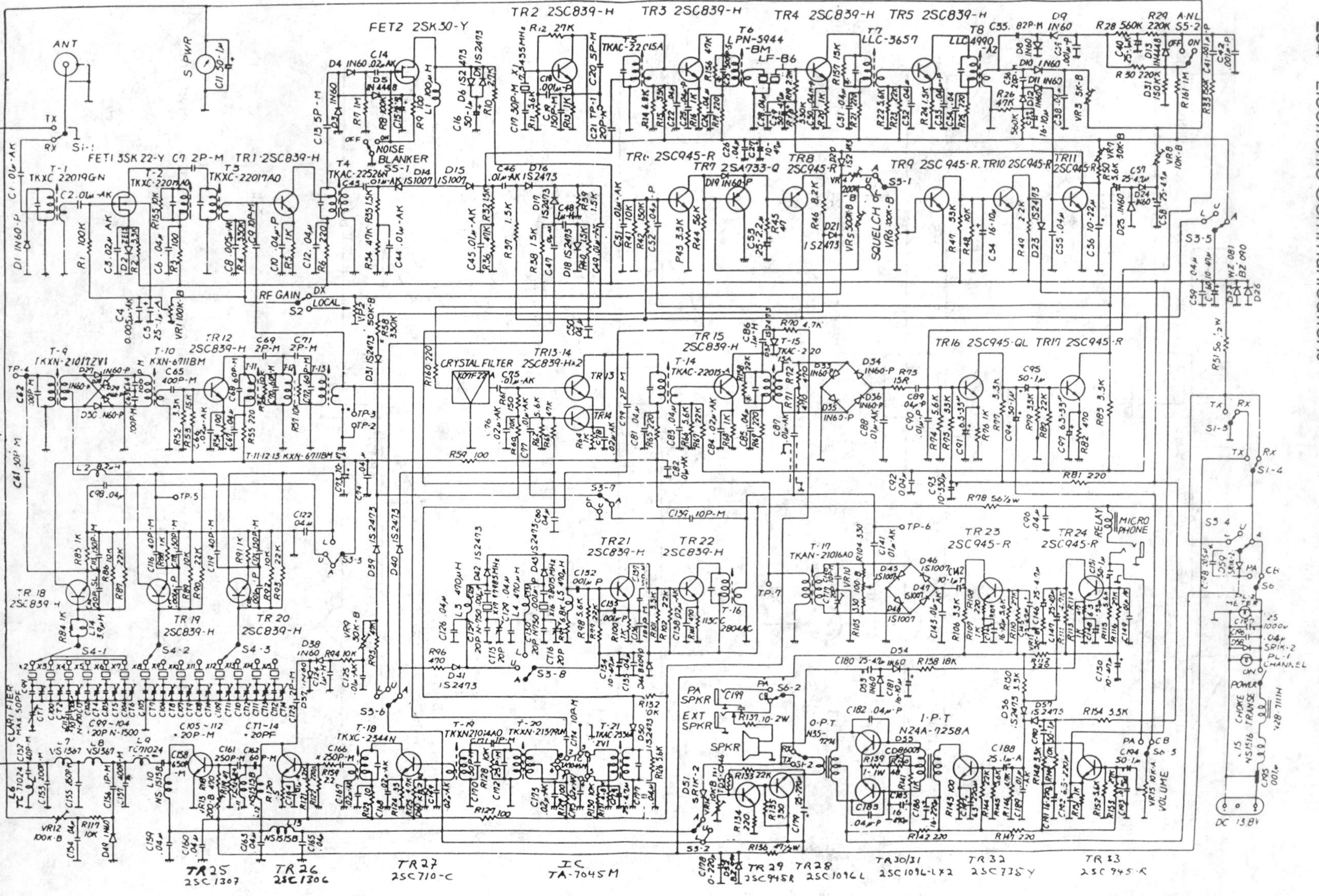

Figure 7-17 Schematic diagram of the Spartan Courier. (Courtesy of Resdel Industries.)

Transmitter (USB) Circuit Diagram

With a microphone connected to the microphone jack, the audio signal is applied to two stages of amplification (T23, T24) before being injected into a balanced modulator (D45 to D48). As the USB crystal oscillator is supplied dc power through switch S3-8, a 7.7985-MHz signal is produced and injected into the balanced modulator. The resultant DSB signal is sent through the crystal filter via diode D17, which is forward biased in the USB or LSB mode (see switch S3-6). It is common practice to use diode switching for directing ac signals. When reverse biased, the diode appears as an open circuit, preventing any signal from passing through. When forward biased, the signal can freely pass through. The crystal filter passes the USB signal (7.7985 to 7.8015 MHz), which is then supplied to the IC mixer TA-7045M via amplifier TR13. The bias to this amplifier is automatically controlled by the automatic load control (ALC) circuit discussed later.

Since both TR18 and TR19 are dc activated, two oscillator frequencies are generated. These are injected into the balanced mixer D27 to D30. All frequencies except the sum frequency are rejected by amplifier TR12 and its associated tuned circuits (T10 to T13). This sum frequency is supplied to the IC mixer TA-7045M.

After mixing with the USB signal, the sum frequencies are chosen by tuned circuits T20 and T19. The resultant frequency upconverted USB signal is then amplifier by the linear preamp, driver, and power amplifier stages. At the output of the Tx driver stage, the signal amplitude is monitored and rectified by the diode detector circuit D38/C124. The ensuing negative dc voltage is superimposed on the normal dc supply voltage and applied as a bias to TR13 via diode D39. If the USB signal is too high, running the danger of generating intermodulation products in the power amplifier stage, the increased negative voltage from the ALC circuit will reduce the forward bias on the SSB amplifier TR13, thereby reducing the drive to the Tx mixer, Tx preamplifier, and so on.

Diode detector D49/C154 monitors the output signal level for the signal power meter. The output π networks match the power amplifier output impedance to the antenna input impedance, and also behave as a LPF to eliminate any spurious high frequencies.

Receiver (USB) Circuit Diagram

After being intercepted by the antenna, the SSB signal is amplified by the relatively low noise FET amplifier (FET 1) and then mixed with the first local oscillator frequency by TR1. Assuming the noise blanker to be inoperative, the difference frequencies from tuned circuit T4 are applied to the crystal filter via diodes D14, D15, and D16. All these diodes are forward biased. If spikes or impulsive noise are present, capacitor C13 delivers the high-frequency components to the noise blanker detection circuitry D4/C14. As only the positive-going spikes are allowed to pass through diode D4 to the FET 2 gate, the FET amplifier is turned on. This grounds the dc drain voltage through resistor R9 if the noise blanker switch is on. This removes the forward bias on diodes D14 and D15, thus preventing any noisy signal from being delivered to the crystal filter.

After selecting the appropriate sideband, USB in our example, the USB is amplified by TR13 and TR15 and then injected into the balanced detector, D33 to D36. When mixed with the X17 frequency (USB), the resultant signal is filtered (LP) and delivered to an audio amplifier (TR16). The audio signal presented to the preamp, driver, and PA stages is attenuated by the volume control VR15. The output push-pull stage can drive either an internal or an external speaker.

This receiver also provides for squelch and AGC. The squelch circuit prevents the receiver from producing an audio-frequency output when very little or no signal energy is being received in the receiver passband. This is accomplished by reducing the gain of the audio preamp TR33. When a signal is absent, transistor TR10 in the squelch circuit is off, as no drive is present from the sound amplifier TR16 via transistors TR17 and TR6, switch S3-1, squelch control VR6, and transistor stage TR9. A rather high dc voltage under these conditions is applied to the emitter of the preamp, via diode D23, causing the preamp to be turned off. On the other hand, when an audio signal is present, transistor TR10 is turned on, reducing the dc voltage to the emitter of the preamp, bringing its gain back up for normal operation.

The AGC voltage is derived from the audio amplifiers TR16 and TR17 and AGC circuitry TR6/7/8. As the audio signal level increases, the dc voltage at diode D21 is reduced, thereby reducing the forward bias and therefore gain of the RF amplifier (FET 1), mixer stage (TR1), and crystal filter amplifier (TR13) via diode D31. The diode detector D24/C58 monitors the audio signal from the audio amplifier TR17. The dc voltage developed is applied to the signal meter indicator, S PWR.

Problems

7-1. Name two methods of obtaining SSB.

7-2. Name three advantages of SSB over AM.

7-3. Name two disadvantages of SSB when compared to AM.

7-4. Show the output frequency spectrum for a double-balanced modulator such as a ring-type lattice modulator (Fig. P7-4). Label the frequencies and show them relative to f_c.

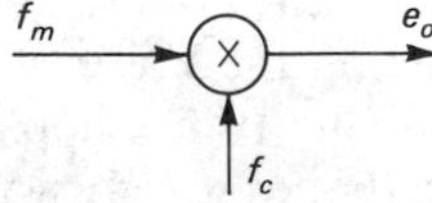

Figure P7-4.

7-5. Why might an integrated circuit balanced modulator be preferred over a discrete-circuit balanced modulator?

7-6. Given the output spectrum for a two-tone test shown in Fig. P7-6, determine the two in-band intermodulation (distortion) product frequencies, and draw them on the graph.

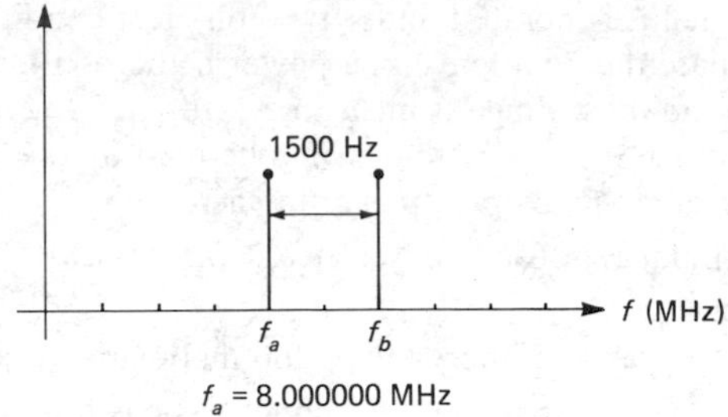

Figure P7-6.

7-7. **(a)** If f_a and f_b are increased 3 dB in amplitude, how much will the intermodulation products be increased?

(b) The intermodulation test is used to test what aspect of a linear amplifier?

7-8. For the transmission types shown in Fig. P7-8, determine the PEP, P_{avg}, and show the frequency spectrum, including the position of the carrier which you will identify. Assume a load of 50 Ω.

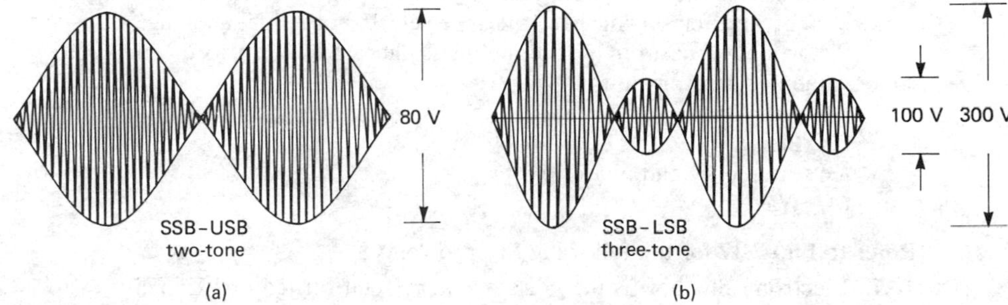

Figure P7-8.

7-9. AM radio station CFRN broadcasts a total signal power of 50 kW with a carrier frequency of 1260 kHz. Assume a modulation index of $m = 1$. *Show the formulas used.*

(a) If this AM signal is fed into a 50-Ω antenna, what minimum rms voltage amplitude must the antenna be capable of withstanding (i.e., PEV_{rms}).

(b) Assume that the station wished to reach the same number of listeners employing SSB-SC operation instead of AM. Find the saving (in watts) in power consumption that would be accomplished.

(c) If a two-tone test (two equal-amplitude tones) were used to check out the new SSB-SC operation assuming that the total output SSB power was 10 kW, what is the PEP of this signal?

7-10. A 100-V_{pp} transmitter output is applied to a 50-Ω antenna. What is PEP represented by this voltage?

7-11. If an AM radio station has a PEP of 100 kW under 100% modulation conditions, what is the average power of the carrier? Assume that $R_L = 50$ Ω.

7-12. A SSB transmitter produces a PEV of 200 V across a 50-Ω load when three equal-amplitude tones are applied. What is the average power output of this transmitter?

7-13. A SSB signal is generated under two-tone test conditions.
 (a) Calculate the peak envelope power if the oscilloscope voltage is 125 V peak-to-peak and the voltage is measured across a 50-Ω load.
 (b) Determine the power that would be read by the Bird 43 wattmeter.
 (c) Determine the average power in each tone.

7-14. What technique is used to select a USB or LSB from the output of a balanced modulator?

7-15. Why must linear amplifiers follow the modulator in an SSB transmitter?

7-16. **(a)** What type of transfer curve characteristic is required for a SSB product detector?
 (b) Give one example of a device that could be used.

7-17. If a frequency shift in the demodulated audio is observed in a SSB receiver, what circuit is adjusted to correct it?

7-18. When demodulating an USB signal, frequency inversion of the baseband is observed. What is the cause of this?

7-19. A SSB transmitter produces no measurable power output on a Bird Thruline wattmeter when no modulation is applied. Is this normal? Explain.

7-20. The power to be transmitted for an amplitude-modulated signal can be greatly reduced without reducing the information content by suppressing the carrier frequency component and reinserting it at the receiver. If both sidebands are transmitted, gross distortion will occur in the output unless the carrier is replaced with *exactly* the correct value of which of the following?
 (1) Amplitude
 (2) Phase
 (3) Frequency and phase
 (4) Waveform

Refer to Fig. 7-17 for the following five problems.

7-21. Electronically, how is the noise blanker circuit turned on? Explain.

7-22. Why are there two capacitors in parallel (C59, C60) on the line with zener diodes D22 and D26 (near S3-5)? Explain.

7-23. Completely explain how diodes D14 and D15 (near TR1) pass an information signal in millivolts.

7-24. How is the receiver protected against overload (too large of received signals)? Identify the protective device and explain how it protects the receiver.

7-25. Refer to the Spartan schematic and identify the major components (transistors, diodes, potentiometers, etc.) that perform the following functions:
 (a) SSB and AM sound amplifier ____________________
 (b) 19-MHz oscillators (LSB) (2) ____________________
 (c) SSB carrier null adjust ____________________
 (d) SSB selectivity ____________________
 (e) SSB receiver IF amplifiers ____________________
 (f) 27-MHz transmitter mixer ____________________
 (g) Ring lattice mixer ____________________
 (h) 27-MHz receiver mixer ____________________
 (i) Modulator for SSB ____________________
 (j) Receiver RF signal overload protection ____________________
 (k) Detector for SSB signal ____________________
 (l) Directing diode for 7.8-MHz LSB oscillator ____________________

7-26. Three of the factors listed below represent the advantages of single-sideband suppressed carrier operation compared to normal amplitude modulation for a given modulating signal. Select the one that is incorrect.
(1) The transmission bandwidth required for a channel is less.
(2) The power required to be transmitted is less.
(3) The signal-to-noise ratio is better.
(4) The receiving equipment required is simpler.

7-27. Refer to Fig. 7-3. Show how the blocks can be arranged to obtain the upper sideband by the phasing method, and write the necessary mathematical equations on your block diagram.

7-28. Name the blocks of the SSB transmitter shown in Fig. P7-28.

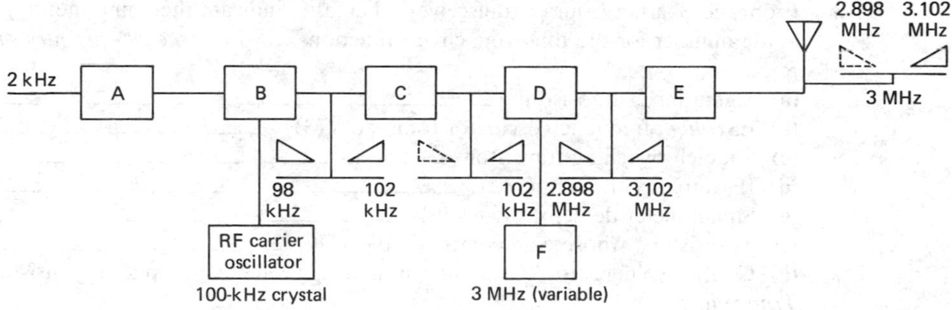

Figure P7-28.

7-29. Refer to the test setup shown in Fig. P7-29(1). Initially, the SSB transmitter is fed by audio oscillator 1. The transmitter is then keyed and the frequency counter reads 6503.00 kHz. Next, audio oscillator 2 is added to the output of oscillator 1 and the resulting oscilloscope and spectrum analyzer displays appear as shown in Fig. P7-29(2).
(a) Calculate the peak envelope power delivered to the dummy load.
(b) Calculate the average power delivered to the dummy load.
(c) Find the carrier frequency of the transmitter.
(d) What is the frequency of audio oscillation 2?
(e) If the transmitter were less linear, third-order intermodulation products would appear in the spectrum. Calculate their frequencies and sketch them on the analyzer display in Fig. P7-29(2). Set their amplitude to 30 dH below the two-tone level.
(f) Why is it not feasible to filter out intermodulation products before they are transmitted?

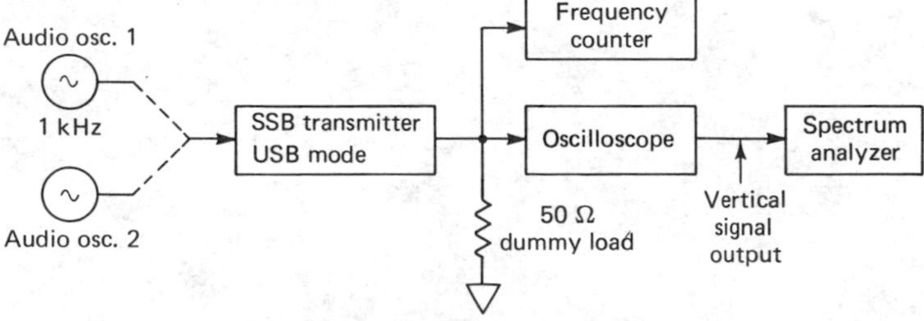

Figure P7-29. (1)

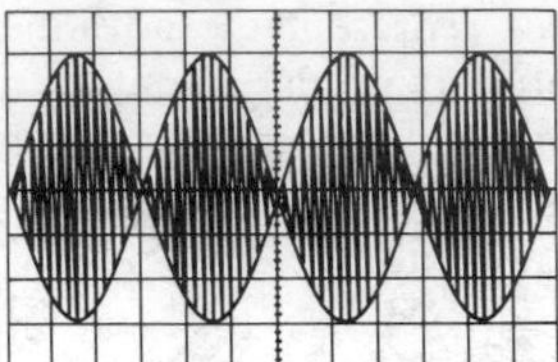

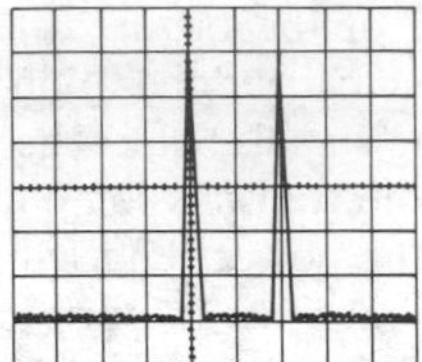

Vertical display 10 dB/div
Center frequency 6.5 MHz
Scan width 1 kHz/div

Figure P7-29. (2)

7-30. From the Spartan Courier transceiver schematic, indicate the component(s) by schematic number for the following circuit functions (*SSB transceiver circuitry only*).

Receiver

(a) Signal mixer transistor __________

(b) LSB *high*-frequency oscillator transistors (2) __________

(c) Squelch switching transistor __________

(d) IF amplifier transistors (2) __________

(e) Signal meter detector diodes (2) __________

(f) Transistors, whose gain controlled by AGC (2) __________

(g) Control voltage (to AGC, squelch, and signal meter) amplifier transistor __________

Transmitter

(a) Modulator carrier null adjustment __________

(b) USB and/or LSB signal directing diodes(2) __________

(c) Forward Tx power detector diode __________

(d) Automatic load control (ALC) detector diodes (2) __________

(e) MIC gain adjustment __________

(f) Reverse battery polarity protection diode __________

Angle Modulation: FM and PM

Wayne Wolinski
William Sinnema

8-1 INTRODUCTION

Information can be superimposed on a sinusoidal carrier wave by varying its amplitude and its phase angle. In AM and SSB the message or modulating signal varies the amplitude of the sinusoid, whereas in angle modulation, whether frequency modulation (FM) or phase modulation (PM), the phase angle of the sinusoid is varied by the modulating signal. Angle modulation results in enhanced noise performance but at the expense of increased bandwidth. It is a compromise of increased spectrum utilization for an improved signal-to-noise ratio. The key difference between these two forms of modulation is that in FM the amount of frequency change of the carrier from some reference frequency is proportional to the message amplitude, while in PM the amount of phase change of the carrier from some reference phase is proportional to the message amplitude.

Unfortunately, the mathematics for angular modulation is somewhat more complex than that for AM or SSB, as the angular modulation is nonlinear. In the case of AM signals, there is a one-to-one correspondence between the modulated signal and the modulating signal, and the modulation is said to be linear. If another modulating tone is added, another corresponding frequency component is added to the output (i.e., superposition holds). In angular modulation when only a single tone is applied as the modulating signal, many sidebands are produced. As a consequence, superposition cannot be applied.

8-2 ANALYSIS OF FM AND PM

The angle of a sinusoidal signal can be described in terms of a frequency and/or a phase angle. When the sinusoid has a constant angular rate, or angular frequency, ω_c, the angular rate of the sinusoid is ω_c radians per second. If the angular rate is not constant, it is more convenient to employ a phasor representation of the signal and to derive the angular rate from it.

Consider the general constant-amplitude, modulated carrier of the form

$$e_c = E_{cm} \cos \theta(t) \tag{8-1}$$

where E_{cm} is the peak value of the carrier sinusoid and $\theta(t)$ represents the phase angle. Its phasor representation, shown in Fig. 8-1, illustrates that if $\theta(t)$ increases linearly with time [i.e., $\theta(t) = \omega_c t$], we say that the phasor has a constant angular rate of ω_c radians per second. If the angular rate varies with time, say $\omega_i(t)$, the relation between the instantaneous angular rate $\omega_i(t)$ and $\theta(t)$ can be expressed as

$$\theta(t) = \int_0^t \omega_i(t)\, dt + \theta_0 \tag{8-2}$$

where θ_0 is the initial angular displacement at $t = 0$. We will assume in our discussions that θ_0 is 0.

Taking the derivative of both sides of equation (8-2), we obtain the angular frequency in terms of the instantaneous phase, that is,

$$\omega_i(t) = \frac{d\theta(t)}{dt} \tag{8-3}$$

Expressions (8-2) and (8-3) are very helpful when relating instantaneous phase to instantaneous frequency, and vice versa when comparing PM to FM.

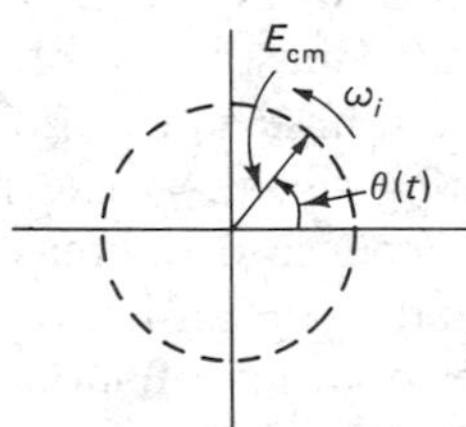

Figure 8-1 Phasor representation of a carrier.

EXAMPLE 8-1

Determine the instantaneous frequency of the signal

$$e_c = E_{cm} \cos(\omega_c t + \theta_0)$$

Solution

$$\theta(t) = \omega_c t + \theta_0$$

$$\omega_i(t) = \frac{d\theta(t)}{dt} = \omega_c \quad or \quad f_i(t) = \frac{\omega_i}{2\pi} = f_c.$$

This result is to be expected, as the angular rate is constant.

EXAMPLE 8-2

Determine the instantaneous angular frequency of the signal

$$e_c = E_{cm} \cos(\cos \omega_c t)$$

Solution

$$\theta(t) = \cos \omega_c t$$

$$\omega_i(t) = \frac{d\theta(t)}{dt} = -\omega_c \sin \omega_c t$$

Note that the angular frequency has the same periodicity as the phase and that it has been "amplified" by ω_c through the differentiation process.

EXAMPLE 8-3

Determine the instantaneous phase of a signal that has a frequency of $f_i = \cos \omega_c t$. Assume that $\theta_0 = 0$.

Solution

$$\omega_i = 2\pi f_i = 2\pi \cos \omega_c t$$

$$\theta_i(t) = \int_0^t 2\pi \cos \omega_c t \, dt$$

$$= \frac{2\pi}{\omega_c} \sin \omega_c t$$

Again note that the phase has the same periodicity as the frequency and that it has been "attenuated" by ω_c through the integration process.

8-2.1 Phase Modulation

In phase modulation the phase angle $\theta(t)$ is varied linearly with the modulating or information signal, which can be expressed as

$$\theta(t) = \omega_c t + k_p f(t) + \theta_0 \tag{8-4}$$

where θ_0 represents the angular displacement at $t = 0$, which we will choose to be 0. k_p is a constant that is dependent on the electrical parameters of the modulation circuit, having the units of rad/V. $f(t)$ represents the modulating signal. The angle by which the phase differs (at any instant in time, t) from the reference value is referred to as the *phase departure*. Thus $k_p f(t)$ represents the phase departure.

Since it is very difficult to deal mathematically with complex signals, consider the case when $f(t)$ is a single tone of the form

$$f(t) = E_m \cos \omega_m t \tag{8-5}$$

where E_m is the peak amplitude of the modulating signal and ω_m is the angular frequency of the modulating signal. Substituting the later expression into (8-4) yields the total phase angle

$$\theta(t) = \omega_c t + k_p E_m \cos \omega_m t \tag{8-6}$$

Substituting (8-6) into (8-1) gives the expression for the single-tone modulated PM signal:

$$\begin{aligned} e_{\text{PM}}(t) &= E_{\text{cm}} \cos(\omega_c t + k_p E_m \cos \omega_m t) \\ &= E_{\text{cm}} \cos(\omega_c t + \Delta\phi \cos \omega_m t) \end{aligned} \tag{8-7}$$

$$e_{\text{PM}}(t) = E_{\text{cm}} \cos(\omega_c t + M_p \cos \omega_m t) \tag{8-8}$$

where $\Delta\phi = k_p E_m = M_p$. M_p (as well as the other equivalents) is referred to as the modulation index for PM and represents the maximum phase departure (called phase deviation).

Employing the relation (8-3), we can now find the corresponding instantaneous frequency departure for the PM signal.

$$\begin{aligned} \omega_i(t) &= \frac{d\theta(t)}{dt} = \frac{d}{dt}(\omega_c t + \Delta\phi \cos \omega_m t) \\ &= \omega_c - \omega_m \Delta\phi \sin \omega_m t \end{aligned} \tag{8-9}$$

Therefore,

$$f_i = \frac{\omega_i}{2\pi} = f_c - f_m \Delta\phi \sin \omega_m t \tag{8-10}$$

Equation (8-10) reveals that the instantaneous frequency is equal to the carrier frequency plus the frequency departure ($- f_m \Delta\phi \sin \omega_m t$).

The frequency departure is seen to be time dependent and the peak value is called the *frequency deviation* or simply the *deviation*. We can also observe that the frequency deviation for PM is proportional not only to the amplitude of the modulating waveform but also to its frequency.

Phase and frequency modulation are closely related; any variation in phase will result in a variation in frequency and vice versa. The essential difference is the nature of dependency on the modulating signal. As a consequence, much of our discussion of FM is equally valid for PM. The differences will be pointed out later.

8-2.2 Frequency Modulation

FM is produced when the instantaneous frequency of a carrier is varied (about some rest value) in response to a modulating signal. The amount by which the instantaneous frequency differs from the carrier (rest) frequency is referred to as the *frequency departure*. Frequency departure is a function of the modulating signal instantaneous amplitude, and the maximum value attained is called the *frequency deviation*, or simply the *deviation*, as in the PM case.

Since the instantaneous frequency departure is to be proportional to the instantaneous value of the modulating signal in FM, we can let

$$\omega_i(t) = \omega_c + k_f f(t) \tag{8-11}$$

where ω_c is the reference carrier angular frequency; k_f is a constant depending on the parameters of the modulation circuit, having units of rad/sec · V; and $f(t)$ is the modulation signal.

For ease of calculation, let $f(t)$ again be the single tone expressed by equation (8-5). Substituting this expression for $f(t)$ into equation (8-11) yields

$$\omega_i(t) = \omega_c + k_f E_m \cos \omega_m t \tag{8-12}$$

$$\omega_i(t) = \omega_c + \Delta\omega \cos \omega_m t \tag{8-13}$$

where $\Delta\omega = k_f E_m$ is the angular frequency deviation.

Dividing both sides of equation (8-13) by 2π, we obtain for the instantaneous frequency:

$$f_i(t) = f_c + \Delta f \cos \omega_m t \tag{8-14}$$

$\Delta f = \Delta\omega/2\pi$ is commonly known as the frequency deviation or simply as deviation.

By employing the relationship of (8-2), the corresponding instantaneous phase may be determined.

$$\theta(t) = \int_0^t (\omega_c + \Delta\omega \cos \omega_m t)\, dt$$

$$= \omega_c t + \frac{\Delta\omega}{\omega_m} \sin \omega_m t \tag{8-15}$$

$$\theta(t) = \omega_c t + \frac{\Delta f}{f_m} \sin \omega_m t \tag{8-16}$$

$$\theta(t) = \omega_c t + M_f \sin \omega_m t \tag{8-17}$$

where

$$M_f = \frac{\Delta f}{f_m} \quad \text{rad} \tag{8-18}$$

The term $\Delta f/f_m$ represents the maximum instantaneous phase departure and is called the *modulation index*, M_f, for FM. Substituting equation (8-17) into (8-1), we obtain for the single-tone modulated FM signal:

$$e_{FM}(t) = E_{cm} \cos(\omega_c t + M_f \sin \omega_m t) \tag{8-19}$$

EXAMPLE 8-4

A 100-MHz carrier is frequency modulated with a 4-kHz 3-V_{peak} sinusoid. If the sensitivity of the modulating circuit is 2 kHz/V, determine:

(a) The maximum frequency deviation of the carrier.
(b) The maximum phase deviation (modulation index) of the carrier.
(c) The expression of the FM signal for a cosine carrier of 5 V_{peak}. Assume that the modulating signal is also a cosine.

Solution

(a) $\Delta f = k_f E_m = 2 \text{ kHz/V} \times 3 \text{ V}_{peak} = 6 \text{ kHz}$.
(b) $M_f = \Delta f/f_m = 6 \text{ kHz}/4 \text{ kHz} = 1.5$ rad.
(c) $e_{FM}(t) = 5 \cos[(2\pi \times 10^8 t) + 1.5 \sin(2\pi \times 4 \times 10^3 t)]$.

Comparing the FM [equation (8-19)] and PM [equation (8-8)] expressions, one notes that basically they are of the same form. Thus the techniques that will be applied to obtain the FM spectrum also apply for PM.

8-3 FM SPECTRUM AND BANDWIDTH

Equation (8-19) reveals that the single-tone modulated FM wave is periodic. Thus the expression should be expandable as a sum of sinusoids from Fourier series theory. From Euler's identity,

$$e^{j\theta} = \cos\theta + j \sin\theta \tag{8-20}$$

we can let

$$\cos\theta = \text{Re}\{e^{j\theta}\} \tag{8-21}$$

where Re stands for "the real part." Thus equation (8-19) can be expressed as

$$e_{FM}(t) = E_{cm} \text{ Re } \{e^{j(\omega_c t + M_f \sin \omega_m t)}\}$$

$$= E_{cm} \text{ Re } \{e^{j\omega_c t} + e^{jM_f \sin \omega_m t}\} \tag{8-22}$$

One can express the last complex term as a Bessel function of the first kind (signified by J) of order n and argument M_f [i.e., $J_n(M_f)$]. The Bessel functions are tabulated in Tables 8-1 and 8-2 and illustrated graphically in Fig. 8-2. Although we do not wish to get enmeshed in Bessel function theory, we can use its following properties:

$$\begin{aligned} &\text{(a) } J_n(M_f) \text{ are real valued} \\ &\text{(b) } J_n(M_f) = J_{-n}(M_f) \text{ for } n \text{ even} \\ &\text{(c) } J_n(M_f) = -J_{-n}(M_f) \text{ for } n \text{ odd} \end{aligned} \tag{8-23}$$

From Bessel function theory,

$$e^{jM_f \sin \omega_m t} = \sum_{n=-\infty}^{\infty} J_n(M_f) e^{jn\omega_m t} \tag{8-24}$$

and equation (8-22) becomes:

$$\begin{aligned} e_{\text{FM}}(t) &= E_{\text{cm}} \operatorname{Re}\{e^{j\omega_c t} \sum_{n=-\infty}^{\infty} J_n(M_f) e^{jn\omega_m t}\} \\ &= E_{\text{cm}} \sum_{n=-\infty}^{\infty} J_n(M_f) \cos(\omega_c + n\omega_m)t \end{aligned} \tag{8-25}$$

Expanding the latter, using properties (b) and (c) of equation (8-23), yields

$$\begin{aligned} e_{\text{FM}}(t) = E_{\text{cm}} \{ & J_0(M_f) \cos \omega_c t \\ & + J_1(M_f) [\cos(\omega_c + \omega_m)t - \cos(\omega_c - \omega_m)t] \\ & + J_2(M_f) [\cos(\omega_c + 2\omega_m)t + \cos(\omega_c - 2\omega_m)t] \\ & + J_3(M_f) [\cos(\omega_c + 3\omega_m)t - \cos(\omega_c - 3\omega_m)t] \\ & + \cdots \} \end{aligned} \tag{8-26}$$

From this result it is evident that the FM waveform with sinusoidal modulation has an infinite number of sidebands. Fortunately, the magnitudes of the spectral components of the higher-order sidebands become negligible and can be neglected.

Once the modulation index is known, the various spectral component amplitudes in a single-tone modulated FM wave can be determined. The first term of equation (8-26) represents the peak carrier amplitude [$(E_{\text{cm}} J_0(M_f)$]; the remaining terms, the sidebands. Adjacent sidebands are spaced by ω_m and are symmetrical in amplitude about the carrier. The even-order sidebands are in phase with each other and with the carrier, whereas the odd-order sidebands are 180° out of phase. One can note from Fig. 8-2 that at modulation indexes of approximately 2.4, 5.5, 8.6, and so on, the carrier amplitude is zero. This fact is helpful in determining the modulation sensitivity, k_f. The carrier frequency is monitored on a spectrum analyzer or narrow-band receiver and the modulating frequency is increased until the first carrier dropout occurs. Measuring both the peak modulating voltage, E_m, and the modulating frequency and knowing that $\Delta f/f_m = M_f$ allows the determination of k_f.

TABLE 8-1 Bessel Function Values for Modulation Indices Less Than Unity[a]

n	$J_n(0.1)$	$J_n(0.2)$	$J_n(0.3)$	$J_n(0.4)$	$J_n(0.5)$	$J_n(0.6)$	$J_n(0.7)$	$J_n(0.8)$	$J_n(0.9)$	$J_n(1.0)$
0	0.998	0.990	0.978	0.960	0.939	0.912	0.881	0.846	0.808	0.765
1	0.050	0.100	0.148	0.196	0.242	0.287	0.329	0.369	0.406	0.440
2	—	—	0.011	0.020	0.031	0.044	0.059	0.076	0.095	0.115
3	—	—	—	—	—	—	—	0.010	0.014	0.020

[a]Note that only values greater than 0.010 are given.

TABLE 8-2 Bessel Function Values for Modulation Indices to 15[a]

n	$J_n(1)$	$J_n(2)$	$J_n(3)$	$J_n(4)$	$J_n(5)$	$J_n(6)$	$J_n(7)$	$J_n(8)$	$J_n(9)$	$J_n(10)$	$J_n(11)$	$J_n(12)$	$J_n(13)$	$J_n(14)$	$J_n(15)$
0	0.765	0.224	−0.260	−0.397	−0.178	0.151	0.300	0.172	−0.090	−0.246	−0.171	0.048	0.207	0.171	−0.014
1	0.440	0.577	0.339	−0.066	−0.328	−0.277	−0.005	0.235	0.245	0.044	−0.177	−0.223	−0.070	0.133	0.205
2	0.115	0.353	0.486	0.364	0.047	−0.243	−0.301	−0.113	0.145	0.255	0.139	−0.085	−0.218	−0.152	0.042
3	0.020	0.129	0.309	0.430	0.365	0.115	−0.168	−0.291	−0.181	0.058	0.227	0.195	0.003	−0.177	−0.194
4	—	0.034	0.132	0.281	0.391	0.358	0.158	−0.105	0.266	−0.220	−0.015	0.183	0.219	0.076	−0.119
5	—	—	0.043	0.132	0.261	0.362	0.348	0.186	−0.055	−0.234	−0.238	−0.074	0.132	0.220	0.131
6	—	—	0.011	0.049	0.131	0.246	0.339	0.338	0.204	−0.015	−0.202	−0.244	−0.118	0.081	0.206
7	—	—	—	0.015	0.053	0.130	0.239	0.321	0.328	0.217	0.018	−0.170	−0.241	−0.151	0.035
8	—	—	—	—	0.018	0.057	0.128	0.224	0.305	0.318	0.225	0.045	−0.141	−0.232	−0.174
9	—	—	—	—	—	0.021	0.059	0.126	0.215	0.292	0.309	0.230	0.067	−0.114	−0.220
10	—	—	—	—	—	—	0.024	0.061	0.125	0.208	0.280	0.300	0.234	0.085	−0.090
11	—	—	—	—	—	—	—	0.026	0.062	0.123	0.201	0.270	0.293	0.236	0.010
12	—	—	—	—	—	—	—	—	0.027	0.063	0.122	0.195	0.262	0.286	0.237
13	—	—	—	—	—	—	—	—	0.011	0.029	0.064	0.120	0.190	0.254	0.279
14	—	—	—	—	—	—	—	—	—	0.020	0.030	0.065	0.119	0.186	0.246
15	—	—	—	—	—	—	—	—	—	—	0.013	0.032	0.066	0.117	0.181
16	—	—	—	—	—	—	—	—	—	—	—	0.014	0.033	0.066	0.116
17	—	—	—	—	—	—	—	—	—	—	—	—	0.015	0.034	0.067
18	—	—	—	—	—	—	—	—	—	—	—	—	—	0.016	0.035
19	—	—	—	—	—	—	—	—	—	—	—	—	—	—	0.017

[a]Note that only values greater than 0.010 are given.

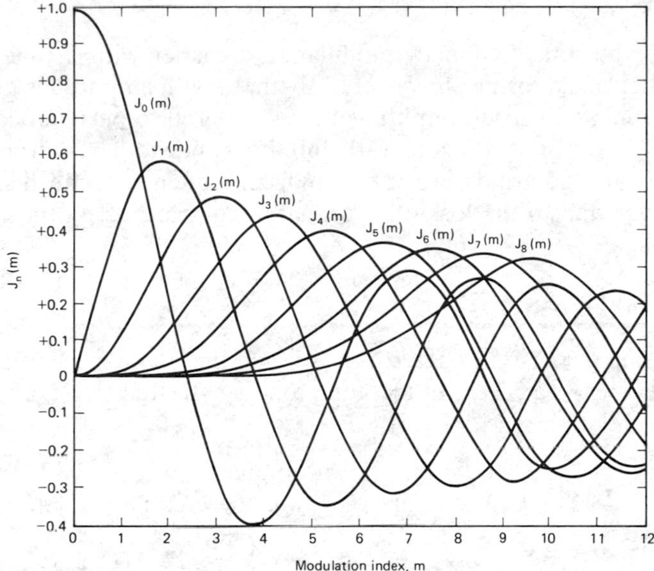

Figure 8-2 Bessel function graph.

EXAMPLE 8-5

Determine the spectrum of a 5 V_{rms} carrier when frequency modulated such that the deviation equals 3 kHz when modulated by a 1-kHz tone.

Solution

(1) $M_f = \frac{\Delta f}{f_m} = \frac{3\text{kHz}}{1\text{kHz}} = 3$

(2) From Table 8-2:

$$J_0(3) = -0.260 \qquad J_4(3) = 0.132$$

$$J_1(3) = 0.339 \qquad J_5(3) = 0.043$$

$$J_2(3) = 0.486 \qquad J_6(3) = 0.011$$

$$J_3(3) = 0.309$$

(3) Multiplying Bessel coefficients by initial carrier amplitude yields the spectrum shown in Fig. 8-3.

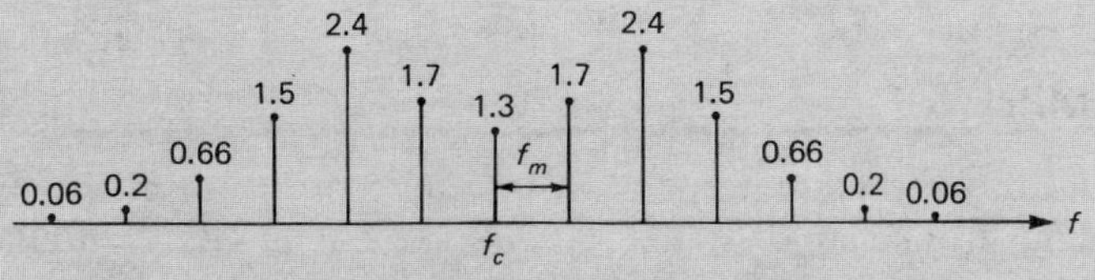

Figure 8-3.

Note that frequency modulating a carrier with a tone produces an infinite number of sidebands; however, only those with amplitudes greater than 1% of the unmodulated carrier amplitude are considered to be significant. Since it is impossible to transmit a signal with infinite bandwidth requirements, only the most significant sidebands are used, and consequently some distortion results at the receiver due to the loss of information contributed by the lower-amplitude sidebands.

EXAMPLE 8-6

Using a spectrum analyzer, it is noted that the carrier frequency first goes to zero when a 5-V_{rms} modulating tone is adjusted to 4.16 kHz. Determine the modulation sensitivity in Hz/V.

Solution At the first carrier dropout $M_f = \Delta f/f_m = 2.405$; therefore, $\Delta f = 2.405(4.16 \times 10^3) = 10$ kHz. Thus the modulating sensitivity is equal to

$$k_f = \frac{\Delta f}{E_m} = \frac{10 \times 10^3}{\sqrt{2} \times 5} = 1.414 \text{ kHz/V}$$

How many sidebands are significant to the FM transmission of a signal? A common rule is to include only sidebands that exceed 1% of the unmodulated carrier. From Table 8-2 one can observe that as M_f is increased, the number of significant sidebands increases. If we maintain a constant frequency deviation, Δf, by maintaining a constant amplitude for the modulating signal, to increase M_f one must have a reduced modulating frequency f_m since $M_f = \Delta f/f_m$. Thus although the number of sidebands is increased, the spacing between sidebands is reduced and an almost constant bandwidth is maintained.

If f_m is increased, the spacing between sidebands is also increased, but the number of significant sidebands decreases due to a reduction of M_f. The net result is a relatively constant bandwidth requirement. Figure 8-4 illustrates this almost constant bandwidth phenomenon. J. R. Carson proposed a general rule of thumb in the 1920s for the bandwidth of an FM waveform which has found common acceptance. It is referred to as *Carson's Rule:*

$$B \approx 2(\Delta f + f_m) \tag{8-27}$$

or substituting $M_f = \Delta f/f_m$ yields

$$B \approx 2f_m(M_f + 1) \tag{8-28}$$

EXAMPLE 8-7

Find the approximate bandwidth required to transmit an FM wave having a carrier frequency of 100 MHz. Assume a maximum modulating frequency of 5 kHz and a modulation index of 5.

Solution $B \approx 2(5 \times 10^3)(5 + 1) = 60$ kHz. The band of frequencies occupied is from 99.97 to 100.03 MHz.

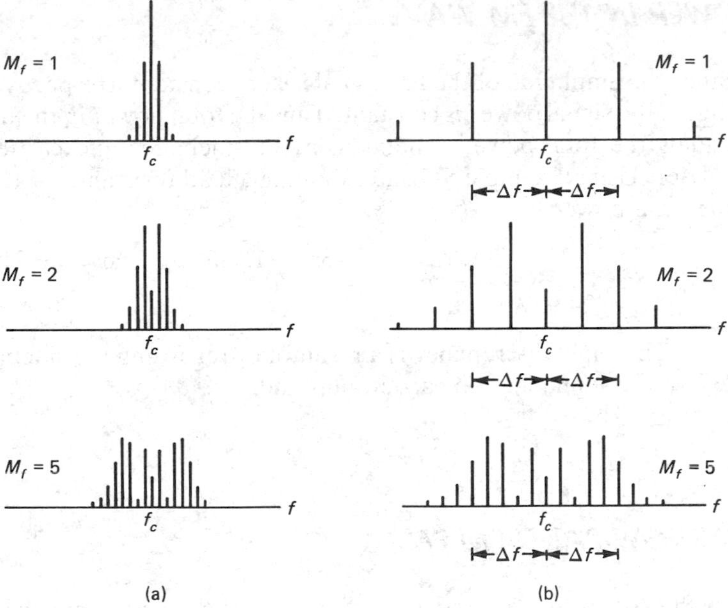

Figure 8-4 Spectrum for sinusoidally modulated FM waveforms for (a) constant f_m and (b) constant Δf.

EXAMPLE 8-8

Find the range of frequencies occupied by the FM waveform

$$e_{\text{FM}} = 5\cos[2\pi \times 10^8 t + 1.5\sin(8\pi \times 10^3 t)]$$

Solution

$$M_f = 1.5$$

$$f_m = \frac{8\pi \times 10^3}{2\pi} = 4\text{ kHz}$$

$$B \approx 2(4 \times 10^3)(1.5 + 1) = 20\text{ kHz}$$

The frequency range occupied runs from (100M − 10k to 100M + 10k) Hz, or from 99.99 to 100.01 MHz.

For commercial FM broadcasting, carrier frequencies are spaced at 200 kHz and have an allowed frequency deviation of 75 kHz. Modulating frequencies up to 15 kHz are permitted.

Depending on the service provided (i.e., stereo TV, mobile radio, subsidiary communication or broadcast FM, and the audio in TV), the allowed frequency deviation varies. Sufficient information to permit bandwidth approximations for these commercial services using FM as well as two-way radio transceivers (such as cellular units) may be obtained by contacting the local governmental agency regulating radio transmission/reception.

8-4 POWER IN THE FM WAVE

Since the amplitude of the FM signal is kept constant irrespective of the modulation index, the signal power is constant. Thus the total power in the unmodulated carrier equals the total power contained in the sidebands and carrier for a modulated carrier. Using Example 8-5 and assuming a load resistance of 1 Ω, we can calculate the total power.

$$5^2 = 1.3^2 + 2(1.695^2 + 2.43^2 + 1.545^2 + 0.66^2 + 0.215^2 + 0.055^2)$$
$$= 24.99\text{W}$$

The slight discrepancy is a result of disregarding the sidebands that fall below 1% of the unmodulated carrier amplitude.

8-5 NARROW-BAND FM/PM

In the following discussion we refer primarily to FM, but the results are equally valid for PM. One should keep this in mind, since FM is often produced by performing a slight modification to modulating signal in a PM transmitter.

Examination of Table 8-1 reveals that for $M_f \leq 0.2$, only the carrier and the first-order sidebands are significant, resulting in a bandwidth of only $2f_m$. Under such conditions the signal is considered to be narrow-band FM (NBFM).

Although the frequency spectrum of AM and NBFM are similar, the sideband phasor relationships are quite different. From the second term in equation (8-26), one notes that the first lower sideband is 180° out of phase compared to the first upper sideband. This results in a 90° shift of the two sidebands relative to the carrier (as compared to AM) as illustrated in Fig. 8-5. The AM case gives amplitude variations with no phase departure, whereas the NBFM case gives rise to phase variations with very slight amplitude change. Figure 8-6 illustrates a simple NBPM modulator.

The output of this circuit will be a phase-modulated signal containing amplitude variations (which can easily be removed), as represented by the resultant in Fig. 8-5(b).

The question can now be raised: How does one generate NBFM? The trick is to convert the term $M_p \cos \omega_m t = k_p E_m \cos \omega_m t$ of the PM equation (8-8) into a form that looks like

$$M_f \sin \omega_m t = \frac{k_f E_m}{\omega_m} \sin \omega_m t$$

of the FM equation (8-19). This is done quite readily by integrating the modulating signal $f(t) = E_m \cos \omega_m t$ before modulating with the phase modulator, that is,

$$\int (E_m \cos \omega_m t)\, dt = \frac{E_m}{\omega_m} \sin \omega_m t$$

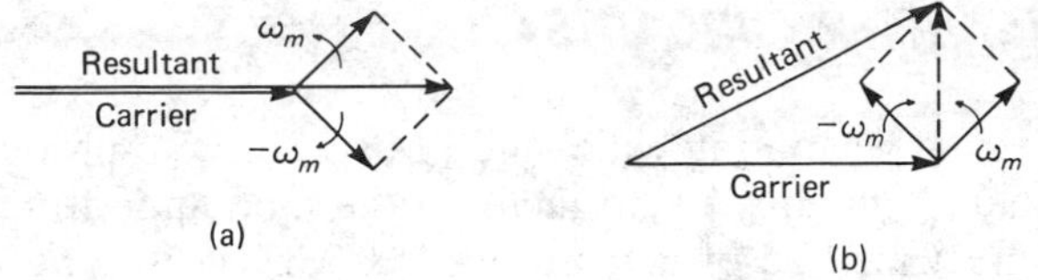

Figure 8-5 Phasor representation of (a) AM and (b) NBFM.

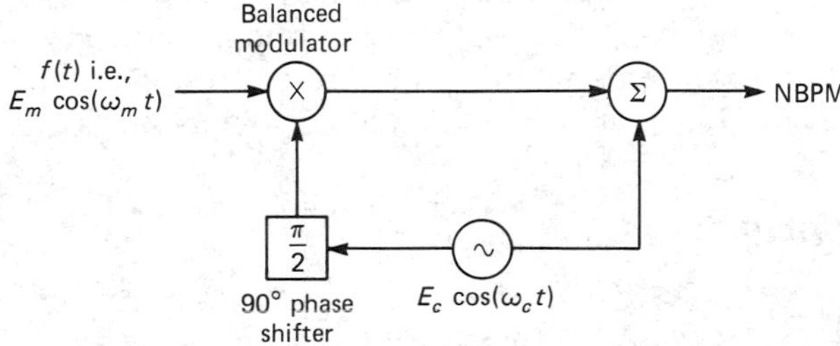

Figure 8-6 NBPM using a balanced AM modulator and a sideband phase shifer.

Thus NBFM may be produced by a NBPM modulator, provided that the modulation signal is first integrated, as illustrated in Fig. 8-7. The integrator can consist of an *RC* network or op-amp integrator with a response similar to that shown in Fig. 8-8.

The Armstrong modulator block diagram of Fig. 8-7 can employ a crystal-stabilized oscillator, ensuring excellent frequency stability at the output. Herein lies the primary advantage of using this "indirect" technique.

To obtain wideband FM (WBFM) that has a high modulation index, frequency multipliers follow the NBFM modulator of Fig. 8-7. This will be followed up in the next section.

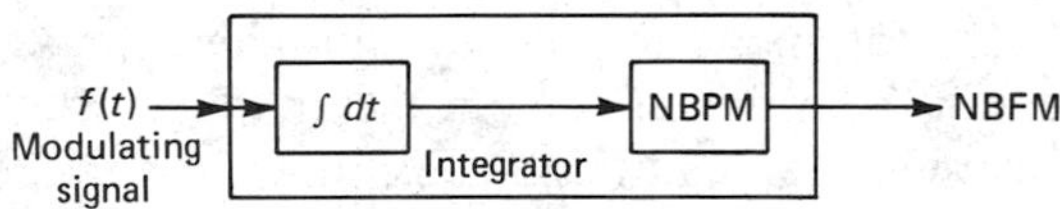

Figure 8-7 Indirect generation of NBFM (Armstrong modulator).

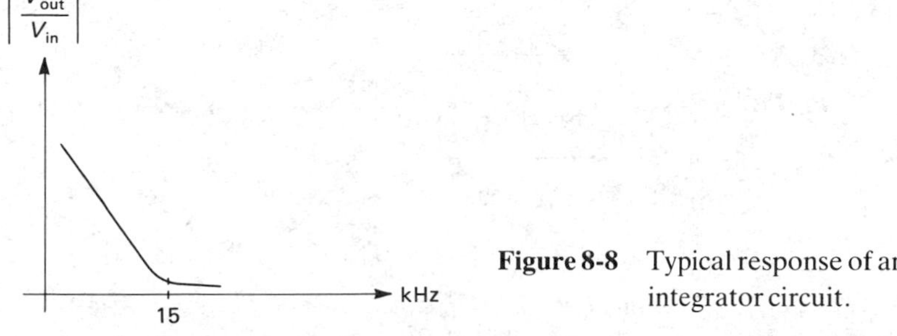

Figure 8-8 Typical response of an integrator circuit.

8-6 FM TRANSMITTER CIRCUITS

One method of generating wideband FM signals is to produce NBFM initially and then frequency multiply up to increase the modulation index. This is referred to as the indirect method of generating WBFM. The second method, known as the direct method, uses a voltage-controlled oscillator (VCO) and has the advantage of producing high-index FM directly. Because of problems with VCO frequency stability, special measures must be taken to minimize frequency drift. Feedback can be used to stabilize the VCO carrier center frequency, a technique referred to as the Crosby AFC (automatic frequency control) method. The final output power amplifiers following their driver amplifiers are operated class C for high efficiency.

8-6.1 Indirect FM

If we apply the FM signal of equation (8-19) to a frequency doubler (i.e., square law device), the resulting signal will be of the form

$$e_0(t) = e_{\text{FM}}^2(t) = E_{\text{cm}}^2 \cos^2(\omega_c t + M_f \sin \omega_m t)$$

Using the trigonometric identity

$$\cos^2\theta = \frac{1}{2}(1 + \cos 2\theta) \tag{8-29}$$

we obtain

$$e_0(t) = \frac{E_{cm}^2}{2} + \frac{E_{\text{cm}}^2}{2} \cos(2\omega_c t + 2M_f \sin \omega_m t) \tag{8-30}$$

Rejecting the dc component through bandpass filtering, we note that the resultant signal is at twice the original center frequency and that the effective modulation index has doubled. Since ω_m or f_m has remained constant, this also means that the deviation Δf has doubled.

In general, for an $\times n$ frequency multiplier, the center frequency, deviation, and modulation index increase by a factor of n. Thus to generate WBFM, the NBFM Armstrong modulator may be followed with frequency multipliers to increase the modulation index as shown in Fig. 8-9. In theory, the modulation index,

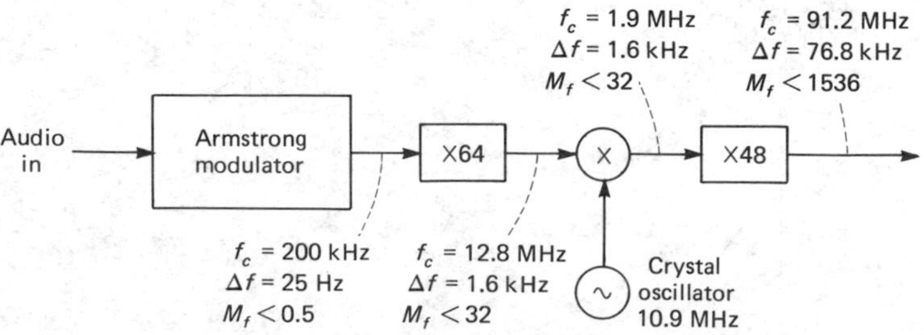

Figure 8-9 Example of WBFM transmitter (Armstrong).

M_f, from the NBFM modulator should be kept below 0.2, but in practice it may reach 0.5. The frequency multiplier blocks ($\times 64$, $\times 48$) are actually made up of several small-order multipliers ($\times 2$, $\times 3$) cascaded, rather than a single high-order stage. This is due to the fact that the output level for a high-order multiplier is very small relative to the input level (in general, harmonics of the input signal decrease in amplitude as frequency increases).

8-6.2 Direct FM

The simplest technique for controlling the frequency of a voltage-controlled oscillator (VCO) is by the use of a varactor diode. The varactor (or varicap) diodes are designed to present a certain range of capacitance values when reverse biased over some voltage range. By employing a varactor in conjunction with the tuned circuit in an oscillator, the carrier frequency can be varied. Note from the capacitance characteristic of Fig. 8-10 the highly nonlinear nature of a typical varactor. Consequently, the voltage variations must be kept small to minimize distortion of the FM signal and frequency multipliers must be employed to increase the frequency deviation.

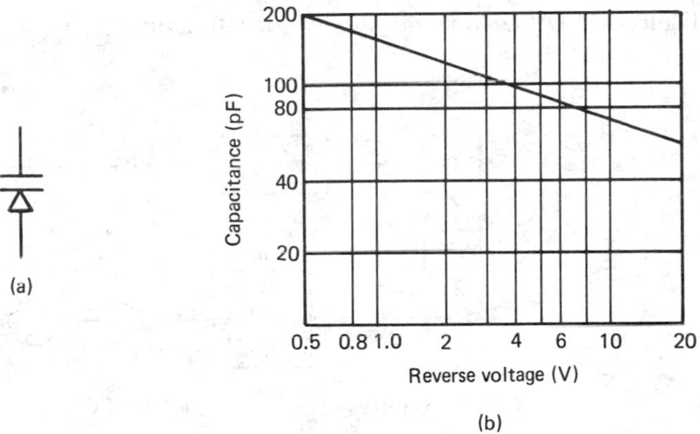

Figure 8-10 Symbol (a) and capacitance versus voltage characteristic (b) of a varactor.

Figure 8-11 gives a simplified circuit of a varactor diode modulator. By applying a modulating signal, the ac bias to the varactor is varied, causing a shift in the VCO frequency. Although the LC tuned VCOs have good deviation sensitivity, they have poor frequency stability due to aging and temperature changes. To improve the frequency stability, a form of frequency control is employed. One of the more common systems is the Crosby system illustrated in Fig. 8-12.

The discriminator in conjunction with the low-pass filter (LPF) provides a slowly varying dc output voltage based on the input frequency. This circuit combination does not respond to the instantaneous frequency inherent in the FM signal but only to carrier frequency drift. The output from the LPF feeds the VCO with the correct polarity dc voltage to force the VCO back to the correct frequency.

The dc voltage would be altering the varactor bias, for instance, of the varactor diode modulator. Since the reference is crystal controlled, the output carrier frequency of the system is very stable.

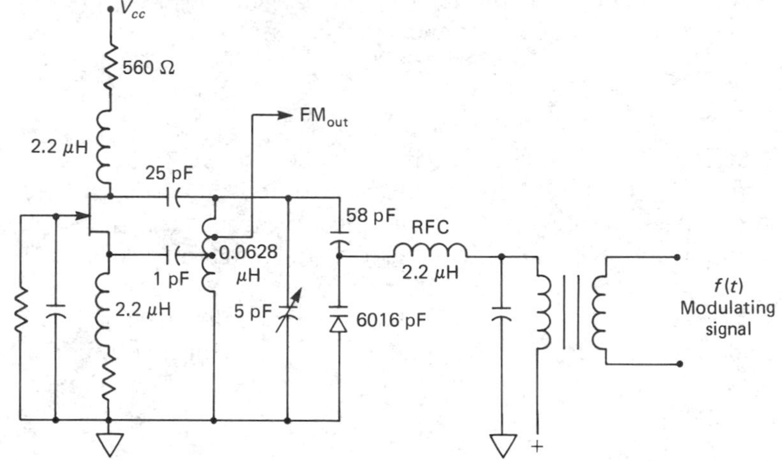

Figure 8-11 Varactor diode modulator with a tuning range of 120 to 200 MHz.

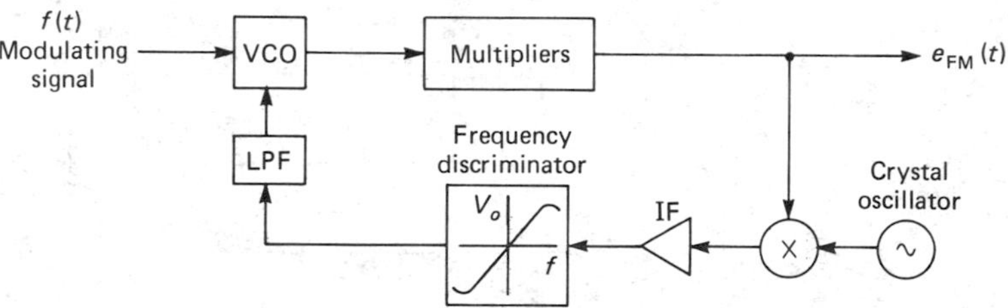

Figure 8-12 Transmitter AFC.

EXAMPLE 8-9

An FM signal has a frequency deviation of 50 Hz for an input 3-V sinusoid. Determine the frequency multiplication required to produce a frequency deviation of 20 kHz for the same modulating tone.

Solution

$$\Delta f_{\text{out}} = 20 \text{ kHz}$$

$$\Delta f_{\text{in}}(\text{to multipliers}) = 50 \text{ Hz}$$

Therefore,

$$n = \frac{\Delta f_{\text{out}}}{\Delta f_{\text{in}}} = 400$$

EXAMPLE 8-10

If the varactor of the circuit in Fig. 8-13 has a capacitance characteristic shown in Fig. 8-10, determine the resonant frequency of the LC circuit when V_{dc} = 7 V. What happens to the resonant frequency when the dc voltage is increased?

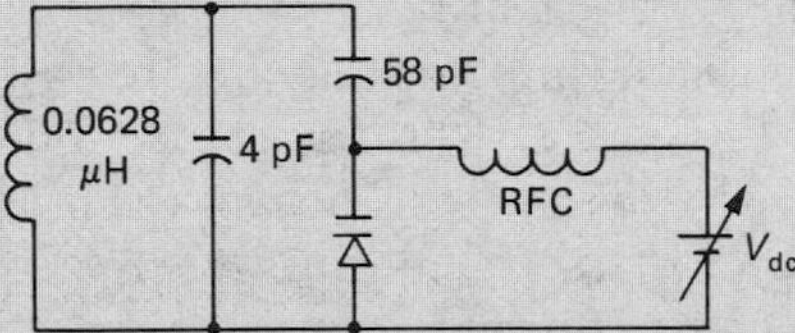

Figure 8-13 Circuit for Example 8-10.

Solution At 7 V, C_{var} (from Fig. 8-10) = 80 pF.

The total series capacitance, C', is given by $1/C' = 1/58\text{ pF} + 1/80\text{ pF}$; therefore, $C' = 33.6$ pF. The net capacitance thus is $c = (33.6 + 4)\text{pF} = 37.6$ pF.

$$f_{res} = \frac{1}{2\pi\sqrt{LC}} = \frac{1}{2\pi\sqrt{0.0628 \times 10^{-6} \times 37.6 \times 10^{-12}}} = 103.6 \text{ MHz}$$

If the dc voltage is increased, the varactor capacitance decreases, causing an increase in the resonant frequency.

8-7 PREEMPHASIS/DEEMPHASIS

When noise is added to a carrier during transmission of an FM signal, the effect on the carrier can be visualized by drawing a phasor representation of random noise added to an unmodulated carrier as shown in Fig. 8-14(a). In an FM receiver, bandpass limiters can be used to remove any amplitude variations on the signal, leaving only the phase variations, $\Delta\phi$, as shown. As a phase variation of $\Delta\phi$ results in an equivalent frequency variation of $f_n\,\Delta\phi$, where f_n represents the noise frequency component f_m in equation (8-10), the FM detector being sensitive to frequency variations will respond with a linear noise voltage output with frequency.

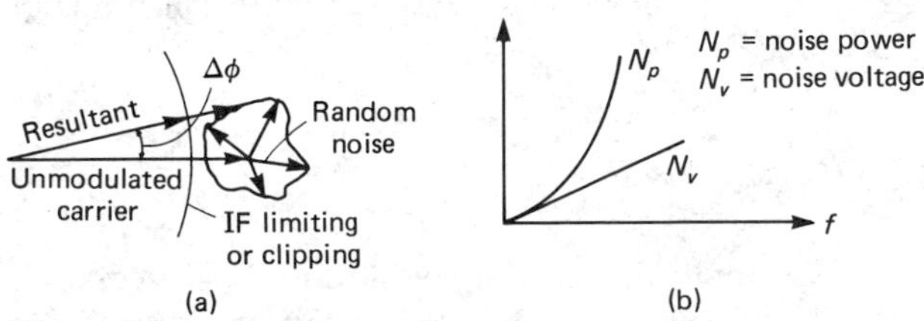

Figure 8-14 Carrier plus noise phasor at limiter output (a) and detector output noise spectrum (b).

Often this is shown as a parabolic noise power spectral density curve, which is the square of the noise voltage curve as shown in Fig. 8-14(b). As a result of this increasing noise with frequency, the higher audio frequencies at the detector suffer a greater signal-to-noise (S/N) degradation than the lower frequencies. Unfortunately, this is further exacerbated by the fact that voice and music contain less energy at the higher frequencies.

To maintain a nearly constant S/N ratio across the audio baseband, the higher frequencies are boosted (preemphasis) before passing the modulating signal to the transmitter. For standard FM broadcast, the 3-dB points occur at the break frequency of 2120 Hz. If a single RC network is used to generate this characteristic, this break frequency is associated with $1/(2\pi RC)$, with $RC = 75\ \mu s$. The net result is an improved noise performance over the voice band of about 12 dB.

The preemphasis network normally precedes the modulator stage in the transmitter, and to restore the correct amplitude-frequency relationship in the receiver, the deemphasis network normally precedes the audio amplifier, as shown in Fig. 8-15. Note that slight architectural variations of that given in Fig. 8-15 may appear, as shown in Fig. 8-16 (preemphasis preceding audio amplifier in Tx). Further discussion of pre- and deemphasis may be found in Appendix B.

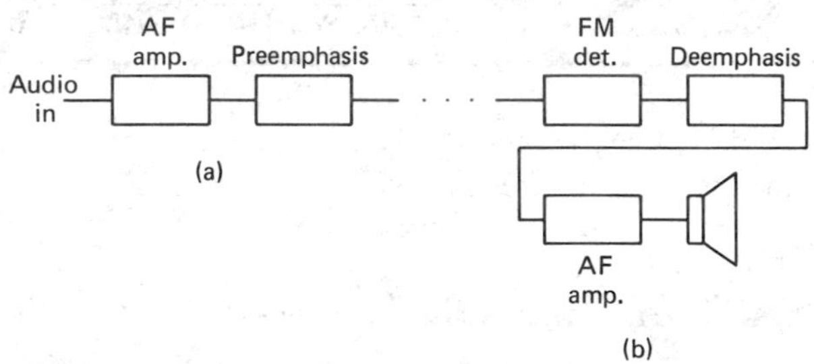

Figure 8-15 (a) Preemphasis; (b) deemphasis.

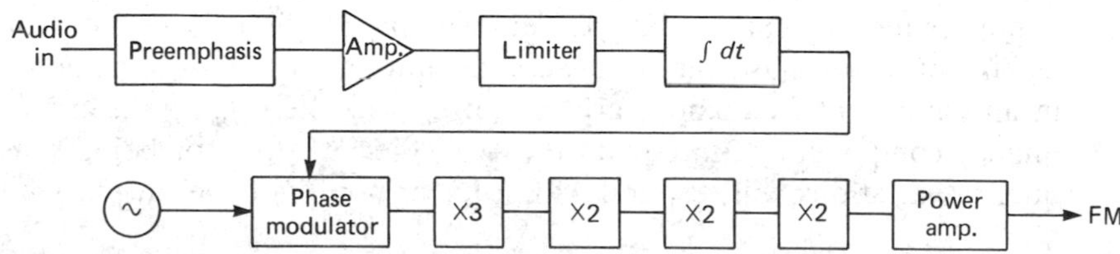

Figure 8-16 FM transmitter with preemphasis.

8-8 FM RECEIVERS

The FM receiver generally follows a double-conversion superheterodyne format similat to the one shown in Fig. 8-17. Automatic gain control (AGC), or circuitry that senses the average signal strength and controls IF and/or RF gain so as to

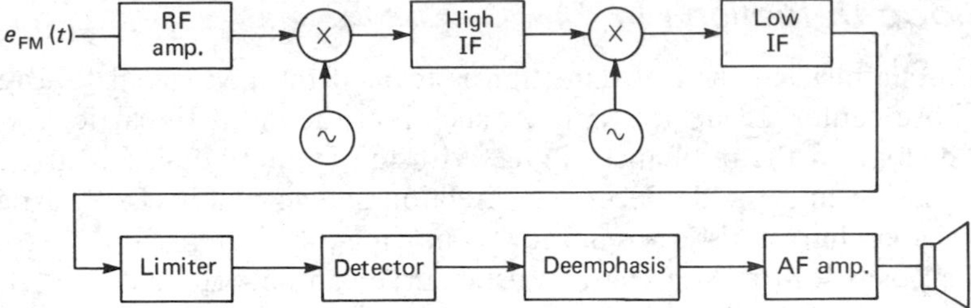

Figure 8-17 Example of a double-conversion superheterodyne FM receiver.

keep the input signal to the limiter of relatively constant amplitude, may or may not be present in a receiver, depending on design. AGC is of particular value in mobile communications, where multipath reception, varying degrees of attenuation, and so on, can result in large signal strength variations at the receiver input. The FM receiver is very similar to those used for other modes of communication (i.e., AM, SSB), the most notable difference appearing after the low IF section.

Many types of FM detectors require a constant-amplitude input signal in order to respond only to frequency variations (amplitude variations, if present, would be due to undesirable factors such as noise, and would produce an interfering output signal from the detector); therefore, a *limiter* is often used (a circuit that produces a constant output amplitude regardless of variations in input amplitude over a limited range of values). If AGC is present, it will ensure the signal fed to the limiter will be within its limiting range.

8-8.1 FM Demodulation

The function of an FM demodulator is to convert instantaneous frequency variations to amplitude variations as shown in Fig. 8-18. Linearity is maintained for reasonable frequency deviations, usually developing a sloppy "∫" transfer characteristic overall. Several categories of circuits exist to accomplish this task, each type having certain distinguishing characteristics. We will first briefly introduce the least popular slope detection/FM discriminator techniques and then discuss the phase-locked loop (PLL) demodulator, the quadrature detector, and the pulse counting FM detector.

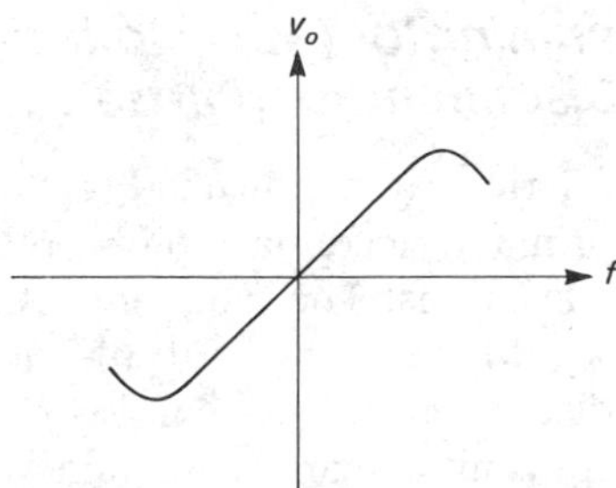

Figure 8-18 Frequency discriminator response.

Slope Detection

The simplest method of detection is to feed the FM signal to a resonant circuit, whose center frequency, f_0, is set such that the input IF carrier frequency falls on the slope of the resonance curve. With the IF unmodulated carrier falling on the high side and the diode in the direction shown in Fig. 8-19, a negative S-curve transfer characteristic is realized. By moving the carrier to the low-frequency slope, the positive S transfer characteristic curve can be obtained.

The slope detector is inefficient, linear only over a very limited frequency range, and is sensitive to amplitude variations (which may be caused by noise, etc.). Therefore, it is rarely used, except in an emergency situation where only an AM receiver is available.

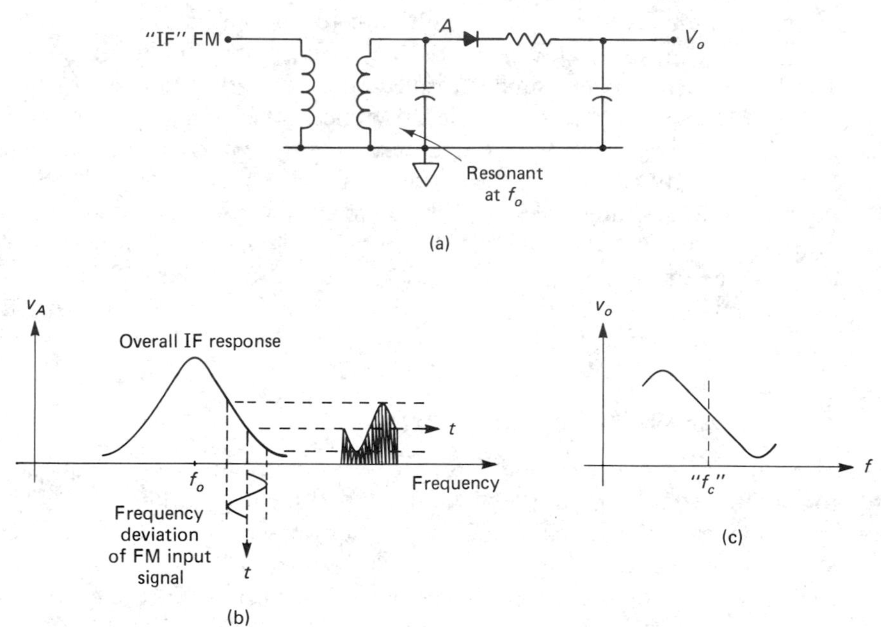

Figure 8-19 Slope detector (a) with IF response (b) and output response (c).

Phase-Shift Discriminator (Also Called Foster–Seely Discriminator, Discriminator, Phase Discriminator)

The Foster–Seely discriminator shown in Fig. 8-20 yields reasonably good linearity, relying mainly on primary–secondary phase relationships, but suffers from sensitivity to amplitude variations. The tank circuits L_pC_p and L_sC_s are tuned to the IF carrier frequency, f_c, which often is at 10.7 MHz in commercial FM broadcast receivers. For our discussion, we will employ f_r as the IF center frequency to avoid confusion with the transmitted carrier frequency, f_c. RFC appears as an open circuit and the "C" capacitors appear as short circuits to the IF carrier frequency.

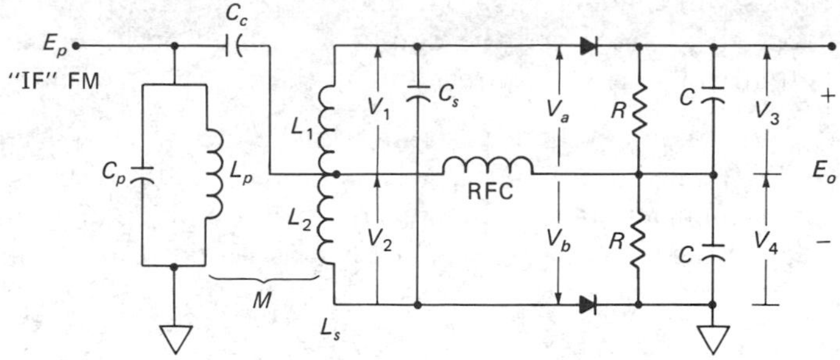

Figure 8-20 Foster–Seely discriminator.

We will analyze the circuit in step form.

1. The primary current in L_p can be expressed as

$$I_p = \frac{E_p}{j\omega L_p} = \frac{E_p}{\omega L_p} \angle -90^\circ$$

2. The induced secondary voltage, described in terms of the mutual coupling coefficient, M, and primary current is

$$E_i = -j\omega M I_p = \frac{-j\omega M E_p}{j\omega L_p} = \frac{-M E_p}{L_p}$$

3. At the unmodulated IF carrier frequency, the circulating L_sC_s tank current I_s will be in phase with E_i since the tank is at resonance: $I_s = E_i/R_s$ at resonance, where R_s is the dc resistance of L_s. If the carrier IF frequency is above resonance, the induced voltage sees an inductive impedance and I_s lags E_i somewhat. Below resonance, I_s will lead E_i as shown in Fig. 8-21(a).

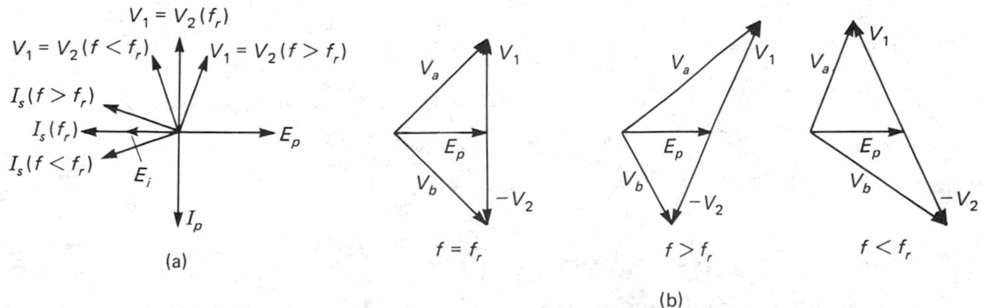

Figure 8-21 Discriminator phase relations.

4. The voltage applied to the diodes V_a and V_b will be the phasor sum of E_p and V_1 and E_p and $-V_2$, respectively [see Fig. 8-21(b)]. At resonance (f_r), the output voltage (E_o) is equal to the sum of V_3 and V_4 and is zero since the

diodes will conduct equal currents at resonance ($|V_3| = |V_4|$). Above resonance, $|V_a| > |V_b|$ and the output E_o will go positive; below resonance, the output will go negative, resulting in the typical "S" characteristic.

If there are amplitude variations in the incoming IF signal, the magnitude ($|V_a| - |V_b|$) will vary, resulting in amplitude variations in the output signal E_o. Therefore, a limiter stage must precede this discriminator to minimize amplitude variations in the incoming IF signal, such as would result from fading, multipath reception, noise, and so on.

Ratio Detector

To reduce sensitivity to input amplitude variations, the Foster–Seely discriminator may be modified to produce the ratio detector of Fig. 8-22. The behavior is quite similar to that of the Foster–Seely, but a very large capacitor C_1 is connected in parallel with $C + C$. Its function is to keep the voltage across the two capacitors constant even for the slowest-frequency deviations of the input signal. The individual capacitor voltages may vary but not the sum of these two voltages. This feature gives rise to the name *ratio detector*. Ignoring the diode forward voltage drop, the output voltage equals $E_o = \frac{1}{2}(|V_a| - |V_b|)$, which is half that of the Foster–Seely output. Thus the ratio detector has the sensitivity of one-half that of the Foster–Seely.

If the carrier signal increases in amplitude, capacitor C_1 puts an extra load on the tank circuit, lowering its Q, and thereby also reducing the signal level to the resonant circuit of the ratio detector.

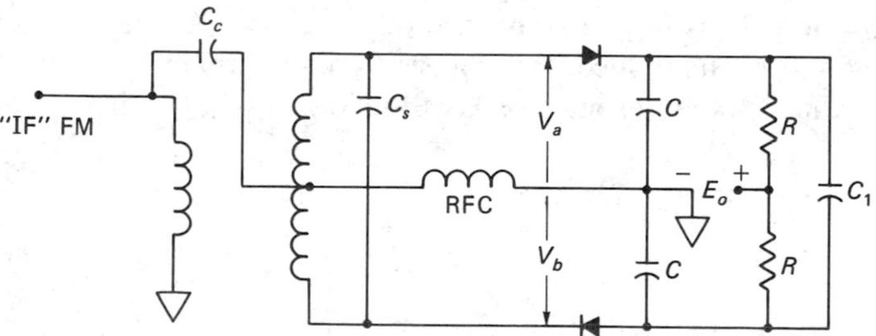

Figure 8-22 Ratio detector.

Quadrature Detector

Quadrature FM detectors rely on a large reactance in series with a tuned parallel resonant circuit to produce two signals in phase quadrature. Frequency changes in the FM signal cause additional leading or lagging phase shift to occur at the *LC* tuned circuit, which is detected with an analog phase detector. Consider the quadrature detector of Fig. 8-23. At any instant of time, the signal at point (b) will be shifted in phase compared to the signal at point (a). The amount of phase shift depends on the frequency.

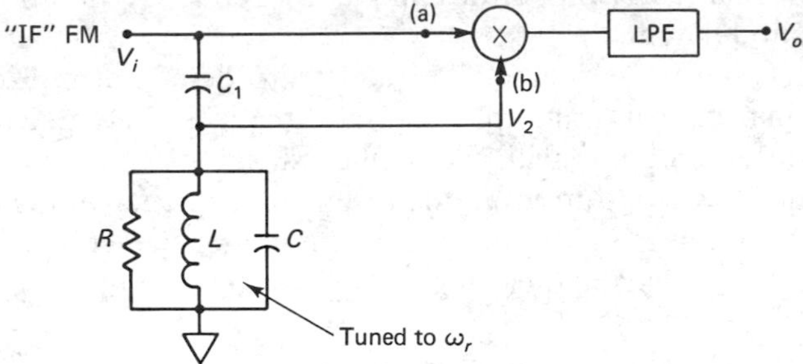

Figure 8-23 Quadrature detector.

At resonance, the parallel tuned circuit appears resistive, R, and under the conditions that $1/(\omega_r C_1) >> R$, the phase of V_2 compared to V_i is $\pi/2$, that is,

$$\frac{V_2}{V_i} = \frac{R}{R + 1/j\omega_r C_1} \approx j\omega_r C_1 R = \omega_r C_1 R \,\angle\, \pi/2 \tag{8-31}$$

V_2 thus leads V_i by approximately 90° and the two signals are said to be in phase quadrature.

Off-resonance, V_2 will lead V_i by angles slightly greater or less than 90°, depending on whether the frequency is less or greater than the resonant frequency of the parallel LC circuit. For small frequencies about ω_r, it can be shown that the phase shift between V_2 and V_i is reasonably linear and expressed approximately by

$$\phi = \frac{\pi}{2} \quad K\,\Delta\omega \qquad \text{rad} \tag{8-32}$$

where K is a constant equal to $2L/R$ and $\Delta\omega = \omega - \omega_r$. Thus if v_i is represented by a sinusoid

$$v_i = V_1 \cos \omega t \tag{8-33}$$

v_2 will be equal to

$$v_2 = V_1 \cos(\omega t + \frac{\pi}{2} - K\,\Delta\omega) \tag{8-34}$$

Multiplying v_i by v_2 will yield a low-frequency component

$$\frac{V_1}{2} \cos\left(\frac{\pi}{2} - K\,\Delta\omega\right) = \frac{V_1}{2} \sin K\,\Delta\omega \tag{8-35}$$

which will pass through the LP filter unattenuated. For small $K\,\Delta\omega$,

$$V_o = \frac{V_1}{2} \sin K\Delta\omega \approx \frac{V_1}{2} K\,\Delta\omega = K'(\omega - \omega_r) \tag{8-36}$$

This demonstrates that the output voltage is proportional to the difference between the incoming frequency and the resonant frequency of the tuned circuit. Hence for an FM signal, consisting of instantaneous frequency variations, the circuit will act as a demodulator.

An alternative method of quadrature detection utilizes a coincidence gate rather than an analog multiplier; this technique is illustrated in Fig. 8-24. The signal at (b) differs in phase with the applied signal at (a) by an amount dependent on the frequency departure from the resonant frequency of the tuned *LC* tank circuit (as described in the preceding analog multiplier detector). The (a) and (b) signals control the opening and closing of the switches.

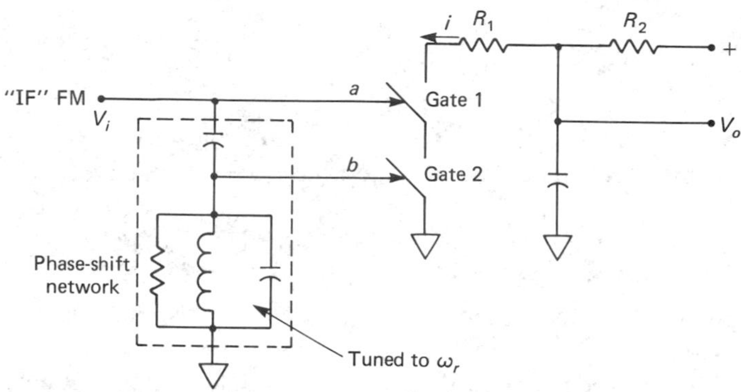

Figure 8-24 Quadrature detector utilizing coincidence gate.

Assuming the applied voltage to be a typical IF FM signal that has been heavily limited and therefore has the appearance of a square wave, the signals at (a) and (b) will appear similar to that shown in Fig. 8-25.

As the incoming frequency varies from the resonant frequency of the *LC* circuit, the phase of V_b will vary, causing the time during which i_1 flows to increase or decrease. The amount by which t_1 will differ from its value at ω_r will be proportional to the difference in frequency between the incoming signal and the resonant frequency of the tuned circuit. Differing values for t_1 will result in differing voltages produced across C_1; hence this circuit may be used for demodulating FM.

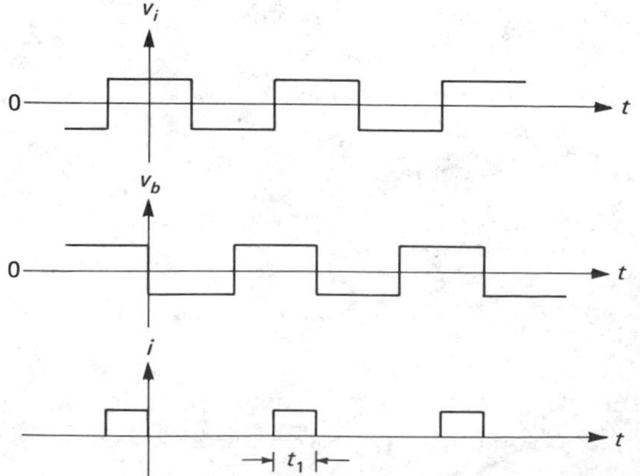

Figure 8-25 Coincidence gate signals at $\omega = \omega_r$.

PLL Demodulator

The PLL, which has been described in some detail in Chapter 4, can also be employed as an FM demodulator. As seen in Fig. 8-26, the PLL consists of a phase detector (PD), low-pass filter (LPF), and a VCO. The phase difference between an input signal, v_i, and the VCO output, produces an error signal which upon filtering produces a voltage, v_e, that controls the output frequency of the VCO. The desirable condition is for the VCO frequency to stabilize at a frequency either equal to, or harmonically related to, the input frequency. Under this condition a constant value for v_e will result, and the PLL is said to be "locked."

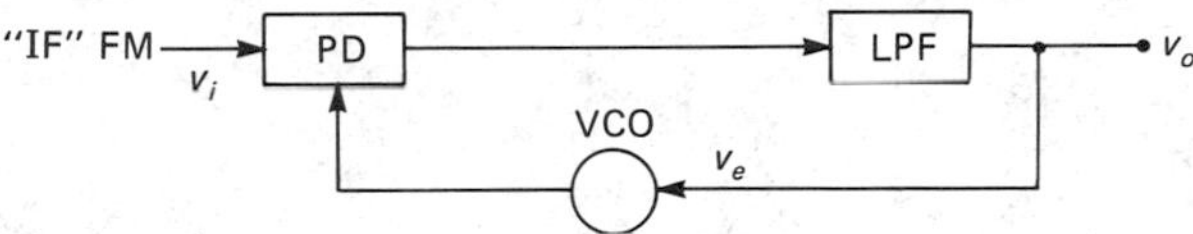

Figure 8-26 PLL FM detector.

As the input signal frequency changes, the PLL will continuously attempt to lock onto the incoming frequency, producing a voltage v_e whose amplitude will depend on the instantaneous incoming frequency. If the applied signal is a frequency-modulated signal, v_e can be considered to be a copy of the original information which had modulated the transmitted FM signal. Therefore, $v_o = v_e \propto v_m(t)$, where $v_m(t)$ is the original message signal.

Advantages of PLLs include the elimination of the need for accurate resonant circuit adjustments (a disadvantage of detectors such as the phase and quadrature detectors) and the ability to provide good FM demodulation in the event of signal carrier drift. Also, the PLL detector can be designed to be relatively insensitive to sudden amplitude variations, such as noise spikes.

Zero-Crossing Detector

The last FM detector we will consider is the zero-crossing detector, which measures the number of zero signal crossings of the FM signal per unit time. From the knowledge of zero crossings, the original message can be reconstructed.

Recall that for an FM signal, the instantaneous radial frequency is given by $\omega_i(t) = \omega_c + k_f f(t)$ [equation (8-11)]. Let t_1 and t_2 be the times associated with two adjacent zero crossings, as shown in Fig. 8-27. If the highest-frequency component in the modulation signal $f(t)$ is much less than the unmodulated carrier radial frequency ω_c, $f(t)$ will vary very little over the time interval (t_1, t_2) and will be essentially constant within this interval.

Since there are π radians between adjacent crossings, under this condition

$$\omega_i(t_2 - t_1) = [\omega_c + k_f f(t)](t_2 - t_1) = \pi$$

$$\omega_i = \omega_c + k_f f(t) = \frac{\pi}{t_2 - t_1} \tag{8-37}$$

or

$$f_i = f_c + Kf(t) = \frac{1}{2(t_2 - t_1)} \tag{8-38}$$

where $K = k_f/2\pi$.

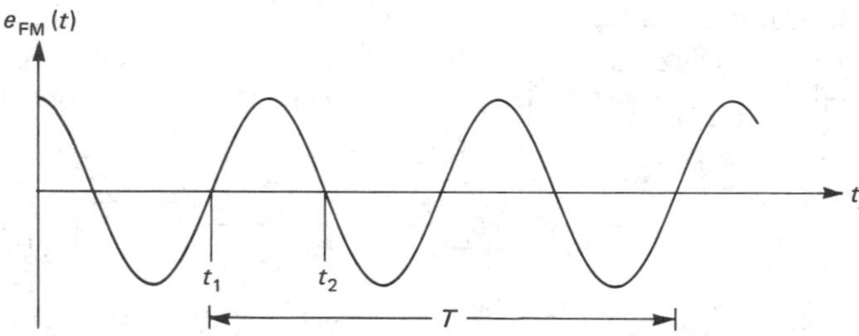

Figure 8-27 Zero crossings of an FM signal.

If we count the number of zero crossings over a larger interval, T, that is still small enough to ensure a reasonable constant $f(t)$ (i.e., T is much less than $1/B$ of the modulating signal), more zero crossings can be detected by the measurement equipment. If n_T denotes the number of crossings in time T, the time between adjacent crossings will be given by

$$t_2 - t_1 = \frac{T}{n_T} \tag{8-39}$$

Thus

$$f_i = f_c + Kf(t) = \frac{n_T}{2T} \tag{8-40}$$

Hence n_T can be used to recover $f(t)$.

An example of a balanced FM zero-crossing detector that eliminates f_c is shown in Fig. 8-28(a). In the accompanying waveform (b), f_i is assumed to be greater than f_c. The monostable vibrator is triggered on the positive slope crossings and produces a pulse of duration $T_c/2$, where T_c is the period of the unmodulated carrier, $1/f_c$. If the applied FM signal is at the carrier frequency, the Q and $\overline{\text{Q}}$ outputs form a square wave and a differential output voltage of zero results. When $f_i > f_c$, the period of f_i (i.e., T_i) is less than T_c and the monostable vibrator will trigger earlier, resulting in an increased net positive voltage at Q and a reduced net positive voltage at $\overline{\text{Q}}$. The low-frequency message signal is extrapolated by filtering and differential amplification. A linear frequency-to-voltage characteristic is obtained for an applied FM signal. The sensitivity of this detector is low and therefore it is seldom used in communication receivers.

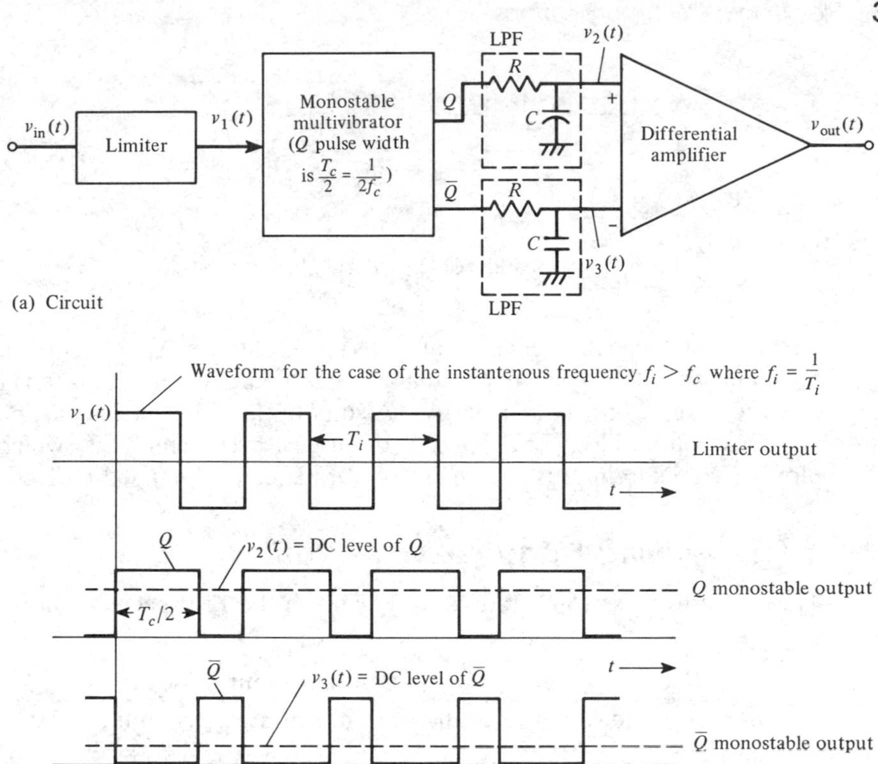

Figure 8-28 A balanced zero-crossing FM detector. (From Leon W. Couch II, *Digital and Analog Communication Systems*, 2nd ed. (New York: Macmillan Publishing Company, 1987), p. 245.

8-9 SIGNAL-TO-NOISE RATIOS

An important factor in quality of reception is the signal-to-noise power ratio present at the receiver output (i.e., at the speaker). Noise, which can be both internally and externally generated (refer to Chapter 9), is always present to some degree within any receiver. For the partial FM receiver given in Fig. 8-29, this noise will manifest itself in the frequency domain as random frequency/amplitude components appearing at the detector input. When summed together, these various components can combine either by themselves, or in the presence of an incoming signal, to produce a total waveform having instantaneous frequency variations, to which the detector responds, producing an output signal. Ultimately, the task at hand is therefore to ensure that the demodulated signal power is sufficient to mask the noise power present at the receiver output in order to recover the original message.

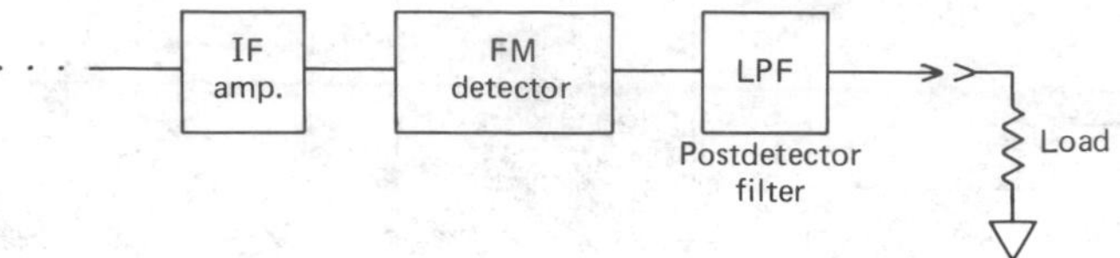

Figure 8-29 Partial FM receiver.

As a detailed mathematical analysis is presented in Appendix B, the following presentation will be primarily factual and based largely on conclusions found there (for the sake of brevity and clarity). Also note that the conclusions in this section are based on an FM system without the use of pre/deemphasis (which when employed, serves to improve the noise performance as discussed in Section 8-7).

8-9.1 Defining Signal-to-Noise Ratio

The signal-to-noise (S/N) ratio at any point in the FM receiver can be defined to be

$$\frac{S}{N} = \frac{\text{signal power without noise}}{\text{noise power in the presence of a carrier (unmodulated)}} \tag{8-41}$$

For the partial receiver given in Fig. 8-29 with only a low-pass filter between the detector and load (which could represent a speaker), Appendix B, equation (B-12), shows the S/N ratio at the load to be

$$\left(\frac{S_o}{N_o}\right)_{\text{FM}} \approx \frac{3\,\Delta f^2\,E_c^2}{4\eta W^3} \tag{8-42}$$

This important result reveals that the best S/N performance will be obtained when the frequency deviation is as large as possible at the input to the FM detector (i.e., given that IF bandwidth is limited, as much as possible should be used, since in general, as deviation increases at the transmitter, so does the bandwidth of the transmitted signal). Also, as E_c is made larger (corresponding to increasing transmitter power, and hence received signal power) the SNR will improve. On the other hand, if η increases (corresponding to an increase of internally or externally generated noise present at the input to the detector), or if the bandwidth of the low-pass filter is increased, the S/N ratio will degenerate. Equation (8-42) demonstrates that actual S/N ratio for a given receiver is a compromise between several factors, as discussed above, and generally 10 log (S/N) should be at least 12 dB or so for satisfactory information recovery.

8-9.2 Output S/N versus Input C/N for FM Detectors

Equation (B-19) (of Appendix B) states that

$$\left(\frac{S_o}{N_o}\right)_{\text{FM}} = 3(M_f)^2\left(\frac{C_i}{N_i}\right)_{\text{FM}} \tag{8-43}$$

This mathematical result is shown graphically in Fig. 8-30.

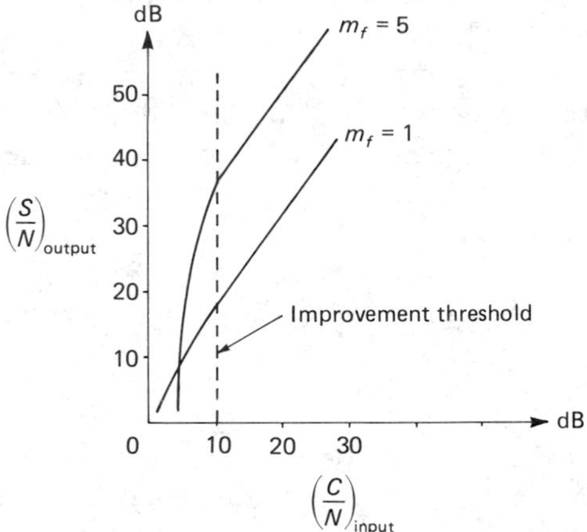

Figure 8-30 Output S/N versus input C/N.

EXAMPLE 8-11

Given that a VHF FM transmitter modulated with a 1-kHz tone, on a carrier frequency of 146.000 MHz, having a deviation of 5 kHz, is used to transmit a tone to an FM receiver:

(a) Use equation (8-43) to find the C/N ratio required at the detector input to produce an output S/N ratio of 40 dB.

(b) Repeat part (a), but use Fig. 8-30.

Solution

(a) (1) $10 \log (S_o/N_o) = 40$; therefore, $S_o/N_o = 10{,}000$.
(2) $10^4 = 3(5)^2\,(C_i/N_i)$; therefore, $C_i/N_i = 133.3$.
(3) $10 \log (C_i/N_i) \simeq 21.2$ dB.

(b) Using Fig. 8-30, an input C/N of ≈ 20 dB is required.

Examination of the plot for $M_f = 5$ in Fig. 8-30 reveals that for C/N decreasing below approximately 10 dB, the output S/N quickly deteriorates. This sharp transition between good and poor output S/N ratios implies a threshold of approximately 10 dB, called the *FM improvement threshold.* As long as the input C/N ratio is greater than this threshold level, reception will be good.

8-9.3 Capture Effect

In Section 8-9.2 we discussed the relationship between detector output S/N and the input C/N ratio. The conclusions drawn were based on the initial assumption that for analytical purposes, the noise spectrum at the detector input was flat and continuous. If instead of a noise spectrum (in addition to a desired signal spectrum) at the detector input, we substitute an interfering signal (such as a second FM signal), also reduced in amplitude compared to the desired signal (as was the noise for analysis), we find that if the main signal to interfering signal ratio is greater than approximately 10 dB, the SNR out of the receiver will be good. This implies a definite suppression of smaller signals in the presence of stronger ones, a phenomenon referred to as the *capture effect.*

If two signals of nearly the same strength are being received, the receiver may fluctuate between stations; this explains the reason behind the distance/local switch on many automotive broadcast FM receivers. When on "local," the receiver sensitivity is reduced to provide a more definite distinction between desired and undesired signals present at the input to the detector.

8-10 GE CENTURY II FM TRANSCEIVERS

As a practical example of an actual commercially available transceiver, the frequency synthesized General Electric Century II will be discussed.

8-10.1 General Description

The Century II is available in various forms: different frequency bands, crystal controlled or frequency synthesized, and tone-coded squelch (optional). A sample specification sheet is given in Table 8-3. The remainder of this section will deal specifically with the frequency-synthesized, 420 to 512 MHz version of the Century II without tone-coded squelch (channel guard).

8-10.2 Block Diagram

The block diagram is given in Fig. 8-31 and follows a more-or-less standard format used by most UHF transceivers: dual-conversion superhet receiver, oscillator/multiplier/amplifier chain for the transmitter, and so on.

Receiver

The incoming UHF signal is amplified (Q401), frequency converted down to a 21.4-MHz intermediate frequency (Q402), and passed to the high-IF section, where selective filtering (crystal filter Z501) and amplification (Q501) occurs. The signal is then once again down-converted, this time to a center frequency of 455 kHz (U501, mixer amplifier), and passed through another very selective filter (ceramic filter Z502). Limiting (to remove amplitude variations) and detection then occurs (U502), after which the audio/squelch circuitry follows. The squelch circuit basically detects high-frequency noise to determine if an incoming signal is present (i.e., noise will be present at the detector output only if a weak signal, or no signal, is present). Detected noise (by D603, D604) produces a more-or-less dc level which feeds a Schmidt trigger (Q603, Q604), whose output will be high or low, resulting in either an enabled or disabled (muted) audio path through audio amplifier U601.

TABLE 8-3 GE Century II System Specifications[a]

Specification	Value
Frequency range	420–512 MHz
Battery drain	
Receiver	
Squelched	200 mA
Unsquelched	650 mA
Transmitter	
KT-179-A	1.8 A at 13.8 V
KT-180-A, KT-198-A	5.5 A at 13.8 V
Frequency stability	0.0005%
Temperature range	−30°C (−22°F) to +60°C (140°F)
Duty cycle	20% transmit, 80% receive
Dimensions, less accessories ($H \times W \times D$)	60 × 180 × 190 mm (2.3 × 7.3 × 7.4 in.)
Weight, less accessories	1.7 kg (3.7 lb)

Transmitter

Power output		
KT-179-A	2–5 W (420–470 MHz)	
KT-180-A	7–20 W (420–470 MHz)	
KT-198-A	6–18 W (470–494 MHz)	
KT-198-A	5–15 W (494–512 MHz)	
Spurious and harmonic emission	−50 dB (5 W) (FCC) −56 dB (20 W) (FCC)	
Modulation	±4.5 kHz	
Audio sensitivity	65–120 mV	
Audio frequency characteristics	Within +1 to −3 dB of a 6-dB/octave preemphasis from 300 to 3,000 Hz per EIA standards; post-limiter filter per FCC and EIA	
Distortion	Less than 3% (1000 Hz) Less than 5% (300 to 3000 Hz)	
Deviation symmetry	0.5 kHz maximum	
Maximum frequency spread	Full Specifications	1-dB Degradation
420–470 MHz	5.5 MHz	10.5 MHz
470–494 MHz	5.5 MHz	7.0 MHz
494–512 MHz	6.0 MHz	7.0 MHz
RF output impedance	50 Ω	

Receiver

ER-116-A (420–470 MHz)
ER-129-A (470–572 MHz)

Audio output (to 4.0-Ω speaker)	3 W (less than 5% distortion) EIA 1.5 W (less than 5% distortion) CEPT	
Sensitivity		
12 dB SINAD (EIA) method)	0.40 μV	
20 dB quieting method	0.45 μV	
20 dB SINAD (CEPT[b])	0.75 μV	
Selectivity		
EIA two-signal method	−85 dB at ±25 kHz (EIA) −75 dB (CEPT)	
Spurious response	−85 dB	
Intermodulation	−75 dB	
Modulation acceptance	±7.0 kHz	
Squelch sensitivity	<8 dB SINAD	
Maximum frequency spread	Full Specifications	3.0-dB Degradation
420–512 MHz	2.0 MHz	3.0 MHz
Frequency response	Within +1 and −1.5 dB of a standard CEPT 6 dB/octave deemphasis curve from 400 to 2700 Hz (1000 Hz reference); also fits +1 to −3 dB from 300 to 3000 EIA	
RF input impedance	50 Ω	

[a]EIA and CEPT unless otherwise noted. These specifications are intended primarily for the use of the service persons. Refer to the appropriate specifications sheet for the complete specifications.
[b]ΔF 60% × ΔF_{max}; F_{mod} = 1 kHz. Measured with psophometric filter.

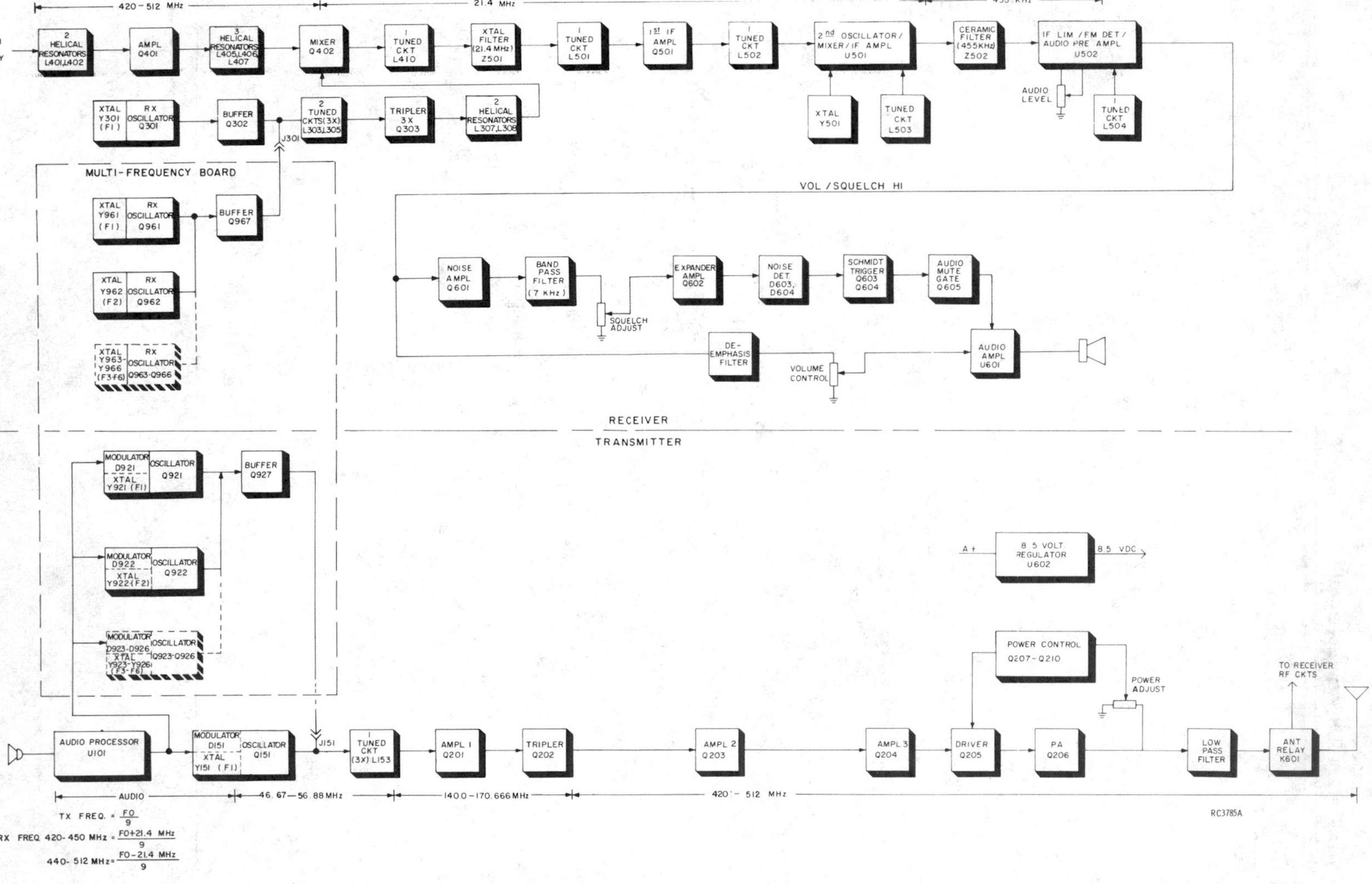

Figure 8-31 GE Century II transmitter/receiver block diagram.

Transmitter

Preemphasis and bandlimiting (to about 3 kHz) occurs in the U101 stage, before passing audio to the synthesizer, where frequency modulation occurs. The FM signal from the synthesizer is then amplified and frequency multiplied as shown in the block diagram. A power control circuit (Q207 to Q210) allows some measure of adjustment on output power (see specifications, Table 8-3). The low-pass filter following the PA (power amplifier) attenuates harmonics produced by class C amplifier Q206 to an acceptable level.

8-10.3 Transmitter Schematic

Figure 8-32 gives part of the circuitry used while in transmit mode. A brief description follows. Audio from the MIC is injected to pin 1 of P101, then passed to audio processor U101, where preemphasis (R102, R103, R104, C104), limiting (D101, D102), and bandlimiting (stage U101A is an active low-pass filter) occurs. Recall that preemphasis is required to improve the received SNR. Limiting is required to prevent overdeviation, and bandlimiting is required to minimize required bandwidth. Output from U101A is then passed to the synthesizer board (via pin 4, P101), where it is used to frequency modulate a carrier.

The FM wave produced by the synthesizer (to be discussed later) is injected to J151 (Q151 collector) and passed to the base of Q201 via tuned circuit L153, C158, C157 (tuned to one-third of the transmitter output carrier frequency). Amplifier Q201 (class C) boosts the signal and passes it to the base of tripler Q202 (also class C), whose output circuits will be tuned to the transmitter carrier frequency (UHF). C213 and C215 will be adjusted to ensure proper tuning of the L208–C213 and L209–C215 tuned circuits. Q203, Q204, Q205, and Q206 are all basically class C amplifiers used to increase power. Power control circuitry provides for adjustment of the effective dc power supply "seen" by driver Q205, thus allowing some control over the peak-to-peak swing present at the collector of Q205. In addition, the power-adjust circuitry stabilizes against power output variations due to temperature changes or component aging: that is, an increase in power output results in increased forward bias on Q210, causing decreased forward bias on Q209, and hence decreased forward bias on Q208, Q207 and smaller dc voltage at the emitter of Q207, which means a lower effective supply seen by Q205. A decrease in power output similarly results in an increase of the effective supply seen by Q205.

8-10.4 Receiver Schematic

Figure 8-33 gives part of the receive mode circuitry: a carrier produced by the synthesizer is injected at J301 (near collector of Q302). (The Rx oscillator will be disabled due to missing crystal X301 in the synthesized version.) The synthesizer-generated carrier is then tripled by Q303 (class C amplifier) to UHF for injection to the source of first mixer (Q402). The incoming signal from the antenna is passed via helical resonators L401 and L402 to the RF amplifier (Q401), where the signal is first amplified. The output of Q401 is passed through more helical resonators (L405 to L407) to achieve front-end selectivity before injection to the first mixer (gate of Q402).

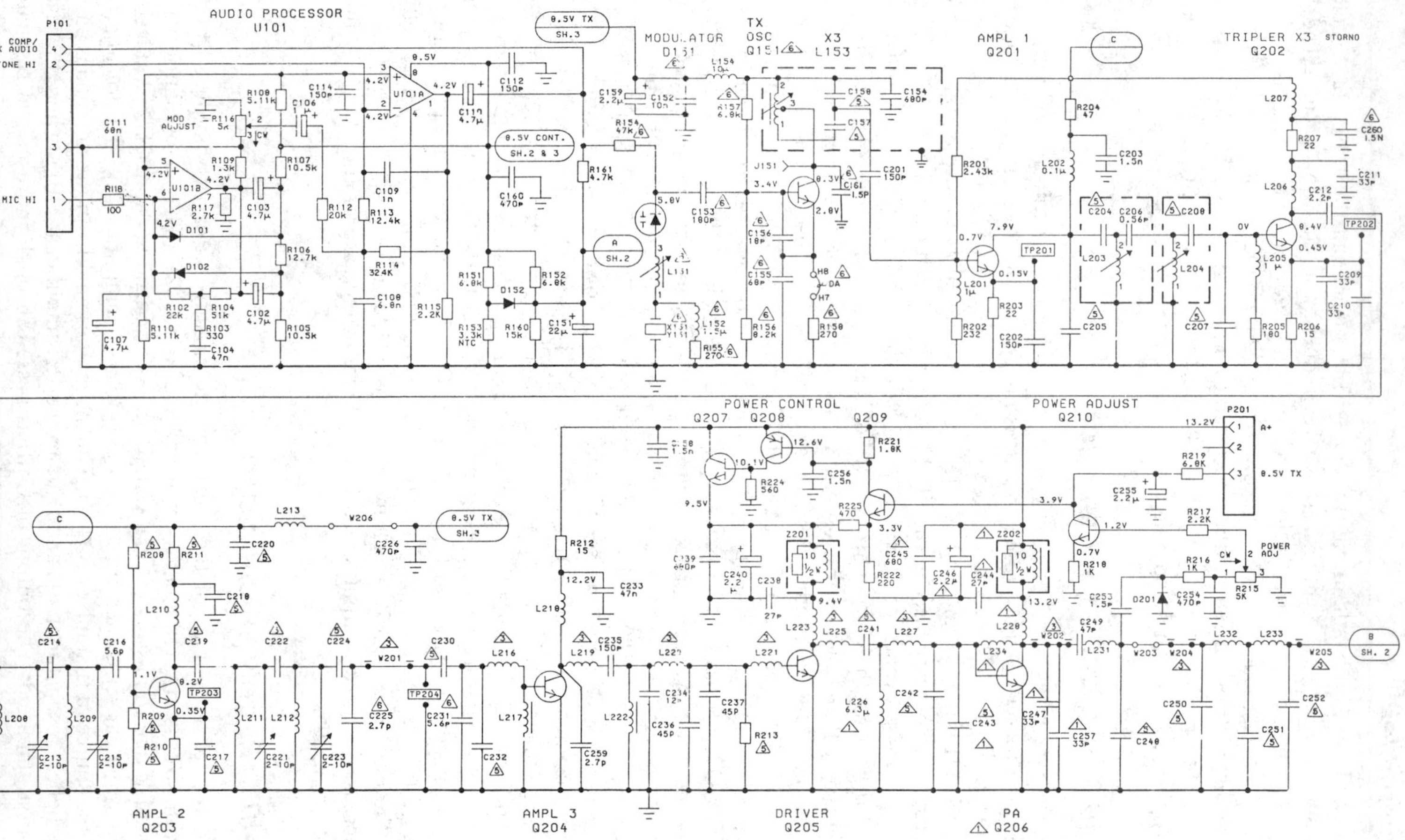

Figure 8-32 Schematic diagram of the 420- to 512-MHz UHF transmitter.

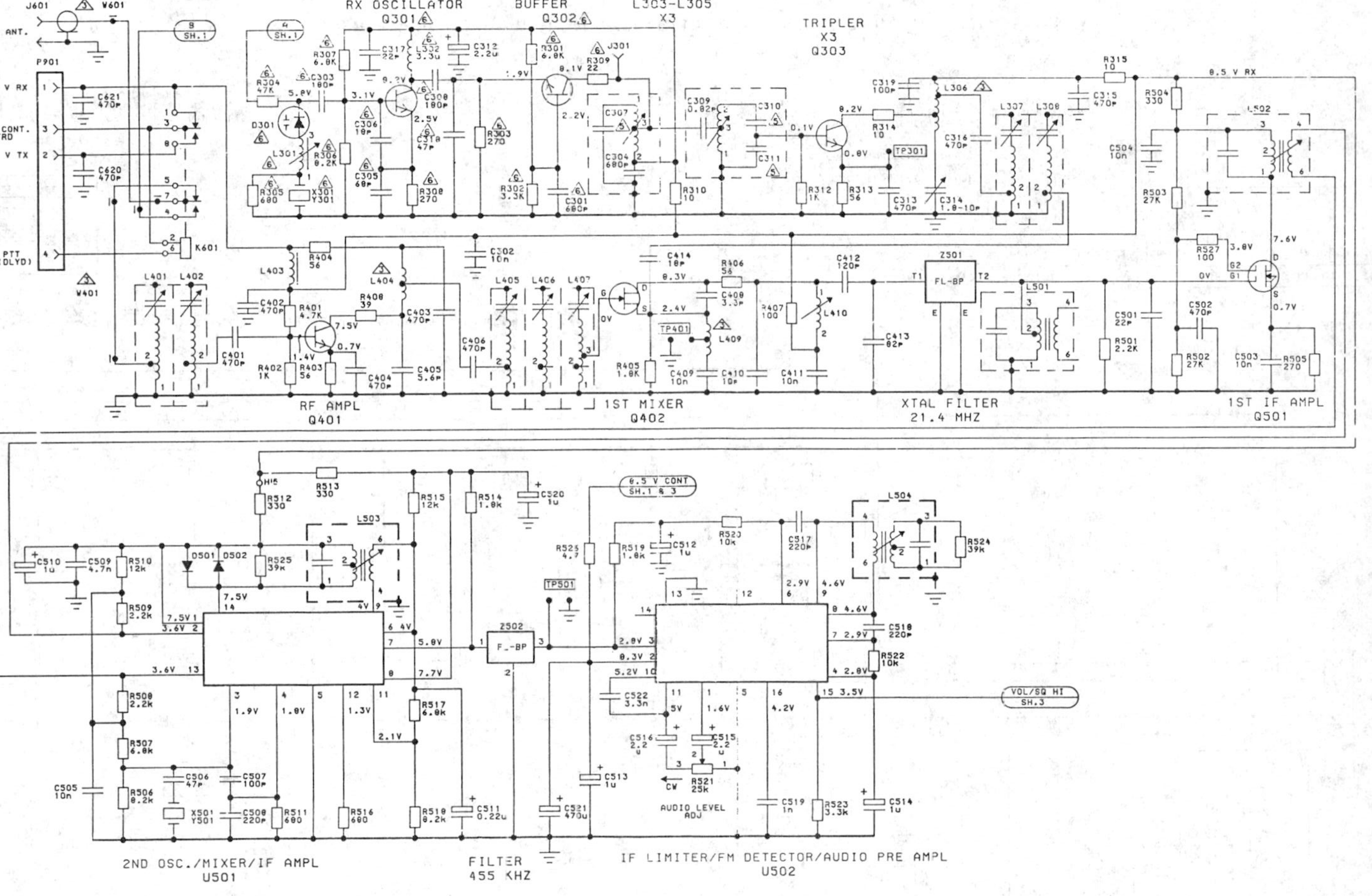

Figure 8-33 Schematic diagram of the 420- to 512-MHz UHF receiver.

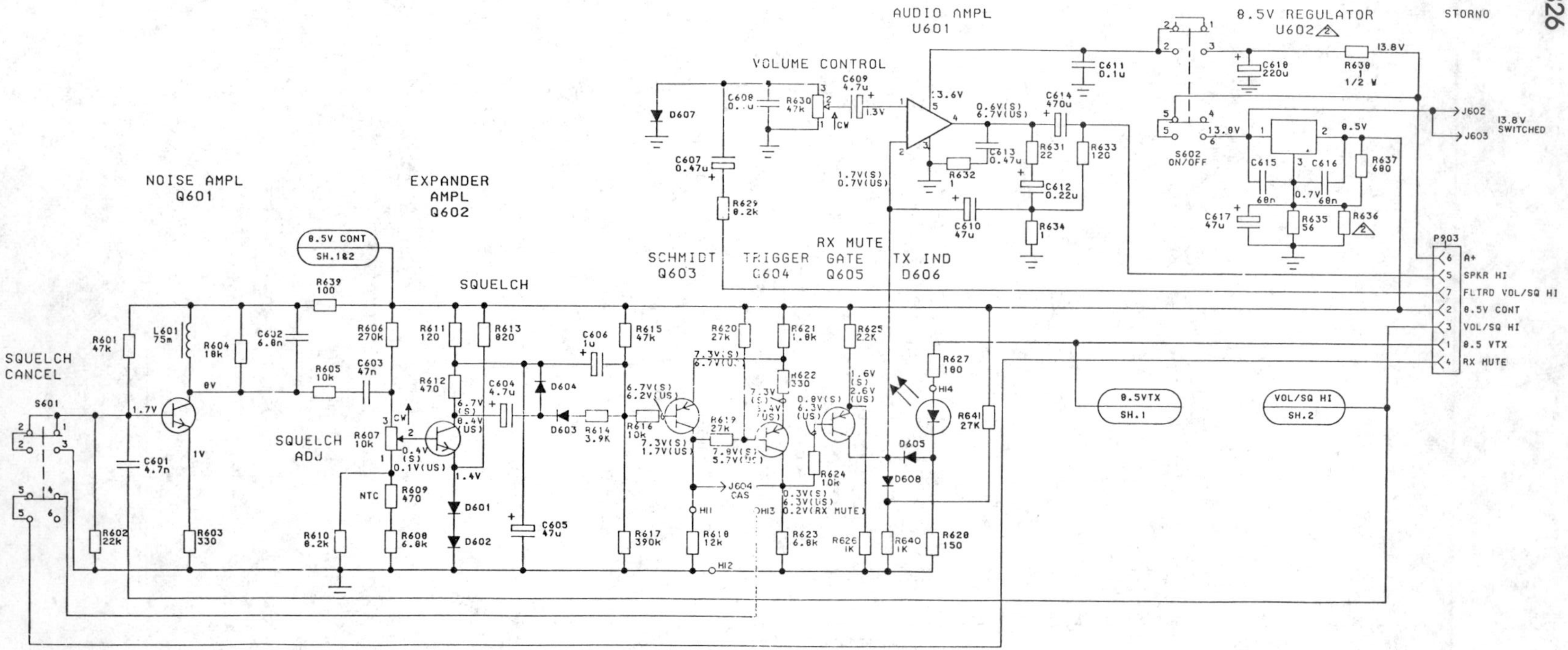

STORNO

MAINTAIN MANUALLY

NOTES:

1. FOR 5 W (G2) AND 3W (G4) TRANSMITTERS REMOVE C243, C244, C245, C246, C247, C257, L226, L228, Q206 & Z202 AND ADD L234 .

2. VALUE OF R636 DEPENDS ON COLOR CODE ON U602.

U602 COLOR CODE	R636 VALUE Ω
BROWN	OMIT R636
RED	270
ORANGE	100
YELLOW	47
GREEN	22
BLUE	6.8

3. PART OF PRINTED CIRCUIT BOARD.

4. TO MODIFY FOR MULTIFREQUENCY, REMOVE R157 (DISABLES TX OSC) AND/OR R309 (DISABLES RX OSC). THIS NOTE DOES NOT APPLY TO UHF-X (GROUP 3 OR GROUP 4).

5. COMPONENT VALUES SEE SHEET 4

6. SEE SHEET 4

ALL RESISTORS ARE 1/4 WATT UNLESS OTHERWISE SPECIFIED.
RESISTOR VALUES IN Ω UNLESS FOLLOWED BY MULTIPLIER k OR M.
CAPACITOR VALUES IN F UNLESS FOLLOWED BY MULTIPLIER u, n OR p.
INDUCTANCE VALUES IN H UNLESS FOLLOWED BY MULTIPLIER m OR u.

VOLTAGE READINGS

VOLTAGE READINGS ARE TYPICAL READINGS MEASURED TO SYSTEM NEGATIVE WITH A 20,000 OHM-PER-VOLT DC VOLTMETER UNDER THE FOLLOWING CONDITIONS:

1. NO SIGNAL INPUT
2. VOLUME CONTROL (R630) SET TO MINIMUM
3. SQUELCH CANCEL (S601) SWITCHED OFF
4. UNSQUELCHED (US)-SQUELCH ADJUST (R607) SET TO MINIMUM (CCW)
5. SQUELCHED (S)-SQUELCH ADJUST (R607) SET TO MAXIMUM (CW)

THIS ELEM DIAG APPLIES TO	
MODEL NO	REV LETTER
19D900158G1	B
19D900158G2	B
19D900158G3	D
19D900158G4	

Figure 8-34 Schematic diagram of the audio/squelch circuits.

Z501 (21.4-MHz crystal filter) follows the mixer to provide high selectivity. First IF amplifier Q501 provides some gain prior to the second frequency conversion by the second mixer in U501 (U501 receives input at 21.4 MHz on pins 2 and 13). Crystal X501 is used by the on-chip oscillator within U501 to provide for mixing down to a center frequency of 455 kHz, which appears at pin 14 of U501, and is then limited by D501 and D502 before injection to the IF amp within U501 (at pin 6). The output of U501 IF amp appears at pin 7 and is then passed through ceramic filter Z502 for additional selectivity. U502 provides limiting, detection (via quadrature detector within U502), and audio preamplification. The recovered audio (pin 15) of U502 is then passed to the audio/squelch circuit given in Fig. 8-34 and is injected at pins 7 and 3 of P903. Pin 7 is the input for deemphasis network C607–R629, which restores the audio to the proper frequency/amplitude relationships before amplification by audio amplifier U601. Pin 3 of P903 provides input to noise amplifier Q601, which is followed by a high-pass filter R605–C603. A signal will be present out of this high-pass filter only when there is either a very weak signal, or no incoming signal, to the receiver (recall the SNR performance for FM).

Noise is detected (converted to dc) by the D604–D603 circuit and passed to Schmidt trigger Q603, which controls muting (disabling) of the audio via Q605 and U601. A Schmidt trigger is used since the received signal often takes sudden dips in strength in a mobile environment, and it is desirable for the receive audio to mute only when the signal gets very weak (weaker then the smallest level that first allows the squelch to open).

8-10.5 Frequency Synthesizer

The frequency synthesizer provides for 16-channel capability without the need for 32 separate crystals (i.e., one Rx, one Tx for each channel). Instead, the frequency corresponding to the channel selected is controlled via the data stored in a 32- by 8-bit PROM (82S23 or TBP 18SA030) mounted in a DIP socket on the synthesizer board. Figure 8-35 illustrates the basic synthesizer structure.

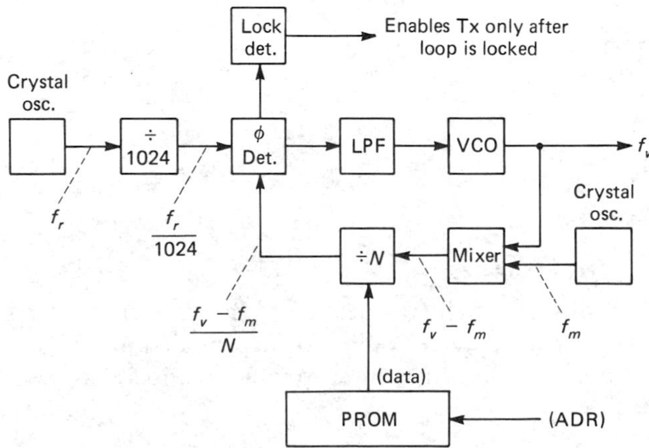

Figure 8-35 Basic synthesizer structure.

Analysis When locked,

$$\frac{f_r}{1024} = \frac{f_v - f_m}{N} \tag{8-44}$$

Therefore,

$$f_v = f_m + \frac{f_r}{1024}(N) \tag{8-45}$$

With both f_m and f_r crystal controlled, excellent frequency stability will be associated with f_v (when locked). Since N is controlled by PROM data, f_v will also be a function of PROM data (up to 32 PROM addresses may be programmed, yielding up to 32 unique frequencies produced).

Receiver/Transmitter Interconnection

To minimize hardware, cost, space, and weight, only one synthesizer is used to provide local oscillator signals for both Rx and Tx. The fundamentals of operation are illustrated in Fig. 8-36; complete information is given in the maintenance manual.

Receive/transmit oscillators triple the crystal frequency and contain temperature-compensation circuitry to avoid excessive frequency error. Only one oscillator is selected at a time, this being controlled by the Rx/Tx switch, which in turn is controlled by push-to-talk (PTT) button on the microphone. The Tx oscillator and VCO are both frequency modulated by incoming audio, as shown in Fig. 8-36. Like the oscillators, only one VCO may be selected, depending on the current mode (Rx/Tx).

Synthesizer IC (U701) (shown within the dashed lines) contains much of the PLL circuitry required. Note that 8 bits of PROM data control M, according to the equation

$$M = 256 + n \tag{8-46}$$

where n is the decimal equivalent of the binary byte corresponding to PROM DATA; for example:

$$\begin{array}{cc} \text{(D7)} \cdots & \text{(D0)} \\ 0\ 0\ 0\ 1 & 1\ 0\ 0\ 0 \\ 2^7 \quad\ 2^4 & 2^3 \quad\ 2^0 \end{array}$$

Therefore, $M = 256 + 2^3 + 2^4 = 256 + 24 = 280$.

U701 also internally divides reference frequency f_r by 1024 (2^{10}). A complete description of lock detector operation is provided in the GE manual, but basically, PTT is supplied to the transmitter only after the synthesizer locks.

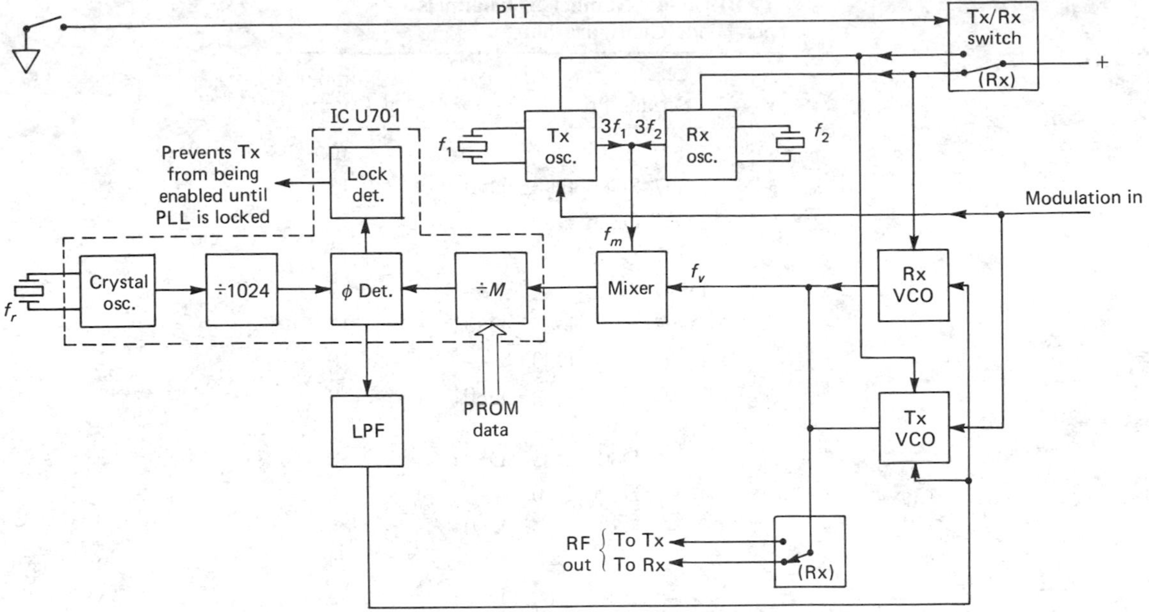

Figure 8-36 Rx/Tx interconnection.

PROM/Display Board

Figure 8-37 shows a block diagram of the PROM/display board, which controls frequency (Rx/Tx) and display information (LED). The channel select switch advances the counter through 16 unique counts. Each count provides a unique address to the two 32 × 8B PROMs (82S23 or TBP 18SA030). Normally, channel numbers 01 to 16 are displayed as a result of the data programmed into the display PROM (if all 16 channels are used). Not all 16 channels need to be addressed. The number of unique counts produced by the counter is dependent on diodes D804 to D808 (see Fig. 8-38). The diode configuration determines the count reached when "reset" occurs to recycle the counter. Table 8-4 illustrates.

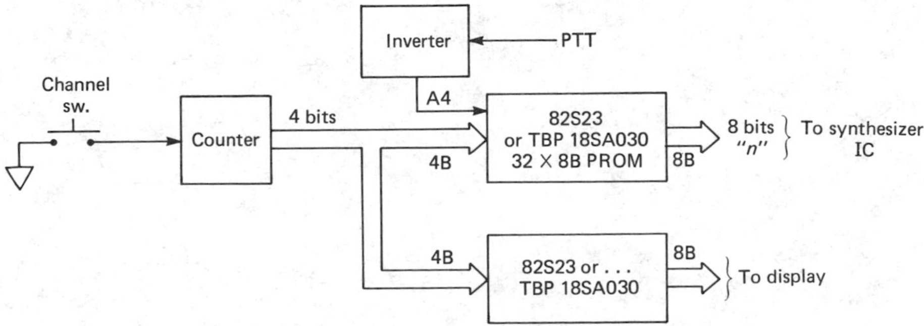

Figure 8-37 Block diagram of PROM/display board.

TABLE 8-4 Number of Channels per Diode Configuration

Diodes Present					Number of Channels
D808					16
	D807	D806	D805	D804	15
	D807	D806	D805		14
	D807	D806	—	D804	13
	D807	D806	—	—	12
	D807	—	D805	D804	11
	D807	—	D805	—	10
	D807	—	—	D804	9
	D807	—	—	—	8
	—	D806	D805	D804	7
	—	D806	D805	—	6
	—	D806	—	D804	5
	—	—	—	—	4
	—	—	D805	D804	3
	—	—	D805	—	2
	—	—	—	D804	1

Note that A4 (address bit 4) of the frequency PROM is controlled by PTT. If the unit is in receive mode, A4 = 0, and the first half of the PROM is addressed by the counter; if transmitting, A4 = 1, and the upper half of the frequency PROM is addressed (see Fig. 8-38).

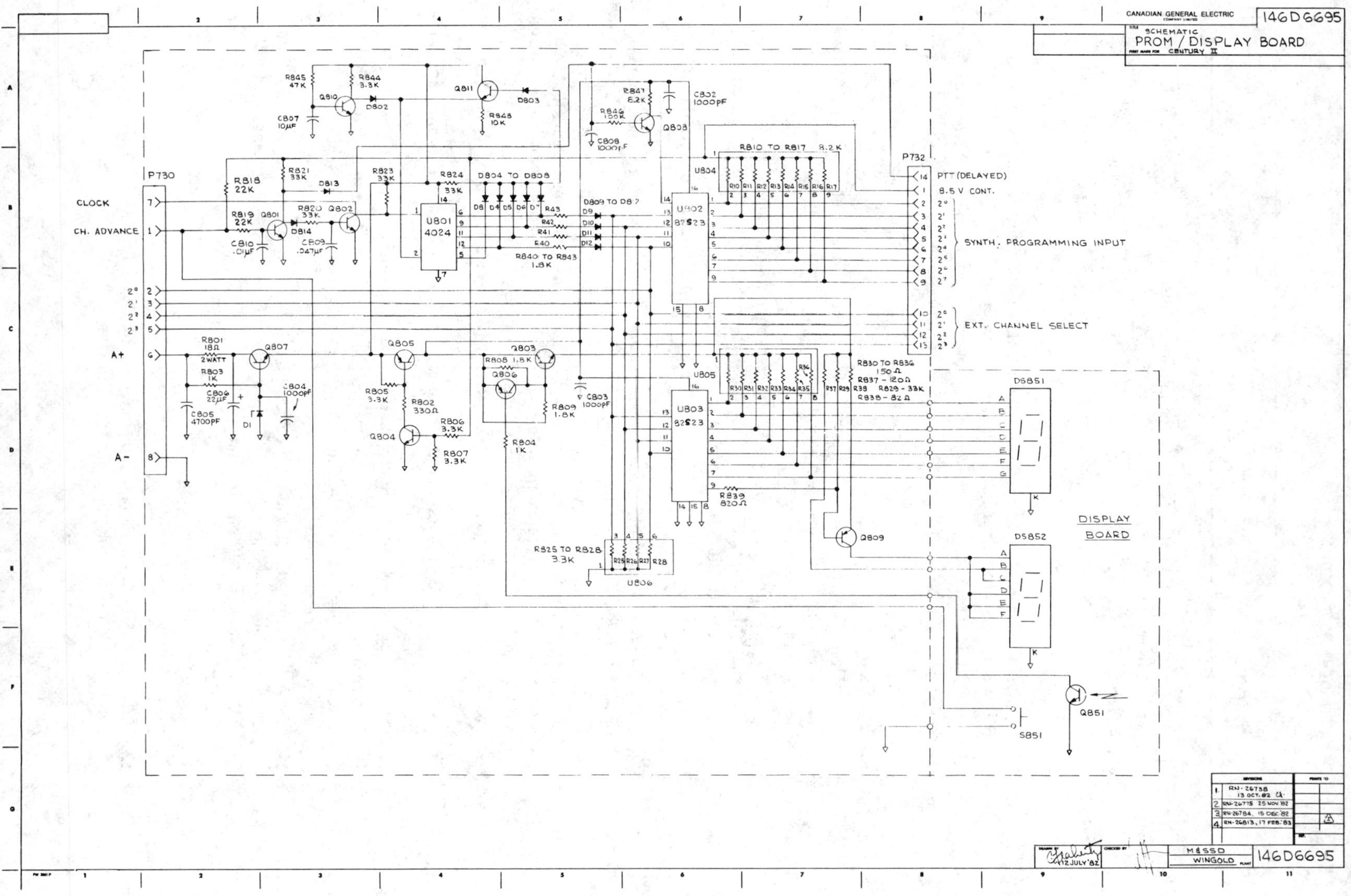

Figure 8-38 PROM/display board.

Problems

8-1. What technical term corresponds to the maximum instantaneous frequency departure for an FM signal?

8-2. Given a wave equation of

$$e(t) = E_c \cos\left(\omega_c t + \frac{\Delta f}{f_m} \cos \omega_m t\right)$$

(a) What is the corresponding expression for the instantaneous frequency?
(b) What expression describes the frequency departure?

8-3. A 100-MHz carrier is frequency modulated by a 2-kHz tone. The positive peak of the audio signal produces a corresponding instantaneous frequency of 100.075 MHz. Find:
(a) The rest frequency
(b) The deviation
(c) The modulation index M_f

8-4. If 1 kHz of deviation is produced out of an ideal FM modulator by a 2-V (rms) modulating tone, what deviation would result if the modulating signal amplitude were increased to 5 V peak?

8-5. A 10-V rms carrier of frequency 100 MHz is frequency modulated by a 15-kHz tone producing a deviation of 30 kHz. Sketch the resulting spectrum showing frequencies and amplitudes (in rms) of all significant components.

8-6. Calculate the power dissipated into a 50-Ω load by the modulated wave described in Problem 8-5.

8-7. Does the power output of an FM transmitter vary with modulating signal conditions, or is it independent of the modulating signal? Explain.

8-8. Explain why an FM signal can be restricted to a finite bandwidth, even when modulating signal conditions (i.e., frequency, amplitude) are constantly changing (such as for voice). Do this by relating to modulation index, deviation (which depends on modulating signal amplitude), and sideband spacing.

8-9. Frequency modulating a 154.0-MHz carrier with a 1-kHz tone produces a deviation of 5 kHz.
(a) Use Table 8-2 to determine the required bandwidth to pass the significant components.
(b) Use Carson's rule to estimate the required bandwidth.

8-10. Determine the total power of the carrier and sidebands transmitted from a 50-Ω antenna if $M_f = 2.0$ and the unmodulated carrier has an amplitude of 1 kV peak (neglecting antenna losses).

8-11. A spectrum analyzer displays a spectrum that has the carrier at the same amplitude as the second set of sidebands. The spectral components have spacings of 10 kHz, and the carrier is at 100 MHz. Find:
(a) The modulation index
(b) The modulating frequency
(c) The frequency deviation

8-12. If the sensitivity of the modulation circuit used in Problem 8-11 is 20 kHz/V, find the amplitude of the modulating input signal in volts peak.

8-13. Under the conditions specified by Problems 8-11 and 8-12, what modulating signal amplitude (in volts peak) will produce the first carrier dropout?

8-14. For the FM signal of

$$e(t) = 100 \cos(2\pi\ 10^8 t + 20 \sin 2\pi\ 10^3 t)$$

(a) Find the approximate bandwidth occupied by the signal.
(b) Is this NBFM or WBFM?

8-15. Determine the amount of frequency deviation produced when a carrier wave is shifted 30° by a 5-V (rms) 10-kHz sine-wave modulating signal.

8-16. An FM transmitter has a deviation of 5 kHz. Determine the deviation when:
(a) The audio frequency is doubled
(b) The audio voltage is doubled

8-17 Why is the indirect method of FM generation often preferred over the direct method when designing a transmitter?

8-18. The preemphasis circuit of Fig. P8-18(a) meeting the standard FM broadcast specifications shown in Fig. P8-18(b) is to drive a load impedance of 10 kΩ. What value for C is required if R = 10 kΩ?

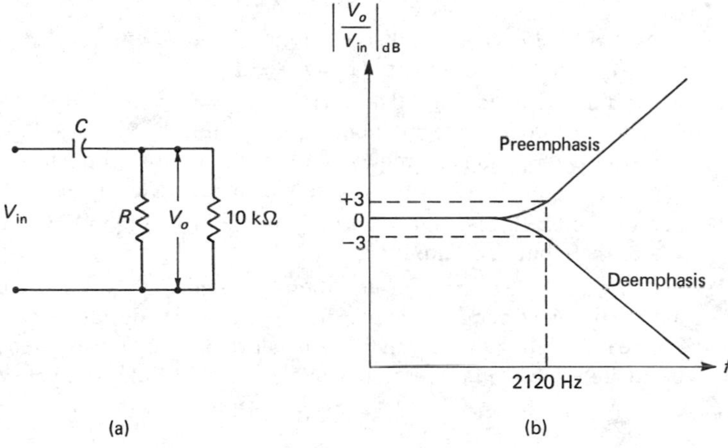

Figure P8-18.

8-19. Explain what the "capture effect" is and the reason for the occurrence of this phenomenon in FM systems.

8-20. The NBFM Armstrong modulator of Fig. P8-20 has a modulating signal of $e_m(t) = E_m \cos \omega_m t$ and an oscillator signal of $E_c \cos \omega_c t$.

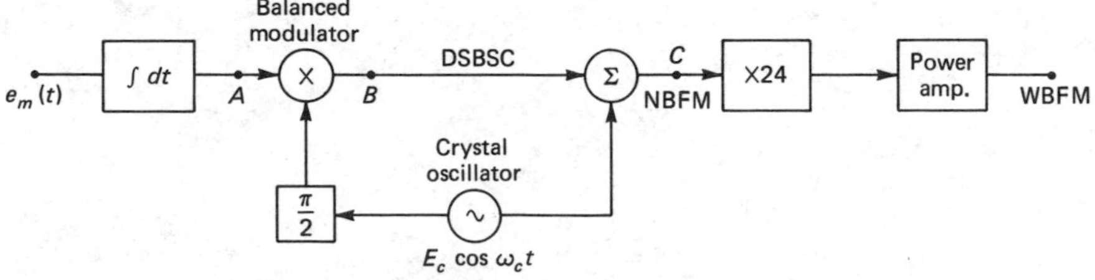

Figure P8-20.

(a) Obtain the expression for the signals appearing as A, B, and C. (*Hint:* Use $\sin x \sin y = \frac{1}{2}[\cos(x - y) - \cos(x + y)]$.)
(b) What frequency components appear at the NBFM output (point C)?
(c) With E_c fixed, what factors determine the amplitudes of the sidebands?
(d) With the phase angle of the carrier at zero, sketch the phasor diagram of the NBFM signal at time t_1 such that $\phi(t_1) = \omega_m t_1 = \pi/4$. Assume that $E_m/\omega_m = E_c/2$.
(e) Sketch the trace of the resultant phasor for various instants in time and note the maximum amplitude and phase variations. How can the amplitude variations be eliminated?

8-21. An FM broadcast station uses the Armstrong method (Fig. P8-20) with "times-24" multiplication as illustrated. If $f_c = 100$ MHz, $\Delta f = 75$ kHz, and the maximum allowed frequency drift at the output is ± 2 kHz, find the following:
(a) The crystal oscillator frequency and its maximum drift.
(b) The modulation index at the WBFM output when $f_m = 15$ kHz.
(c) The modulation index when $f_m = 5$ kHz.

8-22. Given a frequency multiplier with a characteristic of

$$e_o(t) = [e_i(t)]^3$$

and given that $e_i(t) = \cos \theta(t)$, where $\theta(t) = \omega_c t + M_f \sin \omega_m t$, write the expression for $e_o(t)$. [*Hint:* Use $\cos^2\theta = \frac{1}{2}(1 + \cos 2\theta)$.]

8-23. Regarding the quadrature detector:
(a) To what does the term "in quadrature" refer?
(b) Above the resonant frequency for a parallel tuned circuit used in a quadrature detector circuit, does the circuit appear resistive, capacitive, or inductive?

8-24. Sketch the block diagram of a simple phase-locked-loop FM detector. Show clearly where the input and output are.

8-25. Given that a legal limit on deviation is assigned to any particular transmitter (by authorities governing communications) and actual deviation at any instant depends on the modulating signal, can the actual deviation and the allowed maximum deviation be related to obtain a "best" possible S/N ratio at the receiving end?

Noise

Robert McPherson

9-1 INTRODUCTION

As the channel length between a transmitter and receiver is increased, the power level of the desired signal at the receiver input decreases. Thus progressively more gain must be supplied by the receiver in order to recover the information being transmitted. There is, however, an upper limit to this process, as beyond some point the information recovered will be badly degraded or lost entirely, even when more gain is supplied.

The reason for this effect is the presence in the system of signals not due to the desired transmitter. These "noise" signals are introduced in both the channel and the receiver itself. If the desired signal strength is not significantly greater than that of the noise signals present, the composite signal presented to the receiver generates an output that is unintelligible. As a measure of how corrupted the total signal is at a point in a communication system, we are interested in its signal-to-noise power ratio:

$$\frac{S}{N} = \frac{\text{power level of the signal due to the desired source}}{\text{power level of the signals due to all other sources}} \tag{9-1}$$

Noise sources ultimately limit the range of a communication system and are therefore of considerable interest. The examination of noise effects in this chapter is broken into three segments. In the first segment some of the physical phenomena that give rise of noise signals are examined. In the second section a simplified model for the effects of noise sources is presented, and the analysis of various communication systems is undertaken. The reader is also directed to Chapter 18 for further examples of noise analysis in a satellite communication system. The final segment of the chapter deals briefly with some of the measurement techniques used to quantify the noise characteristics of communication system elements.

9-2 SOURCES OF NOISE

For convenience, noise sources may be subdivided into three categories: noise sources in the communication receiver (receiver noise), noise sources in the channel between the transmitter and receiver (system noise), and noise sources that impinge on the system from outside the communication system elements (external noise).

9-2.1 Receiver Noise

The noise generated in a receiver can stem from many causes. Only the more significant ones will be discussed at this time.

Thermal Noise

Any conductive material above absolute zero in temperature will have free charge carriers which are in random thermal motion throughout the material. At any given instant a charge unbalance may exist, resulting in the potential V being developed across the material as shown in Fig. 9-1. The average potential developed must, of course, be zero, but the RMS potential and therefore noise power available is not zero (recall basic ac circuit theory).

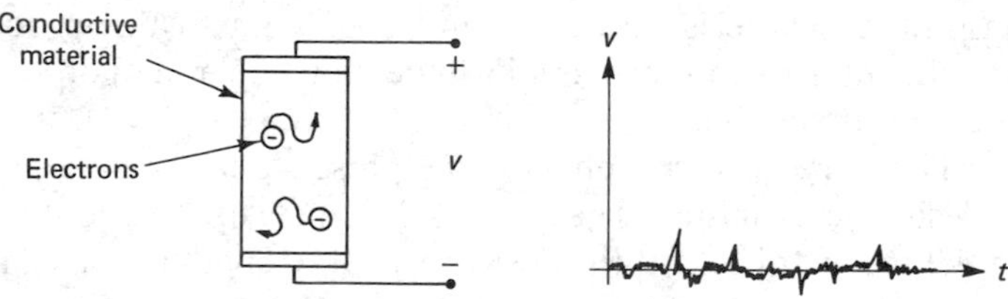

Figure 9-1 Thermal noise source.

An examination of the noise power generated in time domain is quite involved due to the random (probabilistic) nature of the generating mechanism. In the frequency domain, however, the noise power is much easier to describe. It is characterized by a uniform energy distribution over the frequency spectrum out to about 1000 GHz (Fig. 9-2). Such a uniform distribution is called white noise [recall that if all the colors (frequencies) of optical light are combined, the resulting light is white]. Thermal noise is generated by every system element, provided that the temperature is above absolute zero. Thermal noise sets the lower limit for the sensitivity of a system, but is frequently insignificant compared to the noise produced by transistors or tubes.

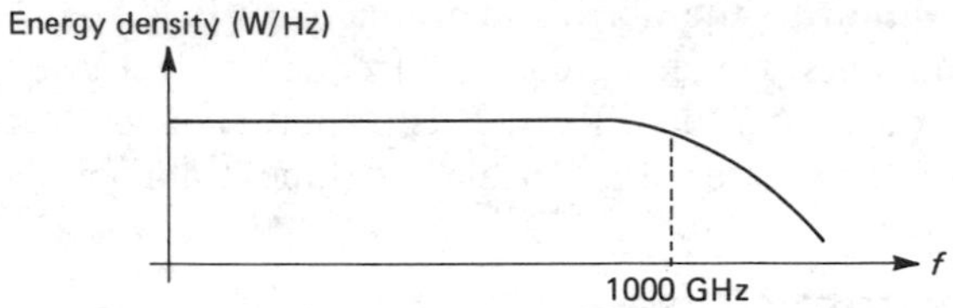

Figure 9-2 Thermal noise power density spectrum.

Receivers generally have multiple filter stages to receive selectively the band of frequencies over which the desired signal is being transmitted. These filters remove thermal noise outside the signal passband. The in-band thermal noise that passes through with the desired signal has a power level given by

$$P_n = kTB \tag{9-2}$$

where P_n is the available noise power, k is Boltzmann's constant $= 1.3803 \times 10^{-23}$ J/K, T is the absolute temperature (Kelvin), and B is the effective bandwidth of the system (hertz). At a standard temperature of 17°C or $T = T_0 = 290$ K,

$$P_n = kT_0B = 4 \times 10^{-21} \text{ W} \tag{9-3}$$

This represents -204 dBW or -174 dBm for a bandwidth of 1 Hz.

Intermodulation Noise

These products are produced because of the nonlinearity of the device or medium. When two signals with frequencies F_1 and F_2 are passed through such a device, the following products are formed which can fall into the passband:

Second-order products: $F_1 \pm F_2$
Third-order products: $F_1 \pm 2F_2$, $2F_1 \pm F_1$
Fourth-order products: $2F_2 \pm 2F_2$, $2F_2 \pm 2F_1$

In multichannel systems carrying complex signals, intermodulation noise appears similar to thermal noise. It may result from driving the system too hard, improper alignment, or nonlinear envelope delay.

Transistor Noise

Noise in transistors stems from three major sources: shot noise, partition noise, and thermal noise. *Shot noise*, which sounds like a shower of lead shot striking a metallic target, is the result of recombination fluctuations of holes combining with the electrons in the base region. The shot effect is dependent on the level of the bias current.

Partition noise is the result of the random fluctuations in the direction of carriers when they have to divide between two or more paths. More electrodes cause an increase in partition noise. In vacuum tubes, pentodes are more noisy than triodes for this reason. *Thermal noise* is due to the resistance of the base region.

In addition to the above, at frequencies below a few kilohertz, *flicker* (or *1/f) noise* appears, which increases as frequency decreases. It is thought to be due to surface recombination effects. It is of particular concern in low-frequency and dc amplifiers encountered in certain instrument and biomedical applications.

9-2.2 System Noise

Crosstalk

On a voice frequency cable system, unwanted coupling or *crosstalk* noise is experienced. It can be caused by coupling between transmission media.

Impulse Noise

Voltage spikes of short duration and of relatively high amplitude are induced on lines by switching and transients on neighboring lines. The presence of these pulse-type signals has a serious degrading effect on data transmission circuits. Impulse noise has only a marginal degradation effect on voice telephony.

9-2.3 External Noise

The most important sources of interfering noise external to the system itself are atmospheric noise, cosmic noise, man-made noise, and sky noise. Figure 9-3 shows the value of some of these noise sources as a function of frequency. A brief description of the physical source of these noise signals follows the figure.

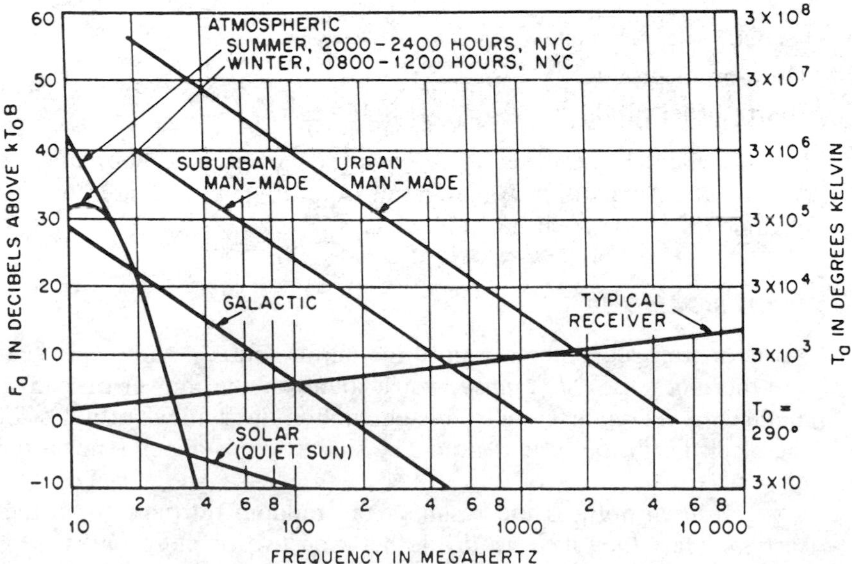

Figure 9-3 Noise power expected from various sources.
(Reproduced with permission of the publisher, Howard W. Sam & Co., Indianapolis, Ind.; Reference Data for Radio Engineers, 5th ed. by Howard W. Sams & Co., copyright © 1968.)

Atmospheric Noise

Atmospheric noise is produced mostly by lightning storms and thus is dependent on the season, time of day, and geographical location. It predominates at frequencies below 20 MHz and is the dominate noise source in the broadcast AM band. In the AM band the external noise dominates over internal circuit noise sources. For this reason, the RF signal is usually brought via some tuned circuit directly to a mixer stage. At higher frequencies, where receiver noise predominates, low-noise tuned RF amplifiers are used ahead of the mixer stage.

Cosmic (or Galactic) Noise

Cosmic noise comes chiefly from the sun and other discrete sources distributed chiefly along the galactic plane. It can be of significance in the FM broadcast band since its limited frequency range is from 18 MHz to about 500 MHz.

Man-made Noise

Man-made noise results from such sources as power lines, electric motors, neon lights, ignition systems, and so on. As shown in Fig. 9-3, the noise level decreases with frequency. These noise sources tend to dominate all others in urban and particularly industrial locations. By locating receivers in quiet remote locations, this noise level can be kept below the galactic noise. Satellite receiving antennas are frequently kept some distance away from urban areas to reduce man-made interference reception.

Sky Noise

Sky noise is due to atmospheric or blackbody radiation from the sky. The noise level is very low, but at frequencies beyond a few gigahertz it can become the primary consideration since the other noise sources decrease to insignificance at very high frequencies.

9-3 MODELING THE EFFECTS OF NOISE

As can be seen from Section 9-2, the sources and characteristics of noise signals are very diverse. Fortunately for most practical communication systems, all the various noise sources can be lumped together and treated as a single thermal (white) noise source which provides the equivalent noise power. The intensity of this equivalent noise source can be measured (some of the measurement techniques are examined later in this chapter) and figures of merit derived which imply how much signal corruption may be expected. The figures of merit most widely quoted are noise figure (or noise factor) and equivalent noise temperature. The following sections define these noise terms. Examples are given to show how these terms may be used to predict the signal-to-noise ratio of a corrupted signal passing through a communication system. Since the thermal noise source is to be our primary noise model, let us digress briefly and examine its representation more closely.

9-3.1 Noise Voltage

The noise voltage shown in Fig. 9-1 has a root-mean-square (rms) voltage value of

$$E_n = \sqrt{4kTRB} \tag{9-4}$$

where R is the end-to-end resistance of the conductive material and the other constants are as defined in equation (9-2). The noisy resistor can be represented symbolically as a noise generator having an emf of E_n in series with a noiseless resistor R, as shown in Fig. 9-4. The maximum thermal noise that can be delivered by such a noise source, called the *available noise power*, occurs when the load impedance is matched to the generator impedance. Under this condition the output rms voltage is half the open circuit value and the available noise power delivered to a matched load becomes $P_n = kTB$, as noted in formula (9-2).

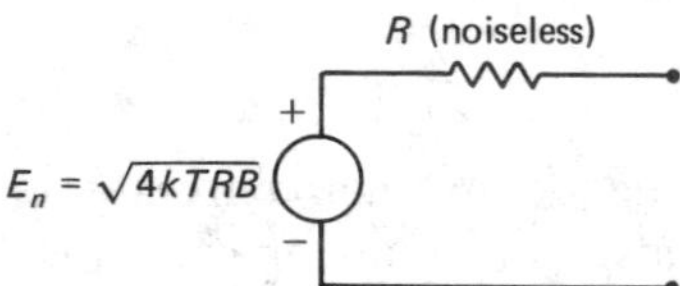

Figure 9-4 Thévenin equivalent of a noisy resistor R.

The available noise power does not depend on the resistance value. As long as the load resistance is matched to the source (generator) resistance, the available noise is dependent only on the temperature and bandwidth. This is normally the condition when connecting communication equipment in tandem. The output impedance of one unit is matched to the input impedance of the following unit. *This is the assumption made in the remainder of this chapter.*

A point that should be noted arises if the case of two resistors connected in series is considered. As shown in Fig. 9-5, the series connection of the two noise models can be reduced to a single equivalent circuit. It should be noted, however, that the two noise voltages, E_{n1} and E_{n2}, cannot simply be added together. The two sources are not simple single-frequency phasor voltages, but represent two independent sources, each containing multiple-frequency components. In this circumstance the total power delivered to a load can be determined by adding together the power provided by each of the independent sources. As a result, the net equivalent voltage is

$$E_{net} = \sqrt{E_{n1}^2 + E_{n2}^2} \tag{9-5}$$

If two resistors R_1 and R_2 are placed in series, the resultant noise voltage will be

$$\begin{aligned} E_n &= \sqrt{4kTR_1B + 4kTR_2B} \\ &= \sqrt{4kTB(R_1 + R_2)} \end{aligned} \tag{9-6}$$

This indicates that the two resistors in series can be replaced by a single resistor of value $R_1 + R_2$.

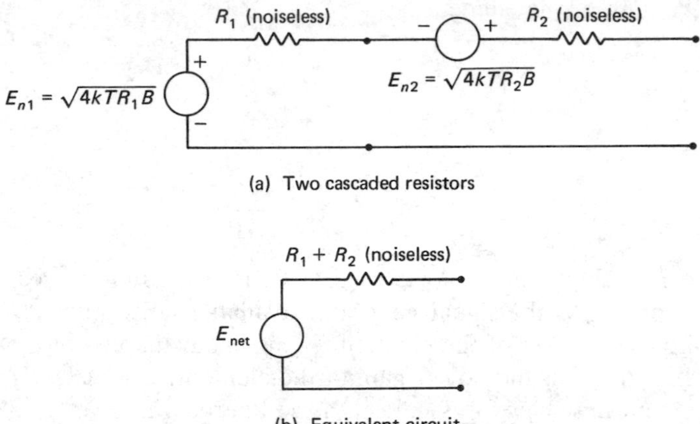

Figure 9-5 Noise model.

9-3.2 Noise Figure and Noise Factor

Let us consider what occurs when a signal is processed by an element in a communication link. Independent of whatever other effects the element has on the incoming signal, some amount of noise will be added to the signal due to the fundamental noise sources within the element itself, as illustrated in Fig. 9-6.

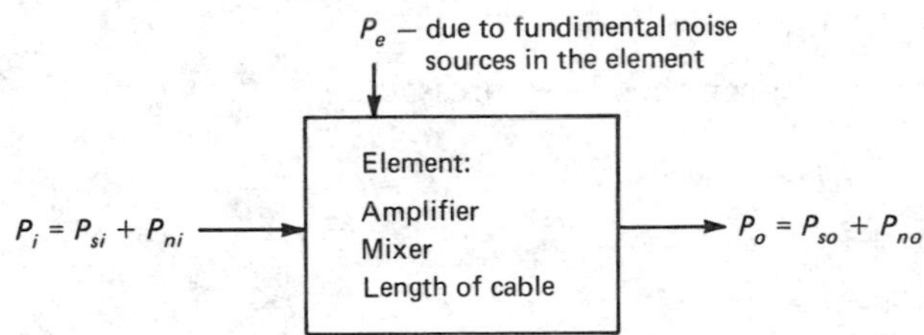

Figure 9-6 Signal contamination in a communication system.

The signal-to-noise (*S/N*) power ratio at the output of the element must therefore be less than the *S/N* power ratio of the incoming signal. The "quality" of the signal is thus degraded.

As a measure of the amount of degradation an element will contribute, H. T. Friis defined the noise factor of an element as

$$\text{noise factor} = \frac{\text{available } S/N \text{ power ratio at the input}}{\text{available } S/N \text{ power ratio at the output}} \tag{9-7}$$

In mathematical form, this can be expressed as

$$F = \frac{P_{si}/P_{ni}}{P_{so}/P_{no}} = \frac{P_{si}}{P_{so}}\frac{P_{no}}{P_{ni}} \tag{9-8}$$

But the gain of the element is

$$G = \frac{P_{so}}{P_{si}} \tag{9-9}$$

Hence

$$F = \frac{P_{no}}{GP_{ni}} \tag{9-10}$$

where F is the element noise factor, P_{si} is the signal power available from the generator, P_{ni} is the signal generator available noise power, P_{so} is the available signal output power of the element, P_{no} is the available noise output power of the element, and G is the power gain of the element. Equivalently the noise factor of an element may be expressed as a noise figure, where

$$\text{noise figure} = F_{\text{db}} = 10 \log F \tag{9-11}$$

Note that F is always greater than 1, and therefore $F_{\text{dB}} > 0$.

As part of the definition, the noise factor is determined using a simple signal source which has an output impedance that matches the element input impedance and is at the same temperature as the element. Specifically, the assumption is that the input noise to the element will be kTB (i.e., $P_{ni} = kTB$). With regard to the temperature T, it is assumed to be 290 K (equal to 17°C). Friis settled on this value because "it makes the value of kT a little easier to handle in computations." We will denote this reference temperature by

$$T_0 = 290 \text{ K} \tag{9-12}$$

Equation (9-10) can thus be expressed as

$$F = \frac{P_{no}}{GkT_0B} \tag{9-13}$$

9-3.3 Noise Factor of an Amplifier

Consider the amplifier with power gain G shown in Fig. 9-7. From equation (9-13) the noise output from the amplifier is

$$P_{no} = FGkT_0B \tag{9-14}$$

This output noise may be factored into two terms as follows:

$$P_{no} = G(kT_0B) + (F - 1)G(kT_0B) \tag{9-15}$$

The first term may be seen to represent the input noise amplified by the gain of the amplifier. The second term represents the noise that would appear at the output even if no noise were injected into the amplifier. This noise is created by the active and passive elements within the amplifier itself. This noise power,

$$P_{no,a} = (F - 1)GkT_0B \tag{9-16}$$

is the noise power at the amplifier output produced by the amplifier only.

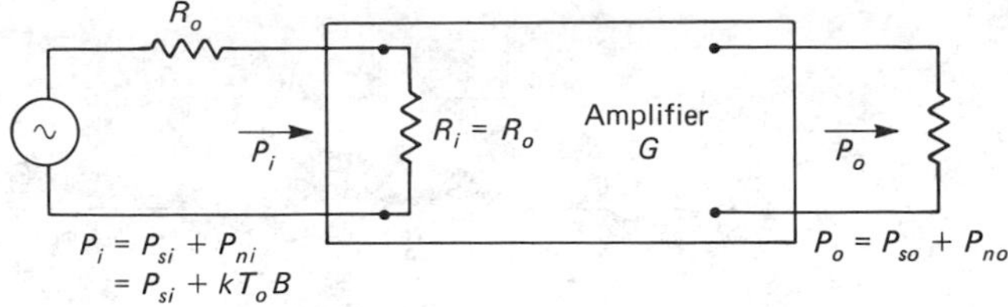

Figure 9-7 Amplifier of power gain G with matched input and output impedances.

The noise contributed by the amplifier itself can be reflected back to the input of the amplifier. In this case it is called the *amplifier effective noise* and is given by

$$P_e = \frac{P_{no,a}}{G} = (F - 1)kT_0B \tag{9-17}$$

An amplifier may thus be modeled as shown in Fig. 9-8.

Note that we are lumping together all the internal noise sources and treating them as a single thermal (white) noise source acting at the input to the amplifier. This representation is not entirely valid, as the noise power densities of the various noise sources are not all uniform with frequency. The model is, however, normally adequate for most receiver systems, since the signal bandwidth of interest will be quite narrow and the power density of all sources may be considered approximately constant within this narrow frequency range.

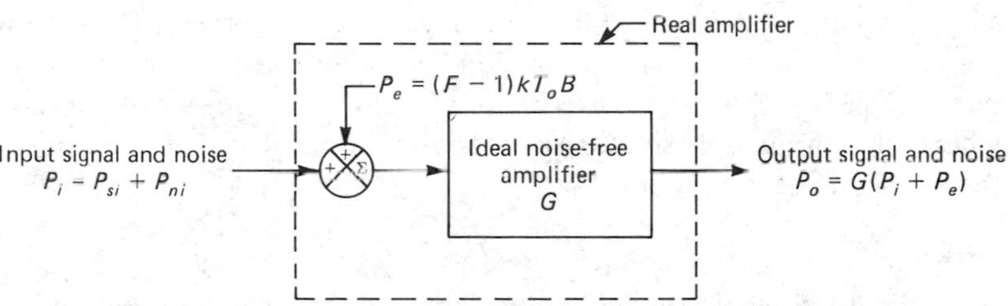

Figure 9-8 Model of an amplifier.

EXAMPLE 9-1

Let us consider how an amplifier, with the characteristics shown in Fig. 9-9, corrupts a signal passing through it. We shall assume an initial signal input that has a signal-to-noise ratio of 60 dB. The output signal and noise power levels can be determined as

$$P_{so} = P_{si} \times G = 10^{-10}\ \text{W}$$

$$P_{no} = (P_{ni} + P_e) \times G = 601 \times 10^{-16}\ \text{W}$$

$$\text{output } S/N = \frac{10^{-10}\ \text{W}}{601 \times 10^{-16}\ \text{W}} = 1.66 \times 10^3 \quad \text{or} \quad 32.3\ \text{dB}$$

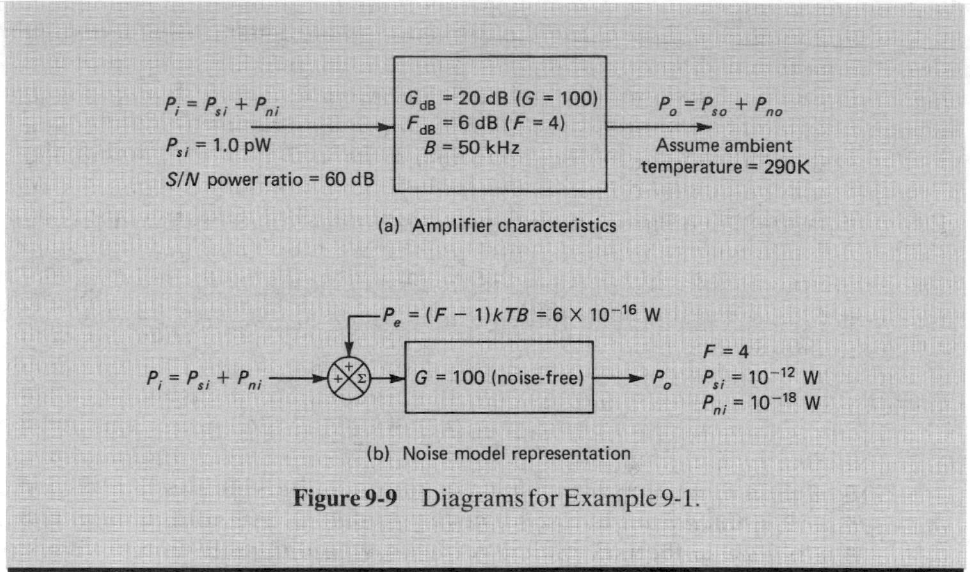

Figure 9-9 Diagrams for Example 9-1.

Note the following with regard to this result:

$$\frac{\text{input } S/N}{\text{output } S/N} = \frac{10^6}{1.66 \times 10^3} = 602 \neq F = 4$$

The reason for the apparent discrepancy comes from the original definition of noise factor. If one reviews the definition, we may see that the ratio input S/N over output S/N will equal the noise factor (F) of the amplifier only in those applications where the input is a simple source with internal impedance and thus an input thermal noise component of

$$P_{ni} = kT_0B$$

In applications where the incoming signal and noise levels are determined otherwise (see the discussions of antennas and cascaded elements in the following sections), the ratio must be determined as shown in the example.

9-3.4 Equivalent Noise Temperatures

Specifying the equivalent noise temperature of an element is simply another manner of describing the added noise power contributed by the element to a signal passing through it. We may note that the equivalent input noise of an element may be represented as

$$P_e = (F - 1)kT_0B = kT_eB \tag{9-18}$$

where

$$T_e = (F - 1)T_0 \tag{9-19}$$

is the equivalent noise temperature and T_0 is some standard temperature (normally, 290 K). Thus for noise calculations an element may be modeled as shown in Fig. 9-10. This form of presenting the noise characteristics of an amplifier is, for example, used when describing the amplifiers used in satellite receiving systems. The first amplifier used in these systems is an LNA (low-noise amplifier) and a typical LNA noise temperature might be 80 K. The 80-K figure provides a direct representation of the noise contribution of the amplifier. An ideal noise-free amplifier would have a noise temperature of 0 K. It should be noted that the equivalent noise temperature value (80 K = −193°C) has little, if anything, to do with the actual physical temperature of the LNA. A low-noise temperature simply means that the amplifier contributes very little distortion and noise to a signal passing through it.

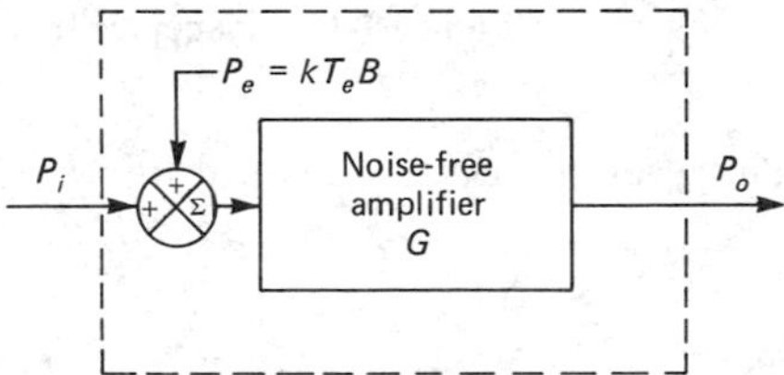

Figure 9-10 Noise model of an amplifier.

The equivalent noise temperature representation is also used when describing antennas. An antenna, as the first element in a receiving system, will contribute some noise to the signal it receives. It is the practice, however, to attribute to an antenna an equivalent noise temperature that represents not only its own internally generated noise (very small), but also includes any noise signals the antenna received along with the desired signal. The noise picked up by the antenna is due to noise sources external to the receiving system. These noise sources were described in Section 9-2.3.

The value of noise temperature attributed to an antenna can fluctuate considerably since the amount of noise picked up at different locations and operating frequencies can vary. Figure 9-3 shows the range of antenna noise temperatures. The lowest possible temperatures are obtained with high-frequency antennas oriented skyward (radio telescopes, satellite receivers, etc.) which largely receive only cosmic noise and thus may have noise temperatures as low as a few kelvin. At the other extreme, lower-frequency antennas in urban areas may have noise temperatures exceeding a hundred million kelvin. This does not imply that the antenna is about to melt down, simply that it is receiving relatively large noise signals from man-made interference sources.

9-3.5 Noise Calculations for Cascaded Amplifiers

Let us consider the front-end sections of a typical receiver system as shown in Fig. 9-11. Noise calculations may be performed by replacing each element by its noise model and cascading the models. Alternatively, the system as a whole may be

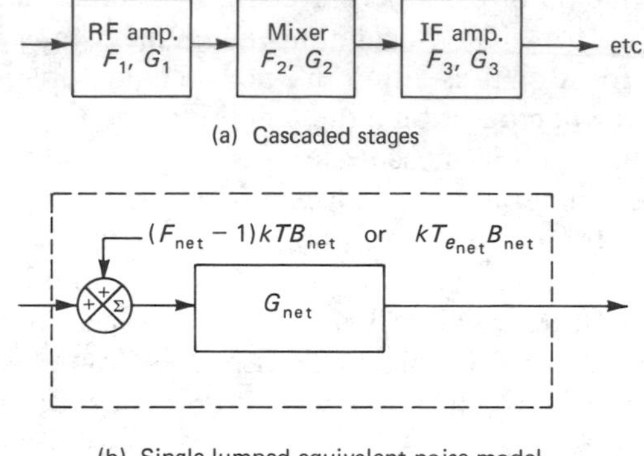

Figure 9-11 Receiver front end.

replaced by a single element, as shown in Fig. 9-11. The net power gain is

$$G_{net} = G_1 \times G_2 \times G_3 \times \cdots \tag{9-20}$$

or

$$G_{dB(net)} = G_{1_{dB}} + G_{2_{dB}} + G_{3_{dB}} + \cdots$$

The net effective bandwidth for series cascaded stages with different individual bandwidths can be calculated; however, typically the IF stage bandwidth will be substantially smaller than the other stages, and as such will establish the overall band of noise frequencies that must be considered *in all stages.*

The net effective noise figure may be established by reflecting the noise contribution of each element back to the input (divide by the intervening gain) and summing them together to find the net effective noise input. This is done in equation (9-21).

$$P_{e_{net}} = (F_{net} - 1)kTB$$

$$= (F_1 - 1)kTB + \frac{(F_2 - 1)kTB}{G_1} + \frac{(F_3 - 1)kTB}{G_2G_1} + \cdots \tag{9-21}$$

which yields

$$F_{net} = F_1 + \frac{F_2 - 1}{G_1} + \frac{F_3 - 1}{G_2G_1} + \cdots \tag{9-22}$$

or similarly, in terms of noise temperature,

$$T_{e_{net}} = T_{e1} + \frac{T_{e2}}{G_1} + \frac{T_{e3}}{G_2G_1} + \frac{T_{e4}}{G_3G_2G_1} + \cdots \tag{9-23}$$

We may note that if the gains are high, the noise figure or temperature contribution of the first one or two elements tends to dominate in the calculation. This effect can be seen to be intuitively correct if we consider that any in-band noise injected near the input will be amplified along with the signal, and thus be much stronger than an equivalent noise signal injected later in the system.

EXAMPLE 9-2

Consider the superhet front end shown in Fig. 9-12. Let us assume that a perfectly noise-free signal of 0.26 μV is applied to the 50-Ω input of the RF amplifier, and then determine the signal and noise power levels leaving the IF amplifier.

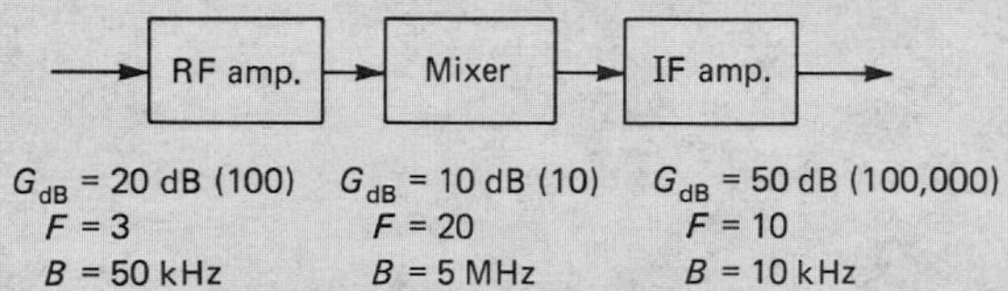

Figure 9-12 First stages in a superhet receiver.

Solution The effective bandwidth of the system is set by the IF amplifier's 10-kHz bandwidth. An overall noise figure can be calculated using equation (9-22).

$$F_{\text{net}} = 3 + \frac{20 - 1}{100} + \frac{10 - 1}{(100)(10)} = 3.199$$

However, care must be exercised in using this figure. The noise entering the system with the signal has been specified as zero rather than the kT_0B value specified in the noise figure definition. The noise figure term is therefore *not* equal to the ratio of output *S/N* to input *S/N*. From Fig. 9-13 the output power levels are

$$P_{no} = (P_{ni} + P_e)G = 8.48 \text{ nW}$$

$$P_{so} = P_{si}G = 135 \text{ nW}$$

$$\text{output } \frac{\text{S}}{\text{N}} = 15.9 = 12 \text{ dB}$$

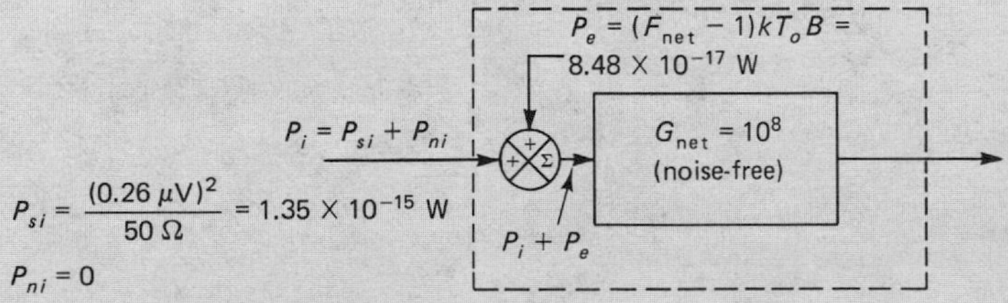

Figure 9-13 Noise model representation of the superhet front end.

EXAMPLE 9-3

Consider the system shown in Fig. 9-14. Assume that the effective noise bandwidth as established by the IF stage is 50 MHz. Calculations using antenna parameters indicate that the received signal strength will be 4 pW. What will the *S/N* power ratio be at the antenna terminals (1) and at the output of the system (2)?

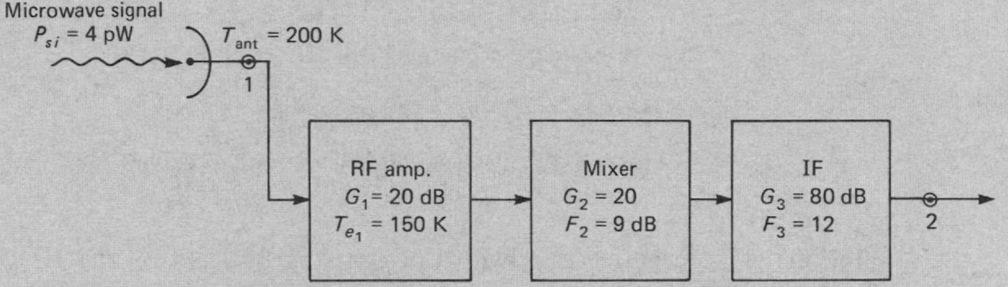

Figure 9-14 Network for Example 9-3.

Solution

(1) *At the input* [see Fig. 9.15(a)]:

$$\frac{S}{N} = \frac{4 \times 10^{-12}\text{ W}}{1.38 \times 10^{-13}\text{ W}}$$

$$= 29.0 \quad \text{or} \quad 14.62\text{ dB}$$

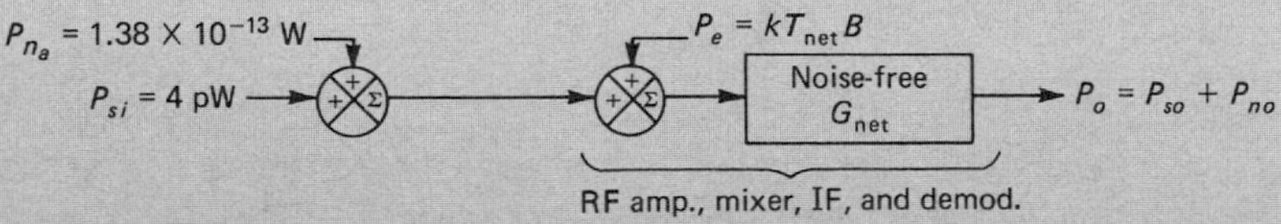

Figure 9-15 Noise model representations.

(2) *At the output:*

$$T_{e_1} = 150\text{ K}$$

$$T_{e_2} = (F_2 - 1)T_0 = (8 - 1)290\text{ K} = 2030\text{ K } (9\text{ dB} \rightarrow 8)$$

(continued)

$$T_{e_3} = (F_3 - 1)T_0 = (12 - 1)290 \text{ K} = 3190 \text{ K}$$

$$T_{\text{net}} = 150 + \frac{2030}{100} + \frac{3190}{(100)(20)} = 171.9 \text{ K}$$

Total effective input noise power:

$$\begin{aligned} P_{\text{op}} &= kT_{\text{ant}}B + kT_{\text{net}}B \\ &= kB(T_{\text{ant}} + T_{\text{net}}) \\ &= (1.38 \times 10^{-23})(50 \times 10^6)(200 + 171.9) \\ &= 2.566 \times 10^{-13} \text{ W} \end{aligned}$$

$$\text{output } \frac{S}{N} = \frac{4 \times 10^{-12} \text{ W} \times G_{\text{net}}}{2.566 \times 10^{-13} \times G_{\text{net}}} = 15.6 \quad \text{or} \quad 11.9 \text{ dB}$$

9-3.6 Lossy Transmission Lines

In many cases the physical length of the feedline between a tower-mounted antenna and the receiver is sufficiently long that signal loss along the line is significant. Let us examine what effect a lossy line has on the signal-to-noise ratio of a signal passing through it. A signal injected into a transmission line will experience a net power loss L, where $L = P_{si}/P_{so}$. The degree of loss is a function of the cable construction and operating frequency. By way of example, let us consider a 200-ft length of RG8/U coaxial cable which is used to carry a signal at 100 MHz. The cable specifications indicated a 2.0 dB loss per 100 ft of length at this frequency. The total loss for 200 ft of cable would be 4 dB and therefore $L = 2.51$ (such that $10 \log L = 4$ dB).

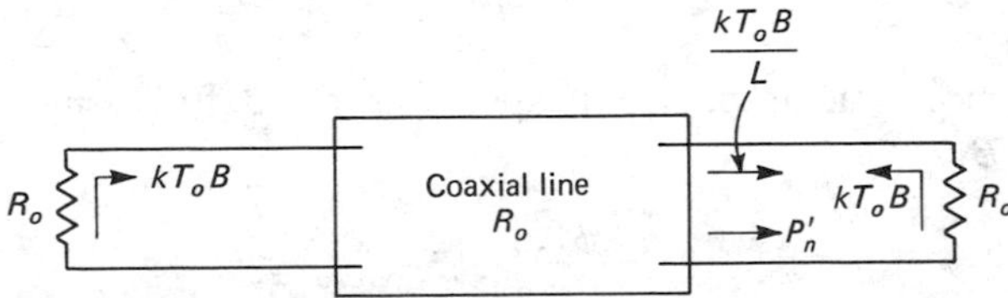

Figure 9-16 Lossy transmission line terminated in matching resistances.

To establish the noise effects of this lossy cable, let us consider what occurs when both ends of the cable are terminated with resistances equal to the characteristic impedance of the transmission line. We will assume that the resistors are at the standard temperature of T_0. This arrangement is illustrated in Fig. 9-16. Note that the resistors, besides acting as matched loads, are also thermal noise sources. Assuming the line and resistors are all at the same temperature (in thermal equilibrium), the total noise power output from the line to the matched load will equal the total power from the noise source (resistor) feeding back into the line. Thus

$$P'_n + \frac{kT_0B}{L} = kT_0B$$

where kT_0B/L represents the noise power transferred from the input and P'_n is the additional noise provided from the line itself. Therefore,

$$P'_n = \left(1 - \frac{1}{L}\right)kT_0B \qquad \text{at the output}$$

or

$$P_e = L\left(1 - \frac{1}{L}\right) kT_0B$$

$$= (L - 1)\, kT_0B \tag{9-24}$$

when reflected back to the input.

Comparing equation (9-24) with equation (9-17), we come to the conclusion that the noise figure of a lossy device is equal to the line loss L

$$F = L \tag{9-25}$$

and therefore

$$T_e = (L - 1)T_0 \tag{9-26}$$

for a lossy device operating at an ambient temperature of T_0. A section of lossy transmission line may thus be modeled for noise analysis using the same representation as was used for amplifiers. The model for 200 ft of RG8/U coax operating at 100 MHz is shown in Fig. 9-17. Note that if the physical temperature of the transmission line and terminating resistors is some value T_L that is different than T_0, then the effective noise temperature is

$$T_e = (L - 1)T_L \tag{9-27}$$

and the noise factor is, strictly speaking, not defined since the input noise level is not kT_0B.

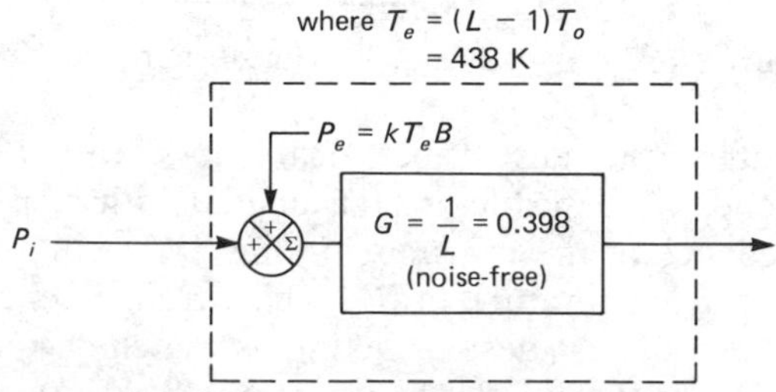

Figure 9-17 Noise model for a lossy line.

EXAMPLE 9-4

For a substantial price premium the RG8/U cable of Example 9-3 can be replaced by Andrew-type HK5 helix cable (0.2 dB/100 ft, attenuation at 100 MHz). What is the equivalent input noise temperature for 200 ft of this cable?

Solution

$$\text{Total loss} = 0.2 \text{ dB/100 ft} \times 200 \text{ ft} = 0.4 \text{ dB}$$

Therefore,

$$10 \log L = +0.4 \qquad \text{and} \qquad L = 1.096$$

For a physical temperature of $T_0 = 290$ K,

$$L = F = 1.096$$

and

$$T_e = (L - 1)T_0 = (1.096 - 1)290$$
$$= 27.98 \text{ K}$$

We see that the use of a low-loss cable not only ensures higher signal strength but will also substantially reduce the total noise added to the system.

EXAMPLE 9-5

Find the noise figure for the three attenuators in cascade as shown in Fig. 9-18. Assume an ambient temperature of T_0.

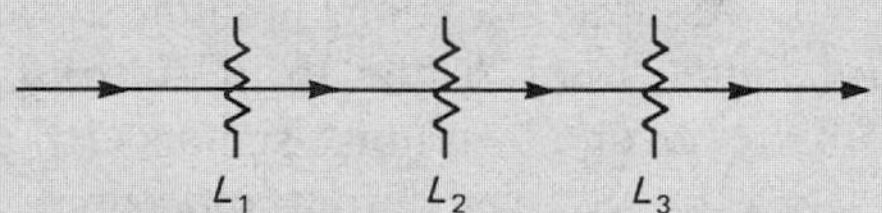

Figure 9-18 Cascaded attenuators.

Solution For three cascaded elements the net effective noise figure is

$$F = F_1 + \frac{F_2 - 1}{G_1} + \frac{F_3 - 1}{G_1 G_2}$$

or specifically for the three attenuators:

$$F = L_1 + (L_2 - 1)L_1 + (L_3 - 1)L_1 L_2 \qquad (9\text{-}28)$$
$$= L_1 L_2 L_3$$

In decibels, the noise figures add, that is,

$$F_{\text{dB}} = L_1 + L_2 + L_3 \quad \text{dB} \qquad (9\text{-}29)$$

9-4 MEASUREMENT OF NOISE FIGURE

The two most common methods of measuring a noise figure value are presented in this section.

9-4.1 CW Signal Generator Method

The CW signal generator method is a very simple method of measuring noise figure but is somewhat inaccurate because the frequency bandwidth of the equipment under test must be measured and the power level of the signal generator is so low that it cannot be directly measured by a standard laboratory power meter. When measuring the noise figure of a receiver, normally only the front-end RF and IF stages up to the detector are taken into account. AGC circuitry must be disabled during the measurement. The minimum test equipment required is shown in Fig. 9-19.

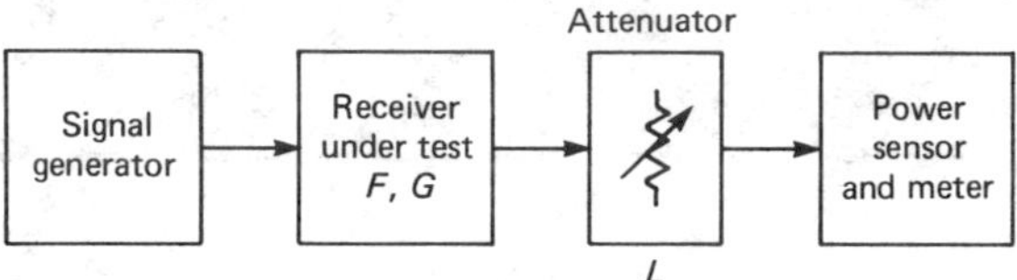

Figure 9-19 Test setup for measurement of noise figure using a CW generator.

The CW generator must have a calibrated signal output level adjustment and have an output impedance which is matched to the input impedance of the receiver under test. If the gain G of the receiver under test is large enough, the adjustable attenuator L will make a negligible contribution to the measured noise figure.

Initially, the receiver is turned on and the signal generator is attached to the input, but left off. Assuming a room temperature of 17°C, the signal generator provides a noise signal of kT_0B due to the output impedance. This noise signal combines with the receiver's own noise signals (P_e) and is amplified. The resulting noise signal reaches the receiver output and passes through the attenuator to the power meter. The attenuator should be adjusted for a convenient power reading.

Now turn the generator on and adjust the output to give twice the power reading. Calibration tolerances of the power meter readings can be eliminated by increasing the attenuator setting by 3 dB and adjusting the generator output to obtain the same power meter reading as was selected originally. With the addition of the input signal, the output now contains signal and noise in equal proportions. From the definition of noise factor [equation (9.7)],

$$F = \frac{(S/N)_i}{(S/N)_o} = \frac{\mathrm{P}_{si}/\mathrm{kT_0B}}{1} = \frac{\mathrm{P}_{si}}{\mathrm{kT_0B}} \tag{9-30}$$

or expressed as a noise figure (in decibels) and assuming that P_{si} is given in dBm:

$$F(\mathrm{dB}) = 173.8 + P_{si}(\mathrm{dBm}) - 10 \log B \tag{9-31}$$

where k and T_0 have been evaluated in the leading constant.

EXAMPLE 9-6

A receiver having a bandwidth of 5 MHz experiences a doubling of its power output when a CW generator is turned on at a power level of $P_{si} = -100$ dBm. Find its noise figure.

Solution

$$F(\text{dB}) = 173.8 - 100 - 10\log(5 \times 10^6)$$

$$= 173.8 - 100 - 67.0 = 6.8 \text{ dB}$$

or a noise factor of 4.8.

9-4.2 Noise Source Generator Method

Noise source generators are usually used to obtain a reliable noise figure for a receiver assembly. This method measures the average noise figure over the bandwidth of the receiver.

Although a simple resistor under controlled temperatures can be used as a noise source, current saturated diodes and gas discharge argon and neon tubes are more prevalent. For argon lamps, the temperature of the discharge noise is around 10,000 K; for neon it is around 20,000 K. The test setup is given in Fig. 9-20. As in the previous method, it is assumed that the receiver gain is high enough that the noise contribution of the variable attenuator is negligible compared to the contributions of the receiver's internal noise sources.

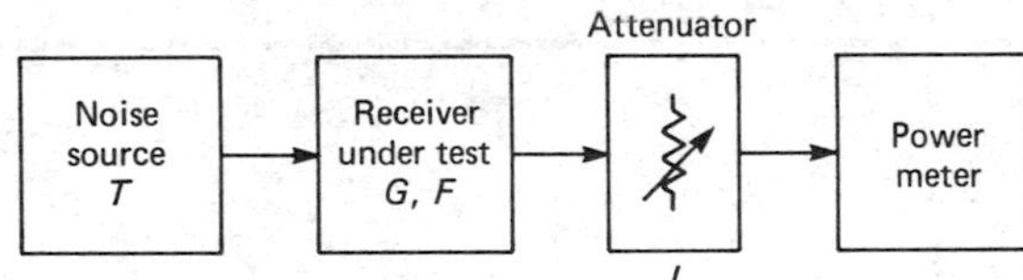

Figure 9-20 Test setup for measurement of noise figure using a noise source.

Noise sources are usually specified in terms of their excess noise, P_{nex}. The excess noise is that noise which is above the matched input noise power $P_{ni} = kT_0B$ which is produced by the noise source when in the off state. The excess noise thus is the total available noise power P_{nT} when the noise source is on and at temperature T, less the source input noise power P_{ni}. Mathematically, this is stated as

$$P_{\text{nex}} = P_{nT} - P_{ni} = kTB - kT_0B \tag{9-32}$$

$$\frac{P_{\text{nex}}}{P_{ni}} = \frac{P_{nT} - P_{ni}}{P_{ni}} = \frac{T - T_0}{T_0} = \frac{T}{T_0} - 1 \tag{9-33}$$

Procedure With the receiver on, initially turn the noise source off and adjust the attenuator setting L to give some convenient power meter reading. The receiver output noise power is (using our noise model of Fig. 9-8)

$$P_{no1} = (P_{ni} + P_e)G \tag{9-34}$$

where $P_{ni} = kT_0B$ and $P_e = (F - 1)kT_0B$. We then switch the noise source on and increase the attenuator setting to give the same power meter reading as was initially obtained. Note the *change in attenuator setting* (Y_{dB} or equivalently a factor of Y). If, for example, the attenuator were increased by 6 dB, the Y factor would be 3.98. The receiver output signal is now

$$P_{no2} = (P_{ni} + P_e + P_{\mathrm{nex}})G \tag{9-35}$$

The Y factor is the ratio of P_{no2}/P_{no1} [equation (9.35) divided by equation (9.34)]. Equations (9.33), (9.34), and (9.35) can thus be combined and manipulated to derive

$$F = \left(\frac{T}{T_o} - 1\right)\frac{1}{Y - 1} \tag{9-36}$$

EXAMPLE 9-7

An argon gas source has a temperature of 10,000 K. How much greater is the excess noise than the thermal noise power of the source (in decibels)?

Solution From equation (9-33):

$$\frac{P_{\mathrm{nex}}}{P_{ni}} = \frac{T}{T_0} - 1 = \frac{10{,}000}{290} - 1 = 33.48 \quad \text{or} \quad 15.25 \text{ dB}$$

EXAMPLE 9-8

Using an argon gas source to measure an FM receiver noise figure, the Y factor was found to be 6 dB (3.98). Find the receiver noise figure.

Solution From the previous example, $(T/T_0) - 1 = 33.48$. Using equation (9-36) yields

$$F = 33.48 \times \frac{1}{3.98 - 1} = 11.23 \text{ (10.5 dB)}$$

9-4.3 Measurement of Antenna Noise Temperatures

A measurement arrangement to determine antenna noise temperature is shown in Fig. 9-21. The current noise temperature of the antenna may be implied by comparison to a calibrated noise source. Tuning the receiver to an unused channel when connected to the antenna, the receiver gain is adjusted to yield a convenient power meter reading. The receiver input is then switched to the variable noise source and the source adjusted to obtain the same power meter reading.

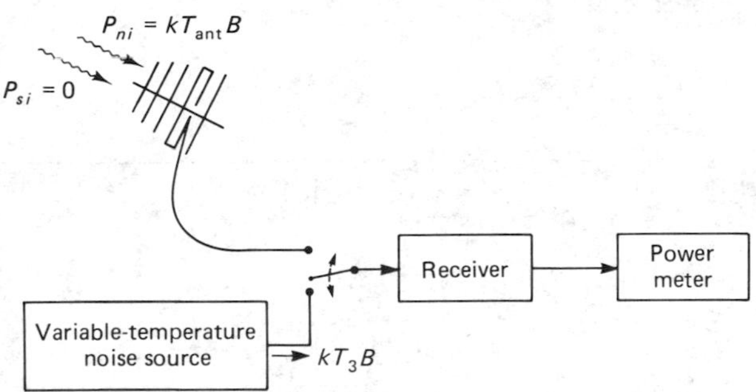

Figure 9-21 Measurement of antenna noise temperature.

A variable-level noise source may be obtained by applying a fixed-temperature noise source to the system via a calibrated attenuator. The noise contribution of the attenuator itself must, however, be included when establishing the effective noise level applied to the receiver by the source/attenuator combination. Figure 9-22 shows the noise model for a noise source feeding an attenuator. The noise contribution of the attenuator (P_{n2}) is a function of its physical temperature (T_2) and current attenuation ratio (L) as outlined in equation (9-27). Transferring the two noise signals to the attenuator output allows an equivalent noise temperature (T_3) to be established.

$$T_3 = \frac{(L - 1)T_2 + T_1}{L} \tag{9-37}$$

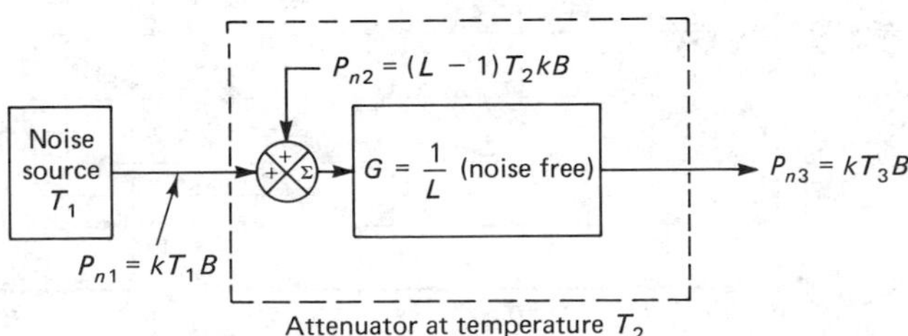

Figure 9-22 Variable noise temperature source.

When the attenuator is zero, $L = 1$, the total noise output temperature will be equal to the noise source temperature T_1. As the attenuation is increased, L becomes large, and the total noise output temperature converges to T_2.

The plots of total noise output temperature for three typical noise sources as a function of attenuation in decibels are given in Fig. 9-23. The attenuator temperature is assumed to be the standard 290 K, and the three sources provide unattenuated noise temperatures as follows. The noise sources consist of a cold source and two hot sources:

1. Liquid nitrogen, $T = 78$ K
2. Argon gas source, $T = 10{,}000$ K
3. Neon gas source, $T = 20{,}000$ K

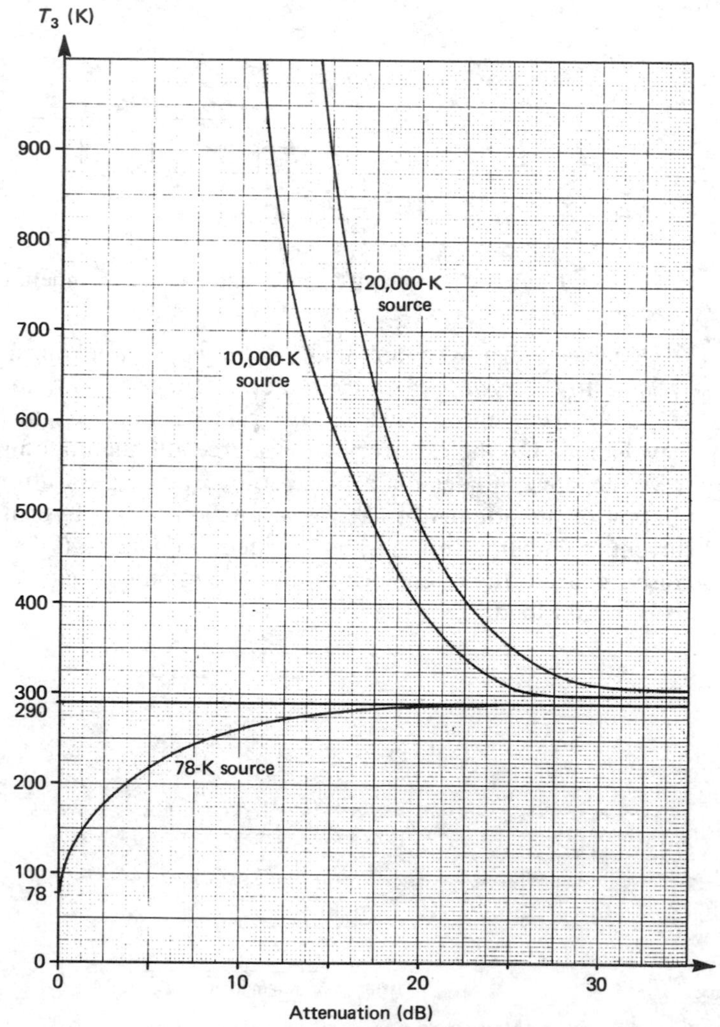

Figure 9-23 Total noise temperature of standard noise sources experiencing an attenuation of L decibels.

EXAMPLE 9-9

A cold load immersed in liquid nitrogen is connected through an attenuator to the input of a LNA. If the ambient temperature (T_2) is 290 K, find the attenuator setting so that the input temperature to the amplifier is 125 K.

Solution Rearranging equation (9-37), we get

$$L = \frac{T_1 - T_2}{T_3 - T_2} \tag{9-38}$$

Thus

$$L = \frac{78 - 290}{125 - 290} = 1.285 \quad \text{or} \quad 1.089\ \text{dB}$$

Problems

9-1. Define briefly noise factor.

9-2. Define briefly noise temperature.

9-3. An amplifier has an equivalent noise temperature T_n. The expression kT_nB is:
  (1) The available noise power at the output of the amplifier
  (2) The total noise power of the amplifier and source referred to the input terminals of the amplifier
  (3) An available noise power, to be added to that of the source at the input, representing the noise produced in the amplifier
  (4) The fraction of the output available noise power that is due to noise produced in the amplifier

9-4. A loss-free capacitor and a resistor have the same impedance magnitude and are at the same temperature. The available noise power from the capacitor is:
  (1) The same as that from the resistor
  (2) Greater than that from the resistor
  (3) Less than that from the resistor
  (4) Zero

9-5. In the equation $P_{no} = G(kTB) + (F - 1)G(kTB)$:
  (a) What does the first term, $G(kTB)$, represent?
  (b) What does the second term $(F - 1)G(kTB)$, represent?

9-6. Consider the system shown in Fig. P9-6. $T_{\text{ambient}} = 290$ K; BW = 50 MHz. Calculations indicate that the received signal strength will be 4.0 pW at 1.
  (a) What will the S/N value be at point 1?
  (b) What is this system's F_{net}?
  (c) What will the S/N value be at point 2?

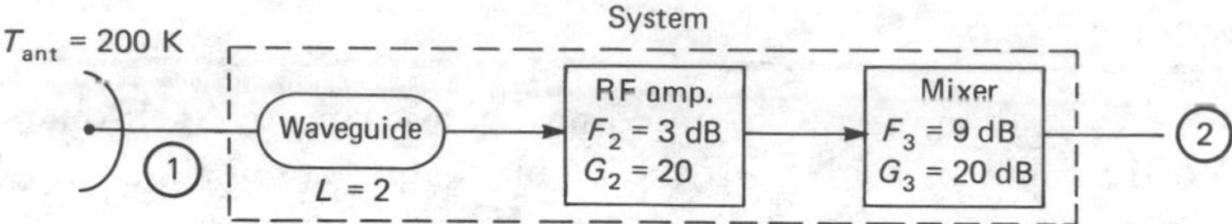

Figure P9-6.

9-7. The receiver system shown in Fig. P9-7 is to be reduced to a single equivalent block for noise calculations. For the equivalent block, determine:
(a) The effective noise bandwidth
(b) The equivalent input noise temperature
(c) The effective power gain (in decibels)

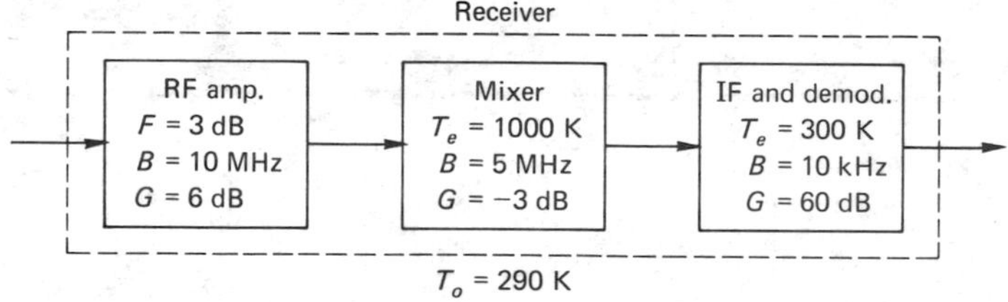

Figure P9-7.

9-8. For the bandwidth of interest in Fig. P9-8, determine:
(a) The signal-to-noise ratio at point A (in decibels)
(b) The signal-to-noise ratio at point B (in decibels)

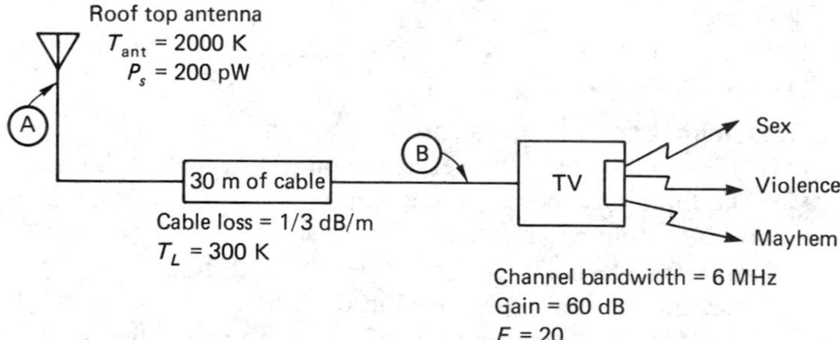

Figure P9-8.

9-9. A receiver with an equivalent input noise temperature of 500 K receives an input signal level of 1.45 pW (Fig. P9-9). The input signal level is slightly corrupted, such that it has a signal-to-noise ratio of 40 dB. The receiver has an effective overall bandwidth of 21 kHz. What will the signal-to-noise ratio (in decibels) be at the receiver output?

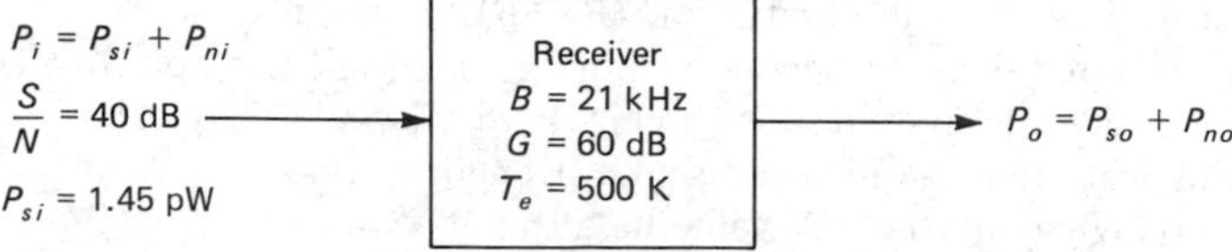

Figure P9-9.

9-10. Consider two possible configurations for the system shown in Fig. P9-10: (1) LNA between antenna and coaxial cable, and (2) LNA between coaxial cable and mixer (as shown).

(a) Which of the two configurations will provide less total noise at the output? Why? Explain.

(b) Calculate the equivalent input noise temperature for both configurations of the above system. *Do not include the antenna.*

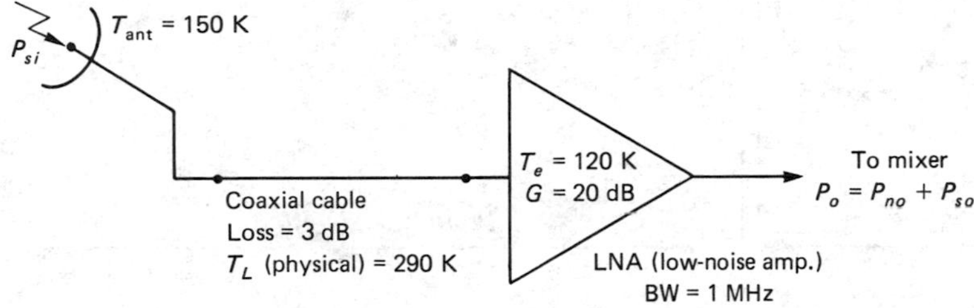

Figure P9-10.

9-11. A 10,000-K argon gas noise source is applied to the input of a receiver through a matched 15-dB attenuator with a physical temperature of T_0. What is the apparent noise temperature seen by the receiver?

9-12. By comparison with a variable noise source, it is established that the apparent noise temperature presented to a receiver input (A in Fig. P9-12) is 240 K. This noise signal is delivered to the receiver by an antenna followed by a 6-dB attenuator. Assuming that the physical temperature of the attenuator is T_0, determine the equivalent noise temperature of the antenna itself (at B).

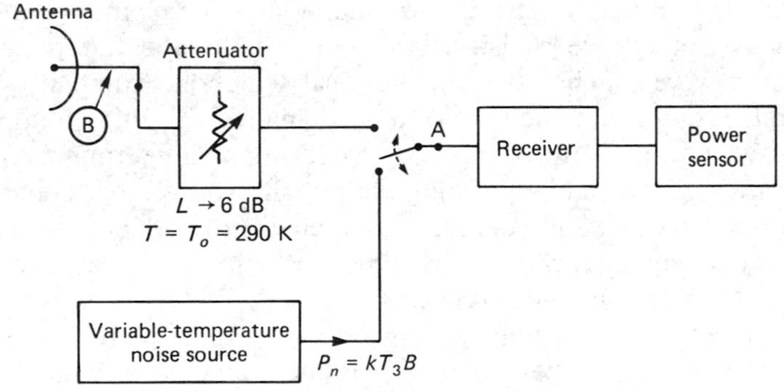

Figure P9-12.

Telephone System Concepts

Peter Hancock

10-1 INTRODUCTION

Deregulation and privatization of telephone systems in recent years have created a highly competitive industry. Prior to these liberalization policies most telephone services were purchased from the local telephone company, which engineered, installed, and maintained the system. In many cases this situation continues. However, we now find interconnect companies supplying terminal equipment and maintenance services and corporations with their own telecom departments and private networks. In addition, computer companies provide local area networks for high-speed data communications, and other specialized companies provide cellular telephones for land mobile service.

The telephone company controls a very valuable resource: the class 5 central office (the local exchange), which is the connection to the telephone network. While government policies have deliberately changed the industry, technological change has been just as dramatic. Digital equipment and fiber optics provide much of the new transmission and switching facilities.

The complexity and reliability of telephone systems are often taken for granted after 100 years of development. The opportunities presented by new technology and political philosophy will provide enhanced telecommunication services, but will require that those involved in telecommunications understand the system capabilities and limitations. To this end the chapter will present a concise treatment of the analog and digital technology in use in present-day telephone systems.

Telephone systems must solve three fundamental technical problems economically. The quantity of equipment to be installed is therefore determined statistically to provide a satisfactory grade of service under normal busy-hour traffic. The three problems are:

1. The *transmission* of voice or data with minimum impairment.
2. A *switching* network which will allow any subscriber to reach any other subscriber.
3. Interfacing the switching and transmission equipment with a suitable *signaling* system.

Our task is to follow the threads of transmission, switching, and signaling through both the exchange area and the toll plant. It is necessary to consider both analog and digital facilities to appreciate the present-day telephone system.

10-2 THE EXCHANGE AREA

10-2.1 Physical Plant

Figure 10-1 looks deceptively simple. Keep in mind that the telephone company has most of its capital tied up in the loop plant. The subscriber's equipment may be a rotary dial or touch-tone telephone and is connected to the house wiring via a RJ-11 modular plug. The modular jack is wired using two-pair JKT house cable to a protector. This device is basically a carbon block spark gap which protects the subscriber from high voltages between each leg and ground. Note that the four-conductor cable contains green, red, black, and yellow wires. The green and red (designated tip and ring, respectively) are used for the first circuit, and in most cases the black and yellow are for future expansion or special applications. If the subscriber's premises are totally "phone centered," a terminating device is placed across the protector so that the central office does not see an open line if all the phones are unplugged. New telephone switching systems have the capability of testing the subscriber's circuit for the presence of the ringer and generating a trouble report. The drop wire provides the connection from the protector in the subscriber's premises to the ready access terminal outside the house on a pole route or to a pedestal with a buried system.

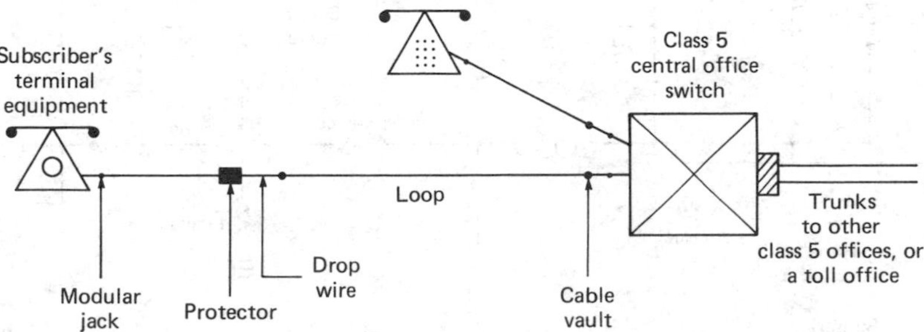

Figure 10-1 The exchange area plant.

At this point a connection is made to the distribution cable. The distribution cable will join a main feeder cable that will eventually enter the exchange cable vault. In the vault, the outside plant cables are spliced to inside cables which go up to a mainframe for cross connection to the switching machine. The mainframe will include protection for the office in the form of heat coils and carbon or gas tube protectors. The heat coils protect against so called sneak currents. It is worth mentioning that the main feeder cables leaving the office are often pressurized with dry air. This prevents the ingress of moisture. An associated alarm system warns of loss of pressure in any section of the feeder so that cable faults can be readily located. Having briefly described the physical loop plant, we must now turn our attention to the associated transmission and signaling problems.

10-2.2 Subscriber Loop

Although the circuit shown in Fig. 10-2 looks straight forward, we will need to spend considerable time examining its performance with respect to the transmission of speech, its susceptibility to induced noise, and its signaling limitations.

When the subscriber goes off-hook (lifts the handset) the loop circuit is closed by the hook switch, thus operating the line relay in the central office due to the negative 48-V battery on the ring and ground on the tip. Note that the relay has two windings and that the loop circuit is balanced; that is, both legs of the loop are electrically the same with respect to ground. This is of great importance and we will return to the subject of noise mitigation in due course.

Now, when the loop is closed, the line relay operates and the central office switching machine effectively gets a "request to send." The central office returns a "clear to send" to the subscriber in the form of dial tone when it is ready to receive signaling information. In common control switching machines this occurs when a digit receiver or register has been attached to the subscriber. Normally,

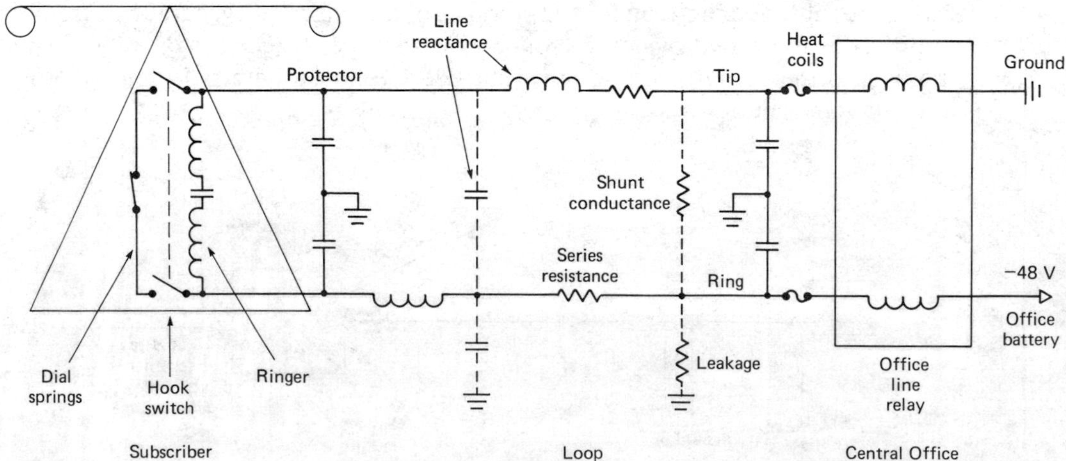

Figure 10-2 The subscriber loop.

sufficient equipment is available so that the provision of dial tone appears to be immediate.

The subscriber now dials the called party and the call is established by the switch; the digit receiver can then be released for use with another call. The switching machine applies ringing to alert the called party. The ringing voltage cycle is usually 2 seconds on, 4 seconds off. The called subscriber may answer the ring during the 2-second ring or during the 4-second silent period. In either case the ringing must be tripped immediately. During the active period of the ring cycle ringing is applied to the subscriber's loop and the central office equipment is disconnected. Hence a ring-trip relay must be provided to trip the ring during the active part of the ring cycle when the central office equipment is isolated from the loop. Figure 10-3 shows the situation in diagramatic form. Note that with standard North American practice, battery biased ringing is applied to the ring conductor. Although the ordinary telephone is not polarity conscious there are a number of cases where the tip/ring integrity of the loop is important (e.g., PABX trunks, Wescom signaling modules and some line drivers).

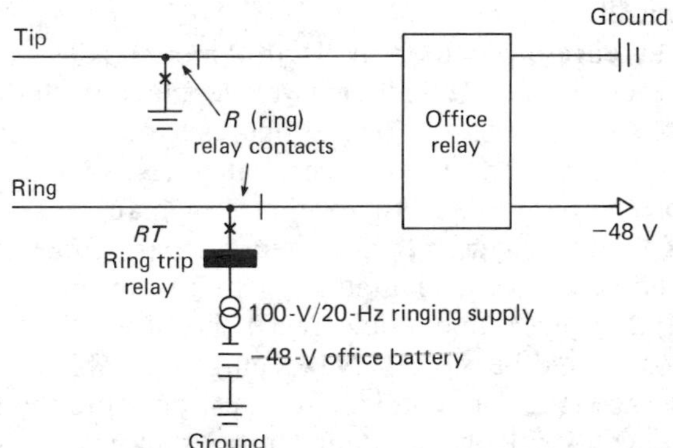

Figure 10-3 Bridged ringing.

It is useful to calculate the peak value of the applied ringing voltage. 100 V rms = 141.4 V peak. Superimposing this on −48 V from the office battery gives a ringing voltage swing between +93.4 and −189 V. It is not surprising that these high voltages have been a major problem for integrated circuits in subscriber line cards. Even modern switching equipments still have a ring relay in each line circuit.

The ringer in the telephone set is tuned to a frequency of 20 to 30 Hz; two windings with a 0.4-μF dc blocking capacitor separating the windings is a common configuration. It is unfortunate that the ringer was not designed for 60-Hz operation, as the provision of a local 20-Hz ringing supply is always expensive. It is no wonder that we refer to 20-Hz, 100-V ringing as power ringing. You will note that numerous extension telephones can be accommodated. Mention should be made of tone ringing, which is a feature on many low-cost telephones. In this case the ringing from the exchange activates a tone oscillator which can drive a speaker or an audio transducer.

The resistance of the loop is critical from both the supervisory and transmission viewpoint. The line relay must respond to the on-hook/off-hook conditions and repeat the dial pulses accurately. However, it is also important that the carbon microphone transmitter operates efficiently. This requires a minimum dc loop current of about 23 mA. It is this transmission limit that controls the length of the bare loop. Let us take an inventory of the resistances in the loop.

Subscriber set[1]	200 Ω
Line relay	400 Ω
Dc loop resistance	?
	600 Ω + dc loop resistance

If 23 mA is the limit with a 48-V office battery, the total resistance can be

$$\frac{48}{23} \times 1000 = 2087\ \Omega$$

Subtracting the 600 Ω allows a maximum loop resistance of 1487 Ω. The office limit is typically 1300 Ω.

The average urban loop in North America is less than 5 km or 2 to 3 miles. How far, then, is 1300 Ω? The answer depends on the gauge of the copper loop plant. Table 10-1 gives some typical figures for twisted-pair cable at 20°C from the finest gauge (No. 26) to the coarsest gauge (No. 19) in general use. There are a number of complications. For example, in many cases the loop is made up of lengths of various gauges. It has been standard practice to use the finest gauge initially, then use a coarser grade to extend the loops. Note that a hypothetical loop using 26-gauge cable would reach the office limit at about 3 miles and the 1000-Hz loss would be 8.6 dB. Nineteen-gauge cable would not normally be used for the average loop for economic reasons, but note the theoretical limit of (1300/85) 15 miles. Most telephone administrations allow a maximum 8 dB of attenuation for urban subscribers and 10 dB for rurals. Hence our hypothetical 19-gauge loop would need loop treatment to meet the transmission standard. The voice-frequency transmission loss of the loop can be reduced by using loading schemes or negative impedance repeaters. The loop current can be maintained at 23 mA by using battery

TABLE 10-1

Gauge	Dc Resistance (per Loop Mile) (Ω)	Loss at 1000 Hz (per Loop Mile) (dB)
26	440	2.85
24	274	2.30
22	171	1.80
19	85	1.27

[1]Note that electronic telephone sets may be as high as 430 Ω.

boost or loop extenders. The twisted-pair loop comprises distributed resistance, inductance, shunt capacitance, and leakage, all of which will affect both the voice and signaling. There is a supervisory limit. That is the loop length at which the on-hook/off-hook supervision and dial pulse signaling becomes unreliable. In practice the transmission is the limiting factor, and setting the office limit at 1300 Ω ensures that the 23-mA limit will be met and that signaling and supervision will be satisfactory.

10-2.3 Subscriber Signaling

Consider the loop from the dc point of view (Fig. 10-4) and its effect on dial pulses (Fig. 10-5). The dial springs break to transmit the dial pulses to a central office relay. The current pulses "seen" by this relay are critical if the relay is to pulse reliably. If a storage scope is attached at the central office end of the loop, the trace would appear as in Fig. 10-5.

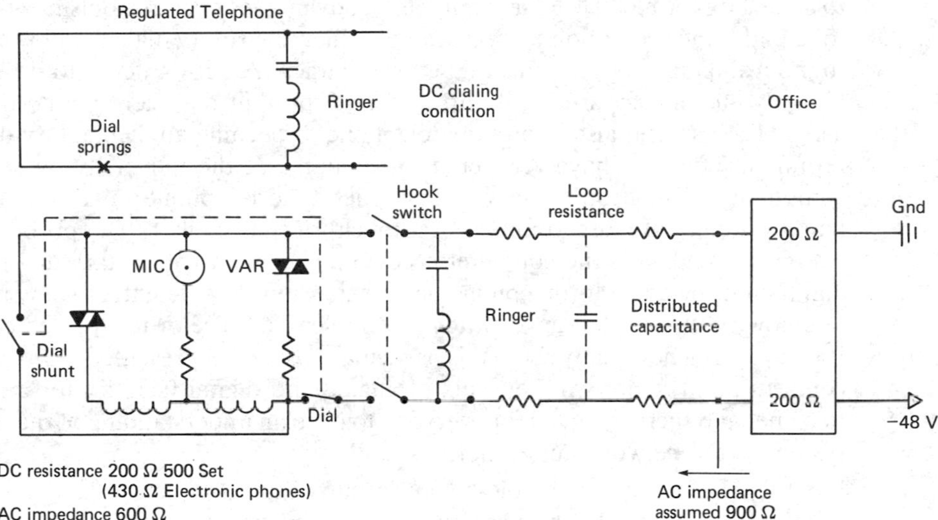

Figure 10-4 Rotary dial signaling.

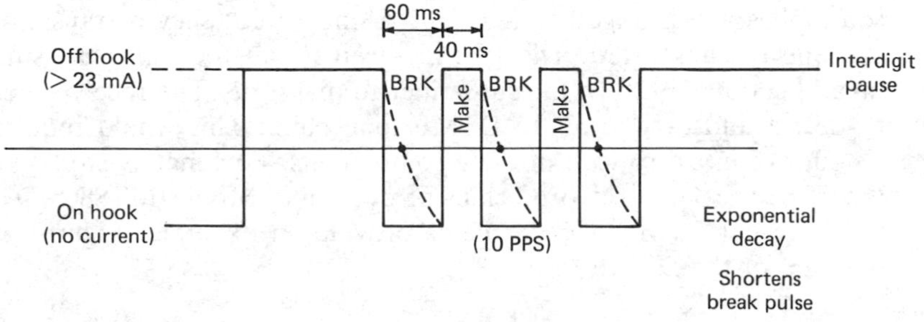

Figure 10-5 Dial pulsing.

The normal pulsing rate is 10 pulses per second for subscriber dialing. This means that the break–make period is 100 ms. Note the effect of the line capacitance (typically, 0.084 μF per loop mile) and the ringer capacitor on the decay of the loop current. When the dial springs open the loop current does not go to zero immediately, resulting in a shortening of the break pulse. To overcome this problem the telephone industry has employed signal conditioning. Simply, the dial springs are operated such that the break pulse is made deliberately longer: typically, 60% break and 40% make. This means that at 10 pulses per second the break will be 60 ms in duration so that the distortion introduced by the loop is counteracted. The interdigit pause is also important, as the switching equipment must recognize the individual digits. In older step-by-step switching the interdigit pause needs to be greater than 600 ms. For later common control equipment 300 ms is typical.

It is interesting to observe that subscriber signaling has been digital (dial pulse) for almost 100 years. Today, modern switching and transmission facilities are digital, but you will note that for subscriber signaling, touch-tone or dual-tone multifrequency (DTMF) telephones are in vogue. Thus the signaling has reverted to an analog form! DTMF has a number of advantages over dial pulsing. The touch-tone pad is more ergonomic (user friendly) than the rotary dial and allows a person to input the digits faster and with better accuracy. A button needs to be depressed for only 40 ms compared to 1 entire second to input digit zero with a dial pulse. The DTMF system also allows the telephone to become an elementary data input terminal. You may have registered for a course at the university by accessing a computer and inputting data via your touch-tone telephone. You may also have had a problem because you are not a touch-tone subscriber. The problem can be overcome by dialing the computer access number with your rotary dial telephone and then using a touch-tone phone on an extension for data entry. Many telephone sets now on the market have switch-selectable dial pulse or touch-tone capability. It is surprising how many times this particular problem is presented. Two comments come to mind; there will be a mix of analog and digital facilities for a long time to come; and there is a need for everyone to have an understanding of the telephone system as the network applications expand.

The DTMF system employs a low group of four frequencies and a high group of four frequencies. Each digit is represented by dual frequencies, one from each group. This gives 16 possibilities for the touch-tone pad. At present a 3 by 4 pad is standard (Fig. 10-6). Depressing button 1 generates the dual frequencies of 697 and 1209 Hz. If the operator's finger slips and two adjacent buttons are pressed, some pads are arranged so that only a single frequency is transmitted. In others, no tone is transmitted. The DTMF receiver can recognize the single tone as an invalid code and return the subscriber to dial tone. The receiver needs the ability to detect and decode all 16 DTMF tone pairs. This would require a significant amount of circuitry with discrete components. Fortunately, today one can implement the receiver in two IC chips or even one. Mitel's MT8865 DTMF filter and the MT8860 DTMF decoder chips provide an example. Figure 10-7 is a circuit diagram of the DTMF receiver.

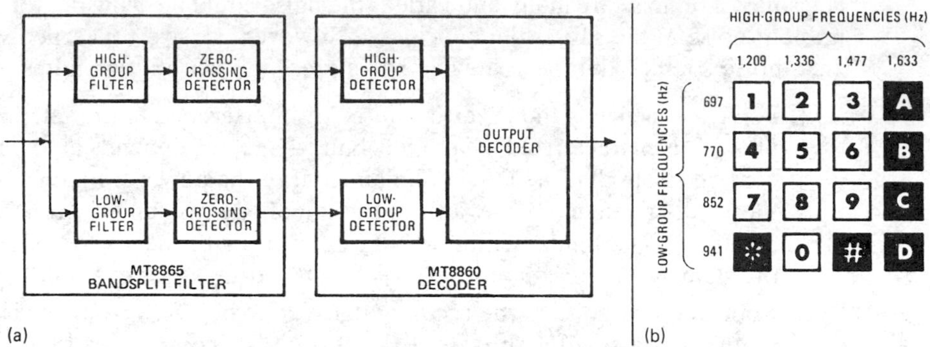

Figure 10-6 DTMF receiver (a); DTMF pad (b).

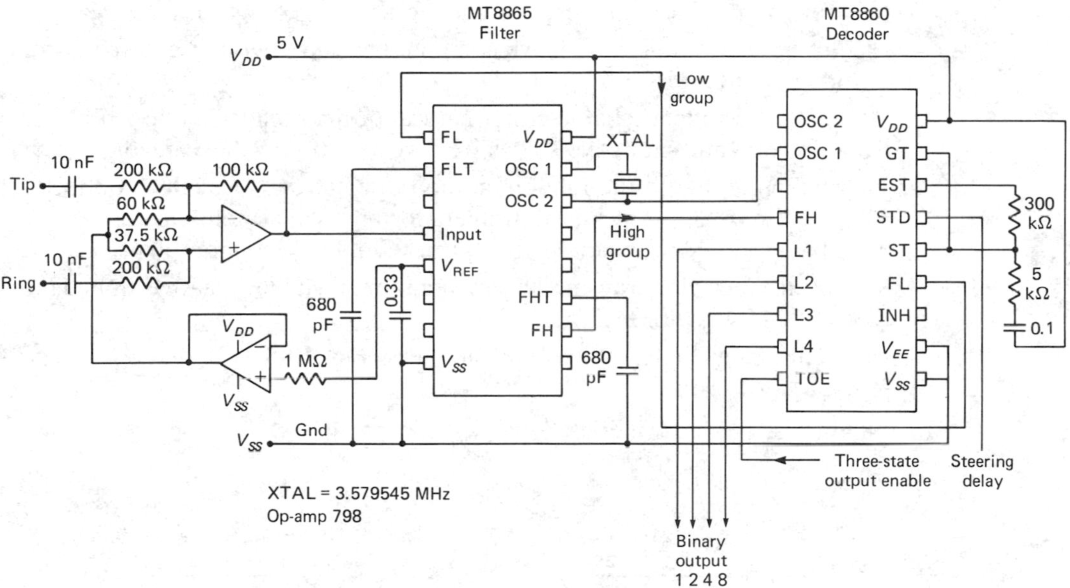

Figure 10-7 DTMF receiver. (*Source*: Mitel Data Book.)

10-2.4 Noise Sources

Transmission and signaling in the loop plant have been examined. It is necessary to return to the transmission viewpoint and consider noise problems in the exchange area. Noise in the telephone context is defined as any extraneous signal. It can be broadly classified into three types:

1. Steady interfering tone or tones
2. Wide-band white noise (random noise)
3. Impulse noise

The sources of noise are many and varied; the noise could be acoustic, for example, ambient noise at the subscriber's location. However, we are concerned with electrical noise sources and the following list is presented with a few notes.

1. Power influence is the greatest source of interference in the telephone loop plant. It is characterized by a steady hum. The pitch depends on the harmonic content. Note that the 60-Hz fundamental of the power frequency is outside the voice-frequency bandwidth of the telephone instrument (roughly 200 to 3400 Hz). The fundamental will not interfere with the voice but may cause problems with dc signaling systems (PABX trunks, for example).
2. Central office noise is the second largest source. Noise from rectifiers (harmonics), combined with degraded filtering, is a common cause.
3. Electromagnetic interference and/or radio-frequency interference (EMI/RFI) can be a problem with electronic devices on a telephone loop. The interference can be introduced directly or via the line cord of an ac-operated device (answering machines and cordless telephones, for example).
4. Electronic device noise: all active components, such as tubes and transistors introduce white noise.
5. Static: electrostatic discharge either natural or man-made can be a problem. Contact noise and inductive spikes from relays fit into this category.
6. Thermal noise due to the random motion of electrons through a conductor. This presents a major technical problem for high-gain satellite receivers with wide bandwidths but is not a problem in the telephone loop plant.
7. Singing in four-wire transmission systems or unstable negative impedance repeaters in two-wire facilities.
8. Crosstalk: interference between cable pairs (Fig. 10-8).
9. Distortion (Fig. 10-9).
10. Echo: a delayed extraneous signal.

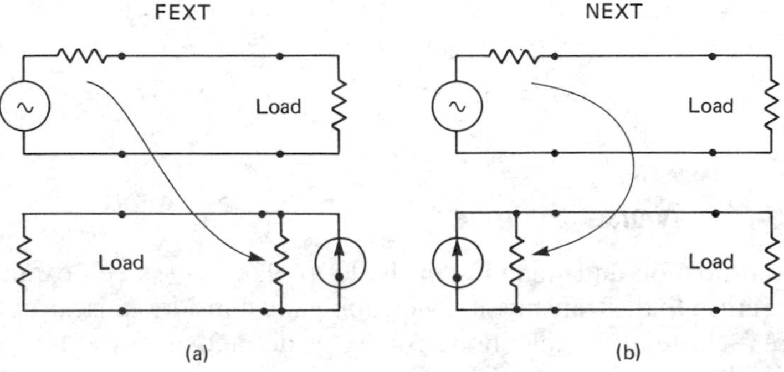

Figure 10-8 Crosstalk.

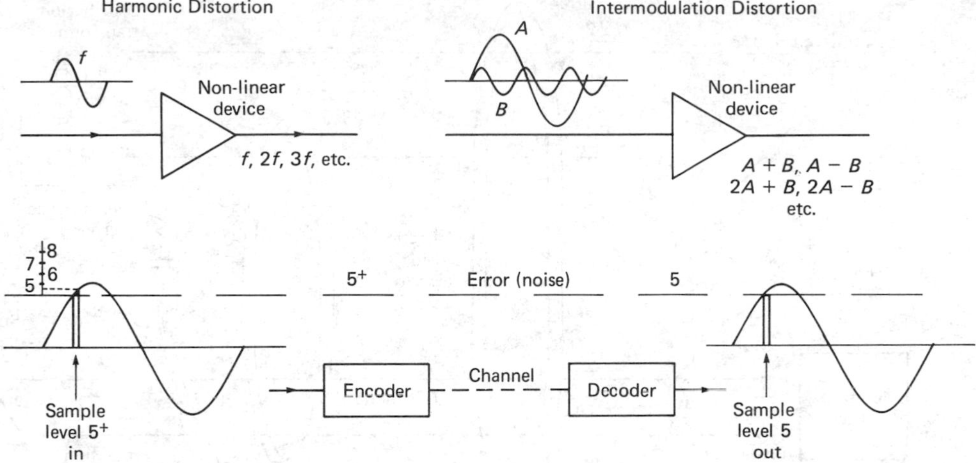

Figure 10-9 Quantization distortion.

10-2.5 Theory of Induction

Power line influence has been identified as the major source of noise in subscriber loops. The noise induction is due to the large ac magnetic field associated with the power-line current (Fig. 10-10). The effect of the electric field is negligible, as the shield around the telephone cable provides an effective electrostatic shield.

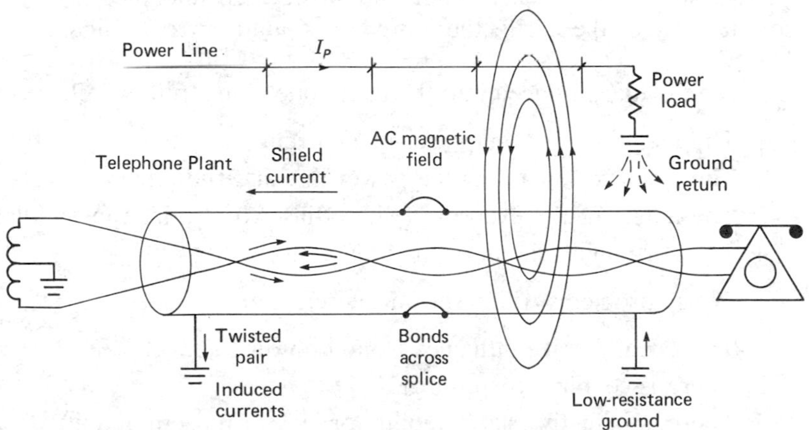

Figure 10-10 Shielding.

You might ask: Should not the magnetic induction cancel out because of the return current? Unfortunately, the return current in many power systems is via the ground (totally or in part), and the canceling effect is lost as the ground return

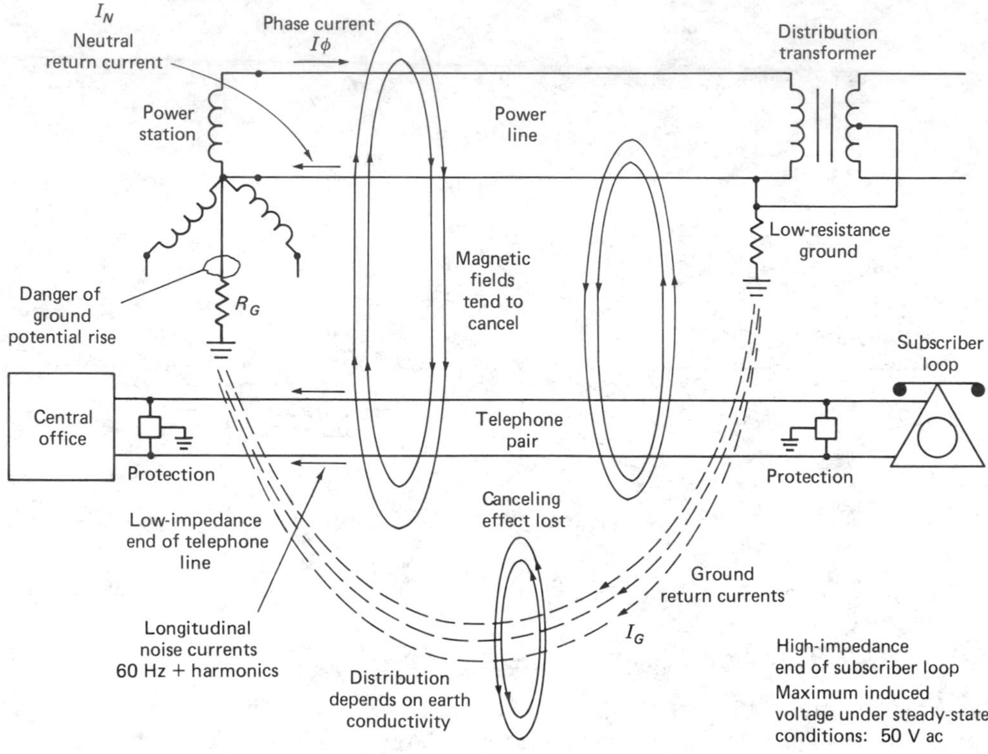

Figure 10-11 Power influence.

current spreads out many hundreds of feet deep into the earth. The poorer the conductivity of the earth, the more the ground current will spread out (Fig. 10-11).

The voltage induced into the telephone plant (power influence) depends on:

1. The magnitude of the power-line current.
2. The distance separating the power line and the telephone cable.
3. The length of the exposure (where the two utilities run parallel).
4. The harmonic content of the induction.

Noise problems will occur on the telephone pair if:

1. There is inductive influence from a power line.
2. There is coupling (exposure).
3. There is inductive susceptability or lack of noise immunity.

The telephone plant has survived in the face of increasing power influence due to balance. That is, since the two legs of the line are electrically equal with respect to ground, the induced noise currents will be equal and opposite and will cancel—in modern terminology this is called common-mode rejection (Fig. 10-12).

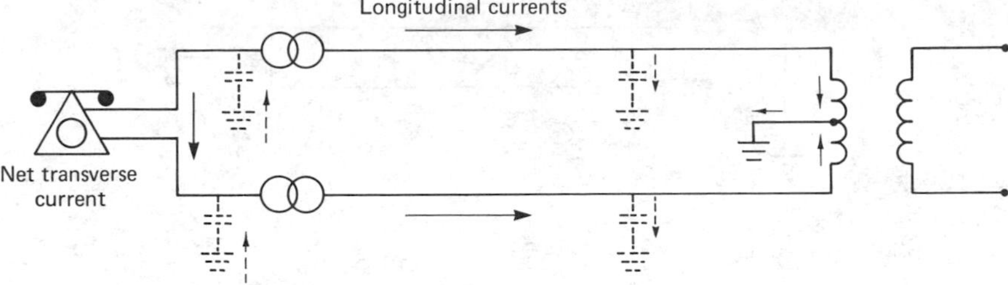

Figure 10-12 Balance.

The major causes of unbalance should be addressed (Fig. 10-13). The series resistance of the tip and ring conductors should be within 5 Ω of each other. The resistance of the individual conductors can be measured relative to ground with a digital meter; however, in the face of high power influence the digital meter may be unusable. In practice a Wheatstone bridge type of instrument is often used. A 5 Ω difference at the low impedance (CO) end of the loop causes a major unbalance. Defective or corroded splices are typical causes. Similarly, a shunt unbalance at the high-impedance (subscriber) end of the loop causes major noise problems. Tracking of a carbon protector, water in the cable, unbalanced foreign equipment, or unbalanced party-line ringers are typical problems. Bridge taps on the cable pair must not be overlooked, as an unbalanced bridge tap will unbalance the loop itself and may present a difficult problem to track down.

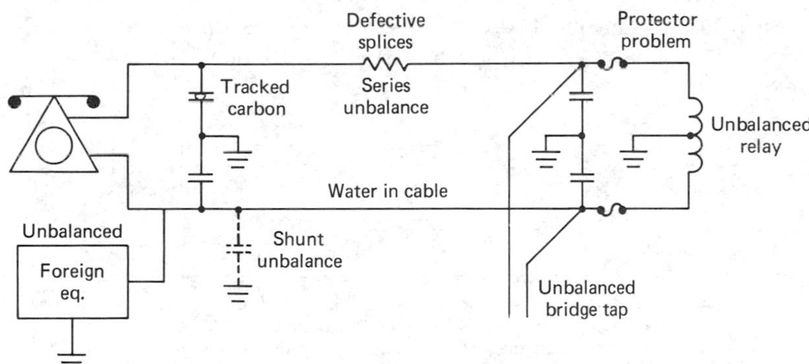

Figure 10-13 Causes of unbalance.

10-2.6 Noise Measurements in the Loop Plant

The power influence on the telephone loop can be measured with a high-impedance ac voltmeter (Fig. 10-14). This can be done by twisting the tip and ring together and grounding the far end. The voltage to ground is then measured at the near end. The voltage will vary from less than 10 V where the power influence is low,

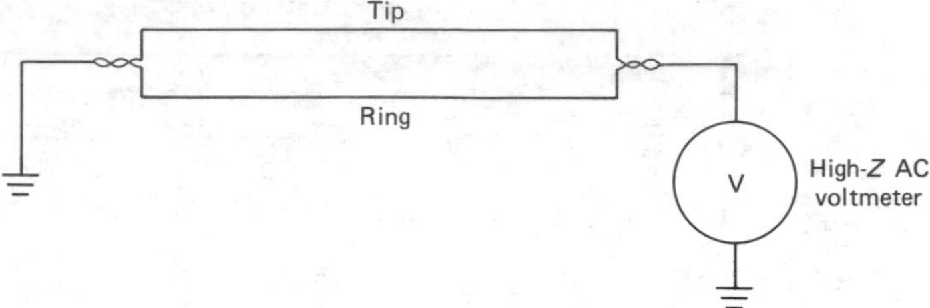

Figure 10-14 Measurement of A.C. induction.

to the steady-state limit of 50 V. The telephone plant is expected to survive in the face of this level of induction. With ground-fault conditions of the power line, the induced ac voltage can be very high for the fraction of a second before the power breakers operate; 1500 V is usually taken as the safety limit under these conditions. It should be obvious that the cable shield and the telephone pairs must be treated as possible live conductors.

The harmonic content of the power influence is also critical. The higher harmonics, especially the odd triples (540 Hz, 900 Hz), lie in the voice band and may be heard by the subscriber. While the fundamental is out of band, it may still provide problems with dc signaling, PCM repeater power feeds, carbon protection, and personnel safety. To measure the power influence and the resulting noise in the subscriber's loop, two measurements are taken, using a noise measuring set that simulates the telephone's response by means of a C-message filter. The two measurements in question are called *noise to ground* (power influence) (Fig. 10-15) and *noise metallic* (circuit noise) (Fig. 10-16).

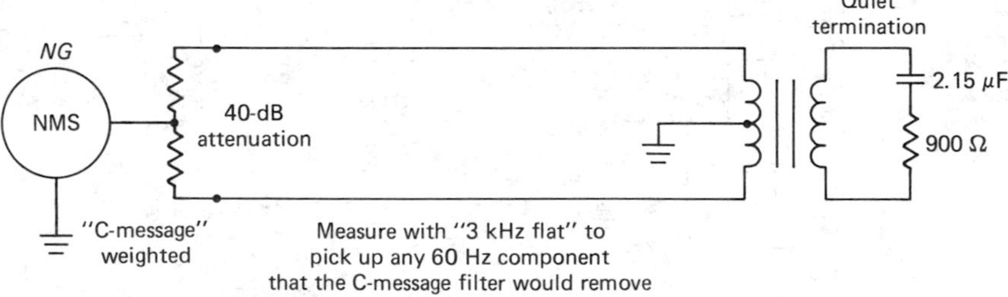

Figure 10-15 Noise-to-ground measurement.

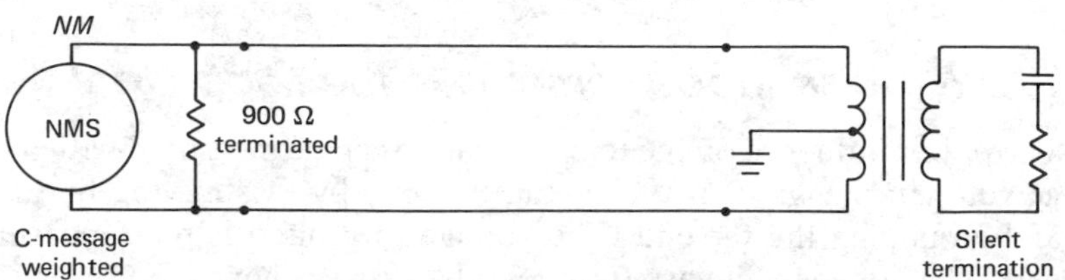

Figure 10-16 Measurement of circuit noise.

Noise-to-ground measurements of between 80 and 90 dBrnc are considered average; above 90 dBrnc the power influence is high. (dBrnc stands for decibels above reference noise, c-message weighted. Reference noise is −90 dBm at 1000 Hz.) The noise metallic or circuit noise is the result of transverse noise currents which flow due to imperfect balance. A terminated (900-Ω) noise measurement is taken at the subscriber's location (Fig. 10-16). The acceptable limit is usually taken to be 20 dBrnc at the subscriber. In practice the telephone company repairman will dial up a silent or quiet termination, which is installed on the central office switch. The transmission loss of the loop can be obtained by dialing up a milliwatt (0-dBm) supply. To do this conveniently requires a transmission measuring set with dial and hold capability (Fig. 10-17). The noise balance (NB) of the loop plant can be computed from the noise ground (NG) and noise metallic measurements (NM).

$$NB = NG - NM$$

In practice we would like the noise balance to be 60 dB or more.

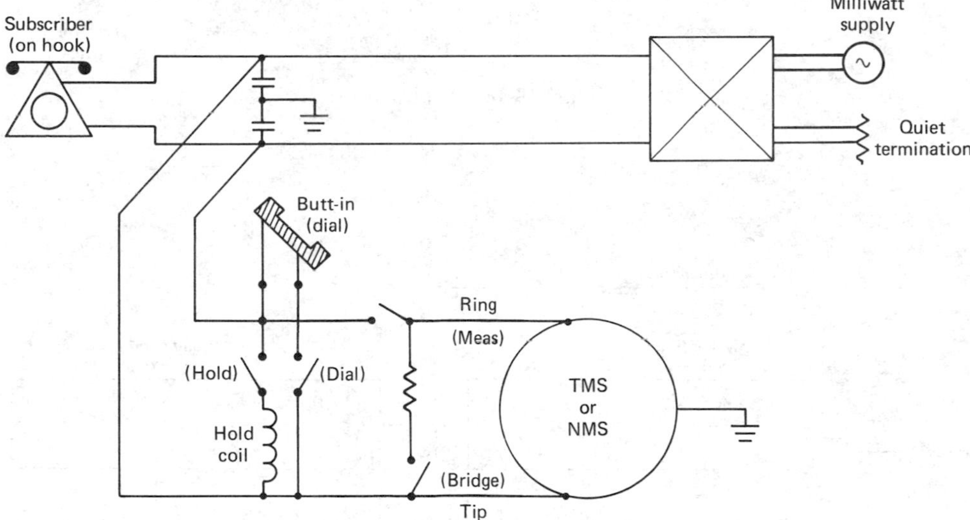

Figure 10-17 Transmission loss/noise measurements.

10-2.7 Trouble-shooting the Unbalance

Locating the unbalance by sectionalizing the loop can be complicated by the fact that the power induction can take place in a part of the cable where no unbalance exists (Fig. 10-18). The Wilcom method for sectionalizing the problem is recommended. Opening the cable at A and taking noise measurements in both directions would not locate the direction of the unbalance, as there is no power influence in section A–C and no unbalance in section A–B. The Wilcom method addresses this problem (Fig. 10-19). Maintaining the circuit so that the longitudinal currents can flow through the unbalanced section will allow the direction of the fault to be readily identified.

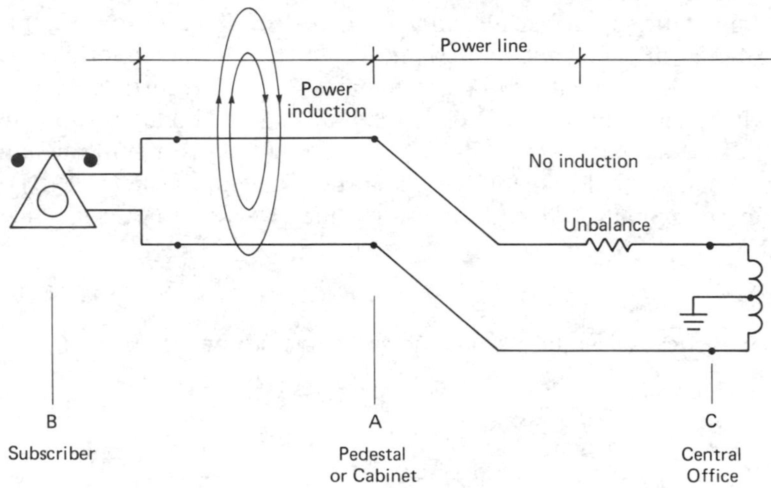

Figure 10-18 Typical noise problem.

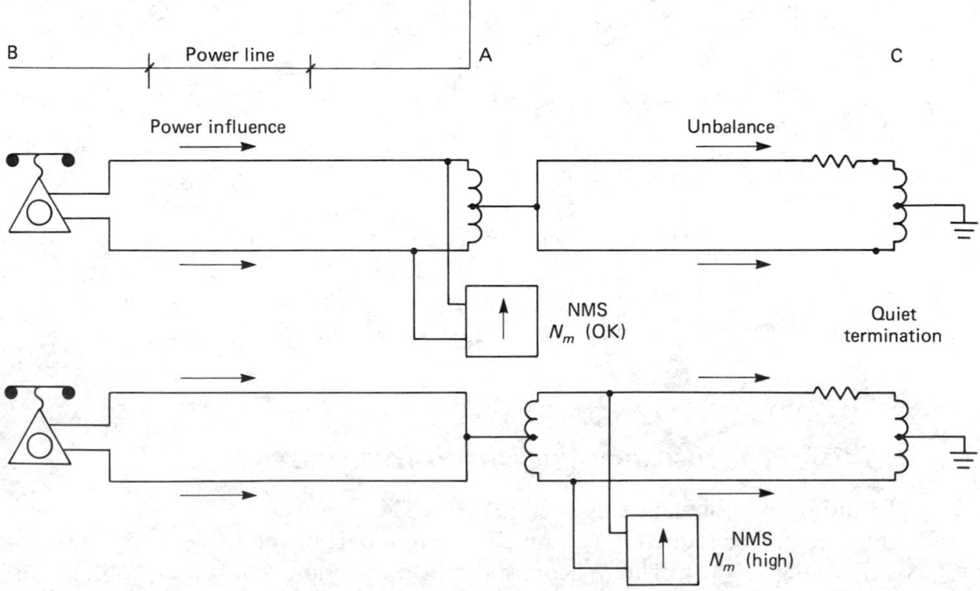

Figure 10-19 Wilcom method.

10-2.8 Shielding

Circuit balance provides much of the noise immunity for the telephone system, but magnetic shielding by the cable shield is also important. Complete magnetic shielding at these low frequencies could only be achieved with a low-reluctance iron shield, which clearly is not economically possible for telephone cables. However, if the extremities of the cable shield are grounded and all the splices are properly bonded, there will be a path for the magnetically induced currents to flow in the aluminum shield that encloses the cable pairs in telephone cables. The shield current will induce a voltage into the cable pairs in a direction that will oppose the primary induction from the power line (Fig. 10-10).

Magnetic shielding obtained by this method is quite effective against the higher harmonics of the power frequency. However, at 60 Hz it is of little value. It is important that the ground connections at the cable ends be very low resistance if the full effect of the shield is to be obtained. Fig. 10-20 shows a situation where intermediate grounding can improve the shielding. In some cases, circuit balance and shielding are unable to solve the problem, or perhaps they cannot be upgraded immediately. Noise mitigation devices can be employed in these situations. The mitigative devices available to the telephone company are as follows:

1. Noise chokes for conditioning single circuits
2. Induction neutralizing transformers, which can be applied to the entire cable and are very effective against both 60 Hz and harmonic induction
3. Drainage coils
4. Reject filters
5. Ringer isolators for party-line ringers

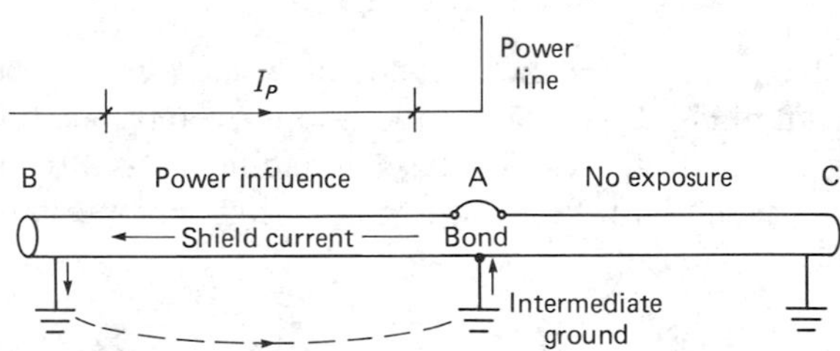

Figure 10-20 Improving the shield effectiveness.

Noise problems will be more pronounced if the transmission levels on the loop are poor. As discussed previously, the dc loop current must be at least 23 mA to ensure efficient operation of the carbon microphone transmitter. The VF loss of the subscriber loop is kept below 8 dB. Eighty percent of the subscribers will be within 2 to 3 miles of the exchange and a bare loop will suffice. The remaining 20% of subscribers may indeed be outside the base rate area (rurals) and the loops will require loop treatment.

10-2.9 Loop Treatment

To reduce the transmission loss and loading, negative impedance repeaters are commonly used. To maintain the dc loop current a device called a loop extender, which provides battery boost, can be installed at the central office. On some special service circuits four-wire repeater circuits may be used (e.g., Wescom package units) in conjunction with SF signaling. It is obvious that conditioned loops are expensive; hence modern practice is to obtain pair gain by use of subscriber loop carriers. Both analog and digital multiplex systems are available (subscriber loop carrier). Fiber optic cable for the feeders from the central office to remote multiplexers is now being deployed. This equipment allows many subscribers to share a common facility. Remote switching units or remote line modules can further increase the economic feasibility of the digital central office.

10-2.10 Loading Schemes

Loading refers to the addition at regular intervals of lumped inductance, series aiding in both the tip and ring conductors. It has been used for many years to reduce the loss on two-wire trunks and loops. Two loading schemes are commonly in use: H88 and D66. H88 involves 88-mH coils every 6000 ft, usually in urban trunks. D66 indicates 66-mH coils every 4500 ft. The D66 loading has a somewhat better high-frequency response and is often used in rural loops. Loading has three major effects on the wire facility:

1. Increases the ac characteristic impedance.
2. Decreases the VF loss within the passband by roughly a factor of two.
3. Increases the dc resistance by approximately 9 to 12 Ω per mile due to the coils.

The frequency response of the loaded line as compared to a nonloaded cable pair is shown in Fig. 10-21. Subscriber loops over 18,000 ft will normally be loaded. The office-end section is made 0.5 section long. Thus if the loop is switched through to another loaded loop or trunk, the loading scheme is maintained (Fig. 10-22).

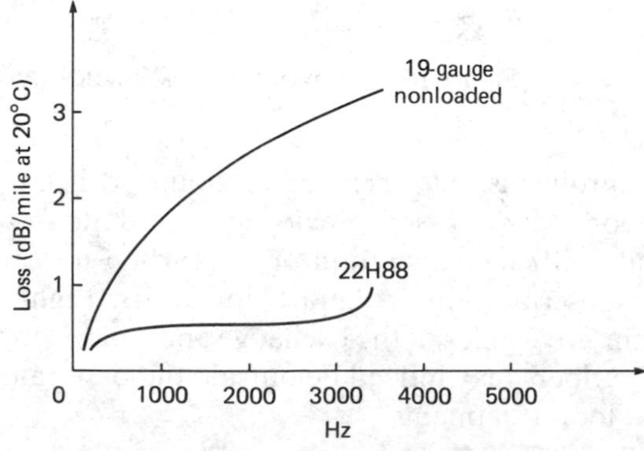

Figure 10-21 Frequency response.

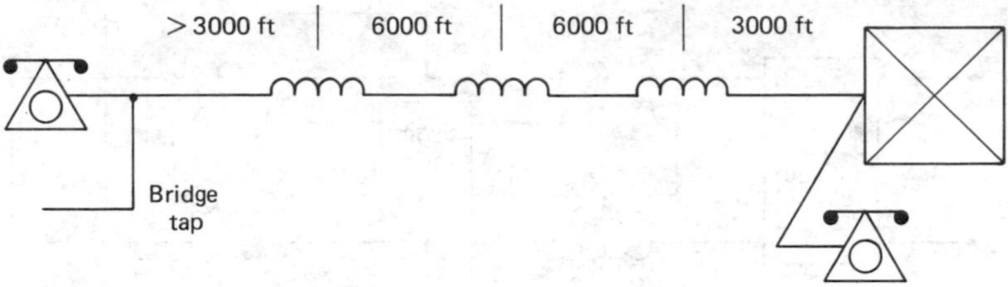

Figure 10-22 Subscriber loop.

The subscriber end section must be greater than 3000 ft and less than 9000 ft. Bridge taps are any branches off the pair. In the construction process, a subscriber may be bridged onto a pair without actually cutting the cable pair going on to a distant terminal box. Electrically, the bridge tap constitutes an added capacitance across the pair. This increases the loss of the facility by some 0.22 dB per 1000 ft. of bridgc tap. Note that any fault or unbalance on the bridge tap will cause problems for the subscriber's loop which can easily be overlooked when troubleshooting. With loaded cable schemes, no bridge taps are allowed between load points. At the end sections any bridge tap must be at least 3000 ft from the last load point. An interesting addition to the bridge tap story is the case of the subscriber having two telephones with the same directory number in different locations. Typically, a doctor may have a home phone and also one in the office in town. If the bridge tap in this case causes significant attenuation, bridge lifters may be installed (Fig. 10-23). These devices are saturable core inductances that cancel the capacity of the tap when the loop is idle. When loop current is drawn, the inductor saturates and is noneffective.

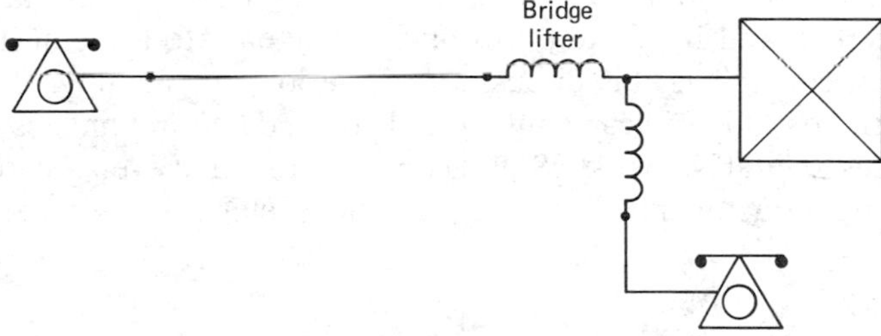

Figure 10-23.

10-2.11 Negative-Impedance Repeaters

These devices are two-wire voice-frequency amplifiers based on the original Bell E6 design. They are widely used in the telephone plant because they maintain dc continuity and provide bidirectional amplification on a two-wire facility. Gain is obtained by reflecting negative resistance into the pair. Obtaining stable gain with a negative impedance device depends very much on the impedance match between the device and the cable plant. Because of this, negative impedance repeaters are

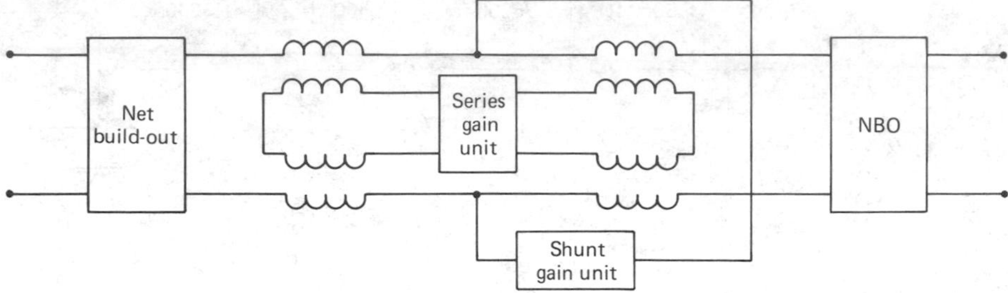

Figure 10-24 Negative impedance repeater.

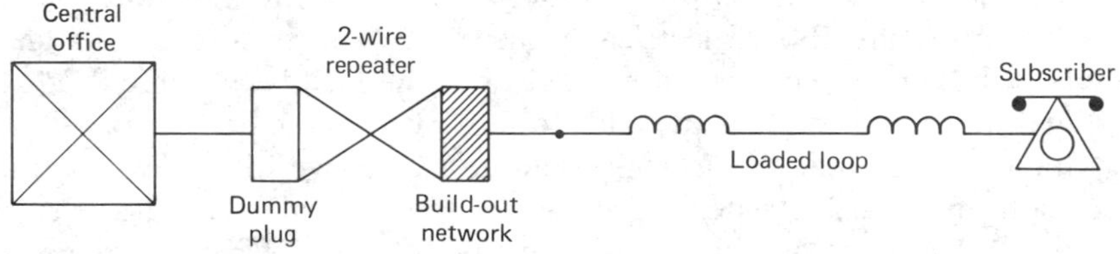

Figure 10-25 Loop treatment.

in general practical only with a loaded facility where the impedance can be stabilized across the voice-frequency band. A block diagram of the device is shown in Fig. 10-24. The unit is usually rack mounted at the central office, where 48-V power and maintenance support are available (Fig. 10-25). A gain of up to 8 dB can be obtained, but 3 to 4 dB is a typical setting. The dc resistance of the repeater is between 65 and 100 Ω, and this must be taken into account in the loop design.

In concluding this discussion of the loop plant, note that in the future the subscriber's loop will carry integrated voice, data, and images using the Integrated Services Digital Network (ISDN) concepts, to make even greater use of the existing copper conductors now in place for analog telephone service.

10-3 THE PUBLIC-SWITCHED VOICE NETWORK

10-3.1 General Toll Switch Plan

For one of the 100 million telephones in North America to reach any other telephone, it is apparent that a sophisticated switching network must be in place with toll transmission facilitites connecting the various switching centers. The General Toll Switch Plan (GTSP) was introduced during the 1950s when direct distance dialing (DDD) replaced operator-assisted long-distance service. In the North American scenario a hierarchy of five classes of switching centers was established. In smaller countries three classes of offices suffice.

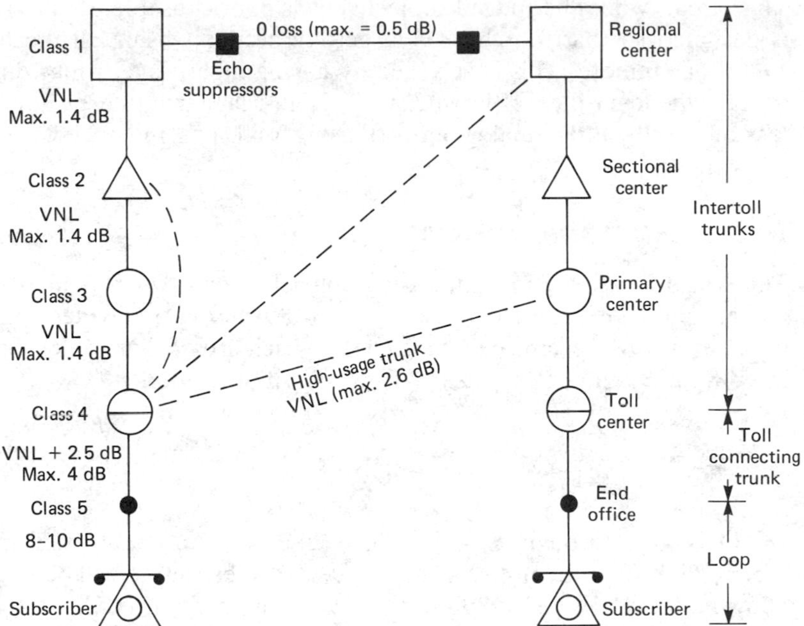

Figure 10-26.

The local exchange or end office is designated as a class 5 office which terminates subscribers and homes on the class 4 toll office. The class 4 office in turn homes on a class 3 primary center, and so on, as shown in Fig. 10-26. Looking at the GTSP from the transmission viewpoint there are a number of features to enlarge upon. The link between the class 5 end office and any higher-order office is referred to as a *toll connecting trunk*. These trunks are engineered to have a loss of 3 to 4 dB to provide stability. We will examine the concept in detail later. The intertoll trunks are designed to have a maximum loss of 1.4 dB, depending on the length and the echo delay of the facility. The concept is to control echo in long trunks by deliberately adding loss. This plan is known as *via net loss design*. Obviously, only so much loss can be added, thus trunks of more than 3000 km (1800 miles) in length are equipped with echo suppressors so that they can be operated at essentially zero loss. The link between regional offices is kept to 0.5 dB loss and would employ echo suppressors. Let us take an inventory of the losses in this analog network for a circuit between subscribers in different regions of the continent using all the links in the hierarchial chain.

2	Toll connecting trunks at 4 dB	8.0 dB
6	Intertoll trunks at 1.4 dB	8.4 dB
1	Regional–Regional link	0.5 dB
2	Sub loops at 8 dB	16.0 dB
	Total interconnection loss (ICL)	32.9 dB

This circuit would be judged subjectively as poor—usable but poor. The network is designed so that the chances of getting a circuit using all the links (the final route) are remote. This is achieved by having high-usage trunks directly between offices whenever the traffic warrants it. High-usage trunks are allowed a maximum loss of 2.6 dB in the analog network using VNL design concepts.

10-3.2 Numbering Plans

The world is divided into nine geographical zones. North America and the Caribbean constitute zone 1. Similarly, North America is divided up into areas by population with specific codes assigned to each area. Thus each subscriber is identified by a *Number Plan Area* (NPA) *code*. For example:

Area code	*Office code*	*Subscriber*
NYX	NNX	XXXX
403	458	7823

where N = 2 through 9, Y = 1 or 0, and X = any number 0 through 9. The second digit (0, 1) identifies an area code. This restriction will be relaxed in future, allowing an NXX XXX-XXXX code. With the present NYX area code there are 160 possible combinations. However, a number are reserved for special services, such as the 911 emergency, 411 information, 611 repair service, or 800 WATS (Wide Area Telecommunications Service) toll-free service. Prefixes must be attached to the NPA code to access the various services:

Prefix	
1	Access to the toll network
0 +	For operator-assisted calls
011	Gives access to the international service

For example, to dial the United Kingdom from North America, one dials 011 44 followed by eight or nine digits for the person being called. England is in zone 4, country code 4.

10-3.3 Toll Transmission Concepts

The telephone instrument and the subscriber's loop are by economic necessity "two wire." However, toll facilities are in general "four wire"; that is, there are separate receive and transmit paths. Hence the need for a two-wire/four-wire interface or hybrid (or four-wire term set). Traditionally, this interface has been located in the toll office, but in the digital world, the hybrid has moved back to the central office either to the line card in the digital end office or to a channel unit in a PCM channel bank in the case of an analog switching office with digital trunks. The interface is the major reflection point which causes echo in the system. The "via net loss" design concept was developed to counter the echo from the terminating hybrid. To understand this problem, it is necessary to present hybrid theory in a qualitative manner.

Consider the hybrid as an ideal four-port device with all four ports correctly terminated (Fig. 10-27). Audio energy from the two-wire divides equally between the transmit and receive ports on the four-wire side. The energy arriving at the receive port is lost; hence the insertion loss from the two-wire to the four-wire transmit port is theoretically 3 dB (Fig. 10-28).

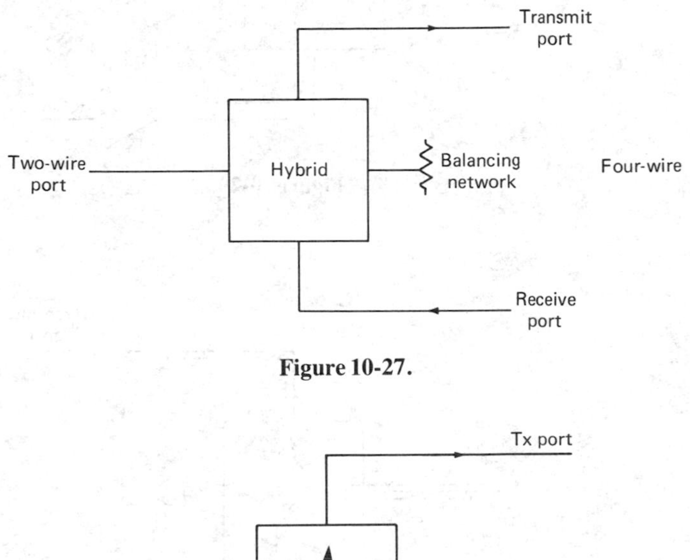

Figure 10-27.

Figure 10-28.

The audio energy arriving at the four-wire receive port splits equally between the balancing network and the two-wire port in the ideal case (a 3-dB insertion loss; see Fig. 10-29). If the input impedance looking into the two-wire from the hybrid does not match the impedance of the balancing network or the termination does not match the line, leakage across the hybrid from the receive port to the transmit port will occur and result in talker echo back to the far end (Fig. 10-30).

Note that if attenuation is placed in the two-wire path, the echo will be attenuated twice. The ratio of the reflected power to the incident power is called the *return loss*. The higher the return loss, the better. In practice, where the switching equipment can connect any subscriber, the minimum return loss objective across the voice frequency band is 10 dB. Where the two-wire termination is variable, the balancing network has to be a compromise, and 2.15 μF plus 900 Ω is typical. Figures 10-31a and 10-31b show these important concepts.

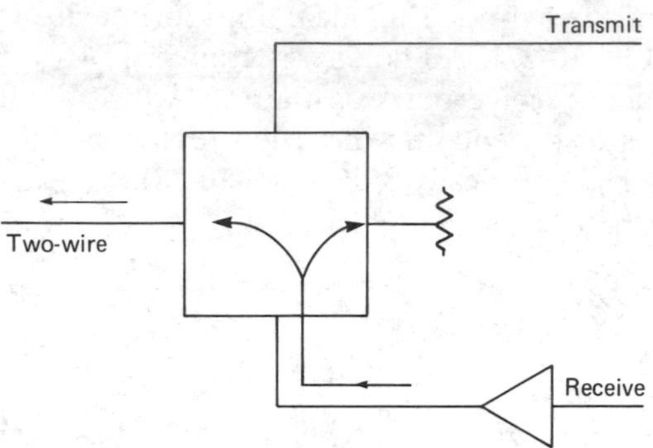

Figure 10-29.

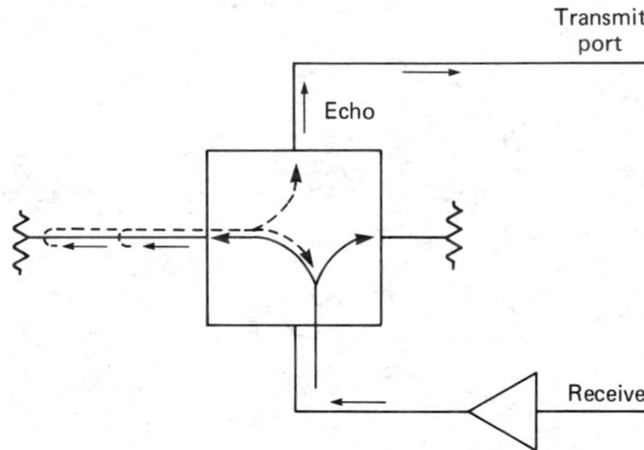

Figure 10-30.

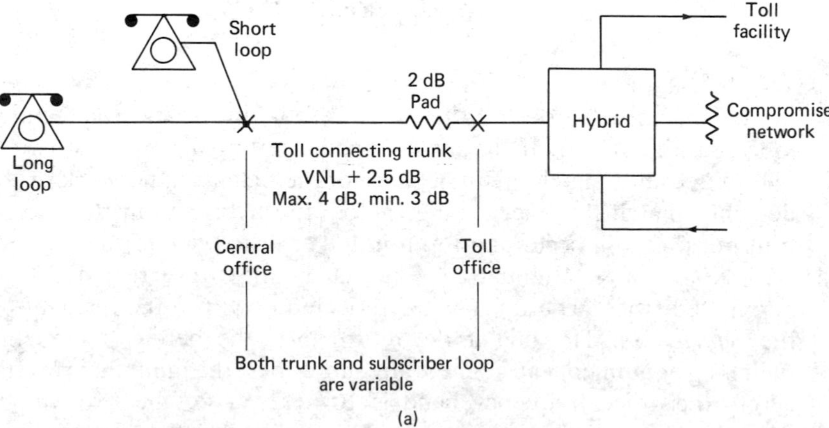

Figure 10-31 (a) Analog network—compromise balance.

(continued)

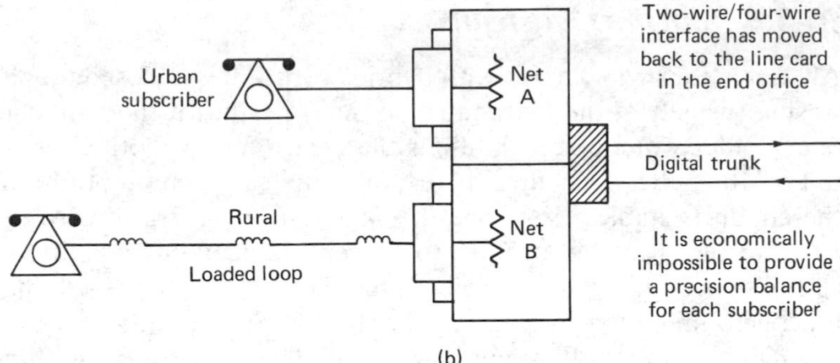

Figure 10-31 (b) Digital end office—compromise balance.

The return-loss measurement is obtained in practice by the following method, as there are usually level coordination pads in the four-wire path as shown in Fig. 10-32. The measurement shown in Fig. 10-32 is called the *trans hybrid loss* (THL). This is not the true return loss, as the measurement includes the 3-dB loss between the four-wire receive and the two-wire and the 3-dB loss between the two-wire and the four-wire transmit port. Also, there are some dissipative losses in the hybrid windings and the two pads. Hence the true return loss (RL) is given by

$$RL = THL - (\text{hybrid loss} + \text{pads})$$

The term in parentheses can be readily determined by placing a direct short temporarily across the two-wire port of the hybrid (RL = 0). The transmission measuring set and oscillator output are adjusted for a reference reading. When the short is removed and the actual termination or an artificial load is connected, the drop in the meter reading in decibels is the true return loss. The worst-case return loss over the whole VF band should be taken. A more dynamic measurement of balance is the echo return loss test, which more closely simulates the working situation. In this case the oscillator is replaced by a noise generator the output of which is shaped for the voice band. The response is obtained using a noise-measuring set with C-message weighting which gives a single-valued average reading called the *echo return loss* (ERL) over the frequency band 500 to 2500 Hz (typically 11 dB).

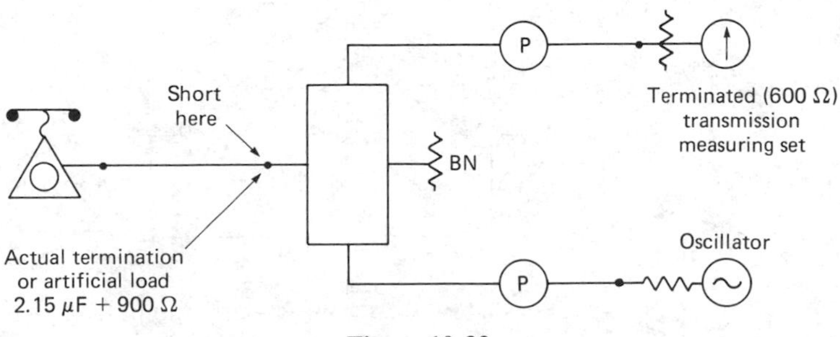

Figure 10-32.

10-3.4 Singing Margin

An active four wire circuit with hybrids constitutes a closed loop that can oscillate or sing (usually at the extremities of the VF band) if the loop gain is greater than one. Under worst-case conditions the return loss at both ends is zero. Referring to Fig. 10-33 the loop gain is 12 dB; the loop losses amount to 12 dB. Theoretically, the circuit is stable. However, the singing margin is also zero. Theoretically, we need a singing margin of 6 dB to prevent a "near-sing condition." This is known as *rain-barrel effect*; the circuit sounds as though one is talking into a pipe. In practice a minimum singing margin of 10 dB is normally specified.

To obtain a 6-dB singing margin under worst-case conditions, the loop gain could be reduced by 6 dB (Fig. 10-34). The same condition could be obtained by placing a 3-dB pad in one of the two-wire circuits (Fig. 10-35). Alternatively, the trunk could be configured with 3-dB pads in both two-wire paths (Fig. 10-36).

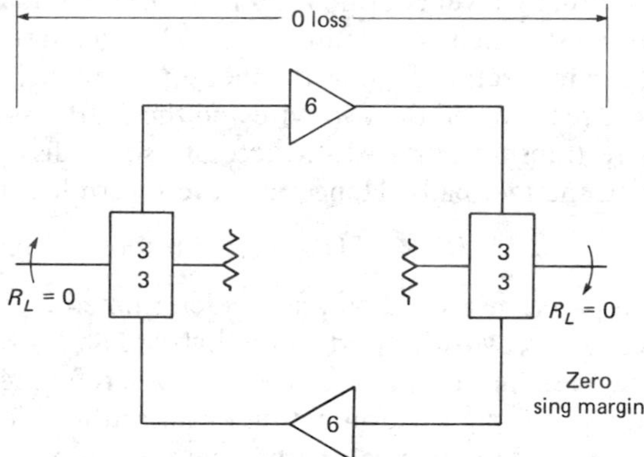

Figure 10-33 2W-to-2W loss is zero.

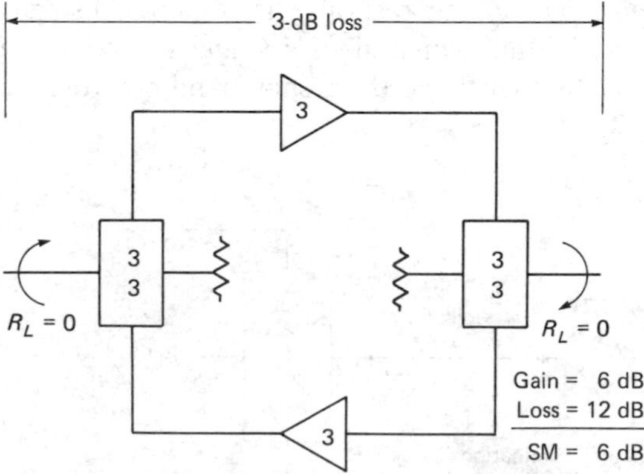

Figure 10-34 Worst-case stability.

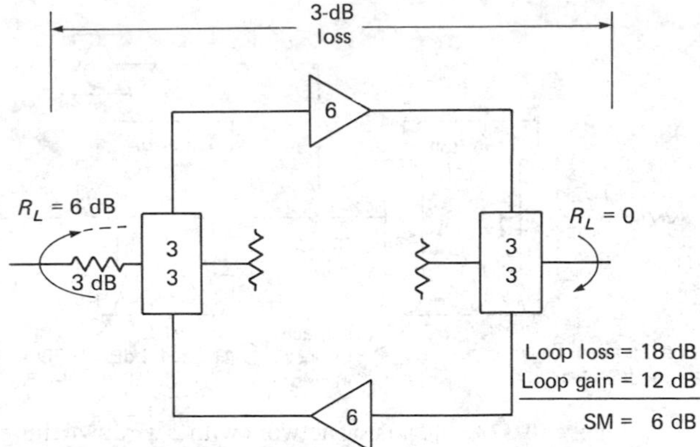

Figure 10-35 Z-wire loss controls stability.

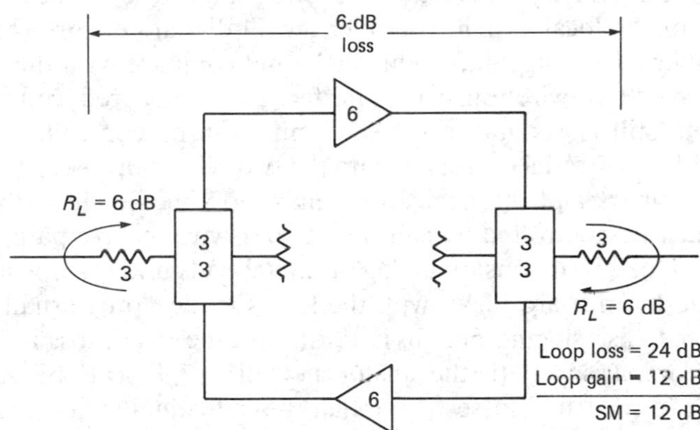

Figure 10-36 Practical solution.

Comparing Figs. 10-34 and 10-35, we see that there is a choice. Either put 6 dB of loss in the four-wire or 3 dB of loss in the two-wire portion. In all cases the loss of the trunk will have to be 3 dB if stability is to be guaranteed. In the analog network the stability of the trunk was achieved by putting a minimum of 3-dB loss in the toll connecting trunks. Originally, all switching was two-wire; at intermediate switching points the return loss could be guaranteed (through balance), but at the terminations the worst-case return loss must be assumed. Putting the required loss in the toll connecting trunks eliminated the need to change the four-wire gain or the external two-wire pads, depending on whether the trunk was terminating or not (Fig. 10-37).

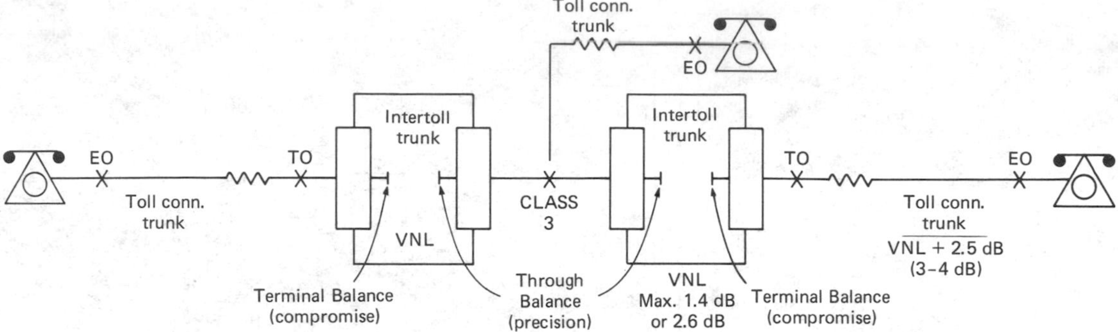

Figure 10-37 Old analog network with 2-wire switching.

Since the introduction of direct distance dialing, the switching equipment has progressed from two-wire to four-wire switching eliminating many of the reflection points for echo generation. The situation has now changed radically, with digital end offices (DEO) where the two-wire/four-wire interface is located in the line circuit of the local switch. Consider the all-digital network (Fig. 10-38). There are a number of major points to highlight and compare with the analog network. The number of two-wire/four-wire interfaces has been reduced to two. The stability problem still exists and talker echo must still be controlled. It is not possible or desirable to introduce a set loss into the two-wire loops because of the large number of subscribers and the need to minimize loss in the subscriber loops. Hence the loss must be controlled by pads in the four-wire receive path. A fixed loss of 6 dB is introduced by means of analog or digital pads under software control. Assuming that the hybrids are ideal, with the losses made up externally, it is apparent that the worst-case singing margin is 12 dB and the trunk loss is 6 dB.

Comparison with the analog network of Fig. 10-37 should be made with respect to the trunk losses. Note that the individual trunk losses in the analog case are measured from the outgoing side of one switch to the outgoing side of the next switch. The loss is VNL + 2.5 dB (minimum 3 dB, maximum 4 dB) for the toll

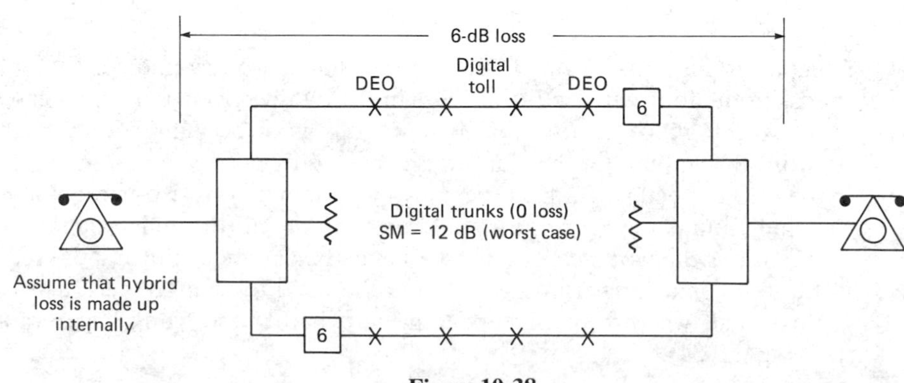

Figure 10-38.

connecting trunk and via net loss for the intertoll. The loss through the switch is negligible. Now in the case of the digital network, the trunk loss must be measured from the analog two-wire to two-wire and the loss of the switch accounted for. Also, digital trunks present zero loss! The 6-dB pads must therefore provide sufficient echo control for long-haul circuits. Compare the built-up analog and digital network overall connection loss:

$$\text{analog network loss} = \text{VNL} + 6 \text{ dB}$$

$$\text{digital network loss} = 6 \text{ dB}$$

The difference of the VNL could provide an improvement of up to 9 dB in the toll network.

There are two additional points to be made. Compare the local–local call with analog and digital end offices (Fig. 10-39). It has previously been shown that to ensure stability a 3-dB loss across the four-wire digital switch will be theoretically necessary. Studies have shown that a worst-case return loss of 1.2 dB can be expected; hence the pads could be reduced to 2 dB. A future solution to this problem is adaptive balancing networks, but cost at this time precludes their use in an ordinary subscriber line circuit. A choice of two different balancing networks to accommodate ordinary loops or loaded loops is another solution. The 2-dB pads are added where necessary to ensure stability. It is interesting to note that any leakage across the hybrids will be encoded and subsequently decoded. This process produces quantization noise, which can build if the circuit is not stable. This results in a loud roar rather than the customary singing tone.

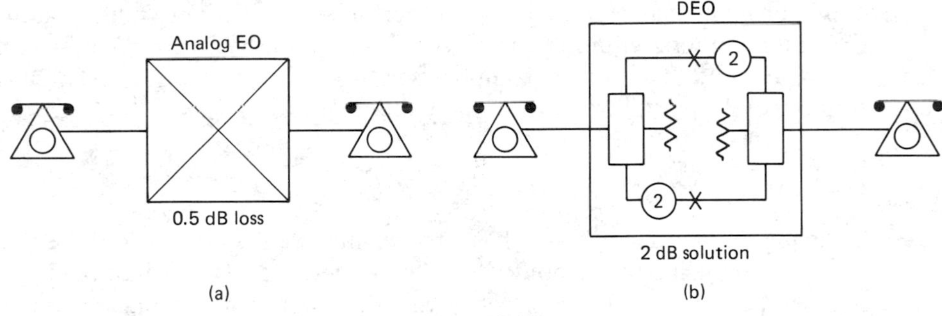

Figure 10-39.

For some time to come, telephone systems will be a mix of analog and digital facilities. Digital trunks that terminate in a digital switch at one end and an analog switch at the other are classified as *combinational trunks*. The digital end office switch must have the capability of switching in 2-, 3-, 5-, or 6-dB pads under software control to accommodate analog, combinational, or digital trunks. Notice that two approaches to the transmission problems of singing and echo control have been presented. The North American analog network is based on the via net loss concept, and the emerging digital network employs the fixed-loss plan. The fixed-loss plan has been used in Europe and Japan throughout the development of the telephone system. In summary of the via net loss concept, there are four types of trunks:

Type	Loss
1. Toll connecting trunk Minimum loss 3 dB Maximum loss 4 dB	VNL + 2.5 dB The 2 dB is obtained with a fixed pad and 0.5 dB is due to the transmission bridge.
2. Intertoll trunk (no echo suppressor) Maximum of 1.4 dB Approximately 700 miles	VNL
3. Intertoll trunk (with echo suppressor) Maximum 0.5 dB	0 dB (nominal)
4. High-usage trunk (no echo suppressor) Maximum of 2.9 dB 1800 miles	VNL

The via net loss for a single trunk is calculated from

$$\text{VNL} = 0.4 + 0.0015 \times \text{trunk length in miles (dB)}$$

or

$$\text{VNL} = 0.4 + 0.00093 \times \text{trunk length in kilometers (dB)}$$

The formulas above assume carrier and/or radio facilities. The 0.4 dB is added to allow for maintenance inaccuracy and performance drift per trunk. Via net loss is the minimum loss required to control echo up to a maximum round-trip delay of 45 ms. Beyond that limit echo suppressors have to be used. A maximum of 22 ms round-trip delay is allowed within any one region.

10-3.5 Signaling in the Network

Signaling in the exchange area has been examined. We now extend the theme and deal with signaling throughout the network (see Fig. 10-40). It should be evident that the human/machine interface is very different from the machine/machine interface. The trunk (relay set) is the signaling interface between the switching equipment and the transmission facility. A trunk is designated by a number of characteristics:

1. Type of signaling employed
2. Method of providing supervision
3. One-way (incoming or outgoing) or two-way signaling capability
4. Function (e.g., local, toll, intertoll, CAMA, ANI)

Let us try to make some sense of what the telephone industry calls a "trunk." It may refer to a pair of wires, a four-wire toll facility, or a trunk relay set (trunk card or trunk module in modern switching plant). In modern parlance, it concerns

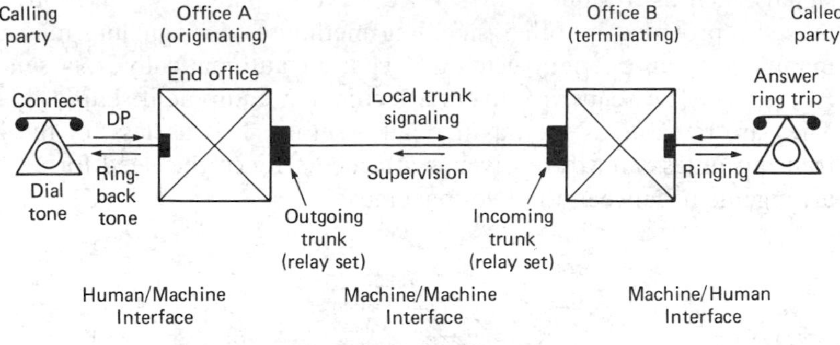

Figure 10-40.

networking. Consider the local trunk between two central offices. Imagine it to be a two-way interlocal trunk (TWILT) consisting of a single pair of conductors. The originating office seizes the trunk simply by closing the loop. The dialed digits are then transmitted by loop pulsing. The terminating switch sets up and applies ringing to the loop of the called party subscriber and ring-back tone is returned to the calling party. When the called party answers, the switch in office B trips the ring. In a toll call, the terminating office must also provide answer supervision to start the billing process. This information is often provided by reverse battery supervision. Office B simply reverses the trip and ring, which causes a polar relay to operate in the toll office. There are a number of points to be made. The system described:

1. Employs dc signaling and loop pulsing.
2. Will be limited in length by pulse distortion (typically) 1200 to 1800 Ω, depending on the switching equipment.
3. Has half-duplex signaling capability.

Half-duplex signaling means that we can signal in only one direction at a time. DC signaling in both directions simultaneously (full duplex) on two wires requires a degree of sophistication. It is, for example, accomplished with duplex signaling (DX) sets and with digital echo cancellors. Loop trunks can be two-way; that is, either office can originate the call. This introduces another problem area, as the trunk can be seized from both ends simultaneously. The problem is referred to as *contention* or *glare*. Trunks are often arranged in groups of (one-way) incoming or outgoing trunks. This eliminates the problem of simultaneous seizure and the trunk groups are also, in general, easier to administrate.

As the network has developed, more sophisticated signaling systems have become necessary. AC systems have provided much of the signaling capability in the analog network. Multifrequency signaling employs two voice-frequency tones out of six to transmit a digit. There are 15 different combinations available allowing 10-digit combinations plus start of signaling (KP) and end of signaling (ST). The frequencies used are in the range 700 to 1700 Hz. The MF system can transmit a 10-digit address within one second, so it is much faster than loop pulsing. Note that MF signaling only transfers address information, so that control and supervisory signals such as a request for service, reverse battery supervision, or disconnect must be provided by other signaling methods. MF signaling may be performed manually by an operator using a keypad or automatically by a sender. Consider Fig. 10-41. The sender/receivers are common equipment shared by many trunks. They are released as soon as the call is set up. It is useless for the MF sender to transmit unless an MF receiver is attached. Hence the need for a "handshaking" arrangement between the two machines.

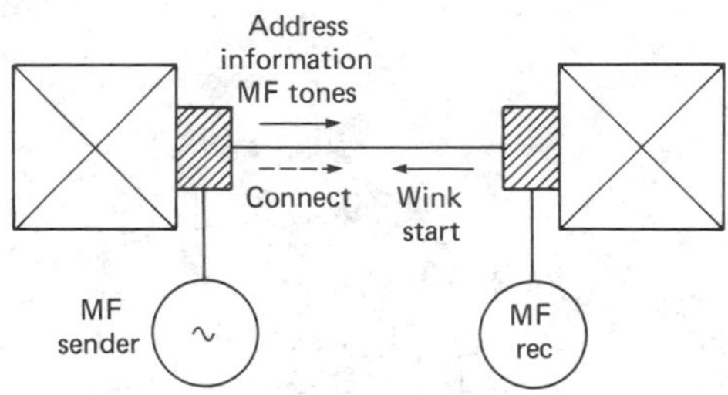

Figure 10-41.

Two methods of controlled outpulsing are in use in North America. They are called *wink start* and *delay dial*. When an MF receiver is attached, the terminating switching machine sends an on hook–off hook–on hook (*wink*). On receipt of this "proceed to send" signal, the originating switch waits 70 ms and sends

KP–ADDRESS DIGITS–ST

The alternative delay dial method is employed by the 4A crossbar switching machine. When the trunk is seized the terminating switch sends an off-hook condition acknowledging the request. Dialing is delayed until an on hook "proceed to send" signal is given to the originating office. The MF signaling system as used in North America is recognized by the CCITT as the R-1 code.

The billing machine for centralized automatic message accounting (CAMA) is usually located in the toll office and services many end offices. It follows that the CAMA machine must receive the directory number of the calling party. In some cases an operator still breaks into the call and asks for the calling party's telephone number. The information is then keyed into the billing machine. This is known as *operator number identification* (ONI). ONI is still a fact of life for small community dial offices. However, the function is now generally automated and is known as *automatic number identification* (ANI). MF signaling is used to transfer the information to the CAMA machine.

10-3.6 E & M Control Leads

So-called E & M signaling is still very prevalent in the telephone system. Control leads designated E (for receive) and M (for transmit) are used in conjunction with specific signaling systems. Figure 10-42 shows the concept. When the switching machine seizes the trunk the loop is closed, causing the trunk relay to operate.

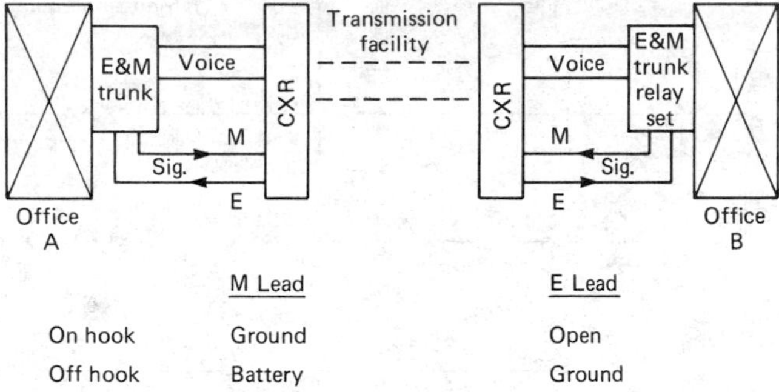

Figure 10-42.

The battery on the M lead operates the transmit signaling relay in the transmission equipment. A current limiter or ballast lamp normally protects the M lead circuit. Note the requirement in this trunk for A and B leads. On the receive side the receive signaling provides a ground on the E lead, which allows a relay in the trunk relay set to operate and seize an incoming trunk. This signaling arrangement, known as a Type 1 interface (Fig. 10-43), is an unbalanced one-wire circuit to a foreign ground. Consequently, Type 1 signaling is susceptible to interference and the range is limited to 25 Ω. The problem becomes especially evident in modern equipment where the relays are replaced with electronic detection circuits. Consequently, the type II interface has been introduced (Fig. 10-44).

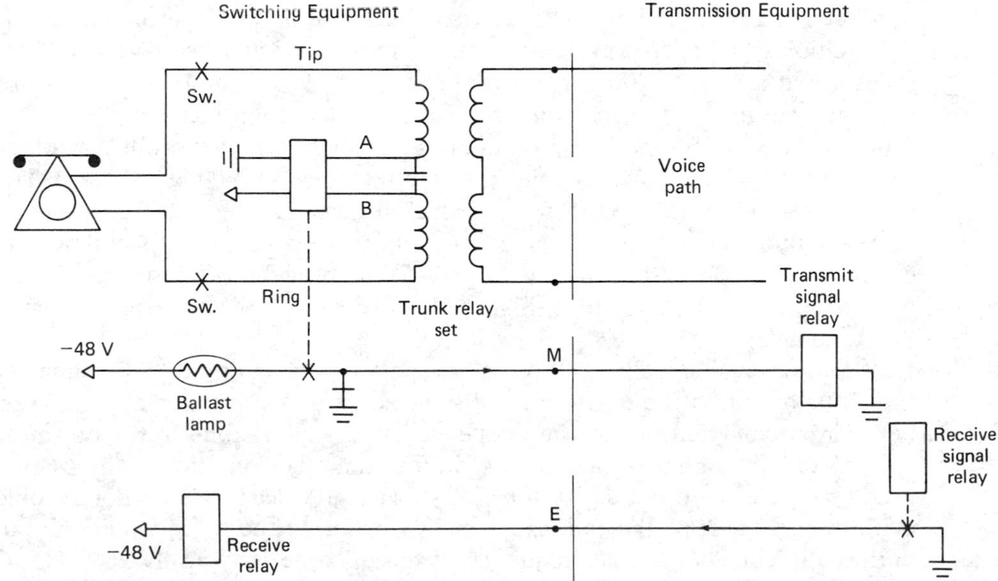

Figure 10-43 Type 1 E & M control.

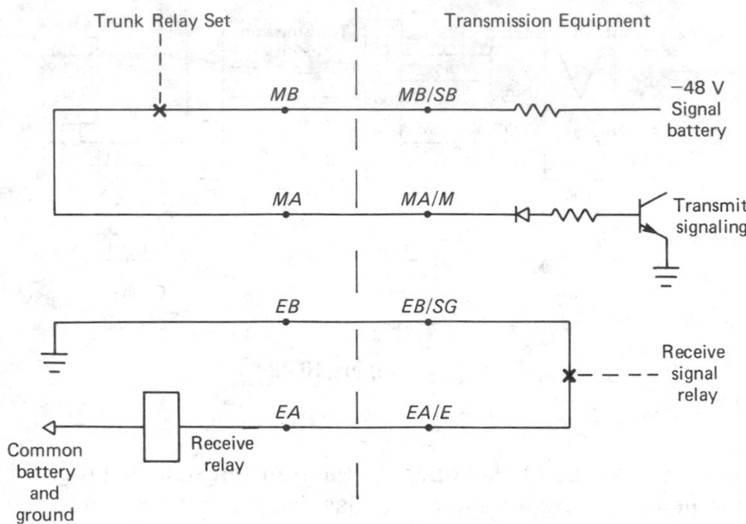

Figure 10-44 Type II interface.

E & M control allows full-duplex signaling. It is used in the toll network in conjunction with single-frequency (SF) signaling systems. Two types of SF signaling are employed in the analog network. In-band signaling uses an external SF set, while out-of-band signaling is normally associated with the multiplex carrier system. SF sets contain a 2600-Hz oscillator to transmit the signaling condition under the control of the M lead. In normal signaling the SF tone is off during the busy condition. Dial break pulses are transmitted as bursts of SF tone. The steady SF tones carrying the signaling and supervisory information would swamp or overload the multiplex and microwave equipment if transmitted at voice levels. Hence SF tones are at a level of 20 dB below the test tone (−20 dBmO). At a −16 transmission level point the tone will be −36 dBm. The tone receiver in the SF set translates loss of SF tone into ground on the E lead. A major problem with early SF sets was "talk-off" due to components of the voice at 2600 Hz. Present-day SF sets have sophisticated guard circuits which desensitize the tone receiver during busy conditions. (Note that singing, crosstalk, or test tones will simulate a busy condition.) It follows that the level of the SF tone at the traditional +7 demod-out of the carrier is −13 dBm. Figure 10-45 shows a typical carrier system with SF signaling.

Out-of-band signaling is normally a part of the multiplex equipment. The E & M leads go via the patch bay to the carrier. There are two major differences between in-band and out-of-band systems. The signaling tone in the out-of-band system is on during busy conditions and off during idle conditions. The frequency in common use in present-day systems is 3825 Hz and voice talk-off is not a problem. With in-band systems the tone is normally off during the busy condition and on during idle conditions. The frequency of the SF tone is typically 2600 Hz. One advantage of in-band SF systems is that the signaling is carried along with the voice, so that patching systems together is simply a matter of patching on a two-wire or four-wire basis.

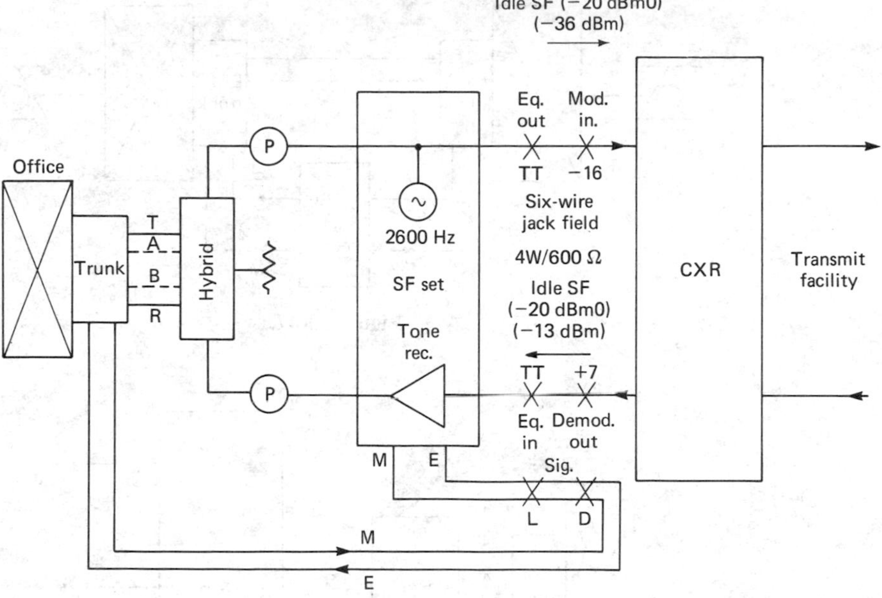

Figure 10-45 SF signalling.

A number of complexities can arise with E & M signaling. For example, if two OOB systems are to be patched in tandem, the signaling must be patched through on a dc basis. Figure 10-46 indicates that on a four-wire basis there must be a level adjustment of 23 dB. Also, the receive E-lead conditions (open and ground) must be changed to M-lead conditions (ground and battery). Modern channel units may have strap options to allow E & M conversion. The alternative is to install a converter known as a *pulse link repeater* (shown in Fig. 10-46). It may not be quite so obvious that we need to change the E-lead conditions for both channel units **or** change both E- and M-lead conditions **on one** channel unit alone (PLR option).

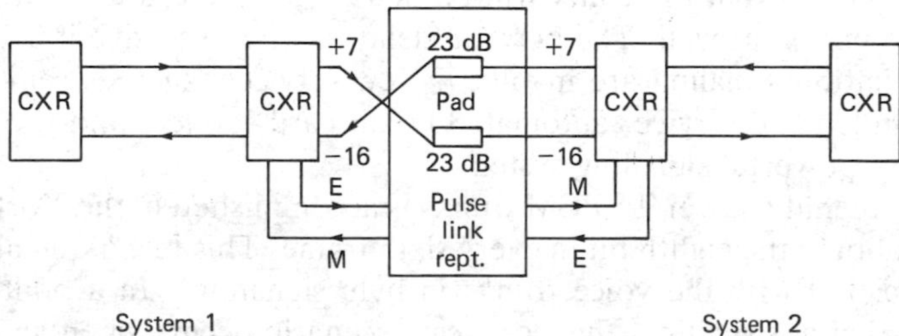

Figure 10-46 Back-to-back operation.

Another conversion comes into play if it is necessary to trunk two nearby switching machines together. In this case we need an M-to-E converter and the commercial unit available is called a *trunk link repeater* (Fig. 10-47). If the switching

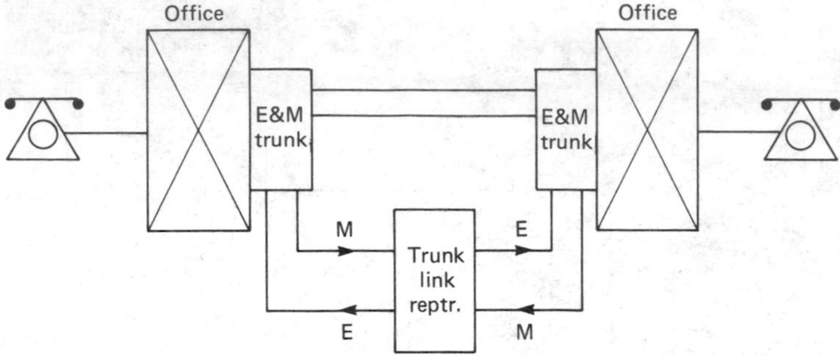

Figure 10-47.

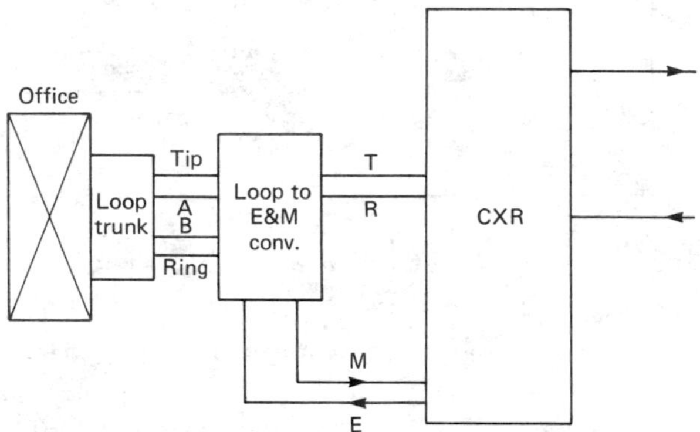

Figure 10-48 Loop-to-E & M conversion.

machine has to work with E & M signaling facilities but has only loop signaling trunks available, another interface unit, called a *loop-to-E & M converter*, has to be installed (Fig. 10-48).

The various signaling systems for the toll network which have been discussed have at least one thing in common: the signaling is associated with the voice trunk. This implies that the trunk will be tied up for the duration of the call and for the setup period as well. The postdial delay can be excessive if a number of trunks or international dialing are involved. New services, such as remote call forwarding, expanded 800 service, automated credit card service, and the ISDN, will require a very powerful signaling system.

Signaling over T1 PCM trunks is accomplished in the North American system by robbing the eighth bit in every sixth frame. This technique also has the signaling associated with the voice trunk (in byte signaling). In a built-up digital network with digital switching, the worst-case scenario occurs when an eighth bit is robbed in every frame somewhere in the system. Consequently, in this case, the performance of the system would be equivalent to 7-bit PCM and would not meet toll noise specifications. All these factors show the necessity for a new concept for signaling, and common-channel signaling in the form of the CCS 7 system is now being introduced in North America.

10-3.7 Common-Channel Interoffice Signaling

The basis for common-channel signaling is readily understood. However, the implementation of CCS 7 is a major undertaking. A number of new concepts must be appreciated to comprehend the system that will play a major role in the future telephone network. The modern switching office is a switching network controlled by a central processing unit. Instead of a signaling path between switches over the voice trunks, it makes sense to provide a data communications path to allow the two processors to talk to each other directly (Fig. 10-49).

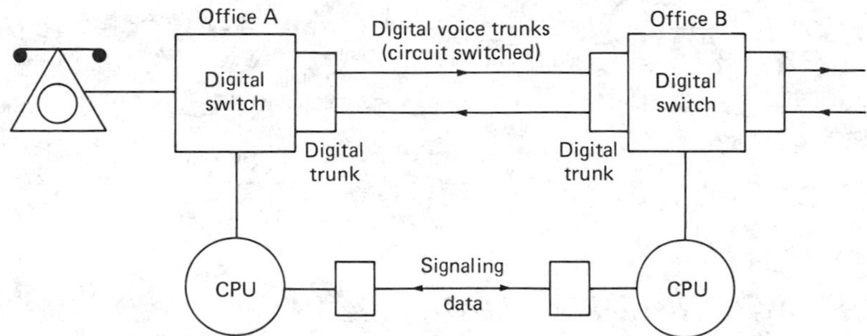

Figure 10-49 Common channel signaling.

Numerous advantages can be listed:

1. Two-way high-speed signaling can reduce the holding time of the trunk and reduce postdial delay to 2 seconds. Also, "look ahead for busy" will be possible so that the busy signal can be returned to the calling party before a connection is made.
2. Enhanced services will be possible: call management for 800 service, special routing for data circuits, and customer identification. Remote call forwarding for an enhanced foreign exchange service and private virtual networks are envisaged.
3. Voice talk-off, test-tone interference, and blue-box fraud will be eliminated.
4. Access to data bases for automated calling card service (AACS) will be possible. The data base is the memory for the future intelligent network. It will allow call processing to be distributed throughout the system.
5. A signaling network with alternate routing can be developed. Integrity of the signaling data can be checked and corrected.
6. Common Channel interoffice signaling will eliminate the need for bit robbing and allow a 64-kbps clear channel to be developed.

The protocol for the packet-switched data network will be similar to the seven-layer Open System Interconnection (OSI) model. Data transfer in the initial CCIS systems was by 2.4/4.8-kilobit modems. CCS 7 will employ 56/64-kilobit signaling over digital facilities. The topology of the network with CCS 7 is shown in Fig. 10-50.

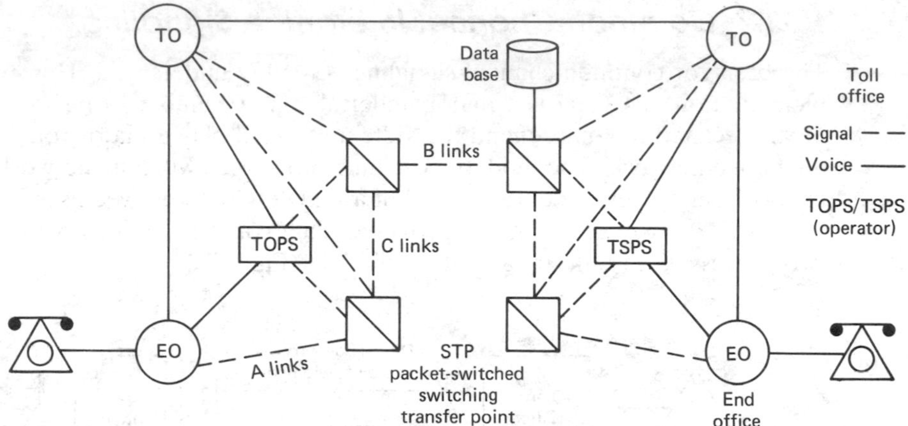

Figure 10-50 CCS 7 network.

10-4 THE SWITCHING PROBLEM

10-4.1 Basic Functions

The switching machine is set up to perform certain basic functions:

1. Provide the connectivity to allow any subscriber or trunk to reach any other subscriber and/or trunk.
2. Scan lines and trunks for requests for service.
3. Allow control of the switching matrix either directly or indirectly from the signaling information.
4. Alert the called party by applying ringing to the subscriber's loop.
5. Provide customer information signals such as dial tone, ringback tone, and busy tone. Generate or pass on signaling information to other offices.
6. Test for busy conditions on lines or trunks.
7. Supervise the call for completion of the call.

The "generic" switching machine is shown in Fig. 10-51.

Before considering digital switching we should examine analog space-division switching. Two types of electromechanical switching are still in use. Step-by-step Strowger switching utilizes direct control, in which the dial pulses from the calling party step the call through a series of selectors to a connector, which completes the call (see Fig. 10-52).

Common control refers to equipment that receives and stores dial pulses, translating them into signals that control the switching network (registers and markers). The switching element is in general a crossbar switch matrix. The wired relay logic of Fig. 10-53 can be replaced today by a processor. The system is then referred to as a *stored program control* (SPC) switching machine (Fig. 10-54).

Generally, class 5 offices and toll offices of this type employ electromechanical crossbar switches (e.g., SP-1). In small-package PABXs, electronic crosspoint switches based on CMOS transmission gates are common (e.g., Mitel SX-200).

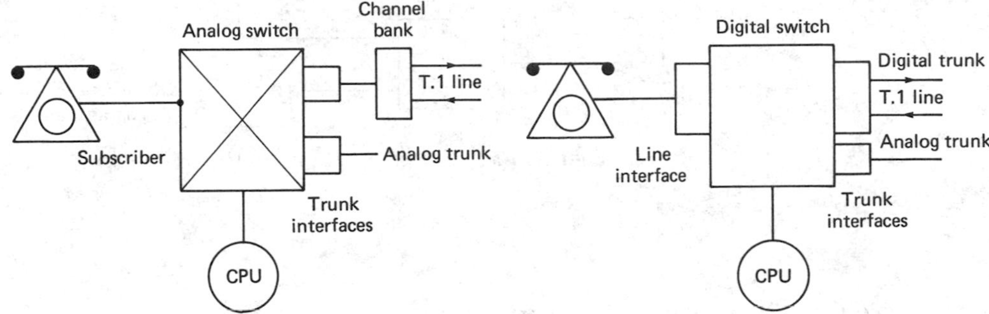

Figure 10-51.

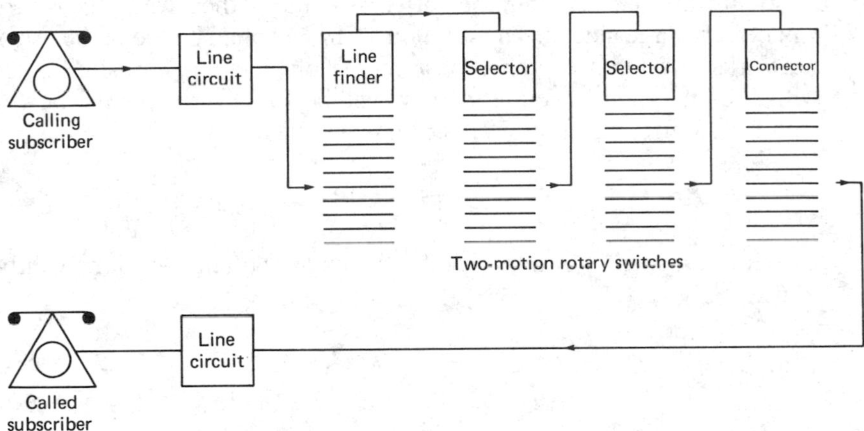

Figure 10-52 Direct control.

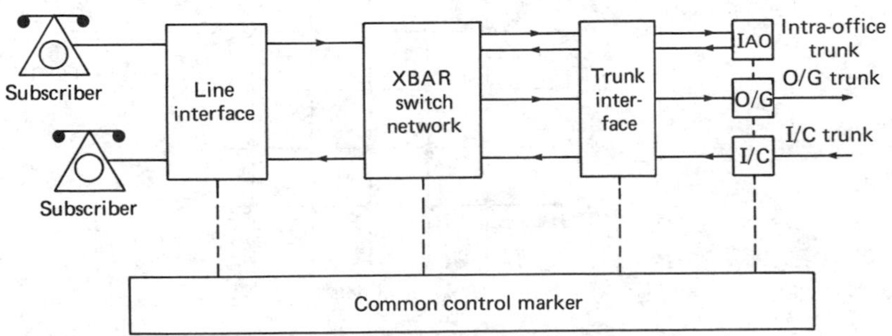

Figure 10-53 Common control concept.

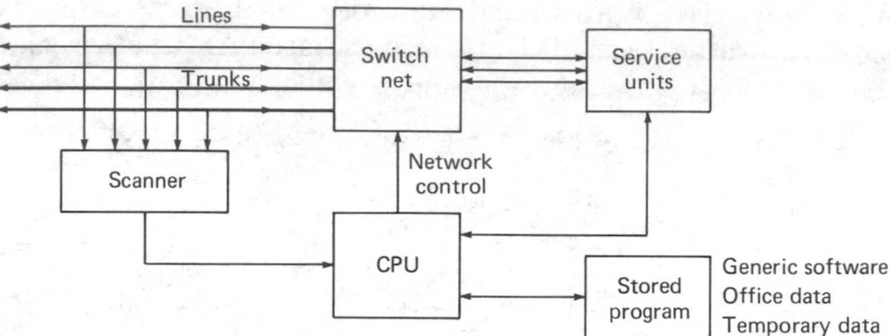

Figure 10-54 Stored program control.

10-4.2 Switching Matrix

Consider a switching matrix where any inlet must be able to reach an outlet (Fig. 10-55). If the matrix is to be nonblocking, there must be as many outlets as inlets (a square matrix). The number of crosspoints is then N^2. If N is large, then N^2 will be very large indeed. The solution to this problem has been the use of multistage networks and the acceptance of a degree of blocking. Consider a square matrix with 100 inlets and 100 outlets. There will be 10,000 crosspoints, of which only 100 could be in use. The use of crosspoints will be

$$\frac{100}{10{,}000} \times 100\% = 1\%$$

Check out a small 8 × 8 matrix, for example. Total crosspoints will be 64, with 8 in use.

$$\frac{8}{64} \times 100\% = 12.5\%$$

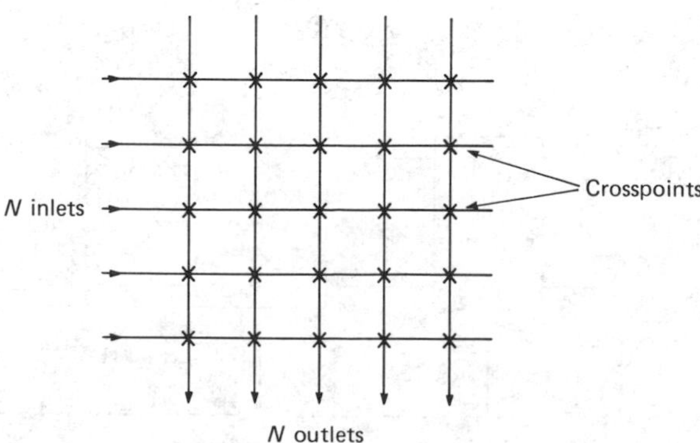

Figure 10-55 Square matrix.

Obviously, the small matrix makes better use of the available crosspoints. Now, a 100-line office would require one of the large switches or 13 of the smaller switches to accommodate subscribers and trunks. Looking at this another way, the 100-line office requires 10,000 crosspoints with the large switch and 832 using the 8×8 matrix. Although this may look encouraging, the connectivity to allow any of the inlets to reach any other inlet still has to be introduced. To increase the number of paths available for both intra- and interswitch links, it is necessary to use a multistage network. Figure 10-56 shows a classical three-stage network where any inlet can reach any outlet. The full link pattern is called *standard link trunking*. It is interesting to note that for the network to be nonblocking, the number of secondary switches has to be $2n - 1$ (see reference 2). This means that the primary stage switches will have more outlets than inlets! The standard North American crossbar switch has 20 inlets and 10 outlets, which introduces a degree of blocking or concentration at the primary stage. A small electromechanical crossbar matrix is expensive on a per crosspoint basis. However, small 8×8 solid-state switches are entirely possible in standard IC packages.

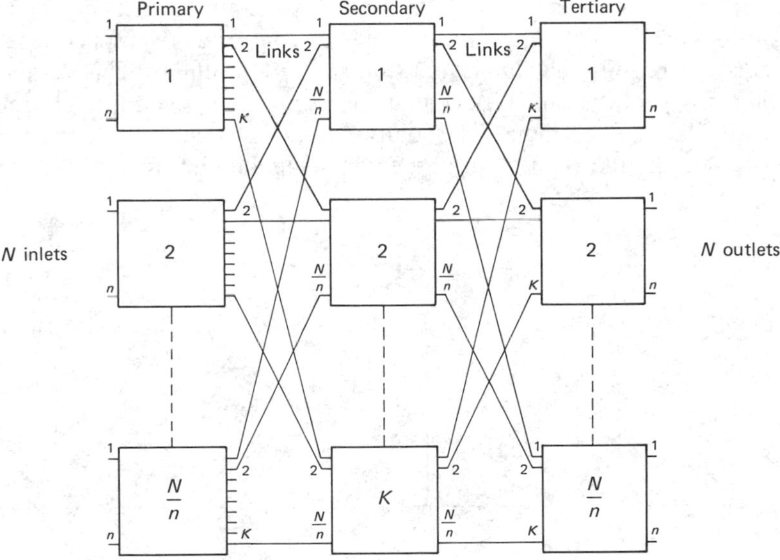

Figure 10-56 3-stage network.

We still have to indicate how any inlet can reach any other inlet. There are two basic architectures, as shown in Figure 10-57, which demonstrate the differences between folded and unfolded networks. To minimize the number of crosspoints, it is necessary to create the switching office by assembling small modules made up of small matrices. This design philosophy has continued with digital switching systems. In analog switchers a degree of blocking is an economic necessity. Non-blocking digital switching networks, on the other hand, are readily implemented.

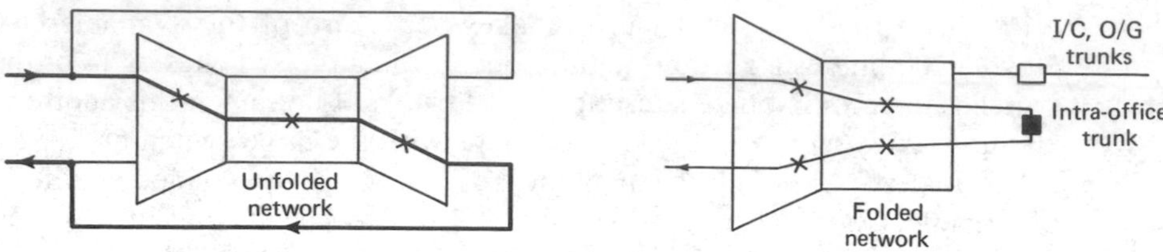

Figure 10-57.

10-4.3 Time-Division Switching

Time-division switching can reduce the number of crosspoints required. Figure 10-58 shows the concept of synchronous time-division switching. The two crosspoints involved with a call between two subscribers close synchronously at the 8-kHz rate. All parties time share the bus. A maximum of about 30 time slots is available; otherwise, the PAM samples become too narrow and have insufficient sound energy to maintain a satisfactory signal-to-noise ratio. Hence, to create a larger switching machine, a number of modules must be assembled. The problem now becomes one of connectivity. The modules must be connected by a space switch. Intermodule calls will require matching time slots on both modules. This system soon runs into blocking problems for lack of matching time slots, even though time slots may still be available on each bus. The solution to this problem would be storage or memory to retain the information until the next available time slot. Storage is practical only if the signal is digital.

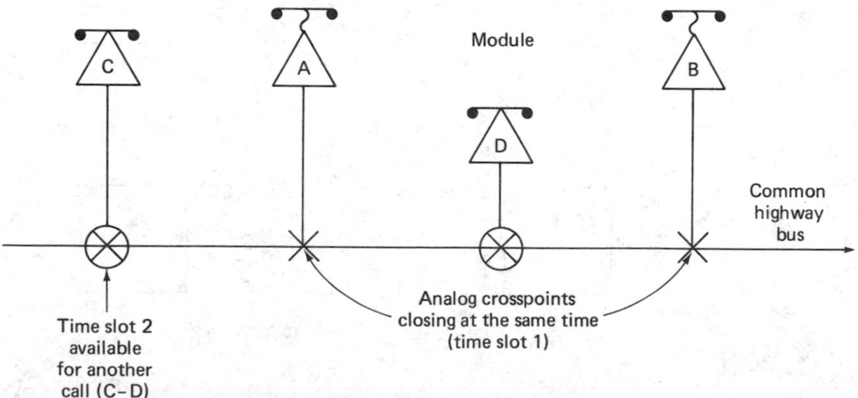

Figure 10-58 Time-division (PAM) switching.

10-4.4 Digital Switching Concepts

If the signal is in digital form, it can be stored in a memory to allow switching through time. Switching in both the time and space domains with integrated circuits provides the technology for large nonblocking networks.

Time-Slot Interchanger

The 32 eight-bit PCM channels are sequentially stored in the information memory. The connection memory is read sequentially and the content of each cell is the address of the information cell to be read. Thus, in Fig. 10-59, the contents of time slot 2 in the information memory are read out at time slot 31. Keep in mind that timing is critical. Three operations have to take place dynamically: (1) read the information memory, (2) write new information into memory, and (3) update the connection memory. Because of timing constraints it is often necessary to write and read the voice data in the parallel mode. Serial-to-parallel conversion will be required at the input and parallel to serial conversion at the output. Note that an 8-bit "byte banger" has been described. If the clocking is speeded up by a factor of 8, a 1-bit time switch can be implemented with a 32 × 1 memory. If the time-slot interchange is fixed, the time switch can reduce to a simple shift register per channel.

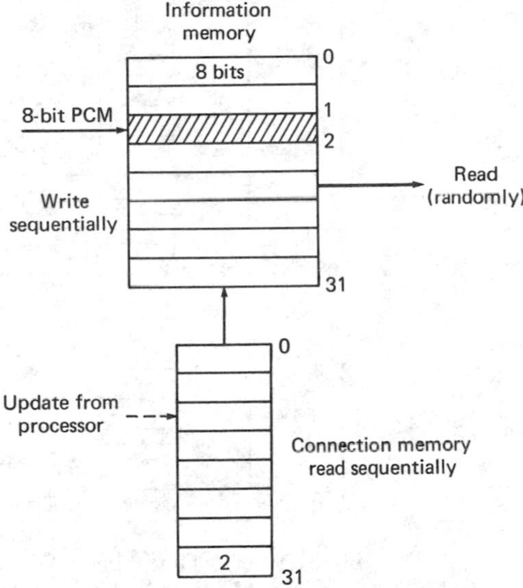

Figure 10-59 The basic time switch.

Consider how the number of inlets can be increased by assembling basic time switches and providing the necessary connectivity. Multistage switching arrays can be configured for digital switching. Let us examine a time–space (T-S) array and point out the conceptual differences between digital and analog switching. The significant features of this digital switching array (Fig. 10-60) are:

1. Digital switching is a four-wire operation.
2. There is modular growth potential with LSI chips.
3. The space switch closes only for the duration of the time slot (not for the duration of the call). The space switch is time-division multiplexed. A 1-of-*N* selector or multiplexer chip can serve as a digital crosspoint.

4. Every crosspoint is a regenerator.
5. The switching network is nonblocking.

Indeed, there is no capability for concentration or expansion in the T-S switch network as shown. A space–time (S-T) array would allow concentration or expansion at the space stage. Current digital switching designs favor a time switch as the primary stage, and as with analog switchers, multistage networks are required. T-S-T, T-S-S-T, and T-S-T-S configurations are commonly used for the networks in large central office or toll switchers.

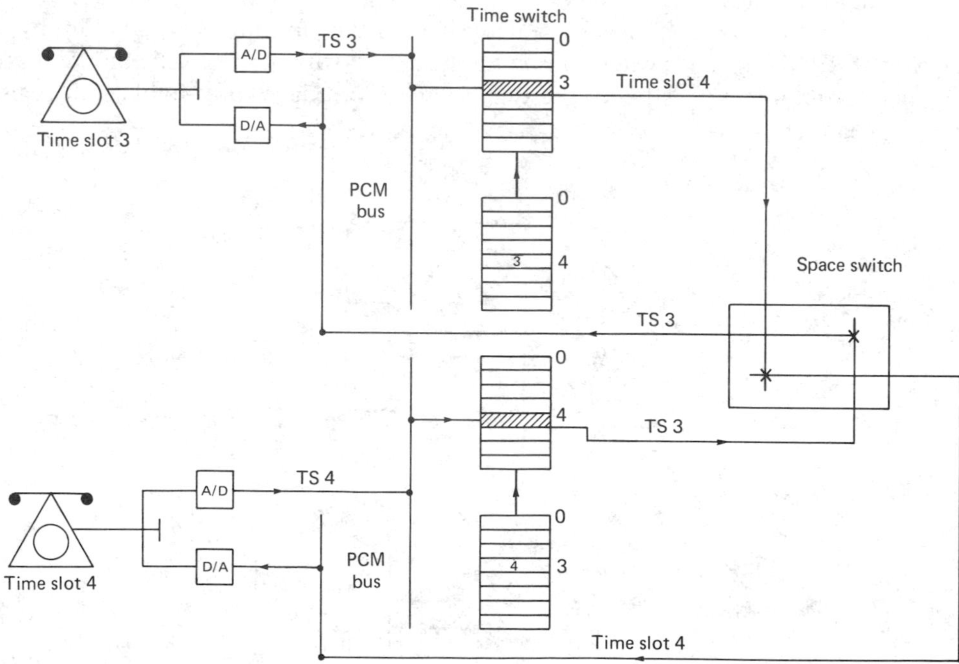

Figure 10-60 Time–space switch.

Time-Multiplexed Space Switch

Consider a digital space switch in which any input stream can reach any outlet (Fig. 10-61). This situation is equivalent to a 256 × 8 crossbar switch of 2048 crosspoints. Now, if a time switch could be placed at each crosspoint, any input channel could then reach any output channel. This is equivalent to a 256 × 256 crossbar switch of 65,536 crosspoints—rather amazing for an 8 × 8 matrix. The concept is implemented in a most elegant manner in the Mitel MT8980 digital crosspoint switch (40-pin DX chip). For a brief description of the chip see Figs. 10-62 and 10-63. The 8-bit word in time slot 31 of stream 2 is stored on page 2 of the data memory in location 31. The information is to be switched to channel 15 in stream 7. The content of cell 15, page 7 of the connection memory low, contains the page and cell to be read. Hence, at time slot 15 the content of location 31 of page 2 will be outputted. The demultiplexer will route the output to stream 7 at that time.

The microprocessor via the control register can change the DX chip into the message mode, and this allows the connection memory low to be updated. The mode control is by bit 3 in the connection memory high (see Fig. 10-63). In addition, in the message mode the microprocessor can read information from the input channels via the data memory, and it can also broadcast data to the output channels via the connection memory low. Mitel refers to the 32-channel serial input and output buses as STi or STo, where the ST stands for serial telephone bus. The format conforms to the (30 + 2)-channel CEPT standard. The bit rate is 2.048 Mbits/s. Internal switching is carried out in the parallel mode.

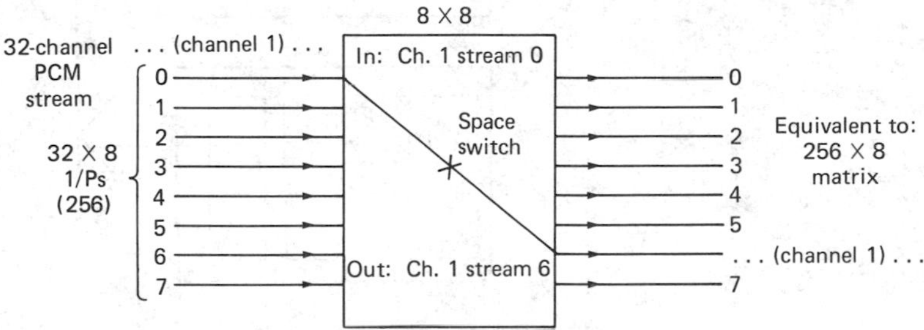

Figure 10-61 Time-multiplexed space switch.

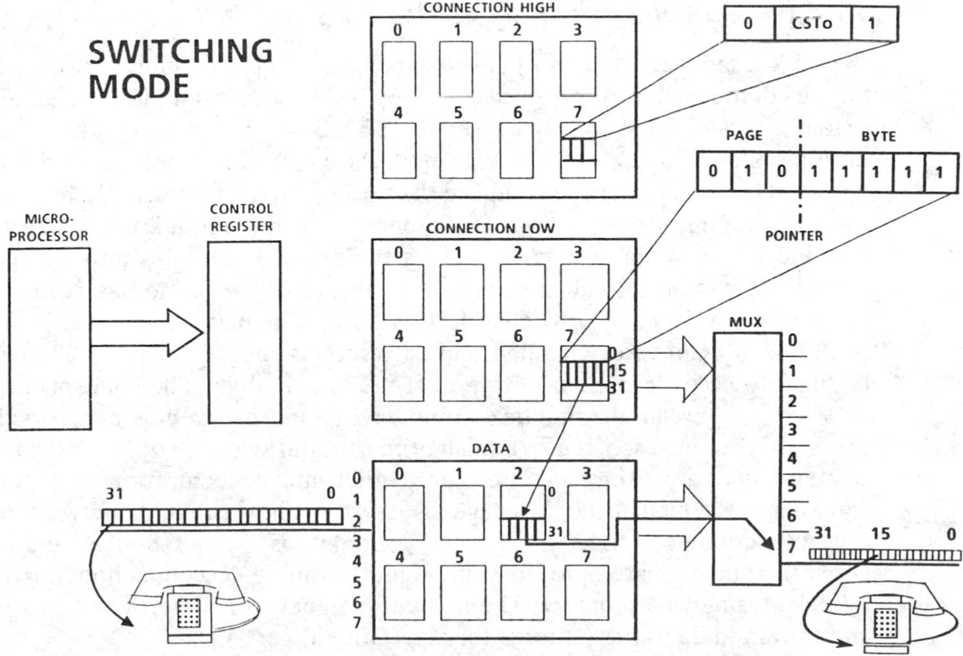

Figure 10-62 ISO-CMOS MT8980 digital crosspoint switch.

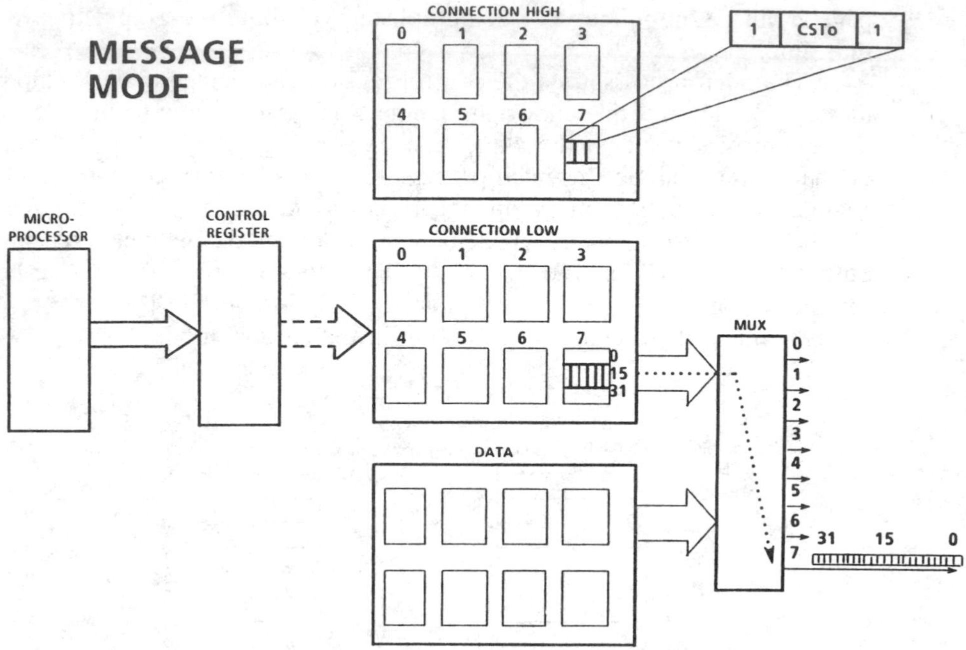

Figure 10-63 DX chip (message mode).

10-4.5 Digital Switching System

We have examined the concepts and topology of the switching network. We will now examine the Northern Telecom DMS-100 as an example of a large digital switching system or switching office (see Fig. 10-64). Subscribers terminate on the line modules. The size of the line modules was 512 subscribers in the original design. The line modules concentrate the traffic onto PCM links. 512 lines compete for a maximum of two links or 60 channels. The line modules compare to small 500 line offices which are assembled by the network modules into a large office capable of handling 100,000 lines and trunks. A network module is made up of small 8 × 8 arrays and provides 64 PCM ports. The network module handles (64 × 30) 1920 PCM voice channels and corresponds roughly to a 10,000-line office, taking into account the concentration at the line module. The concepts presented previously regarding the minimization of crosspoints are well demonstrated by the DMS-100. For instance, it uses a small primary input switch (8 × 8) for the network, small modules are assembled into a large office, and concentration with about 10% trunking is accepted at the line stage. The system allows up to 32 network modules in a fully equipped office. Each network module is a T-S-T-S array and a folded network topology is employed. Inter- and intramodule connections provide the foldback using wire *junctors*. The number of junctors is determined by the inter- and intramodule traffic patterns for a specific office.

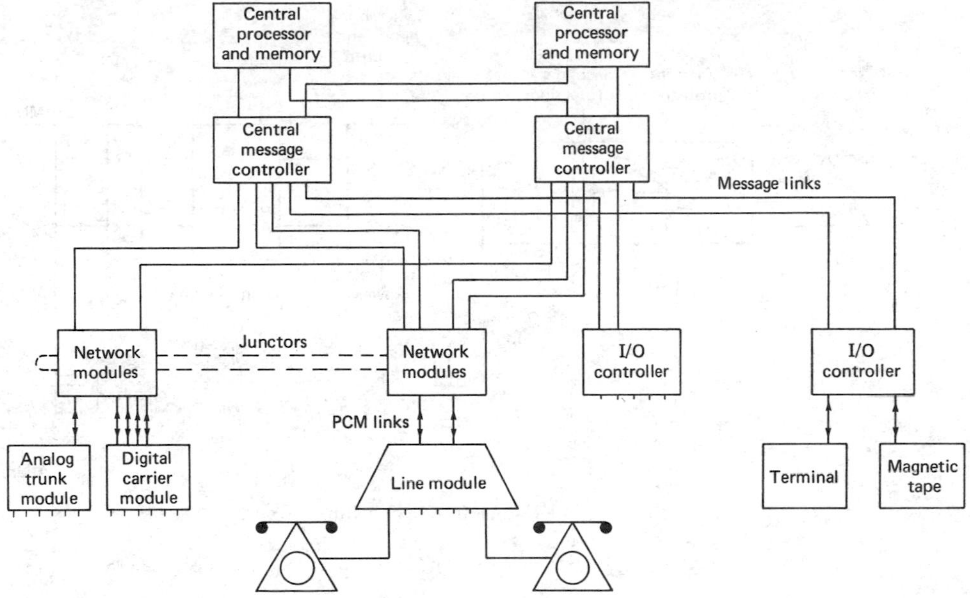

Figure 10-64 DMS-100 system.

The DMS-100 system diagram of Fig. 10-64 can provide the vehicle for a discussion of many digital switching concepts. Line modules consist of line circuits which support the subscriber telephones. Consider the services the line circuit must provide:

1. *B*attery to supply the dc loop current, preferably with constant current feed and overcurrent protection.
2. *O*vervoltage protection for IC components from lightning or power faults.
3. *R*inging: 105 volt/20 Hz. Off-hook detection during ringing to provide ring trip.
4. *S*upervision: to detect seizure and disconnect.
5. *C*odec: for nonlinear A/D and D/A conversion.
6. *H*ybrid: for the two-wire/four-wire interface.
7. *T*est: facilities to check the analog and digital paths through the line circuit.

The *BORSCHT* functions can be provided today using hybrid IC chips. A subscriber line interface chip (SLIC) can provide all the functions except the PCM codec and the associated transmit and receive filters. The Mitel 88610 is a good example of present SLIC technology. The filters and codec are also available in one integrated 18-pin package (e.g., Mitel 8960). The hybrid transformer is replaced in the SLIC by an op-amp circuit. The 8960 allows the transmit and receive filter gains to be adjusted up to 7 dB (in 1-dB steps) under software control. Chip testing and analog and digital loopback is also featured. Study the line circuit of Fig. 10-65.

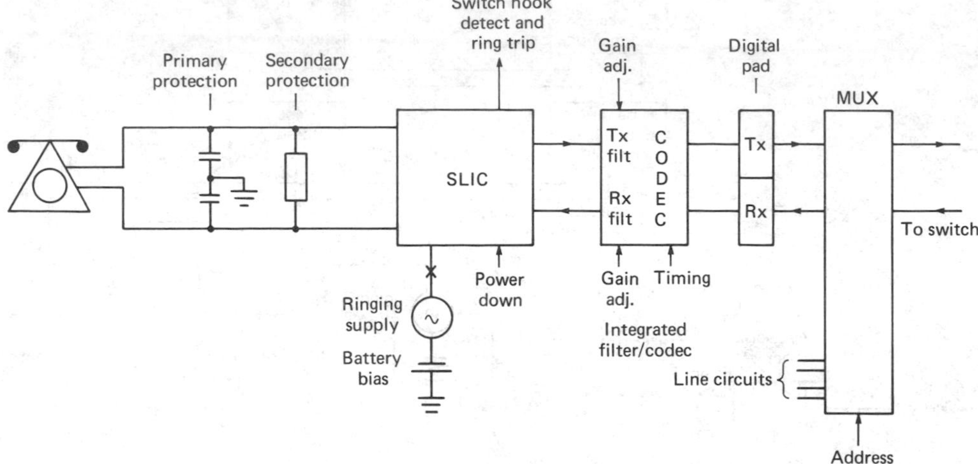

Figure 10-65 The line circuit.

Returning to the DMS-100 (Fig. 10-64), we see that the line module is connected to the network module by 30 + 2 PCM links. It seems logical to anticipate a remote line module (RLM) off the digital office. T1 lines provide trunks between the remote and the digital carrier module. Remote line modules/remote switching units and digital concentrators play an important role in rural telephone service with the modern digital switch as the host.

The digital carrier module (digital trunk) provides the interface between the North American T1 repeatered lines and the switching system. The functions the digital trunk must perform are:

1. Bipolar conversion of the incoming T1 line signal to TTL logic levels.
2. Recovering the clock from the incoming PCM.
3. Recovering the framing pattern.
4. Recovering the signaling from the sixth frames and formatting the signaling for the internal PCM links.
5. Providing "out-of-frame" alarms for the maintenance position.
6. Looking for bit 2 and activate the remote alarm if every bit 2 is a logic 0.
7. Buffering the incoming PCM to allow for varying bit rates and delay.

On the transmit side the trunk unit must:

8. Format the signaling from the internal PCM links and insert it into the DS-1 signal.
9. Insert the framing sequence.
10. Force all bit 2's to zero during a local alarm. Provide for simultaneous restoral.

11. Provide a loopback feature for testing. Insert a "keep alive" signal for the T1 line when in a loopback condition.
12. Convert TTL to the DS-1 bipolar AMI line signal.

To this sizable list must be added the requirement to change 24 DS-1 time slots into the (30 + 2)-channel format. In some cases the digital trunk is referred to as the DS-1 formatter. The channel difference is taken care of in the DMS-100 by formatting 5 DS-1 streams (120 channels) into four (30 + 2) links (120 channels).

The analog trunk module will provide A/D and D/A conversion for 30 analog trunks. All types of signaling, including MF, must be accommodated. Echo suppressors are also a part of the analog trunk interface.

10-4.6 Processor Organization and Redundancy

Taking the DMS-100 as our example, the control philosophy of switching machines can be explored. Reliability of service is paramount with large digital switching offices: hence the level of duplication throughout the system is very high. In our example, referring to Fig. 10-64 we can identify distributed control with duplicated microprocessors in the line, trunk, and network modules. The network switching modules have duplicate switch planes. The PCM links are also fully duplicated. The peripheral processors have specific tasks related to the module involved. The central control oversees the entire system and communicates with the lower-order processors over duplicate message links. The central processor and the central message controller are both duplicated. The control philosophy for the main processor in this example is duplicate synchronous control. Both processors perform the same task at the same time and their outputs should agree. If there is a discrepancy, a decision has to be made as to which processor is "insane." The control capacity is twice the office requirement. Alternative philosophies are:

1. *Duplicate hot standby*. Twice the control capacity is required.
2. *Dual-load sharing*. In the event of a failure, one machine carries the full load. System performance could be reduced.
3. *Multiprocessing*. The trend is toward multiprocessors instead of one large main processor. This should not be confused with distributed control in the peripheral processors. These processors will still be in place. In multiprocessing a number of smaller computers carry out the central control tasks. Each of these processors is able to communicate directly with each other and have access to all the associated memories.

Memory provided for the main central computer appeared to be very large when stored program control (SPC) was first introduced. In light of the memory capacity now available with personal computers, the size does not seem so unreasonable. For example, the original DMS-100 central processor had a program store of approximately 1 million bytes of 17-bit words. The data store was 16 million bytes. Backup memory is often provided by an extra memory card, which can be a rover. This is referred to as $N + 1$ redundancy.

10-4.7 Synchronization of Digital Offices

PCM transmission and switching has solved many of the problems associated with analog systems. However, it brings a new problem of synchronization of switching networks. Digital data is a transient event and must be captured or it will be lost. The data must be clocked into a memory in step with the frame timing in order to extract each 8-bit word. The problem is accentuated by the fact that there is one exchange clock but there are many digital trunks with different and varying delays. The exchange clock will vary due to aging and temperature. The problem is overcome by two techniques: elastic buffering and clock control. Elastic buffers are used to counter the short-term variations (Fig. 10-66). If the exchange clock is faster than the recovered write clock, the buffer will empty. If the data is coming in faster than it is being transferred, data will be overwritten. Either case is referred to as a *slip*. When a slip occurs, the elastic store is designed to fill or drop data to the halfway mark. This is a controlled slip. As the elastic store has 386 memory locations, the system loses one frame, but framing is maintained. Only one 8-bit sample is repeated or lost from each voice channel.

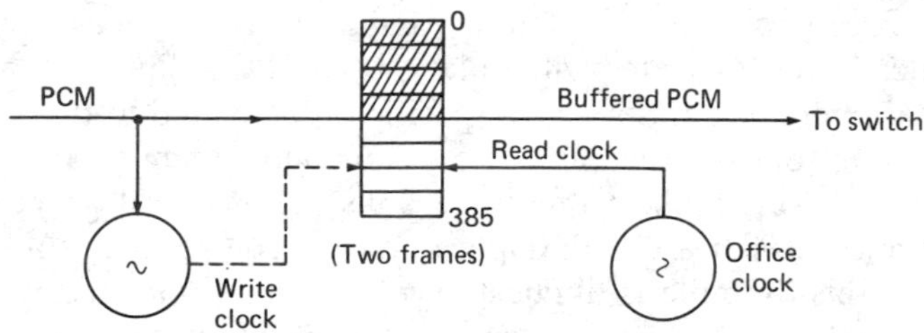

Figure 10-66 Elastic store.

It has been established that only one out of 25 slips results in an audible click, and that a slip rate of 300 per hour can easily be tolerated. Analog and digital data, however, can be adversely affected, depending on the data system. Analog VF data can lose synchronization. Digital data will be lost unless the protocol has the retransmit feature. This would also apply to PCM signaling.

The network objective is one slip in 5 hours on an end-to-end basis and not more than one slip in 20 hours for any one trunk. Consider a built-up circuit of six offices and assume an average clock variation of ± 2 parts in 10^7. The frame slip rate would be

$$5 \text{ offices} \times 2 \times 10^{-7} \frac{1}{\text{office}} \times 1.544 \times 10^6 \frac{\text{bits}}{\text{sec}} \times 3600 \frac{\text{sec}}{\text{hour}} \times \frac{1 \text{ slip}}{193 \text{ bits}} = 28.8 \text{ slips/hour}$$

That is, a "noticeable" slip would occur about once per hour on a voice circuit. If we consider one trunk between two digital offices, with the same clock variation the slip rate would be roughly 6 slips per hour. To meet the network objective of 1 slip in 20 hours would require an improvement of 120 times in the variation (two orders of magnitude). Thus with two free-running clocks, the difference in frequency must be better than $2 \times 10^{-9} \times$ clock frequency to meet the objective. Over the long term, this type of relative clock accuracy or stability can be achieved only with cesium atomic clocks (better than 1 part in 10^{11}). Cesium beam atomic clocks are expensive and need periodic maintenance. Hence digital switches in the network are usually synchronized in a master–slave arrangement with the atomic clock at the regional master switching node. The free-running mode (called plesiochronous) may occur if the master reference clock is lost under fault conditions.

The synchronous master–slave mode has a number of topologies as shown in Fig. 10-67. In the synchronous method each digital switch has its own duplicated voltage-controlled crystal oscillator (temperature controlled), which due to aging may drift in the order of 1×10^{-10} parts per day. The VCXO is synchronized to the master clock from an incoming digital trunk. Digital radio or DS-4 fiber optic trunks are preferred. The VCXO stability over the short term (1 or 2 days) is excellent and loss of the master reference will not be catastrophic. The block diagram of the frequency locked VCXO is given in Fig. 10-68.

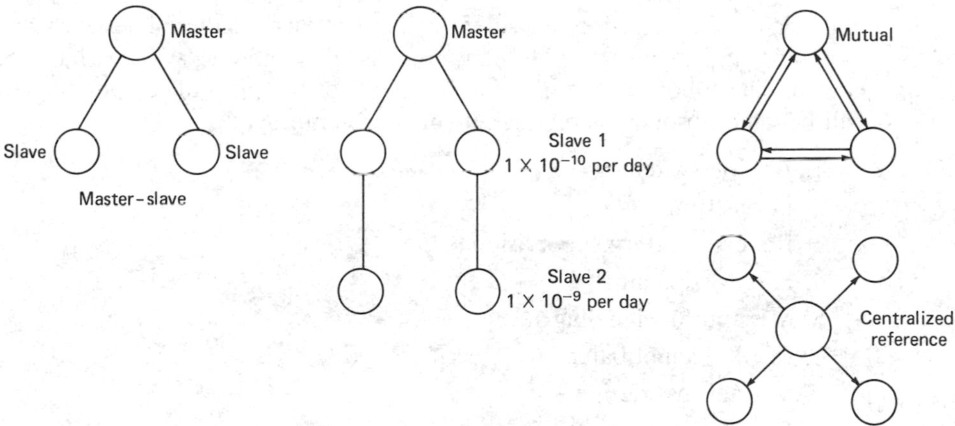

Figure 10-67.

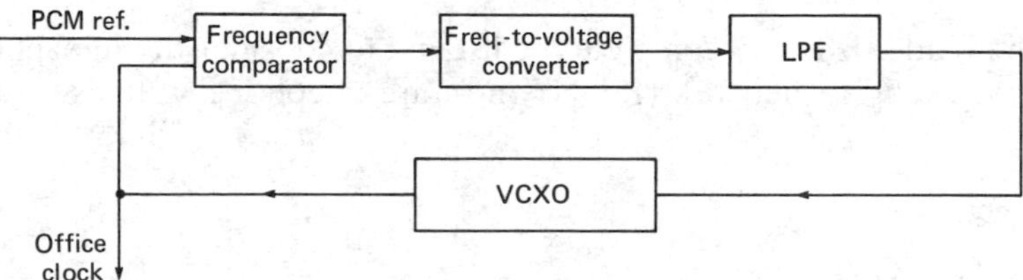

Figure 10-68.

Digital channel banks have a stability in the order of 3×10^{-5}; hence with combinational trunks the timing of the channel bank must be derived from the digital switch. This is achieved by looping the receive clock at the channel bank. Figure 10-69 illustrates the concept. The channel banks are therefore bit synchronized with the switch, although channel 1 of the channel bank will not necessarily be time slot 1 for the switch. D3 frame synchronization is established by the buffer in the digital switch (D3/D4 channel banks must be used when interfacing with a digital switch).

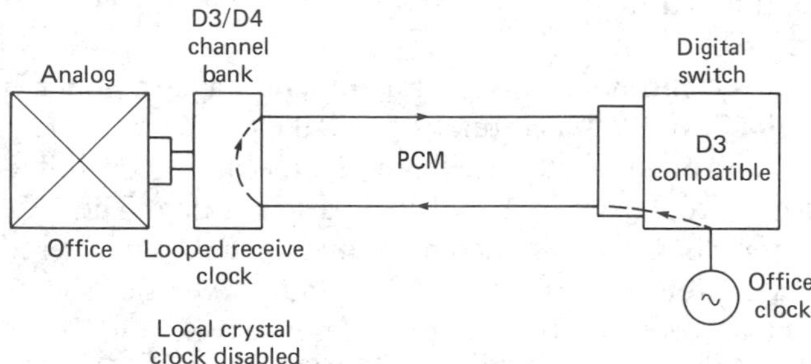

Figure 10-69.

To close this section, consider the implications of a digital network with digital switching and digital trunks. Many aspects of telephone systems have been discussed. The following is a list of components of the present telephone system that will become obsolete as digital networks are employed:

1. Analog switching
2. Loop trunks
3. Reverse battery supervision
4. E & M signaling
5. MF and SF signaling
6. PCM channel banks
7. Via net loss design
8. Echo suppressors
9. Transformer hybrids
10. etc.

The reader is left to complete the list and to consider the impact of the integrated services digital network (ISDN) on future telephone systems.

Problems

10-1. A telephone repairman goes to a subscriber's protector and dials up the milliwatt supply (test tone). He measures the loop loss as 10 dB. He knows that the loop is around 10,000 ft of 26 gauge cable. Is the measurement within reason?

10-2. Modern test sets for checking subscriber loops often assume that the telephone instrument has a dc resistance of 430 Ω. Older instruments measure the dc loop current through a standard telephone butt-in. The loop current measurements were found to agree reasonably well on long loops, but close to the office a major discrepancy between the two instruments was noted. Explain the difference in loop current readings. Why would the new instrument assume a dc resistance of 430 Ω?

10-3. List the problems to be expected with low loop current.

10-4. Suggest a method to determine the noise balance of a telephone circuit when insufficient power influence is present.

10-5. A touch-tone telephone is used to access information on a computer data bank via the public switched network. A dial pulse to DTMF converter was installed at the computer to allow rotary dial telephones to access the system. Explain simply why the installation was not successful.

10-6. Near-end crosstalk tests were performed on a large cable to determine the worst-case situation. It is not practical to measure all possible pair combinations; hence a sample number of tests were performed and statistical methods were used to calculate the lowest crosstalk coupling loss. A mean value of 70 dB with a standard deviation of 1.9 dB was obtained. Determine the lowest crosstalk coupling loss to be expected.

10-7. What is the purpose of the hold coil in a transmission measuring set? Does it modify the reading significantly?

10-8. Transmission and noise measurements taken at the subscriber's premises showed an 8-dB loss on the loop and 20 dBrnc of noise. Calculate the S/N ratio at the subscriber.

10-9. The transmission measuring set indicates a difference of 4 dB between a telephone in the off-hook condition and a short-circuit termination at the subscriber. Calculate the return loss (Fig. P10-9).

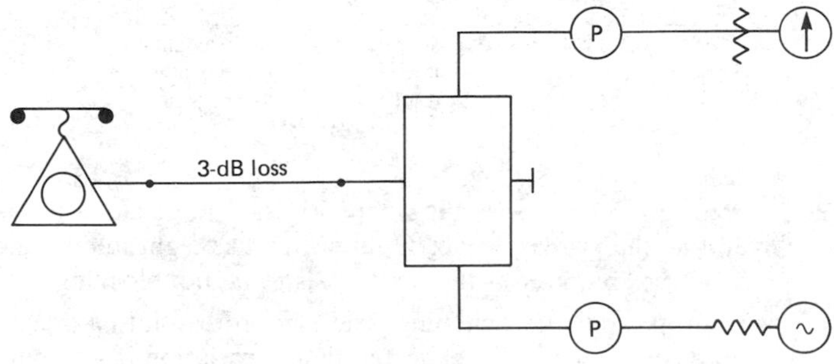

Figure P10-9.

10-10. A toll connecting trunk has a maximum loss of 4 dB, which is obtained by a combination of VNL and 2.5 dB of fixed loss. The transmission facility is loaded cable with a via net loss factor of 0.03 dB/mil. Calculate the length of the trunk in miles.

10-11. Consider the signaling between the central office and a PABX as shown in Fig. P10-11. The PABX is seized by a normal ringing signal from the central office. The locals on the PABX have direct outdialing capability and access the trunks by dialing 9. Discuss any problems you anticipate with two-way trunks using this system. Include possible solutions in your discussion.

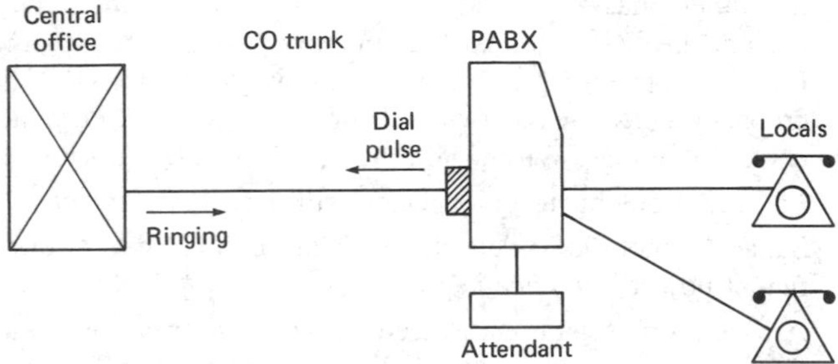

Figure P10-11.

10-12. A foreign exchange subscriber is connected to the central office by means of a four-wire transmission system. Standard single-frequency signaling is employed for the transmission of dial pulse information **from the subscriber**. It is also used to control the ringing **to the subscriber**. Discuss a problem that will occur if the transmission path is lost. Refer to Fig. P10-12 and propose solutions to this problem.

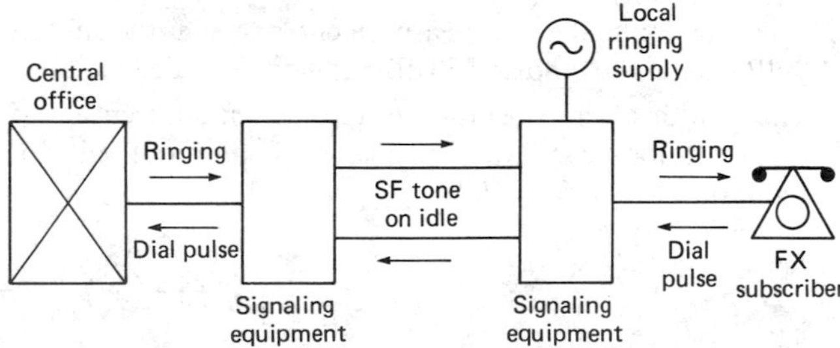

Figure P10-12.

10-13. A three-stage digital switching network is of the space–time–space configuration. Given that the primary space input switch has eight inlets, calculate the number of time switches required if the network is to be nonblocking.

10-14. Two European digital switching offices are forced into a nonsynchronous mode due to equipment failure. The expected timing variation is $\pm$ 3 parts in 10^7. The CCITT digital trunks have a capacity of 32 channels and employ 8-kHz sampling and a 2.048-Mb/s line signal. Calculate the frame slip rate.

10-15. Compare a folded and nonfolded switching network. Consider the number of crosspoints involved in an intraoffice call and the trunk requirements of the two networks.

10-16. Comment on the architecture and design philosophy of the digital switching systems shown in Fig. P10-16.

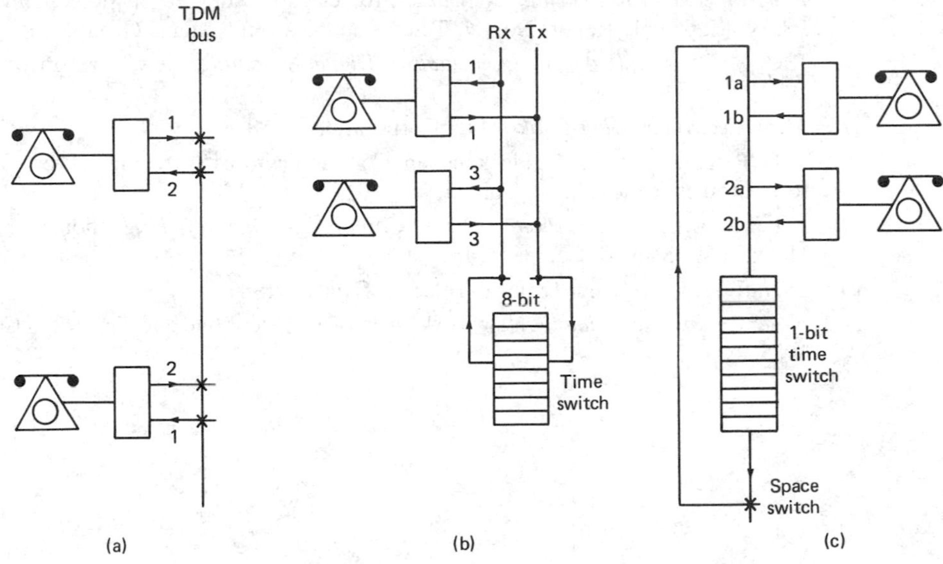

Figure P10-16.

10-17. Design a system to provide outgoing trunks between an old analog switching office and a new digital toll office. Digital transmission facilities will be employed. Indicate how the incoming PCM will be synchronized with the digital office clock.

10-18. The cross-office (local–local) loss between subscribers on a modern digital PABX was found to be 6 dB. Explain why this is necessary. Indicate how the loss could be inserted for intraoffice calls but removed for incoming calls from the central office.

10-19. Common channel interoffice signaling will allow 64-kb/s "clear channels" on T.1 PCM trunks. How true is this statement? What advantages can you identify for a 64-kb/s clear channel?

10-20. List the problems you foresee for digital transmission right from the subscriber's telephone. Give two methods of allowing two-way digital communications on a single metallic pair.

References

1. M A. Clement, *Transmission*. Chicago: Telephony Publishing Corp., 1969.
2. C. Clos, "A Study of Non-blocking Switching Networks." *Bell Syst. Tech. J.*, Vol. 32, 1953.
3. R. L. Freeman, *Telecommunication System Engineering*. New York: John Wiley & Sons, Inc., 1980.

4. R. Gundrum, *Power Line Interference: Problems and Solutions. Lee's abc of the Telephone.* Vol. 14. Geneva, Ill.: Lee's abc of the Telephone, 1982.
5. P. B. Hancock, *Trunking Design of a P.A.B.X. Employing a Computer Controlled Electromechanical Switching Network*. Project paper, degree of M.Sc. University of Essex, May 1971, Report No. 46, Telecommunication Systems Group.
6. J. C. McDonald, (Ed.), *Fundamentals of Digital Switching*. New York: Plenum Press, 1983.
7. *Microelectronics Data Book*. Mitel Corporation, 1988.
8. S. D. Personick and W. O. Fleckenstien, Communications Switching—From Operators to Photonics. *Proc. IEEE*, Vol. 75, No. 10, October 1987.
9. J. P. Ronayne, *Introduction to Digital Communications Switching*. Indianapolis, Ind.: Howard W. Sams & Company, Inc., 1986.
10. *Digital Network Notes*. Ottawa: Telecom Canada, 1983.
11. *Notes on Distance Dialing*. New York: American Telephone and Telegraph Company, 1975.

Digital Communications

Henry Edwards

11-1 OVERVIEW

Digital communication has been around for a lot longer than is generally realized. Ever since human beings began to signal beyond the voice range, it has, for the most part, been the form of signaling used. The term *digital* itself is the adjective form of "digits" and involves a means of displaying something by the use of digits. One early example of digital communication was the display of two flags held at various angles to indicate numbers and letters of the alphabet as the transmitted digits. This type of communication system involved more than two-state information processing, and one of its major problems was having a clear line of sight in daylight and good weather.

One of the early ways of processing two-state information was the use of the heliographic system, which was in common use, prior to the telegraph system, mainly in India. This involved the reflection of light from a mirror, or polished surface, to indicate digits transmitted. Ideal for signaling from mountaintops but limited again by clear line of sight, daylight, and weather patterns. With the invention of the telegraph system the sender and receiver had finally no need to worry about line of sight, weather, or time of day, as, in most cases, information could be transmitted in two-state information over long distances at any time. This was obviously a great improvement on earlier systems, but the speed of information still depended on the skill of the operators at either end of the line. With the standardization of the Morse code (1865) the speed of information became universal with the further introduction of a telegraph machine that interpreted received information at approximately 60 words per minute.

The telegraph system was the last major information-processing system that used two-state information prior to the introduction of modern digital communications. The two-state signals of the telegraph system were basically current or no current on a timed basis. This timing varied according to the speed of the operator or machine on the line, and a series of currents and no currents indicated letters of the alphabet and numbers—exact timing was not crucial. The telegraph system was subsequently replaced by the telephone system, which used a voice analog signal that did not consist only of two-state information.

Exact timing of two-state information became very crucial with the advent of present digital communication systems. Two-state information is again being sent, but the synchronization of this information is highly important, as more than one transmitted message is being sent on any one circuit. The two-state information no longer represents letters of the alphabet and numbers, but synchronized pieces of analog signals contained in a binary code. The general term for this type of transmission is *pulse code modulation*.

Pulse code modulation (PCM) is information in the form of combinations of two electrical states over a given time period. The two electrical states can be considered very similar to turning a light switch on and off. One moment you have electrical energy flowing, followed by none at all. The time period will depend on the speed of your finger action. Generally, because of this two-state operation, PCM is considered to be a digital signal, and as a digital signal it is unidirectional and requires four-wire switching: two wires for transmit and two wires for receive. This type of system lends itself to the use of binary mathematics, which is based on the number 2 with units 1 and 0.

11-2 COMPARISON BETWEEN ANALOG AND DIGITAL (PCM) SIGNALS

An analog signal is too rigid for use as a digital signal because (1) it is continuous in time, (2) the amplitude varies with time and, (3) it cannot be stored easily except by cassette or magnetic tape (Fig 11-1 illustrates the first two factors).

A digital signal has a set time period and discrete amplitude for its two states. With today's technology the digital signal can be stored very easily and in large quantities on microchips such as those used in solid-state circuitry.

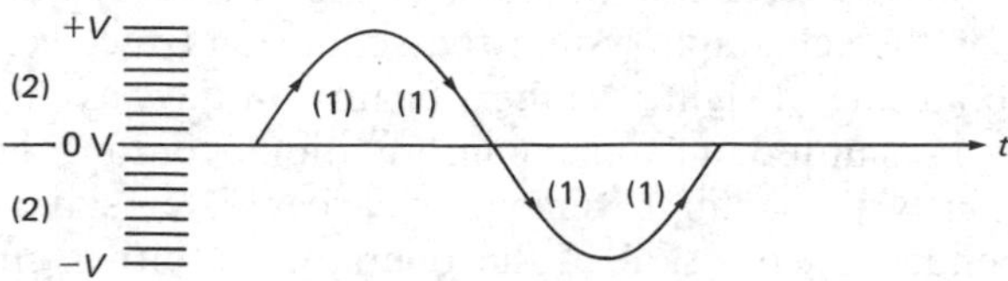

Figure 11-1 Analog signal problems.

11-3 TYPES OF DIGITAL SIGNALS USED AS LINE CODES

11-3.1 Non-Return to Zero (NRZ)

The two states of this signal are shown as 0 = 0 V or 1 = + *V* and are set over a time period *t*. Consecutive 1's do not return to zero but remain at the + *V* amplitude until a 0 appears. The problem with this type of digital signal is that in reality it is not as clear cut as shown in Fig. 11-2 because a signal does not rise to maximum potential or return to zero potential instantaneously. The situation shown in Fig. 11-3 gives a more realistic indication of a particular level of voltage that is induced throughout the signal, due to the rise and fall problems, which becomes an average dc value over a period of time and is commonly known as the *dc component.*

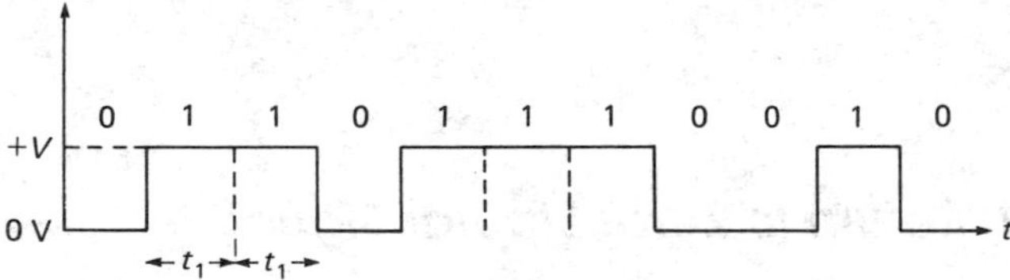

Figure 11-2 Non-return-to-zero signal.

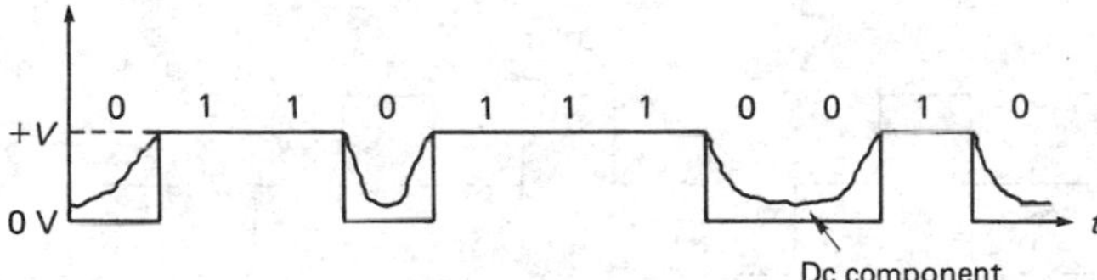

Figure 11-3 Realistic signal showing the dc component.

11-3.2 Return to Zero (RTZ)

This type of digital signal (Fig. 11-4) reduces the dc component and gives a steadier stream of pulses. The consecutive 1's return to zero potential, but the disadvantages are the creation of problems with attenuation, noise, distortion, and as the frequency increases, so do the losses.

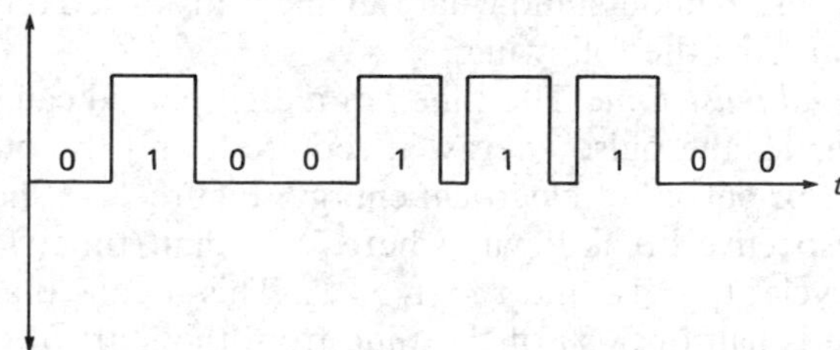

Figure 11-4 Return-to-zero signal.

11-3.3 Non-Return to Zero: Bipolar Signal

In this signal (Fig. 11-5) the two states are represented by $1 = +V_{max}$ and $0 = -V_{max}$. Consecutive 1's and 0's do not return to zero potential and the dc component is vastly reduced.

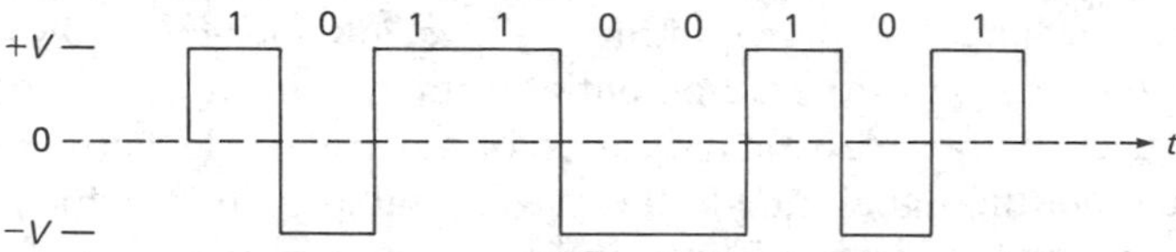

Figure 11-5 Non-return-to-zero: bipolar signal.

11-3.4 Return to Zero: Bipolar Signal

In this signal consecutive 1's and 0's return to zero potential (Fig. 11-6), which gives a further reduction of the dc component.

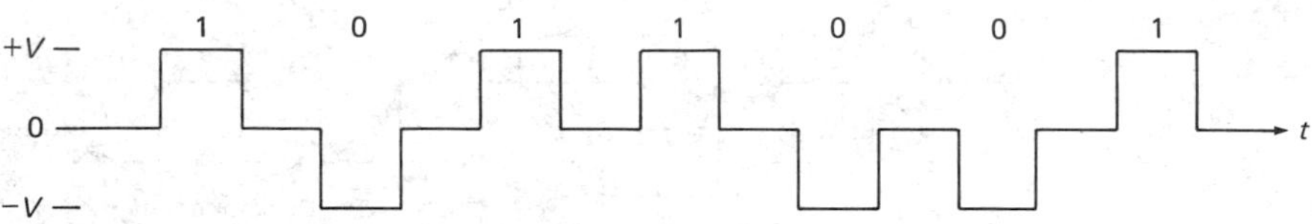

Figure 11-6 Return-to-zero: bipolar signal.

11-3.5 Return to Zero: Bipolar Signal with Alternate Mark Inversion

This signal is derived from that shown in Fig. 11-6 and is the actual PCM signal used. However, to understand what is being transmitted on the line requires further development from the following:

1. *Actual pulse time.* The time t of a digit period can be split into two halves. One half carries the pulse energy, if required, and the other half is considered a quiet period or devoid of electrical energy. Figure 11-7 shows a pulse as one cycle with the respective breakdowns where, t = digit time, t_1 = pulse energy, t_2 = 50% duty cycle, $t_1 = t_2$, and $t = t_1 + t_2$. This means that the length or duration of the pulse is half or 50% of the time from the start of one pulse to the start of the next one.

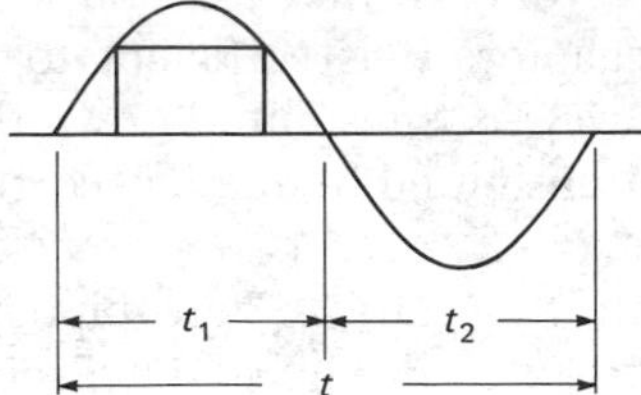

Figure 11-7 Pulse cycle.

2. *Alternate mark inversion.* This is a further improvement of the return to zero bipolar signal and is used as a maintenance check for receiving equipment on incoming logic 1's which are transmitted with alternate 1's at opposite polarity, as shown in Fig. 11-8. In being able to achieve this, the logic 0's must be at zero potential or low energy level. The advantage with this method becomes fairly obvious on considering incoming PCM pulses. If two 1's consecutively are at the same potential, there must be an incoming error which is termed a *bipolar violation* as shown (Fig. 11-9). The receiving equipment will then take the necessary steps to rectify the situation.

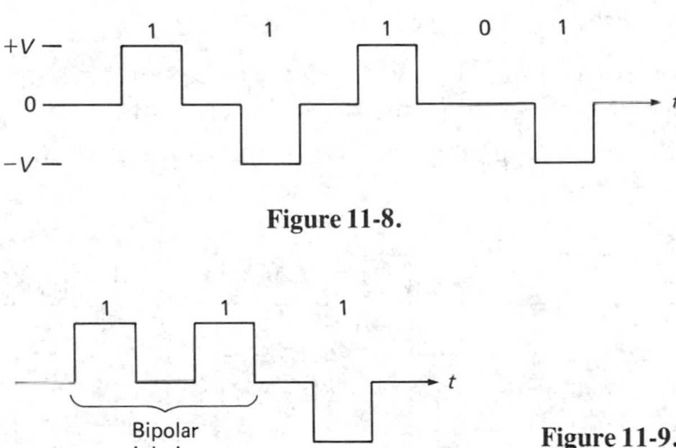

Figure 11-8.

Figure 11-9.

3. *Threshold level.* When looking at a series of incoming pulses the receiving equipment has to decide which is a logic 1 and which is a logic 0. Over a long distance the amplitude level decreases due to line effects, and therefore a level is set, with a logic 1 being above this level and a logic 0 below it. Generally, this level is set at half the peak-to-peak levels entering the receiving equipment, an example of which is shown in Fig. 11-10.

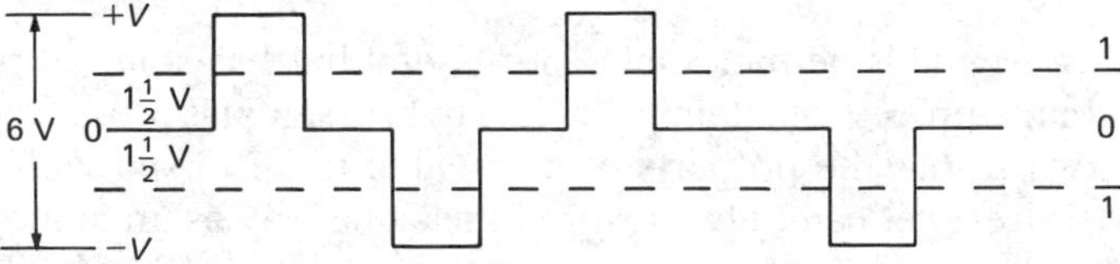

Figure 11-10 Threshold levels of bipolar AMI signal.

4. *Frequency*. This is set according to the PCM pulse rate used, which will be explained fully later in the chapter. It is sufficient to say that the discrete amplitude level of the frequency cycle is transmitted down the line, the energy of which is constituted in half of the transmitted frequency, corresponding to the logic 1's.

These four improvements on the return to zero bipolar signal produce our standard PCM signal, which is illustrated in Fig. 11-11. Producing this final PCM signal has the following advantages:

1. No dc component.
2. Lots of pulses.
3. Error detection (bipolar violation).
4. It will allow the use of repeat coils (inductor coils) used in the regeneration of the applied PCM line signal when the signal energy level is low due to cable effects.
5. Energy is concentrated at half the applied frequency, and therefore there is less attenuation and likelihood of transfer of energy to another close-running circuit.

and some disadvantages:

1. A three-level signal (+V, 0 −V) is more noise susceptible, but as long as the noise peaks are below the threshold level, the right decisions will be made.
2. A three-level signal gives a lower threshold margin.

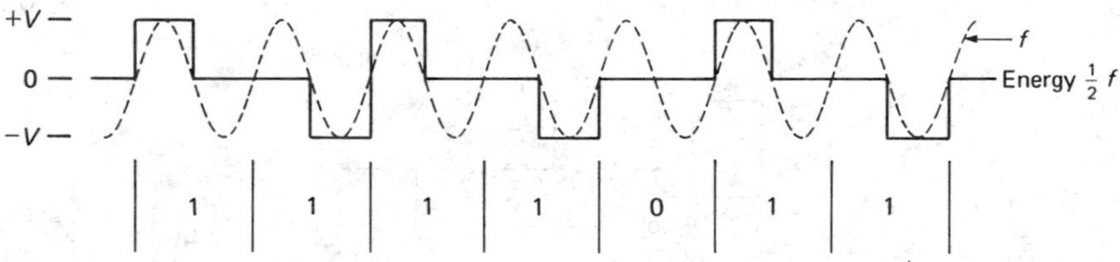

Figure 11-11 Frequency of bipolar AMI signals.

11-4 HYBRID SIGNALS

A hybrid signal is neither analog nor digital but has some of the characteristics of both. In the process of attaining PCM a hybrid signal is produced at the intermediate stage, which has the property of being able to be mixed with similar signals on a pair of wires, yet can only be stored the same way as an analog signal. This signal is commonly known as *pulse amplitude modulation* (PAM). The following shows how it is derived, along with two other hybrid signals.

11-4.1 Pulse Amplitude Modulation

This is a generation of a train of pulses with heights proportional to the sampled point of the analog sine wave at the same instant in time (Fig. 11-12). By producing this signal the amount of speech information carried by the original analog is drastically reduced, but the original speech is still well represented. By keeping the time width of each sample small and a set time period between these samples, we can interleave other analog samples to time share a common transmission medium. This is an example of time-division multiplex (TDM).

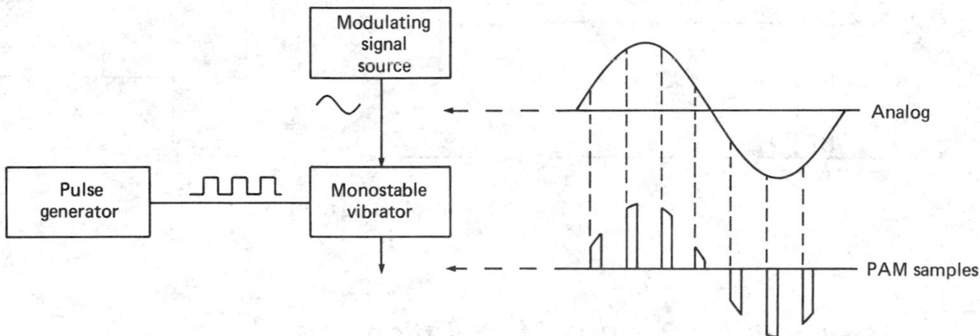

Figure 11-12 Analog to PAM.

To reproduce the original analog signal from the PAM sample requires only the use of a low-pass filter, as shown in Fig. 11-13. Generally, the signal reproduced is never exactly the same as the original but certainly of good-enough quality for human understanding.

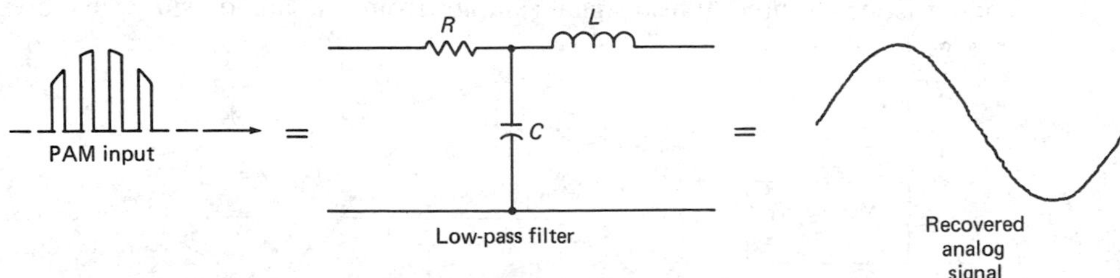

Figure 11-13 PAM to analog.

11-4.2 Pulse Duration Modulation

Pulse duration modulation (PDM) consists of sample pulses of constant amplitude that vary in time duration but are proportional to the point of sampling on the original analog signal (Fig. 11-14). The trailing edge of the sample occurs at a fixed time, but the leading edge varies according to the amplitude of the analog signal at point of sampling. This produces two main difficulties, which make it unsuitable for PCM processing purposes:

1. To interleave other PDM signals will require considerable guard times due to the varying time lengths of individual samples.
2. The smallest or thinest samples are more susceptible to becoming lost or garbled on the transmission media.

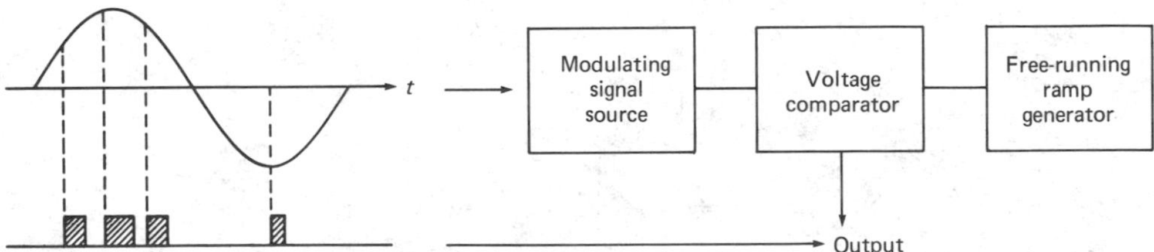

Figure 11-14 Analog to PDM.

11-4.3 Pulse Position Modulation

The PDM pulses contain the information in their leading edges, and therefore the power to transmit them can be high and wasteful due to the varying times between leading and trailing edges. To overcome this, pulses can be produced of equal time and amplitude but can vary their position in the individual time intervals assigned to them (Fig. 11-15). This is called pulse position modulation (PPM). As can be seen, this is just a further development of PDM, with the improvement of generating constant energy pulses. However, it still has the same time-interleaving problem for adding other samples and is therefore of little use in processing PCM. For comparison, all three hybrid signal samples from one analog sine wave are shown in Fig. 11-16.

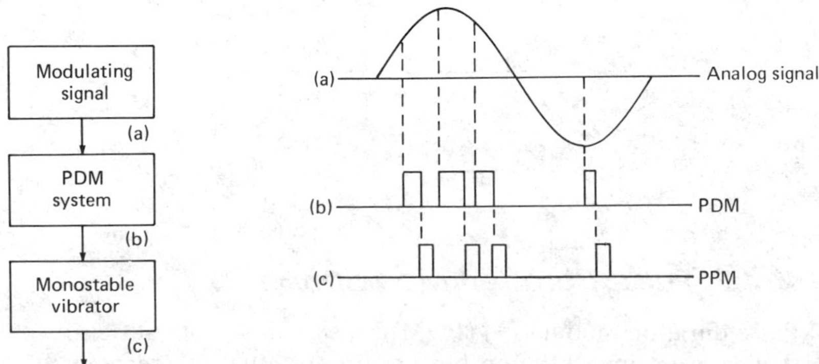

Figure 11-15 Analog to PPM.

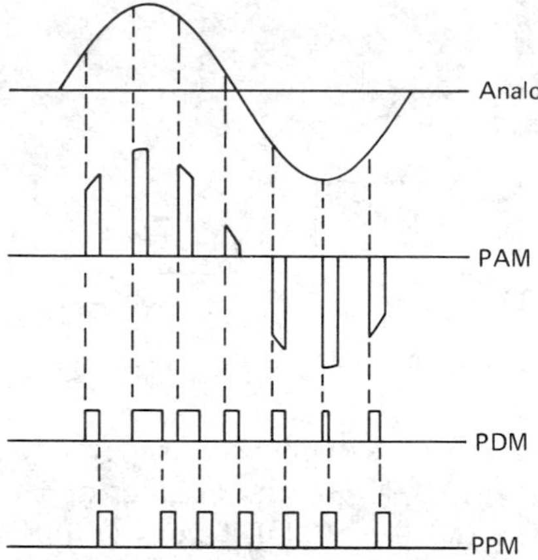

Figure 11-16 Comparison of analog, PAM, PDM, and PPM.

11-5 PRINCIPLES OF PCM

11-5.1 Sampling

As shown previously, the first process of PCM is to sample the analog signal to the PAM format. What has not been stated is at what rate the analog signal should be sampled. Based on the theorem of H. Nyquist of the Bell Telephone Laboratories, called the *Nyquist rate*, the sampling is set at twice the highest frequency used. In the speech band the frequency is generally stated as between 300 and 4000 Hz; therefore, the sampling frequency is 2×4000 Hz $=$ 8000 samples per second. Most manufacturers of PCM equipment set the speech band prior to sampling at 180 to 3400 Hz by using bandpass filters.

By sampling at the 8-kHz rate and using a maximum of 3400 Hz in the speech band, the upper and lower sidebands of the sampling signal do not overlap each other, as shown in Fig. 11-17, which is one reason for using it. If the sampling rate was reduced to 6 kHz, the situation totally changes when sampling a speech signal, as seen in Fig. 11-18. In this situation the upper and lower sidebands of the sampling signal overlap and give rise to the term *aliasing*, which is imposing one sideband partly on top of another or making one frequency look like another. The 8-kHz sampling rate then becomes the standard for sampling of speech signals, not only to conform to the North American (NA) PCM standard but also throughout the world no matter what PCM system is used.

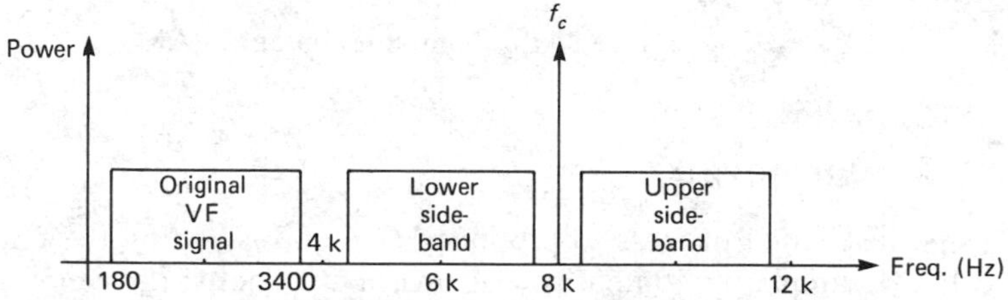

Figure 11-17 Correct sampling rate.

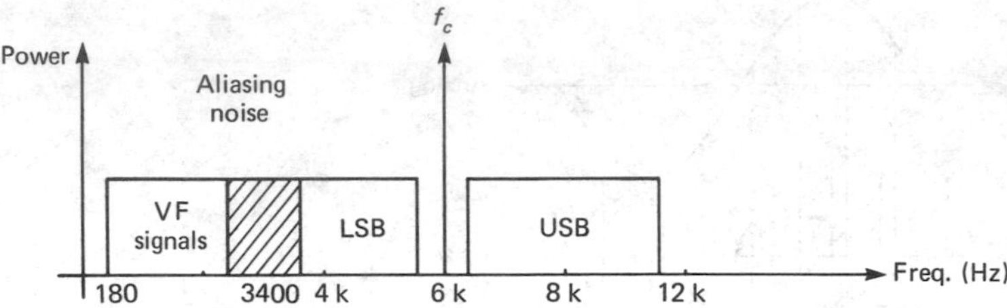

Figure 11-18 The aliasing problem.

For practical purposes, if a single transmission line is considered to be sampled at the 8-kHz rate and the analog signal is 1000 Hz, the number of samples taken of this signal over one cycle would be 8 (8000 ÷ 1000). Over the total 1000 cycles it would be 8000. Each sample would be taken every 1/8000 sec = 125 μs, and if only one transmission is taken, this would be uneconomical (as normal transmission needs only one signal and is a lot cheaper) for transmitting further as a single message, but if other transmission lines were sampled and interleaved with this signal, it becomes a more economic proposition. In North America the standard number of transmission lines or channels to be combined in this manner is 24. This means that for the same time period needed to sample one channel (125 μs), 23 others can also be sampled, and this will give the individual sample a time period of 125/24 = 5.2 μs. A good illustration for this sampling rate is shown in Fig. 11-19.

The total of 24 channels sampled over a period of 125 μs is known as a *frame*. The frame time is important because of the two PCM systems in general use in the world where the time of the frame is the same. The only difference is that the other system will sample more channels for the same period (32). However, as an example in the North American system, the first line, or channel, will be sampled for a period of 5.2 μs, and then after 119.8 μs (125 − 5.2) it will begin sampling again on channel 1.

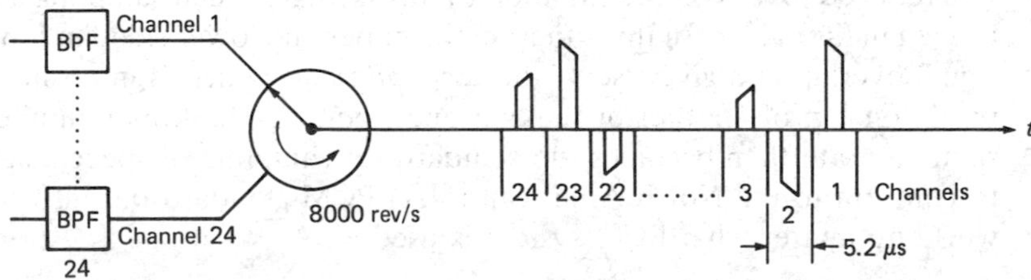

Figure 11-19 Sampling analog to PAM.

11-5.2 Quantizing

This is the second process to produce PCM and is setting the amplitude levels of the PAM sample by comparison against a step network over a range of voltage amplitudes. The PAM sample itself is subject to a signal-to-noise ratio as the slope of the sampled curve varies in magnitude over the various samples taken. The samples then have to be "squared off," in proportion, over the total range of samples. This squaring off of the signal can be made at a point where the signal energy is at its weakest, namely at the slope, and if this is so, the resultant squared-off sample is not a true representation of the signal energy. If part of the slope is incorporated into the resultant signal sample, it is said to ride along with the required signal as *quantization noise*—it is the unwanted energy of the quantized sample.

For the samples to be squared off a nonlinear step scale has been developed, which consists of 256 voltage levels (128 positive, 128 negative) over the voltage range of the original transmitted speech. As these voltage levels approach zero potential, they come closer together, and as they approach maximum potential they become farther apart. This results in an equal ratio between the signal (wanted energy) to the noise (unwanted energy) over all the samples taken (Fig. 11-20).

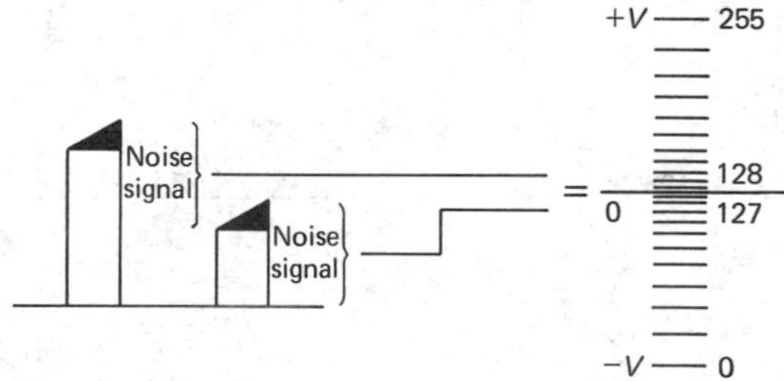

Figure 11-20 Quantizing.

The steps are derived from the μ-law companding curve, which is logarithmic in form and gives a steep-enough slope for the signals to approach equality over all ranges offered to it. The μ is the compression of the curve, and in the PCM case it is equal to 255. This is illustrated in Fig. 11-21 and it can been seen that for low-level input signals, the curve is practically linear, and for higher input signals the curve increases proportionally nonlinear. This is shown only in the positive part of the curve, but the negative side is exactly the same. The curve is segmented into steps for input signal application. (These steps are separate from the compression $\mu = 255$ of the curve.)

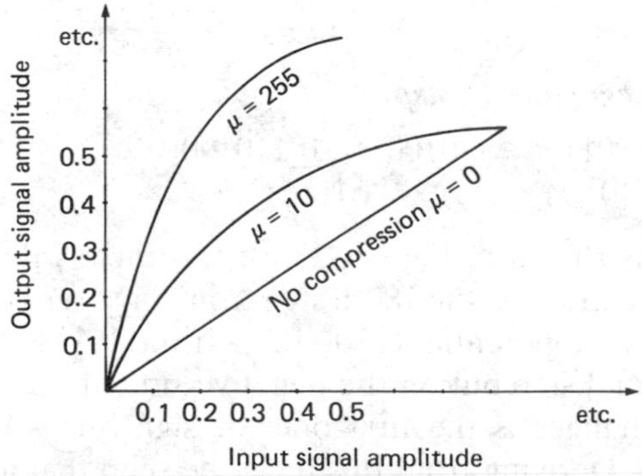

Figure 11-21 μ curve comparisons.

11-5.3 Encoding

To apply a binary code to the original PAM it is first applied to the compressed μ curve in steps, as stated, but the steps are derived from further compressing the curve into eight nonlinear segments each of which contains 16 equal steps within that segment (0 to 15). This gives a total of 128 (8 × 16) steps to the positive side of the curve and 128 steps to the negative side of the curve. Each segment step is twice the amplitude within that segment as in the preceding smaller segment. Figure 11-22 shows the segmented curve with the total 256 steps.

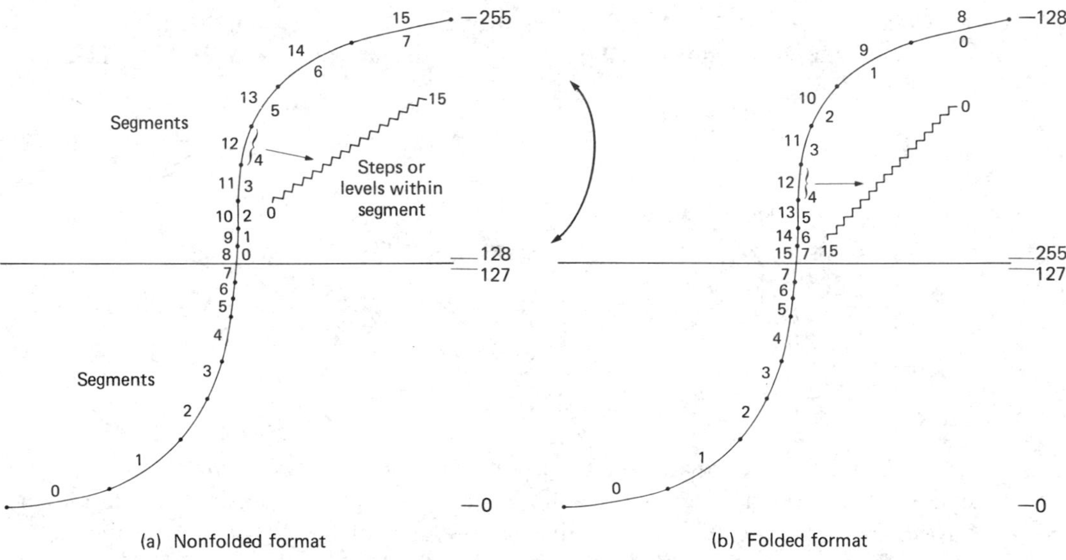

Figure 11-22 Segmented curve.

The binary code being produced from the sample consists of eight bits which are made up in order of sign, segment, and step. For instance, if a PAM sample is quantized at the fourth step in segment 3 on the positive side, the PCM code would be as follows:

Sign	*Segment*	*Step*	
1	011	0100	= <u>10110100</u>
(0–1)	(0–7)	(0–15)	

In the nonfolded format as shown in Fig. 11-22(c), this means that the amplitude of the PAM sample has been set on the 180th step from logic zero (maximum negative) or the 53rd step from zero potential (180–127). If the PAM sample was quantized in the same segment and step but on the negative side of the sine wave, the only binary number that changes is the first one, or sign bit, which would become a logic zero (00110100). Deriving from this it can be seen that all samples in the positive half have a sign bit of 1, and all samples in the negative half have a sign bit of 0. Understanding this, it is easier now to consider the segments from 0 to 7 in each positive and negative curve, with the 0 being the largest or higher-amplitude samples, and the 7 the smallest or low-amplitude samples. This then gives our total of 256 steps as mentioned.

Producing PCM binary codes in this way creates a problem in indicating idle channels to the distant end. An idle channel does not exhibit a PAM sample, and therefore the condition of 0 V is sent in the respective time-slot period. The step indicating this based on Fig. 11.22(a) is 128 or 10000000 (positive, segment 0, step 0) and as seen produces a multitude of 0's. Unfortunately, the transmission line does not like a stream of 0's, as the clocks in the repeater units along the line can only be kept synchronized to the logic 1's and will drift if too many 0's are sent. There is also the chancc that 0's will be interpreted as 1's if excessive noise is encountered. To overcome this the folded binary format is used as shown in Fig. 11-22(b). In this case an idle channel will produce a PCM code of all 1's, as the top half of the curve has been folded over and the lowest step becomes 255 (1111111), with the highest now being 128 (10000000). The negative side of the curve remains the same. This is generally in use and known as the *folded binary format* for producing PCM binary codes.

Examples of Encoding

1. (a) PAM signal quantized on step 206_{10} in nonfolded format [Fig. 11.22(a)]. Sign = positive = 1 (above 127). Segment = (206 − 127)/16 = 4 segments + 14 steps in positive curve. Segment = 4 = 100, step = 14 = 1110. Therefore,

$$\text{PCM code} = \underset{+\text{ve}}{1} \quad \underset{\text{seg. }4}{100} \quad \underset{\text{step }14}{1110} = 11001110$$

   or simply $206_{10} = 11001110_2$

   (b) PAM signal quantized on step 206 in folded format [Fig. 11.22(b)]. Sign = positive = 1 (above 127). Segment = (256 − 206)/16 = 50/16 = 3 segments + 2 steps in positive curve. This places the quantized level at segment 4, step 14. Therefore,

$$\text{PCM code} = \underset{+\text{ve}}{1} \quad \underset{\text{seg. }4}{100} \quad \underset{\text{step }14}{1110} = 11001110_2$$

   or simply $206_{10} = 11001110_2$ *Note:* In either case the resulting PCM format is the same.

2. PAM sample quantized on step $76._{10}$. Sign = negative = 0 (below 127). 76/16 = 4 segments + 12 steps = segment 4 = 100, step 12 = 1100. Therefore,

$$\text{PCM code} = \underset{-\text{ve}}{0} \quad \underset{\text{seg. }4}{100} \quad \underset{\text{step }12}{1100} = 01001100_2$$

3. A positive-going sample is quantized in segment 7, step 5. Positive = 1, segment 7 = 111, step 5 = 0101. Therefore, PCM code = 11110101 (step 245_{10})

11-6 PCM STANDARDS

There are two PCM systems in use in the world today, recommended by the Consultive Committee International Telephone and Telegraph (CCITT) world governing body. Each is dealt with below.

11-6.1 North American 24-Channel System (CCITT Recommendation G733)

This is the system that has been explained earlier in deriving PCM from analog signals containing 24 channels or transmission circuits. Each channel, as stated, has a time duration of 5.2 μs, and 24 channels, which make up the frame with a total time duration of 125 μs. The channels contain a PAM sample in PCM eight-bit coded format as described. Figure 11-23 shows the layout of a typical frame.

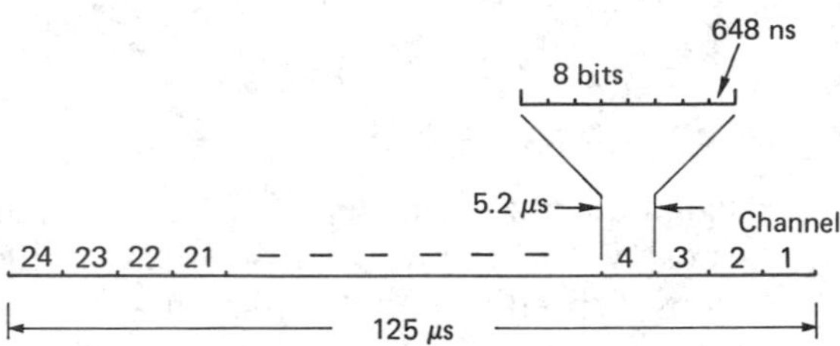

Figure 11-23 N.A. PCM system.

The number of bits on a per frame basis is $24 \times 8 = 192$ with an extra bit added for frame synchronization at the receiving equipment. This extra bit is variously known as the 193rd bit, framing bit, or S bit. It is a pattern of 1's and 0's and for synchronization purposes the receiving equipment aligns a 2-kHz wave on the 1's as shown in Fig. 11-24. This extra bit also allows the receiving equipment to discriminate between the sixth frames, which carry the signaling information, and alternate sixth frames when A and B signaling is used.

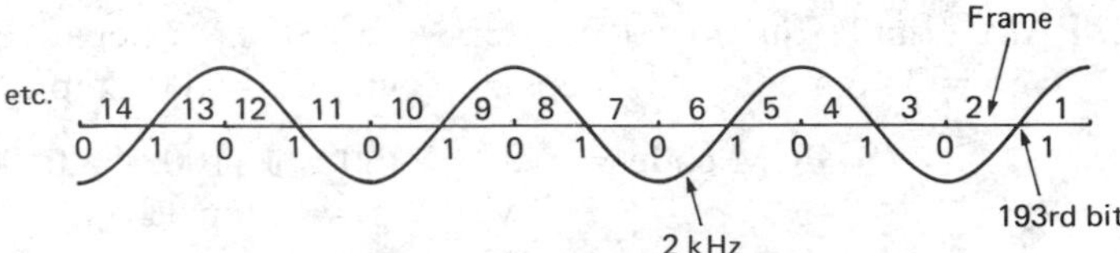

Figure 11-24 Basic alignment wave.

Signaling

This system carries the signaling for the channels as the lowest significant bit of each channel within the sixth frame. By using this method of signaling, the bit signifies a possibility of only one of two states, indicating the particular channel that is in use or is idle. As was shown using the folded binary format, an idle channel will produce eight logic 1's at all times of sampling, but to indicate the channel being idle in the sixth frame the last logic 1, or lowest significant bit of the channel, is suppressed to a logic 0 when transmitted. Therefore, the busy indication will always be a logic 1 in the lowest significant bit of the channel in the sixth frame. Figure 11-25 shows a typical example of this. The loss of the lowest significant bit from every sixth frame would not be critical in voice reception.

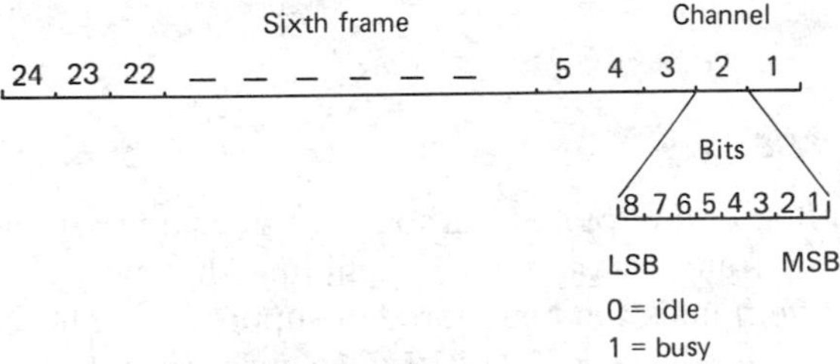

Figure 11-25 Sixth-frame signaling.

Bit Rate

Each frame consists of 24 eight-bit channels, with an extra bit added to the frame for alignment purposes. This gives a total of 193 bits, as stated previously. Using a sampling rate of 8000 samples per second, the bit rate can be calculated as 193 $\times$ 8000 = 1.544 Mbits/s. This is standard for this system and is known as the T1 or DS-1 rate.

A and B Signaling

By using two signaling systems, additional signals can be sent over one transmission line. A typical use for this is between digital remote and a class 5 central office and is generally known as a *superframe*. All conditions apply for the standard 24-channel system except for the pattern of the frame alignment bits (S bits). This has to change to be able to indicate the signaling frame for either signaling system. It is done by combining the signaling bit pattern of 101010 . . . with a framing bit pattern of 111000111000111000. . . . They are arranged in this way to form the combined 193rd-bit pattern of 01110$\underline{1}$10001$\underline{0}$ continually repeated. The sixth and twelfth bits in this pattern indicate the A and B signaling frames, respectively (Fig. 11-26). Every sixth frame still indicates the signaling as before except that in this type of arrangement the 24 extra bits can make up control words for transmitting from, or to, a remote office.

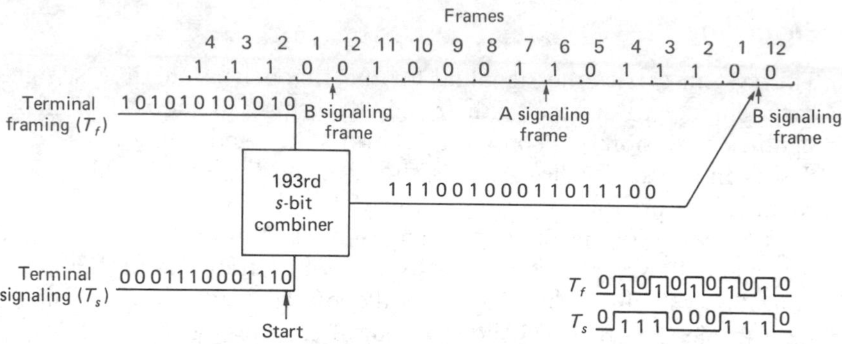

Figure 11-26 A and B signaling.

Dial Pulses

The sampling of dial pulses, although at the same rate as speech, is required only in the sixth frame as signaling. Sampling that would normally be done for the previous five frames can be ignored or suppressed. The sampling for the dial pulses can be considered every 750 μs or once every sixth frame. At this rate a single dial pulse (100 ms) is sampled approximately 133 times, of which 77 times is the break pulse (58 ms) and 56 times the make pulse (42 ms). As an example, to send the code 424-9168 would require 4522 samples $[(4 + 2 + 4 + 9 + 1 + 6 + 8) \times 133]$.

Bit 2

There must be a logic 1 in bit 2 of at least one channel in a frame. If it is not present in this form, the receiving equipment assumes that the information in the frame is erroneous and will reject the frame. This is a transmission integrity check, commonly called the *remote alarm*.

Zero Code Suppression

To prevent the repeater clocks from drifting on a PCM transmission line the number of consecutive 0's sent is limited to 13. If a fourteenth zero appears, it is forced to a logic 1 by the transmitting equipment.

Loss of Frame Alignment

If the 193rd bits (S bits) begin to drift out of alignment, the frame synchronization is lost and the receiving equipment will attempt to resynchronize using the 2kHz wave and try to realign on logic 1's over a quantity of frames. Until successful realignment, the information in the frames is ignored, but if the equipment fails to resynchronize in a set period, an office alarm is activated.

Signaling Systems

The first PCM signaling unit was the D1 channel bank. This subsequently evolved to a series of channel banks known as the D2, D3, and D4 channel bank signaling systems. Although these D-coded systems are slowly fading from general use, they are still to be found in various PCM installations and the following gives a brief description of their operational characteristics.

D1 Channel Bank System This was the first system and only 128 steps were used for quantizing, which resulted in 7-bit voice encoding. The eighth bit (LSB) was used for signaling on every channel in every frame. This was a single system of 24 channels or *digroup*.

D2 Channel Bank System This system uses 256 quantizing steps and is the standard 24-channel system described previously. It uses a controlled random sampling method and the signal-to-noise ratio (quantizing) was improved by approximately 6 dB in range of higher-level signals. It is used with A and B signaling systems and also standard E and M interface signaling, where transmission of signaling information originally comes into the channel bank from a trunk circuit on an M lead and terminates at a distant trunk circuit on an E lead.

D3 Channel Bank System This is exactly the same as the D2 system, with the exception of using a logical order of PAM sampling (channels 1 to 24). Note: The system in general use is the D2/D3 channel unit, which, as the code implies, is a combination of the D2 and D3 channel bank systems.

D4 Channel Bank System This system uses 48 channels, or two digroups, and consequently, the bit rate is increased to 3.152 Mbits/s; that is, two frames are sent where before only one was transmitted. This is known as the T1C rate.

11-6.2 European (CEPT) System (CCITT Recommendation G732)

This system is more commonly known as the 30 + 2 system, because it contains 30 voice channels and two signaling and alignment channels per frame. This makes it a better binary number to handle for office equipment and provides a higher bit rate due to 32 eight-bit channels operating in the same frame time that 24 eight-bit channels operate in the North American system. Figure 11-27 shows the makeup of the frame. As can be seen, the signaling in this system is dedicated to channel 16 only, but the signaling appears in this channel on every frame, as shown in the following pattern:

Frame 1 signaling in channel 16 for channels 1 and 17.
Frame 2 signaling in channel 16 for channels 2 and 18.
Frame 3 signaling in channel 16 for channels 3 and 19.
etc.
Frame 15 signaling in channel 16 for channels 15 and 31.

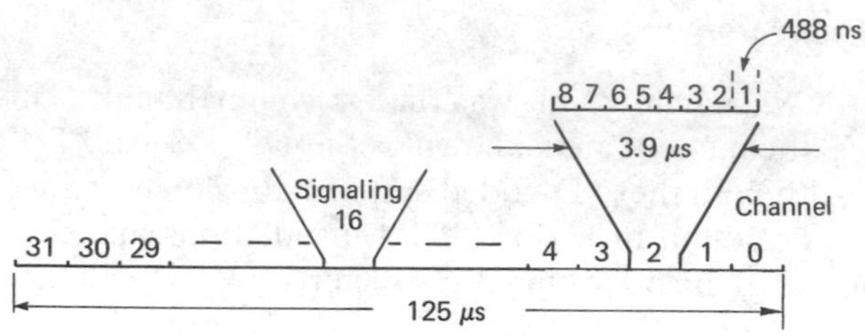

Figure 11-27 CEPT PCM system.

The signaling for all voice channels is thus carried over 15 frames. The absence of channel 0 as a voice channel allows it to be available for use as frame alignment, remote alarms, or similar uses.

The use of channel 16 as signaling for only two channels means that the channels have a choice of four bits each for denoting their signaling information. Only one bit is normally needed for busy or idle status, so other conditions can also be sent regarding the channel included in the other three bits. In Fig. 11-28 a typical frame structure as used in Europe is shown. For even-number frames (i.e., frames 2, 4, 6, 8, . . . , 14) channel 0 is the same as in frame 0. For odd-number frames (i.e., frames 3, 5, 7, 8, . . . , 15), channel 0 is the same as in frame 1. The 30 + 2 system is now popular in North America for use with digital switching offices.

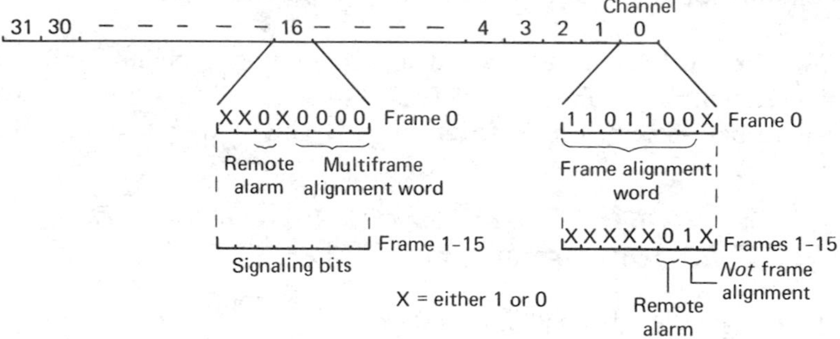

Figure 11-28 Channel 0 and 16 format.

11-7 DIGITAL HIERARCHY

With the advent of the 24-channel PCM system in North America and the advance of technology, it became advantageous to increase the amount of information transmitted between switching centers by increasing the bit rates and thus increasing the quantity of PCM channels carrying information. The frame time was left the same (125 μs), but the channels within it were increased in proportion to the new bit rates. The outcome of all this was that a hierarchical system of four basic bit rates was produced known as the DS (data stream) or T rates. The basic 24-channel 1.544 Mbits/s described earlier become the foundation for the developing multiplexing hierarchy and thus became known as the T1 or DS-1 rate. The following table designates the present digital hierarchical system and Fig. 11-29 shows how they are developed from the basic T1 rate.

Digital Signaling Level	Transmission Rate (Mbits/s)		Number of Voice Channels (per frame)
DS-1	T1	1.544	24
DS-1C	T1C	3.152	48
DS-2	T2	6.312	96
DS-3	T3	44.736	672
DS-3A	T3A	91.04	1344
DS-4	T4	274.176	4032

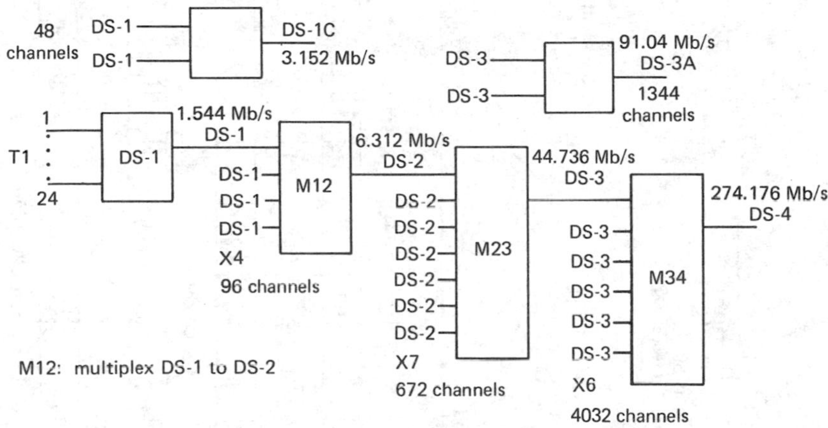

Figure 11-29 Progression of DS rates.

It can been seen from Fig. 11-29 that each upgading of the DS-1 rate does not produce a correspondingly equal increased bit rate. The number of bits required to increase the speed of information, and hence channels, will always be more, as with the faster speeds more forms or synchronization and framing alignment patterns are required. This means increasingly extra bits. For example, in going from the DS-1 to DS-2 rates, the four incoming DS-1 or T1 lines are sampled faster than the 1.544-Mbit rate. This means that one bit on a T1 line will be sampled twice every 49 times, and to prevent it from being transmitted further as two separate bits the 49th bit, one of the two sampled, is suppressed and another bit added, which is known as a *housekeeping bit*. A cumulation of housekeeping bits is then used to indicate subframes of the transmitted DS-2 rate (M bits), the framing pattern (F bits), and stuffed bits that are used to raise the bit rate to the desired bit rate of the DS-2 system (C bits). For transmitting from the DS-2 to DS-3 and DS-4 bit rates a similar process is followed, but the number of extra bits added to meet the new rates increases proportionally.

In Fig. 11-30 a typical illustration is shown of four DS-1 rates becoming a DS-2 rate and Tables 11-1 to 11-3 show the makeup of the DS-2, DS-3, and DS-4 bit rate frames.

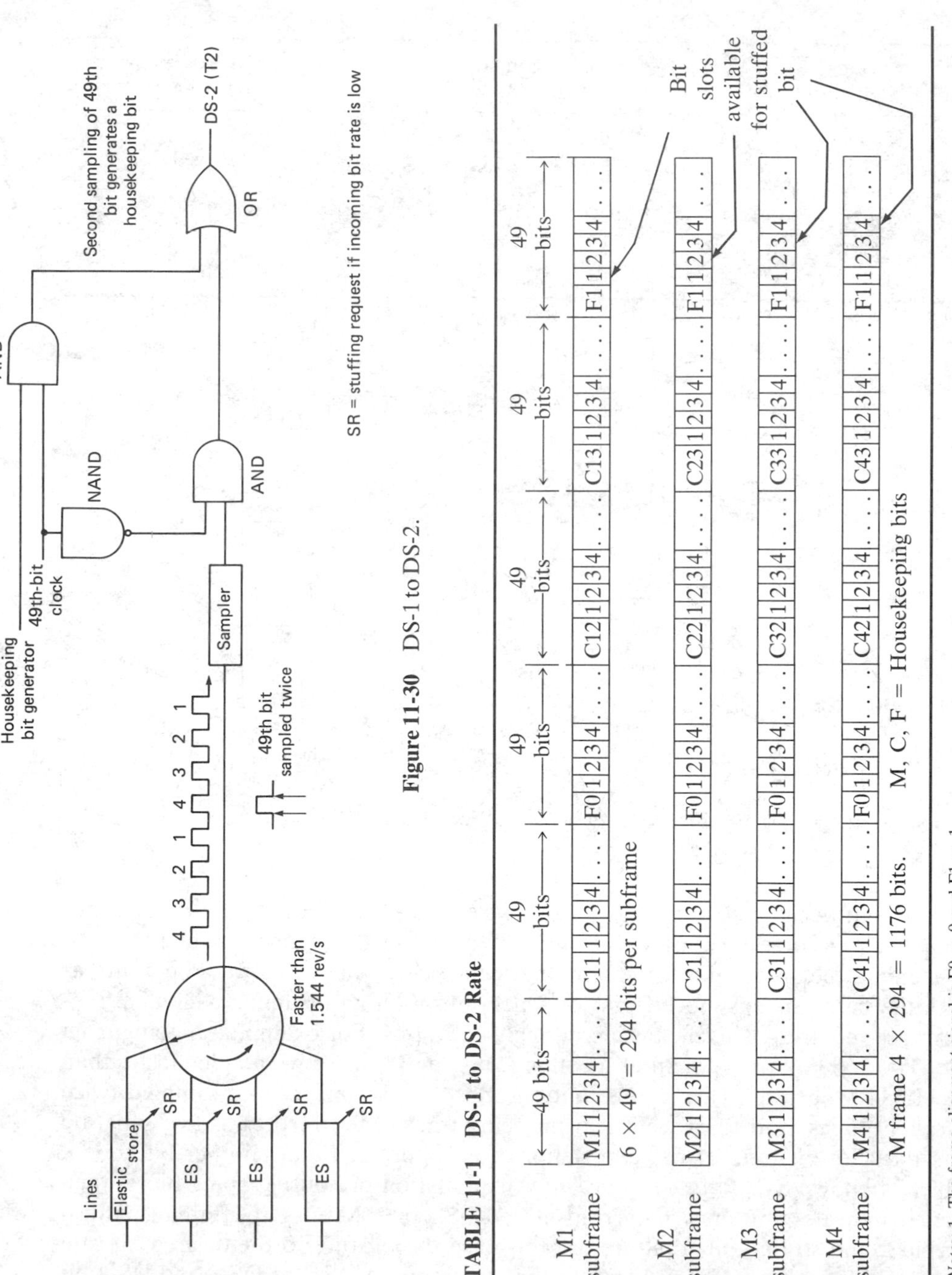

Figure 11-30 DS-1 to DS-2.

TABLE 11-1 DS-1 to DS-2 Rate

	49 bits	49 bits	49 bits	49 bits	49 bits	49 bits
M1 subframe	M1 1 2 3 4	C11 1 2 3 4	F0 1 2 3 4	C12 1 2 3 4	C13 1 2 3 4	F1 1 2 3 4 . . .
M2 subframe	M2 1 2 3 4	C21 1 2 3 4	F0 1 2 3 4	C22 1 2 3 4	C23 1 2 3 4	F1 1 2 3 4 . . .
M3 subframe	M3 1 2 3 4	C31 1 2 3 4	F0 1 2 3 4	C32 1 2 3 4	C33 1 2 3 4	F1 1 2 3 4 . . .
M4 subframe	M4 1 2 3 4	C41 1 2 3 4	F0 1 2 3 4	C42 1 2 3 4	C43 1 2 3 4	F1 1 2 3 4 . . .

$6 \times 49 = 294$ bits per subframe

Bit slots available for stuffed bit

M frame $4 \times 294 = 1176$ bits. M, C, F = Housekeeping bits

Notes: 1. The frame alignment signal is F0 = 0 and F1 = 1.
2. M1, M2, M3, and M4 represent the multiframe signal, which is 011X, where X may be used for an alarm service digit.
3. C11, C12, and C13 represent the stuffing indicator word for DS-1 input; 000 indicates no stuffing, and 111 indicates stuffing done.
4. Stuffing bit slots are indicated following F1 in each subframe.
5. The maximum stuffing rate per DS-1 input is 5376 bits/s.
6. The nominal stuffing rate per DS-1 input is 1796 bits/s.
7. The first bit slot before each 1-bit slot is a control bit time slot.
8. 1 designates a bit time slot belonging to DS-1 #1 input; 2 designates a bit time slot belonging in DS-1 #2 input; and so on.
9. Stuffing is only requested when the incoming bit rate on the DS-1 lines is out of synchronization with the 1.544-Mbit rate.

TABLE 11-2 DS-2 to DS-3 Rate (44.736 Mbits/2)

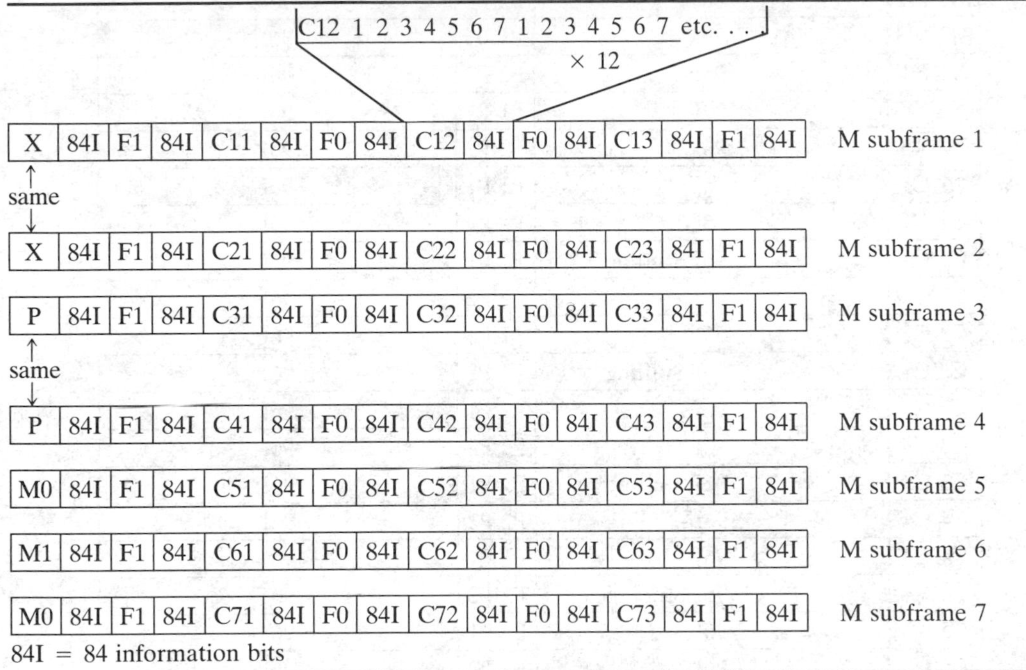

X	84I	F1	84I	C11	84I	F0	84I	C12	84I	F0	84I	C13	84I	F1	84I	M subframe 1
X	84I	F1	84I	C21	84I	F0	84I	C22	84I	F0	84I	C23	84I	F1	84I	M subframe 2
P	84I	F1	84I	C31	84I	F0	84I	C32	84I	F0	84I	C33	84I	F1	84I	M subframe 3
P	84I	F1	84I	C41	84I	F0	84I	C42	84I	F0	84I	C43	84I	F1	84I	M subframe 4
M0	84I	F1	84I	C51	84I	F0	84I	C52	84I	F0	84I	C53	84I	F1	84I	M subframe 5
M1	84I	F1	84I	C61	84I	F0	84I	C62	84I	F0	84I	C63	84I	F1	84I	M subframe 6
M0	84I	F1	84I	C71	84I	F0	84I	C72	84I	F0	84I	C73	84I	F1	84I	M subframe 7

84I = 84 information bits

Notes: The stuffing bits are placed in the seventh bit slot immediately following the F1 alignment signal at the end of each M subframe. The pattern of the preceding C bits in that subframe indicates if stuffing has been done (111 = stuffing done, 000 = no stuffing done). Each individual DS-2 line where stuffing was done uses the same bit number after the F1 alignment signal: for example, stuffing for DS-2 line 1, where C11, C12, C13 = 111:

C13 84I F1 1 (bit slot available for stuffing), 2 3 4 5 6 7

and stuffing for DS-2 line 2, where C21, C22, C23 = 111:

C23 84I F1,2 1 bit slot available for stuffing 3 4 5 6 7

1. The frame alignment signal is F0 = 0 and F1 = 1 (1001).
2. M0, M1, and M0 represent the multiframe alignment signal, which appears in the fifth, sixth, and seventh subframes.
3. PP is parity information taken over all information time slots in the preceding M frame. PP = 11 if the digital sum of all information bits is 1 and PP = 0 if the sum is 0. These two parity bits are in the third and fourth M subframes.
4. XX may be used for an alarm service channel. In any one M frame the two X-bits must be identical.
5. The maximum stuffing rate per 6.312 Mbits/s input is 9,398 bits/s. Minimum rate for the DS-2 line is 6.216671 Mbits/s.
6. The nominal stuffing rate per 6.312 Mbits/s input is 3671 bits/s.

TABLE 11-3 DS-3 to DS-4 Rate (274.176 Mbits/s)

M1 $\overline{\text{M1}}$	1 2 3 4 5 6 1 2 3 4 5 6 . . . (96I)	P1 P1	96I
M2 $\overline{\text{M2}}$	96I	P2 P2	96I
M3 $\overline{\text{M3}}$	96I	P1 P1	96I
X1 $\overline{\text{X1}}$	96I	P2 P2	96I
X2 $\overline{\text{X2}}$	96I	P1 P1	96I
X3 $\overline{\text{X3}}$	96I	P2 P2	96I
C1 $\overline{\text{C1}}$	96I	P1 P1	96I
C1 $\overline{\text{C1}}$	96I	P2 P2	96I
C1 $\overline{\text{C1}}$	96I Stuffing	P1 P1	96I
C2 $\overline{\text{C2}}$	96I	P2 P2	96I
C2 $\overline{\text{C2}}$	96I	P1 P1	96I
C2 $\overline{\text{C2}}$	96I Stuffing	P2 P2	96I
C3 $\overline{\text{C3}}$	96I	P1 P1	96I
C3 $\overline{\text{C3}}$	96I	P2 P2	96I
C3 $\overline{\text{C3}}$	96I Stuffing	P1 P1	96I
C4 $\overline{\text{C4}}$	96I	P2 P2	96I
C4 $\overline{\text{C4}}$	96I	P1 P1	96I
C4 $\overline{\text{C4}}$	96I Stuffing	P2 P2	96I
C5 $\overline{\text{C5}}$	96I	P1 P1	96I
C5 $\overline{\text{C5}}$	96I	P2 P2	96I
C5 $\overline{\text{C5}}$	96I Stuffing	P1 P1	96I
C6 $\overline{\text{C6}}$	96I	P2 P2	96I
C6 $\overline{\text{C6}}$	96I	P1 P1	96I
C6 $\overline{\text{C6}}$	96I Stuffing	P2 P2	96I

96I = 96 Information bits
Stuffing = stuffing bit slot available within these 96 Information bits

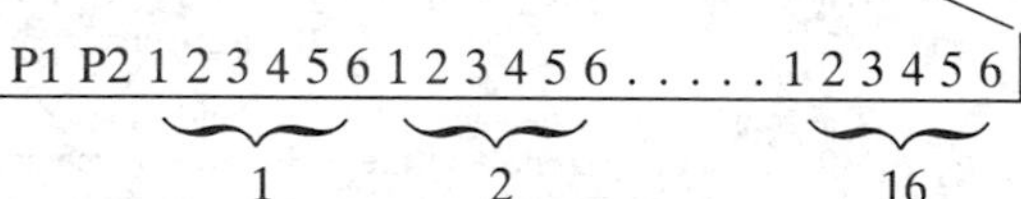

e.g. 1) C1 C1 C1 = 111 = stuffing done for DS-3 line 1.

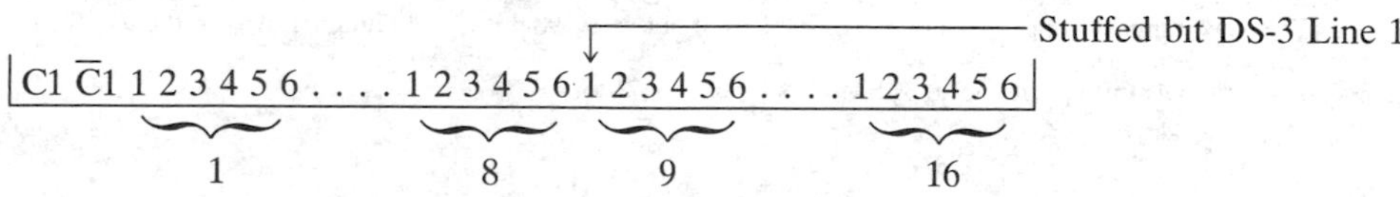

2) C2 C2 C2 = 111 = stuffing done for DS-3 line 2.

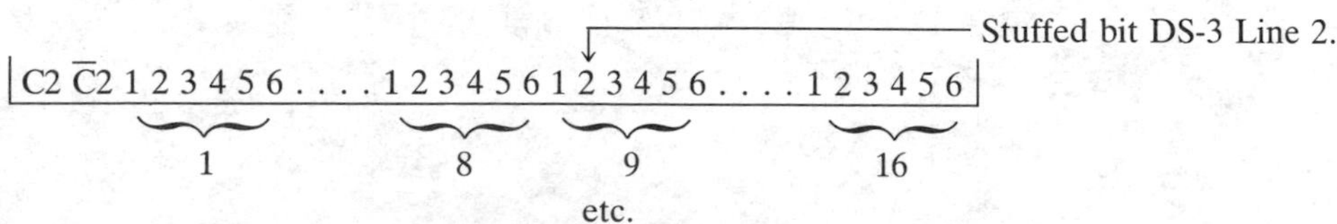

etc.

Notes: 1. The two supervisory bits are logically complementary, with the first bit carrying normal supervisory information.
2. In each pair of parity bits the two bits are logically the same.
3. M1, M2, and M3 (101) indicate the start of a superframe.
4. Superframes occur at the rate of 58.2857 kHz.
5. X1, X2, and X3 (000) may be used for DS-4 restoration.
6. An indicated stuff occurs in the 96 data bits following the third-stuff supervisory bit pairs, with the eighth and ninth bits of the stuffed channel carrying the same information.
7. Maximum stuff rate is 58.2857 kbits/s per DS-3 input.

11-8 CHANNEL BANK

Channel bank is the term given to the originating and terminating equipment on a PCM transmission line. It contains two major portions of circuitry known as the channel unit and common equipment. Normally, the equipment for these operations is contained on the same PWC card as a single unit, but for explanation purposes it is explained and shown separately (Fig. 11-31).

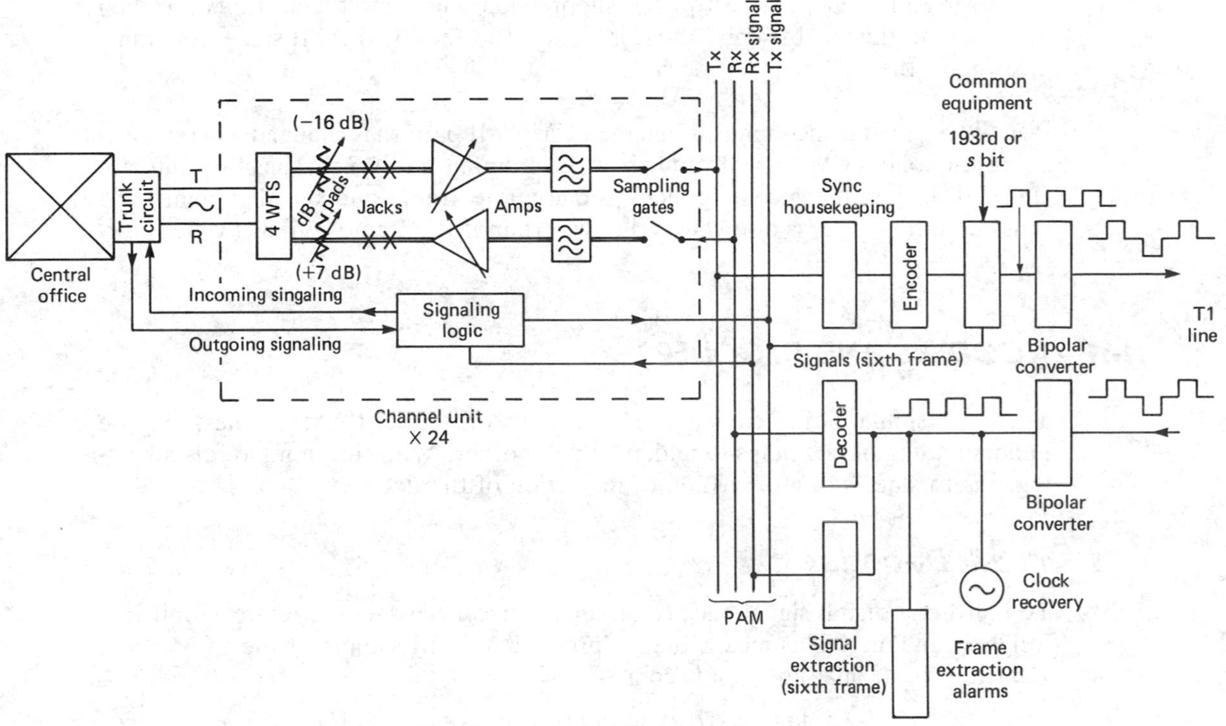

Figure 11-31 Channel bank layout.

11-9 CHANNEL UNIT

The purpose of the channel unit is to produce PAM samples from incoming analog signals, on the transmit side, and to produce analog signals from the incoming PAM samples on the receive side. This involves changing two-wire analog circuits to four-wire PAM (and PCM) circuits. To accomplish this a four-wire terminating set is required to establish the transit and receive paths.

On the transmit side the amplitude decibel levels have to be set, and prior to sampling the signal must be locked at a set frequency (180 to 3400 Hz). The sampling switch is under the control of clock pulses (8000 per second), which sample the frequency into the relevant channel or time slot. The same is required for the reverse situation on the receive side.

One channel unit is required per channel or time slot, and in the North American system 24 channel units are required to make up a frame.

11-10 COMMON EQUIPMENT

This consists of two parts: the transmit and the receive side.

Transmit This equipment encodes the PAM samples from the channel units into unipolar PCM 8-bit formats. It then inserts these eight formats into the required channels or time slots and makes up the frame in so doing. Added to this result is the 193rd framing bit pattern, and signaling, regarding each channel, is placed in every sixth frame. The zero code suppression and remote alarm bits are then inserted and this final frame format is changed to the PCM AMI signal for transmission on the T1 line.

Receive On the incoming T1 line the PCM AMI pattern is changed to a positive-going unipolar PCM format, and then the framing and synchronization bits are checked against the incoming clock for accurate frame and bit alignment. The remote alarm (bit 2) is also checked to ascertain if the frame contains valid information.

11-11 ENCODERS AND DECODERS

There are similarities in circuits used to encode or decode PAM samples, and the understanding of one helps to understand the other. With this in mind it is advantageous to start with a discussion of the action of the decoder.

11-11.1 Decoder

To convert a digital signal back to an analog it is necessary to treat each bit in a weighted manner. This means that to produce a PAM sample of the PCM code, 11001011 in decimal notation becomes

$$1(128) + 1(64) + 0(32) + 0(16) + 1(8) + 0(4) + 1(2) + 1(1)$$
$$= 128 + 64 + 8 + 2 + 1$$
$$= 203$$

For ease of examination and illustration, a four-digit code will be used, but the principle of adding the binary weighting factors together remains the same. A simplified circuit to show this method is indicated in Fig. 11-32.

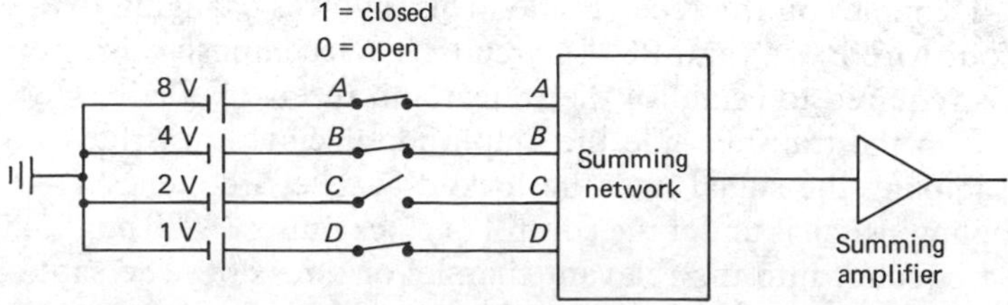

Figure 11-32 Decoder.

The switches can be relay or semiconductor devices, but the reference voltages are various amplitudes, corresponding to the weighting of the binary code. In this example the binary code 1101 is shown by the closed switches ABD. This gives a resulting voltage added in the summing network as X = 8 + 4 + 0 + 1 = 13 V.

The varying reference voltages do not make the circuit of Fig. 11-32 very practical, so only one reference voltage is generally used, which requires a special circuit to be able to weight and sum the results from the voltage switches. This is known as the resistive *ladder network*. An example of this network using our selected binary code is shown in Fig. 11-33.

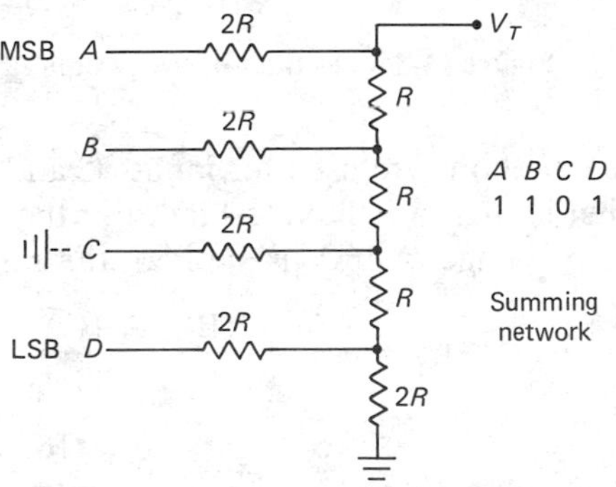

Figure 11-33 Ladder network.

There are just two values of resistors, R and 2R, and each input voltage source should be considered separately. By using the superposition theorem for each of these inputs, the rest being replaced by short circuits, a total can be made when all input results are added together.

Considering the most significant first (A), the equivalent circuit reduces to that shown in Fig. 11-34. If the next bit is analyzed, it produces the resultant circuit shown in Fig. 11-35. By using Thévenin's theorem and breaking the circuit at point Y, an equivalent circuit is shown.

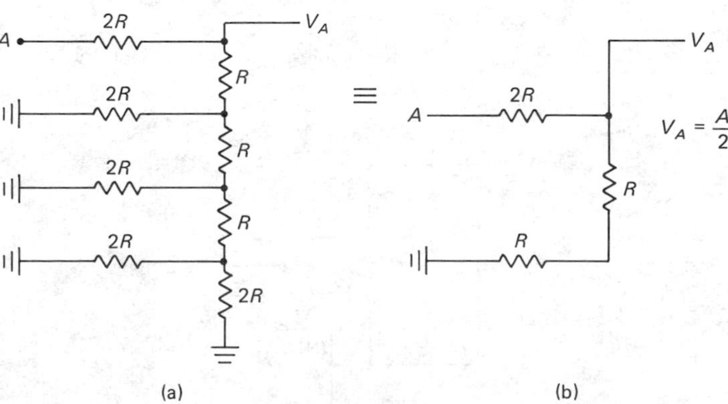

Figure 11-34 Ladder network equivalent for A.

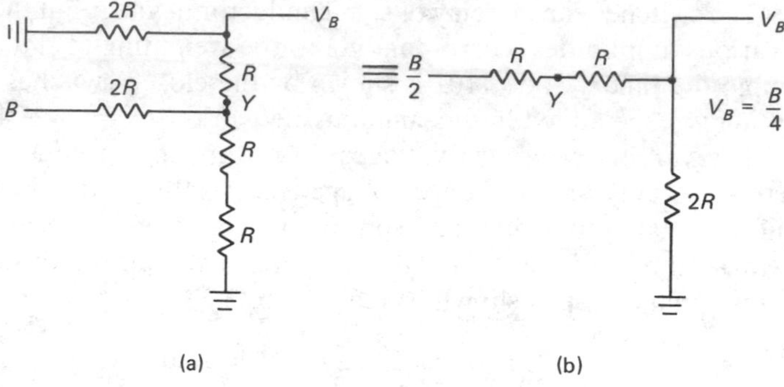

Figure 11-35 Ladder network equivalent for B.

If the same methods are used for inputs C and D, the resultant voltages can be added together in their weighted fashion. For the code that we are using (1101), and if a reference voltage of 16 V is applied, the resultant equation becomes

$$
\begin{aligned}
V_T &= \frac{1A}{2} + \frac{1B}{4} + \left(\frac{1C}{8}\right) + \frac{1D}{16} \\
&= \frac{16}{2} + \frac{16}{4} + 0 + \frac{16}{16} \\
&= 8 + 4 + 0 + 1 \\
&= 13 \text{ V}
\end{aligned}
$$

Voltage Switches

The voltage switch must be of a high-speed operation and be able to input a logic 1 or 0 and output a ground or reference voltage. There are various devices that can accomplish this and the one chosen here shows a logic gate and two MOSFETs, one being a P-channel device and the other an N-channel device. No output circuit exists until one of them is switched on. This gives rise to the term *complementary*, and the two devices together are known as CMOS (Fig. 11-36).

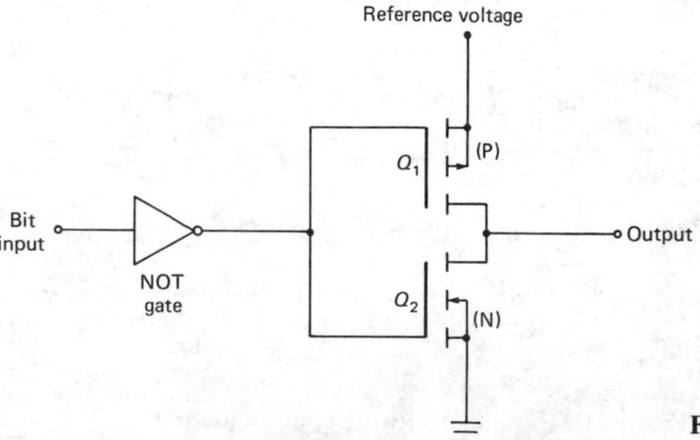

Figure 11-36 Voltage switch.

Operation

When a logic 1 is presented at the input it is inverted at the NOT gate to a logic 0 (ground), which is then applied to the common-gate terminal of the CMOS circuit. The P-channel device Q1 is biased on and the N-channel Q2 is off. The reference voltage then shows at the output terminal, with the ground being disconnected by Q2. When a logic 0 is presented at the input it is inverted to a logic 1 (+5 V), and this arriving at Q1 and Q2 will turn off Q1 and turn on Q2, thereby disconnecting the reference voltage and allowing the ground of Q2 to appear at the output terminal.

The advantage of using a CMOS circuit in this adaptation is that the power dissipation and voltage drop across the turned-on transistor Q1 and Q2 is extremely small compared to similar devices. If there are eight similar circuits, their outputs can be timed into the ladder network to produce the PAM sample, as mentioned previously.

Summing Amplifier

The purpose of this unit is to produce a PAM sample from the summing network with a discrete quantized voltage level. To do this requires the following: (1) infinite voltage gain, (2) infinite input impedance, (3) zero output impedance, and (4) input voltage at virtual ground (high input resistance). The unit that meets these requirements is the operational amplifier. The inputs to it are the resistors used in the ladder network for summing, and it is basically an extension of that circuit (Fig. 11.37).

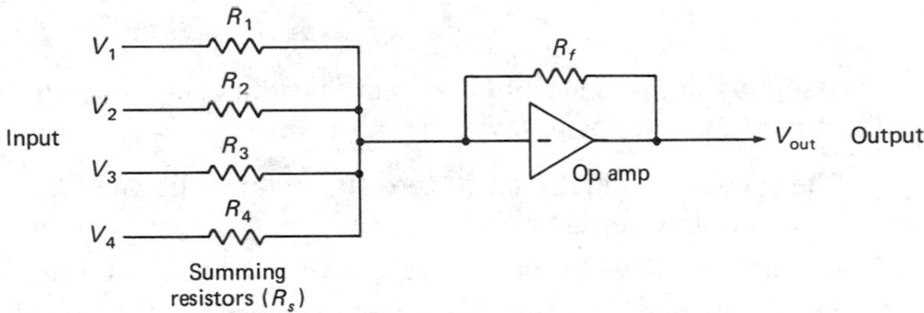

Figure 11-37 Summing amplifier circuit.

If, in this circuit, $R_s = 10\ \text{k}\Omega$ and $R_f = 100\ \text{k}\Omega$, the amplifier would have a gain of 10:1. In practice, the gain of the amplifier will be related to the desired number of levels of quantization. For example, in the circuit given,

$$V_{\text{Out}} = -\left(\frac{R_f V_1}{R_1} + \frac{R_f V_2}{R_2} + \frac{R_f V_3}{R_3} + \frac{R_f V_4}{R_4}\right)$$

and

$$I_{\text{Out}} = -\left(\frac{V_1}{R_1} + \frac{V_2}{R_2} + \frac{V_3}{R_3} + \frac{V_4}{R_4}\right)$$

11-11.2 Encoder

The general method now used to convert PAM samples to an encoded format is called *successive approximation*, where the PAM sample is compared to particular voltage levels until a comparative level has been reached. Each of the voltage levels compared in this fashion will produce either a logic 1 or a 0, and to produce a PCM code of eight bits will require eight voltage comparisons. The equipment that does the level comparisons is known as a *comparator*, and the circuit that provides the comparing voltage levels is the now familiar *ladder network*, which is itself controlled by a register and control logic. One conception of this system is shown in Fig. 11-38.

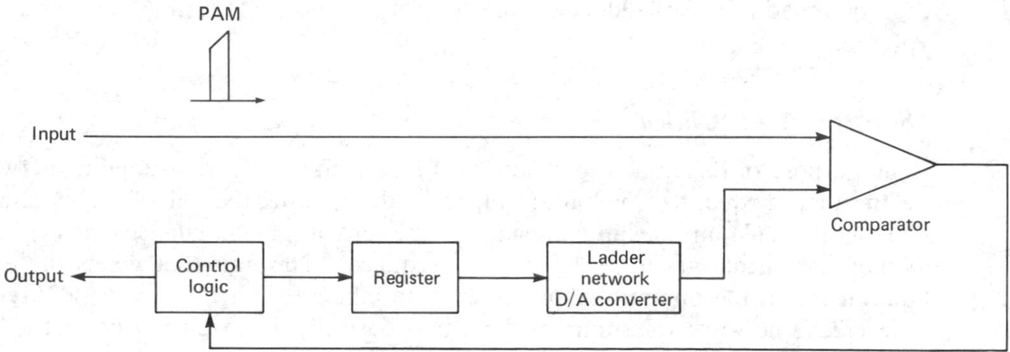

Figure 11-38 Encoder.

Assuming in this conceptual system that the comparator has a range of ∓24 V, the circuit operates as follows:

1. The register is initially set to zero. It contains the eight bits to indicate sign, segment, and step number.
2. An incoming PAM sample is received at the comparator.
3. The comparator is automatically set to transmit a logic 0 unless the incoming PAM sample is positive. In this case a latch within the comparator will clock and produce a logic 1 at the transmit lead to the control logic.
4. The control logic will output the incoming logic 1 or 0 from the comparator as the first bit of the PCM word, which now becomes the sign bit.
5. The control logic will also set the most significant (MSB) bit of the register the same as the sign bit received (−ve = 0, +ve = 1). The second MSB of the register will then be set to a logic 1 and the total code will be forwarded to the ladder network (01000000 or 11000000).

6. The ladder network acts on the incoming code and either halves the voltage potential upward (12 V) or downward (−12 V) for comparison with the PAM sample at the comparator.
7. The comparator now does the first comparison, and if the PAM sample voltage level is higher than the compared level, it transmits a logic 1, but if lower, it transmits a logic 0 back to the control logic.
8. The logic 1 or 0 received at the control logic is then transmitted out as the second bit of the PCM code.
9. According to this logic bit, the register is again set. If a logic 1, the third MSB becomes a 1, giving the code 11100000 or 01100000 to the ladder network. If a logic 0, the second MSB is reset to 0 and the third bit is set at logic 1, giving the code 10100000 or 00100000 to the ladder network.
10. The ladder network then sets the voltage level again by halving the remaining scale either to ∓18 or ∓6 V and the comparator then compares this to the incoming PAM sample.
11. The result of the comparison will produce a logic 1 or 0 as the third output of the eight-bit PCM word and set the register again by either retaining or resetting its third MSB and setting the fourth MSB as a logic 1.
12. The entire process is repeated until all eight bits of the PCM word have been transmitted.

Figure 11-39 shows two examples that illustrate this procedure. The actual PAM sample level may not be reached exactly but will be reproduced within ∓ one-half of the lowest significant bit (LSB). The resulting error is called *quantization error.*

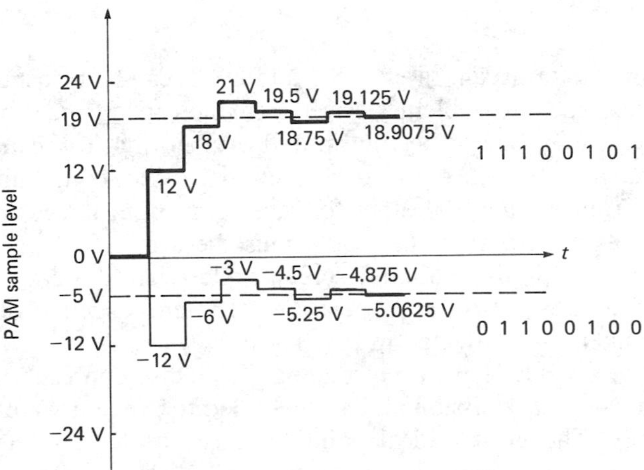

Figure 11-39 Comparison of voltage levels to PAM sample.

EXAMPLE 11-1: PAM = 19 V

Register Setting	Ladder Network O/P (V)	Comparator Result	PCM 8-bit Word Format
00000000	0	1	1
11000000	12	1	11
11100000	18	1	111
11110000	21	0	1110
11101000	19.5	0	11100
11100100	18.75	1	111001
11100110	19.125	0	1110010
11100101	18.9375	1	11100101

EXAMPLE 11-2: PAM = −5 V

Register Setting	Ladder Network O/P (V)	Comparator Result	PCM 8-bit Word Format
00000000	0	0	0
01000000	−12	1	01
01100000	−6	1	011
01110000	−3	0	0110
01101000	−4.5	1	01101
01101100	−5.25	0	011010
01101010	−4.875	1	0110101
01101011	−5.0625	0	01101010

11-12 PCM CABLES

The lower data stream rates, DS-1 and DS-2, can be transmitted over cable routes that overcome the problems derived from their speed and synchronization qualities. However, they cannot be transmitted over normal toll cables due to their unidirectional nature, so have to be transmitted in cables containing two basic sections—one for transmit and the other for receive. Furthermore, to prevent crosstalk and induction, each part of the cable must be screened from the other. For these requirements the D, Z, and T screen cables were designed for transmitting PCM over long distances in large quantities. The cross section of these cables showing their makeup is illustrated in Fig. 11-40.

These cables have an insulating metallic screen separating pairs for the two directions of transmission and are gel filled to prevent water absorption over long distances. The most critical point for these cables is at the termination to the repeater stages, where the cable pairs for each section exit from their screened areas and crosstalk coupling loss could result, although it is kept to a minimum by terminating at the last possible point.

These cables are designed to carry large volumes of PCM traffic (e.g., 100 pairs and up), but for smaller and shorter PCM routes the PIC cables have become

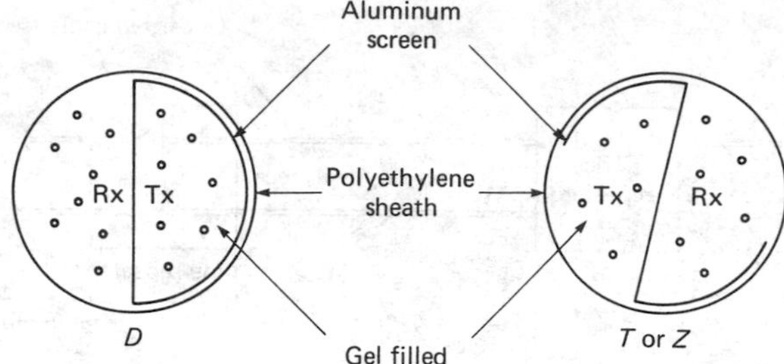

Figure 11-40 PCM cables.

popular. They consist of fewer pairs laid out in binder groups and are even count cables. One binder group is used in one dirction of transmission and another binder group is used in the other direction. Each of these groups is recognized by a different-colored binder twisted around it (hence its name). Figure 11-41 shows an example of these cables.

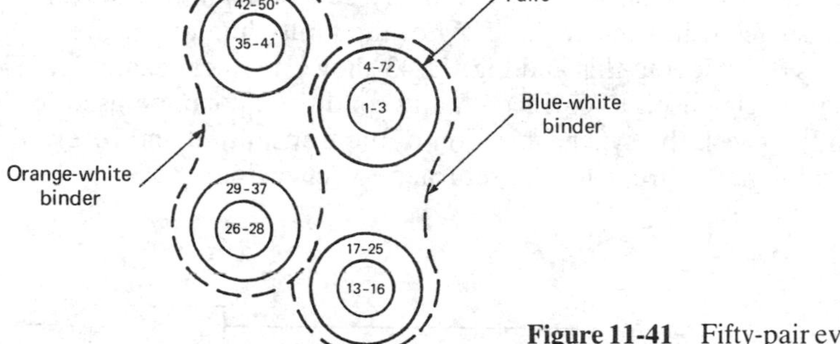

Figure 11-41 Fifty-pair even-count PIC cable.

All PCM transmission cables carrying the DS-1 rate are generally known as T1 cable routes, or with the DS-2 rate as T2 cable routes. For the higher rates (DS-3, DS-4) the synchronization problems become astronomical, and coaxial, fiber optics, or digital radio have to be employed.

11-13 T1 REPEATERED LINE

For connecting the T1 cable route from office to office the following equipment is required: (1) protection switch, (2) simplex power feed, (3) ALBO line repeater, (4) line repeater housing (pedestal), and (5) office repeater. The layout of these is shown in Fig. 11-42. The channel bank and screened cable have already been dealt with, but in regarding the screened cable the only exception shown here is that inside the central office the cable becomes "dry" (i.e., not gel filled), to prevent a possible flammable situation, as the gel is considered a fire hazard. An explanation of the operation of the other equipment follows.

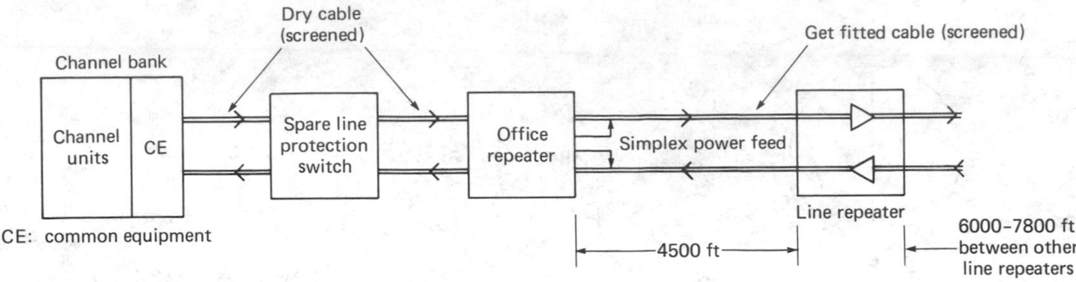

Figure 11-42 T1 repeatered line.

11-13.1 Protection Switch

The repeatered outside line is the more likely cause of failure, and if a fault is detected, the switch will transfer the PCM of the faulty line to a spare line. There are many ways of achieving this and Fig. 11-43 shows a typical example. When the spare line is not being used to correct a faulty condition, it can be used for looping at the far end to check the synchronization of the transmitted and received pulses. In this case all logic 1's are sent and returned for checking.

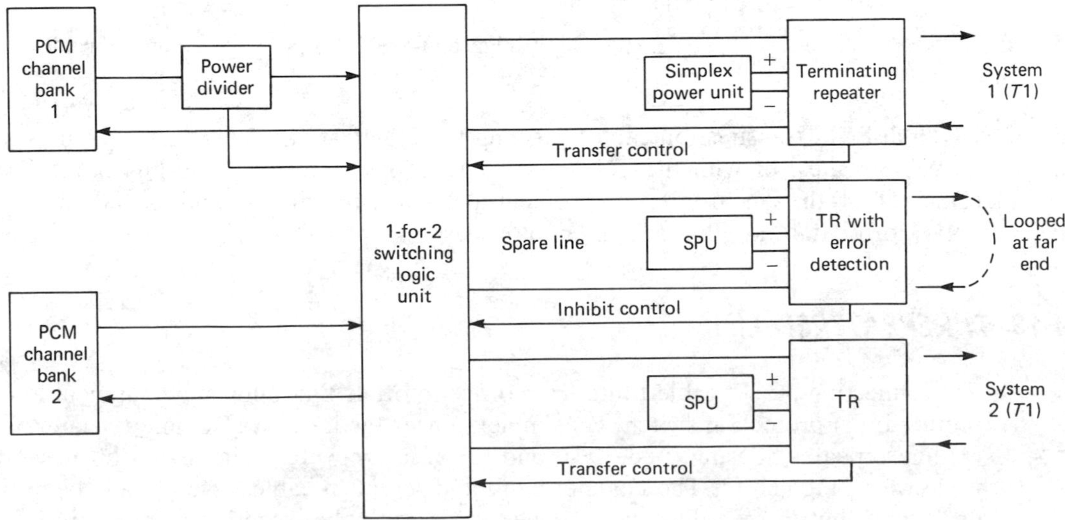

Figure 11-43 Protection switch.

11-13.2 Simplex Power Unit

To be able to apply power to each repeater used along the line, this unit is required to output the necessary voltage supplies over a loop formed by the simplexing of two cable pairs, as shown in Fig. 11-44(a). The distant office loops the current back if within the range of 10 line repeaters, but if there are more line repeaters, the tenth repeater will have to provide the looping conditions [Fig. 11-44 (b)]. The distant office or another location will have to provide similar power for the next 10 repeaters. This distance between power points is known as a *section*, and between offices as a *span*

The supply voltage must be able to provide at least 150 mAmps through the loop, and to ensure this, calculations must be carried out based on cable used, distance between repeaters, and type and number of line repeaters. The required voltage is then selected from combinations of ∓130 V, −48 V, and ground.

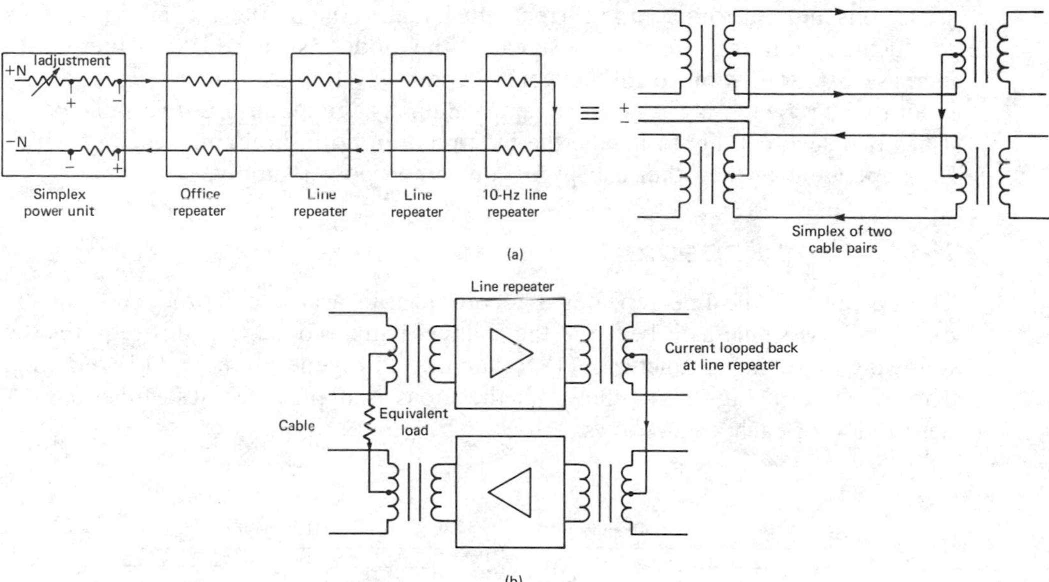

Figure 11-44 Powering of the T1 repeatered line.

11-13.3 Automatic Line-Build-Out Network (ALBO)

The ALBO circuit is an integral part of each line repeater and is designed to maintain the appearance of constant equal length between line repeaters. As shown in Fig. 11-45, an example of this circuit produces a variable resistance by varying the current in silicon diodes. This is controlled by the repeater and is proportional to the input signal.

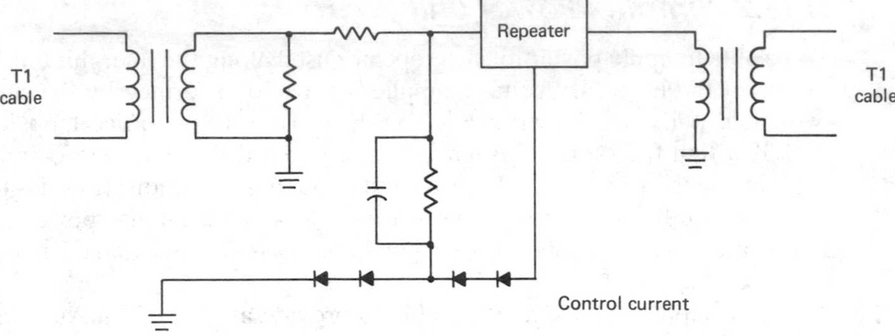

Figure 11-45 Automatic line-build-out network.

If a cable pair is short, the input signal will be strong and the peak detector senses this and puts out a large current which makes the diode resistance low, thus producing a shunting effect of the signal. If the diode resistance is low, there is an increase of attenuation to the network. In this way the level of the signal at the input to the repeater always has the same value. As a repeater uses two cable pairs, it has two such circuits to handle the bit stream in both directions, each of which is independent of the other except for a common power supply.

11-13.4 Line Repeater

The purpose of the line repeater is to produce an accurate reproduction of the original PCM signal as it began at the channel bank. To achieve this requires the following three major functions: (1) reshaping, (2) regeneration, and (3) retiming. The repeater in Fig. 11-46 shows all the areas that provide these functions. A description of each area follows.

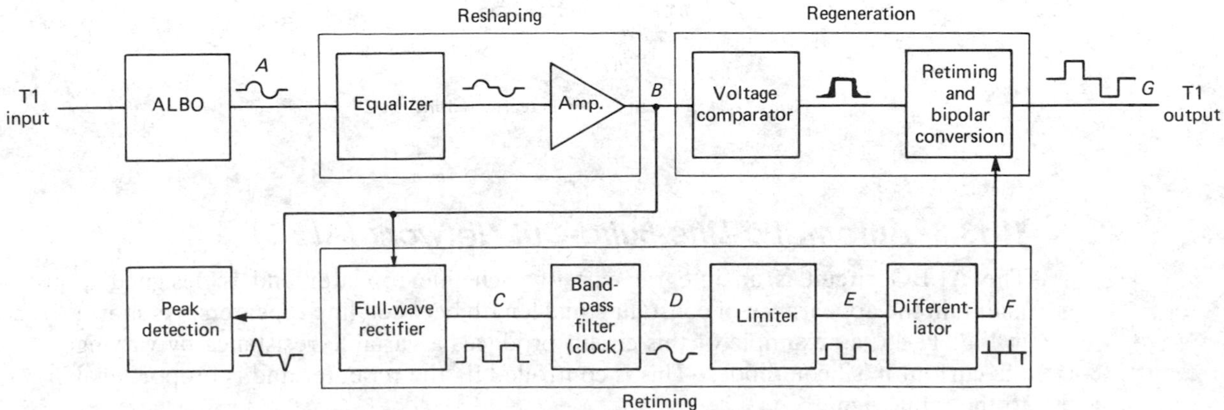

Figure 11-46 Line repeater.

Reshaping

Due to cable and line conditions, a pulse becomes distorted and at 6000 to 7800 ft (approximately 1818 to 2365 m), which is a standard for repeater spacing, may appear as shown in Fig. 11-47(a). The pulse is spread out in time and the task of the equalizer is to restore the pulse to its original width, although not to its original band. A rounded pulse appears at the equalizer output [Fig. 11-47(b)] and is applied to a predetermined level before forwarding to the next two functions.

The incoming pulse, which has widened in time, must also be checked to see that it has not overlapped into the next time frame and prevented the detection of a "no pulse" condition. One of the functions of pulse reshaping is to ensure that the reshaped pulse has returned to zero potential before the next pulse time period appears, when a sample is taken.

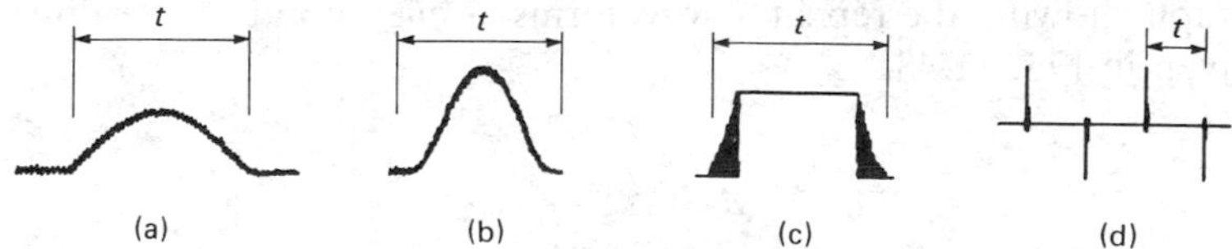

Figure 11-47 Pulse progression in line repeater.

Regeneration

The amplified pulse enters a voltage comparator circuit which has a threshold such that if any incoming pulse does not exceed 50% of the nominal pulse height, it will be forwarded from the comparator as a logic 0. If it does exceed the 50% threshold level, it is forwarded as a logic 1. This circuit thus becomes a pulse detector, and it is imperative that the threshold level be measured accurately.

An output pulse from the comparator is a square wave which can be carrying *timing jitter*, as shown in Fig. 11-47(c). All pulses forwarded will to outputted at the correct time set by the retiming function at the bipolar converter.

Retiming

The incoming pulse train carries its own timing information and the retiming starts after equalization and amplification of the signal. The incoming signal splits in two directions as shown in Fig 11-46 and the retiming signal is passed into a full-wave rectifier. This produces a unipolar bit stream of equal polarity which contains the timing information that must be extracted to provide the repeater timing. This extraction is accomplished by passing the unipolar pulse train through a bandpass filter to reproduce the original 1.544 MHz sine wave. The output is then fed to an amplitude limiter, which changes this to a set square wave with a constant voltage level. This wave then becomes differentiated and timing spikes are produced to provide the correct points of timing for the outgoing pulses. The timing circuit will also generate a timing spike for the termination of a new pulse [Fig. 11-47(d)].

The retiming function must correct any timing jitter coming from the comparator by producing a sample pulse which is lined up with the nominal center of the incoming comparator pulse to minimize effects of the timing jitter at the pulse edges. The new pulse is started at this nominal center.

The number of logic 1's in the incoming bit stream is kept fairly constant by using zero code suppression, thereby keeping the timing accurate at 1.544 MHz, but other problems, such as noise riding on the signal, cable impulse noise, or crosstalk, can influence the retiming function and produce timing jitter. If it affects one repeater, this problem will be produced arithmetically by all subsequent repeaters and is the chief cause of the limitation of length of transmission of the PCM cable system. The timing jitter will produce the timing spikes at the incorrect times and consequently the synchronization of the bit stream is in jeopardy. A graph showing the repeater waveforms at each point of the functions described is given in Fig. 11-48.

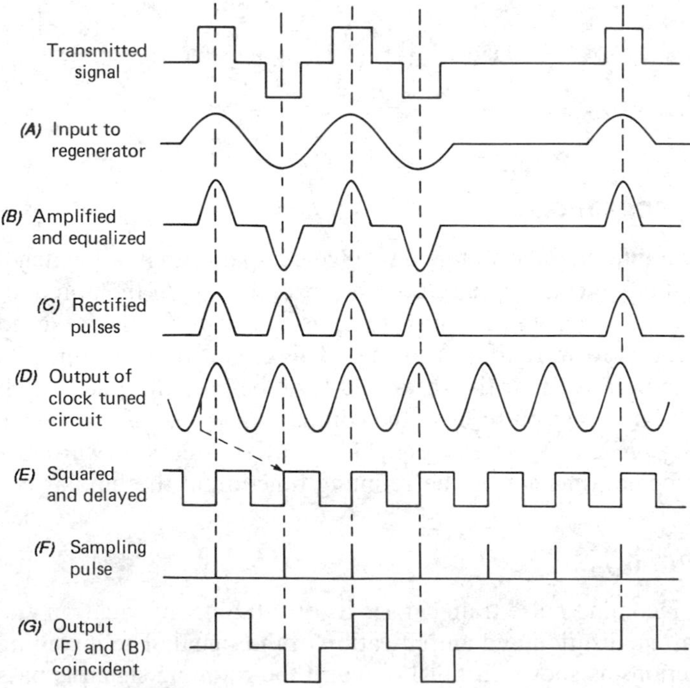

Figure 11-48 Line repeater waveforms.

11-13.5 Office Repeater

This is an originating and terminating repeater for the T1 line which (1) provides a transmitted PCM signal of sufficient energy to be acceptable at the first repeater on the line (4500 ft); (2) restores the incoming PCM signal to full amplitude; (3) provides a point for testing or patching the transmit or receive lines; and provides (4) monitoring of the incoming bit stream for errors (e.g., 1 to 10^5 errors is acceptable, 2 in 10^5 errors is unacceptable).

Order Wire

One circuit in a PCM cable route is set aside as a subscriber line to allow maintenance personnel access between line repeater locations and the terminating offices. This is a practical circuit to allow personnel to be able to call the various test positions at each office.

11-14 USES OF PCM

11-14.1 Digital Offices

Only with the advent of the digital offices in the late 1970s did PCM virtually replace the old analog system, not only on cable routes but throughout the switching system, the only exception being that the subscriber hears and speaks in analog; therefore, all transmissions must always start and finish with the human voice.

The digital office has some advantages and disadvantages compared with the analog system. A comparison of these is shown in Table 11-4. The analog office referred to here is of the electromechanical or step-by-step type.

All offices do not use the same PCM system. Some use the North American system and others the CEPT system (30 + 2, as described). Even the ones using the same system add extra bits to the word format and switch in serial or parallel within the office. Serial format is no change from that described earlier in this chapter, but parallel switching moves all the information of a word at the speed of one bit (i.e., eight bits transmitted in one bit time). This obviously requires eight links where before only one was needed, but has savings in speed and reduction of some logic circuits.

Problems arise with synchronization of the bit streams, and this is critical in modern digital offices; therefore, each office is equipped with the most accurate "clock" possible to produce the pulses. The acceptable rate of error inside the office is a minimum of 1 in 10^6 pulses, which is less than the acceptable rate on the T1 line outside.

All PCM passed into the switching network is in the form of unipolar pulses and a type of interface is used to change the incoming PCM rate from a T1 line to the switching network to allow this. When subscribers line circuits or analog trunks are involved, codecs (encoder/decoder) are required to produce the necessary PCM from analog, and vice versa. Table 11-5 indicates some digital offices with their respective PCM systems.

TABLE 11-4 Comparison of Digital and Analog Offices

Digital Office	Analog Office
Faster call processing	Slower
More traffic capacity	Limited
Space saving by using small solid-state circuitry	Larger switching components require more floor space
Modular concept allows shorter installation time	Hard cabling required throughout
Changes can be made in software	All changes require hard wiring
Excessive duplication for reliability	Limited duplication or none at all
Quiet	Noisy
Centralized intelligence (computer runs system)	Distributed intelligence (e.g., S × S)
Stray currents or static electricity can affect components	No effect
Temperature intolerable (requires large air-conditioning plants)	Temperature tolerable
Continuous power consumption for solid-state circuitry	No calls being processed; little current drain on equipment
Four-wire switching for conversation	Two-wire switching for conversation
Less chance of blocked or lost calls	More chance of blocked or lost calls
More custom calling features	Some features require too much hard wiring

TABLE 11-5 PCM of Various Digital Offices

Office	Class	Size	PCM	Bits per Word	Parallel or Serial
Northern Telecom					
DMS-1	Remote	256 lines	NA	8	S
DMS-10	5	7,500 lines	CEPT	8	S
DMS-100	5	120,000 lines	CEPT	10	S
DMS-200	4	60,000 trunks	CEPT	10	S
SL-1	PBX	2,000 lines	CEPT	8	S
GTE AE Microtel					
3-EAX	4	60,000 trunks	NA	9	P
5-EAX	4/5	145,000 lines	NA	12	P
GTD-120	PBX	100 lines	NA	8	P
Western Electric					
4-ESS	1–4	107,500 trunks	CEPT	8	S

The actual switching networks are made up of combinations of time and space switches. The time switch has the ability to hold up a PCM word in time, and the space switch is a point of distribution between various time switch networks.

The layout of a typical digital office is shown in Fig. 11-49, and it highlights the duplication used and the fact that all PCM information is now carried on buses (highways, links) and all equipment can be directly accessed by a main processor via smaller processors generally called *peripheral processors*. The smaller processors do repetitive functions (e.g., pulse storage, off/on-hook transitions, supply tones, etc.) and relieve the main processor of some of the burden. This is an example of an attempt at *distributed processing*.

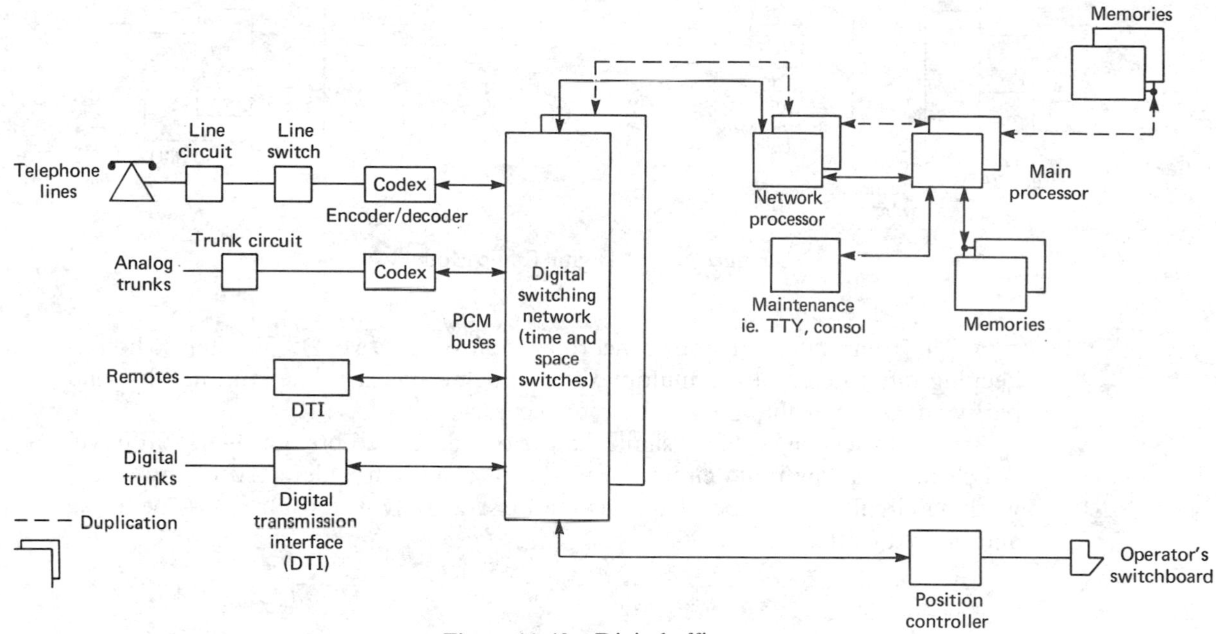

Figure 11-49 Digital office.

11-14.2 Digital Radio

Using the DS-3 rate two systems can be combined or multiplexed into a DS-3A system at 91.04 Mbits/s and sent over long-haul routes operating in the 8-GHz band. The rate used combines 1344 telephone conversations. A block diagram of this system using the Northern Telecom DRS-8 digital radio system is shown in Fig. 11-50. There are 10 DS-3 inputs into the system, giving a total of five DS-3A outputs from the multiplexers (transmitters) through the protection switch to the microwave radio, and vice versa at the receiving end. Of the six microwave radio sites shown, five are active and one is a spare. The range of this system is 6565 km, with hops of 50 km between microwave radio towers.

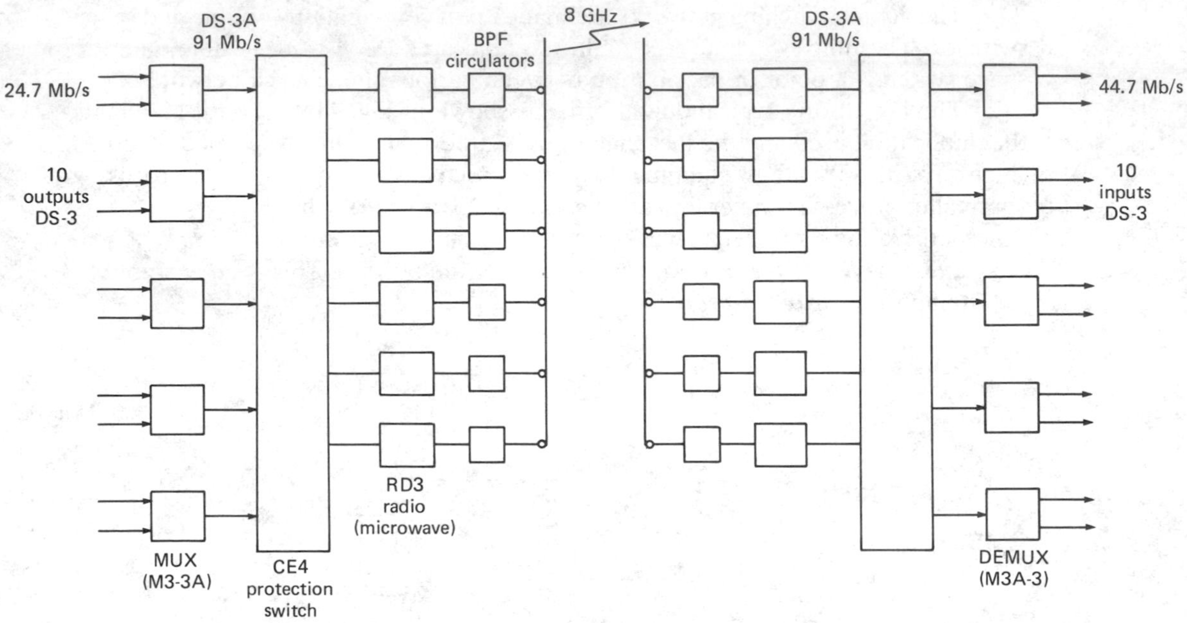

Figure 11-50 Digital Radio Model.

The transmitter provides synchronization of the two DS-3 systems, housekeeping bits to control the multiplexer and demultiplexer, line stuffing bits, and parity checks to enable greater bit rate accuracy.

The protection switch is similar to the one described previously in that it will switch to a standby radio circuit when a fault condition is detected in any of the working circuits. The type of fault condition generally tested for is loss of frame and bit error rate.

Problems

11-1. What does the term *modulation* mean?

11-2. Why is an analog signal unsuitable for use as a digital signal?

11-3. What is meant by the terms *dc component, threshold level, bipolar violation*, and *alternate mark inversion* with respect to digital signals?

11-4. Why use the AMI format on a digital line?

11-5. Show the binary code 11010110 in the AMI format.

11-6. What are the disadvantages of using PDM and PPM as digital signals?

11-7. What are the three processes for producing PCM? Explain one of them in your own words.

11-8. What influence did H. Nyquist have on producing PCM?

11-9. What is aliasing, and why is it unaccpetable to PCM sampling?

11-10. How many samples of a 3.3-kHz signal would be taken every cycle?

11-11. Why is a nonlinear format used in the encoding process? Explain its connection with the μ curve?

11-12. Explain in our own words what is meant by the term *folded binary format*. Why is it used?

11-13. Explain the binary code 00111101 in terms of sign, segment, and step. On a linear scale of +24 to −24 V, what would the voltage of the PAM sample be?

11-14. What are the codes of PAM samples quantized on the following steps in the folded binary format?
(a) 103
(b) 232
(c) 54
(d) 177

11-15. Compare and contrast the North American and CEPT PCM systems. What advantages/disadvantages does each have?

11-16. Explain the following terms with respect to North American PCM.
(a) Basic alignment wave
(b) S bit
(c) Sixth frame signaling
(d) A and B signaling
(e) Zero code suppression
(f) Remote alarm

11-17. What are the times for the following in the NA and CEPT PCM systems?
(a) Bit rate **(b)** Frame
(c) Channel **(d)** Pulse

11-18. How many times would the dialed pulse digit 6 be sampled for inclusion into the sixth frame, channel 10, signaling of the NA PCM system? Over what time period would all six digits be sent?

11-19. How many times would four dial pulses be sampled for inclusion in channel 19 of the CEPT PCM system?

11-20. Explain the different signaling systems used in the NA and CEPT PCM systems.

11-21. What is the difference between the D1 and D3 channel bank systems?

11-22. List the digital systems in order of their bit rates (lowest to highest). Where would each of these systems be used?

11-23. Show by the use of a diagram how a DS-1 system can be sent over a DS-3 fiber optic link and returned to a DS-1 system at the receiving end.

11-24. How many T1 lines are required to produce the DS-2, DS-3, and DS-4 rates?

11-25. Explain the following terms with regard to the DS-1 to DS-2 rate.
(a) Subframe
(b) Stuffing

11-26. Of what equipment does a channel bank consist? Explain the functions of one of them.

11-27. Explain the operation of the decoder. Why does it require a ladder network? What other equipment requires a ladder network for its operation?

11-28. Using diagrams, show how the binary code 1010 can be decoded to a voltage potential level.

11-29. Why does an encoder need the use of a comparator circuit? Explain how this comparator circuit is used in producing a binary code from an incoming PAM sample.

11-30. Using graphs and tables, show the binary codes of the following PAM samples (+24 to −24 V range).

(a) −21 V
(b) 7.5 V
(c) −11.5 V
(d) 15 V

11-31. Explain the terms *successive approximation, quantization error, summing amplifier*, and *CMOS* and give an example of where each occurs.

11-32. What design characteristics are required in a cable to effectively transmit 150 pairs using the T1 PCM rate?

11-33. State what units are used in the T1 repeatered line. Explain the operation of one of them.

11-34. What does the term *ALBO* mean, and why is it used?

11-35. Explain the terms *reshaping, retiming*, and *regenerating* with regard to the T1 repeatered line.

11-36. What is timing jitter, and what effect does it have on the range of a T1 repeatered line?

11-37. If an outside plant person wanted to test a PCM circuit from a line repeater, how would contact be made with the testing personnel inside the central office?

11-38. What are the major advantages/disadvantages of a digital central office?

11-39. What differences are there between the PCM systems used in a digital office compared with those used outside?

11-40. How many DS-1 lines would be required in six microwave radio sites operating in the DRS-8 digital radio mode?

Data Communications Techniques

Tom McGovern

12-1 INTRODUCTION

Telecommunications may be defined as the process of sending and receiving information. The principal sources of the information may be people, computers, or cameras. The telephone system was created for the transmission of voice, but because of the cost of building communication links on a national and international scale, the telephone system has been adapted for data communication between computers and terminals. Although voice is the major form of communication between business organizations, there is a growing need for data communication within and between organizations. Data processing has become a central controlling function within business organizations. The distribution of this information, both locally and to remote geographical locations, is a prime consideration in the efficient functioning of any business. In most organizations data processing leads to a requirement for data transmission. The future of information communication lies in the integration of facilities for voice, data, and pictures for facsimile transmission, graphics, freeze-frame television, and video conferencing.

12-1.1 Data Communications Networks

The word *network* is a broad term similar in meaning to *system*. Both are used to describe organized complexity. Data communications networks can be defined as a collection of computers, terminals, and related devices connected by communications facilities. Data communication networks are concerned principally with the transfer of information between computers and terminals. As the number of terminals and computers in a network increases, problems concerned with the design, efficiency, cost, and control of such networks become paramount.

To begin with, a network may be described as a group of nodes interconnected by media. These nodes may be terminations or junctions. *Terminations* are devices, including computers and computer terminals, which are connected to the network for the purpose of sending or receiving information. *Junctions* are devices that perform such functions as completing a link between two terminals (switching), deciding on which order the various information transfers should take place (scheduling), and allocating temporary storage of information along the route of its travel (buffering). The connecting medium may come in the form of an electrical conductor, a light-beam channel, or even free space. Figure 12-1 illustrates the basic components of a network and the interrelationship of the nodes.

Networks may be described in many different ways: for example, in terms of their components (terminals, personal computers, mainframe computers, multiplexers, front-end processors), their connections (bus, tree, ring, multipoint, packet switching), and in terms of their protocol (X.25, CSMA/CD, token passing).

In this chapter we discuss all of the foregoing facets of networks, as well as the major network types (wide area networks, local area networks, and public data networks) and the main data communication applications (distributed data processing, distributed data base, micro/mainframe links, and system network architectures).

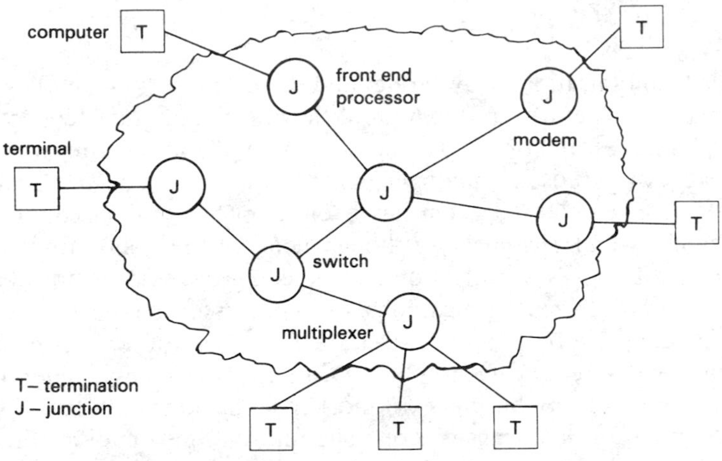

Figure 12-1 Network nodes.

12-2 FUNDAMENTALS

There are several reasons for establishing a common vocabulary. First, it is important to establish good communication between people as well as machines. Data processing people find it necessary to converse with technical personnel, read and evaluate performance specifications for new hardware, use communications software ranging from simple terminal emulation to sophisticated multilevel protocols, as well as use a variety of unfamiliar data bases using unfamiliar networks. For these reasons it is essential to become familiar with the basic data communication terms and concepts from a descriptive and functional viewpoint rather than mathematically or electrically.

Data transmission involves the production, movement, and delivery of information over a transmission medium between a transmitter and a receiver (Fig. 12-2) and includes consideration of (1) the type of signal or information to be sent, (2) conversion of data to a form suitable for the medium, (3) the quality of information received, (4) the number of users, and (5) the cost of transmission.

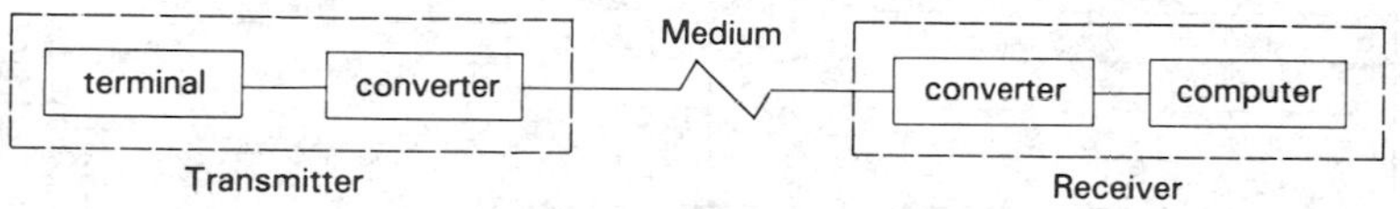

Figure 12-2 Data transmission.

12-2.1 Codes

Information is converted by terminals and computers into digital pulses. Codes are used to give meaning to a group of pulses (1's and 0's); for example in one code the letter 'A' is represented by 1000001.

Historically, codes developed in two mainstreams: telecommunications and computers. In telecommunications, the Morse code and the Baudot code preceded the now internationally accepted American Standard Code for the Interchange of Information (ASCII). In the early days of computers, each manufacturer developed their own code, but soon this was realized to be contrary to the best interests of the manufacturer, the user, and the computer industry as a whole. Because of the predominance of IBM, its favorite, the Extended Binary-Coded Decimal Information Code (EBCDIC) emerged as a computer industry standard. The structures of the Baudot, ASCII, and EBCDIC codes will now be discussed.

The Baudot Code (Fig. 12-3)

The Baudot code is a five-bit code used mainly for telegraphy and for some keyboards, printers, and readers. Although five bits can accommodate only 32 unique codes, two of the codes, "figures" (FIGS) and "letters" (LTRS), extend the capacity of the Baudot code. Preceding the other bit combinations by either FIGS or LTRS permits dual definition of the remaining codes. In other words, each bit combination could represent either a particular letter or a particular number or special character, depending on whether a LTRS character or a FIGS character precedes it. A substantial amount of international data communications traffic uses the Baudot code.

CHARACTER		BIT POSITION				
LOWER CASE	UPPER CASE	1	2	3	4	5
A	.	•	•			
B	?	•			•	•
C	:		•	•	•	
D	S	•			•	
E	3	•				
F	!	•		•	•	
G	8		•		•	•
H	•			•		•
I	8		•	•		
J	'	•	•		•	
K	(	•	•	•	•	
L	)		•			•
M	.			•	•	•
N	,			•	•	
O	9				•	•
P	0		•	•		•
Q	1	•	•	•		•
R	4		•		•	
S	BELL	•		•		
T	5					•
U	7	•	•	•		
V	;		•	•	•	•
W	2	•	•			•
X	/	•		•	•	•
Y	6	•		•		•
Z	"	•				•
LETTERS (SHIFT TO LOWER CASE)		•	•	•	•	•
FIGURES (SHIFT TO UPPER CASE)		•	•		•	•
SPACE				•		
CARRIAGE RETURN					•	
LINE FEED			•			
BLANK						
PRESENCE OF • INDICATES MARK BIT ABSENCE OF • INDICATES SPACE BIT						

Figure 12-3 Baudot code.

MOST SIGNIFICANT DIGIT (HEX)

LEAST SIGNIFICANT DIGIT (HEX)	0	1	2	3	4	5	6	7
0	NUL	DLE	SP	0	@	P	`	p
1	SOX	DC1	!	1	A	Q	a	q
2	STX	DC2	"	2	B	R	b	r
3	ETX	DC3	#	3	C	S	c	s
4	EOT	DC4	$	4	D	T	d	t
5	ENQ	NAK	%	5	E	U	e	u
6	ACK	SYN	&	6	F	V	f	v
7	BEL	ETB	'	7	G	W	g	w
8	BS	CAN	(	8	H	X	h	x
9	HT	EM	)	9	I	Y	i	y
A	LF	SUB	•	:	J	Z	j	z
B	VT	ESC	+	;	K	[	k	{
C	FF	FS	,	<	L	\	l	|
D	CR	GS	-	=	M	]	m	}
E	SO	RS	.	>	N	^	n	~
F	SI	US	/	?	O	–	o	DEL

Figure 12-4 ASCII code.

The ASCII Code (Fig. 12-4)

The ASCII code is a seven-bit-plus-parity code. Most communication terminals on the market today are designed to conform to ASCII format. Some manufacturers, however, use minor variations of the ASCII code to make it more applicable to special-purpose terminals. For example, a card reader/line printer terminal that communicates using asynchronous transmission might use the SYN code (normally used for establishing synchronization during synchronous transmission) as a control character to indicate that the printer is out of paper. The parity bit is another change that many manufacturers make. Table 12-1 identifies the purpose of the control characters. There are three categories of control character: communication control (CC), format effector (FE), and information separator (IS). We shall be interested primarily in the communication control characters.

TABLE 12-1 Control Characters

Code	Mnemonic	Meaning	Category
0	NUL	Null character	
1	SOH	Start of header	CC
2	STX	Start of text	CC
3	ETX	End of text	CC
4	EOT	End of transmission	CC
5	ENQ	Enquiry	CC
6	ACK	Acknowledge	CC
7	BEL	Bell	
8	BS	Backspace	FE
9	HT	Horizontal tabulation	FE
10	LF	Line feed	FE
11	VT	Vertical tabulation	FE
12	FF	Form feed	FE
13	CR	Carriage return	FE
14	SO	Shift out	
15	SI	Shift in	
16	DLE	Data link escape	CC
17	DC1	Device control 1 (XON)	CC
18	DC2	Device control 2	
19	DC3	Device control 3 (XOFF)	CC
20	DC4	Device control 4	
21	NAK	Negative acknowledge	CC
22	SYN	Synchronization character	CC
23	ETB	End of transmission block	CC
24	CAN	Cancel	
25	EM	End of medium	
26	SUB	Substitute	
27	ESC	Escape	
28	FS	File separator	IS
29	GS	Group separator	IS
30	RS	Record separator	IS
31	US	Unit separator	IS
127	DEL	Delete	

The EBCDIC Code (Fig. 12-5)

IBM's EBCDIC code has its roots in data processing. It began when Herman Hollerith created the 12-bit Hollerith code for punched cards before computers were invented. The computer version of this was the binary-coded decimal (BCD) code, a six-bit compression of the Hollerith code. The EBCDIC code is the BCD code extended to eight bits. The extra capacity is used for additional control codes as well as providing spare capacity for future expansion. The main problem with this code for communications purposes is the lack of a parity bit for error checking. The solutions are to convert back and forth between ASCII and EBCDIC, and in the IBM world, to incorporate other forms of error checking.

LEAST SIGNIFICANT DIGIT (HEX)

MOST SIGNIFICANT DIGIT (HEX)	0	1	2	3	4	5	6	7	8	9	A	B	C	D	E	F
0	NUL	SOH	STX	ETX	PF	HT	LC	DEL			SMM	VT	FF	CR	SO	SI
1	DLE	DC_1	DC_2	DC_3	RES	NL	BS	IL	CAN	EM	CC		IFS	IGS	IRS	IUS
2	DS	SOS	FS		BYP	LF	EOB	PRE			SM			ENQ	ACK	BEL
3			SYN		PN	RS	UC	EOT					DC_4	NAK		SUB
4	SP										¢	.	<	(	+	|
5	&										!	$	•	)	;	¬
6	-	/										,	%	–	>	?
7											:	#	@	'	=	"
8		a	b	c	d	e	f	g	h	i						
9		j	k	l	m	n	o	p	q	r						
A			s	t	u	v	w	x	y	z						
B																
C		A	B	C	D	E	F	G	H	I						
D		J	K	L	M	N	O	P	Q	R						
E			S	T	U	V	W	X	Y	Z						
F	0	1	2	3	4	5	6	7	8	9						

Figure 12-5 EBCDIC code.

12-2.2 Computer–Communications Interface

Data move within a computer along parallel paths. The number of parallel lines is governed by the computer's architecture and is typically 16 for microcomputers and 32 for mainframe computers. The external communication path is normally a single line, so that an interface is required to convert between the two systems. In addition, there are two ways of transmitting computer information between computers. Asynchronous communication is common between micros and mainframes, and synchronous communication is more common between mainframes.

Parallel Transmission

To transmit a code character, the bits can be transmitted in either parallel or serial fashion. In parallel data transmission, all code elements are transmitted simultaneously. That means that for the five-level Baudot code, five pairs of lines must interconnect the receiver and transmitter (Fig. 12-6). Parallel transmission is only used over short distances and is therefore restricted to communication between devices in the computer center: for example, between the processor and the printer.

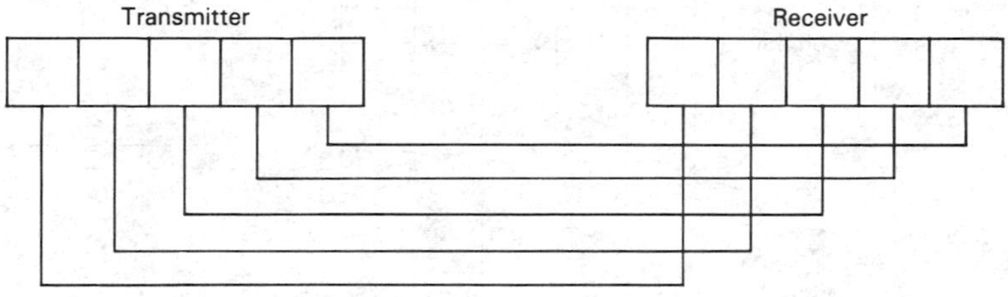

Figure 12-6 Parallel data transmission.

Serial Transmission

In serial data transmission each code element is sent sequentially (Fig. 12-7). This requires only one pair of wire conductors for interconnecting the receiver to the transmitter. Serial transmission is the basis for network communication. Whether the network nodes are geographically remote, as in wide area networks, or quite close together, as in local area networks, serial communication is obviously more economical although slower than parallel transmission.

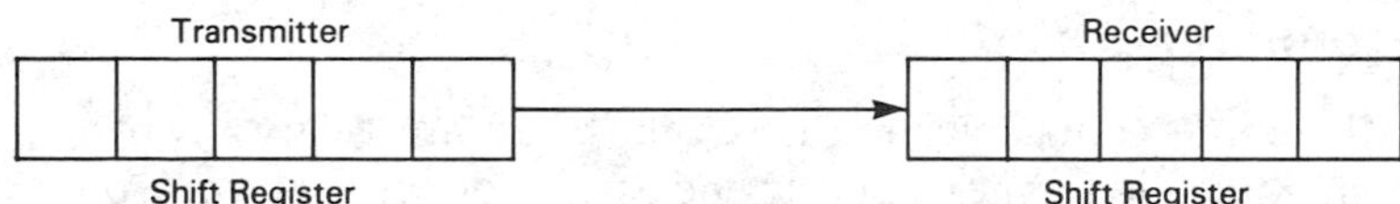

Figure 12-7 Serial data transmission.

Parallel/Serial Transmission

If we can visualize characters moving around the computer on an eight-lane highway which must be reduced to a single-lane bridge to reach another computer or a terminal, we can see the need for a device that acts as a traffic policeman. This device must convert the parallel traffic to serial for external transmission and convert the incoming serial traffic to parallel for computer processing. Such a device is the universal synchronous/asynchronous receiver transmitter (USART; Fig. 12-8). The USART is available as an integrated circuit or *chip*, which permits the selection of a number of parameters, including baud rate and modem control.

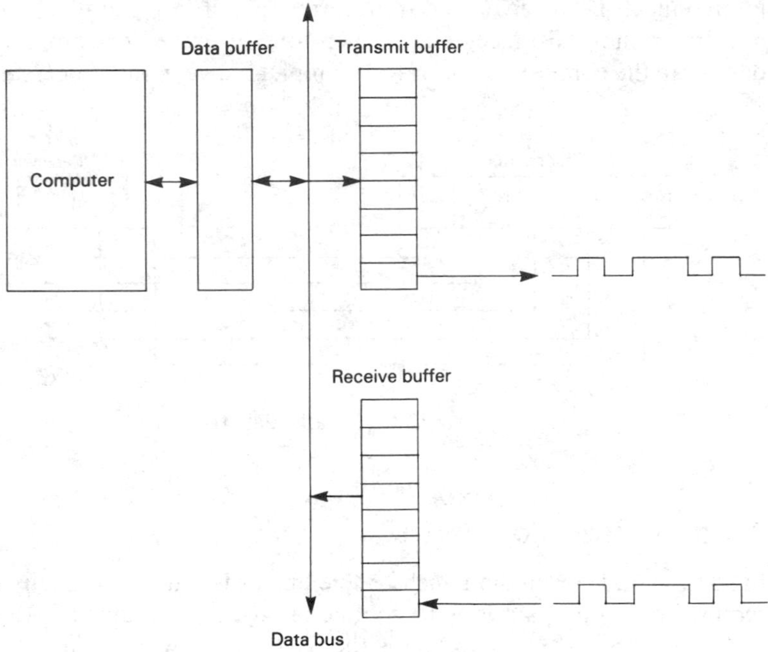

Figure 12-8 USART.

Synchronous and Asynchronous Serial Communication

All remote communication is serial. There are two ways of implementing serial communication: synchronously and asynchronously. *Asynchronous communication* is the simpler of the two, in that each character is transmitted independently of all other characters. In order to separate the characters and permit the receiving device to recognize them correctly, a start bit precedes the character and a stop bit follows it. Information can be transmitted at the speed of a typist or, in the case of micro to mainframe, at speeds up to 56,000 bits per second (56 kbits/s). As an example of asychronous serial transmission, consider the transmission of the letter E in the ASCII code (Fig. 12-9). The data bits of each character are preceded by a start bit, which is always a zero, and followed by an optional parity bit and one or more stop bits.

In the idle condition the transmission remains in the 1 state. As soon as a zero is detected, the receiver begins to detect the incoming character. After seven bits (or eight bits if there is a parity bit), a stop bit should be detected and the

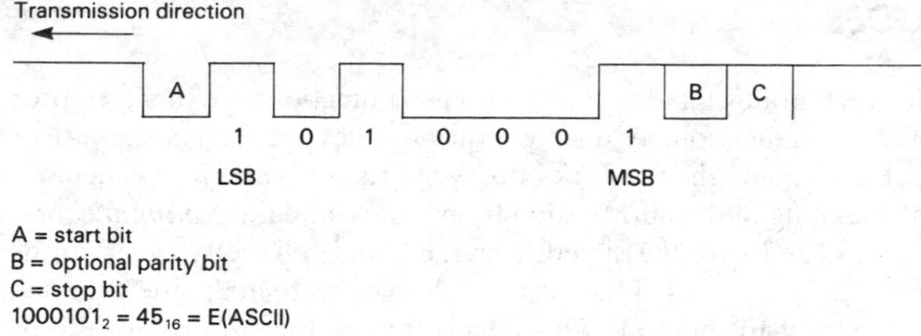

Figure 12-9 Asynchronous transmission of an "E."

receiver is ready for the next character. Because the start and stop bits take up a significant amount of time, the information throughput is less than for the case where they are not present.

Synchronous transmission (Fig. 12-10) is used for the high-speed transmission of a block of characters. The sending and receiving modems must operate in conjunction with one another (i.e., they use a common clock pulse at both the transmitting and receiving ends). Start and stop bits for each character are not required. Synchronization is established by preceding the data block with SYN control characters. The block is also terminated by SYN characters.

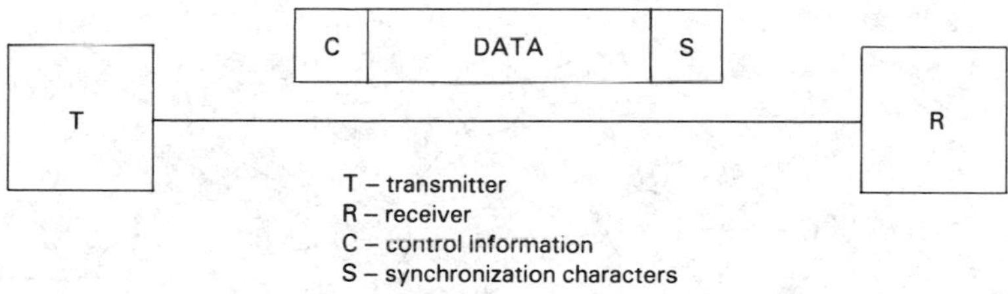

Figure 12-10 Synchronous transmission.

Interface Standards

Early in the history of planning data communication networks, the Electronic Industries Association (EIA) accepted a set of standards for the interconnection of DTEs (data termination equipment) or terminals and DCEs (data communication equipment) or modems, known as the EIA RS-232-C specification. This standard contains (1) the electrical signal characteristics, (2) the interface mechanical characteristics, and (3) a minimal amount of control information.

New standards continue to appear as both computer and telecommunication technology advance. Any attempt to upgrade operational equipment to a new standard is seriously undermined by the costs involved. For this reason and because it has been so successful, the RS-232-C specification will continue to be a major communication standard.

12-3 NODES

Network nodes may be terminations or junctions. A junction provides the facility for communication. A termination provides the user access to the network. It is useful to divide the network into two parts: the *computer network*, which interprets the meaning of the information transferred, and the *communication network*, which is concerned with the speed, accuracy, and reliability of the transmission process (Fig. 12-11). A goal of the network designer is to make the communication network transparent to the user. This means that neither the route, the distance traveled, nor the method of handling the data affect the user's ability to process the data. The computer network contains the terminations.

Terminations are user network nodes. They include terminals, personal computers (PC), minicomputers, and mainframes. They may be classified by their degree of intelligence. The primary interface between a termination and the rest of the network requires special consideration. Networks may be constructed privately for a particular user. Most often, a large number of users share a public network. The most common of these is the packet-switching network. Some terminations require an extra interface for use on packet-switching networks. PCs are converted to terminals by software or by a combination of hardware and software. This is referred to as *terminal emulation*.

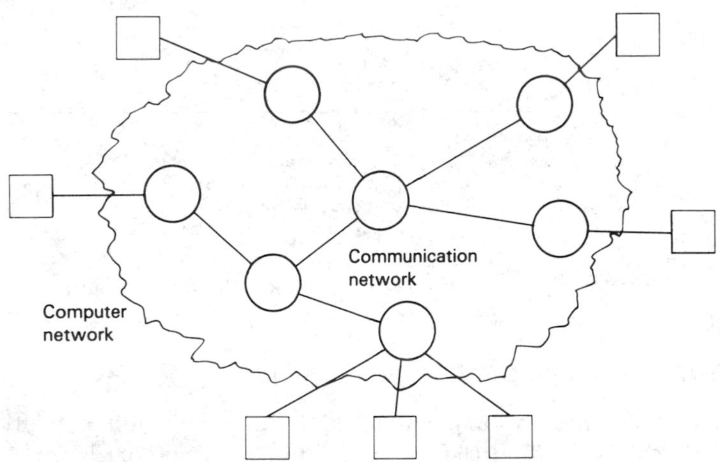

Figure 12-11 Network parts.

12-3.1 Terminations

A *termination* is a node with one connection to one network though it may be used to access other networks at different times. A termination may be a terminal used to transmit instructions or data to a remote computer; a terminal that receives output from a remote computer; a PC operating as a terminal; or a microcomputer, minicomputer, or mainframe computer. Major factors that influence the selection of terminations are method of data collection, form and quality of output, processing power, transmission rates, and price. Terminations may be classified in various ways. The approach that has maximum implication for the network is the processing power or "degree of intelligence." This determines the level of sophistication of equipment used to connect the termination to the network. An equally important consideration is the relationship between the intelligence of the terminal and the trade-off between processing costs and communication costs.

Data Termination Equipment

A major network interface occurs at the boundary between the computer network and the communication network. This is described as the interface between the termination or DTE and the modem or data communication equipment (DCE). The modem is a special node in the network that handles the physical conversion of the data for transmission (modulation). An example of this interface was referred to as the RS-232-C specification. It should be noted here that the DTE/DCE interface is also found within the communication network between modems and other junctions. This is discussed in Section 12-3.2.

Packet Assembly/Disassembly Modules (PAD)

Networks must be controlled. Control requires the establishment of a set of rules called a *protocol*. Numerous wide area networks have been developed in recent years with a variety of protocols. There has been a slow but discernible gravitation toward an international standard for protocol, based on a recommendation by the Consultative Committee for International Telephony and Telegraphy (CCITT), called the *X.25 protocol*. X.25 provides a set of guidelines that describe the physical interface and logical interface between the DTE and the DCE as well as the movement of data through the network. The X.25 specification requires that the terminal use point-to-point synchronous circuits. This requirement is not severe since most terminal manufacturers include it as a standard or as an option.

Terminations not capable of directly interfacing to a network using X.25, such as the simple and preprogrammed terminals, may be connected to a network through a PAD. A *PAD* is a shared communications controller which enables nonuser-programmable terminals access through synchronous or asynchronous connection. X.3 and X.29 are recommendations defined by the CCITT which specify the standards for PADs.

Terminal Emulation

Self-reconfiguring terminals have the ability to behave as a variety of different terminals. They do this by means of programs stored in EEPROMs. This feature may be termed *hardware emulation*. User programmable terminals are microcomputers, which may also be made to behave like a variety of different terminals.

They accomplish this by means of a program called a terminal emulator package. This feature may be termed *software emulation*. Software emulators provide many additional features, including the ability to up-load and down-load files to and from a mainframe computer.

Terminals

Ever since the appearance on the data processing scene of the microcomputer, observers have been predicting the demise of the terminal. Since the price of microcomputers has been falling rapidly while their performance has been improving dramatically, including their ability to emulate any popular terminal on the market, it is not surprising that users requiring personal computers for their workplace also appreciate their potential as terminals. In reaction to this, terminal manufacturers have begun to offer competition by incorporating microprocessors in their terminals. This allows them to decrease prices while increasing flexibility and functionality. Our classification of terminals as simple, preprogrammed, and self-reconfiguring is being obscured by the incorporation of features from sophisticated terminals into simple terminals. Developments include:

1. Special graphics
2. Multinational character sets
3. Standard sophisticated editing facilities, including:
   (a) Character and line insertion and deletion
   (b) Smooth and incremental scroll
   (c) Paging modes
   (d) Protected fields
   (e) Tab functions
   (f) Split screens (windows)
4. Low-profile, detachable keyboards
5. Nonglare screen
6. Tilt/swivel monitors
7. 60-Hz refresh rates
8. CRT saver function
9. User-friendly menu-driven setup procedures

A comparison of specific display, edit, and communication parameters for current typical self-reconfiguring terminals provides a clear indication of the power and versatility of these machines (Fig. 12-12).

The competition for market share between terminals and microcomputers is not over. As usual, success will depend not only on price and performance but also on market strategy.

12-3.2 Junctions

Network junctions provide an interface between network terminations, usually between terminals and computers, but often between computers. Two kinds of junction will be discussed here: modems and control nodes.

Parameters	CIT-101e	DT-1	WY-75	VC4604
DISPLAY				
screen diagonal (in)	12	12	14	12
tilt/swivel	×	×	×	×
24 lines × 80 columns	×	×	×	×
24 lines × 132 columns	×	×	×	×
matrix	7 × 9	-	7 × 13	7 × 10
ASCII characters	96	96	128	128
EDITING				
autorepeat	×	×	×	×
scroll/page	×	×	×	×
split screen	×	×	×	×
cursor control	×	×	×	×
blink	×	×	×	×
half/full intensity	×	×	×	×
reverse video	×	×	×	×
underscore	×	×	×	×
character/line insert/delete	-	×	×	×
protected fields	-	×	×	-
COMMUNICATIONS				
max. speed	19.2 kbps	s	s	s
half-duplex	×	×	×	×
full-duplex	×	×	×	×
variable parity	×	-	-	×
variable stop bits	-	-	×	-
RS-232-C	×	×	×	×

× available - unavailable s same

Figure 12-12 Comparison of performance characteristics.

Modems

The *modem* is used by data networks to convert data signals into signals that can be carried by the telephone system, and vice versa. It provides the terminal with its initial connection to the network, modulates and demodulates the data signals going to and coming from the network, and supplies the appropriate control signals to effect data transmission.

Control Nodes

Control nodes are responsible for line control, message buffering, and error detection, as well as optimizing the efficiency of the network. Each node has its own function related to its position in the network. This category of junction includes multiplexers, concentrators, front-end processors, and switches.

Multiplexers and *concentrators* are used by groups of remote terminals to provide a cost-effective way of transmitting and receiving data via a shared high-speed line.

Front-end processors are used to off-load the communications-handling chores from a mainframe computer. The function of the mainframe computer is to run applications programs that help to solve a wide variety of business problems. Direct interaction with a network could downgrade the mainframe's ability to handle its application work load.

Switches act as traffic police in networks by providing a variety of routes between terminations. They have the intelligence to schedule transfers and optimize routings.

Because of the impact of IBM on computers, data processing, and therefore data networks, it is important to note IBM's unique terminology. Front-end processors are referred to as *communications controllers* and concentrators are *cluster controllers*. Because IBM networks operate principally in synchronous mode, while microcomputers and many terminals are asynchronous devices, the protocol converter is a junction used to connect asynchronous terminals and microcomputers to IBM networks.

Public networks require monitoring and control of the complete network, which is often being shared by many unrelated users. This overall supervision of the network is performed by the *network control center* (NCC).

Like terminals, junctions may be classified in terms of their level of sophistication. They range from simple physical switches to full-fledged computer systems with specialized instruction sets dedicated to communications handling.

Network junctions connect computers and terminals efficiently. Modems transmit data accurately and efficiently. The multiplexer, the concentrator, and the switch improve efficiency by providing line sharing. The front-end processor permits the host processor to focus on data processing. The network control center is responsible for the overall performance of the network. Perhaps the most significant development in communication network nodes parallels that of terminal development, namely, that the functional demarcation lines of different kinds of equipment are becoming less well defined. We may consider that a communications processor is a general-purpose computer adapted to perform as a front-end processor, a multiplexer, a concentrator, or an intelligent switch. The adaptation consists of configuring the input–output capability and modifying the software to specific purposes. Many of the software functions are standard: for example, transmission and reception of data, buffering, message assembly and disassembly, error handling, routing, and scheduling. It may be that in the future the function of a communications processor will depend solely on its position in the network.

12-4 CONNECTIONS

Computer networks (terminations) and communication networks (junctions) are connected together to form data networks. Because the physical shape of wide area networks is dictated by geography, we can consider their development in terms of how they are interconnected to optimize performance (logical connection) and how their irregular shape has placed an emphasis on routing (physical connection). Local area networks are characterized as having simple physical shapes such as the ring network, in which each node is connected to the next node in line until a ring is formed. Since their logical interconnection is governed by these simple physical shapes, we need only consider their physical connections.

Logical connection deals with the best way to connect the nodes of a network to optimize termination interaction. As the number of terminations increases, so does the need to maximize throughput, minimize network overhead, maintain flexibility of routing, ensure data integrity and security, and of course minimize costs. Methods of logical connection of network nodes have developed from simple physical switching, through polling and interrupt systems, to a sophisticated multilayer protocol environment employing virtual circuit connections.

Topology is defined as the physical description of a network. Examination of the simple shapes a network may take helps us understand the nature of the information flow in the network. For example, a fundamental approach to problem solving is the use of hierarchies to break the problem up into manageable pieces. So too a hierarchical network design lends itself to levels of control, overall at the top, detailed and specific at the bottom.

Wide area networks are shaped by geography and tend to be highly irregular. The need for intermediate nodes depends on the number of users, the volume of data transmitted, and the flexibility of network routing. Local area networks are confined to one or more contiguous buildings and have simple topologies with a high degree of symmetry.

12-4.1 Logical Connections

The need for sophistication in protocols developed as networks grew in complexity. The first connections between terminations were made point to point (Fig. 12-13). In point-to-point connections, each termination is connected to any other termination by an individual and permanent line. The obvious disadvantages of this system include cost and underutilization of lines; that is, the user pays for the line whether it is being used or not.

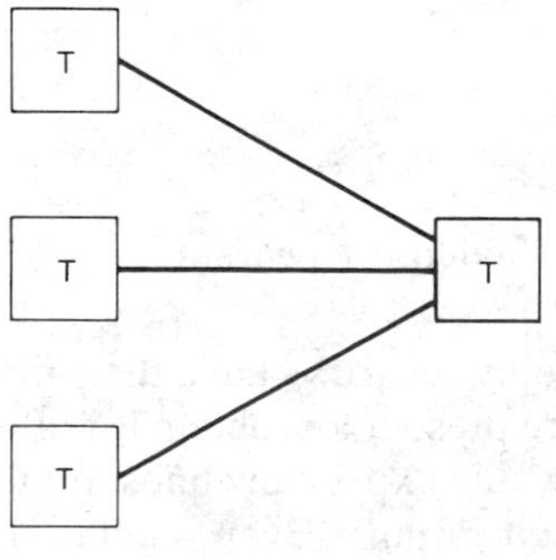

Figure 12-13 Point-to-point connection.

Circuit and Multipoint Switching

One solution found to the drawbacks of point-to-point switching was circuit switching (Fig. 12-14). In this system, a switching center is created to provide a path between two terminations for the duration of the communication. The user pays only for the time connected and for sharing the lines and switching centers with other users.

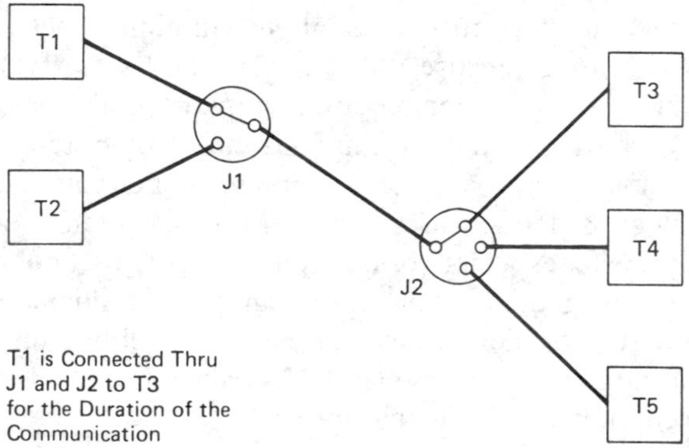

Figure 12-14 Circuit switching.

An alternative improvement over point-to-point connection is a multipoint, or multidrop, network, which facilitates a one-to-and-from-many termination connection (Fig. 12-15). One termination controls the data transfer between itself and all other terminations by means of polling. *Polling* describes the process by which the controlling node checks, in a predefined sequence, whether any of the other nodes needs to establish a communication dialogue. The other nodes in the network must wait for the completion of this dialogue before they, in turn, are polled.

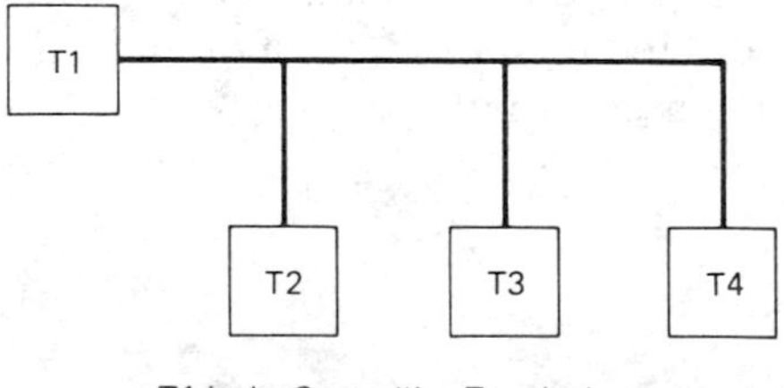

Figure 12-15 Multidrop network.

Both circuit switching and multipoint networks have their drawbacks. *Circuit switching* times can be high and therefore present a problem for interactive systems. Most users of a remote enquiry system would expect an almost immediate response, for example, in checking customer credit ratings. Even a delay of a few seconds, when experienced on a regular basis, becomes intolerable.

In *multipoint systems*, polling implies a programming overhead; that is, time must be spent checking each terminal to find out if there is a requirement for data transfer from a particular terminal. This is accomplished by the execution of a program in the controlling termination. The cost of polling is normally distributed among the terminal users. Another problem of multipoint systems is underutilization of terminals due to the fact that there is only one connecting line.

Packets

Two factors in computer communication systems that were fundamental to the development of data networks are the creation of international standards and the idea of packet switching. The establishment of standards within countries and across continents is essential to the development of such data-processing applications as electronic funds transfer. The network of the future must be accessible to a wide variety of users, and it must perform its own housekeeping functions (i.e., it must be intelligent and use its resources efficiently).

Packet switching is a refinement of message switching. *Message switching* involves transferring messages between terminations as units of information (Fig. 12-16). The message travels through the network with a header that precedes the message and identifies the destination. Subsequent messages may take entirely different routes.

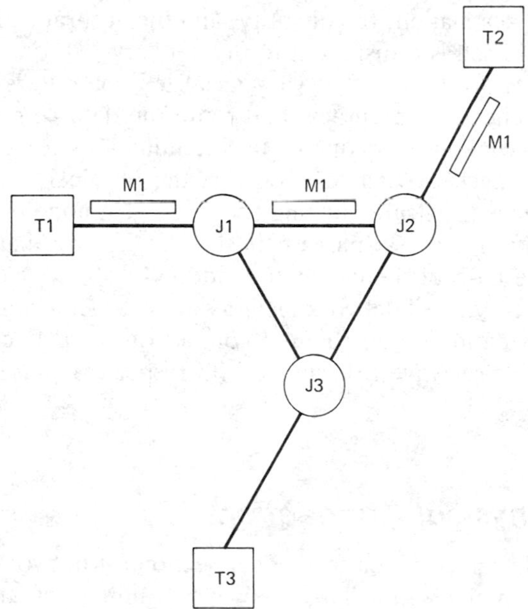
T2
M1
M1
M1
T1
J1
J2
J3
T3
Message M1 is Transmitted from Terminal T1
to Terminal T2 via Junctions J1 and J2

Figure 12-16 Message switching.

In *packet switching*, information is transferred between terminations in discrete, fixed-length units called packets. Figure 12-17 shows that each packet consists of: a header that contains control information and the destination address, and the message or data field, which is transparent to the network. A frame check sequence (FCS) is attached to the packet for error control at the local level.

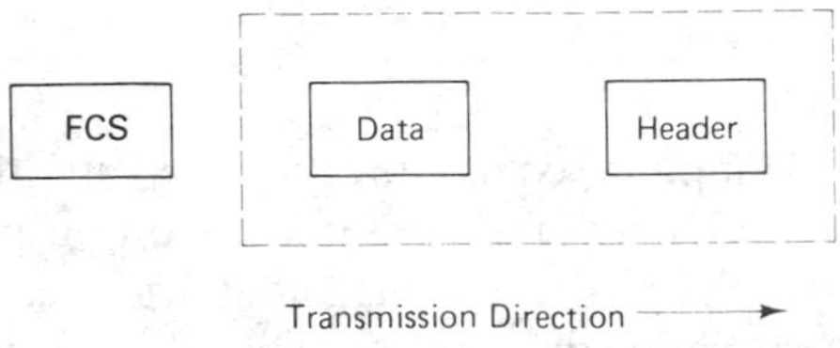

Figure 12-17 Packet.

Transparency is a term frequently used in computer systems to describe a lack of awareness. For example, the user is often unaware of how much the operating system does to satisfy a request made in a high-level language. It can be said that the operating system is transparent to the user. In this regard, it can be claimed that the network and the software that drives the network have no interest in the actual information being transmitted. The network is concerned about the source, the destination, the quantity, and the accuracy of the information, but the information itself is transparent to the network.

Packets are transmitted via specialized computers that interpret header information, check for accuracy, and route the data (Fig. 12-18). Circuit switching describes the physical linking of two terminations for the period of the communication. In packet-switched networks, the physical interconnections are permanently in place. Logical switching occurs at each node of the network by means of the destination address (part of the header). A routing decision is made by the node on the basis of the destination address.

Packet-switched network facilities are charged primarily on the basis of quantity of information rather than on connection time. Packet-switching systems are becoming less sensitive to distance with respect to charging than they were in the past.

12-4.2 Physical Connections

Topology describes the pattern of connection for network nodes. It determines the layout of communication links between terminations and junctions. Topological design algorithms must select links and link capacities on the basis of message transmission delay, cost, network traffic, and perhaps most importantly, future expandability.

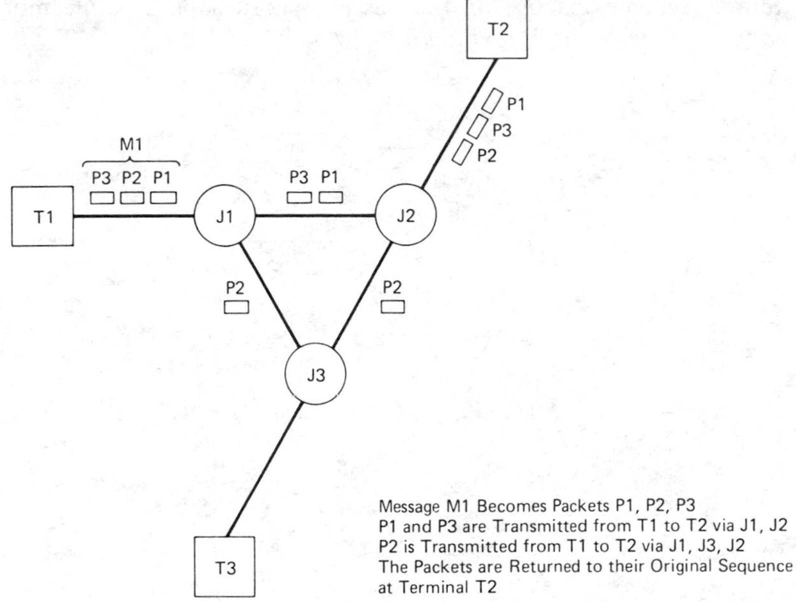
T2
P1
P3
P2
M1
P3 P2 P1
P3 P1
T1
J1
J2
P2
P2
J3
T3
Message M1 Becomes Packets P1, P2, P3
P1 and P3 are Transmitted from T1 to T2 via J1, J2
P2 is Transmitted from T1 to T2 via J1, J3, J2
The Packets are Returned to their Original Sequence
at Terminal T2

Figure 12-18 Packet switching.

Wide Area Network Topologies

A fully connected mesh network (Fig. 12-19) is one in which each junction has a direct link to every other junction and has a routing table defining the unique link to be used to reach each destination. The routing algorithm consists of a simple table search. A more realistic mesh network is called the irregular mesh (Fig. 12-20). Junction node location is governed by user requirements. Node interconnection depends on route availability, line bandwidths, and costs. Route selection is much more sophisticated than in the fully connected mesh. The public telephone system is the largest irregular mesh in the world.

In a circuit-switched network the routing algorithm operates during call setup, while the route is being selected. In a packet-switched network the algorithm may either determine individually the routing of each data packet, or set up a route for a sequence of packets. For an irregular mesh-connected packet-switching network it may not be possible to apply a routing algorithm because the topology at specific nodes may be unknown or may be subject to change. In this case, packets may be transmitted by *flooding*, where multiple copies of each packet are transmitted to all connecting nodes. An alternative is to use *random routing*, in which single copies of a packet are transmitted to a connecting node chosen at random. In modern packet networks *directory routing* is used. The destination address obtained from the packet header provides access to a routing table to determine the best route.

Simple schemes choose on the basis of shortest path, while more sophisticated schemes take network loading into account.

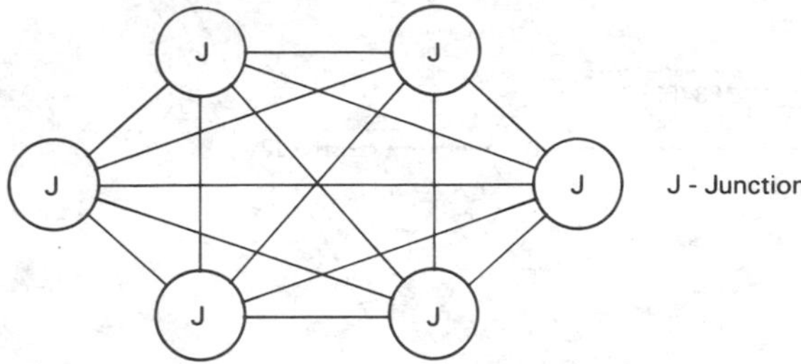

Figure 12-19 Fully connected mesh.

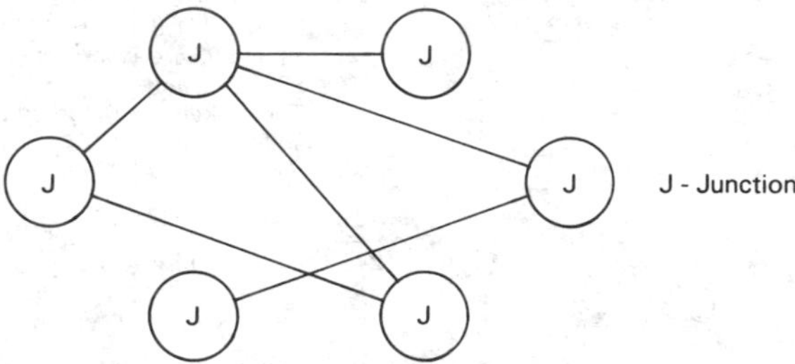

Figure 12-20 Irregular mesh.

Local Area Network Topologies

Local area networks (LANs) were created to share resources within an organization's local environment. As personal computer usage grew in organizations both for management applications and for office automation, it became clear that dramatic saving in disk storage, printing facilities, and software purchases could be made by connecting these personal computers together. The result was the creation of networks, within and between buildings, called local area networks. LAN topologies are known principally for their simplicity and symmetry. These factors, together with their restricted physical boundaries, account for their low operating system overhead and high performance.

In the *star topology*, each node is connected via a point-to-point link to a central control node (Fig. 12-21). All routing of network traffic, from the central node to outlying nodes and between outlying nodes, is performed by the central node. The routing algorithm is a simple matter of table lookup. The star formation works best when the bulk of communication is between the central node and outlying nodes. When the message volume is high among outlying nodes, the central switching feature may cause delays. The central control node is the most complex of the nodes and governs the success, size, and capacity of the network. Star networks require greater emphasis on reliability for this reason.

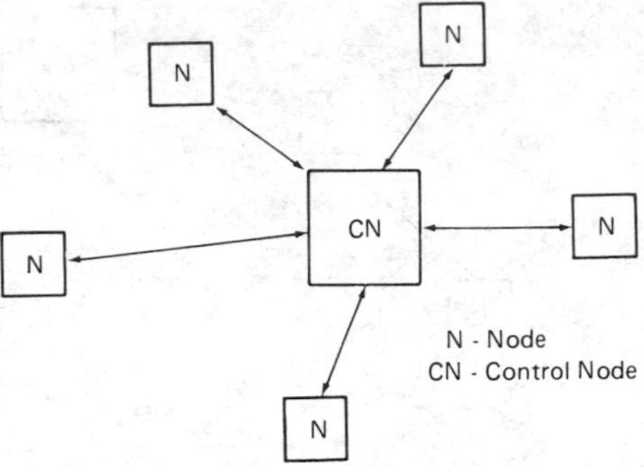

Figure 12-21 Star topology.

In the *ring topology*, each node is connected to two other nodes in a closed loop (Fig. 12-22). Transmitted messages travel from node to node round the ring in one direction. Each node must be able to recognize its own address as well as retransmit messages addressed to other nodes. Since the message route is determined by the topology, no routing algorithm is required; messages travel automatically to the next node on the ring. This is achieved by circulating a bit pattern, called a *token*, to facilitate sharing the communications channel. A node gains exclusive use of the channel by "grabbing" the token. It passes the right to access the channel on to the other nodes when it has finished transmitting. This is the basic protocol used in ring topologies. When control is distributed, each node can communicate directly with all other nodes under its own initiative.

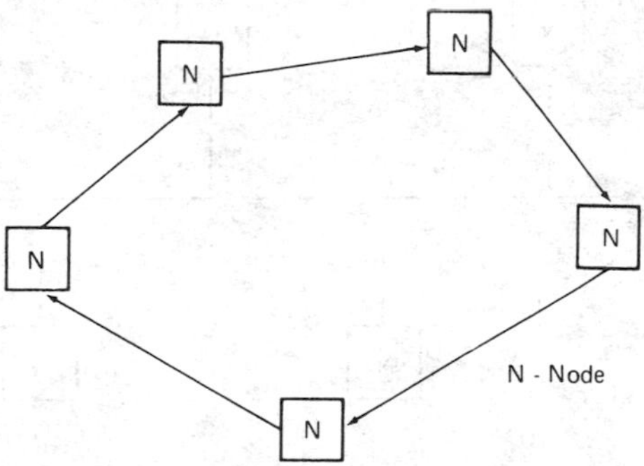

Figure 12-22 Ring topology.

Ring networks with centralized control are often referred to as *loops* (Fig. 12-23). The control node permits the other nodes to transmit messages and acts like the central control node of a star network. Since ring communication is unidirectional, failure of one node brings the network down. However, simple bypass mechanisms may be employed to minimize downtime.

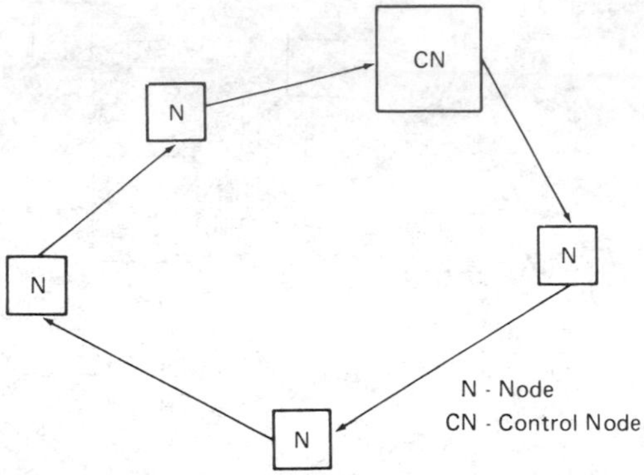

Figure 12-23 Loop topology.

In *bus topology*, unlike the star topology, no switching is required and there are no repeating messages, as they are passed round the ring. The bus is simply the cable that connects the nodes (Fig. 12-24). The message is "broadcast" by one node to all other nodes and recognized by means of an address by the receiving node. This broadcast of the message propagates throughout the length of the medium. One advantage over the ring is that delays due to repeating and forwarding the message are eliminated. Since the nodes are passive the system is fail-safe. A faulty node will not affect the other nodes. Bus networks are also easily installed and expanded.

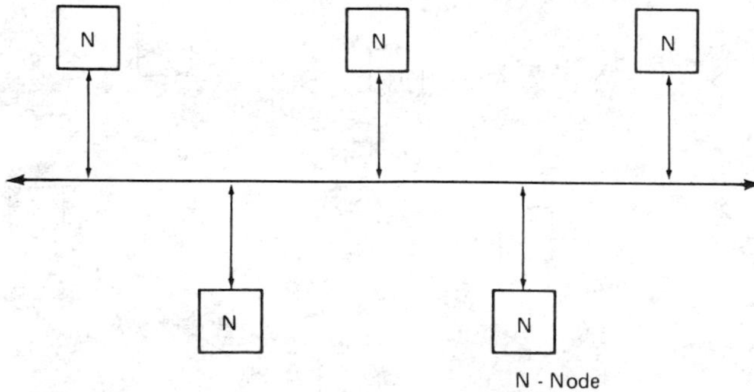

Figure 12-24 Bus topology.

A *tree* is a generalization of bus topology in which the cable branches at either or both ends, but which offers only one transmission path between any two nodes (Fig. 12-25). As with the bus, any node "broadcasts" its message, which can be picked up by any other node in the network.

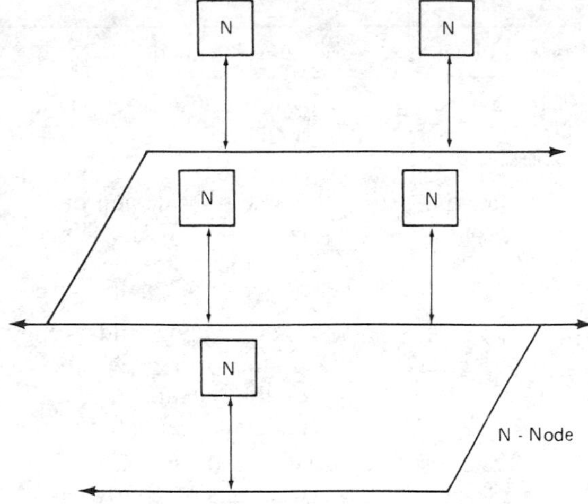

Figure 12-25 Tree topology.

12-5 PROTOCOL

As networks become more complex and wide reaching, and as the requirements for interlinking grow, the need for international protocol standards grows proportionately. A *protocol* is defined as a "set of rules." Most familiarly used in a military or diplomatic sense, it is now applied to data communication systems. There are protocols for making connections, transferring messages, and performing enough other functions that they may be considered as a hierarchy of levels or layers. The establishment of a connection between two terminals would obey the lowest level of protocol. The transfer of a file of information to solve a specific problem would follow the protocol of a higher level in the hierarchy, and so on.

Protocol may be defined in terms of its level of "closeness" to the hardware in the same way that computer languages are categorized as high or low level. The International Standards Organization (ISO) has proposed a seven-level architecture called the *Reference Model of Open Systems Interconnection*, which may be remembered by the palindromic acronym ISO-OSI (Table 12-2). The ISO-OSI is a set of guidelines, not a precisely defined standard.

12-5.1 Physical Level

Protocols at the physical level involve such parameters as the signal voltage swing and bit duration; whether transmission is simplex, half-duplex, or full-duplex; whether synchronous or asynchronous; and how connections are established at each end.

TABLE 12-2 ISO-OSI

Level	Function	Examples
1: Physical	Physical transmission via a medium	RS-232-C RS-449 X.21
2: Data link control	Reliable transmission of messages across a single data link	HDLC SDLC BISYNC ASYNC
3: Network control	Packet handling, addressing, routing, call establishment, maintenance, and clearing	X.25
4: Transport level	End-to-end control, transmission control, and efficiency	
5: Session control	System-to-system control, session monitoring, and management	
6: Presentation control	Library routines, encryption, compaction, and code conversion	
7: Application	Network management and network transparency	

To interconnect data termination equipment (DTE) such as a computer, terminal, word processor, or printer to data communication equipment (DCE), commonly known as a modem or data set, an interface standard is required. The EIA RS-232-C, the EIA RS-449, and the CCITT X.21 are three prominent physical level standards.

EIA RS-232-C

This standard is used extensively in North America and contains detailed specifications for a 25-pin connector in terms of electrical signal characteristics, mechanical characteristics, and a functional description of the interchange circuit specifications for particular applications.

The electrical characteristics for RS-232-C specify that in addition to the protective and signal grounds, all circuits carry bipolar voltage signals. Voltages at the connector pins with respect to the signal ground may not exceed +25 or −25 V. Any pin must be able to withstand a short circuit to any other pin without sustaining damage to the equipment. With a 3kΩ to 7kΩ load, the driven output represents a logic 0 for voltages between +5 and +15 V and a logic 1 for voltages between −5 and −15 V. A voltage at the receiver from +3 to +15 V represents a 0, while a voltage between −3 and −15 V represents a 1. Voltages between +3 and −3 V lie in the transition region and are not defined.

The most significant standards for establishing the protocol for communication between a terminal and a modem are contained in the functional description of the interchange circuits. The most important of these are:

Request to Send (RTS): an indication by the DTE to the DCE that it is ready to transmit.

Clear to Send (CTS): an indication by the DCE to the DTE that it is ready to receive and to retransmit the data.

Transmitted Data: serial data are transmitted from the DTE to the DCE to be modulated.

Received Data: demodulated data are received serially by the DTE from the DCE.

Data Carrier Detect (DCD): an indication by the remote DCE to the remote DTE that it is receiving a carrier signal.

Recall that the two main problems associated with serial transmission are the conversion between serial and parallel data and mutual synchronization. The solution to the synchronization problem leads to two major classes of communication systems: asynchronous and synchronous communication.

Asynchronous Communication

The proliferation of personal computers has increased the demand for a relatively low speed, unsophisticated method of communication between the microcomputer and a mainframe installation. The most commonly used system to date is an asynchronous full-duplex link running at 300 to 2400 bits/s.

Figure 12-26 shows the signal sequence for a half-duplex communication at level 1. The RTS initiates the carrier, which turns on DCD and CTS; then the data are transmitted. When transmission is complete, RTS is turned off, which turns off the carrier and CTS. Carrier off turns off DCD.

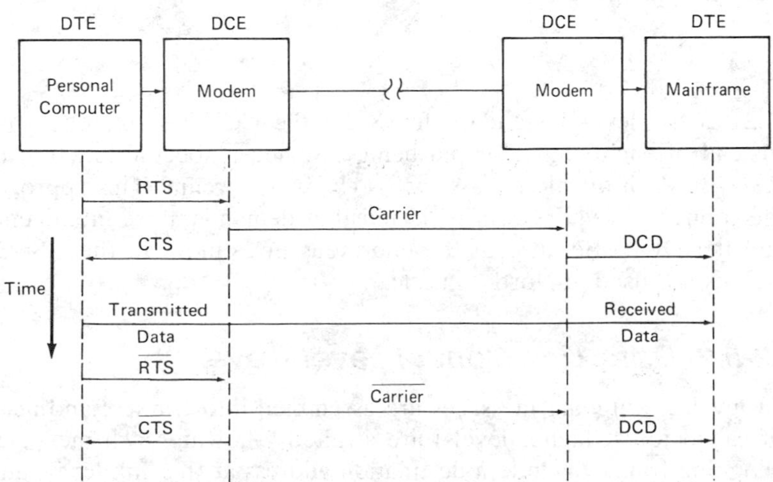

Figure 12-26 Level 1 communication.

In full-duplex mode both DTEs are transmitting and receiving, so that Transmitted Data and Received Data are active at the same time. This means that RTS and DCD are always on.

We have assumed here that there is no dialing or switching involved. On a switched line, manual or autodialing would precede these steps. On a leased line the modem power switch also initiates the communication sequence.

Synchronous Communication

For large volumes of data transmitted between mainframes, a higher transmission rate reduces line costs. Synchronous transmission is more efficient in this context, particularly in conjunction with a large number of remote sites.

At level 1, the protocol established between the DTE and the DCE is similar to that of the asynchronous system. The main difference is that the data are sent in synchronism with timing signals generated by a clock in the DCE. Synchronous modems communicate with one another to synchronize their clocks. Two additional signals on the RS-232-C are therefore required:

Transmit Timing: a signal from the DCE to the DTE to "clock" the data being transmitted from the DTE to the DCE.

Receive Timing: a signal from the DCE to the DTE to enable the incoming data to be sampled at the correct speed. This clock is derived from the carrier by the DCE.

EIA RS-449

An EIA upgrade of the RS 232-C, the RS-449 consists of a 37-pin connector containing additional functions, such as diagnostic circuits, and a 9-pin connector for secondary channel circuits. The RS-449 has been designed to accommodate a growing variety of communication needs, including increased cable length and transmission speeds.

X.21

X.21 is the level 1 standard for X.25, the CCITT's packet-switching protocol. Instead of each connector pin being assigned a specific function (RS-232-C and RS-449), each function is assigned a character stream. This approach has reduced the connector requirement to 15 pins but demands more intelligence of the DTE and the DCE. Because of the enormous investment in the RS-232-C interface, X.21 is not used in North America.

12-5.2 Data Link Control Level (Level 2)

At level 2, outgoing messages are assembled into frames, and acknowledgments (if called for at higher levels) are awaited following each message transmission. Outgoing frames include a destination address at the link level, and if the higher levels require it, a source address as well, plus a trailer containing an error-detecting or error-correcting code. The data portion of the frame is whatever comes down to this level from level 3, without reference to its significance. Correct operation at this level assures reliable transmission of each message. Data link control is the most interesting level of protocol because it occurs at the interface between hardware and software. It is the most important level of control because it includes consideration of protocols ranging from the simple microcomputer/printer interface to the most sophisticated high-speed synchronous packet-switched networks. It is the most difficult level because of the number of different facets involved and their interrelationships. A map of the data link control world would be helpful for understanding this level (Fig. 12-27 on p. 484). The major characteristics of data link control that must be considered are (1) whether the protocol is half- or full-duplex, (2) how the units of information are laid out, (3) how nodes establish and relinquish control of the line, (4) how errors may be detected and corrected, and (5) how continuous sending and receiving can be controlled.

Error Handling

Errors incurred in transmitting messages or frames or packets or blocks or any other unit of information transfer over noisy lines are usually handled by the error check character(s). The error check character(s) are calculated by the transmitter and appended to the message. The receiver repeats the calculation and compares check character(s).

Checksum This error detection method adds together the numerical values of all the characters in the block and takes the least significant n bits of the result, where n is the number of bits in the checksum character(s) (Table 12-3). This is a relatively poor method of error detection, used primarily with asynchronous half-duplex communication packages.

TABLE 12-3 Checksum

	Hex		ASCII Code
	C 43		1 0 0 0 0 1 1
	H 48		1 0 0 1 0 0 0
	E 45		1 0 0 0 1 0 1
	C 43		1 0 0 0 0 1 1
	K 4B		1 0 0 1 0 1 1
	b̸ 20		0 1 0 0 0 0 0
	S 53		1 0 1 0 0 1 1
	U 55		1 0 1 0 1 0 1
	M 4D		1 0 0 1 1 0 1
Sum		=	1 0 0 1 1 1 0 0 1 1
Checksum		=	1 1 1 0 0 1 1

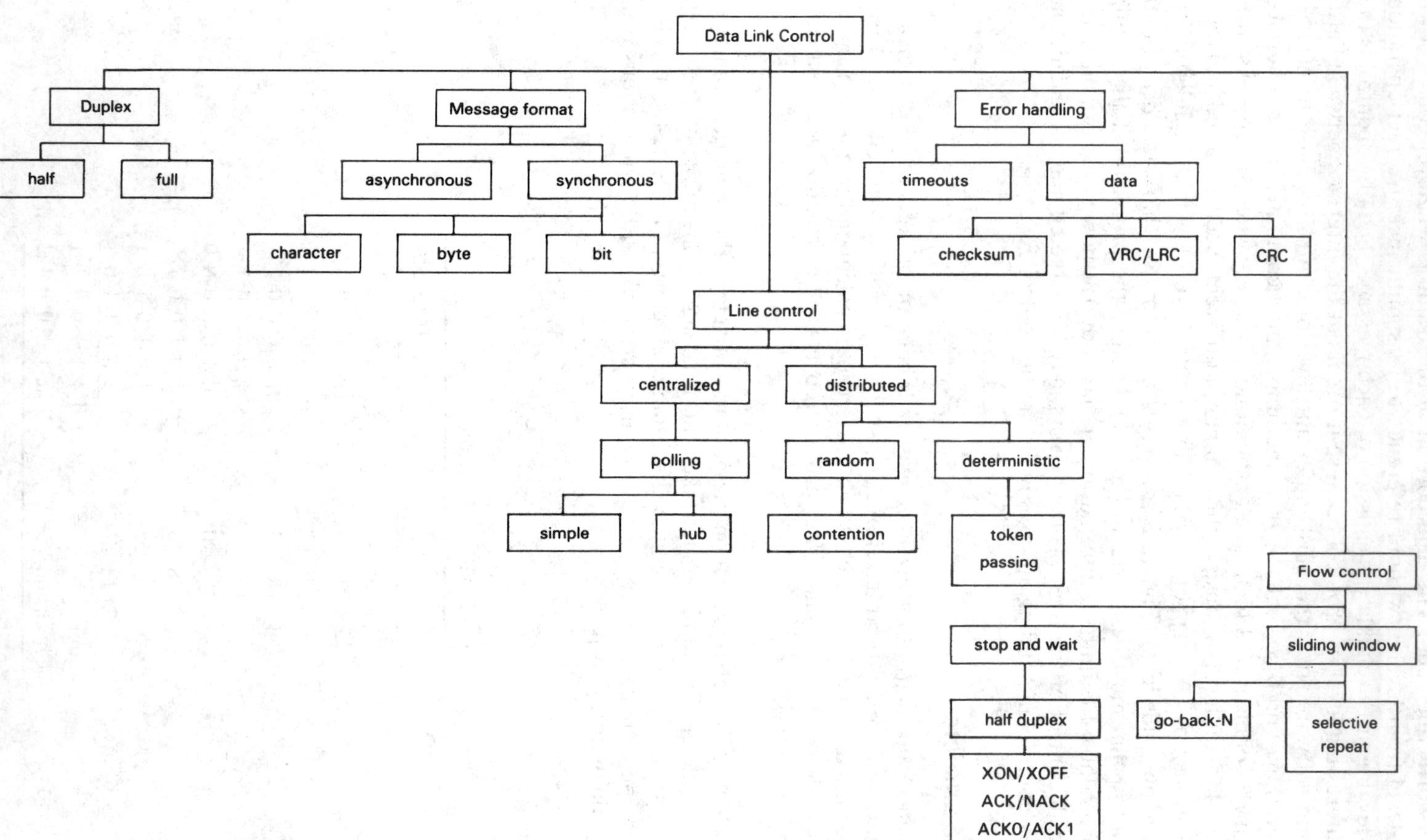

Figure 12-27 Data link control map.

VRC/LRC This error detection method is a combination of the traditional parity addition to each character transmitted (vertical redundancy check) and a horizontal parity bit generated from each row of the block (longitudinal redundancy check). VRC/LRC is used by IBM's BISYNC protocol (Table 12-4).

TABLE 12-4 VRC/LRC (Used to Create BCC for BISYNC)

Bit Position	Character							Block Check Character
	V	R	C	/	L	R	C	
1	0	0	1	1	0	0	1	0
2	1	1	1	1	0	1	1	1
3	1	0	0	1	1	0	0	0
4	0	0	0	1	1	0	0	1
5	1	1	0	0	0	1	0	0
6	0	0	0	1	0	0	0	0
7	1	1	1	0	1	1	1	1
Odd parity	1	0	0	0	0	0	0	0

CRC The cyclic redundancy check is a more accurate way of checking a block of characters than is checksum or VRC/LRC. This method divides the data stream by a preset binary number. The transmitted data are the dividend and the check data appended to the block are the remainder. CRC is used by bit-oriented protocols such as X.25. The CCITT-recommended divisor is 10001000000100001, which produces 16 bits of check data. The derivation of the check data for CRC is illustrated in Fig. 12-28 using a five-bit divisor. The information plus the number of bits in the frame check sequence (FCS) as zeros (in this example, four) is divided by the generator polynomial to produce the real FCS. At the receiving end the division of information and FCS should produce a zero remainder, signifying no errors in transmission.

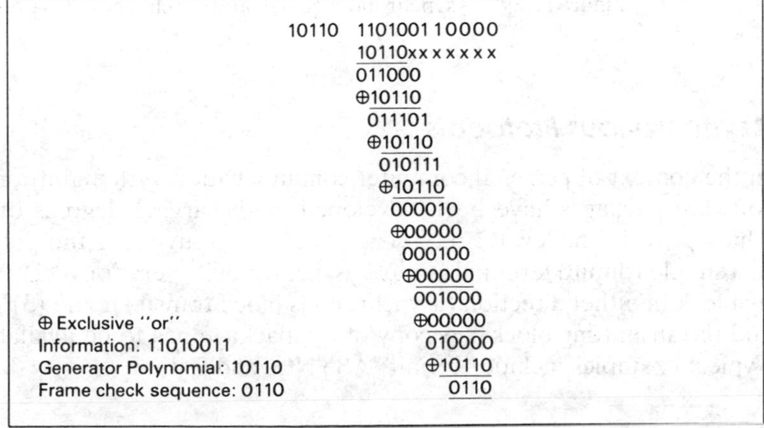

Figure 12-28 CRC (illustrated using a five-bit polynomial).

Applications

Level 2 protocol may be subdivided into two categories: synchronous and asynchronous protocols. The synchronous protocols can be further classified into character, block count, and bit protocols. The asynchronous protocols may be subdivided into the simple character-at-a-time transfers that are used to transmit information via a keyboard, and asynchronous block transfer used by asynchronous communication packages for file transfer. Asynchronous block transfer is a combination of the start/stop character used for simple asynchronous operation and the header/data/trailer layout of synchronous protocols (Fig. 12-29).

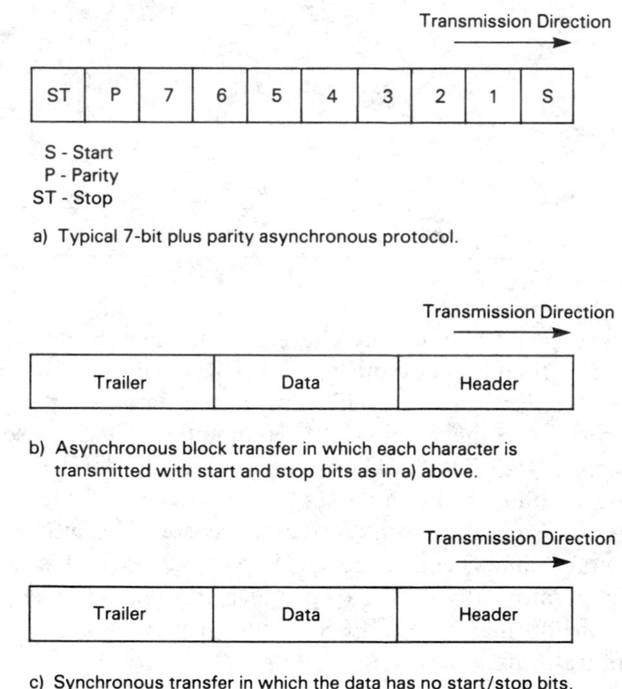

Figure 12-29 Asynchronous (a) (b) and synchronous (c) protocols.

Asynchronous Protocols

In the context of personal computer communication with mainframes, a variety of software packages have been developed, with varying degrees of sophistication, which provide the level 2 requirements for (1) converting the personal computer to a simple (dumb) terminal (simple asynchronous operation), (2) transmitting files by block in either direction (asynchronous block transfer), and (3) detecting errors and retransmitting blocks. A copy of the package has to be available at each end. Typical examples include Kermit, ASYNC, and Zstem.

Synchronous Protocols

A Character-Oriented Protocol: BISYNC IBM's Binary Synchronous Communications, a more traditional approach to protocol, is still in common use. The format of the BISYNC message is shown in Fig. 12-30.

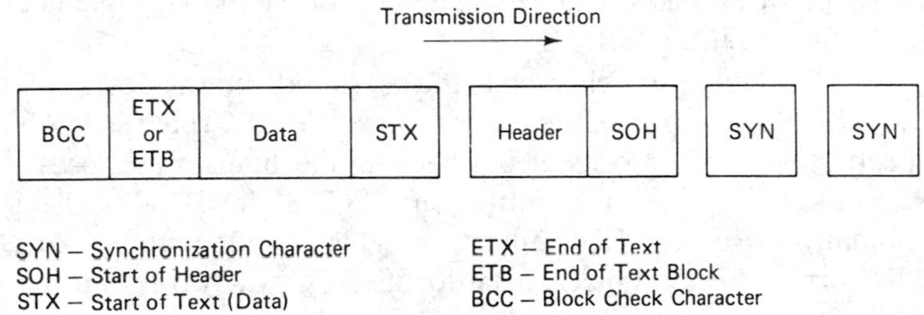

Figure 12-30 BISYNC (IBM's Binary Synchronous Communications Protocol).

SYN is a synchronization character that is used to keep the sending and receiving terminals in step. SYN is added by the sender and removed by the receiver.

SOH (Start of Header) tells the receiver that the information following is control information which relates to addressing and sequencing of user data.

STX (Start of Text) identifies the start of the user's information.

There are two kinds of information that may be transferred:

1. Characters from a standard character set, such as ASCII or EBCDIC. The user's information is then terminated by EXT (End of Text) or ETB (End of Text Block) (Fig. 12-31).

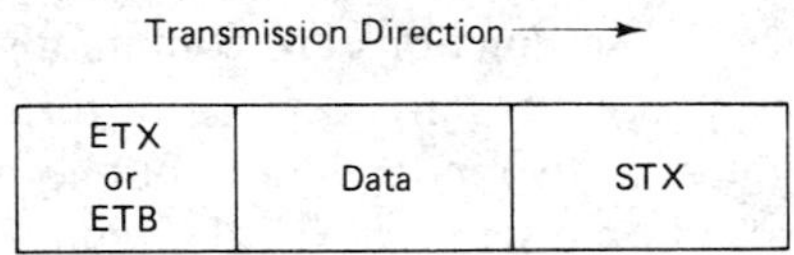

Figure 12-31 Standard character set.

2. Data that may contain control characters, from an analog-to-digital converter, for example. To prevent user data from being "recognized" by the receiver as control characters, BISYNC uses DLE (Data Link Escape) (Fig. 12-32).

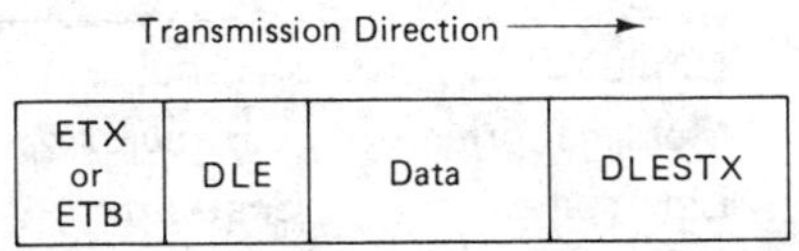

Figure 12-32 Nonstandard text.

When DLE is first encountered by the receiver, the data that follow are treated as user data; that is, they are not tested for control characters. Any control characters present on the data are "transparent" to the receiver. The only problem, of course, is that DLE is also used to signal termination of user's data. If a pattern of bits occurs on the user's data that is equivalent to a DLE, the transmitter adds a second DLE to indicate that the first one is data. The receiver removes the second occurrence of all pairs of DLEs. The end of user's data is detected by means of DLE ETX or DLE ETB.

BCC (block check character) is used for detection of errors that have occurred during the transmission of the user's data. Since ASCII is a seven-bit code, BISYNC incorporates parity into its error-checking mechanism and uses a combination of LRC and VRC when transmitting in ASCII. The BCC is the LRC part of the combination. Because EBCDIC is an eight-bit code with no parity, BISYNC uses a CRC error-checking mechanism for its BCC when transmitting EBCDIC.

A Bit-Oriented Protocol: X.25 The prime function of level 2, the data link, is the transfer of data between the user (DTE) and the network (DCE). This transfer must include control transmissions for initiating, sequencing, checking, and terminating the exchange of user information. For this level of protocol, the CCITT X.25 recommends a procedure compatible with the High-level Data Link Control (HDLC) procedure standardized by the International Standards Organization (ISO). IBM's Synchronous Data Link Control (SDLC) is a variant of HDLC used within IBM's own networks.

The control procedures adhere to the principles of a new ISO class of procedures for a point-to-point balanced system. (A balanced electrical circuit is more tolerant of electrical noise than is an unbalanced circuit. Because of this the speed and distance characteristics of balanced circuits are superior to unbalanced circuits.) The link configuration is a point-to-point channel with two stations. Each station has two functions. The primary function is to manage its own information transfer and recovery. The secondary function is to respond to requests from the other stations. In essence, these functions are commands and responses, respectively.

The data link procedures are defined in terms of commands and responses and may be thought of as two independent but complementary transmission paths that are superimposed on a single physical circuit. The DTE controls its transmissions to the DCE; the DCE controls its transmissions to the DTE.

Table 12-5 provides a summary and a comparison of BISYNC and X.25 based on the facets of data link control discussed earlier in the chapter.

TABLE 12-5 BISYNC versus X.25

Factor	BISYNC	X.25
Duplex	Half-duplex	Full-duplex
Message format	Character	Bit
Line control	Centralized	Centralized
Error handling	VRC/LRC	CRC
Flow control	ACK0/1	Go-back-N

12-5.3 Network Control Level (Level 3)

At level 3, outgoing messages are divided into packets. Incoming packets are assembled into messages for the higher levels. Routing information, included in the packet, defines the destination of the packet and indicates the order of transmission. (The packets are not necessarily received in the same order in which they were sent when a packet network is used.) The header usually includes a source address.

This level of protocol specifies the way in which users establish, maintain, and clear calls in the network. It also specifies the way in which user's data and control information are structured into packets. The relationship between the user data packet and the frame structure of the X.25 data link control protocol is shown in Fig. 12-33.

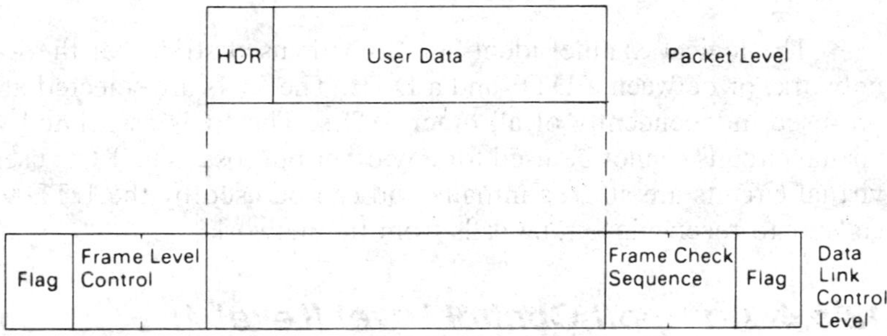

Figure 12-33 Frame packet relationship. (Reproduced courtesy of Telecom Canada.)

A fundamental concept of packet switching is that of the virtual circuit. Just as circuit switching permits a physical link between two stations, so does a virtual circuit permit a logical bidirectional association between two stations (DTEs). This logical link is assigned only when packets are being transferred. A permanent virtual circuit is a permanent association between two DTEs, much like a private line. A switched virtual circuit is a temporary association between two DTEs and is initiated when a DTE sends a call request packet to the network.

At the packet level, DTEs may establish simultaneous communication with a number of other DTEs in the network. This is accomplished in a single physical circuit by means of asynchronous time-division multiplexing (ATDM). This type of multiplexing is similar to the straightforward time-division multiplexing concept of improving the efficiency of information transfer by interleaving messages into gaps of other messages. Dynamically allocating the bandwidth to active, virtual-circuits further improves the transfer rate. The time previously allocated to non-active terminals is now used on a first-come, first-served basis.

The interleaved packets are transferred from the DTE to the DCE using the frame format of the data link control protocol. Each packet contains a logical channel identifier (LCI) that is assigned, at installation time, to a switched or permanent virtual circuit. Figure 12-34 illustrates a typical packet transfer between the DTE and the DCE.

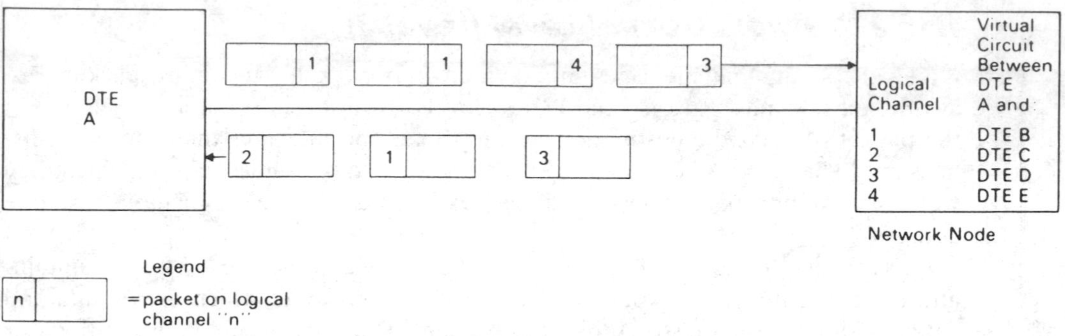

Figure 12-34 Packet flow. (Reproduced courtesy of Telecom Canada.)

The logical channel identifier (LCI) is used strictly at the local level of the network, or between a DTE and a DCE. The LCIs are selected at each DTE and are used independently of all other DTEs. The LCIs associated with permanent virtual circuits cannot be used for any other purpose. The LCIs used with switched virtual circuits are all free initially and can be used by the DTE to originate new calls or to receive incoming calls from the network.

12-5.4 Transport Control Level (Level 4)

Level 4 may be the busiest of all the architectural levels. Its protocol establishes network connections for a given transmission—for example, whether several parallel paths will be required for high throughput, whether several paths can be multiplexed onto a single connection to reduce the cost of transmission, or whether the transmission should be broadcast. This is the lowest level of strictly end-to-end communication, where the involvement or even the existence of intervening nodes is ignored.

The transport level provides the facilities that allow end users to transmit across several intervening nodes. Of the upper four levels, the transport level is the one that is relatively well defined; for example, the European Computer Manufacturers' Association ECMA-72 transport protocol standard. The services provided by this level include optimizing costs, quality of service, multiplexing, data unit size, and addressing. The main purpose of the transport level is to free the higher levels from cost-effective and reliability considerations.

12-5.5 Session Control Level (Level 5)

At this level the user establishes the system-to-system connection. It controls logging on and off, user identification and billing, and session management. For example, on a data base management system, a failure of a transmitting node during a transaction would be calamitous, because it would leave the data base in an inconsistent state. Level 5 organizes message transmissions in such a way as to minimize the probability of such a mishap—perhaps by buffering the user's inputs and sending them all in a group more quickly than they could be sent under control of a higher level.

This level provides the end users with the means of organizing and synchronizing their dialogue as well as managing their data exchange and includes (1) initiation of the session, (2) management and structuring of all session requested data transport actions, and (3) termination of the session.

12-5.6 Presentation Control Level (Level 6)

The presentation level provides for the representation of selected information. Any function that is requested often enough to justify a permanent place is held in the presentation level. Such functions include library routines, encryption, and code conversion. Three protocols are being developed for the presentation level: virtual terminal, virtual file, and job transfer and manipulation.

12-5.7 Application Control Level (Level 7)

This is the level seen by individual users. At this level network transparency is maintained, hiding the physical distribution of resources from the human user, partitioning a problem among several machines in distributed-processing applications, and providing access to distributed data bases that seem, to the user, to be concentrated in his CRT terminal.

The application level is the highest level of the ISO-OSI and is normally developed by the user. The user determines the messages used and the actions to be taken on receipt of a message. Other features include: identification of partners, establishment of authority, network management statistics, network transparency, and network monitoring.

12-6 PUBLIC NETWORKS

A major factor in the development of networks has been the public telephone system. Data network technology inherited a connection system that had been developed for voice communication. Voice signals are continuous, low frequency, and do not require a high degree of accuracy for the communication to be acceptable. Computer-generated data signals are discrete, are most efficiently transmitted at high frequencies, and must be transmitted accurately. A variety of measures may be taken to make public telephone voice-grade lines more suitable for data transmission. An alternative is to consider replacement of the telephone connection system with a more suitable one. Telephone companies are implementing both approaches.

The scale of the problem is an important consideration for both approaches. For example, in Europe a telephone exchange may contain transistor and integrated-circuit switches, as well as switches using large-scale integration and microprocessors. The interfacing, maintenance, and upgrading problems are considerable. It is somewhat ironic to consider that the highly populated highly developed countries may be hindered by their own initiative when they attempt to improve voice-grade lines for high-speed data communication. The lead in developing new international data networks comes from countries like Canada. With less commitment to older technology, vast distances to span, and a national coordinating group, Telecom Canada, Canada is in a good position to contribute to international network development. In the United States, the free-enterprise system encourages technological development through competition. At the same time, the creation of incompatible systems retards the development of national and international networks. The solution is to incorporate the standards of organizations such as the International Standards Organization (ISO) and the Consultative Committee on International Telegraphy and Telephone (CCITT) into the competitive process.

12-6.1 Network Types

An organization requiring a data communications network has a choice; it can select from a variety of leased communication facilities to build a private network or it can use a public network. Whichever choice is made, the connection between two DTEs is complicated by the inclusion of the carrier facilities (Fig. 12-35). This brings us to the heart of communication systems—shared responsibility. As networks become more complex, the problems that occur may be the responsibility of the user, the computer system supplier, or one of a number of public networks through which the data pass. Thus network management, monitoring, and control become premium considerations for all concerned.

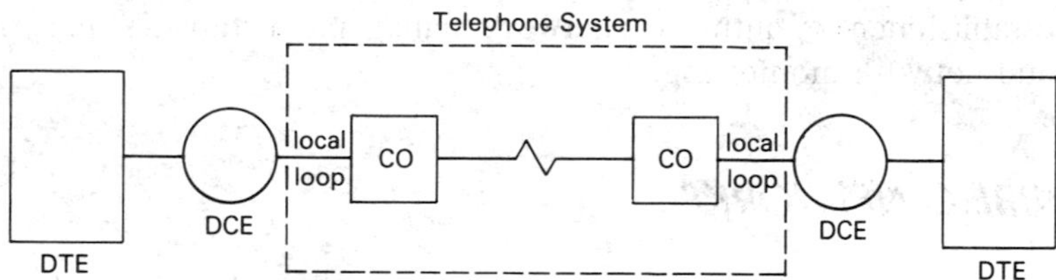

Figure 12-35 DTE to DTE using the telephone system.

As well as the standard dial-up and private analog services that may be used for data transmission, there are three main categories of public digital data networks: circuit switching, packet switching, and dedicated digital transmission. All are available from common carriers. Circuit switching is economical in low traffic volumes and batch-oriented applications. Packet switching is more appropriate for efficient, flexible, higher-volume transmission as well as interactive applications (Table 12-6). With the growth of satellite transmission and the conversion of higher-level telephone switching offices to digital switches, dedicated digital data transmission is one of the fastest-growing areas in data communications.

TABLE 12-6 Circuit Switching versus Packet Switching

	Circuit Switching	Packet Switching
1	Physical connection	Logical connection
2	Messages not stored	Messages stored
3	Statically established path for complete message	The route is established dynamically for each packet
4	Setup time	Negligible setup time, variable transit time
5	Busy signal if receiver not available	Busy message if receiver not available
6	Switching offices	Switching computers
7	Any length of message accepted	Packet length
8	Low-volume data	High-volume data
9	No speed or code conversion	Speed and code conversion
10	Fixed bandwidth	Variable bandwidth

Future Development

There are three ways in which data transmission services are improving:

1. The traditional approach is to adapt analog telephone networks for data transmission. In particular, the development of modulation techniques is significantly improving the performance characteristics of voice-grade lines.
2. The main thrust of development is toward special networks to handle data, such as packet-switching networks and dedicated digital data transmission networks.
3. The long-term solution appears to be an integrated network for voice, data, and video. The Integrated Services Digital Network (ISDN) is often referred to as "the network of the future." It is not a separate new network but will emerge from the gradual transformation and enhancement of existing networks and services. The concept of ISDN has gained wide support. Its implementation will include:
   a. A single interface between the user and the telephone company for all services
   b. A gateway to route information between the user and other networks (circuit switched, packet switched, digital) via the ISDN network
   c. The ISDN core network, digital in nature
   d. Network control for the interface, the gateway, and the core network

12-7 LOCAL AREA NETWORKS

A local area network (LAN) is a system for interconnecting data communicating components within a relatively confined space. A system is a group of interrelated parts with the focus on the interrelationship. LANs are concerned principally with methods of communication among their components. The components are mutually compatible devices and include microcomputers, disk storage, and printers. The emphasis on compatibility is purely practical. Incompatible microcomputers may be able to share a printer but not programs or data. The term "a relatively confined space" describes the distinction between LANs and wide area networks. LANs are most commonly contained within one building but may spread to contiguous buildings. Wide area networks usually operate between cities, countries, and continents. LANs are restricted to intra- and interbuilding communication, the maximum distance dependent on the medium used to connect components.

Other significant characteristics of LANs, which do not appear in the definition, include ownership, speed, and availability. LANs are privately owned and therefore not subject to regulation by public bodies or networks. This fact, coupled with the simple, symmetrical topologies used in LANs, facilitates most acceptable speeds of between 1 and 10 Mbits/s. The speed of a LAN is limited by the way the software interacts with the network and the access speed of shared storage rather than the speed of the medium that connects the nodes. Although a large number of LANs are now commercially available, the variety of technical options is limited. There are three major media, three topologies, and two protocols.

LANs are, by definition, distance limited. The access potential of a LAN node is greatly increased by gateways and by connection to other LANs through a metropolitan area network. An alternative approach to using a LAN is to incorporate data transmission facilities into a private branch exchange, the switchboard that provides telephone facilities within most organizations.

The terminal nodes of a LAN are usually either special or general-purpose microcomputers. The compatibility of these terminal nodes is crucial to the maximization of resource sharing. The junctions on a LAN are interface boards which are capable of generating and receiving control and data signals for the network. These interface boards are housed in the workstation (microcomputer) and file server (hard disk control unit). In addition, a junction box is used at specific points in the network to physically connect workstations to the network. The most significant component is the line, that is, the medium of communication. There are three popular media: twisted pair, coaxial cable, and fiber optics. The technology of LANs, as distinct from distance networks, is simple and contains a high degree of symmetry. There are three principal topologies: bus, ring, and star.

LAN protocol, the network's way of controlling traffic, is also distinguishable from wide area networks, which normally use some form of polling, in that the workstations have an equal say in controlling the traffic. There are two main protocols or access methods of LANs: carrier sense multiple access/collision detection (CSMA/CD) and token passing.

12-7.1 Protocol

The International Standards Organization's reference model of Open Systems Interconnection (ISO-OSI) may also be applied to LANs. Since LANs are concerned primarily with the transmission of information over a physical medium, only the first two levels, physical and data link control, need be addressed.

The physical level has already been defined as consisting of an interface card coupled to an appropriate medium. The data link control level may be discussed in terms of message format (the frame), error handling (CRC), flow control (buffering), and line control.

Line control may be subdivided into two categories: polling (for networks with a control node) and distributed access methods (for LANs in which each node has equal control). Two distributed access methods have emerged to dominate the LAN market: contention, in which any node has the ability to initiate transfer at any time, and token passing, in which each node must wait its turn.

Contention

The most common form of the contention protocol is carrier sense multiple access with collision detect (CSMA/CD), which is usually associated with bus/tree topology. *Carrier sense* is the ability of each node to detect traffic on the channel by "listening." Nodes will not transmit while they "hear" traffic on the channel. *Multiple access* lets any node send a message immediately upon sensing the channel is free of traffic. This eliminates the waiting that is characteristic of noncontention protocols. One problem that may arise is that because of propagation delays across the network, two nodes may detect that the channel is free at the same time, since each will not have detected the other's signal. This causes a collision. *Collision detect* is the ability of a transmitting node to "listen" while transmitting, identify a collision, abandon transmitting, wait, and try again. Figure 12-36 shows the sequence of events for this protocol. CSMA/CD is a highly efficient form of distributed access.

Token Passing

Token passing is most often used with ring topologies, although it can be applied to bus/tree topologies by assigning the nodes logical positions in an ordered sequence, with the last member followed by the first—in other words, a logical ring.

Tokens are special bit patterns that circulate from node to node around the ring when there is no traffic. A node may take possession of the token. This gives the node exclusive access to the network for transmitting its message. This technique eliminates the possibility of conflict among nodes.

For example, node A holds the token and sends a message specifying node D. All the nodes on the ring check the message as it passes. Each node is responsible for identifying and accepting messages addressed to it, as well as for repeating and

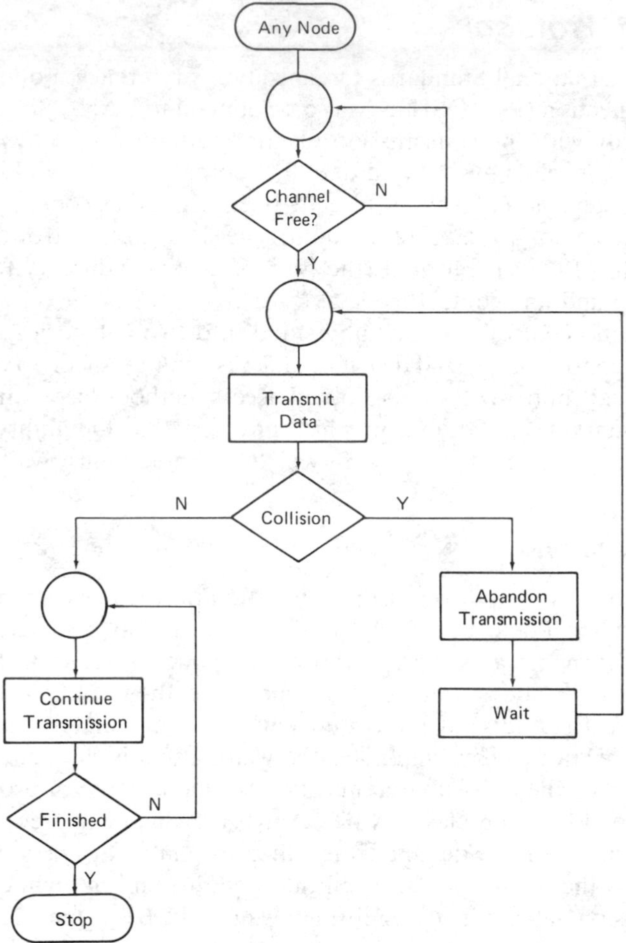

Figure 12-36 Contention protocol.

passing on messages addressed to other nodes. Node D accepts the message and sends the message to node A to confirm its receipt. When node A receives confirmation that the message arrived and was accepted, it must send the empty token round so that another node will have a chance to take over the channel. Figure 12-37 shows the sequence of events for both the transmitting and receiving nodes.

A variation of token passing, known as *slotted access*, is found exclusively in ring topology. Instead of a token, the nodes circulate empty data frames, which a node may fill in its turn. Usually, a fixed number of slots or frames of fixed size circulate. Each frame consists of source and destination addresses, parity and control, and data. Note that the slotted access technique limits not only when a node transmits but how much at a time.

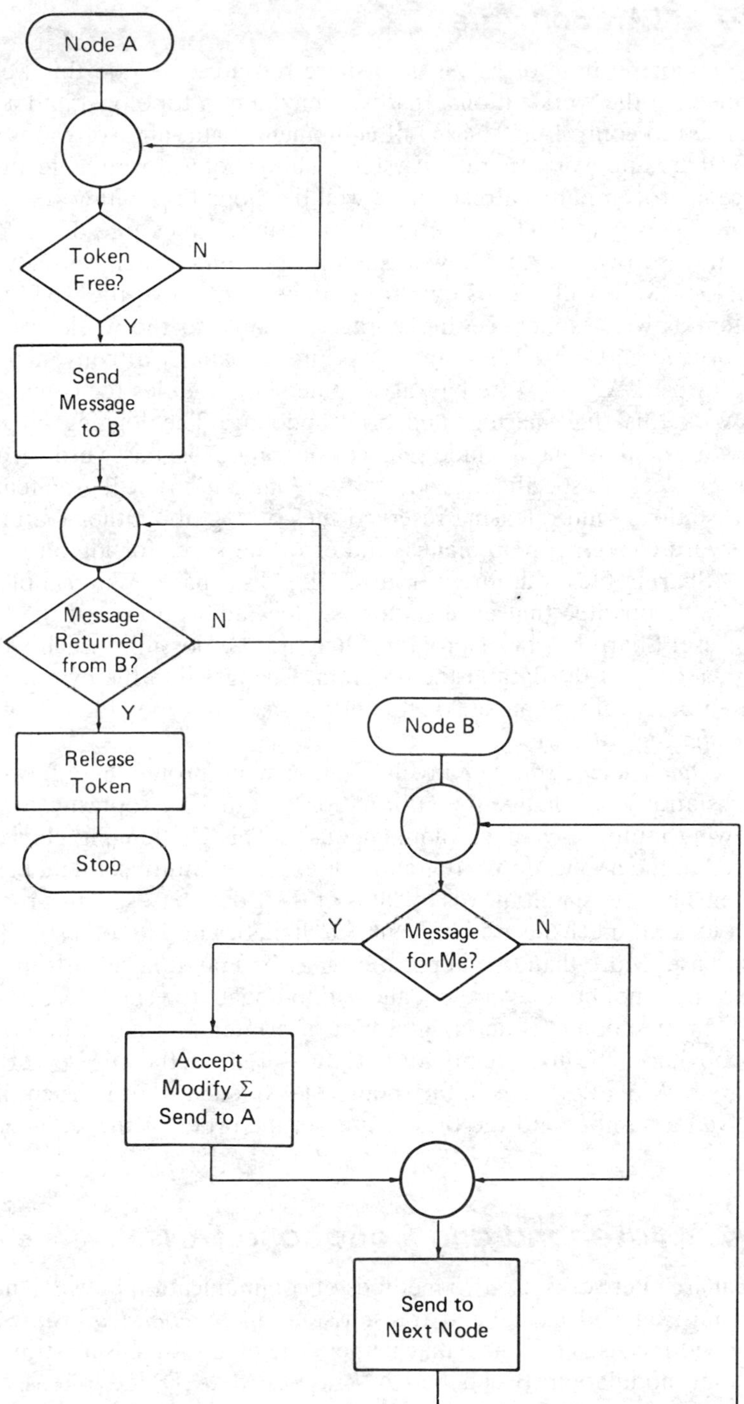

Figure 12-37 Token protocol.

12-7.2 LAN Software

The major purpose of LANs is to share resources. To do this we need a way of connecting the workstations: that is, a medium, a topology, and a protocol. Then we need to control the sharing of equipment, data, and software, in other words, a local area network operating system. This is not as formidable a task as it sounds since microcomputers already have well-developed operating systems, such as MS-DOS, OS/2, and UNIX. The problem then becomes one of interfacing network function software with the workstation operating system already available. The LAN software will add its own restrictions to those of the workstation operating system as well as increase file storage available to the workstation, and provide for private and shared files and for printer capability at convenient locations.

Most LAN software packages come with modules for logging on and off the network, disk/file sharing, and print spooling. The logging on and logging off network module may include such considerations as password security, validation of user access to specific files and software, an automatic log-on feature for specific workstations, and releasing reserved files on log-off. Other features may include password changing, help menus, and error messages for log-on problems.

Sharing files is different from sharing disk space. With simple LAN software it may be possible that each station is allocated a portion of the hard disk for its own use. Sharing data or programs may not be possible. Each workstation takes its own copy of the data or the program. Changes to a file by one workstation will not automatically be passed on to other users. This type of LAN software is termed *disk-sharing software*.

Data-sharing software involves either write protecting a file so that only one workstation may change the information at a time or sophisticated LAN software allowing multiple changes simultaneously. This challenging chore is often transferred to the applications program. For example, multiuser data base management systems handle simultaneous updates of their data bases. This process is restricted by a lockout mechanism to prevent simultaneous update of a specific record in the data base. More than one application may be updating records in the data base at the same time, but they are not allowed to update the same record simultaneously.

Print spooling is an integral part of printer sharing. It describes the process of copying a file to a temporary storage area on the file server until it can be printed. While the file is in the spooler it is queued; that is, each file, as it arrives, is given a number and the files printed in the order of the assigned numbers.

12-7.3 Baseband and Broadband LANs

Local area networks use two methods of communication between nodes: baseband transmission and broadband transmission. In *baseband transmission*, the entire bandwidth is used for one data channel. The digital signals are transmitted directly without modulation. Baseband LANs use CSMA/CD and token-passing protocols on a variety of topologies.

In *broadband transmission*, a single physical channel, usually coaxial cable, is frequency divided into a number of independent channels. These independent channels are allocated different bandwidths and may be used for voice and video, as well as data transmission. Since this is an analog signal technique, the digital signals must be modulated. High-frequency modems are required to provide channel carrier frequencies that do not interfere with each other.

One distinction between broadband and baseband systems is that signals in baseband systems must travel in one direction, whereas on broadband the signal travels in both directions. Both use the modified bus or tree topology. In broadband LANs, separate send and receive channels allow two-way exchange of information.

Broadband LANs use frequency-division multiplexing (FDM) to subdivide the coaxial cable into many channels, each of which can have the capacity of a baseband LAN. Since this is an analog signal technique, the digital data signals must be converted. The main drawback of broadband LANs is the cost of modems and FDMs. The major benefit lies in the fact that the network can transmit data, voice, and video signals simultaneously. The coaxial cable used in broadband LANs is the same as that used in cable television (75 Ω). Its capabilities include a bandwidth of 300 MHz and a transmission rate of up to 5 Mbits/s per channel.

To accommodate transmissions of data, voice, and video, the cable is divided into bands that are multiplexed into many subchannels. Figure 12-38 shows a possible allocation of bands for low-speed data, switched voice and data, and video channels, with capacity reserved for future expansion.

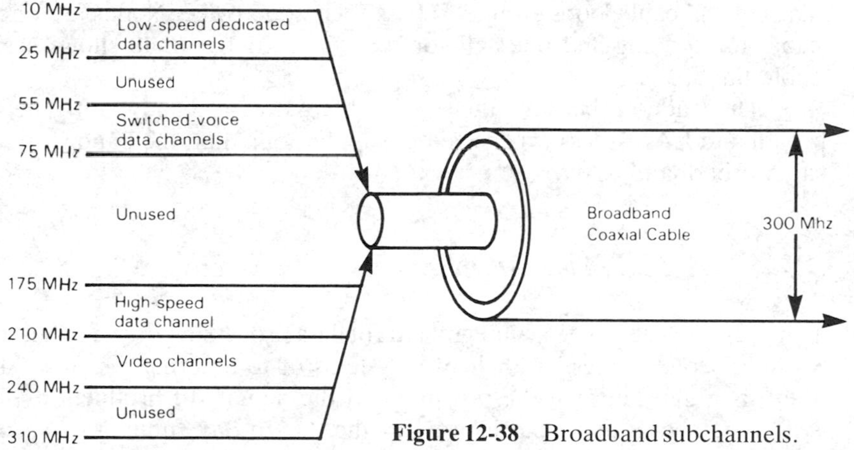

Figure 12-38 Broadband subchannels.

On single-cable systems, compatible with cable TV, this is accomplished by halving the capacity of the cable. A central retransmission facility (CRF) remodulates lowband (send) to highband (receive) signals. Figure 12-39 shows the IEEE 802 working group recommendation for a broadband standard and establishes a 108 to 162 MHz guardband between high- and lowband nodes.

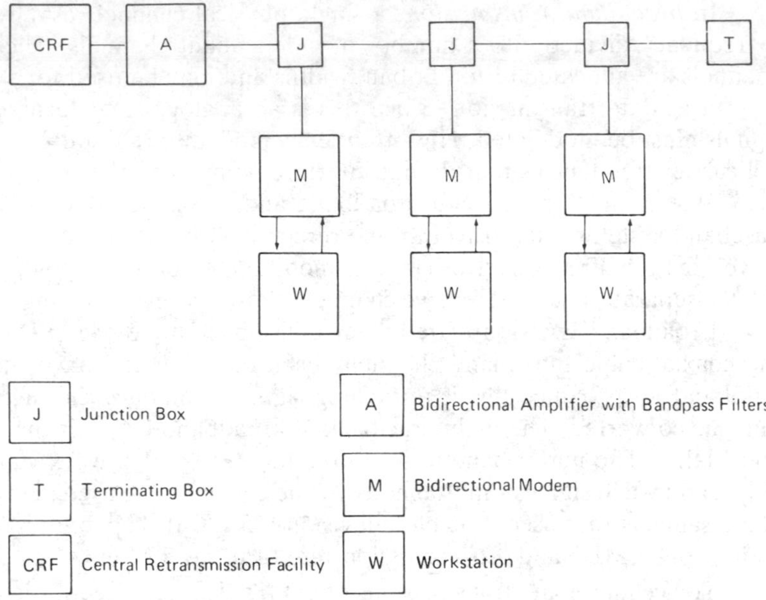

Figure 12-39 Single-cable broadband LAN.

In a dual-cable system the entire bandwidth is available in both directions. The coaxial cable loops around to pass each node twice. Nodes use one-half of the cable for sending and one-half for receiving. Figure 12-40 shows a simple dual-cable linkup.

The trade-off between the systems is cost versus bandwidth. The single-cable broadband LAN is less expensive but supplies only half the bandwidth of the dual-cable broadband LAN.

12-8 COMPUTER APPLICATIONS

Networks have become an integral part of data processing. Few installations operate without remote access to their own systems or to external systems. As is normal, there is a gap between the possible and the actual. In production environments where data processing participates in the day-to-day running of an organization (on-line systems), there is a natural tendency toward caution in implementing new technologies. This tendency is offset by rapid changes in technology and the need to stay competitive.

The main thrust in the application of networks is toward distributed data processing (DDP). This is a simple concept involving the decentralization of data processing resources but it is a complex and difficult system to implement. Because of the complexities and the philosophical change in an organization's perspective, from centralized to decentralized, progress has been slow in the development of

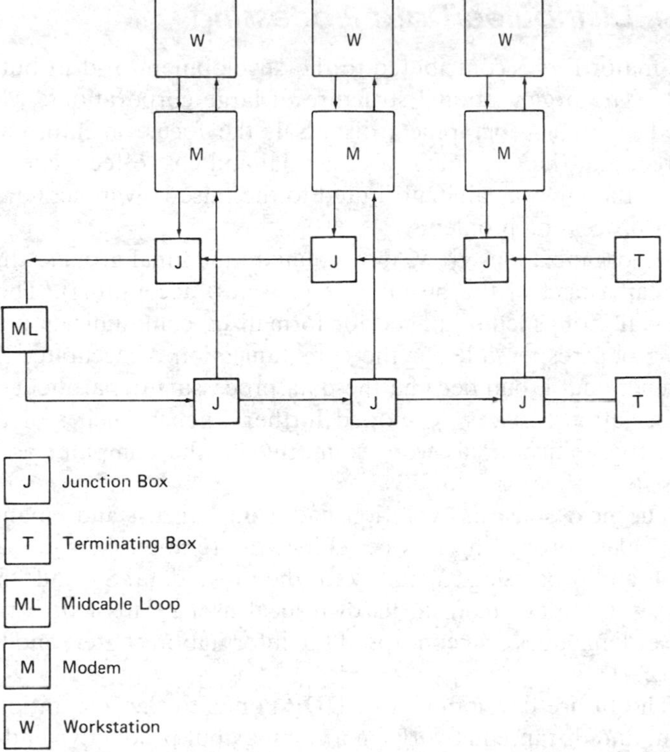

Figure 12-40 Dual-cable broadband LAN.

wide area networks for DDP. A more practical example of DDP occurs in the field of local area networks, where shared software, hardware, and data are basic to the design of such networks. Although there are a few distributed data base management systems (DDBMS) available for commercial use, these applications remain rooted in the research/university arena of activity.

A more practical example of network development lies in the micro/mainframe link. Because of the microcomputer's increasing power and decreasing price, it is possible to solve sophisticated data processing problems on the microcomputer using productivity tools such as spreadsheets, local data bases, and graphics. When the microcomputer can also be used to access corporate data bases, locally and remotely, and upload and download files, a powerful combination is created.

The possibility of user implementation of a DDP has decreased in recent years as computer manufacturers concentrate on developing network architectures based on their own hardware. It has become critical for them to provide interfaces to public packet-switched networks, local area networks, and other manufacturers' networks. This is the most realistic way of developing DDP systems on wide area networks.

12-8.1 Distributed Data Processing

Three major forces contributed to the development of distributed data processing (DDP). The organizational structure of large corporations, which becomes fragmented through geographical dispersal, the focus on information as a valuable resource, and the technological potential and cost-effectiveness of handling information in a new way all contributed to the raised awareness of DDP as a solution to data processing problems.

As organizations grew, they created functional groups which specialized in a particular aspect of the business (e.g., sales, accounting). The creation of these functional groups led to a need for formalized communication and then to a functional group responsible for the communication of accurate, timely information. This functional group became the data processing department. The centralization of information services developed further when branches of the organization in other geographical areas were connected to the computer as remote access terminals.

The next stage in evolution had a dual thrust and emanated from the end users of data processing services. First, the remote users demanded local access, control, and processing capability for their own data. Second, the large centralized system was under fire from its own local users. This led to the development of end-user languages, the concept of the information center, and local area networks (Fig. 12-41).

The future development of DDP points to the integration of LANs, metropolitan, and distance networks in a complex but productive end-user environment.

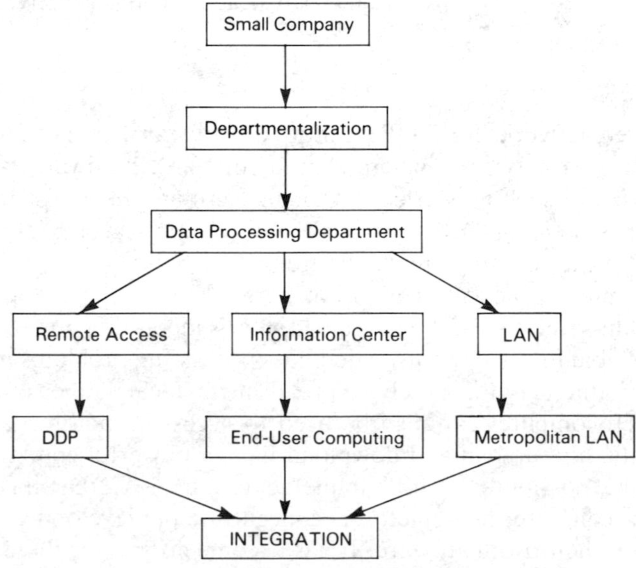

Figure 12-41 Evolution of DDP.

Structure

There are two facets to the issue of centralization versus decentralization. One relates to organizational philosophy, the other to information handling. The hierarchical nature of organizations and the need for company-wide goals leads to the idea of centralized control. The head-office mentality has come under criticism, yet it has its strengths: relatively simple decision-making paths, consistent company-wide policy implementation, and a high degree of organizational control. At the other end of the spectrum, in organizations with geographically dispersed sites, there are good arguments for local control and autonomy: familiarity with local conditions, faster reaction to changing conditions, and the motivational need for control over one's own destiny. The conflict is often viewed as a power struggle. Certainly, there is evidence at all levels of society of the growing awareness of individuals and groups to the possibility of influencing their own lives.

In terms of decentralizing information services, the first solution was to access a centralized system by remote terminal. This was to provide decentralized decision making (optimum solutions to local problems are achieved by those who are there) as well as overall monitoring and control by the central site. This did not satisfy the remote end user's information needs (nor, for that matter, the local end user's needs). In addition, the perception that the head office can operate on the basis of summaries or summaries of summaries is open to question.

The end user may be defined as the person who needs specific information to make specific decisions to fulfill the organizational goals of profit and service in a productive manner. Whether located at the head office or in a branch office, the end user is a manager who directly affects organizational performance. All other groups in the organization provide support, whether they are in personnel, data processing, or senior management. This perception of the end user as the driving force in an organization has led to an emphasis on providing the information and productivity tools necessary for the end user to do the job wherever he or she is located.

DDP is not a process or a technology; it is a concept that enables hardware, software, and communication choices to meet the needs of the end user. A centralized information system reflects a shared approach to information resources. A decentralized system tends toward the dedicated, autonomous use of information. A distributed system lies somewhere in between, where location and processing of information may be decentralized to a much greater degree than overall organizational control.

Definition

DDP has evolved from an attempt to satisfy organizational information systems goals to an emphasis on end-user productivity. Because of its conceptual simplicity but complex and varied implementation, DDP has been defined in many ways. Consideration of several definitions (Table 12-7) may provide insight into the breadth of meaning, the variety of emphasis, and the changing perception.

TABLE 12-7 DDP Definitions

1. The concept of arranging hardware, software and networks to meet organizational needs
2. The philosophy of placing computers geographically and organizationally close to the application
3. Putting resources where the people are
4. The use of multiple computers in a cooperative arrangement to support common systems objectives
5. The transfer of data processing functions from a central system to small dedicated processors where the information is used
6. Programs and data reside in processing nodes and are integrated by networks
7. The interconnection of components of a system, permitting shared resources and data bases
8. Placement of intelligent devices at multiple sites to execute user-written applications to update and store local files and access remote data bases
9. A local site connected to a remote host via a network in order to access the corporate data base and provide local processing, data entry, and printing
10. The provision of an integrated computer network for end-user productivity

The first three definitions are broad and emphasize the relationship between goals and resources. Definitions 4 and 5 emphasize the processing needs of distributed systems, 6 and 7 focus on the software and data aspects of DDP, and 8 and 9 are very specific. Eight defines DDP in terms of the three words *distributed*, *data*, and *processing*. Definition 9 attempts to identify functional responsibility. Number 10 encompasses both local and remote processing and reflects the more recent preoccupation with productivity tools for the end user.

Advantages and Disadvantages

The decision to implement a distributed data-processing system usually involves a series of trade-offs or compromises. The factors on which the decision is based must be clearly identified and carefully weighed. The prior commitment to one strategy (a centralized information system) requires that the case for change significantly outweigh the decision to stick with the present system. The nature of the organization (conservative or risk taking), the application, the potential rewards, and the present performance of the organization are all factors that may swing the decision one way or the other. The choice is such a fundamental and important one, heavily influenced by world economy and market strategy, that a relatively cautious, planned, phased implementation may be appropriate. The advantages and disadvantages, often different sides of the same coin, are summarized in Table 12-8.

TABLE 12-8 Advantages and Disadvantages of Distributed Systems

Advantages	Disadvantages
Response time	Maintenance
End-user applications control	Loss of data control
Flexibility	Complexity
Costs	Costs

Classification

Network topology describes the relationships of communication nodes. In distance networks the topology is dictated by geography; in LANs, by the network manufacturer (usually characterized by simplicity and symmetry). The termination nodes of a network may also be classified by different topologies, the most common of which are vertical, horizontal, and functional.

Vertical (Hierarchical) Topology (Fig. 12-42) This topology parallels the classical organizational structure. For that reason, it is the most popular topology. Because the vertical model pervades human thinking, it may be overutilized. Typically, we have three levels of processing: at the host, at the remote site, and in intelligent terminals. Transactions enter and leave at the lowest levels. Processing not possible at the intelligent terminal is passed to the remote-site processor. Processing not possible at the remote site is passed to the host or primary site. The lower levels may perform partial processing before passing the data up the hierarchy. Each level may support its own data base, which may or may not be shared with other levels. IBM's Systems Network Architecture (SNA) provides this environment.

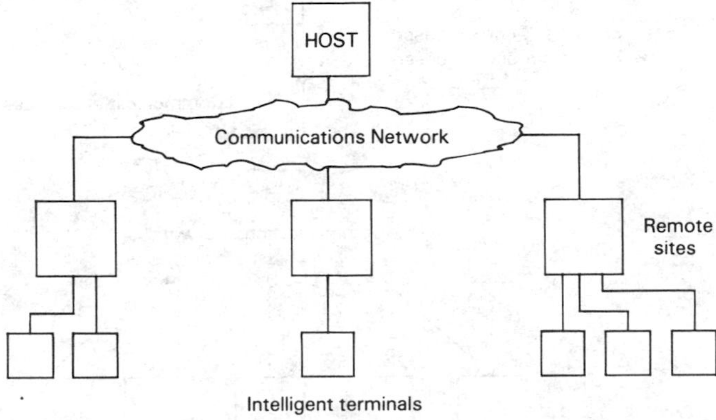

Figure 12-42 Vertical topology.

Horizontal Topology (Fig. 12-43) Horizontal topology is characterized by peer relationships among nodes. Every node in the network can communicate with every other node without having to consult a central node. The advantage of such a configuration is that each node can be equally responsive to user requests. A network user at one node can easily gain access to applications and facilities at other nodes. Since every data exchange and remote-access operation does not have to pass through a central node, communication overhead is reduced and network performance increases in efficiency. The Digital Equipment Corporation's DECnet network is a good example of horizontal topology for distance networks. Both ring and bus LAN topologies are also examples of this form of distributed processing.

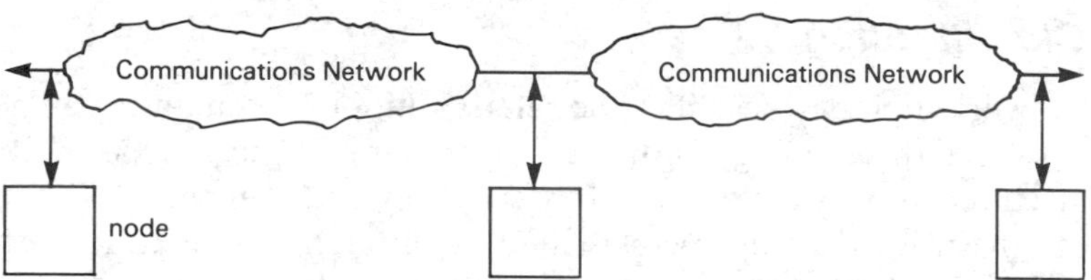

Figure 12-43 Horizontal topology.

Functional Topology (Fig. 12-44) This involves the separation of processing functions into separate nodes. Typical functional separations are (1) data base processing, (2) application processing, (3) communication processing, and (4) transaction processing. The major advantage of this type of topology at the host end is

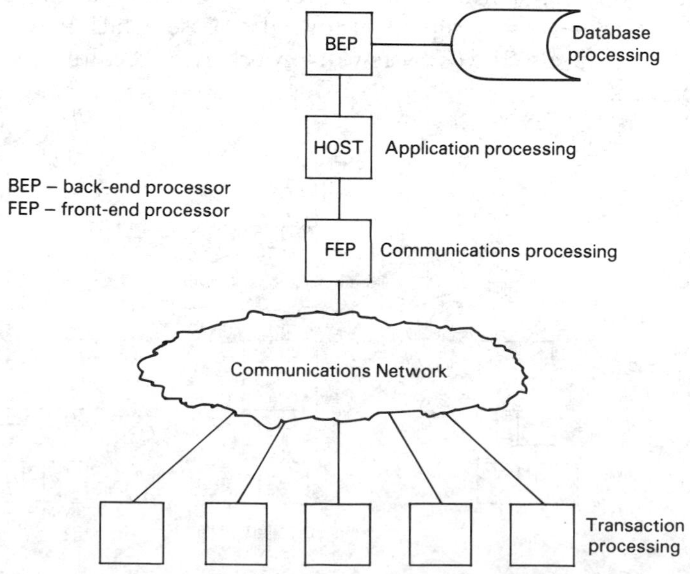

Figure 12-44 Functional topology.

the offloading of data base searching and retrieval and communication handling of large numbers of terminals. This improves the application throughput on the host significantly.

Another aspect of computer network configuration is the compatibility or incompatibility of the machines. A homogeneous network avoids the problems of protocol conversion, the need for gateways from one manufacturer's system to another, emulation software, and multivendor servicing problems. The heterogeneous network takes advantage of manufacturer strengths and price breaks.

12-8.2 Distributed Data Base

A distributed data base (DDB) is a collection of data that belongs logically to the same system but which is spread over the computer nodes of a network. Distributed applications are applications that access data from more than one DDB: for example, the transfer of an account from one bank branch to another. A DDB can also be implemented on a LAN where the nodes share not only server data but also each other's files. Rather than emphasize the geographical dispersion of data that apply to distance networks, we may broaden our definition to include LANs as follows: A *distributed data base application* is one in which a node accesses data bases connected to other nodes in the network. The characteristics of a centralized data base are well catalogued. They are compared with the requirements of a distributed data base in Table 12-9.

TABLE 12-9 Centralized versus Distributed Data Base

Characteristic	Centralized Data Base	Distributed Data Base
Centralized control	Data base administrator (DBA)	Global DBA, local DBAs
Data independence	Schema	Distributed transparency
Reduction of redundancy	Avoid inconsistencies, share data	Redundancy desirable, replication, failsafeness
Complex physical structures for efficient access	Indexes, chains, two-way circular linked list	Different solution distributed access plan
Integrity, recovery, concurrency control	Transactions must be completed or cancelled	More difficult in DDB
Privacy and security	Authorized access through DBA	Local protection schemes; network is the weak link

Why Distribute Data Bases?

It turns out that the reasons for distributing data bases are similar to those used for DDP. This is not surprising since DDB is a major application of DDP. The main difference lies in the emphasis on data access rather than on distributed processing. The factors are listed here for completeness:

1. Fits naturally into the decentralized structure of many organizations.
2. Provides a natural solution when several data bases already exist in the organization.
3. Supports a smooth incremental growth with minimum impact on already existing nodes.
4. Reduces communication costs.
5. Improves performance through parallelism (true for any multiprocessor system).
6. Provides a higher degree of availability and reliability (fail-safe).

The development of commercially available DDBs is dependent on the continuing low cost of microcomputers and minicomputers, on network technological developments which facilitate DDB communication and on DDBMS software developments to accommodate the problems of concurrency and recovery.

Problems in Distributed Data Base Management

The distributed data base faces the same kinds of control problems associated with the management of a centralized data base. These problems are complicated by the fact that the data are located at different sites.

Recovery. A transaction is a program unit whose execution preserves the consistency of the data base. It must be executed to completion or not at all. In a distributed data base it is more difficult to ensure that this happens. The failure of one site may result in erroneous computations. It is the function of the transaction manager to ensure that transactions are executed to completion or cancelled. The Transaction Manager maintains a log for recovery purposes and participates in concurrency control.

Concurrency Control Most computer installations operate in a multiprogramming mode in which several transactions are executed at the same time (i.e., concurrently). There are several schemes for controlling the interaction of concurrent transactions to prevent them from destroying the consistency of the data base in a centralized data base. These schemes involve a transaction that locks portions of the data base so that other transactions may not access specific data until they are unlocked. These schemes can be modified for use in a distributed environment.

Deadlock Handling A serious problem that can occur in concurrent processing is called *deadlock*. A system is in a deadlock state if there exists a set of transactions such that every transaction in the set is waiting for another transaction in the set. There are two ways of dealing with the deadlock problem. A deadlock prevention protocol ensures that a deadlock state is never entered. Alternatively, a deadlock detection and recovery scheme allows the system to enter a deadlock state and recover from it. Deadlock prevention and recovery schemes can be implemented in a distributed system.

12-8.3 Micro/Mainframe Links

DDP and distributed data base are major network applications that require considerable additional resources and reorganization of a centralized data processing environment. These areas are still relatively undeveloped technologically (except for LANs) and still face numerous design and implementation problems. The proliferation of powerful low-cost microcomputers has encouraged many organizations to seek ways to interface these devices to their mainframes both locally and remotely.

The main thrust for the development of micro/mainframe facilities came from the end users who already had personal computers on their desks for use with productivity tools such as spreadsheets, data management programs, and graphics. The simplest request was to use the personal computer as a terminal for separate access to mainframe data. It was recognized that it would be very useful for end users to be able to switch from using a productivity tool on their personal computers to, for example, using a mainframe program directly to access the corporate inventory control system, without changing machines. Terminal emulation has become a major concern of software suppliers. An important feature is the ability to emulate a variety of terminals and gain access to different in-house mainframes as well as national financial and statistical data bases. These applications led quickly to the demand for access to mainframe data for incorporation into end-user solutions and for transmission of results to the mainframe. There are two ways of exchanging data: downloading and uploading. Downloading a segment of data from the mainframe to the micro for reprocessing, integrating, and incorporating into a report is straightforward. It has thc advantagc of dcaling with current information. Uploading data to the mainframe from the micro is useful for concatenating reports; for example, department reports could be combined to produce a company-wide report with appropriate summaries and graphs.

In addition, the data processing department saw the potential and the advantage of using the micro as a development tool for mainframe applications, principally for saving mainframe resources. The possibilities range from editing on the micro and uploading to the mainframe for compilation and debugging, to applications development. The micro can be used to build a prototype of a system with a fourth-generation language for conversion to another language on the mainframe or for use with the mainframe version of the productivity tool. A number of organizations are finding the combination of PC FOCUS and mainframe FOCUS a way of offloading the mainframe during project development while using the multiple-user capability of the mainframe version for running the finished product. Since the two versions of FOCUS are compatible the exchange is simple:

Downloading ON TABLE PCHOLD AS d:filename.ext
Uploading XFER ddname FROM d:filename.ext

The micro/mainframe link improves end-user productivity by making data more easily available and by providing local processing capability. The incorporation of the micro/mainframe link into a distributed data base application is a good possibility for the future where local and remote data access may be combined in a way that is transparent to the user.

Classification

There is some confusion about the options available for micro/mainframe connections. Contributing factors include:

1. Microcomputers support asynchronous transfer as their standard user terminal interface.
2. IBM supports synchronous communication with its standard 3270 series terminals.
3. Asynchronous terminal communication uses ASCII, as do micros and minis.
4. IBM uses EBCDIC (as do some other mainframe manufacturers).
5. Mainframe computers often unload display processing onto the terminal (or the terminal emulating micro).
6. The ANSI terminal standards are interpreted in different ways.

The characteristics of micro/mainframe tools may be classified in the following way. Simple programs that are used to allow the microcomputer to behave like a terminal are called *terminal emulation software*. Asynchronous communications packages include terminal emulation facilities, file transfer capabilities, and other features (Fig. 12-45). They usually have to be present at both ends of the link for full operation. A combination of hardware and software tools is used to supply the very important synchronous terminal world. The products available on the market add to the confusion by supplying overlapping features for a variety of applications.

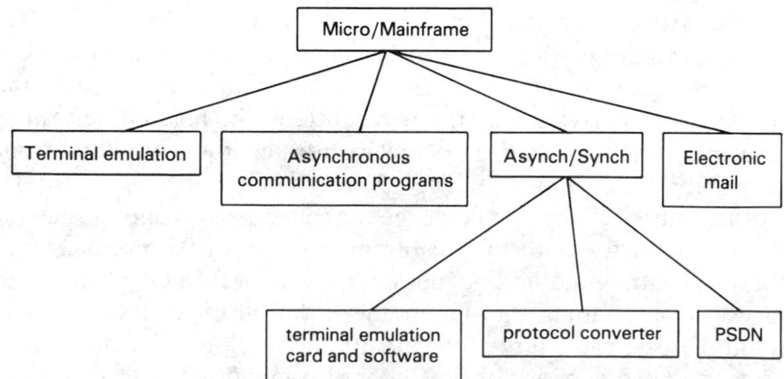

Figure 12-45 Micro/mainframe classification.

Terminal Emulation

This is a relatively simple aspect of micro/mainframe communication. PCs may emulate simple (dumb) terminals using software to act like a specific asynchronous terminal; for example, TTY, VT52, or the very popular VT100 terminal specifications. A simple terminal emulation program is normally used to communicate with a host mainframe or mini computer. Many packages have the ability to emulate more than one terminal.

Asynchronous Communication Programs

These communications programs have the following features:

Terminal Emulation This is as described above for terminal emulation.

File Transfer This is the ability of the program to transfer (upload and download) files between the micro and another micro, a mini, or a mainframe.

Configuration Setting This is the ability to create communications sets, macros, or script files. These names describe the files that provide initializing information for the communications programs, such as a table of telephone numbers for dial-up routines, setting parity, the number of bits per character, the number of stop bits, and the transmission rate.

Soft Key Programming This is the ability to program function keys so that a single key may be used to sign on to a specific mainframe, set up the micro for unattended access and file transfer, or provide help facilities. These procedures are sometimes called *command* or *script files*.

Error Detection The methods range from simple checksum in hobbyist environments to the standard CRC-16 of commercial environments.

Editing Some programs provide insertion, deletion, and creation of data, as well as copying, renaming, and deleting of files, without having to exit the program to an editor or the operating system.

A number of modem manufacturers (e.g., Hayes), are now supplying communication software as an option with the modem. This guarantees compatibility between the software and the modem and may be cheaper than separate purchases.

Async/Sync

Synchronous techniques are used in direct, coaxial cable connection to an IBM mainframe or remote controller (Fig. 12-46). Since the micro market is dominated by PC-DOS and MS-DOS, there are many vendors who produce hardware and software products for linking PCs as IBM 3270 emulators.

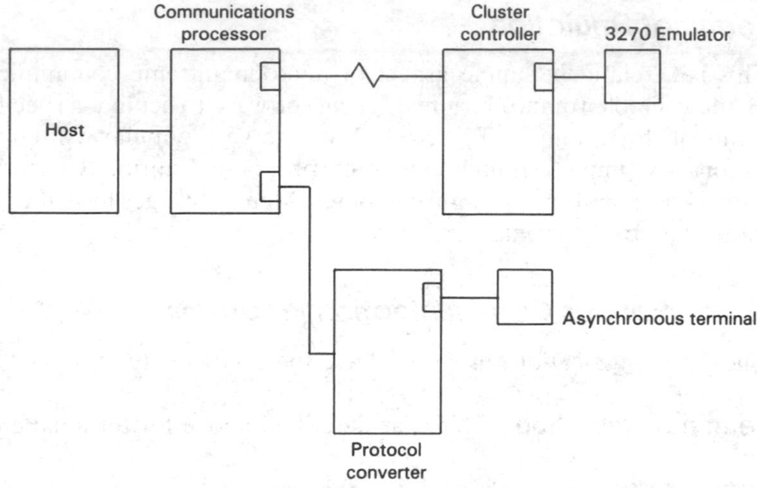

Figure 12-46 Async/sync connections.

The IBM 3270 terminals are the most widely used synchronous devices on the market. The emulation of these devices is a major market for many vendors that produce hardware and software to perform this task. One problem relates to the different keyboards. PC keys must do double duty to represent all the 3270 keyboard keys. The video control is different; PCs are limited in what they can simulate. The IBM mainframe expects to treat a printer attached to a PC as a separate device. These difficulties must be addressed by the hardware and software of the emulator.

The main problems in using a PC for 3270 emulation are:

1. The async to/from sync protocol conversion:

$$\text{TTY} \longleftrightarrow \text{SDLC}$$

$$\text{TTY} \longleftrightarrow \text{BISYNC}$$

2. The code conversion:

$$ASCII \longleftrightarrow EBCDIC$$

These problems are usually solved in one of three ways: by the terminal emulation card, the protocol converter, or the public switched data network.

Terminal Emulation Card

A terminal emulation card with accompanying software allows the PC to communicate with an IBM mainframe directly, via a communications controller or via a remote cluster controller. A number of variations in terminal emulators are available:

1. Emulation of the 3278/9 terminals. The 3278 is a monochrome display; the 3279 is the color graphics version. The card is designed to communicate with the 3274 (32 device cluster controller) or the 3276 (a single integrated control unit and display unit).
2. Emulation of the 3274 controller; or emulation of the 3276 controller. These cards, with the accompanying software, provide 3270 emulation, file transfer, and windows for multiple 3278 sessions.

Typical packages include Data Communication Associates IRMA board plus the IRMAlink software, Blue-Lynx 3270 Advanced Coax, and IDEACOMM 3278.

Protocol Converter

These are intelligent devices that convert ASCII to/from EBCDIC, twisted pair to/from coax, and the PC screen to/from the 3270 screen. Most protocol converters have a single port attachment to the mainframe (replacing the 3274) and multiple ports that can be attached to PCs, modems, terminals, printers, LANs, and statistical multiplexers. Despite having a variety of features, their principal function is to convert asynchronous to synchronous communication.

Public Switched Data Networks

When there are many remote users, the mainframe can be equipped with a direct link to a public packet-switched network. Asynchronous micros can dial into the network with a local number and be connected to a remote host. The network can treat the microcomputer as a simple terminal or provide protocol conversion for 3270 operation.

Electronic Mail

Designed specifically for message transmission, electronic mail systems are used for accessing on-line services such as the Source, Dow Jones News/Retrieval, and Compushare. Another facet of this application is the interface with TELECOM Canada's ENVOY 100, which offers customers access to each other via PC, through DATAphone, Datapac, TWX, Telenet and Tymnet, Telex, and International Networks.

12-8.4 Network Architectures

Network architectures facilitate operation, maintenance, and growth of the communication and processing environments by isolating the user and the application programs from the details of the network. Many network architectures are based on the ISO-OSI seven-level protocol discussed earlier. The main producers of network architectures are public packet-switching networks and computer manufacturers. We have discussed the X.25 architecture developed by the CCITT for the first three levels of the ISO-OSI model and used by most public packet-switching networks. Two widely used network architectures are IBM's System Network Architecture (SNA), based on a hierarchical computer topology; and Digital Equipment Corporation's (DEC) DECnet, based on a horizontal topology.

Systems Network Architecture

Because IBM has a large share of the mainframe market, SNA is a de facto standard for network architectures. SNA formally defines the responsibility of communication system components in terms of nodes or devices and paths. Both nodes and paths are arranged hierarchically in several levels. The nodes provide functional distribution under a central control. A typical IBM network configuration (Fig. 12-47) consists of hosts (370 series), front-end processors or communications controllers (3705/3725), cluster controllers (3274/3276/3174), and terminals (3270) or personal computers. Each device controls a specific part of the network at its level of the hierarchy and operates under the control of a device at the next level. The paths provide flexibility and redundancy in routing.

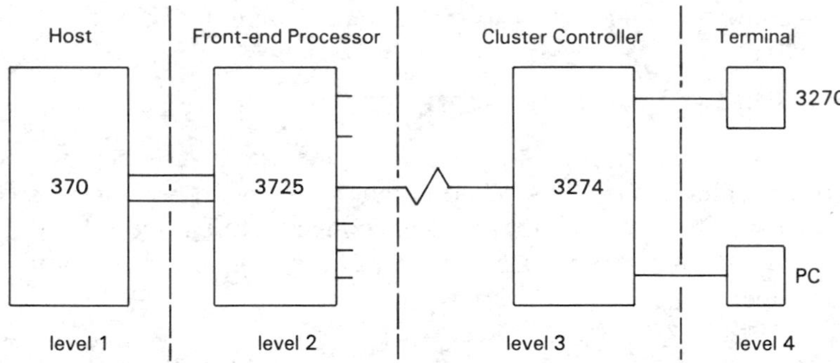

Figure 12-47 Typical network configuration.

Digital Network Architecture

Digital Network Architecture (DNA) is the model of structure and function, "the framework of specifications," on which DECnet is based. DECnet is a family of hardware and software communication products that form a network environment for Digital Equipment Corporation's (DEC) computers. The DNA model is closely based on the ISO-OSI model. The protocol layers have the same function. This means that the DECnet/X.25 interface is relatively straightforward.

Network architectures are at the heart of most data networks at the present time. Most computer manufacturer architectures and public packet-switched networks were developed from the ISO-OSI standard. IBM's SNA architecture is unique. The various architectures continue to develop, to provide more services, and to become more flexible and integrative. The main user problems continue to be with the incompatibilities of different networks.

12-8.5 NAIT Network

The Northern Alberta Institute of Technology (NAIT) network is a small communications network contained within the city of Edmonton. Nevertheless, it contains many of the hardware elements discussed previously. It is a realistic example of the way in which many networks operate. The prime consideration is to share local and remote computing facilities among a large number of users (students, faculty, and administration). As a result, software requirements are not complex and consist of terminal emulation (VT100 and 3270), resource sharing (printer spooling), file transfer (Kermit), and micro/mainframe links (PC plus IRMA boards).

The NAIT network consists of two campuses and three satellites (Fig. 12-48). The main campus contains the major components of the network: IBM4381 for academic use, VAX8200 for student records, MAI8030 for financial systems, and PACX1V for switching. The Plaza users connect directly to one of three VAX750 systems via a Develcon switch primarily for CML applications and indirectly through 56 lines to the main campus. The satellite communities are Patricia Campus (Automotives), the Kennedy Building (Purchasing), and the Kingsway Building (Accounting).

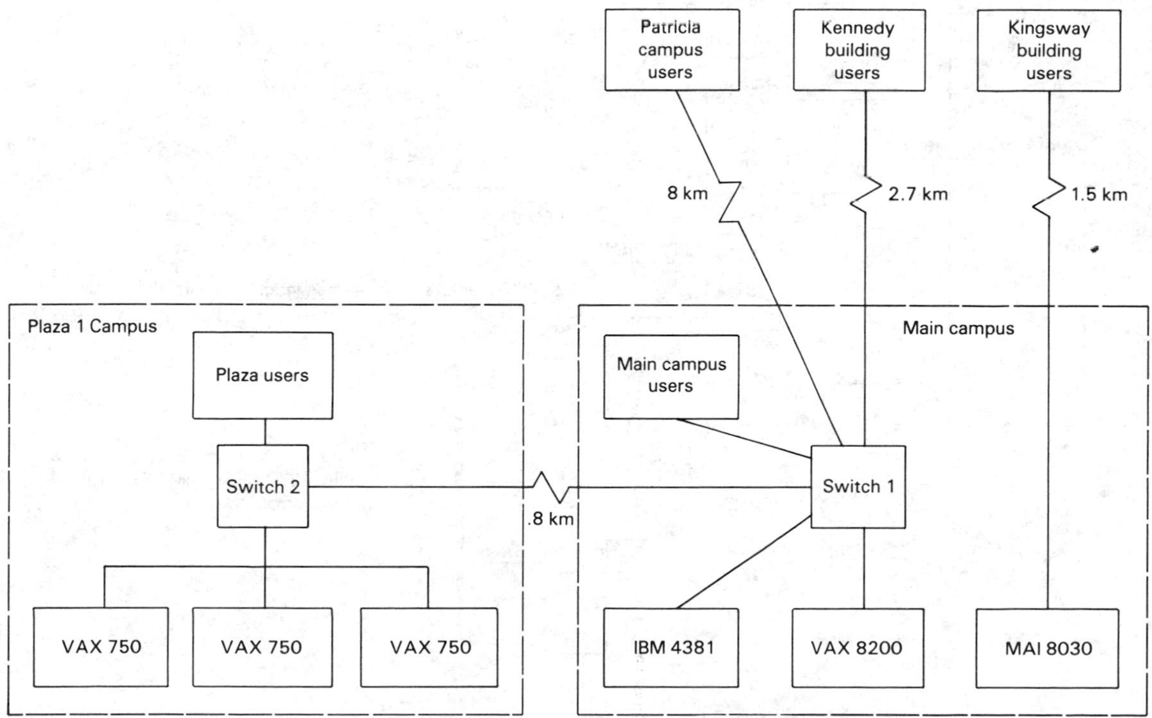

Figure 12-48 NAIT network geography.

Academic Applications

The heaviest users are Computer Systems Technology students, who program applications in COBOL, Pascal, SQL, and FOCUS. Business Administration, Mathematics, Earth Resources, and Chemistry students use application packages and program in BASIC.

One of the main problems associated with the academic applications is the increasing demand for hardware resources by both users and applications (ISPF, PROFS, SQL, and FOCUS) in the context of budgetary constraints. Solutions include running noninteractive applications in batch mode and deferring them to off-peak periods, and restricting access to applications for parts of the day.

The network nodes for academic applications include a communications controller (3725), cluster controllers (a 3274 and the 3174/3299 used to extend the coax range), and three protocol converters (7171s used for asynchronous to BISYNC conversion) (Fig. 12-49).

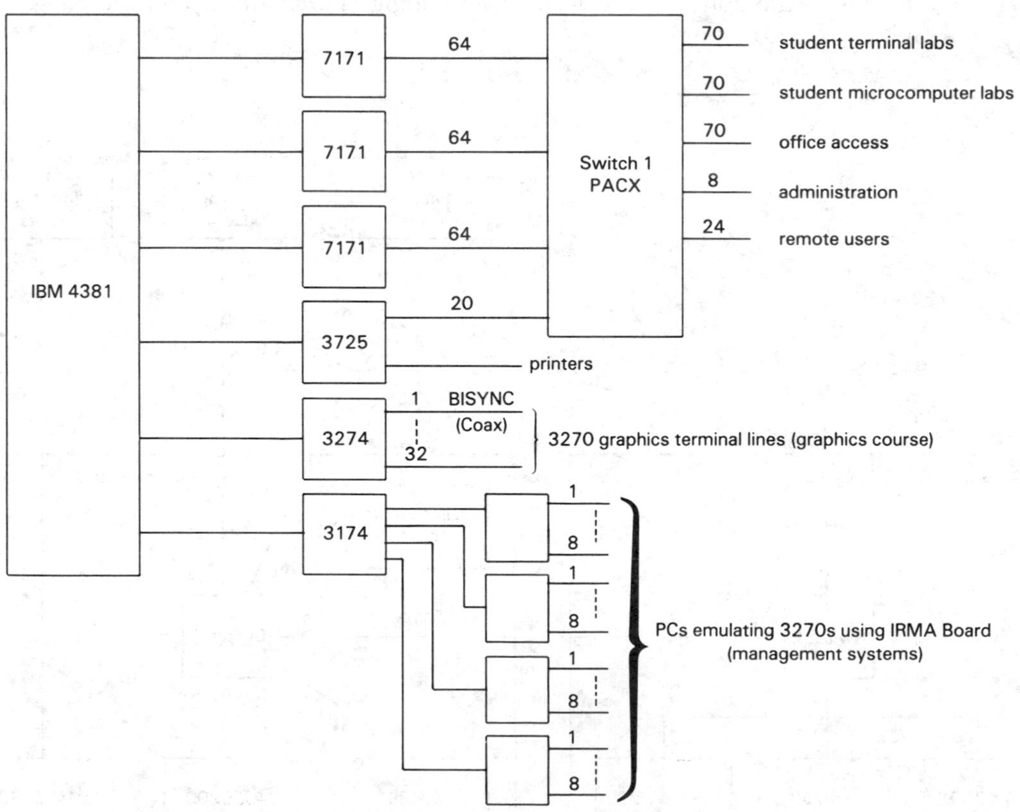

Figure 12-49 NAIT network academic system.

Financial Systems

These applications are typical of most educational organizations and include general ledger and inventory control systems. Remote access from the Kingsway Building is handled using statistical time-division multiplexers on two 4800-bits/s leased lines (Fig. 12-50). The statmuxes are:

1. A Gandalf PIN9103 which has 32 asynchronous lines converted to one synchronous line using HDLC protocol with dynamic line allocation up to 9600 bits/s.
2. The Gandalf SWITCHMUX 2000 converts 16 asynchronous lines to a synchronous line operating at speeds up to 19.2 kbits/s using HDLC protocol. (A similar configuration is used for the Kennedy Building to Main Campus interface.)

The terminals are asynchronous devices emulating VT100 operation.

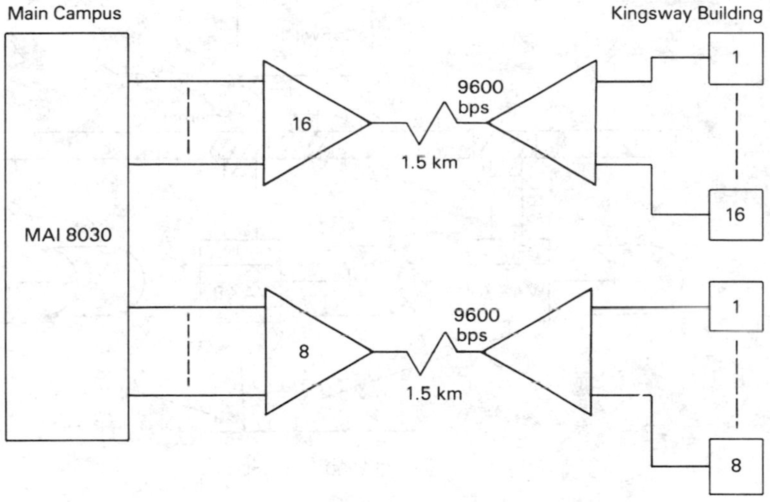

Figure 12-50 NAIT network: financial system.

Student Records

Principally used by the Registrar's Office and the Continuing Education Department, student records are available to other departments on a limited access basis. The main applications are in admission processing, registration, and mark processing. Communication is via the PACX switch. The system is based on the POISE Data Management System tailored to NAIT's requirements. The applications operate in an on-line enquiry and update mode.

CML

Computer Managed Learning (CML) is growing into a major application at NAIT. Many departments make use of the software by providing their own test banks. These include Power Engineering, Electrical Engineering, and Health Sciences. The communications facilities are comprised of 56 lines to the main campus, 14 dial-up lines, and a VAX/Ethernet cluster to optimize resource usage (Fig. 12-51). The single-entity concept, of sharing storage among several processors, is used to smooth user demand for access. The Develcon switch allocates users to each VAX on a rotational basis, in increments of four. The terminals are asynchronous devices emulating VT100 operation.

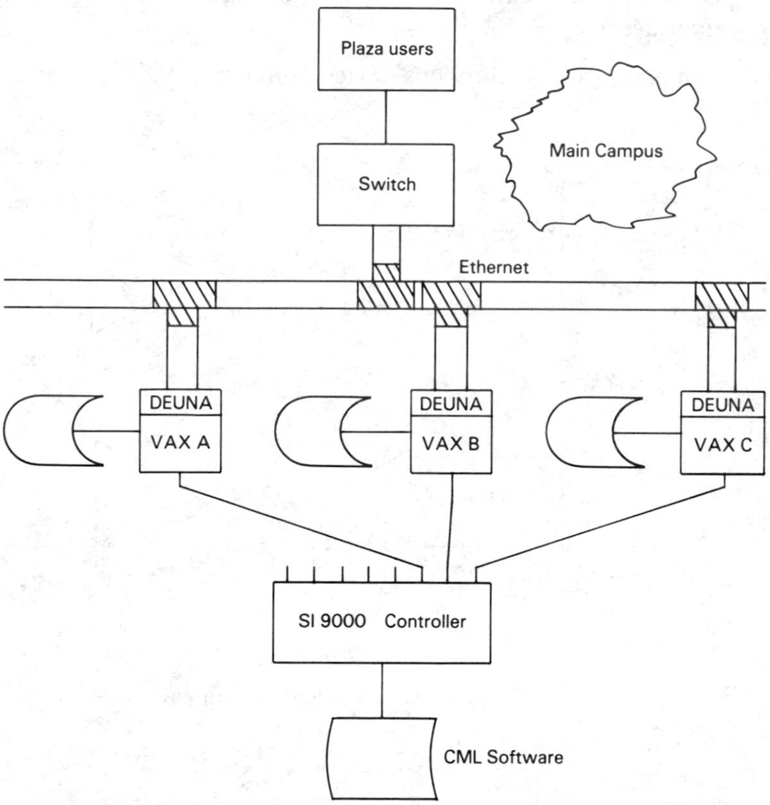

Figure 12-51 NAIT network: CML system with SILINK cluster.

The application is an on-line update system since students are evaluated as they enter results. Hard-copy terminals are provided to optimize terminal availability, to enable students to consult reference materials (e.g., steam tables), and for backup in the event of machine downtime. An optical scanner input device is provided for high-speed multiple-choice examinations.

The VAX cluster should not be confused with a network. A cluster has an integrated file and record system, can use common batch and print queues, and is confined to a small area. A cluster offers resource sharing and a common security data base. It is often part of a network.

In a SILINK cluster all nodes are connected to a shared disk using a Systems Industries 9900 disk controller and are interconnected through a DEC Ethernet system. Each node has a DEC Unibus Network Adapter (DEUNA). Up to eight SI disks may be shared in a single cluster. The Plaza-to-Main Campus interface running between the Develcon and the PACX switches is a statmux configuration using Gandalf SWITCHMUX 2000s (Fig. 12-52).

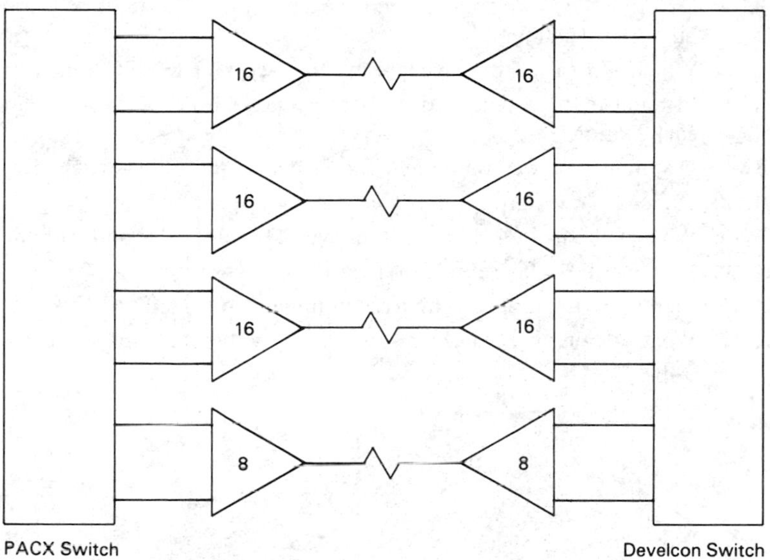

Figure 12-52 NAIT network: main campus/plaza interface.

Future

The NAIT network is expanding and evolving. There are already problems developing in line connections. Twisted-pair connection conduit is reaching maximum capacity. Consideration is being given to the use of thin coaxial cable.

The main issue being addressed is that of connectivity between network nodes. Microwave links are ruled out because of the closeness of the Edmonton Municipal Airport. In the near future, fiber optic trunks may be installed. The software interfaces for noncompatible hosts are available from computer manufacturers, independent software companies, and telephone companies. As with most telecommunication systems, the only constant in the evolution of the NAIT network is change.

Problems

12-1. How many unique information codes can the Baudot code accommodate?

12-2. Are asynchronous transmissions clocked?

12-3. With reference to Fig. 12-12, answer the following questions:
- **(a)** Which terminal has the clearest character set?
- **(b)** Which terminals have more control capability?
- **(c)** Which terminal has the poorest editing capability?
- **(d)** Which terminals have flexible error-checking capability?
- **(e)** Which terminal has flexible asynchronous character length?

12-4. Describe the routing algorithm for each of the following LAN topologies: ring, star, bus, tree, and loop.

12-5. Which error-checking mechanisms are used by BISYNC and X.25 protocols?

12-6. In point-to-point balanced four-wire systems using X.25, what are the functions of each station?

12-7. Distinguish between a permanent virtual circuit and a switched virtual circuit.

12-8. Under what conditions is circuit switching superior to packet switching?

12-9. What is the main technical difference between LANs and wide area networks?

12-10. Why can DDP be defined in a variety of ways?

12-11. Identify the main problem areas in implementing DDP.

12-12. What are the three problems associated with connecting a micro to an IBM mainframe?

Transmission Lines and Waveguides

William Sinnema

13-1 INTRODUCTION

Transmission lines and waveguides are structures that guide electric signals from one point to another. Transmission lines consist of two or more conductors separated by a dielectric, whereas waveguides are formed by a single hollow conductor. Two common forms of transmission lines are the two-wire line and the coaxial line, shown in Fig. 13-1. To maintain separation of the two electrical conductors, a nonconducting dielectric is provided. The more commonly used dielectrics are polyethylene, polystyrene, and Teflon. Since all dielectrics are somewhat lossy, minimizing the volume of dielectric is desirable if small attenuation of the signal is required. Periodic dielectric spacers, foamed polyethylene, and helically wound dielectrics give the lowest loss characteristics.

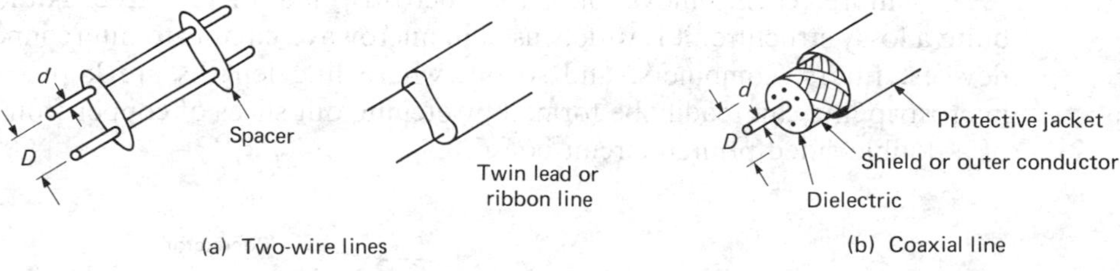

Figure 13-1 Transmission lines.

Coaxial cable produced under the Andrew trademark Heliax consists of a hollow corrugated center conductor, separated from an outer corrugated conductor by a helically notched ribbon of polyethylene. The corrugations permit slight flexing and bending of the cable. An outer jacket reduces the risk of damage due to crushing and abrasion and also prevents galvanic corrosion where the cable makes contact with metal supporting structures. Frequently used jacket materials consist of polyvinyl chloride (PVC), polyethylene, and Teflon. Shields can be found made of aluminum–polyester foils, braided copper, spiral copper, and semiconductive textile tapes. The attenuation of a few types of 50-Ω coaxial lines is given in Table 13-1. Most of those listed are identified by their RG/U (Radio Guide Universal) number.

TABLE 13-1 Attenuation of 50-Ω Coaxial Cable

Cable Type	Dielectric	Attenuation (dB/100 m)				
		50 MHz	100 MHz	400 MHz	1000 MHz	4000 MHz
RG-58	Polyethylene	10.2	14.8	32.8	55.8	
RG-8	Polyethylene	5.5	7.2	15.4	29.2	70.5
RG-58	Cellular polyethylene	10.5	14.8	29.5	47.6	
RG-8	Cellular polyethylene	3.9	5.9	13.8		59.1
Heliax	½-in. HJ4	1.8	2.5	4.0	8.1	19.0

The two-wire line is a balanced line in that both conductors have equal impedance with respect to ground, whereas the coaxial line usually has its outer sheath or braid at ground potential. The coaxial line has the advantage of confining the electromagnetic (EM) field within the dielectric and therefore is not subject to radiation losses or interference. The open-wire EM field extends well away from the structure and is therefore not used extensively above a few hundred megahertz. Radiation is discussed in more depth in Chapter 15.

A more recent innovation is the microstrip line of Fig. 13-2. Although it is quite a lossy structure, it is widely used in microwave circuits to interconnect active devices, filters, combiners, and so on, where line lengths are kept short. The microstrip line can readily be formed by etching out slices of copper from one side of a double-sided printed circuit board.

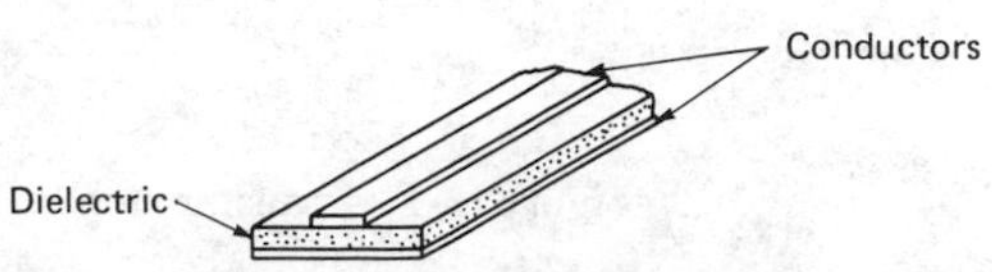

Figure 13-2 Microstrip line.

The electrical characteristics of transmission lines are determined chiefly by the capacitance and inductance of lines. The capacitance of a line is directly related to the dielectric contant (ε) of the insulating medium. For the coaxial line of Fig. 13-1(b), the capacitance per meter of line is given by

$$C = \frac{2\pi\varepsilon}{\ln(D/d)} \quad \text{F/m} \tag{13-1}$$

where ε is the dielectric constant or permittivity or capacitivity (F/m) of the insulating medium, d is the diameter of the inner conductor, and D is the inner diameter of the outer conductor. In free space, the dielectric constant is equal to $1/36\pi \times 10^{-9}$ F/m and is symbolized by ε_0.

Manufacturers usually specify the relative dielectric constant (ε_r) of materials, which is defined as the ratio of the dielectric constant of the material to the dielectric constant of free space: that is, $\varepsilon_r = \varepsilon/\varepsilon_0$. Table 13-2 gives the relative dielectric constant of some materials. It should be noted that ε_r can vary slightly with frequency and temperature.

TABLE 13-2 Relative Dielectric Constants

Material	ε_r
Air	1
Aluminum oxide	8.8
Polystyrene	2.55
Polyethylene	2.26
Teflon	2.1
Germanium	16
Silicon	12

The inductance per unit length of the coaxial line of Fig. 13-1(b) is given by

$$L = \frac{\mu}{2\pi} \ln\frac{D}{d} \quad \text{H/m} \tag{13-2}$$

where μ is the permeability of the insulating medium. For dielectrics it is equal to the permeability of free space: $\mu_0 = 4\pi \times 10^{-7}$H/m.

In addition to the distributed capacitance and inductance, a transmission line also has conductive losses due to the finite resistivity of the dielectric, resulting in a distributed shunt conductance, and ohmic losses due to the resistance of the wire conductors, resulting in a distributed series resistance. The net circuit model can be depicted as sketched in Fig. 13-3.

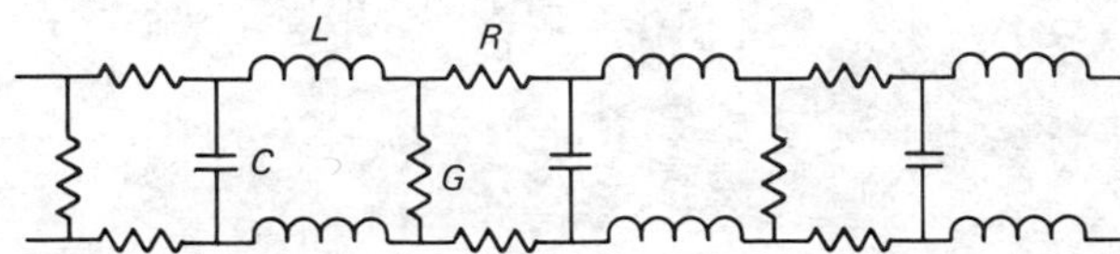

Figure 13-3 Distributed circuit model of a transmission line.

The distributed conductance current, which permits a leakage current between the conductors, is normally small for most dielectrics and can often be neglected. The series resistance can, however, be significant, particularly at the higher frequencies, due to skin effect. At high frequencies the current is forced to flow near the conductor surface due to the larger inductive reactance caused by the greater flux linkages toward the center of a current-carrying conductor.

As shown in Fig. 13-4(a), the current density in a conductor decays in amplitude exponentially with the depth of penetration from the surface. At sufficiently high frequencies, essentially all the current is carried near the surface and none in the central regions. If the conductor diameter is large compared to the skin depth δ parameter, where δ is defined by

$$\delta = \frac{1}{\sqrt{\pi f \mu \sigma}} \quad \text{m} \tag{13-3}$$

where f is the frequency, σ the conductivity of the conductor, and μ the permeability of the conductor, the total current is the same as if a uniform current density existing between the surface and a depth of one skin depth and the current was zero beyond that. This is shown in Fig. 13-4.

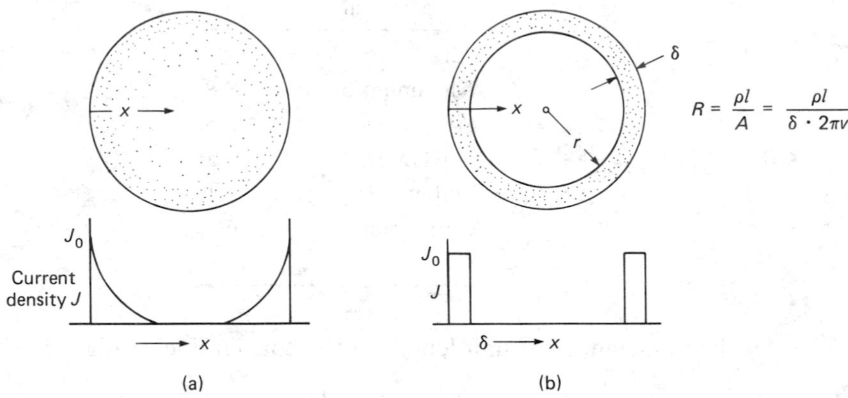

Figure 13-4 Current density distribution in a cylindrical conductor (a) and it's equivalent uniform current density flow (b).

EXAMPLE 13-1

Find the skin depth and the resistance of a 1-m length of copper coaxial cable at 4 GHz having the dimensions shown.

$$\sigma = \frac{1}{\rho} = 5.81 \times 10^7$$

$$\mu = \mu_0$$

(continued).

Solution

$$\delta = \frac{1}{\sqrt{\pi(4 \times 10^9)(4\pi \times 10^{-7})(5.81 \times 10^7)}}$$

$$= 1.044 \times 10^{-6} \text{ m} = 1.044 \times 10^{-3} \text{ mm}$$

For a conductor of length 1 m,

$$R = \frac{\rho}{A} = \frac{1}{\sigma A} = \frac{1}{\sigma\delta 2\pi r}$$

The resistance of the inner conductor (R_i) and outer conductor (R_o) is

$$R_i = \frac{1}{5.81 \times 10^7(1.044 \times 10^{-6})2\pi(1.08 \times 10^{-3})} = 2.43\ \Omega$$

$$R_o = \frac{1}{5.81 \times 10^7(1.044 \times 10^{-6})2\pi(3.62 \times 10^{-3})} = 1.38\ \Omega$$

$$R_{\text{total}} = 3.81\ \Omega$$

13-2 TEM WAVES

When considering guided-wave structures, the energy transmitted along the structures can be described in terms of traveling voltages and currents, or alternatively, in terms of electric and magnetic field intensities. For transmission lines, voltage and current descriptions are most commonly employed, but as these are difficult to define for waveguides, the electromagnetic wave description is found to be more convenient for the latter case. Within either structure, different field configurations or modes can be formed depending on the conductor spacing and size and frequency of operation. When employing transmission lines, the transverse electromagnetic wave mode is commonly assumed, in which case the electric field and the magnetic fields are entirely transverse (perpendicular) to the direction of propagation, as well as transverse to one another. However, if the conductor spacing exceeds approximately one-half a wavelength, higher-order modes can be formed. This can occur at the higher frequencies. Figure 13-5 shows the field vectors for a TEM wave in a coaxial line. The field intensities are greatest at the surface of the inner conductor.

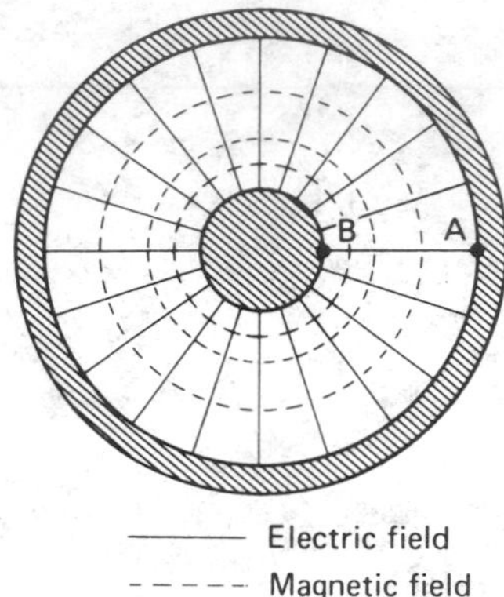

Figure 13-5 TEM fields in a coaxial line.

13-3 CHARACTERISTIC IMPEDANCE, PROPAGATION CONSTANT, AND PHASE VELOCITY

When waves or currents and voltages travel down a transmission line, traveling waves are encountered. The wave moving toward the receiving end is called an *incident wave* and the wave moving away or reflected from the receiving end is called a *reflected wave*.

These traveling waves experience an impedance called the *characteristic impedance* of the transmission line. That is, the ratio of incident voltage to incident current or the ratio of reflected voltage to reflected current experience the characteristic impedance Z_0. In general, this impedance is expressed by

$$Z_0 = \sqrt{\frac{R + j\omega L}{G + j\omega C}} = \sqrt{\frac{Z}{Y}} \tag{13-4}$$

where R is the series resistance per unit length (Ω/m); L is the series inductance per unit length (H/m); G is the shunt conductance per unit length (S/m); C is the shunt capacitance per unit length (F/m); $\omega = 2\pi f$, where f is the frequency of operation; $Z = R + j\omega L$ is the series impedance per unit length (Ω/m); and $Y = G + j\omega C$ is the shunt admittance per unit length (S/m). If losses are ignored (i.e., $R = G = 0$), the characteristic impedance is reduced to

$$Z_0 = \sqrt{\frac{L}{C}} \tag{13-5}$$

which for a coaxial transmission line equates to [substitute expressions (13-1) and (13-2) into (13-5)]

$$Z_0(\text{coax}) = \frac{1}{2\pi}\sqrt{\frac{\mu}{\varepsilon}}\ \ln\frac{D}{d} = \frac{60}{\sqrt{\varepsilon_r}}\ \ln\frac{D}{d} \tag{13-6}$$

This is the impedance given by manufacturers when they specify the characteristic impedance of coaxial cable. From the latter expression it should be observed that Z_0 (1) does not depend on the line termination and (2) does not depend on the line length, but (3) depends only on the spacing and size of the conductors and the type of dielectric used. As we shall see later, for lossy lines, Z_0 also depends on the conductor and dielectric losses as well as frequency.

As a wave travels down a transmission line, the wave experiences both phase delay and attenuation. These can be expressed in terms of the propagation constant, γ, a complex quantity that may be written as

$$\gamma = \alpha + j\beta = \sqrt{(R + j\omega L)(G + j\omega C)} = \sqrt{ZY} \tag{13-7}$$

where R, L, G, C, and ω are defined as in equation (13-4).

The real part of this expression (α) is called the *attenuation constant*, expressed as nepers per meters. If, for example, an incident voltage is traveling toward a load in the x direction, it exponentially decays as

$$|V(x)| = |V_s|e^{-\alpha x} \tag{13-8}$$

where V_s is the sending or reference voltage.

To calculate the total attenuation over a distance l, solve $|V| = |V_s|e^{-\alpha l}$ by taking the natural logarithm of both sides: $\alpha l = -\ln|V/V_s|$ nepers. To obtain the corresponding attenuation in decibels, note that

$$\begin{aligned} \text{dB} &= 20 \log\frac{V_{\text{in}}}{V_{\text{out}}} \\ &= 20 \log e^{\alpha l} \\ &= (\alpha l)20 \log(e) = 8.686(\alpha l) \end{aligned}$$

Thus

$$1 \text{ neper (Np)} = 8.686 \text{ dB} \tag{13-9}$$

β in expression (13-7) is called the phase constant (rad/m). This constant indicates the phase delay of the incident voltage (or current) as it travels down a distance of 1 m. Thus, including any attenuation that the signal might suffer, the voltage expression can be expressed as

$$\begin{aligned} V(x) &= V_s e^{-\alpha x} \angle -\beta x \\ &= V_s e^{-(\alpha + j\beta)x} \end{aligned} \tag{13-10}$$

The phase velocity of the wave can be expressed in terms of β by

$$v_P = \frac{\omega}{\beta} \tag{13-11}$$

Again, let us consider a loss-less coaxial line (i.e., $R = G = 0$). In this case, equation (13-7) simplifies to

$$\gamma = \alpha + j\beta = \sqrt{(j\omega L)(j\omega C)} = j\omega\sqrt{LC} \tag{13-12}$$

Thus $\alpha = 0$ and the signal experiences no attenuation;

$$\beta = \omega\sqrt{LC} \tag{13-13}$$

For a coaxial line this can be expressed as [substitute expressions (13-1) and (13-2) for L and C]

$$\beta = \omega\sqrt{\mu\varepsilon}$$

Since for dielectrics $\mu = \mu_0$, this can be expressed as

$$\beta = \omega\sqrt{\varepsilon_r}\sqrt{\varepsilon_0\mu_0}$$

Substituting the later expression for β into (13-11), we obtain

$$v_p = \frac{1}{\sqrt{\varepsilon_r}\sqrt{\mu_0\varepsilon_0}}$$

Now

$$\frac{1}{\sqrt{\mu_0\varepsilon_0}} = \frac{1}{\sqrt{4\pi \times 10^{-7} \times \dfrac{1}{36\pi} \times 10^{-9}}} = 3 \times 10^8 \text{ m/s}$$

which is equal to the velocity of light in free space and denoted by c. Thus

$$v_p = \frac{c}{\sqrt{\varepsilon_r}} \tag{13-14}$$

Manufacturers call $1/\sqrt{\varepsilon_r}$ the *velocity factor*. It determines how much the velocity is reduced from the velocity of light in a cable. For example, in a polyethylene cable where $= 1/\sqrt{\varepsilon_r} = 1/\sqrt{2.3} = 0.66$, the velocity of propagation if $0.66c = 0.66 \times 3 \times 10^8$ m/s $= 2 \times 10^8$ m/s.

One can also establish a relationship between the wavelength and the phase constant by noting that a *wavelength* (λ) is defined as that distance a wave travels in one cycle or through an angle of 2π radians. If the phase constant or phase shift per unit distance on the line is given by β, then

$$\beta\lambda = 2\pi \qquad \text{or} \qquad \lambda = \frac{2\pi}{\beta} \tag{13-15}$$

Equivalently,

$$\lambda = \frac{2\pi}{\omega} v_p = \frac{v_p}{f} \tag{13-16}$$

Since Z_0 and C are often specified for a transmission line, it may be useful to express β and v_p in terms of these constants. Substituting equation (13-13) into (13-11), we find that

$$v_p = \frac{1}{\sqrt{LC}} \tag{13-17}$$

By combining equation (13-5) with equations (13-13) and (13-17), β and v_p can be found in terms of Z_0 and C:

$$v_p = \frac{1}{Z_0 C} \tag{13-18}$$

and

$$\beta = \omega Z_0 C \tag{13-19}$$

As an incident wave moves along a transmission line toward a load, the wave will "observe" the characteristic impedance, Z_0 of the line. If the line is terminated with a load equal to the characteristic impedance of the line, the wave continues to "see" Z_0 when it reaches the load. A line terminated with its characteristic impedance is called a *matched line*. The wave energy is thus dissipated in the load and no reflected energy occurs. If the line is lossy, some of the energy is also lost as it moves along the line.

If the line is not terminated in Z_0, a reflected wave is set up. Such a mismatched line will be considered further later. Let us now consider several examples that should help elucidate these concepts.

EXAMPLE 13-2

A telephone paper-insulated exchange cable has the following characteristics: wire gauge 24 AWG; $\gamma = 0.355 \angle 45.53^0\ \text{mi}^{-1}$, and $Z_0 = 778 \angle 44.2^0\ \Omega$. Find the attenuation (dB/mi), wavelength (mi), and the velocity of propagation (mi/s) if the frequency of operation is 1000 Hz.

Solution

$$\gamma = \alpha + j\beta = 0.355 \angle 45.53^0 = 0.249 + j0.253\ \text{mi}^{-1}$$

$$\alpha = 0.249\ \text{Np/mi} = 8.686 \times 0.249 = 2.2\ \text{dB/mi}$$

$$\beta = 0.253\ \text{rad/mi}$$

$$v_p = \frac{\omega}{\beta} = \frac{2\pi(1000)}{0.253} = 24{,}835\ \text{mi/s}$$

$$\lambda = \frac{2\pi}{\beta} = \frac{2\pi}{0.253} = 24.8\ \text{mi}$$

or

$$\lambda = \frac{v_p}{f} = \frac{24{,}835}{1000} = 24.8\ \text{mi}$$

EXAMPLE 13-3

An 8/U JAN-C-17/A polyethylene coaxial cable has the following constants: $Z_0 = 52\ \Omega$, $C = 96.8$ pF/m, and attenuation $= 15.4$ dB/100 m at 400 MHz. Find the attenuation and phase constants per meter of line, and the velocity factor.

Solution

$$\alpha = \frac{15.4}{(100)(8.686)} = 0.01773 \text{ Np/m}$$

$$\beta = \omega Z_0 C = (2\pi)(400 \times 10^5)(52)(96.8 \times 10^{-12})$$

$$= 12.65 \text{ rad/m}$$

$$v_p = \frac{\omega}{\beta} = \frac{1}{Z_0 C}$$

$$= \frac{1}{52(96.8 \times 10^{-12})} = 1.987 \times 10^8 \text{ m/s}$$

$$\text{velocity factor} = \frac{v_p}{c} = \frac{1.987 \times 10^8}{3 \times 10^8}$$

$$= 0.662 = 66.2\%$$

For polyethylene, $1/\sqrt{\varepsilon_r} = 1/\sqrt{2.26} = 66.5\%$. The velocity factor is not quite equal to $1/\sqrt{\varepsilon_r}$ because of the effect of the resistive and conductive losses in the line.

EXAMPLE 13-4

Consider a 0.5-m length of coaxial line having the same constants as that given in Example 13-3 connected to a matched load (i.e., $Z_R = Z_0$; Fig. 13-6). The generator voltage is $V_g = 10\ \angle 0°$ and the frequency of operation is 400 MHz. Find the load voltage.

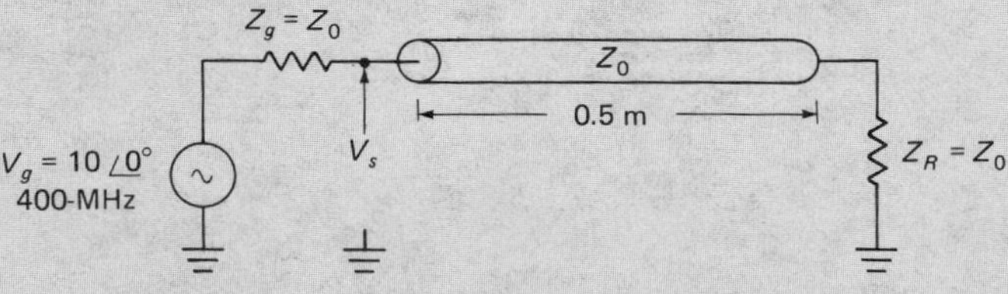

Figure 13-6.

Solution Since the line is terminated by Z_0, there is no reflected voltage at the load. At the generator end, the generator "sees" the Z_0 of the line and thus the equivalent circuit at the generator end is as shown in Fig. 13-7. Thus

$$V_s = 10 \angle 0° \times \frac{Z_0}{Z_0 + Z_0} = 5 \angle 0°$$

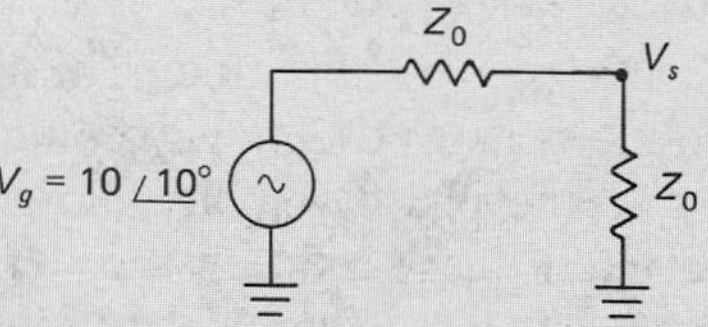

Figure 13-7.

Employing equation (13-10) yields

$$V(1) = V_s e^{-\alpha l} \angle -\beta l$$

$$V(0.5) = 5 \angle 0° \, e^{-(0.01773)(0.5)} \angle -(12.65)(0.5)$$

$$= 4.956 \angle -6.325 \text{ rad} = 4.956 \angle -362.4^0 = 4.956 \angle -2.4°$$

EXAMPLE 13-5

The cable of Example 13-3 is to be cut to a length of $\lambda/4$ at 100 MHz. What is the physical length of this cable?

Solution

$$\beta = \omega Z_0 C = (2\pi)(100 \times 10^5)(52)(96.8 \times 10^{-12}) = 3.163 \text{ rad/m}$$

$$\lambda = \frac{2\pi}{\beta} = \frac{2\pi}{3.163} = 1.987 \text{ m}$$

$$\frac{\lambda}{4} = 0.497 \text{ m} = 49.7 \text{ cm}$$

Since both the voltage and the current decay exponentially as they move down a matched transmission line, the power decays as $e^{-2\alpha l}$ with distance. The received power therefore is

$$P_R = P_s e^{-2\alpha l} \tag{13-20}$$

The power that is dissipated in the line is therefore

$$P_{\text{diss}} = P_R - P_S = P_S(1 - e^{-2\alpha l}) \tag{13-21}$$

$$P_{\text{diss}} = P_S(1 - 1 + 2\alpha l) = P_S 2\alpha l \qquad \text{for } \alpha l << 1 \tag{13-22}$$

From this it can be observed that the greater the loss, the greater the power dissipated in the line. Since α increases with frequency due to skin effect, the power-handling capacity P_s of the line must be reduced.

13-4 DISTORTIONLESS TELEPHONE LINE

To pass voice through a cable pair without undergoing distortion, the attenuation of the message, which contains many frequency components, should be constant regardless of frequency, and the phase velocity should also be constant over the voice band to maintain a fixed phase relationship between all frequency components. The received message will be identical to the transmitted message, but somewhat attenuated and delayed in time. If the phase velocity must be constant with frequency, the phase constant must be proportional to the frequency as observed from relation (13-11). Therefore, for distortionless transmission,

$$\alpha = \text{constant}$$

$$\beta = \omega \times \text{constant}$$

If a line is loss-less where $R = G = 0$, $\alpha = 0$ and $\beta = \omega\sqrt{LC}$ and we have distortionless transmission. Also, at very high frequencies, using the binomial expansion on equation (13-7), it can be shown that α and β can be approximated by

$$\alpha \approx \frac{R}{2}\sqrt{\frac{C}{L}} + \frac{G}{2}\sqrt{\frac{L}{C}} \tag{13-23}$$

$$\beta \approx \omega\sqrt{LC}$$

and we again have distortionless transmission.

In the voice-frequency range, however, these approximations do not hold and the line is generally not distortionless. A distortionless line can still be obtained by forcing the line constants to bear the following relationship:

$$\frac{R}{L} = \frac{G}{C} \tag{13-24}$$

Substituting the condition into (13-7), we obtain

$$\alpha \approx \sqrt{RG} \tag{13-25}$$

$$\beta \approx \omega\sqrt{LC} \tag{13-26}$$

and thus $v_p = 1/\sqrt{LC}$.

For a typical nonloaded telephone cable, the ration R/L is usually much greater than G/C. The most cost-effective method of forcing the distortionless condition is to increase L by inductive loading. Reducing R or C results in large-diameter conductors or large conductor spacings, respectively, and increasing G causes increased attenuation of the signal.

Loading can be achieved by wrapping a highly permeable material, such as iron wire, helically around the conducting core. This method is used on transatlantic submarine cables where the high expense can be justified. On land-line systems, lumped inductances, as shown in Fig. 13-8, are inserted in the line at regular intervals, causing the line to act much like a low-pass filter. The effect of lumped loading is similar to the continuously loaded line up to a critical cutoff frequency. Beyond the cutoff frequency, the attenuation increases sharply, as shown in Fig. 13-9. Using 88-mH loading coils at 1.6-km spacing results in a cutoff frequency of around 3.5 kHz. A certain amount of distortion is usually tolerated and the inductance does not arrive at the values required by the ratio given in equation (13-24).

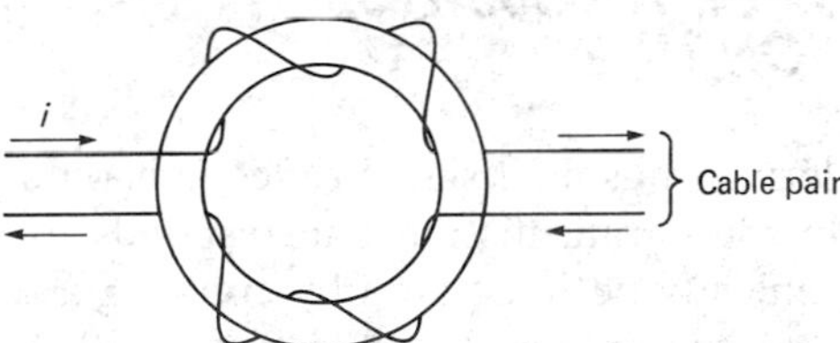

Figure 13-8 Loading coil winding.

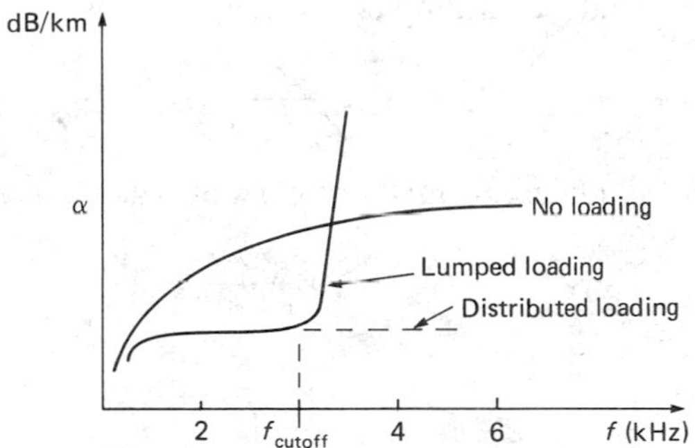

Figure 13-9 Effect of loading coils on attenuation.

The propagation velocity of the signal on a loaded line also tends to be more constant than for the unloaded line, but has the disadvantage of being much lower. This flatness results in very little delay distortion, but the longer delay is of real concern when reflections or echoes are present on the line. The latter tend to be very annoying if the delay is in the order of several milliseconds. Typical transmission velocities along a loaded wire pair are around 30,000 km/s at voice frequencies. This results in a delay of 0.03 ms/km or 30 ms for a 1000-km run. Loading also causes the characteristic impedance to increase and become real and equal to $\sqrt{R/G}$. This permits excellent matching over the voice band range.

For local telephone loops, the line linking the subscriber to the local exchange, 19H88 and 22D66 cable is most frequently used. The first two digits represent the wire gauge, the letters H and D indicate loading-coil spacings of 6000 and 4500 ft, and the last two digits indicate the inductance in millihenries. The loading coils increase the magnetic field strength about the conductors, which results in more crosstalk between adjacent lines.

The sharp cutoff of high frequencies on lumped loaded cable precludes the transmission of the high frequencies necessary for multiplexing or data transmission. Therefore, loading coils must be removed from a line if these higher frequencies are to be transmitted and amplifiers used at regular intervals to prevent signal degradation on long lines. Baseband data transmission rates of 300 kbits/s are generally obtainable on subscriber lines as long as all loading coils are absent.

13-5 TRANSIENTS FOR A LOSS-LESS TRANSMISSION LINE

Before considering sinusoidal signal reflections on a transmission line, let us consider how a loss-less transmission line responds to a voltage and current surge. This can conveniently be illustrated by applying a dc voltage through a switch to the input of a coaxial line as shown in Fig. 13-10. When the switch closed at $t = 0_S$, the incident wave initially sees Z_0 of the line. The wave does not know what is at the receiving end until it reaches that point at a time $t_1 = l/v_p$, where v_p is the velocity of propagation and l is the line length. Just after the switch closes at $t = 0^+$ s, the incident sending end voltage, v^+, will be, by voltage-divider action,

$$v^+ = V_g \frac{Z_0}{Z_g + Z_0} = V_g \frac{Z_0}{2Z_0} = \frac{V_g}{2} \tag{13-27}$$

Similarly, the sending end incident current will be

$$i^+ = \frac{V^+}{Z_0} = \frac{V_s}{2Z_0} \tag{13-28}$$

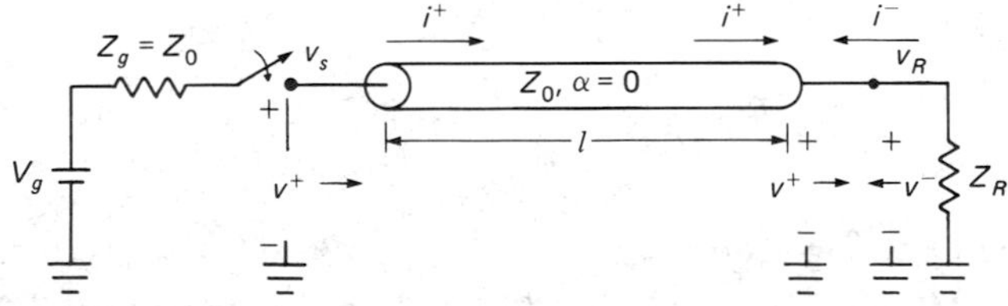

Figure 13-10 Step wave applied to a transmission line.

When the incident voltage and current arrive at the receiving end, the voltage and current must be related by Ohm's law (i.e., $v/i = Z_L$). If $Z_R = Z_0$, the incident voltage-to-current ratio is equal to Z_0 and the total voltage or current can be taken to be the incident voltage and current. In other words, no reflections occur and Ohm's law holds: that is, $v^+/i^+ = Z_R = Z_0$. If however, $Z_R \neq Z_0$, another wave must be established to assure that Ohm's law is obeyed. The total voltage and current consist of incident and reflected waves. Thus

$$\frac{v}{i} = \frac{v^+ + v^-}{i^+ + i^-} = Z_R \tag{13-29}$$

where v^- and i^- are the reflected voltage and current. Since the reflected current must travel in the opposite direction,

$$i^- = -\frac{v^-}{Z_0} \tag{13-30}$$

Substituting equations (13-28) and (13-30) into (13-29), we have

$$Z_R = Z_0 \frac{v^+ + v^-}{v^+ - v^-} \tag{13-31}$$

From this equation, the relation between the reflected voltage and the incident voltage can be found:

$$\frac{v^-}{v^+} = \frac{Z_R - Z_0}{Z_R + Z_0} = \Gamma_R \tag{13-32}$$

The constant Γ_R is called the *voltage reflection coefficient*. By equation (13-32) the reflected voltage will be

$$v^- = \Gamma_R v^+ \tag{13-33}$$

It can be noted that if $Z_R = Z_0$, Γ_R goes to zero and no reflected wave is present.

From equations (13-28) to (13-30) the current reflection coefficient can be found to be

$$\frac{i^-}{i^+} = \frac{Z_R - Z_0}{Z_R + Z_0} = -\Gamma_R \tag{13-34}$$

These reflected signals now travel back toward the generator end, and since the generator end is matched to the transmission line (i.e., $Z_g = Z_0$), no further reflections are encountered. If $Z_g \neq Z_0$, multiple reflections occur, gradually descending in magnitude. For the example of Fig. 13-10, the resultant voltages at the receiving and sending end with time are those shown in Fig. 13-11.

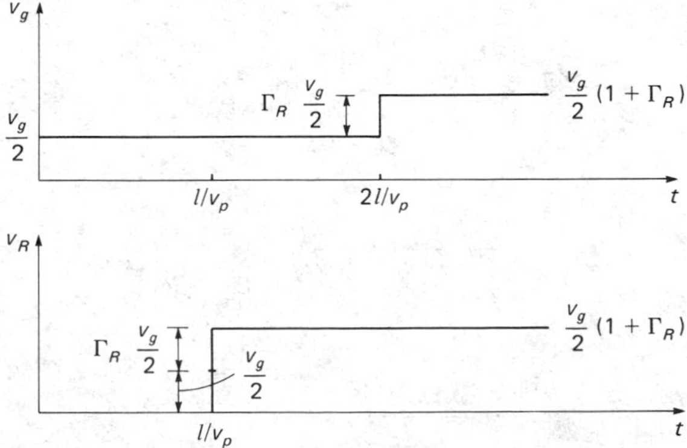

Figure 13-11 Waveforms at the sending (a) and receiving (b) end of Figure 13-7.

EXAMPLE 13-6

A step input is applied to a 20-m length of coaxial cable that is terminated with its characteristic impedance of 50 Ω, but that has a 25-Ω fault 15 m from the input. If the velocity factor is 66%, find the voltages at the input and output points of the cable if a step voltage is applied to the line by a 10-V 50-Ω generator. Assume no attenuation.

Solution The problem can be depicted as shown in Fig. 13-12. To travel 15 m takes $15/(0.66 \times 3 \times 10^8) = 75.75$ ns. To travel 5 m takes $5/(0.66 \times 3 \times 10^8) = 25.25$ ns. When the 10 V is applied, $10[50/(50 + 50)] = 5$ V begins to move down the line. When this incident wave reaches point B, 75.75 ns later, it experiences a mismatch of 25 Ω in parallel with the 50 Ω of the continuing section of transmission line: that is,

$$\Gamma_B = \frac{Z_B - Z_0}{Z_B + Z_0} = \frac{16.67 - 50}{16.67 + 50} = -0.5$$

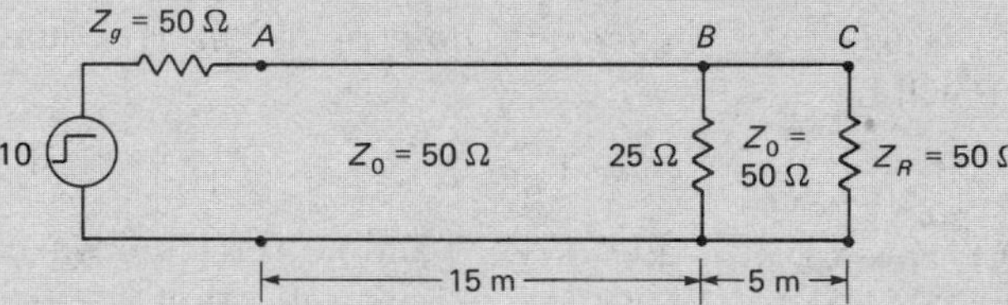

Figure 13-12
Circuit for Example 13-6.

The reflected voltage from point B will be

$$v_B^- = \Gamma_B v^+ = (-0.5)(5) = -2.5 \text{ V}$$

This reflected wave of −2.5 V will travel back to the generator and will not experience another reflection at the generator, as it is matched.

The net voltage at B, $v_B^+ + v_B^- = 5 - 2.5 \text{ V} = 2.5$ V, will be the reduced incident wave marching toward the load. As the load is also matched, there will be no further reflection at the load end. The resulting voltage waveforms are shown in Fig. 13-13.

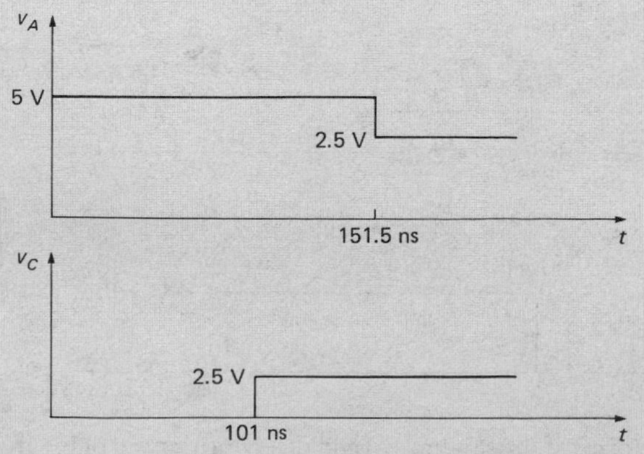

Figure 13-13
Waveforms on the cable shown in Figure 13-12.

As a check, the final voltages in the cable under dc conditions should be (see Fig. 13-14)

$$V = 10\text{ V}\,\frac{16.67}{50 + 16.67} = 2.5\text{ V}$$

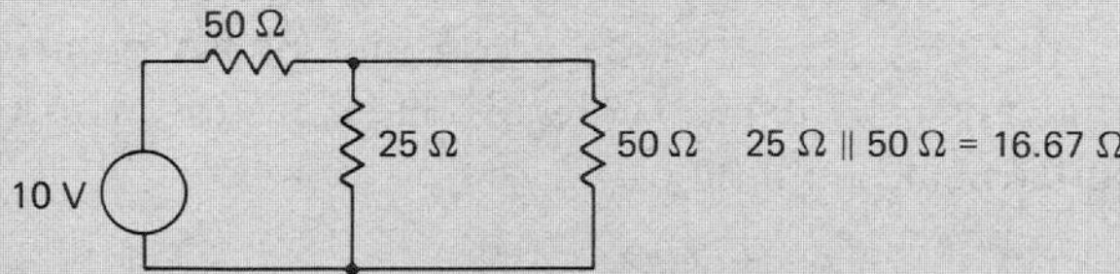

Figure 13-14 Dc equivalent of Figure 13-12.

13-5.1 Time-Domain Reflectometer

The time-domain reflectometer (TDR) is an instrument that applies a step voltage or a bell-shaped pulse to a transmission line while simultaneously monitoring on a CRT both the ensuing incident and reflected waves. From the display, the location and nature of any discontinuities can be determined. The shape and magnitude of the reflected wave indicate the nature and magnitude of the mismatch, while the location of the mismatch is determined by the elapsed time between application of the step voltage and the reflected signal. The elapsed time viewed on the oscilloscope is double the time for the incident wave to arrive at the discontinuity. If the reflected voltage causes an increase in voltage display, $Z_R > Z_0$, and if it causes a decrease, $Z_R < Z_0$.

Figure 13-15 shows some typical TDR displays. If the cable is rather lossy, the rise time and shape of the reflected voltage can be severely degraded.

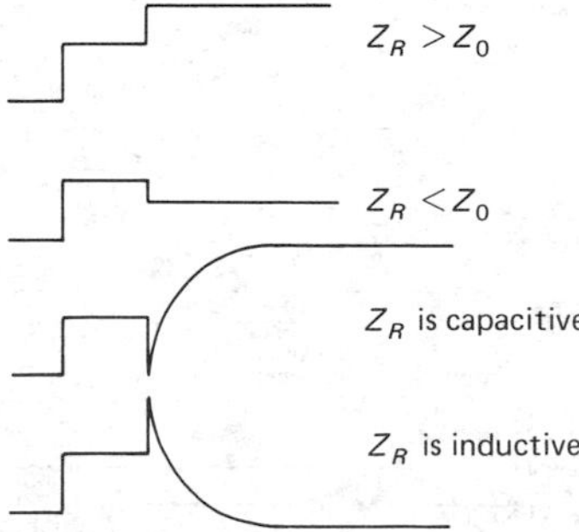

Figure 13-15 TDR display for various line terminations.

13-6 SINUSOIDAL TRAVELING AND STANDING WAVES

Just as with transient signals, sinusoidal signals also travel along a transmission line with finite velocity. Figure 13-16 shows a typical plot of a voltage along a lossy transmission line for three different instances of time. This is equivalent to taking three flash pictures of the voltage along the line for three different times. One can note that the wave appears to move in the $+x$ direction. The phase velocity of the wave (i.e., the velocity at which a constant phase point on the wave travels) is equal to that given by expression (13-11). If no loss is present, the sinusoids will all have equal peak amplitudes.

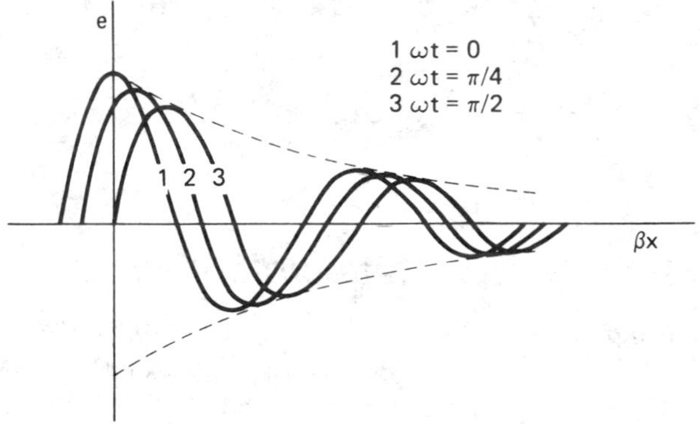

Figure 13-16 Traveling wave on a lossy line.

If the line is mismatched, a similar decaying traveling wave would move in the opposite direction. When both incident and reflected waves are present on a line, the interaction results in the creation of a standing-wave pattern. The name *standing wave* is given to these waveforms because they appear to be motionless, varying only in amplitude. Figure 13-17 gives an example of such a standing wave on a loss-less line.

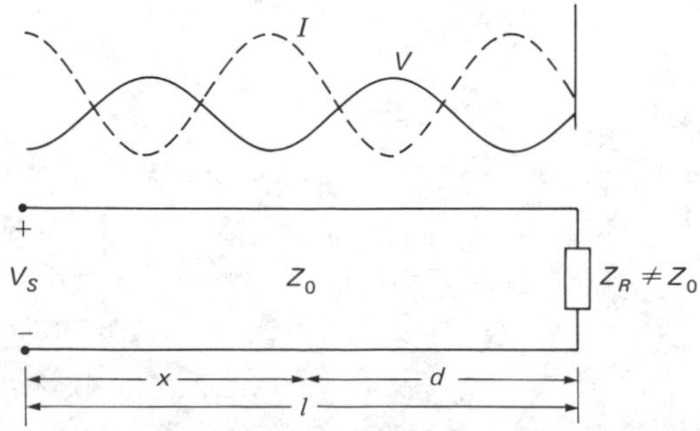

Figure 13-17 Standing wave on a lossless mismatched line.

Short-Circuited Transmission Line

Let us consider the particular case of the frequently employed short-circuited lossless transmission line where $\alpha = 0$ and $Z_R = 0$ and therefore $\Gamma_R = -1$. It can be shown that the voltage and current along the short-circuited transmission line are given by

$$V(d) = jI_R Z_0 \sin \beta d = jI_R Z_0 \sin \frac{2\pi}{\lambda} d \tag{13-35}$$

$$I(d) = I_R \cos \beta d = I_R \cos \frac{2\pi}{\lambda} d \tag{13-36}$$

and therefore,

$$Z(d) = jZ_0 \tan \beta d = jZ_0 \tan \frac{2\pi}{\lambda} d \tag{13-37}$$

where I_R is the line current at the short circuit and d is the distance from the load.

When comparing $I(d)$ and $V(d)$, we can note that they are always 90° out of phase. This could have been expected since no power can be absorbed in a lossless shorted line. Plotting these expressions in graphical form (Fig. 13-18), we can note that a standing-wave pattern results, with distance between nulls being a half-wavelength. This phenomenon is frequently used to measure high frequencies (e.g., light waves).

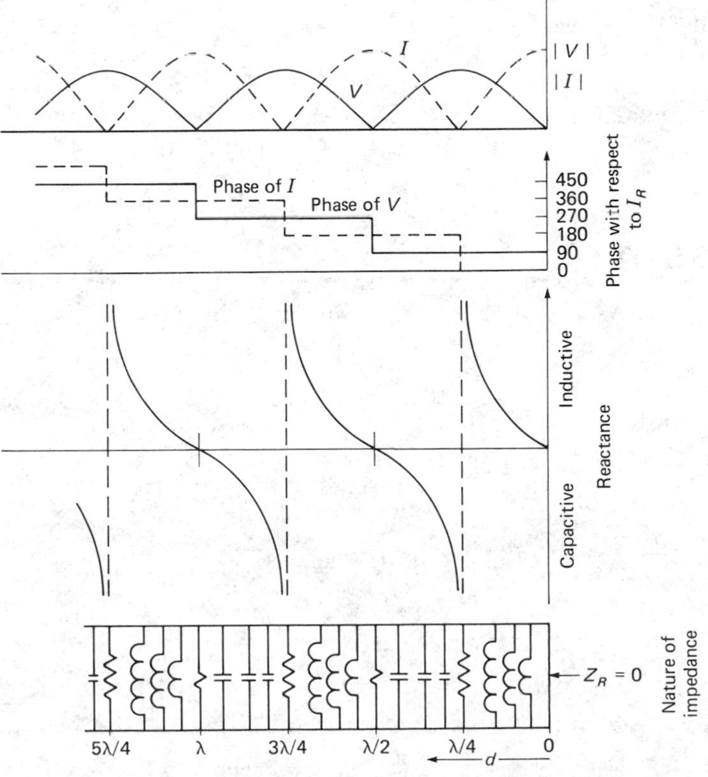

Figure 13-18 Voltage, current, and impedance along a short-circuited transmission line.

It should also be noted that a shorted line produces a purely reactive circuit, depending on the line length and frequency of operation. The shorted line is frequently used to form a series or a parallel resonant circuit, depending on whether the line length is at around a half-wavelength or a quarter-wavelength, respectively.

If the voltage reflection coefficient at the load is Γ_R [see equation (13-28)], it can be shown that the reflection coefficient anywhere along the line is given by

$$\Gamma(d) = \frac{V^-(d)}{V^+(d)} = \Gamma_R e^{-2\gamma d} = \Gamma_R e^{-2\alpha d} e^{-j2\beta d} \tag{13-38}$$

Considering only the magnitude of the voltage reflection coefficient, the reflection coefficient diminishes in magnitude as line positions farther from the load are considered (larger d values). [i.e., $|\Gamma(d)| = |\Gamma_R| e^{-2\alpha d}$]. The reason for this phenomenon is that the reflected signal undergoes more attenuation that the incident signal component, as it must travel an extra distance of $2d$ from the reflection coefficient measurement point.

The *voltage standing-wave ratio* (VSWR), which is defined as the ratio of the maximum voltage along the line to the minimum voltage along the line, is convenient to measure. The VSWR is related to the reflection coefficient by the expression

$$\text{VSWR} = \frac{|V(d)_{\text{max}}|}{|V(d)_{\text{min}}|} = \frac{1 + |\Gamma(d)|}{1 - |\Gamma(d)|} \tag{13-39}$$

In decibels, this is expressed as

$$\text{VSWR (dB)} = 20 \log \text{VSWR} \tag{13-40}$$

For a matched line, where $\Gamma(d) = 0$, the VSWR has a minimum value of 1.

EXAMPLE 13-7

A 50-Ω loss-less cable is terminated with an impedance of $Z_R = 50 - j50\ \Omega$. Find Γ_R and the VSWR on the line.

Solution

$$\Gamma_R = \frac{Z_R - Z_0}{Z_R + Z_0} = \frac{50 - j50 - 50}{50 - j50 + 50} = \frac{j1}{2 - j1}$$

$$= \frac{1 \angle 90^\circ}{2.236 \angle -26.57^\circ} = 0.447 \angle 116.57^\circ$$

$$\text{VSWR} = \frac{1 + 0.447}{1 - 0.447} = 2.6 = 20 \log 2.6 = 8.35 \text{ dB}$$

EXAMPLE 13-8

The magnitude of the voltage reflection coefficient 2 m from the shorted end of a lossy transmission line is 0.8. Determine the VSWR at the load and at the 2-m point.

Solution At the load, $|\Gamma_R| = |\Gamma(0)| = 1$. Substituting into expression (13-39), the VSWR is found to be infinity. At the 2-m point, $|\Gamma(2)| = 0.8$. The VSWR here is $(1 + 0.8)/(1 + 0.8) = 9$.

Thc VSWR always improves as one moves away from the load on a lossy line. An attenuator or "pad" between the signal generator and a transmission system under test is often inserted in practice to make the system appear as a good match.

13-7 DIRECTIONAL COUPLER

The directional coupler is a passive device that couples the signal to an auxiliary arm, separating the incident and reflected waves. The microstrip-line dual directional coupler shown in Fig. 13-19 causes a portion of the incident wave to appear at port D and a portion of the reflected wave to appear at port C. The ratio, expressed in decibels, of the forward power in the main arm P_A^+ to the power coupled to port D (P_D^+), is called the *coupling factor* of the directional coupler. If

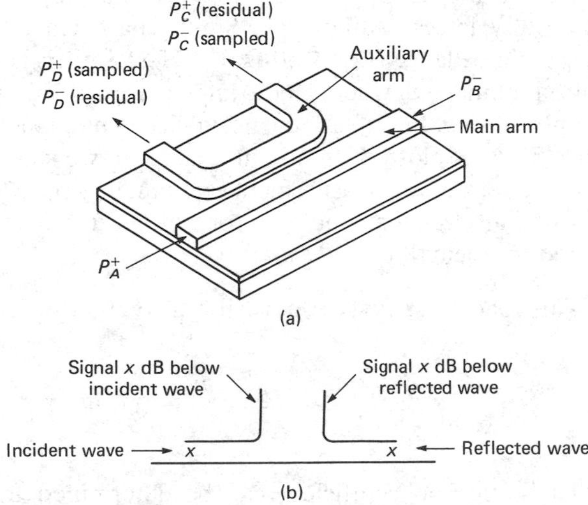

Figure 13-19 Microstrip-line dual directional coupler (a) and the corresponding symbol (b).

port C is not well matched, some portion of the reverse power may also find its way to port D. For the same power applied in the reverse direction on the main arm as applied previously in the forward direction, the ratio, expressed in decibels, of the forward sampled power (P_D^+) to the undesired portion of the reverse signal power (P_D^-) is called the *directivity* of the coupler. The directivity indicates how well a coupler can isolate two signals, and therefore sets the limit on the accuracy of a specific measurement. Usually, both arms track with frequency if symmetrically designed. Often, one auxiliary arm of the directional coupler is internally terminated with its characteristic impedance, permitting the measurement of either the incident or reflected wave, depending on the orientation of the coupler.

In equation forms, these definitions can be expressed as

$$\text{coupling factor} = 10 \log \frac{P_A^+}{P_D^+} \quad \text{dB} \tag{13-41}$$

$$\text{directivity} = 10 \log \frac{P_D^+}{P_D^-} \quad \text{dB} \tag{13-42}$$

where $P_B^- = P_A^+$. Directional couplers can be purchased with various coupling factors and the directivity should exceed 30 dB for the more precise measurements.

13-8 SMITH CHART

To calculate the input impedance of a length of transmission line terminated in a complex load, or to determine a load impedance when the impedance is measured some distance from a load, can be a very tedious affair. For convenience, P. H. Smith devised a chart, now called a *Smith chart*, which graphically represents the line impedance at any point along a loss-less line. Modifications can be made to incorporate any losses. Although a Smith chart can be developed for a single characteristic impedance (i.e., a 50-Ω or a 75-Ω Smith chart), usually the universal Smith chart is employed, where all the impedances (or admittances) are normalized. On the Smith chart of Fig. 13-20, the normalized impedance Z/Z_0, or the normalized admittance Y/Y_0, is plotted. In all our examples, we shall use a Z_0 of 50 Ω ($Y_0 = 1/50 = 0.02$ S), as most measurement equipment employs 50-Ω coaxial lines.

Before considering a concrete problem, let us consider a load of $Z_R = 30 + j70$ attached to a length of 50-Ω line.

1. This load must first be normalized to 50 Ω before plotting on the Smith chart.

$$\frac{Z_R}{Z_0} = \frac{30 + j70}{50} = 0.6 + j1.4$$

2. The reflection coefficient can be determined from equation (13-32) or graphically from the Smith chart. The magnitude of the reflection coefficient is

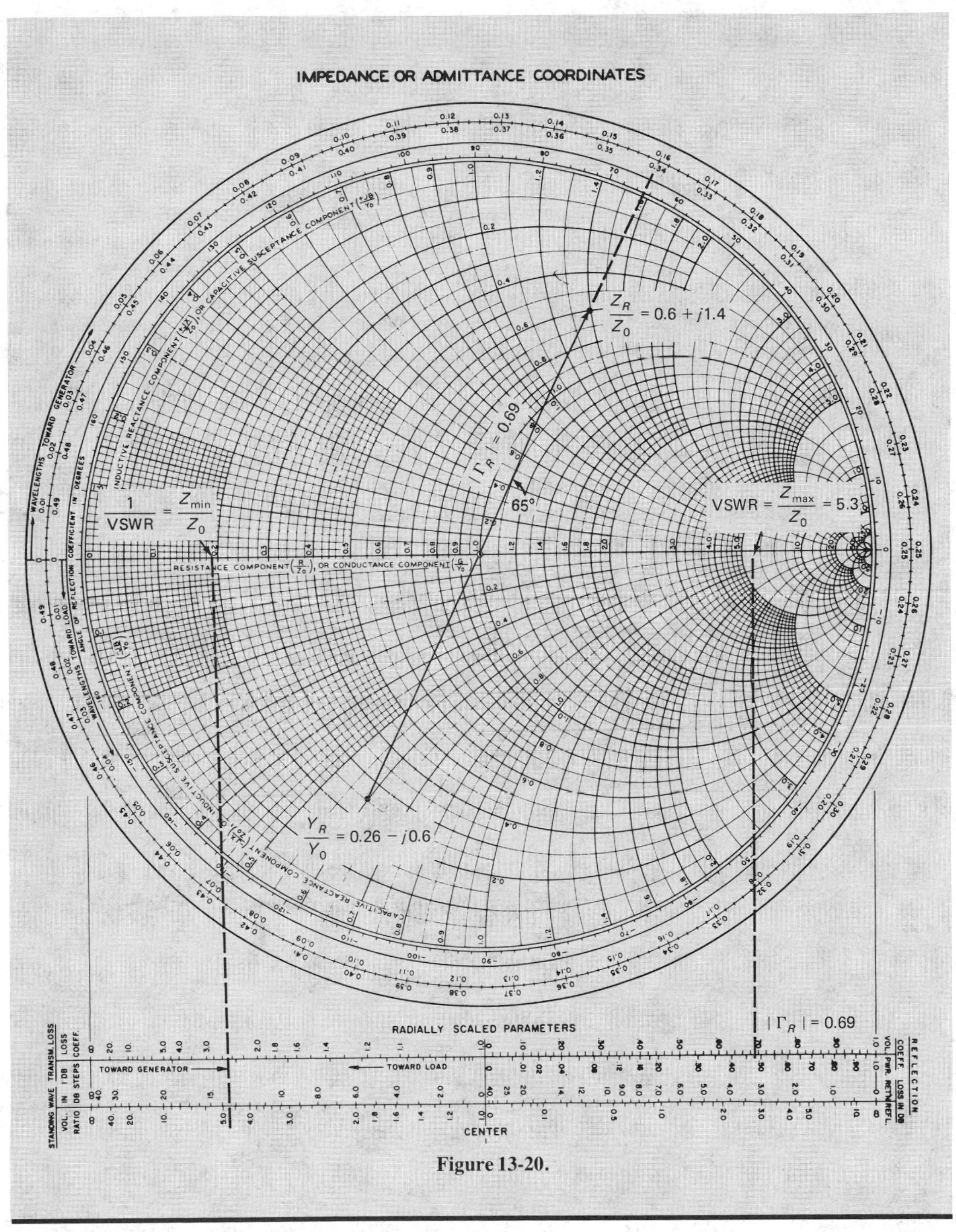
IMPEDANCE OR ADMITTANCE COORDINATES
$\frac{Z_R}{Z_0} = 0.6 + j1.4$
$|\Gamma_R| = 0.69$
65°
$\frac{1}{VSWR} = \frac{Z_{min}}{Z_0}$
$VSWR = \frac{Z_{max}}{Z_0} = 5.3$
$\frac{Y_R}{Y_0} = 0.26 - j0.6$
RESISTANCE COMPONENT, OR CONDUCTANCE COMPONENT
INDUCTIVE REACTANCE COMPONENT, OR CAPACITIVE SUSCEPTANCE COMPONENT
CAPACITIVE REACTANCE COMPONENT, OR INDUCTIVE SUSCEPTANCE COMPONENT
WAVELENGTHS TOWARD GENERATOR
WAVELENGTHS TOWARD LOAD
ANGLE OF REFLECTION COEFFICIENT IN DEGREES
RADIALLY SCALED PARAMETERS
$|\Gamma_R| = 0.69$
TOWARD GENERATOR
TOWARD LOAD
CENTER
STANDING WAVE
TRANSM. LOSS
REFLECTION

Figure 13-20.

found by measuring the radial distance from the center of the Smith chart to the normalized impedance. This radial distance is linear, having a maximum reflection coefficient magnitude of 1 when at the chart circumference. For convenience a "refl. coeff." scale is provided below the Smith chart. The phase angle of the reflection coefficient can be determined by extending the radial line to the protractor scale located outside the circumference of the Smith chart. In this example, $\Gamma_R = 0.69 \angle 65°$.

3. The apparent impedance seen looking into the input terminals of a network consisting of a transmission line terminated in a complex load Z_R is not in general the same as Z_R. In fact, the impedance varies depending on how long a length of transmission line is inserted between the load and the point of impedance measurement. The line length is measured in wavelengths and the impedance changes occur as a result of the phase shifts that are present between the input and load terminals of the transmission line.

The Smith chart provides a convenient method for predicting the impedance seen looking into a transmission line some distance from the actual load. As noted previously, the first step is to plot the normalized impedance Z_R/Z_0 on the chart. Then at this point from the load, one can visualize walking down the transmission line toward the generator and observing the changing impedance you see looking back toward the load.

4. Since the magnitude of the reflection coefficient does not vary as one moves along a loss-less transmission line [see equation (13-38)], only its phase angle changes; a circle passing through the *normalized* impedance forms the locus of possible impedance seen along the transmission line.

5. The VSWR on the line can be determined by noting where the constant $|\Gamma|$ circle intercepts the right half of the centerline of the chart. In this case VSWR = 5.3. This also happens to be the maximum impedance point (or maximum voltage point) along the line. Please note that this is a purely resistive impedance: that is, $Z_{max}/Z_0 = 5.3$ or $Z_{max} = 365\ \Omega$. The opposite side or minimum impedance point is the 1/VSWR point.

6. The point directly opposite any point on the constant $|\Gamma|$ circle gives the corresponding inverse normalized impedance or admittance. For the $30 + j70$ load,

$$1 \Big/ \frac{Z_R}{Z_0} = \frac{Y_R}{Y_0} = \frac{1}{0.6 + j1.4} = 0.26 - j0.60$$

$$Y_R = 0.02(0.26 - j0.6) = 0.0052 - j0.012\ \text{S}$$

7. One complete revolution around the Smith chart represents a $\lambda/2$ distance change on the transmission line. From equation (13-38), a $\lambda/2$ change in d represents a rotation of $-2\beta d = -2(2\pi/\lambda)(\lambda/2) = 2\pi$ rad.

EXAMPLE 13-9

The input impedance seen looking into a 2-m length of RG8/U 50-Ω line is measured to be $50 - j30\ \Omega$ at 1 GHz (Fig. 13-21). The line has a capacitance of 96.8pF/m. If losses are ignored, what is the load impedance?

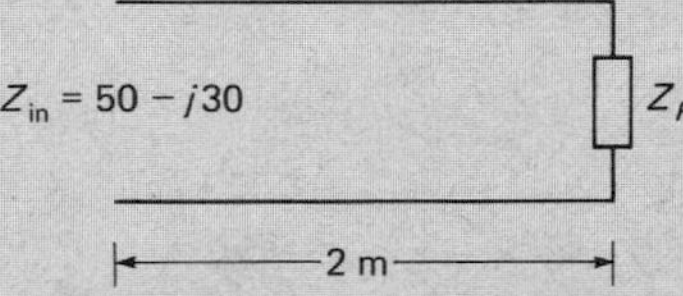

Figure 13-21 Transmission line used in Example 13-9.

Solution

$$\beta = \omega Z_0 C = (2\pi \times 10^9)(50)(96.8 \times 10^{-12}) = 30.4 \text{ rad/m}$$

$$v_p = \frac{\omega}{\beta} = \frac{2\pi \times 10^9}{30.4} = 2.067 \times 10^8 \text{ m/s}$$

$$\lambda = \frac{v_p}{f} = \frac{2.067 \times 10^8}{10^9} = 0.2067 \text{ m}$$

Length of line in wavelengths = 2/0.2067 = 9.676λ. To find the load impedance, plot

$$\frac{Z_{in}}{Z_0} = \frac{50 - j30}{50} = 1 - j0.6$$

on the Smith chart of Fig. 13-22 and rotate toward the load 9.676λ on the $|\Gamma| = 0.29$ locus. Since one rotation around the Smith chart is equivalent to a distance of λ/2, it is necessary to rotate 19 times around the chart [9.5λ = 19(λ/2)] plus an additional 0.176λ.

From the "wavelengths toward load" scale reading of 0.149, it is necessary to effectively move to the 0.146 + 0.176 = 0.325 reading on the same scale. The normalized load impedance is read as

$$\frac{Z_R}{Z_0} = 1.22 + j0.62$$

$$Z_R = 50(1.22 + j0.62) = 61 + j31\ \Omega$$

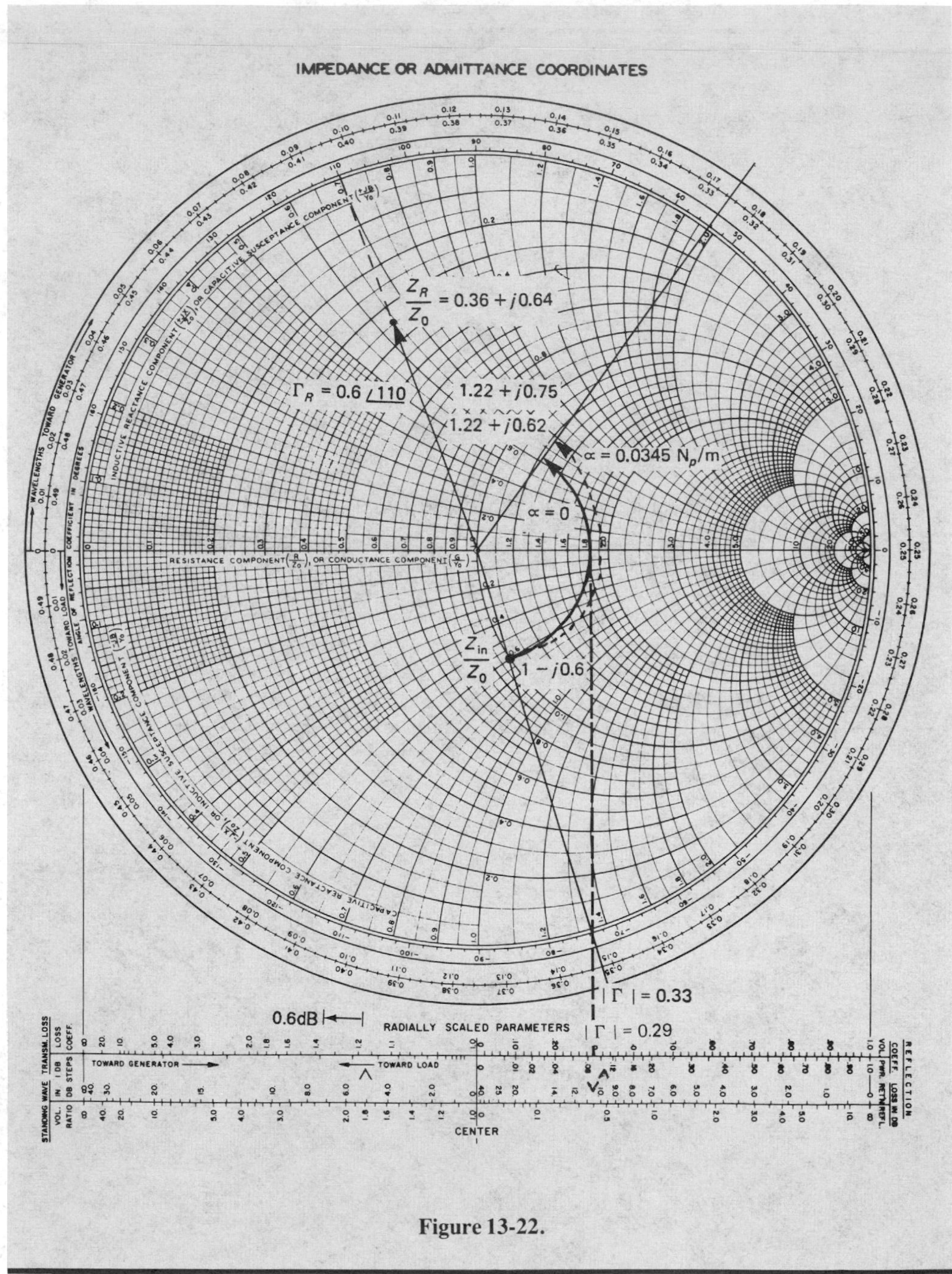
IMPEDANCE OR ADMITTANCE COORDINATES
$\frac{Z_R}{Z_0} = 0.36 + j0.64$
$\Gamma_R = 0.6 \angle 110$
$1.22 + j0.75$
$1.22 + j0.62$
$\alpha = 0.0345\ N_p/m$
$\alpha = 0$
$\frac{Z_{in}}{Z_0}$
$1 - j0.6$
$|\Gamma| = 0.33$
$|\Gamma| = 0.29$
0.6dB
RADIALLY SCALED PARAMETERS
TOWARD GENERATOR
TOWARD LOAD
CENTER
RESISTANCE COMPONENT, OR CONDUCTANCE COMPONENT

Figure 13-22.

EXAMPLE 13-10

If the line of Example 13-9 has a loss of 0.3 dB/m at 1 GHz, what would the load impedance be for the same measured input impedance?

Solution From equation (13-38) we note that when moving toward the generator, the magnitude of the reflection coefficient decays by $e^{-2\alpha d}$. Therefore, when moving toward the generator, the reflection coefficient must increase by a factor of $e^{2\alpha d}$. For a loss of 0.3 dB/m, $\alpha = 0.3/8.868 = 0.0345$ Np/m.

Since the reflection coefficient has a magnitude of 0.29 at the generator, it will have a magnitude of $\Gamma_R = 0.29e^{2(0.0345)2} = 0.29 \times 1.148 = 0.33$ at the load. Locating this on the same radial line as for Example 13-9, we obtain $Z_R/Z_0 = 1.22 + j0.75$ or $Z_R = 61 + j37.5\ \Omega$.

A loss scale in 1-dB steps has been provided below the Smith chart to minimize the calculations. In our example, a total of $0.3 \times 2 = 0.6$ dB loss is experienced, so when rotating toward the load, we must increase $|\Gamma|$ by this amount. The arrows on the scales below the Smith chart indicate the sequence to be followed.

EXAMPLE 13-11

Find the length of 50-Ω short-circuited transmission line required to obtain an inductive reactance of $j100\ \Omega$ at a frequency of 500 MHz. Assume a polystyrene dielectric ($\epsilon_r = 2.55$) and zero loss.

Solution

$$\lambda = \frac{v}{f} = \frac{3 \times 10^8}{\sqrt{2.55} \times 500 \times 10^6} = 0.3757 \text{ m} = 37.57 \text{ cm}$$

$$\frac{Z_R}{Z_0} = 0$$

$$\frac{Z_{in}}{Z_0} = \frac{j100}{50} = j2$$

To find the length of line required, plot $Z_R/Z_0 = 0$ on the Smith chart (Fig. 13-23) and rotate toward the generator on a constant $|\Gamma|$ locus ($|\Gamma| = 1$) until $j2$ is intercepted. Thus $l = 0.176\lambda$ or 6.61 cm.

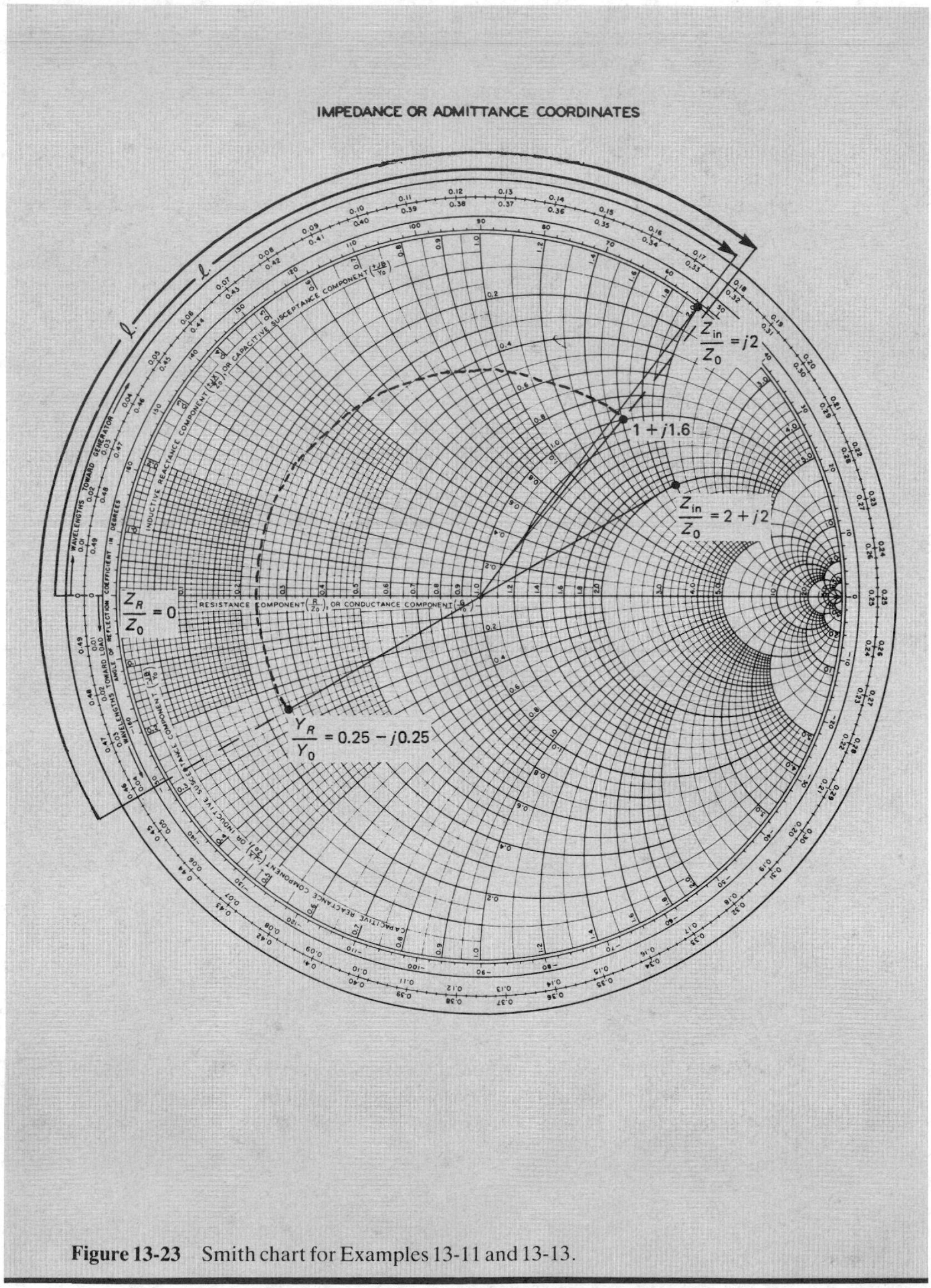

Figure 13-23 Smith chart for Examples 13-11 and 13-13.

EXAMPLE 13-12

Determine the expression for the input impedance to a $\lambda/4$ length of transmission line terminated with Z_R. How can such a "quarter-wave transformer" be used for matching a 75-Ω load to a 50-Ω generator?

Solution The normalized impedance a quarter-wavelength away from a load appears on the opposite side of the Smith chart and therefore is equal to the inverse of the normalized load impedance. Thus $Z_{in}/Z_0 = 1/(Z_R/Z_0)$ or $Z_{in} = Z_0^2/Z_R$. To match a 75-Ω load to a 50-Ω source, a quarter-wavelength of transmission line having a $Z_0 = \sqrt{Z_{in}Z_R} = \sqrt{50 \times 75} = 61.2\ \Omega$ is inserted at the load (Fig. 13-24). The characteristic impedance of a coaxial line can be altered by changing the diameter of the inner or outer conductors, or by selecting a different dielectric.

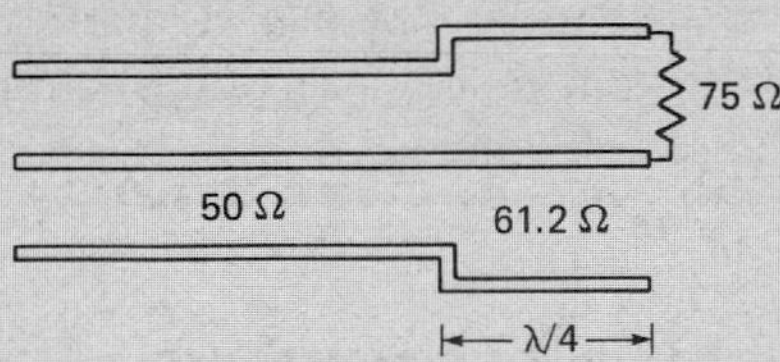

Figure 13-24.

EXAMPLE 13-13

A $100 + j100\ \Omega$ load is to be matched to a 50-Ω transmission line by placing a purely reactive element in parallel with the line at the appropriate location (Fig. 13-25). Determine the location, l_1, and the reactance required.

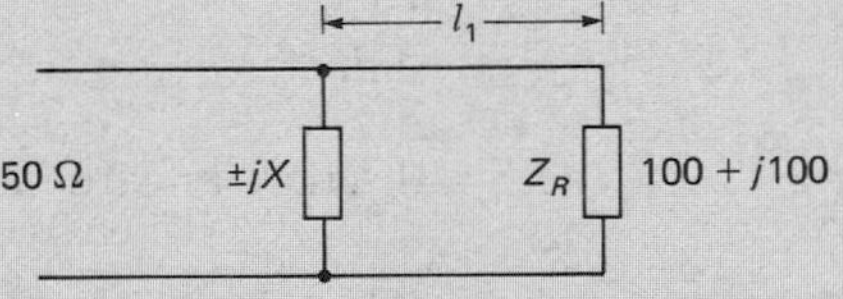

Figure 13-25.

Solution As the reactance is to be placed in parallel with the line, it is convenient to employ admittances since parallel admittances add. The procedure is as follows (refer to Fig. 13-23).

1. Plot $Z_R/Z_0 = (100 + j100)/50 = 2 + j2$ on the Smith chart. Once the normalized impedance is plotted a constant radius (constant $|\Gamma|$) circle can be drawn through the normalized impedance point. The normalized admittance, $Y_R/Y_0 = 0.25 - j0.25$, can be obtained by rotating 180° from the impedance point around the circle. The normalized admittance seen looking back into the line at a particular distance back from the load may

be determined by rotating around the circle from the Y_R/Y_0 point, the corresponding distance on the outside wavelength scale.

2. Since the normalized admittance at the input is to be $Y_{in}/Y_0 = 1 + j0$, rotate along a constant $|\Gamma|$ locus until the normalized conductance is equal to 1. This can always occur at two points and we will select the first intersect. Whatever susceptance is obtained at that location can be canceled out by the reactive component. Thus at a point $l_1 = 0.219\lambda$ from the load, the normalized admittance is equal to $1 + j1.6$.
3. The normalized susceptance of $j1.6$ can be canceled out with a susceptive element having a normalized susceptance of $-j1.6$. The net normalized admittance at the generator side of this junction then is $1 + j0$ and a match is obtained. The susceptance, B_C, of the element is thus capacitive in nature and equal to $1.6Y_0 = 1.6/50 = 0.032$ S or $X_C = 1/B_C = 31.25\ \Omega$. If the frequency and line constants are known, C and l_1 could be calculated. Rather than a lumped capacitance, a section of short-circuited line called a *stub* could also be used to emulate X_C.

13-9 IMPEDANCE MEASUREMENT TECHNIQUE EMPLOYING THE DUAL-DIRECTIONAL COUPLER

In modern impedance measurement equipment such as the Hewlett-Packard network analyzer, the dual-directional coupler of Fig. 13-19 is employed together with a voltmeter that is able to measure the voltage from each of the two pairs of input probes as well as the phase difference between these two input channels. By measuring the incident and reflected voltages at the unknown load terminals, the load reflection coefficient can be calculated (i.e., $\Gamma_R = E_R^-/E_R^+$), and from this the load impedance can be calculated from equation (13-32), rewritten as

$$Z_R = \frac{1 + \Gamma_R}{1 - \Gamma_R} Z_0 \tag{13-43}$$

or determined by plotting Γ_R on the Smith chart and reading the corresponding normalized impedance.

As it is impractical to monitor the incident and reflected voltages right at the load terminals, measurements must be taken some distance from the load. Since impedance measurements are repetitive every half-wavelength, one could place the dual-directional coupler $n\lambda/2$ back from the load (n being an integer). However, this has the disadvantage of needing physical readjustment for every change in frequency as the wavelength changes with frequency.

A practical measurement system is shown in Fig. 13-26. The 50-Ω adjustable line stretcher is adjusted initially to equalize the incident and reflected signal path lengths from some common point near the dual-directional coupler to the sampling probes of the voltmeter ($l_1 = l_2$). This adjustment is initially made by replacing the load by a short circuit ($\Gamma_R = -1$) and adjusting the line stretcher until a phase difference of 180° is read on the phase indicator of the voltmeter. The voltage

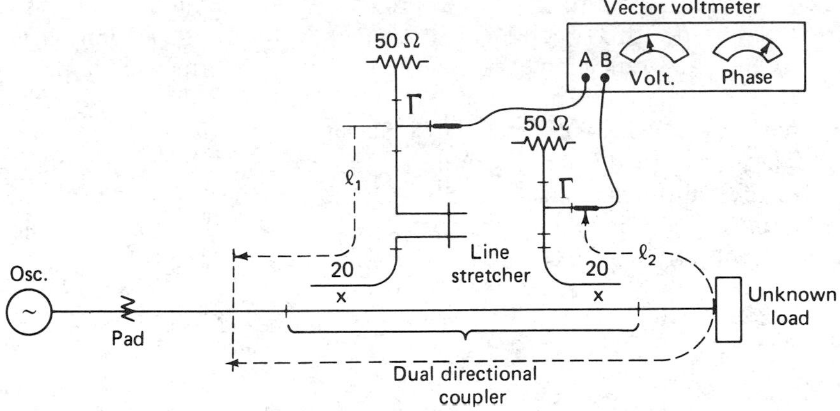

Figure 13-26 Impedance measurement system.

magnitudes should be identical for both channels, barring differences in connector losses, and the like since for a short circuit, the magnitude of reflected voltage is equal to that of the incident voltage. The path length l_1 may now be an integral number of half-wavelengths longer or shorter than the path length l_2, or it may be identical. To assure the latter, vary the frequency of the generator and observe whether or not the phase angle remains at 180°. If not, the stretcher should be readjusted until the 180° phase difference holds steady as the frequency is varied.

With the sampling points now being electrically speaking at the same point as the load terminals, replace the short circuit with the unknown load. Measure the incident and reflected voltage and determine Z_R.

EXAMPLE 13-14

After proper calibration with the short circuit, the following measurements are noted on the voltmeter with the unknown load in place:

$$|V^+| = 60 \text{ mV (channel A)}$$

$$|V^-| = 36 \text{ mV (channel B)}$$

and channel B leads channel A by 110°. Thus

$$\Gamma_R = \frac{V^-}{V^+} = \frac{36 \times 10^{-3}}{60 \times 10^{-3}} \angle 110° = 0.6 \angle 110°$$

Plotting this on the Smith chart of Fig. 13-22, we obtain

$$\frac{Z_R}{Z_0} = 0.36 + j0.64$$

$$Z_R = 50(0.36 + j0.64) = 18 + j32 \ \Omega$$

The HP network analyzer provides the complete system just described, and when used with a sweep oscilliator, can quickly plot the magnitude and phase angle of the load's reflection coefficient or display the mormalized impedance on a Smith chart overlay, depending on the display option chosen. In the latter case, the network analyzer is calibrated by replacing the load temporarily with a short circuit. The scope trace is initially adjusted to form a circular path around the periphery of the Smith chart ($|\Gamma| = 1$) and the line stretcher then adjusted to form a single point at the $0 + j0$ impedance location ($\Gamma_R = -1$). With the load attached, the normalized impedance curve is viewed directly on the Smith chart.

13-10 POWER AND RETURN LOSS

When a signal hits a discontinuity or a mismatched load, a portion of the signal is reflected with the rest passed on or absorbed by the load. The return loss at a discontinuity, such as that illustrated in Fig. 13-27, is defined as the ratio in decibels of the power incident on the discontinuity to the power reflected from the discontinuity. If we let the incident power be P^+ and the reflected power be P^-, the return loss is given by

$$\text{return loss} = 10 \log \frac{P^+}{P^-} = 10 \log \frac{|E^+|^2/Z_0}{|E^-|^2/Z_0}$$

$$= 20 \log \left| \frac{E^+}{E^-} \right| = 20 \log \frac{1}{|\Gamma|} \qquad (13\text{-}44)$$

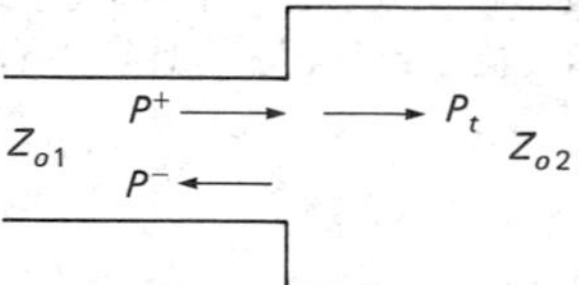

Figure 13-27 Reflections at a discontinuity.

In terms of the VSWR, this can be expressed as [see equation (13-39)].

$$\text{return loss} = 20 \log \frac{\text{VSWR} + 1}{\text{VSWR} - 1} \qquad (13\text{-}45)$$

The power loss at a discontinuity is defined as the ratio in decibels of the incident power to the output or transmitted power P_t. Since $P_t = P^+ - P^-$, the power loss is given by

$$\text{power loss} = 10 \log \frac{P^+}{P_t} = 10 \log \frac{P^+}{P^+ - P} = 10 \log \frac{1}{1 - |\Gamma|^2} \qquad (13\text{-}46)$$

$$\text{power loss} = 10 \log \frac{(\text{VSWR} + 1)^2}{4\ \text{VSWR}} \qquad (13\text{-}47)$$

EXAMPLE 13-15

Find the power reflected, power transmitted, return loss, and power loss when 100 mW of power is incident on the discontinuity shown in Fig. 13-28.

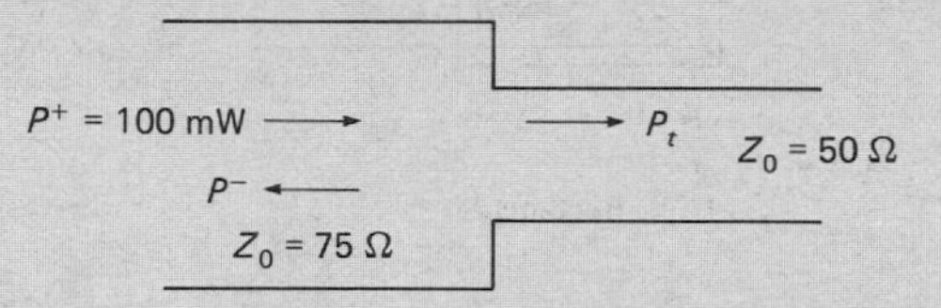

$P^+ = 100$ mW → ; → P_t ; $Z_0 = 50\ \Omega$; P^- ← ; $Z_0 = 75\ \Omega$

Figure 13-28 Discontinuity of Example 13-12.

$$\Gamma = \frac{50 - 75}{50 + 75} = -0.2$$

$$|\Gamma| = 0.2 \qquad |\Gamma|^2 = 0.04$$

$$P^- = |\Gamma|^2 P^+ = 0.04(100 \times 10^{-3}) = 4 \text{ mW}$$

$$P_t = P^+ - P^- = (100 - 4) \text{ mW} = 96 \text{ mW}$$

$$\text{return loss} = 20 \log \frac{1}{0.2} = 10 \log \frac{100 \text{ mW}}{4 \text{ mW}} = 13.98 \text{ dB}$$

$$\text{power loss} = 10 \log \frac{1}{1 - 0.04} = 10 \log \frac{100 \text{ mW}}{(100 - 4) \text{ mW}} = 0.18 \text{ dB}$$

13-11 WAVEGUIDES

A single hollow conductor, commonly called a *waveguide*, cannot support the TEM wave. In such structures, many possible field configurations can propagate. For the rectangular and circular waveguides of Fig. 13-29, the field configurations are described in terms of a TE or TM wave. Although both types may exist simultaneously, for efficient coupling and minimum multimode distortion only one type should exist in the guide. Furthermore, only one mode within each of these types should be permitted to exist for the same reasons.

The TE (transverse electric) wave has its E field entirely perpendicular to the direction of propagation with a component of the H field in the direction of propagation, whereas the TM (transverse magnetic) wave has its H field intensity wholly perpendicular to the direction of propagation, with some of the E field existing in the direction of propagation. The various field configurations within these two types are further specified by the subscripts m and n, where m and n are integers. For the rectangular waveguide, the first subscript m denotes the number of half-wavelength variations of the transverse field in the x-direction, and the second subscript denotes the number of half-wavelength variations of the transverse field in the y-direction.

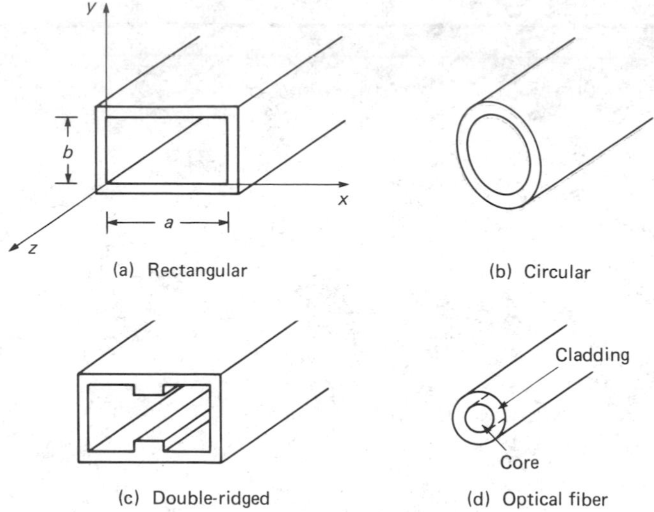

Figure 13-29 Waveguides.

The frequency of operation, the method of excitation, and the waveguide internal surfaces determine the modes that can propagate in a waveguide.

At the boundary of a good conductor, no tangential (parallel) component of the electric field can exist, as the voltage gradient in a good conductor must be zero. In addition, no component of the magnetic field intensity, H, can exist normal to the surface. Stating these boundary conditions in a positive manner, the E field must be normal at the surface and the H field must be tangential to the surface.

Because the optical fiber has no metal boundaries, a hybrid set of modes exist. These modes have all components of the E and H fields present.

13-12 MODES IN A RECTANGULAR WAVEGUIDE

For hollow waveguides, each mode has a specific cutoff frequency, below which the mode is attenuated. The mode with the lowest cutoff frequency is called the *dominant mode*. For the $TE_{m,n}$ or $TM_{m,n}$ modes in the rectangular waveguide of Fig. 13-29(a), the cutoff frequency is given by

$$f_c = \frac{C}{2\sqrt{\mu_r \varepsilon_r}} \sqrt{\left(\frac{m}{a}\right)^2 + \left(\frac{n}{b}\right)^2} \quad \text{Hz} \tag{13-48}$$

where m and n are the mode numbers, a and b are the inside dimensions (m), and μ_r and ε_r are the relative permeability or dielectric constants of the material filling the guide. The corresponding cutoff wavelength is then

$$\lambda_c = \frac{c}{f_c} = \frac{2\sqrt{\mu_r \varepsilon_r}}{\sqrt{(m/a)^2 + (n/b)^2}} \tag{13-49}$$

If we consider a typical rectangular waveguide with an aspect ratio of $b/a = 0.5$, the mode that has the lowest cutoff frequency of the TE modes is the TE_{10} mode. Assuming the waveguide to be filled with air or nitrogen, the cutoff frequency for the TE_{10} mode is

$$f_{cTE10} = \frac{c}{2a} \tag{13-50}$$

and

$$\lambda_c = 2a \tag{13-51}$$

or the width of the waveguide is exactly a half-wavelength (free space).

The ridge waveguide of Fig. 13-29(c) lowers the cutoff frequency of the dominant mode, thus increasing the operation frequency bandwidth. It has the disadvantage of reducing the maximum power-handling capability.

To obtain a clearer picture of how the mode might be configured, consider the two plane TEM waves traveling at slightly different directions of Fig. 13-30(a). Only the E fields are indicated with solid lines (positive crests) and dashed lines (negative crests), but corresponding to these should also be the perpendicular magnetic field intensity lines. Note that where these solid and dashed lines meet, the net electric field is zero. It is at these locations that a vertical conductive wall can be located without interfering with the wave propagation. Such walls can only be placed at distinct separations. For the narrowest separation, only the TE_{10} mode exists, followed by the TE_{20} mode, and so on, for the larger spacings. By completing the enclosure with top and bottom plates, the rectangular waveguide is formed. In actual practice there is but one wave, each being a reflection of the other.

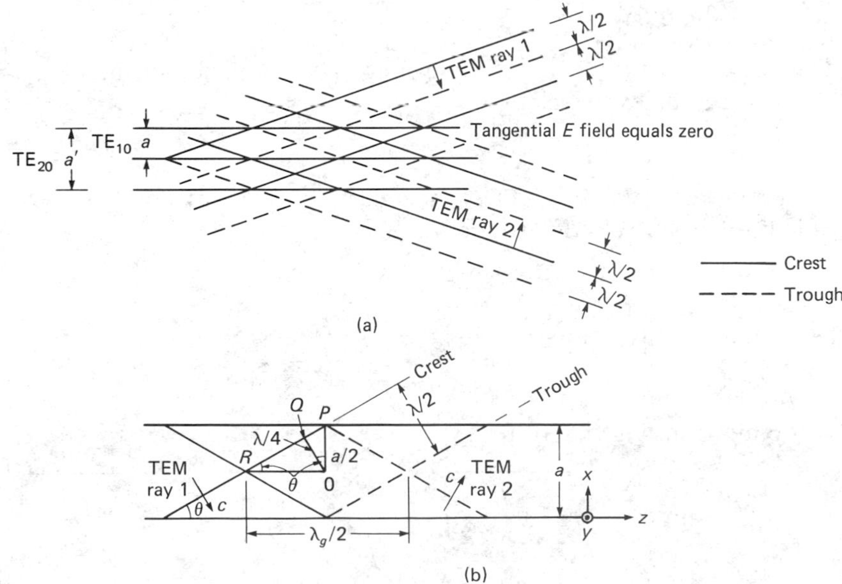

Figure 13-30 Production of the TE_{10} mode by two intersecting plane waves (a) and the geometry used in determining TE_{10} wavelength relationship (b).

Figure 13-30(b) is a magnification of the individual TEM waves forming the TE wave. An observer monitoring the signal along the waveguide would note the waveguide wavelength λ_g as shown in Fig. 13-30(b). This wavelength is longer than the wavelength, λ, of the transmitted signal, which can be obtained by means of the right-angled triangles *OQP* and *RQP*, that is,

$$\cos\theta = \frac{\lambda/4}{a/2} = \frac{\lambda}{2a}$$

$$\sin\theta = \frac{\lambda/4}{\lambda_g/4} = \frac{\lambda}{\lambda_g}$$

Now $\sin^2\theta + \cos^2\theta = 1$; therefore, $(\lambda/\lambda_g)^2 + (\lambda/2a)^2 = 1$ and

$$\lambda_g = \frac{\lambda}{\sqrt{1 - (\lambda/2a)^2}} \tag{13-52}$$

From equation (13-52) it is seen that when $\lambda = 2a$, the waveguide wavelength goes to infinity and the individual TEM waves just bounce back and forth between the walls with no wave motion along the Z axis. The longest wavelength TEM wave that can produce a wave traveling down the waveguide, referred to as the *cutoff wavelength*, λ_c, is equal to $2a$.

$$\lambda_g = \frac{\lambda}{\sqrt{1 - (\lambda/\lambda_c)^2}} \tag{13-53}$$

13-12.1 Phase and Group Velocity of the TE_{10} Mode

The velocity of the plane wave traveling in the direction of the ray shown in Fig. 13-30(b) can be expressed as

$$v = c = \lambda f \tag{13-54}$$

The velocity at which the signal energy propagates along the waveguide axis, called the *group velocity*, is therefore given by

$$v_g = c \sin\theta = \lambda f \sin\theta \tag{13-55}$$

Using the identity

$$\sin\theta = \sqrt{1 - \cos^2\theta} \tag{13-56}$$

we obtain from Fig. 13-30(b)

$$\sin\theta = \sqrt{1 - \left(\frac{\lambda/4}{a/2}\right)^2} = \sqrt{1 - \left(\frac{\lambda}{2a}\right)^2} = \sqrt{1 - \left(\frac{\lambda}{\lambda_c}\right)^2} \tag{13-57}$$

Therefore,

$$v_g = \lambda f \sqrt{1 - \left(\frac{\lambda}{\lambda_c}\right)^2}$$

$$= c\sqrt{1 - \left(\frac{\lambda}{\lambda_c}\right)^2} \tag{13-58}$$

$$= c\sqrt{1 - \left(\frac{f_c}{f}\right)^2} \tag{13-59}$$

An observer, however, sees a wavelength λ_g along the axis of the guide, suggesting that the phase velocity is given by

$$v_p = \lambda_g f = \lambda_g \frac{c}{\lambda} \tag{13-60}$$

which when substituting equation (13-53) gives

$$v_p = \frac{c}{\sqrt{1 - (\lambda/\lambda_c)^2}} \tag{13-61}$$

$$v_p = \frac{c}{\sqrt{1 - (f_c/f)^2}} \tag{13-62}$$

From equation (13-62) note that v_p is greater than the velocity for light for an air-filled guide, whereas the group velocity [equation (13-59)] is always less than c. Note also that

$$v_p v_g = c^2 \tag{13-63}$$

If the waveguide is filled with a dielectric the velocity of light, c, in equations (13-54) through (13-63) should be replaced by $c/\sqrt{\mu_r \varepsilon_r}$.

Figure 13-31 indicates graphically the variation of phase and group velocity with the frequency of operation. From the figure it should be noted that the velocity of propagation varies with frequency. For any signal comprising a band of frequencies, distortion or dispersion occurs in the group velocity of a propagating node. It is the only distortion, usually termed *intramodal distortion*, occurring in single-mode waveguides. The region over which the waveguide should be operated is delineated in Fig. 13-31. At frequencies above this region, higher modes can be set up, causing multimode distortion; that is, different modes have varying group velocities. The propagation constant for the TE_{10} mode can be expressed as

$$\beta_g = \frac{\omega}{v_p} = \frac{\omega}{c}\sqrt{1 - \left(\frac{f_c}{f}\right)^2} \tag{13-64}$$

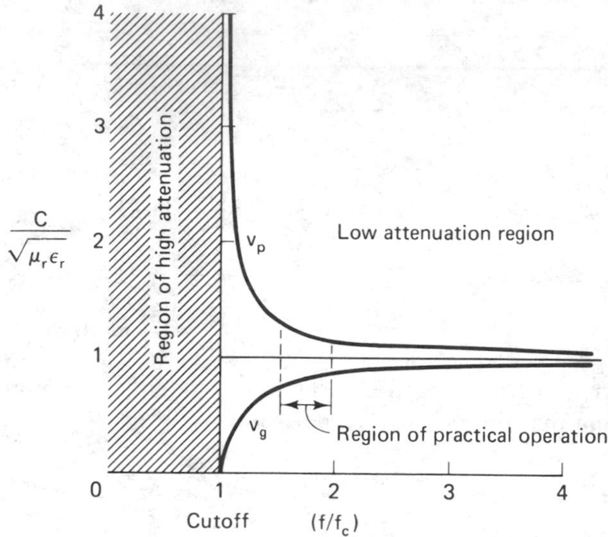

Figure 13-31 Variation of phase and group velocity with frequency.

13-12.2 Higher-Order TE and TM Modes

In addition to the TE_{10} mode just described in some detail, many other modes may exist. The field configurations for some of the lower-order $TE_{m,n}$ and $TM_{m,n}$ modes in rectangular waveguides are shown in Fig. 13-32. The cutoff frequencies for the $TM_{m,n}$ modes are also given by equation (13-48). Because the magnetic field lines must be continuous, there must be at least one-half-wave variation of the magnetic field in both transverse directions, disallowing the existence of the TM_{01} and TM_{10} modes. For an aspect ratio of $b/a = \frac{1}{2}$, the TE_{01} and TE_{20} modes both have cutoff frequencies equal to c/a, which is twice the TE_{10} cutoff frequency. Figure 13-33 shows the relative cutoff frequencies for several rectangular waveguide modes ($b/a = \frac{1}{2}$).

In principle the mode set up in the waveguide is dependent on the method by which the signal is coupled into the waveguide. In practice, irregularities in the waveguide surface and at discontinuities such as bends, matching devices, and so on, cause other modes to be excited.

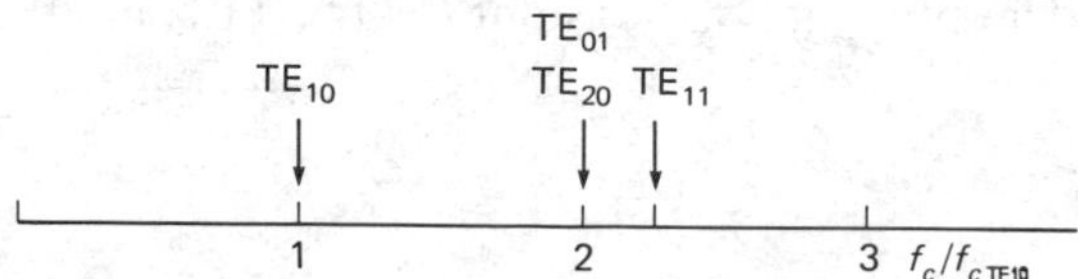

Figure 13-33 Cutoff frequencies of several rectangular modes relative to the dominant, TE_{10} mode cutoff frequency ($b/a = 1/2$).

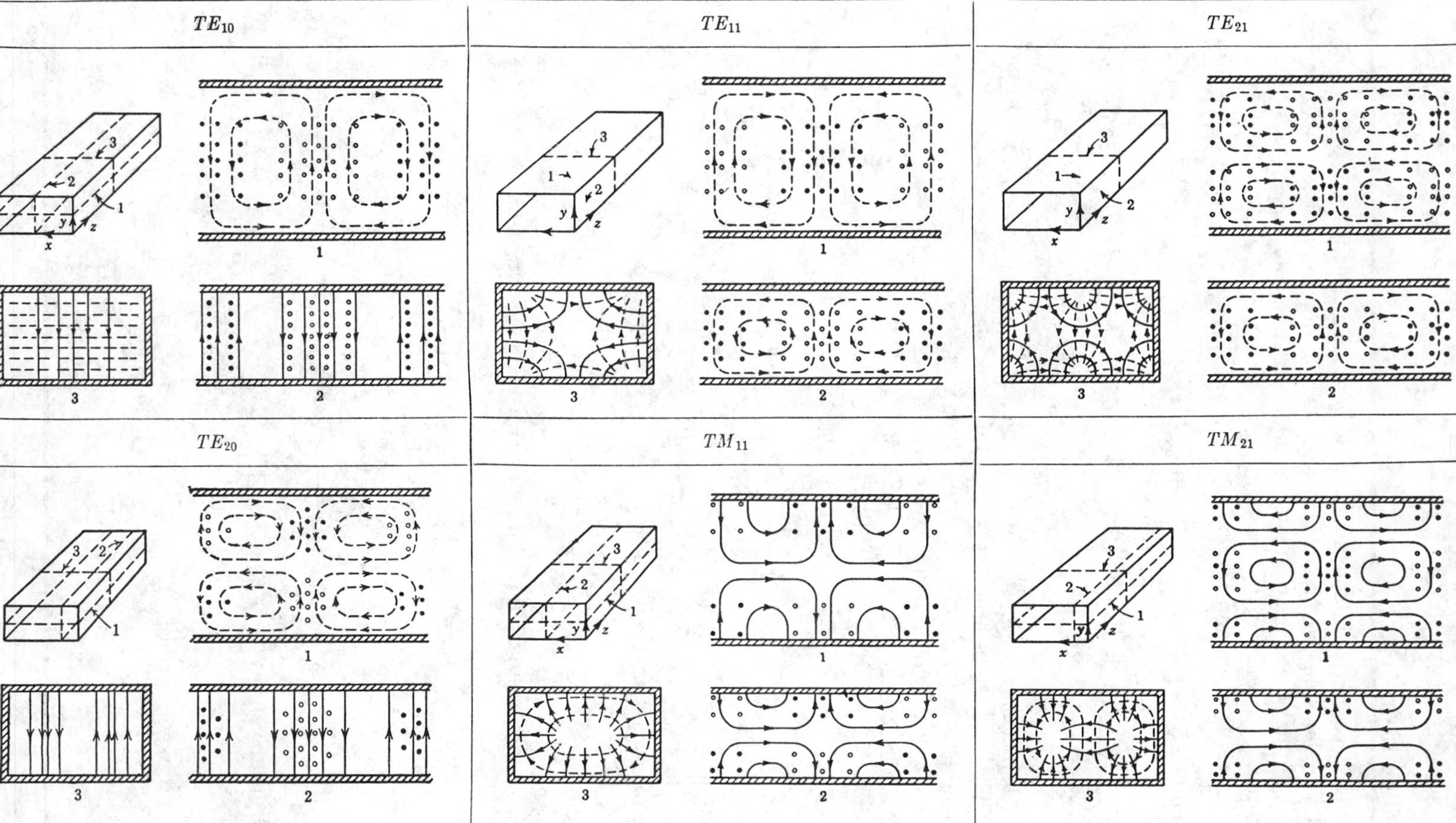

Figure 13-32 Some field configurations for rectangular waveguides. The closeness of the line spacing indicates the strength of the fields. Dots represent field lines coming out of the plane of the paper. Small circles represent lines going into the paper. (From S. Ramo, J. R. Whinnery, and T. VanDuzer, *Fields and Waves in Communication Electronics*, John Wiley & Sons, Inc., New York, 1965, Table 8.02.).

EXAMPLE 13-16

A WR90 X-band air-filled rectangular waveguide has the internal dimensions, width 22.86 mm and height 10.16 mm. Find the cutoff frequency of the dominant mode and the next highest mode. If the signal frequency is 10 GHz, find the guide wavelength, phase and group velocities, and the propagation constant for the dominant mode.

Solution By equations (13-50) and (13-51),

$$f_{c(\mathrm{TE}10)} = \frac{c}{2a} = \frac{3 \times 10^8}{2(0.02286)} = 6.56 \text{ GHz}$$

$$\lambda_c = 2a = 2(2.286) = 4.57 \text{ cm}$$

If the aspect ratio was $\frac{1}{2}$, the next-higher-order modes would be the TE_{01} and TE_{20} as seen from Fig. 13-33. In this example, the aspect ratio is 1/2.25, indicating that the TE_{20} would have the next-lowest cutoff frequency.

Applying equation (13-48) for $m = 2$, $n = 0$ yields

$$f_{c(\mathrm{TE}20)} = \frac{3 \times 10^8}{2} \sqrt{\left(\frac{2}{0.02286}\right)^2} = 13.1 \text{ GHz}$$

The cutoff frequency for the TE_{01} mode would be

$$f_{c(\mathrm{TE}01)} = \frac{3 \times 10^8}{2} \sqrt{\left(\frac{1}{0.01016}\right)^2} = 14.8 \text{ GHz}$$

The free-space wavelength, λ, is

$$\lambda = \frac{c}{f} = \frac{3 \times 10^8}{10 \times 10^9} = 3 \text{ cm}$$

For the dominant TE_{10} mode, the guide wavelength [from equation (13-53)] is

$$\lambda_g = \frac{3}{\sqrt{1 - (3/4.57)^2}} = 3.98 \text{ cm}$$

The phase velocity [from equation (13-61)] is

$$v_p = \frac{3 \times 10^8}{\sqrt{1 - (3/4.57)^2}} = 3.98 \times 10^8 \text{ m/s}$$

The group velocity [from equation (13-58)] is

$$v_g = 3 \times 10^8 \sqrt{1 - (3/4.57)^2} = 2.26 \times 10^8 \text{ m/s}$$

The propagation constant [from equation (13-64)] is

$$\beta_g = \frac{2\pi(10 \times 10^9)}{3 \times 10^8} \sqrt{1 - \left(\frac{6.56}{10}\right)^2} = 158 \text{ rad/m}$$

13-12.3 Coupling to Waveguides

The two commonly used methods of coupling fields from a coaxial line to a waveguide are by means of the probe coupler and the loop coupler. The probe is placed at the peak of the electric field for launching the TE_{10} mode as shown in Fig. 13-34(a). This coupler is usually used as a transition from a coaxial line to a waveguide or as a coupler from a klystron source.

By noting the field configurations of Fig. 13-32, it is quite evident which mode the various probe couplings will excite in Fig. 13-34.

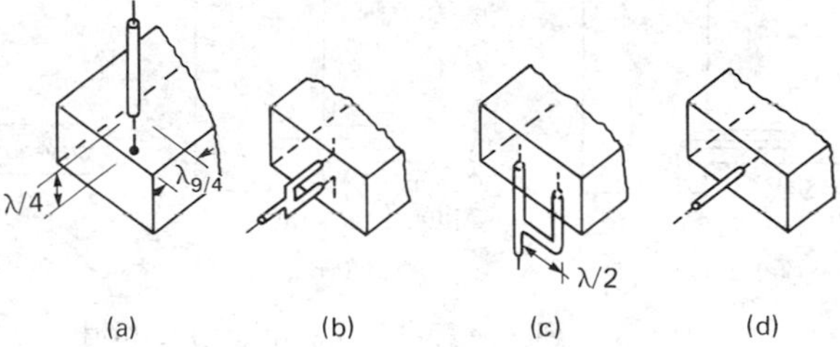

Figure 13-34 Probe coupling to the TE_{10}(a), TE_{11}(b), TE_{20}(c), and TM_{11}(d) modes.

Loop coupling couples to the magnetic field rather than to the electric field. It is used for example, for coupling to one of the magnetron tube cavities. Figure 13-35 indicates loop coupling for launching the TE_{10} mode.

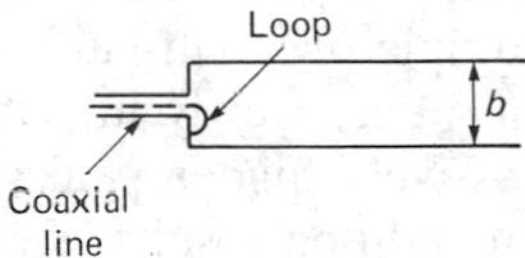

Figure 13-35 Loop coupling to the TE_{10} mode in a rectangular waveguid

13-13 WAVEGUIDE CAVITIES

Let us consider what occurs in a length of waveguide when shorting plates are applied to the input and output (see Fig. 13-36). The resulting physical form is that of a closed cavity. An electromagnetic wave can be established in the closed cavity by attaching it to a waveguide carrying a signal and then drilling a coupling hole to allow some of the fields to leak into the cavity. Alternatively, a signal on coax can be coupled into a cavity using a probe or loop arrangement (see Fig. 13-37) to launch the electromagnetic wave. When a wave traveling in this closed guide impinges on the shorting plate at the end of the guide it is reflected and reverses its direction of propagation in the guide. Each portion of the wave injected thus undergoes multiple reflections between the end plates. In general, subsequent

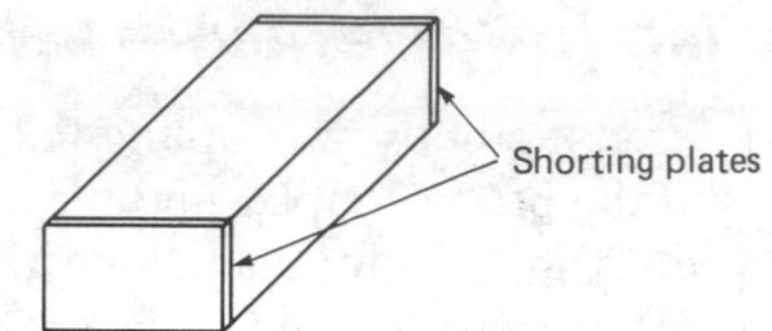

Figure 13-36 Rectangular waveguide cavity.

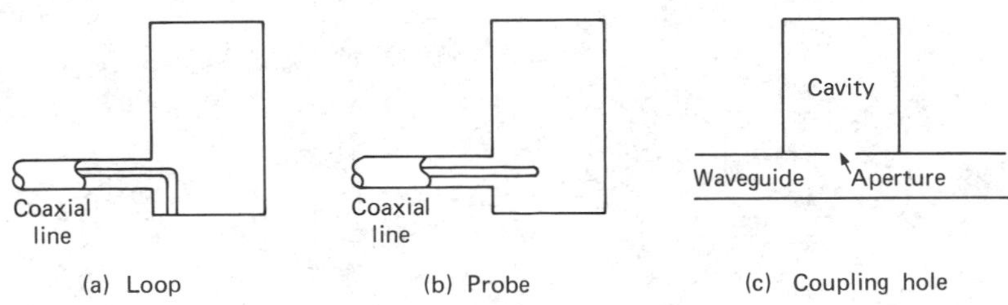

Figure 13-37 Coupling methods.

signals are injected in random phase with the existing waves, and this dissonant combination of waves leads to a net field strength approaching zero (therefore, zero stored energy). The cavity thus presents an apparent load to the signal source that is large and reactive. If, however, the frequency of the injected signal is such that the cavity is some multiple of a half-guide-wavelength long, quite a different situation arises. In this case each newly injected wave will be in phase with the existing waves, and successively higher peaks in the electromagnetic fields are developed. This resonant condition results in extremely intense field occurring in the cavity (large stored energy).

Resonant cavities are used in microwave circuits in much the same way that resonant *LC* circuits are used in low-frequency circuits. The principal difference occurs in the resonant circuit *Q* value that can be obtained. In *LC* circuits it is difficult to reduce the circuit losses sufficiently to obtain *Q* values greater than a few hundred. Waveguides, however, are very low loss devices, and as a result, *Q* values in the order of several thousand are possible.

With some understanding of the field configurations in a waveguide, the functions of various waveguide components can generally be comprehended. Table 13-3 lists some of the more common waveguide components, with a brief indication of their applications.

TABLE 13-3 Waveguide Components

Physical Shape	Type	Application or Function
Junctions		
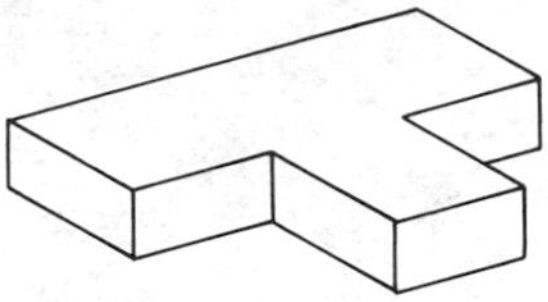	H-plane tee	Combining or splitting signals
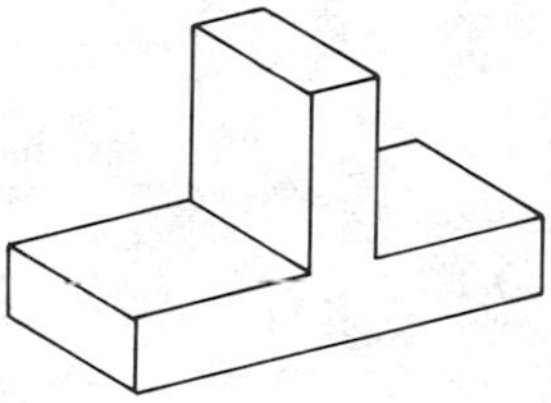	E-plane tee	
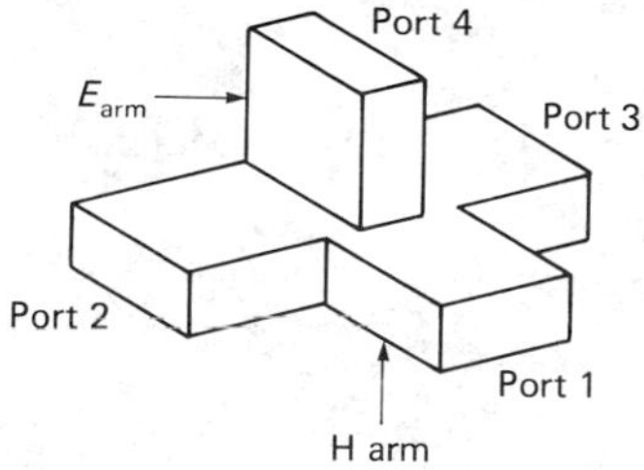 Port 4, E_{arm}, Port 3, Port 2, Port 1, H arm	Magic tee or hybrid tee	A wave incident at port 4 divides equally but 180° out of phase at ports 2 and 3 A wave incident at port 1 divides equally and in phase at ports 2 and 3 Excellent isolation between ports 1 and 4 Used as a mixer in a microwave receiver 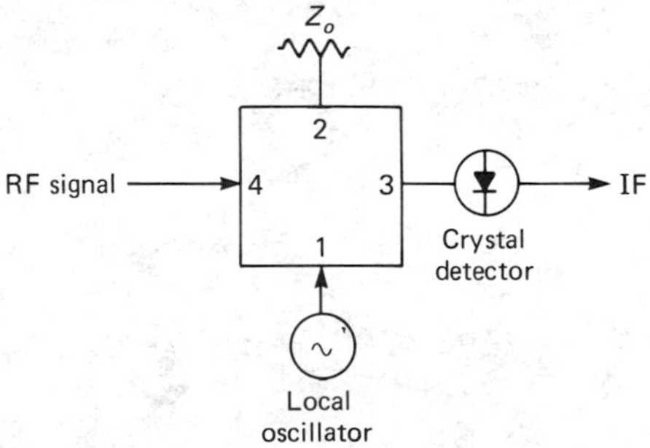
		Z_o, 2, RF signal, 4, 3, IF, 1, Crystal detector, Local oscillator
Attenuators and Matched Load		
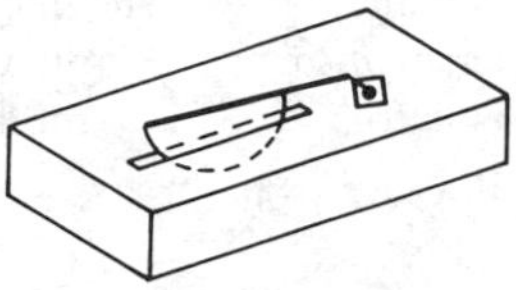	Flap	Variable attenuator

(continued)

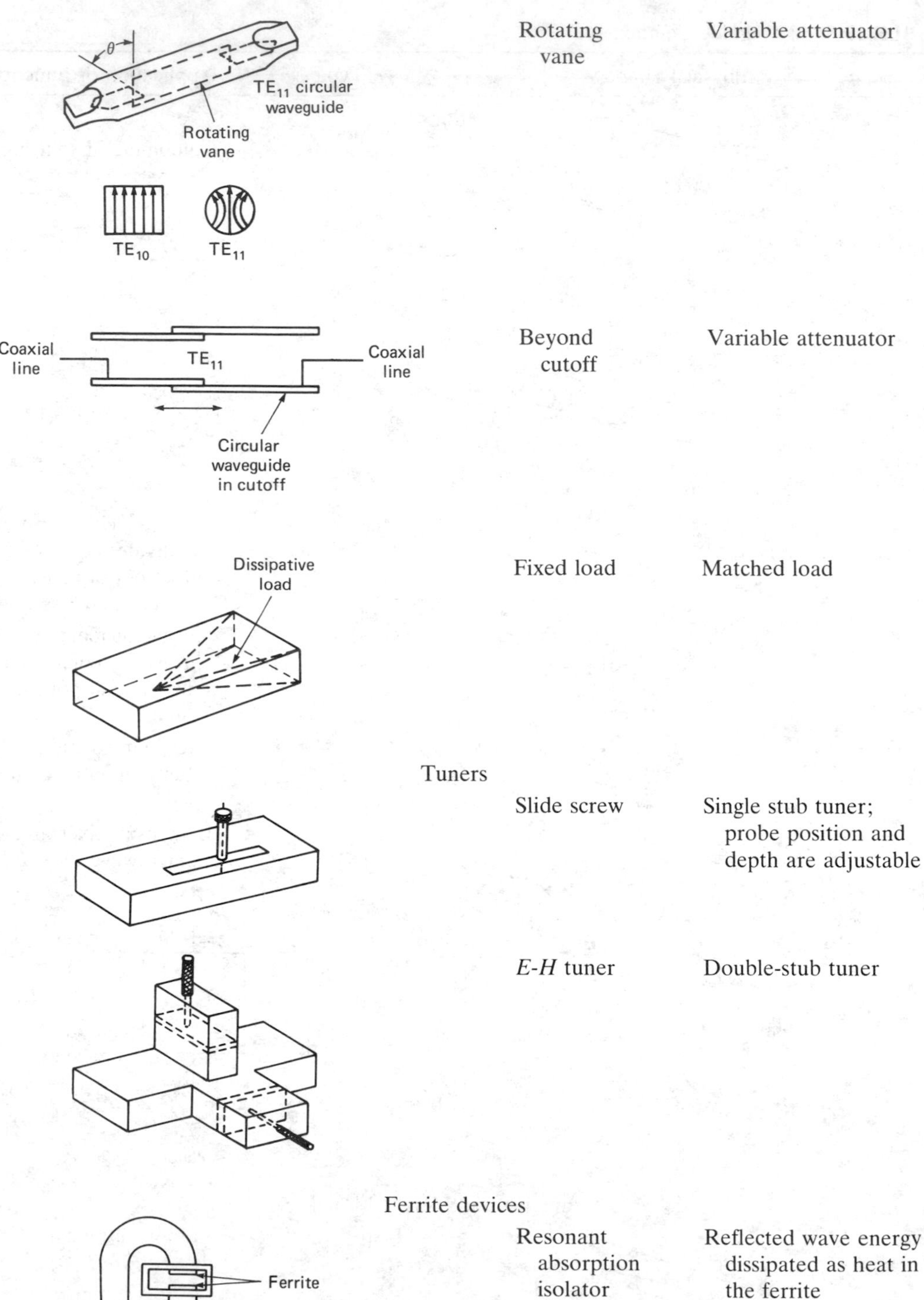

	Rotating vane	Variable attenuator
	Beyond cutoff	Variable attenuator
	Fixed load	Matched load
Tuners		
	Slide screw	Single stub tuner; probe position and depth are adjustable
	E-H tuner	Double-stub tuner
Ferrite devices		
	Resonant absorption isolator	Reflected wave energy dissipated as heat in the ferrite

Circulator

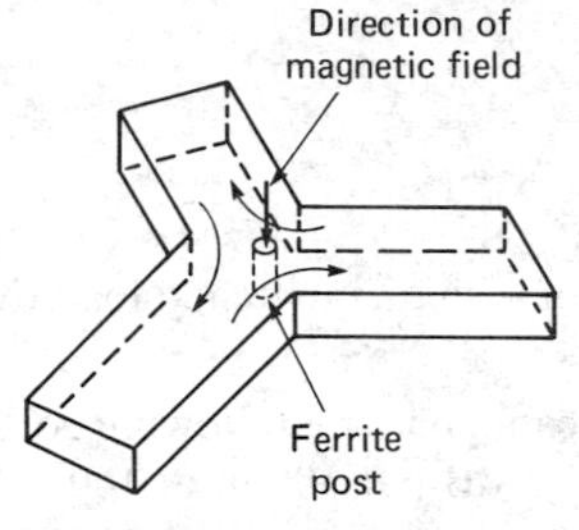

Used in a duplexer

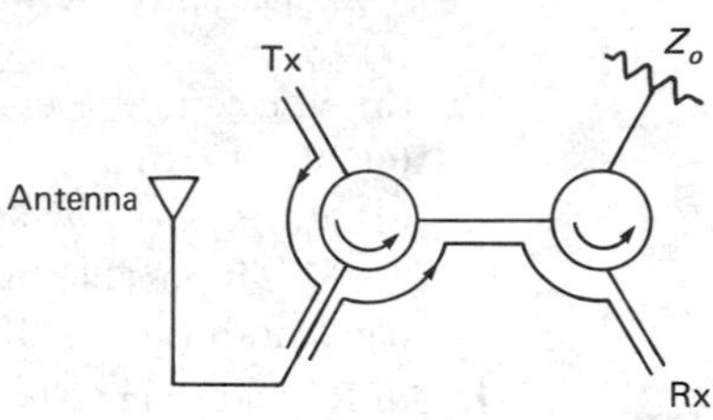

Coupler

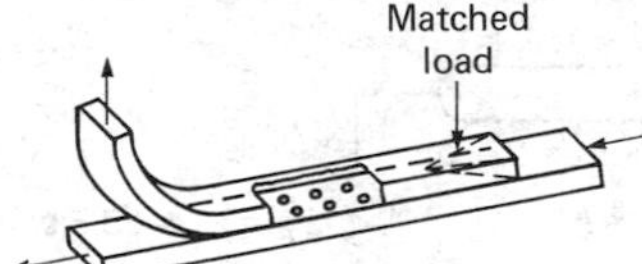

Broadband directional coupler

To extract incident or reflected wave

Slotted line

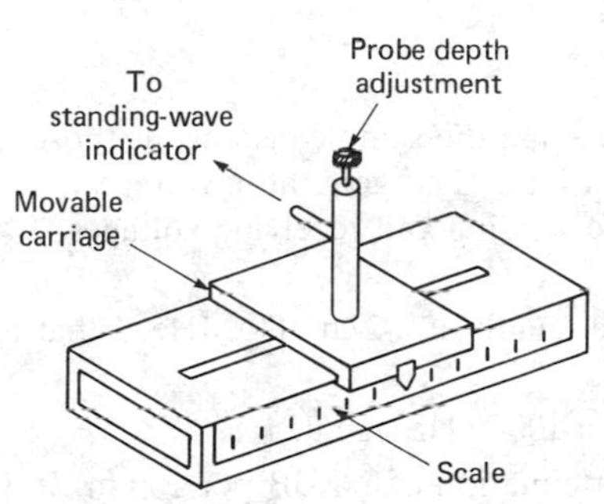

Waveguide slotted line

For impedance measurement; an RF diode detector is usually placed in the output arm

---

Problems

13-1. A polyethylene coaxial cable has an inner conductor diameter of 0.078 in. and an outer conductor with an inner diameter of 0.285 in. Find:

(a) The capacitance per meter
(b) The inductance per meter
(c) The velocity factor for this cable

13-2. Find the skin depth of a copper conductor at 100 MHz.

13-3. A AWG 26-gauge paper-insulated exchange telephone cable has the following line constants at 1 kHz. $R = 440\ \Omega$/mi, $G = 1.6\ \mu$S/mi, $L = 0.45$ mH/mi, and $C = 0.069\ \mu$F/mi. Determine:

(a) Z_0

(b) γ, α, and β (per mile)

(c) λ

(d) v_p

Is it possible to terminate this cable with a unique matched load over the voice bandwidth (i.e., 300 to 3300 Hz)?

13-4. An RG58/U Jan-C-17A polyethylene coaxial cable has the following constants: $Z_0 = 53.5\ \Omega$ and $C = 93.5$ pF/m. Attenuation is 13.5 dB per 100 m at 100 MHz. Find:

(a) The attenuation constant

(b) The phase constant

(c) The velocity factor of the cable at 100 MHz.

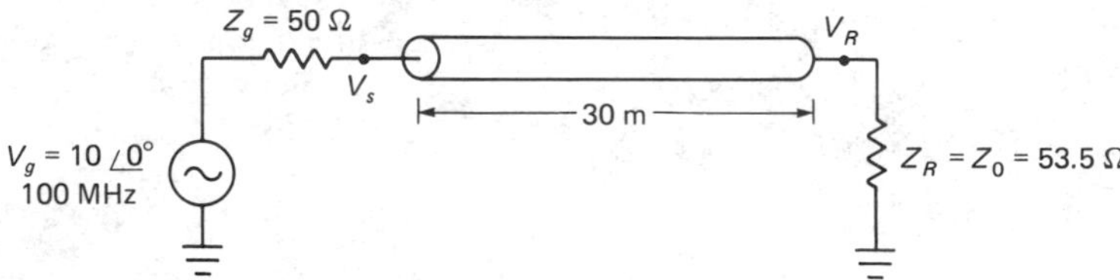

Figure P13-4.

13-5. A 30-m length of coaxial line (Fig. P13-4) having the same constants as those given in Problem 13-4 is connected to a matched load. The generator voltage is $10\ \angle 0°$ and its internal impedance is $50\ \Omega$. Find the sending and receiving voltages, V_S and V_R.

13-6. The cable of Problem 13-4 is to be cut to a length of $\lambda/2$ at 100 MHz. What is the physical length of this cable?

13.7. Determine the wavelength in free space of a 100-MHz signal.

13-8. A 7/8-in HJ5 Heliax coaxial cable has an attenuation of 1.1 dB per 100 m. If 4 kW of power is applied to a 100-m length of this matched cable, how much power is dissipated in the cable?

13-9. Determine the voltage reflection coefficient for a 50-Ω line terminated with the following resistances:

(a) Short circuit

(b) 25 Ω

(c) 50 Ω

(d) 100 Ω

(e) Open circuit

13-10. A lossless line with a $2Z_0$ terminating load is connected to a Z_0-ohm generator at time $t = 0$ having a voltage of 20 V_{dc}. If the velocity factor is 66%, sketch the waveforms at the sending and receiving ends of a 100-m length of this line.

13-11. The voltage waveform shown in Fig. P13-11 is seen on a 50-Ω TDR display when connected to the input of a length of faulted 50-Ω line. Determine the fault location and the fault impedance. Assume that the velocity factor is 66%, and ignore any losses.

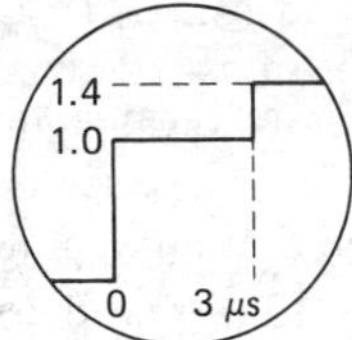

Figure P13-11.

13-12. The voltage reflection coefficient at a load, as measured on a 50-Ω system, is 0.5 $\angle 30°$. Determine the load impedance.

13-13. Find the VSWR on the line for the various terminations given in Problem 13-9.

13-14. A directional coupler having a coupling factor of 30 dB is used to sample a signal on a 1-kW transmitter. How much power is present at the matched sampling arm?

13-15. The input impedance to a 24.26-m length of *R*G 58/U line is measured to be 35 + *j*20 Ω at 100 MHz. The 50-Ω line has a velocity factor of 66.6%.

(a) If losses are ignored, find the load impedance.

(b) If the line attenuation is 13.5 dB per 100 m at 100 MHz, what is the load impedance? Assume a 66.6% velocity factor.

13-16. **(a)** Explain the proper calibration procedure for the measurement system shown in Fig. 13-26.

(b) After proper calibration, with an unknown load attached to the 50-Ω measurement system of Fig. 13-26, the vector voltmeter reads the following: channel A = 100 mV and channel B = 60 mV. Channel B lags channel A by 30°. Find the load impedance.

13-17. Find the power reflected, power absorbed, return loss, and power loss when 1 W of power is incident from a 50-Ω line to a 75-Ω load.

13-18. A directional coupler having a 20-dB coupling factor monitors 50 mW of incident power at the matched output of the sampling arm. When the directional coupler main arm is reversed, 5 mW of reflected power is obtained at the matched sampling arm. Determine the VSWR and reflected and incident power in the main arm.

13-19. The voltage standing wave shown in Fig. P13-19 is observed on an air dielectric transmission line. Determine:

(a) The VSWR on the line

(b) Frequency of operation

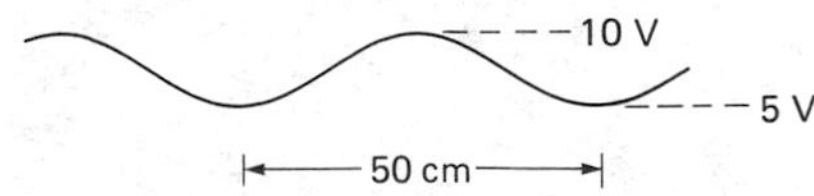

Figure P13-19.

13-20. A hollow brass C-band (4.64 to 7.05 GHz) rectangular waveguide designated as WR159 has dimensions of a = 40.39 mm and b = 20.193 mm. The number 159 notes the waveguide width in hundredths of inches.

(a) Determine the cutoff frequency for the TE_{10}, TE_{20}, and TE_{11} modes.

(b) Which modes can propagate if f = 8 GHz?

(c) Determine the cutoff wavelength of the dominant mode.

(d) Determine the free-space wavelength and waveguide wavelength if the frequency of operation is 6 GHz.

13-21. A hollow brass X-band (8.2 TO 12.5 GHz) rectangular waveguide designated as WR90 has dimensions of $a = 22.86$ mm and $b = 10.16$ mm.

(a) Determine the cutoff wavelength and frequency of the dominant mode.

(b) Determine the free-space wavelength, the waveguide wavelength, and the phase and group velocities at an operating frequency of 10 GHz.

13-22. Prove from equation (13-3) that the resistance for a cylindrical conductor of radius r and length l, in meters, at a frequency f is

$$R = \frac{l\sqrt{\mu\pi f\rho}}{2\pi r} \quad \Omega$$

Wave Propagation

Dennis Morland

14-1 INTRODUCTION

Radio-wave propagation occurs when signals travel through space from one antenna to another. This chapter deals mainly with the nature of radio waves and factors that affect their propagation in the atmosphere. Some of the general behavioral characteristics of electromagnetic waves, such as speed of propagation, wavelength, and characteristic impedance, are included. The wave phenomena refraction and diffraction are also described, with special reference to their effect on radio waves. Antennas, devices that perform the conversion between signals traveling in transmission lines or waveguides and signals traveling through space, are dealt with in Chapter 15 and so are not discussed here except in very general terms. From the early experiments with "wireless' communication in the nineteenth century, radio has developed rapidly and is used now in many different forms. Public broadcasting stations, mobile two-way radios, and microwave point-to-point communication links are all possible due to the ability of waves to travel through space.

Radio waves offer the obvious advantage of being able to allow communications between very distant points without the need for a direct connection, such as a wire or waveguide, but there are disadvantages, too, such as:

1. The radio spectrum is a limited resource that must be shared among users (refer to Table 1-1), so only a limited number of transmitters can be "on the air" at any one time, since each transmitter requires its own slice of the spectrum. Anyone wishing to transmit a signal must obtain permission from the government regulatory agencies that manage spectrum usage.

2. The receiving antennas always pick up interfering signals and noise along with the desired signals. There is no way to prevent this, so reliable communication is possible only when the ratio of signal to noise plus interference at the receiver is good enough.

Figure 14-1 shows a sketch of a radio communication system and an equivalent model. This chapter deals with the characteristics of the propagation path in such systems.

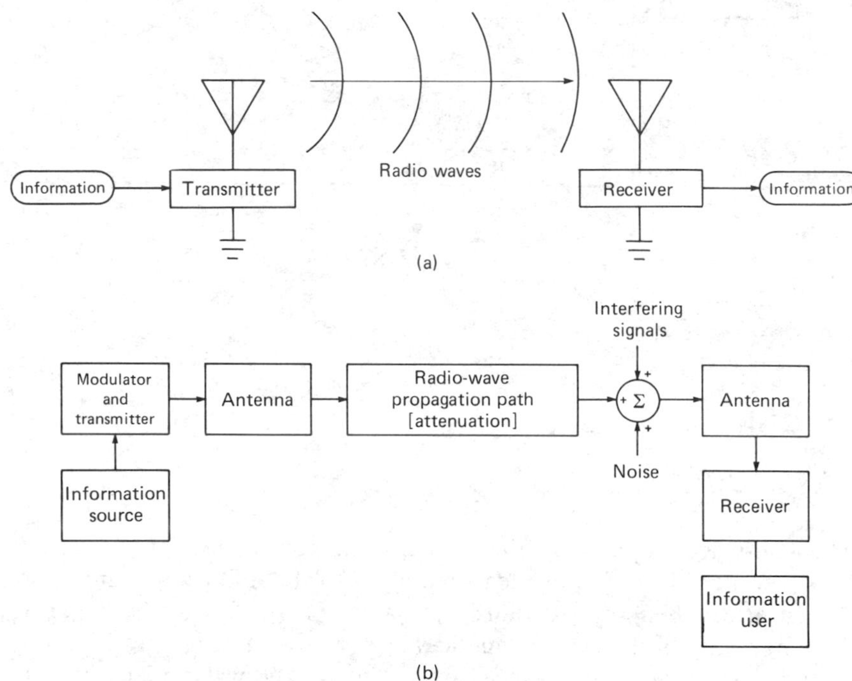

Figure 14-1 (a) Radio link; (b) systems model of radio link.

14-2 TRAVELING ELECTROMAGNETIC WAVES

Radio waves propagate through space as traveling electromagnetic (EM) waves. That is, the energy of the signals exists in the form of electric and magnetic fields that are varying sinusoidally with time and are also in motion. The two types of fields always exist together because a changing electric field will generate a magnetic field and a changing magnetic field will create an electric field. In fact, the reason that EM waves are able to travel through space is that there is a continual flow of energy from one field to the other. Since these fields are four-dimensional quantities, involving the three physical dimensions and time, they are difficult to depict both mathematically and graphically. We will start with the simplest EM wave, the transverse electromagnetic (TEM) wave. Equations (14-1) and (14-2) describe the behavior of the electric (E) and magnetic (H) fields of the TEM wave represented in Fig. 14-2.

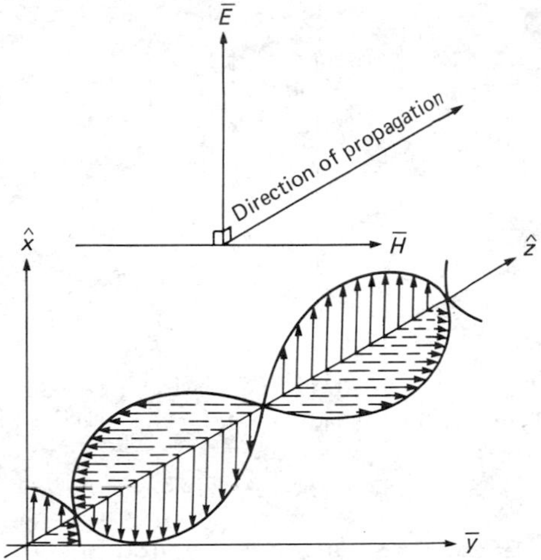

Figure 14-2 Transverse electromagnetic wave.

$$\overline{E}(x, y, z, t) = Ex \cos(\omega t - \beta z)\hat{x} + 0\hat{y} + 0\hat{z} \tag{14-1}$$

$$\overline{H}(x, y, z, t) = Hy \cos(\omega t - \beta z)\hat{y} + 0\hat{x} + 0\hat{z} \tag{14-2}$$

The symbols $\hat{x}$, $\hat{y}$, and $\hat{z}$ refer to the x, y, and z directions of the Cartesian coordinate system. The electric field lines described by equation (14-1) are always in the x direction, since the y and z components are zero. The field has a maximum strength of Ex and varies sinusoidally. At any given point, its phase is changing at the rate of ω radians per second and at any given instant of time, its phase is changing at the rate of β radians per meter.

Note that the magnetic field lines described by equation (14-2) are varying in exactly the same manner as the E field but they are oriented in the y direction. This is a characteristic of waves in free space. Because the electric and magnetic field lines are at right angles to each other and to the direction of propagation, this type of wave is called a transverse electromagnetic wave.

In a TEM wave, the direction of propagation is related to the directions of the electric and magnetic field vectors by the right-hand rule. Curl the fingers of your right hand from the electric field direction to the magnetic field direction and your thumb will point in the direction that the wave is moving.

14-3 POLARIZATION

The polarization of an EM wave is specified by the direction of the electric field lines, the x direction in Fig. 14-2. This direction is important to know because it indicates the direction in which electric currents can be induced in any obstacle that the wave meets. For example, if a metal antenna rod is oriented parallel to the electric field lines of an incoming wave, as in Fig. 14-3(a), currents will be induced in the antenna and can be fed to a receiver through a transmission line. If the antenna is oriented perpendicular to the incoming electric field lines, as in

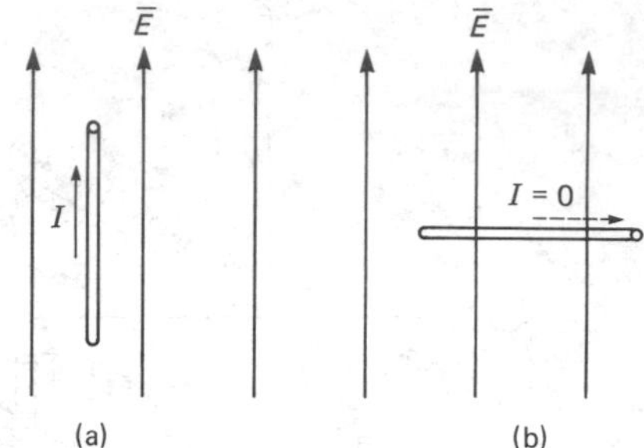

Figure 14-3 (a) Currents are induced in a wire oriented in the same direction as the electric field; (b) currents are not induced when the wire is at right angles to the polarization.

Fig. 14-3(b), no currents can be stimulated and the antenna will not be able to receive the signal. The polarization of a wave is important when considering the propagation of signals from one antenna to another. Radio waves are usually described as being vertically polarized or horizontally polarized, depending on the orientation of electric field lines with respect to the surface of the earth. A vertically polarized transmitting antenna should be paired with a vertically polarized receiving antenna. (Circular or elliptical polarization can also exist but this topic will be left to Chapter 15.) Reflection or scattering of radio waves by the terrain or other large objects such as buildings can shift the polarization of the total field in some cases.

14-4 CHARACTERISTIC IMPEDANCE OF FREE SPACE

The magnitudes of the electric and magnetic field strengths in a traveling electromagnetic wave bear a fixed relationship to each other for any given material. The ratio of the magnitudes of the electric field to magnetic field is known as the *intrinsic impedance* of the media.

$$\eta = \frac{|E|}{|H|} \qquad \text{units} = \frac{\text{V/m}}{\text{A/m}} = \frac{\text{V}}{\text{A}} = \text{ohms} \tag{14-3}$$

By analogy with transmission lines, this relation is also called the characteristic impedance or the wave resistance and has units of ohms. As in the case of transmission lines, this quantity depends on the parameters (permittivity and permeability) of the material through which the wave is moving.

The exact value for the characteristic impedance for a TEM wave in a lossless, isotropic, unbounded region is given by

$$\eta = \sqrt{\frac{\mu}{\varepsilon}} = \frac{\sqrt{\mu_0\mu_r}}{\sqrt{\varepsilon_0\varepsilon_r}} \qquad \textit{ohms} \tag{14-4}$$

where μ is the absolute permeability of the medium and ε is the absolute permittivity. The symbols μ_r and ε_r refer to relative permeabilities and permittivities, where μ_0 and ε_0 are the absolute permeability and permittivity of a vacuum (free space). The characteristic impedance of free space is therefore

$$\eta_0 = \frac{\sqrt{\mu_0}}{\sqrt{\varepsilon_0}} = \frac{\sqrt{4\pi \times 10^{-7}}}{\sqrt{\frac{1}{36\pi} \times 10^{-9}}} = 120\pi \text{ ohms} \simeq 377\ \Omega$$

The characteristic impedances of lossy dielectrics also depend on the conductivity (σ) of the medium and the operating frequency.

$$\eta = \sqrt{\frac{j\omega\mu}{\sigma + j\omega\varepsilon}} \tag{14-5}$$

For other loss-less dielectrics, the value above must be divided by the square roots of the relative permittivity and relative permeability. For most dielectrics, the relative permeability is very close to 1, but the relative permittivity can vary considerably. The characteristic impedance for most loss-less dielectrics therefore is given by

$$\eta = \frac{120\pi}{\sqrt{\varepsilon_r}} \quad \text{ohms} \tag{14-6}$$

EXAMPLE 14-1

Find the characteristic impedance of polyethylene, which has a relative permittivity of 2.25. Then find the magnetic field strength of an EM wave that has an electric field strength of 10 V/m.

$$\eta = \frac{120\pi}{\sqrt{2.25}} = 251.3\ \Omega$$

$$|H| = \frac{|E|}{\eta} = \frac{10 \text{ V/m}}{251.3\ \Omega} = 39.8 \text{ mA/m}$$

14-5 WAVEFRONTS AND RAYS

Wavefronts and rays are often used to represent the fields in electromagnetic waves in order to simplify their description. Usually, only the electric field is shown, since the magnitude of the coexistent magnetic field can be found by using the characteristic impedance of the medium. *Rays* are lines that show the direction of propagation of the EM wave. In Figure 14-4, which shows a radiating point source, the rays point outward in all directions. *Wavefronts* are surfaces perpendicular to the direction of propagation. They are surfaces of constant phase of the electric field. In Fig. 14-4, the wavefronts are spherical surfaces centered on the point source. Wavefronts are usually drawn a fixed distance apart—often a wavelength or a half of a wavelength.

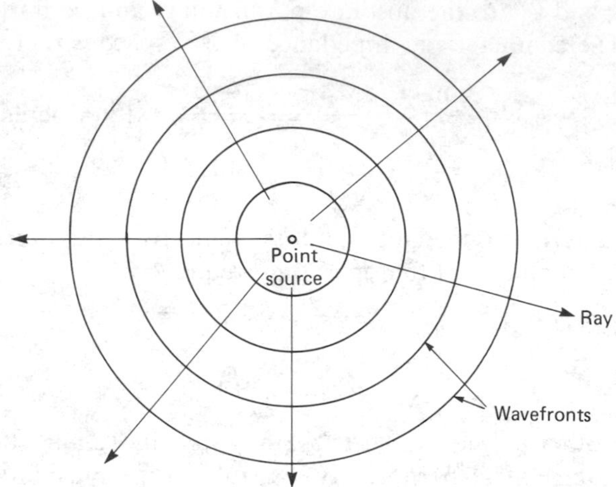

Figure 14-4 Radiating point source of electromagnetic waves.

14-6 ISOTROPIC RADIATION

An isotropic radiator is an EM source that radiates equally in all directions. In other words, it is a point source as depicted in Fig. 14-4. As the waves move outward from the source, their power is spread out over the surface of an expanding sphere. Assuming that the total power being radiated is P and the medium around the radiator (air) is loss-less, the power density in watts per square meter (W/m^2) at any radius R_1 is

$$U_1 = \frac{P}{4\pi R_1^2} \qquad \text{W/m}^2 \tag{14-7}$$

since the area of the surface of a sphere is $4\pi R^2$. At some other radius, R_2,

$$U_2 = \frac{P}{4\pi R_2^2} \qquad \text{W/m}^2 \tag{14-8}$$

Note the ratio:

$$\frac{U_2}{U_1} = \frac{P/4\pi R_2^2}{P/4\pi R_1^2} = \frac{R_1^2}{R_2^2} \tag{14-9}$$

This indicates that the power density is inversely proportional to the square of the distance from the source. This is the inverse square law, which applies to all EM radiation and describes how radio waves get weaker as they move away from a source. This concept is developed further in Chapter 15, where formulas will be developed to predict the amount of coupling between two antennas. A true isotropic radiator cannot be constructed physically, but the power density in the far field of any antenna will decrease with an inverse-square-law relationship with respect to distance.

EXAMPLE 14-2

The power density of a radio wave at a distance of 10 km from a transmitter is measured to be 8 μW/m^2. What would the power density be at a distance of 80 km from the transmitter? Assume free-space (square-law) propagation.

Solution:

$$\frac{U_2}{U_1} = \left(\frac{R_1}{R_2}\right)^2$$

$$U_2 = U_1\left(\frac{R_1}{R_2}\right)^2 = (8.0 \times 10^{-6}\ \text{W/m}^2)\left(\frac{10}{80}\right)^2$$

$$= 0.125\mu\text{W/m}^2 = 125\ \text{nW/m}^2$$

Note: Every time the distance is doubled, the power density is reduced by $\frac{1}{4}$ (6 dB).

At distances that are large (in terms of wavelengths) from the source, the curvature in the spherical wavefronts is very small (Fig. 14-5). Eventually, the wavefronts become almost totally flat, and then we essentially have plane waves. In a true plane wave, the electric and magnetic field lines exist as straight lines at right angles to the direction of propagation and to each other (Fig. 14-6). This kind of transverse electromagnetic wave is the simplest to represent [see equations (14-1) and (14-2), for example] and is generally used in the analysis of wave phenomena such as reflection and refraction (which will be dealt with in a later section).

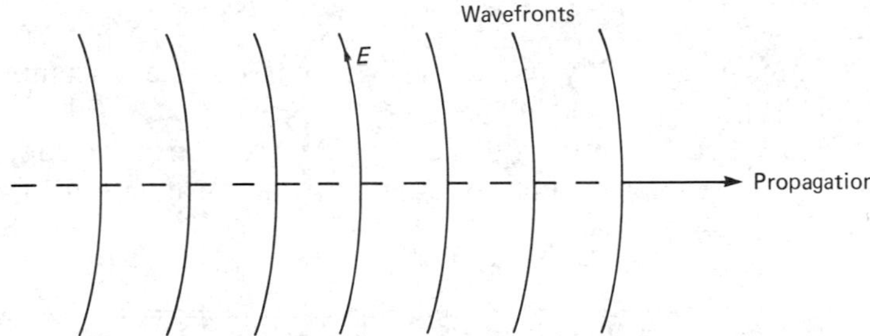

Figure 14-5 Large-radius spherical wave.

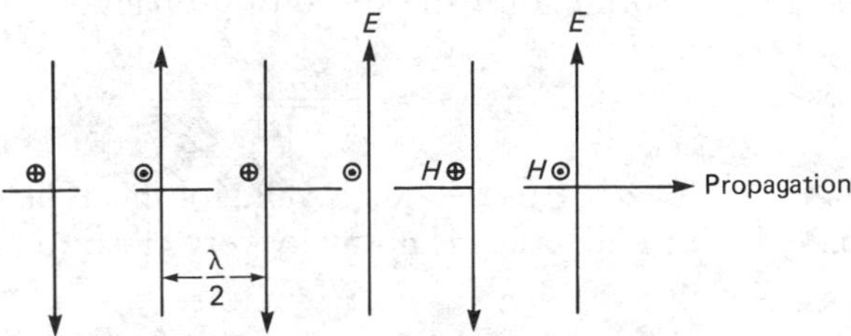

Figure 14-6 Wavefronts in a plane TEM wave.

14-7 SPEED OF PROPAGATION

The speed with which a TEM wave moves is dependent on the type of material the wave is moving through. For a loss-less dielectric, this speed will be the same for all frequencies and can be calculated by the relation

$$v = \frac{1}{\sqrt{\varepsilon\mu}} \tag{14-10}$$

where ε is the absolute permittivity and μ is the absolute permeability of the medium.

A more convenient form of this relation involves using relative parameters instead of absolute parameters. We can use the following relation:

$$\varepsilon = \varepsilon_r \varepsilon_0 \tag{14-11}$$

where ε is the absolute permittivity, ε_r the relative permittivity, and ε_0 the absolute permittivity of a vacuum.

$$\varepsilon_0 = \frac{1}{36\pi} \times 10^{-9} \qquad \text{farads/meter} \tag{14-12}$$

where μ is the absolute permeability, μ_r the relative permeability, and μ_0 the absolute permeability of a vacuum.

$$\mu_0 = 4\pi \times 10^{-7} \qquad \text{henries/meter} \tag{14-13}$$

Using these relations, the speed of propagation becomes

$$v = \frac{1}{\sqrt{\varepsilon_r \varepsilon_0 \mu_r \mu_0}} = \frac{1}{\sqrt{\varepsilon_r \mu_r}\,\sqrt{\varepsilon_0 \mu_0}} \tag{14-14}$$

The relative permeability and permittivity of free space (vacuum, air) are both very close to 1.

The speed of propagation of an EM wave in a vacuum can be found by substituting the known values of ε_0 and μ_0 into formula (14-14) with $\varepsilon_r = \mu_r = 1$ (vacuum).

$$v = \frac{1}{\sqrt{(1)\,(1)}\,\sqrt{(4\pi/36\pi) \times 10^{-9} \times 10^{-7}}} = 3 \times 10^{8} \text{ m/s}$$

This is the well-known value c, the speed of light in a vacuum. Using this quantity allows the general formula for the speed of propagation to be further simplified.

$$v = \frac{c}{\sqrt{\varepsilon_r \mu_r}} \simeq \frac{c}{\sqrt{\varepsilon_r}} \tag{14-15}$$

where $c = 3 \times 10^{8}$ m/s. This relation is valid for most dielectrics, since the relative permeability of such materials is normally very close to 1.

14-8 WAVELENGTH AND PHASE CONSTANT

The wavelength (λ) of a signal is the distance it travels during the period of one cycle. This quantity depends on the speed of propagation of the wave and on the frequency of the signal. If a wave travels a distance λ meters each cycle and there are f cycles per second (hertz), the wave must be moving at a rate of

$$v = \lambda f \quad \text{m/s} \tag{14-16}$$

so

$$\lambda = \frac{v}{f} \tag{14-17}$$

Clearly, then, the wavelength is inversely proportional to the frequency and also depends on the velocity of propagation, which in turn depends on the relative permittivity and permeability of the medium. Table 1-1 illustrates this relationship for free-space waves.

The phase constant (β) of a wave is the rate with which phase is changing with respect to distance. This quantity can be found from the wavelength. During one cycle, the phase of an alternating signal changes by 2π radians or 360°. Since a wavelength is the distance the wave travels during one cycle, then

$$2\pi = \beta\lambda \tag{14-18}$$

or

$$\beta = \frac{2\pi}{\lambda} \quad \text{rad/m} \tag{14-19}$$

$$\beta = \frac{360°}{\lambda} \quad \text{deg/m} \tag{14-20}$$

EXAMPLE 14-3

Calculate the velocity of propagation, the wavelength, and the propagation constant of a 100-MHz signal traveling through Teflon, which has a relative permittivity of 2.1.

$$v = \frac{c}{\sqrt{\varepsilon_r \mu_r}} = \frac{3 \times 10^8 \text{ m/s}}{\sqrt{(2.1)(1)}} = 2.07 \times 10^8 \text{ m/s}$$

$$\lambda = \frac{v}{f} = \frac{2.07 \times 10^8 \text{ m/s}}{100 \times 10^6} = 2.07 \text{ m}$$

$$\beta = \frac{2\pi}{\lambda} = \frac{2\pi}{2.07} = 3.035 \text{ rad/m}$$

EXAMPLE 14-4

Knowing the speed of propagation in a material and the characteristic impedance of it, we can solve a number of problems by using the methods developed for transmission-line analysis. For example, consider a plane TEM wave meeting a slab of polyethylene ($\varepsilon_r = 2.25$) 8 mm thick (Fig. 14-7). The incoming ray is normal (perpendicular) to the surface. If the operating frequency is 10 GHz, find:

(a) The reflection coefficient of the face of the slab
(b) The percent reflected power
(c) The percent transmitted power

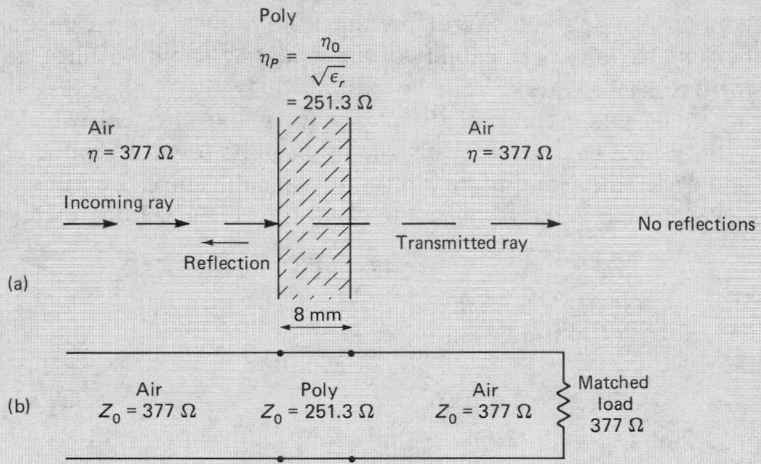

Figure 14-7 (a) Wave meeting a slab; (b) transmission-line analogy.

Outside the slab:

$$\text{characteristic impedance} = \eta = 377\ \Omega$$

$$\text{speed of propagation} = c = 3 \times 10^8\ \text{m/s}$$

Inside the slab:

$$\text{characteristic impedance} = \eta_p = \frac{377}{\sqrt{\varepsilon_r}} = 251.3\ \Omega$$

$$\text{speed of propagation} = v = \frac{c}{\sqrt{\varepsilon_r}} = 2 \times 10^8\ \text{m/s}$$

$$\text{wavelength (at 10 GHz)} = \lambda_p = \frac{v}{f} = \frac{2 \times 10^8}{10 \times 10^9} = 0.02\ \text{m} = 2\ \text{cm}$$

The 8-mm thickness therefore represents an electrical length of

$$\frac{L}{\lambda} = \frac{0.8\ \text{cm}}{2\ \text{cm}} = 0.4\lambda$$

If we presume that no returning signals come back to the slab from the region to the right of it, the air on the right represents an impedance of 377 Ω at the interface. This would represent a normalized impedance of 377/251.3 = 1.5 + j0. This normalized load impedance can be plotted on a Smith chart, and then the normalized impedance at the left side of the slab can be found by moving clockwise from this point by a distance equal to the slab thickness (0.4λ). Figure 14-8 shows that this normalized impedance is 1.05 + j0.42. The actual impedance that the left slab face presents to the incoming wave is therefore $Z = (\eta_p)(z) = (251.3)(1.05 + j0.42) = 263.9 + j105.56\ \Omega$.

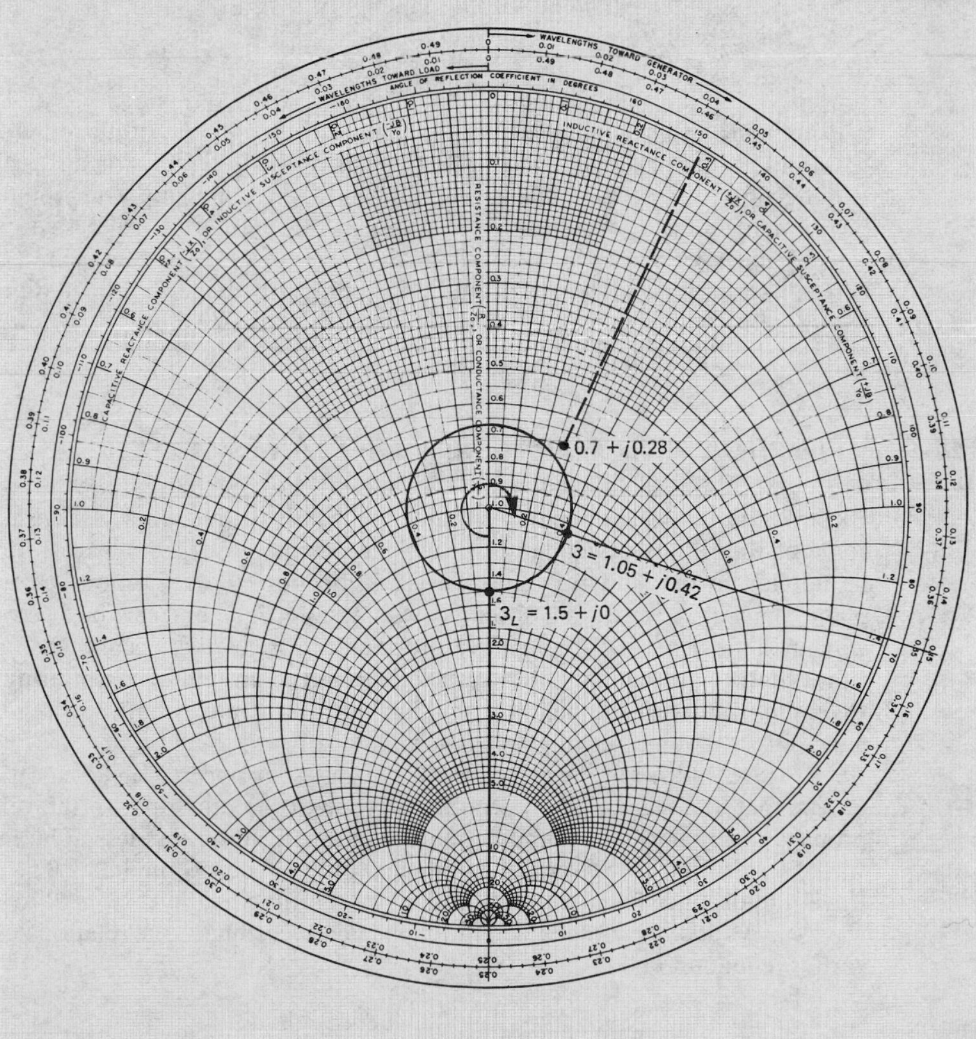

Figure 14-8.

The reflection coefficient for the wave meeting the left face of the slab can be found by normalizing this impedance to 377 Ω and plotting it on the Smith chart:

$$\text{normalized impedance} = \frac{263.9 + j105.56}{377} = 0.7 + j0.28$$

(a) From the Smith chart, the reflection coefficient for the incoming wave is

$$\Gamma = 0.23 \angle 129^\circ$$

(b) The percent reflected power

$$|\Gamma|^2 \times 100\% = (0.23)^2(100) = 5.33\%$$

(c) The percent transmitted power

$$1 - |\Gamma|^2 \times 100\% = 94.67\%$$

Repeating this problem with a thickness of 1 cm (one-half wavelength), we would get an impedance at the left interface of 377 Ω. This means that the reflection coefficient would be 0, and 100% of the incident power would pass through the slab. This principle is used in the design of protective covers (radomes) for antennas. Also, an example involving a very thin film ($L \ll \lambda$) of a dielectric would result in a very small percentage of reflected power, regardless of the exact dielectric constant of the material.

14-9 REFLECTION AND REFRACTION OF PLANE WAVES

Electromagnetic waves behave the same regardless of frequency. Because of this, some of the results developed in the science of classical (geometrical) optics can be applied to the study of radio waves. Reflection and refraction are two such phenomena that were first observed with visible light but can be used to help explain some aspects of radio-wave propagation. *Reflection* occurs when an electromagnetic wave bounces off a surface such as the ground or a scattering object. *Refraction* refers to the bending of a light ray as it moves between regions that have different velocities of propagation.

Figure 14-9 shows a plane wave moving from one region to another. The second material has a higher permittivity constant than the first, so the velocity of propagation is slower. Both materials are assumed to be loss-less. The incoming wave is refracted (bent) at the interface with the second medium. The time to travel the distance from A to A' must equal the time to travel from B to B'. The wave travels at different speeds in the two media according to the relation developed earlier [equation (14-15)]:

$$v = \frac{c}{\sqrt{\mu_r \varepsilon_r}}$$

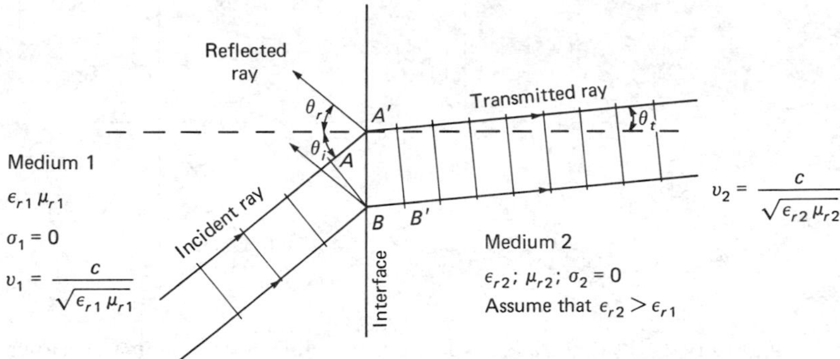

Figure 14-9 Plane-wave refraction at a dielectric interface.

To maintain the equiphase fronts as plane waves in the second medium, a shift in the direction of propagation must occur. The amount of refraction may be evaluated by considering the wavefront $A-B$ traveling to $A'-B'$.

$$\frac{\text{distance } A \text{ to } A'}{\text{distance } B \text{ to } B'} = \frac{A-A'}{B-B'} = \frac{v_1 t}{v_2 t} = \frac{A-A'/A'-B}{B-B'/A'-B} = \frac{\sin\theta_i}{\sin\theta_t}$$

$$\frac{\sin\theta_i}{\sin\theta_t} = \frac{v_1}{v_2} = \frac{c/\sqrt{\varepsilon_{r1}\mu_{r1}}}{c/\sqrt{\varepsilon_{r2}\mu_{r2}}} = \frac{\sqrt{\varepsilon_{r2}\mu_{r2}}}{\sqrt{\varepsilon_{r1}\mu_{r1}}} \tag{14-21}$$

This relationship is called *Snell's law:*

$$\frac{\sin\theta_i}{\sin\theta_t} = \sqrt{\frac{\mu_{r2}\varepsilon_{r2}}{\mu_{r1}\varepsilon_{r1}}} = \frac{n_2}{n_1} \tag{14-22}$$

where n is defined as the *index of refraction:*

$$n = \frac{c}{v} = \sqrt{\frac{\varepsilon_r}{\mu_r}} \tag{14-23}$$

With respect to the reflected ray, note that it travels away from the interface at the same speed as the incident wave, so there is no bending effect. Therefore, $\theta_r = \theta_i$. When a wave moves from a less dense (lower permittivity) medium to a higher-density one, the propagation angle will always be closer to a line normal to the interface (i.e., $\theta_t < \theta_i$). Conversely, if a wave moves from a region with a higher permittivity to a lower-permittivity region, the ray will be refracted farther away from the normal ($\theta_t > \theta_i$). As Fig. 14-10 depicts, an angle for θ_i will exist where θ_t becomes 90° and where $\sin\theta_t$ becomes equal to 1. The incident angle that results in this situation is called the *critical angle* and is given by

$$\theta_c = \sin^{-1}\sqrt{\frac{\varepsilon_{r2}}{\varepsilon_{r1}}} \tag{14-24}$$

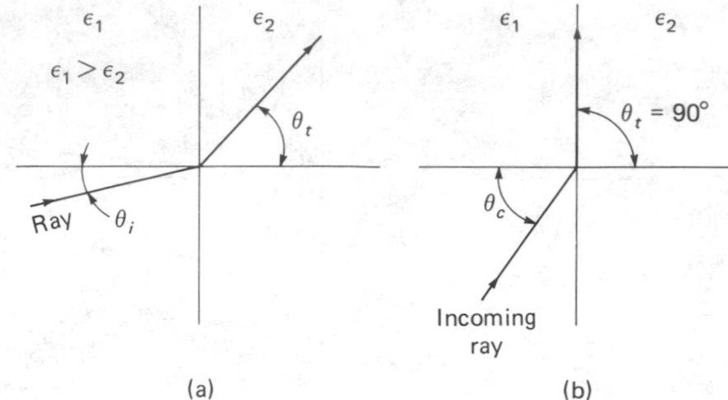

Figure 14-10 (a)Refraction of a wave traveling into a region with a lower permittivity constant; (b) showing critical angle.

At incident angles greater than or equal to the critical angle, no wave energy remains in the second medium; all of the wave energy is reflected. This behavior is of extreme importance in the analysis of propagation in fiber optics.

14-10 PROPAGATION PATHS BETWEEN TWO ANTENNAS

Propagation can occur between a pair of antennas via more than one path at the same time, as shown in Fig. 14-11. Usually, one of these paths will have a much stronger signal than the others and so predominate. A number of factors, such as frequency, distance, and terrain, determine which path is the predominant one.

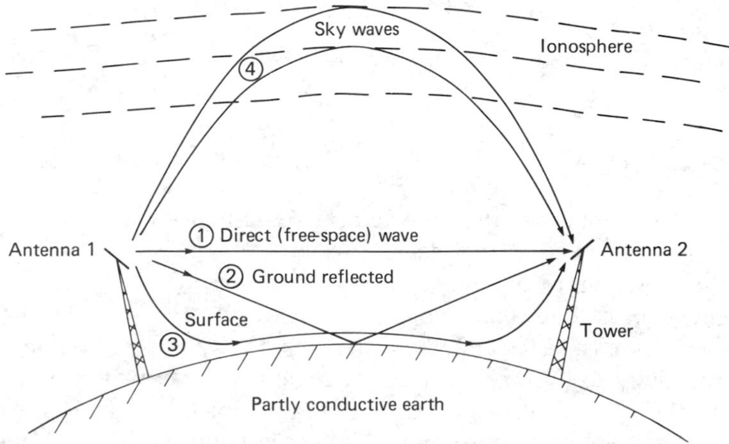

Figure 14-11 Normal propagation paths.

The *direct* or *free-space wave* is shown as path 1; a wave reflected from the ground is path 2. These two paths together are sometimes referred to as the *space wave*. EM waves can also be guided along the surface of the earth like waves in a transmission line. This *surface wave*, path 3, is significant at lower frequencies where strong coupling can exist between the transmitting antenna and the ground. The surface wave and the space wave together make up the *ground wave*.

The *sky wave* of path 4 exists when some of the energy in a wave are reflected back to the ground by ionized regions in the upper atmosphere. This propagation mechanism can result in communication over very long distances but is the least reliable because the ionosphere is constantly changing. Additional coupling mechanisms can exist if the two antennas are very close together, but these *near-field effects* are negligible if the antennas are more than a few wavelengths apart. Since antennas are usually much farther apart than this, we will ignore these effects.

At frequencies less than about 1500 kHz, the surface wave provides the primary coverage. The attenuation of this path increases rapidly with frequency because of skin-effect losses in the earth. The sky wave is important in the approximate frequency range 3 to 30 MHz, along with the direct wave. At frequencies above this, only the space wave remains.

These propagation mechanisms are examined in more detail in the next sections.

14-11 GROUND WAVE AT LOW FREQUENCIES (SURFACE WAVE)

For frequencies less than 3 MHz, the wavelengths will be greater than 100 m. Since the wavelengths are so large, most antennas can be considered to be close to the ground in terms of wavelength. This results in a strong coupling to the surface wave by a transmitted signal. At the same time, the direct wave and ground reflected waves tend to cancel each other out in this frequency range, as depicted in Fig. 14-12, leaving the surface wave to predominate.

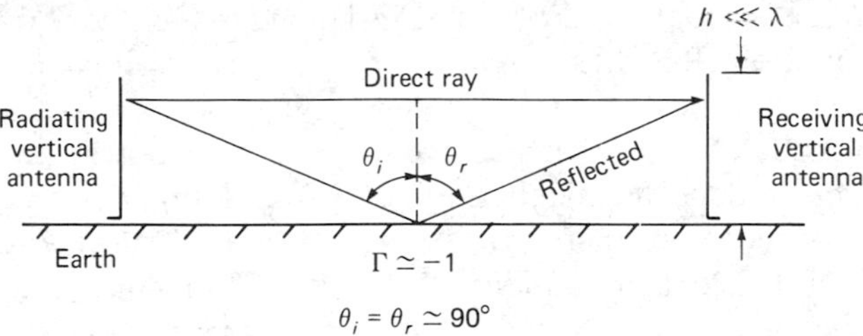

Figure 14-12 Direct and reflected waves at low frequencies. Antenna heights are exaggerated.

If two antennas are at moderate heights above the ground, the difference between the direct and ground reflected waves will be small compared to the wavelengths of low-frequency signals. Since the earth acts as a reasonably good conductor at low frequencies, the reflection coefficient of the earth will be approximately -1. That is, the reflected wave will be of almost the same amplitude as the direct wave but will be 180° out of phase with it. When the direct and the reflected waves recombine at the receiving antenna, they therefore tend to cancel each other out.

Surface-wave coverage varies with frequency and with the type of terrain. Skin-effect losses decrease with decreasing frequency since the electric field can penetrate farther into the ground, into more conductive (wetter) regions. Vertical polarization will give lower attenuation than horizontal polarization because, with the latter, horizontal currents are induced in the upper most layer of the earth. This upper layer is the driest and therefore the most lossy. Vertical polarization stimulates vertically oriented currents that penetrate into wetter regions where the conductivity is higher; hence losses are reduced. Typical attenuation rates for the surface wave at 20 kHz is about 6 dB/km over land and 3 dB/km over seawater. At the higher frequencies of the AM broadcast band (520–1640 kHz) the attenuation rates are much higher and very large amounts of power (such as 50 kW) are needed to allow surface-wave coverage of an area large enough for broadcast purposes.

The ground wave is significant only at low frequenties for a number of reasons. Frequencies higher than a few megahertz are very weakly coupled to the surface wave because the wavelengths are shorter, so the antennas are separated from the ground by significant distances in terms of wavelengths. The ground waves that are stimulated are very strongly attenuated because the skin-effect forces the ground-induced currents to flow in the uppermost layer of the earth, which is the driest and the most lossy. Hence propagation occurs mainly by the space wave and the sky wave for signals above the medium-frequency range.

14-12 LINE-OF-SIGHT PROPAGATION: DIRECT WAVE

This type of propagation path, used by VHF, UHF, and microwave communication link attenuation of the radio signal, is dependent on the universal square law and on precipitation. Reliable terrestrial communication links using less than 10 W of power are possible over typical distances of 20 to 60 km. Satellite links over distances of about 37,000 km are also achieved with similar power levels.

Distance depends mainly on line-of-sight limits due to the earth's curvature. This is affected by refraction in the troposphere (lower atmosphere) due to the decrease of density and humidity of the air with increasing height (see Fig. 14-13). This refraction is usually modeled by representing the earth as having less curvature than it really has and by drawing the transmitted rays as straight lines. This is easier to do than drawing curved rays.

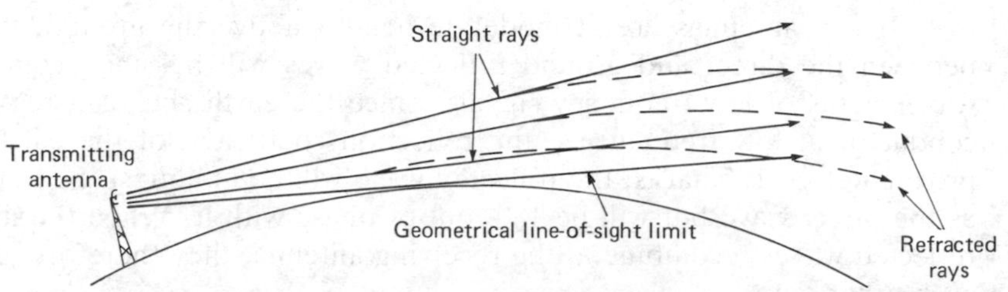

Figure 14-13 Line-of-sight limit and refraction.

The effective refraction is represented by a K factor, where

$$K = \frac{\text{effective earth radius}}{\text{actual earth radius}} \tag{14-25}$$

The most typical value for K is $\frac{4}{3}$, but the value depends on local conditions, especially humidity, and can vary considerably. The effect of the earth's curvature can be represented by plotting the apparent earth bulge, as shown in Fig. 14-14. The apparent earth bulge (h) at a point of distance d_1 away from one antenna and a distance d_2 from another is given by the formula

$$h = \frac{d_1 d_2}{2Ka} \tag{14-26}$$

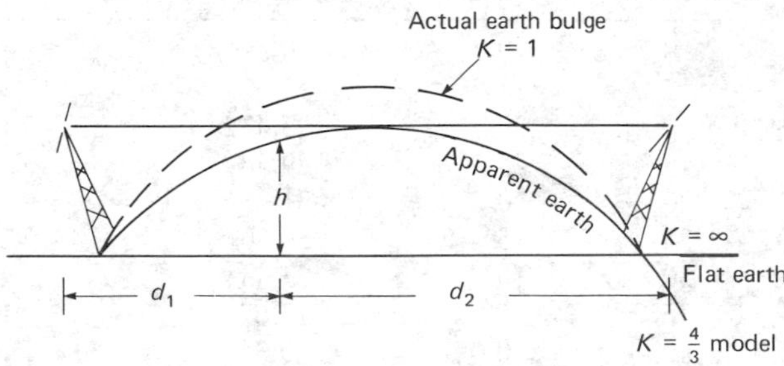

Figure 14-14 Apparent earth bulge for different K factors.

where d_1 and d_2 are the distances to the antennas, K is the refraction factor, and a is the radius of the earth $\simeq$ 6400 km. The value for h will be in the same units as the lengths d_1, d_2 and a. If the values $K = \frac{4}{3}$ and $a = 6400$ km are assumed, the formula can be written as

$$h = (0.0586)(d_1)(d_2) \qquad \text{meters} \tag{14-27}$$

if d_1 and d_2 are in kilometers.

EXAMPLE 14-5

Two antenna towers are to be constructed 40 km apart. Plot the actual earth bulge ($K = 1$) and the apparent earth bulge (assuming ($k = \frac{4}{3}$) every 5 km (Table 14-1 and Fig. 14-15).

In practice, some clearance between the direct ray and the ground must be maintained. This is further complicated by the possibility of a ground reflected wave being a problem, in combination with diffraction effects (see Fig. 14-16). The phase of the reflected wave will change by 180° (about) at the reflection point, but if the path length is $\lambda/2$ longer than the direct wave, it will add rather than subtract. The necessary clearance is called the first Fresnel zone radius and is given by

$$F_1 = \left(\frac{\lambda d_1 d_2}{d_1 + d_2}\right)^{1/2} \tag{14-28}$$

The result will be in the same units as λ, d_1, and d_2. Other Fresnel zones are defined for clearances that result in multiples of half-wavelength path-length differences.

The general formula for Fresnel zone clearance is

$$F_n = \left(\frac{n\lambda d_1 d_2}{d_1 + d_2}\right)^{1/2} = \sqrt{n}\, F_1 \tag{14-29}$$

where F_1 is the first Fresnel zone radius.

TABLE 14-1

d_1 (km)	d_2 (km)	h ($K = 1$) Meters Actual	h ($k = \frac{4}{3}$) Meters Apparent
0	40	0	0
5	35	13.67	10.255
10	30	23.43	17.58
15	25	29.30	21.975
20	20	31.25	23.44
25	15	29.30	21.975
30	10	23.43	17.58
35	5	13.67	10.255
40	0	0.0	0.0

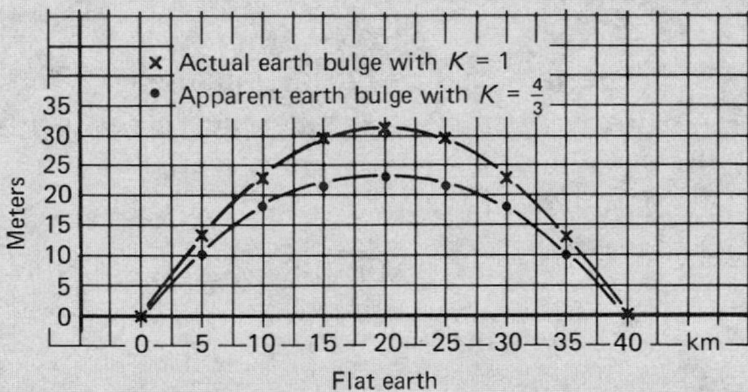

Figure 14-15 Plot of earth bulge.

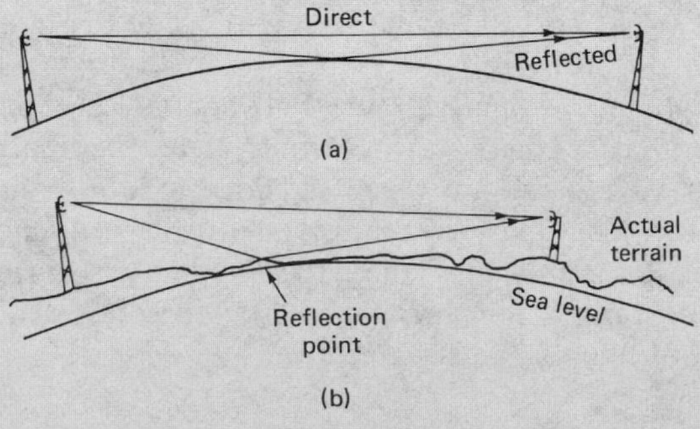

Figure 14-16 Reflections.

There are successive zones of alternating maxima and minima. Reflection from any point in odd zones will increase the signal (up to +6 dB, in-phase addition). Reflection from even zones will decrease the signal (down to $-\infty$ dB, out-of-phase cancelation). Due to the high directivity of microwave antennas, only the first few zones are significant. Figure 14-17 shows the effect of reflections on the strength of the received signal in terms of the clearance between the line of sight and the reflecting surface, for three different kinds of reflecting surface. The plane earth curve represents reflection from a flat surface. The smooth sphere curve represents reflection from a curved surface without any surface roughness. The knife-edge curve represents the effects of a sharp edge, such as the top of a mountain or building. These three simplified models of actual terrain are depicted in Fig. 14-18. For the plane earth and smooth sphere cases it is assumed that the ground will act as a good reflector, so a reflection coefficient of -1 is used. The knife edge is assumed to act as a blockage, or absorber, so a reflection coefficient of zero is used. This makes more of a difference for negative clearances than for positive clearances, as shown by Fig. 14-17.

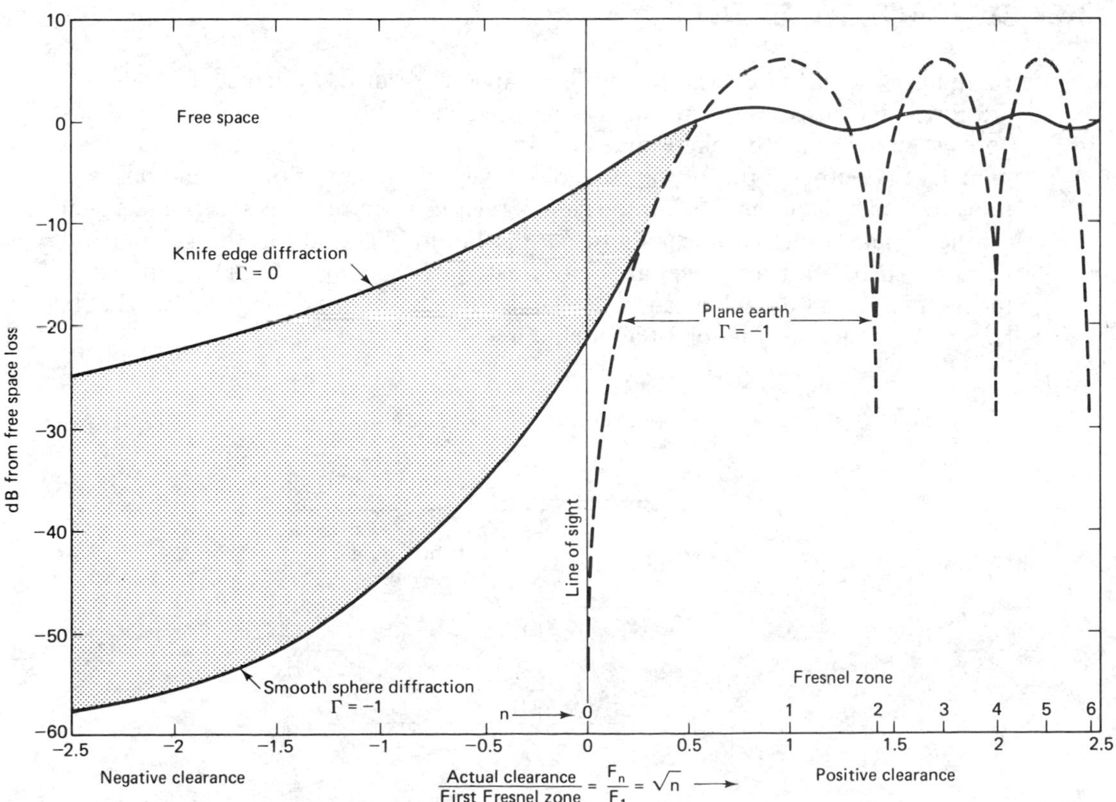

Figure 14-17 Effect of path clearance on radio-wave propagation. (Adapted from Sinnema, *Electronic Transmission Technology*, 2nd ed., Prentice Hall, Inc., Englewood Cliffs, N.J., 1988.)

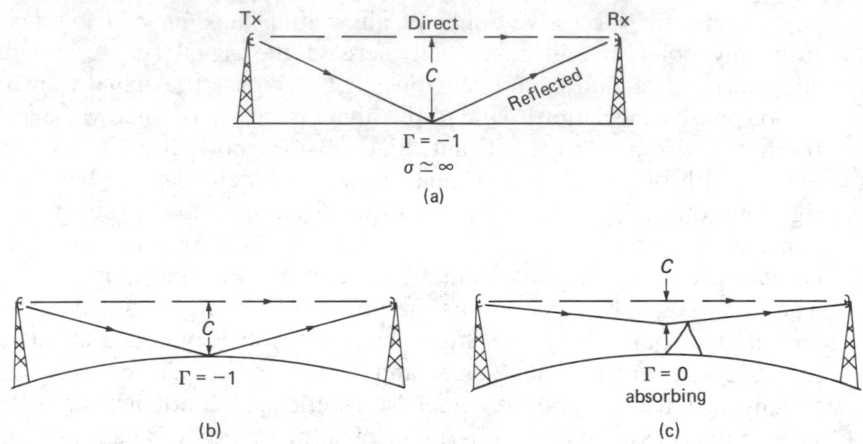

Figure 14-18 Simplified terrain models.

14-13 DIFFRACTION EFFECTS

In addition to the cases where there is positive clearance, as depicted above, we are interested in cases where there is negative clearance, as in Fig. 14-19. (This applies to the smooth sphere and knife-edge models only, not the plane-earth model.) Like other forms of waves, radio waves will tend to flow around obstructions to some extent and leak into shadow regions. The result is that some signal reaches points which are actually below the line of sight and in the geometrical shadow. Such diffracted waves are quite weak (large loss in decibels) compared to the nonobscured waves, but can still be large enough to, for example, allow reliable communication over the top of a mountain.

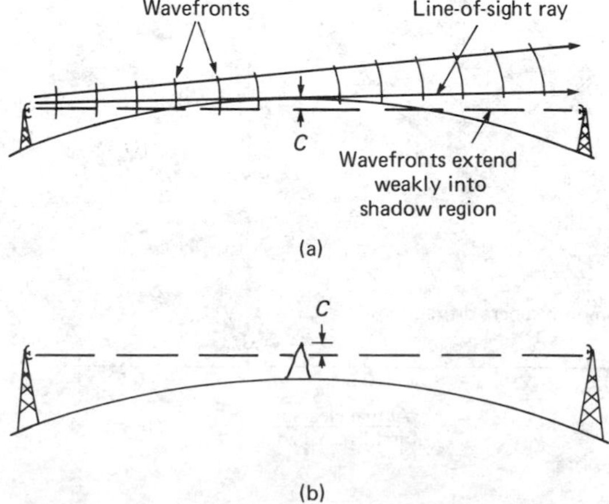

Figure 14-19 Diffraction of obstructed wavefronts.

EXAMPLE 14-6

Two microwave towers are to be set up for a microwave link operating at 6.25 GHz. The distance between the towers is 48 km. A survey shows that only five points are potential obstructions.

TABLE 14-2

Distance from Left Tower (km)	Obstruction Elevation above Sea Level (m)	Distance from Right Tower (km)
13.6	118	34.2
19.76	104.6	28.24
24.0	102	24.0
29.4	104	18.6
36.6	116	11.4

(a) Find the required elevation of the two towers assuming they have the same height if a minimum clearance of $0.6F_1$ at $k = \frac{4}{3}$ is desired.
(b) Find the effect of the reflection, with this minimum clearance, on the strength of the received signal.

Solution **(a)** First find the wavelength of the signal.

$$\lambda = \frac{c}{f} = \frac{3 \times 10^8 \text{ m/s}}{6.25 \times 10^9 \text{ Hz}} = 0.048 \text{ m}$$

Then construct a table using the formulas for the apparent earth bulge (h) and for the Fresnel radius. The answer is the highest total elevation, which in this case is 158.38 m. That is, both antennas should be this height above sea level (smooth earth) to achieve the required minimum clearance.
(b) Figure 14-18 shows that a $0.6F_1$ clearance would change the received signal strength by 0 dB relative to a free path signal. By comparison if the clearance was increased to $1F_1$, the received signal would increase by +6 dB. Clearance less than $0.6F_1$ would decrease the received signal.

TABLE 14-3

d_1 (km)	d_2 (km)	Earth Bulge (h) $\frac{d_1 d_2}{2Ka}$	Minimum Clearance $0.6F_1$:[a] $0.6\left(\frac{d_1 d_2 \lambda}{d_1 + d_2}\right)^{1/2}$	Obstruction Elevation (m)	Total Elevation Distance above Flat Earth (m)
13.60	34.4	27.41	12.97	118.0	158.38
19.76	28.24	32.70	14.17	104.6	151.47
24.0	24.0	33.75	14.40	102.0	150.15
29.4	18.6	32.5	14.03	104.0	150.08
36.6	11.4	24.45	12.26	116.0	152.71

[a] F_1 is the first Fresnel zone radius.

14-14 IONOSPHERIC PROPAGATION

The earth's atmosphere above about 30 km exists as a partly ionized plasma (free electrons and positively charged gas ions) as a result of radiation from space. Ultraviolet rays and particle emissions from the sun are the main cause, although cosmic rays play some part, too. This plasma acts as a partial conductor and is capable of reflecting radio transmissions back toward the earth.

The ionosphere is in several layers at different heights, which can all be reflecting signals at the same time, as in Fig. 14-20. The degree of ionization increases with height, partly because the incoming radiation is stronger and partly because the recombination time is longer since the air is less dense. At lower levels, a free electron has a much better chance of bumping into a positive gas ion and recombining than in the upper regions, where the air is very thin. The degree of ionization also varies considerably from night to day. Strong radiation from the sun during the day causes a high level of ionization, which tends to fade away at night except at the higher levels where the recombination time is very long. A wave passing through an ionized region tends to be refracted or bent back toward the earth. The amount of bending increases with the level of ionization and decreases as the frequency gets higher.

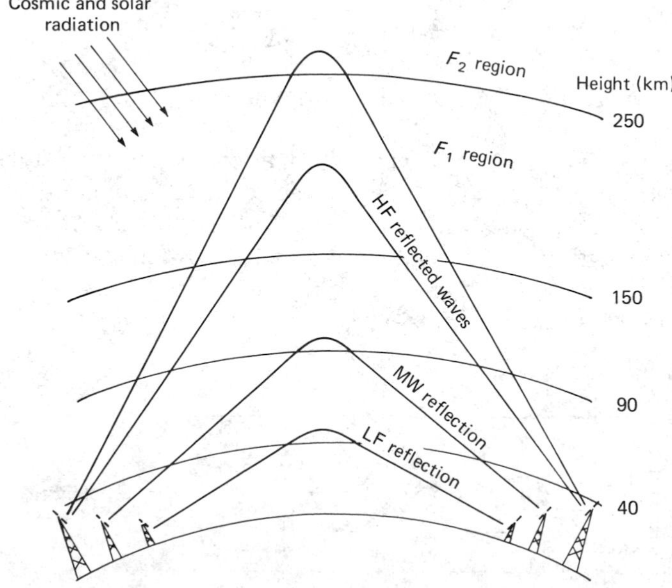

Figure 14-20 Waves in the ionosphere.

Another factor that strongly influences the degree of ionization is the behavior of the sun's sunspot cycle. Sunspots tend to send out huge bursts of gas, which, when encountering the ionosphere, cause greatly increased levels of ionization.

The visible effect of this is the Aurora Borealis or "northern lights" (in the north) and the Aurora Australis (in the south). Long-distance communications via the ionosphere can be greatly enhanced during periods of high sunspot activity. Unfortunately, this activity is unpredictable except for a long-term cyclical variation. Sunspot activity seems to vary in an 11-year cycle. A peak should occur in 1991, according to the Space Environment Sciences Center of the U.S. National Bureau of Standards. Activity will probably taper off until reaching a minimum in about 1998. Another maximum will occur perhaps in 2002. The number of sunspots during a maximum is about six times the number during a minimum.

14-15 IONIZATION LAYERS

The *D layer* ranges from about 40 to 90 km and has a relatively weak amount of ionization because most of the incoming radiation is absorbed by the higher layers. Only low-frequency (LF) waves are bent sufficiently by this layer to return to earth. Because it is in a relatively dense part of the atmosphere, it disappears entirely during the night.

The *E layer* ranges from about 90 to 145 km and can reflect signals in the medium-frequency (MF) range and above (up to about 20 MHz). It also disappears at night. It is sometimes called the *Kennelly–Heaviside layer*.

The *F1 layer* ranges from approximately 145 to 250 km during the day, while the *F2 layer* exists from about 250 to 400 km during daylight. At night, the rate of recombination is too slow, due to the rarity of the air, to allow this layer to disappear. However, the boundary between the two layers becomes indistinct, so a single F layer seems to exist, ranging from 150 to 350 km. It is capable of reflecting high-frequency (HF) signals to about 30 MHz because the degree of ionization is very high.

14-16 CRITICAL FREQUENCY

This is the highest frequency, for any layer, that would be reflected back to the earth if the transmitted ray traveled straight up from the antenna. Although these frequencies will vary with conditions in the ionosphere, typical values are 300 kHz for the D layer, 4 MHz for the E layer, 5 MHz for the F1 layer (daytime), and 8 MHz for the F2 layer (daytime). At night the combined F layer has a critical frequency of (typically) 6 MHz.

14-17 MAXIMUM USABLE FREQUENCY AND OPTIMUM FREQUENCY

Frequencies higher than the critical frequency can be refracted back to earth if they leave the transmitting antenna at an angle close to the horizontal. The maximum usable frequency (MUF) is the highest frequency that will be able to return to the earth at a given distance from the antenna (see Fig. 14-21). Government

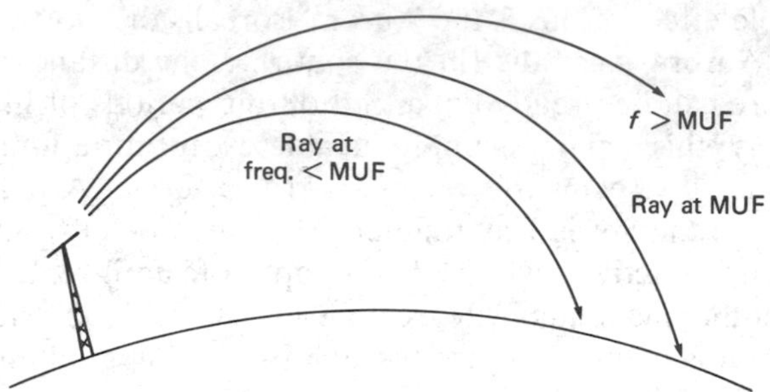

Figure 14-21 Maximum usable frequency.

agencies that monitor the ionosphere publish estimates of the MUF. Typical behavior is shown in Fig. 14-22. Operation at the MUF is not very reliable because of constantly changing atmospheric conditions. So the recommended highest operating frequency for any given distance is the optimum working frequency (OWF), which is 85% of the MUF.

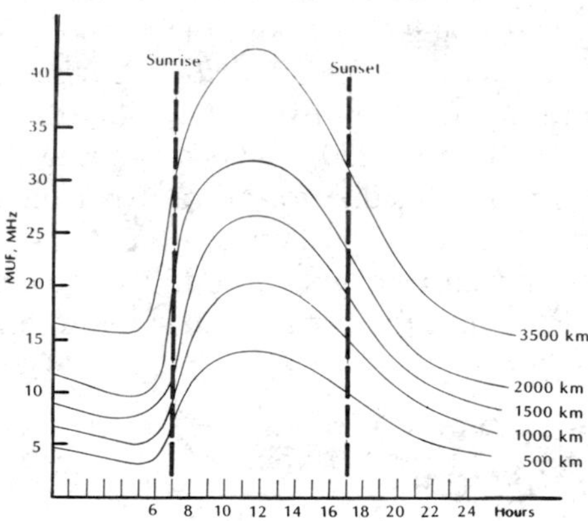

Figure 14-22 Typical plot of maximum usable frequency with distance and time. (From Sinnema, *Electronic Transmission Technology*, 2nd ed., Prentice Hall, Inc., Englewood Cliffs, N.J., 1988.)

14-18 SKIP DISTANCE AND SKIP ZONE

For frequencies above the critical frequency, there will be a region that is too close to the transmitting antenna to receive reflected waves. The closest distance to the transmitting antenna that will receive ionospheric waves is called the *skip distance*. There is a *quiet zone* or *skip zone* extending from the limit of the ground wave to the skip distance (see Fig. 14-23).

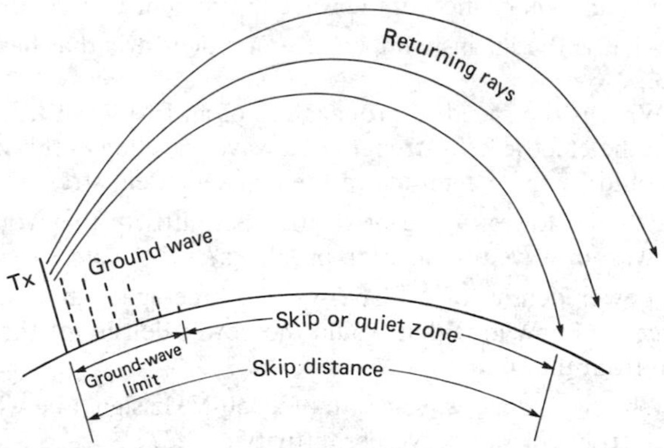

Figure 14-23 Skip zone and skip distance.

14-19 TROPOSPHERIC SCATTER

Radio communication into the skip zone is difficult but not impossible. When a wave is traveling through the lower atmosphere, or troposphere, which extends to about 10.5 km, some of the energy is scattered by the small variations in density and humidity that always exist in air. The amount of reflection is very small, but if enough power is used (often many kilowatts), reliable communication can occur. Often, the frequencies that are used for this purpose are in the microwave range. Sometimes tropospheric scatter, especially sporadic *E*-layer reflections, can cause problems with VHF point-to-point and mobile communications.

Problems

14-1. The vectors shown in Fig. P14-1 represent the electric and magnetic fields in a traveling plane wave. Is the wave traveling to the right or the left?

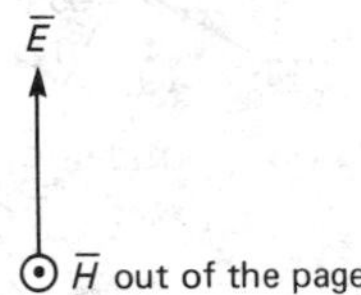

Figure P14-1.

14-2. What is the polarization of the wave in Problem 14-1? (The top of the page is "up.")

14-3. **(a)** What is the intrinsic impedance of a dielectric that has a relative permittivity of 3.2?
(b) What is the speed of propagation of an EM wave traveling in this material?
(c) If the electric field strength of a wave traveling in this material is 50 mV/m, what could be the magnitude of the magnetic field strength?

14-4. What is the lowest value of relative permittivity that you will reasonably have to deal with in wave propagation problems?

14-5. The power density of a plane wave in free space at a distance of 4 km from the source is 12 μW/m. What would the power density of the wave be at a distance of 1 km from the source?

14-6. **(a)** What would the wavelength of a 150-MHz signal be when it is traveling through a dielectric with a relative permittivity of 3.2?
(b) What would the propagation constant (β) be in this material, for 150 MHz?

14-7. A plane wave in air impinges normally on a sheet of glass 3 mm wavelengths thick (which is 0.10 wavelength thick at the operating frequency: see Fig. P14-7). Assume the glass to be loss-less and to have a relative permittivity of 4.00.
(a) What is the frequency of the wave?
(b) Sketch a transmission-line equivalent circuit for this case.
(c) Using a Smith chart, find the reflection coefficient at the incident face of the sheet.

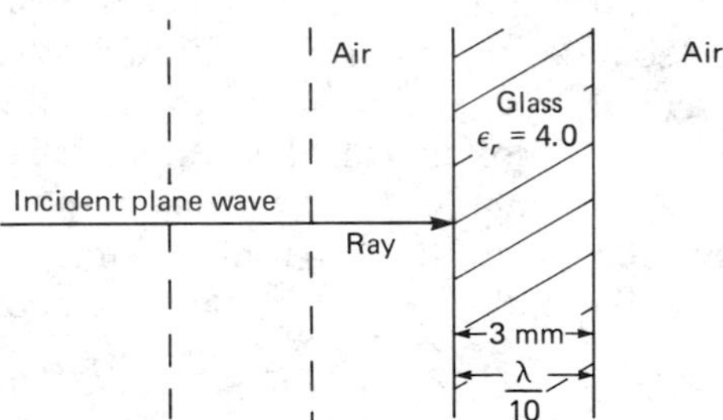

Figure P14-7 Wave incident on a glass sheet.

14-8. Light is incident normally on the short face of a 30°–60°–90° prism as shown in Fig. P14-8. A drop of liquid is placed on the hypotenuse of the prism. If the index of refraction of the prism is 1.50, find the maximum index of refraction the liquid may have if the light is to be totally reflected.

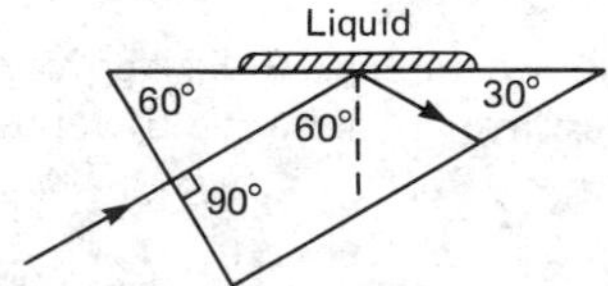

Figure P14-8 Total internal reflection in a prism.

14-9. A uniform plane wave approaches from the left (Fig. P14-9) and has an electric field strength of 10 V/m.

(a) What is the electric field strength of the reflected wave? (A Smith chart may be used if desired.)

(b) If the frequency of the wave is 10 MHz, how thick is the plastic in meters for a half-wavelength thickness.

(c) The plastic is replaced with a good conductor in the same position. With the incident field as in part (a), what is the electric field strength of the reflected wave in this case?

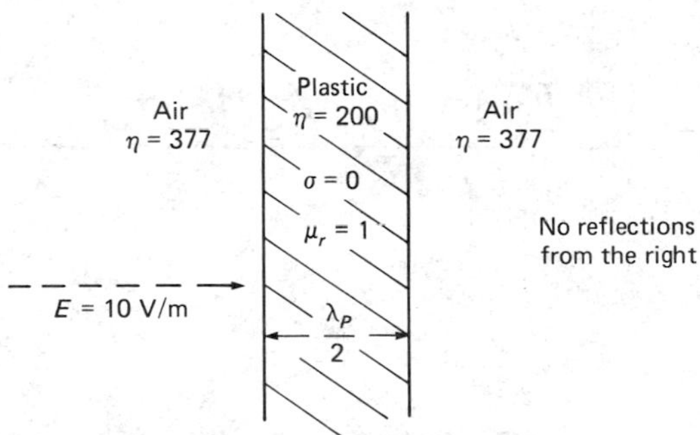

Figure P14-9 Reflection from a slab.

14-10. Two microwave towers of equal heights are to be installed between two points 50 km apart. The K factor for this location is 1.2 and the frequency of operation is 4 GHz. For a clearance of 0.75 of the first Fresnel zone, above a smooth earth, find the required antenna elevation.

14-11. Why is the surface wave not used for VHF communications?

14-12. Two microwave towers are set up 48 km apart and operate at a frequency of 6 GHz. Both towers have an elevation of 67.7 m above sea level. The appropriate K factor is $\frac{4}{3}$. Two obstructions exist: (1) at 8.1 km from the left tower, a hill with an elevation of 27.7 m; (2) at 16.2 km from the left tower, a building with an elevation of 20 m. To ensure that there will be no interference with the beam, a clearance of 0.6 of the first Fresnel zone is desired.

(a) Is the $0.6F_1$ clearance condition met by the obstructions?

(b) A contractor wishes to erect a new building 40.5 km from the left tower. It is to have an elevation of 41.5 m. Should the communications company oppose the building permit application? That is, will the building meet the clearance criterion?

14-13. A receiving antenna is mounted below the crest of a ridge that produces knife-edge diffraction. The amount of negative clearance is equal to $0.5F_1$. What would be the expected attenuation relative to free-space propagation?

14-14. What, according to Fig. 14.22, is the maximum usable frequency if we want to communicate via the ionosphere with a location 1000 km away at 8 A.M.?

Antennas

Dennis Morland

15-1 INTRODUCTION

An antenna is a device that couples energy between waves in free space and waves in guiding structures (transmission lines or waveguides). To achieve this function, the antenna generally has to provide impedance matching and directional radiation or reception. Communication via radio waves requires antennas at both the transmitting and receiving ends of the propagation path.

Antennas exist in a vast variety of shapes and types, from simple straight wires to complicated structures of metal and dielectrics. Since antennas are three-dimensional objects and the fields surrounding them are four-dimensional (involving time as well), the detailed analysis of antennas is very difficult mathematically. Fortunately, the overall behavior of antennas can be summarized in a small number of terms, such as input impedance, gain, and polarization, which are often fairly easy to measure. This chapter is concerned mainly with such antenna fundamentals. Some of the more popular antenna types are introduced. Sample calculations will show how these antenna parameters relate to practical problems.

15-2 RECIPROCITY THEOREM

An antenna has the same characteristics regardless of whether it is being used to transmit or receive radio waves. This relates to antenna parameters such as gain, impedance, beamwidth, and radiation pattern. For example, consider the impedance of an antenna. An antenna connected to a transmission will appear as a complex impedance to the transmission line [see Fig. 15-1 (a) and (b)]. The real part of this impedance represents the "radiation resistance" of the antenna, which accounts for the radiated signal. That is, to model the radiation process using simple circuit elements, we pretend that the power radiated from the antenna is dissipated in the radiation resistance. This allows us to treat the antenna as a normal circuit component for the purposes of circuit analysis. The same antenna impedance would represent a generator output impedance when this antenna is used for receiving signals [see Fig. 15-1 (c) and (d)].

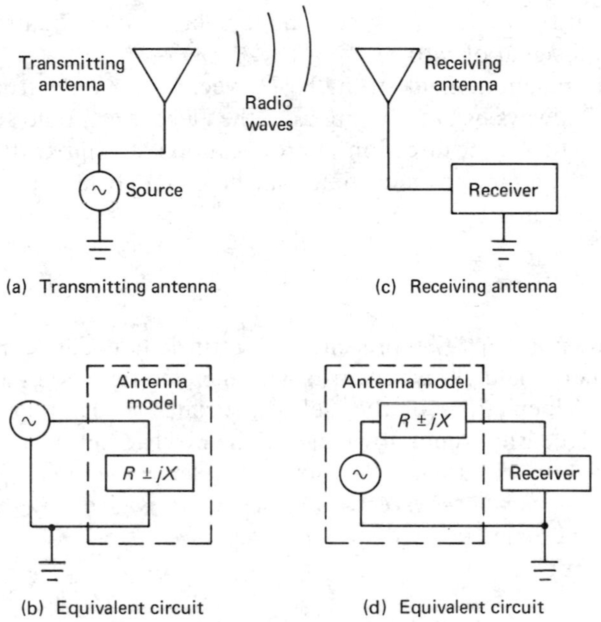

Figure 15-1 Antennas and equivalent circuits.

The reciprocity theorem simplifies things a bit, since we do not have to worry about whether an antenna is to be used for transmission or reception when stating its characteristics. When measuring antenna parameters, the antenna can be placed at either the transmitting or receiving end, whichever is more convenient.

15-3 POLARIZATION OF ELECTROMAGNETIC FIELDS

The electromagnetic (EM) field radiated from (or impinging on) an antenna is usually specified in terms of the electric field strength, which has units of volts per meter. For example, if an incoming EM wave with an electric field strength of 10 μV/m meets a wire antenna with an effective length of 1 m, the voltage induced in the wire will be (10 μV/m)(1 m) = 10 μV if the wire is oriented parallel to the electric field vector. [This is actually the open-circuit or Thévenin equivalent voltage as it would relate to Fig. 15-1(d).] If the wire and the electric field vectors were not in the same direction, the induced voltage would be less. In fact,

$$V = [L][E] \cos \theta \tag{15-1}$$

where θ is the angle between these two directions, E the magnitude of the electric field, and L the effective length of the antenna. The term *polarization* is used to describe the direction of the electric field lines in an electromagnetic wave. This is an important parameter to know when dealing with antennas, since the amount of energy coupled into an antenna by a wave is affected strongly by the relation between the antenna orientation and the wave polarization. If the antenna were oriented at right angles ($\theta = 90°$) to the polarization, equation (15-1) would predict a received signal of zero.

The magnetic field strength (H) vector of an electromagnetic wave in free space will always be at right angles to the electric (E) field strength vector and also at right angles to the direction of propagation (remember the right-hand rule). The magnitude of the magnetic field can be found from the electric field using the simple relation

$$|H| = \frac{|E|}{377} \quad \text{A/m} \tag{15-2}$$

where the number 377 represents the intrinsic impedance of free space. Although the magnetic field always coexists with the electric field in a radio wave, it is often ignored in depictions of the fields associated with antennas because it can be deduced so easily from knowledge of the electric field. Thus subsequent drawings in this section will usually show only the electric field component.

The power density vector ($\overline{S}$) of a wave is given by the cross product of the electric and magnetic field strength vectors:

$$\overline{S} = \overline{E} \times \overline{H} \tag{15-3}$$

That is, the magnitude of the power density is given by the product of the magnitudes of the electric and magnetic field strengths, and the direction associated with this value is given by the right-hand rule and points in the direction that the energy of the wave is traveling (see Fig. 15-2). Since the E and H values are related by the intrinsic impedance, it can also be stated that

$$|S| = \frac{|E|^2}{377} \quad \text{W/m}^2 \tag{15-4}$$

For example, the 10-μV/m wave mentioned earlier has a power density of 0.266 pW/m².

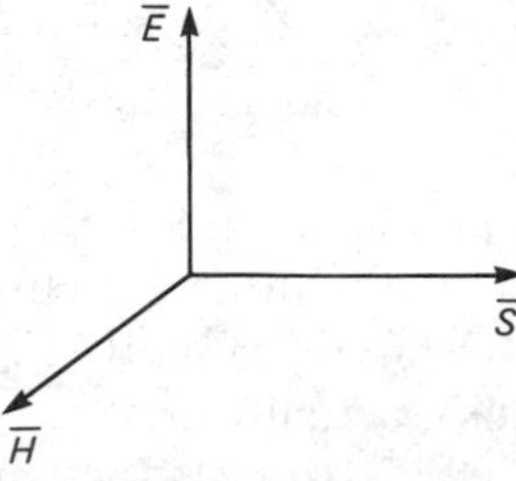

Figure 15-2 Power density cross product (right-hand rule).

15-4 RADIATION PATTERN OF A SHORT DIPOLE (HERTZIAN DIPOLE)

A Hertzian dipole is a conducting rod or wire which is much shorter than a wavelength of the operating signal. Currents on such an antenna are assumed to be uniform along its length, although the currents on longer antennas are tapered. Although the Hertzian dipole is not a practical antenna, it is useful to consider at this point for two reasons: first, it has a very simple mathematical description that will make it useful in illustrating fundamental antenna concepts, and second, longer antennas can be modeled by assuming that they consist of a collection of several Hertzian dipoles in series. Since antennas radiate time-varying vector fields into three-dimensional space, graphical and mathematical descriptions of the radiation process are complicated. It is usually convenient to use a spherical coordinate system instead of a Cartesian coordinate system to represent the radiated fields. (A quantity in space can be described using either of these coordinate systems, or a cylindrical system.) Any point in space can be described by giving the X, Y, and Z values or by giving the radius from the center and the two angles, theta (θ) and phi (ϕ). Theta is the angle, relative to the Z-axis, in the Z-Y plane. Phi is the angle, relative to X, in the X-Y plane. Figure 15-3 shows the spherical coordinate system, and Fig. 15-4 shows a small dipole antenna placed at the center of a spherical coordinate system.

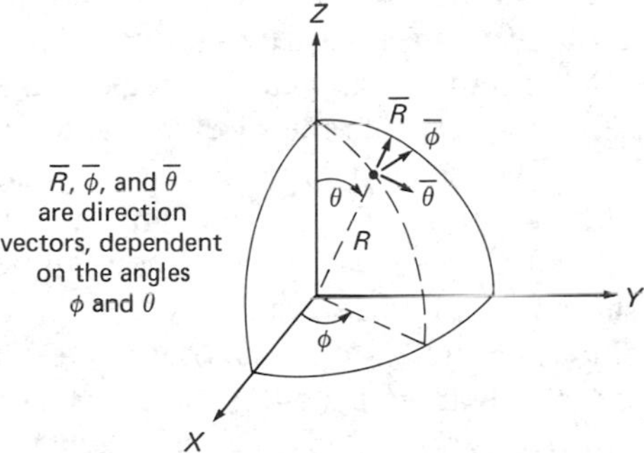

Figure 15-3 Spherical and cartesian coordinate systems.

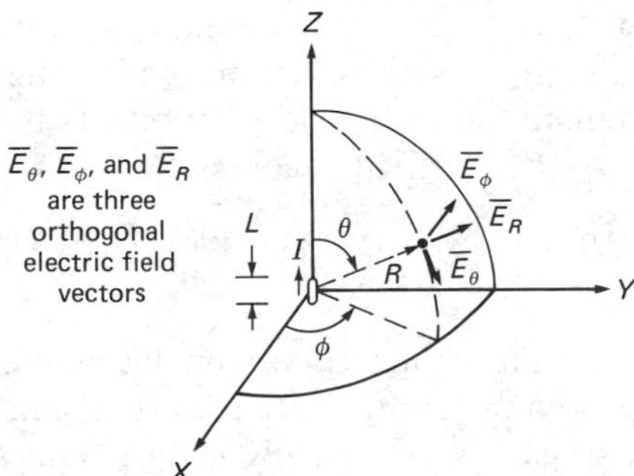

Figure 15-4 Hertzian dipole oriented along the Z-axis, at the center of a spherical coordinate system.

If a short (elemental) dipole has an alternating current (I) flowing on it, the fields produced can be described as follows.

$$E_\phi = 0 \qquad \text{(i.e., no "sideways" electric field)} \tag{15-5}$$

$$E_\theta = \frac{\eta IL \sin\theta}{4\pi}\left(\frac{j2\pi}{\lambda R} + \frac{1}{R^2} - \frac{j\lambda}{2\pi R^3}\right) e^{j(2\pi ft - \beta R)} \tag{15-6}$$

$$E_R = \frac{\eta IL \cos\theta}{4\pi}\left(\frac{2}{R^2} - \frac{j\lambda}{\pi R^3}\right) e^{j(2\pi ft - \beta R)} \tag{15-7}$$

where I is the magnitude of the current (rms) (A), L is the length of the conductor (m), θ is the angle toward the observer position, R is the distance from the antenna to the observer, λ is the wavelength (L, R, and λ must be in same units), f is the frequency (Hz), β is the propagation constant (radians per unit length), and η is the intrinsic impedance (ohms) (377 for air).

Before proceeding further, we will consider the three terms that involve R inside the parentheses in equation (15-6). These terms describe how the field strength decreases as the wave moves away from the source. Since there are three terms, there are three components to the total field. The first term, involving $1/R$, decays the most slowly of the three and is known as the *radiation term*. The second term, involving $1/R^2$, is the induction component, which relates to transformer-type coupling between antennas. The third term, involving $1/R^3$, is called the *electrostatic term* and relates to a capacitive-like coupling between antennas. The field strength due to these last two terms drops off much more quickly than that due to the radiation term and can be neglected if the transmitting antenna is a reasonable distance away from the receiving antenna. If the receiving antenna is close enough that all three terms are significant, it is in the *near-field* or *Fresnel region*. If only the radiation term is significant, it is in the *far-field* or *Fraunhoffer region*. Since antennas are usually used to communicate over long distances, the near-field effects are usually irrelevant, but a word of caution is in order. When measuring antenna parameters on a laboratory test range, it is often difficult to separate the antennas by large distances. There is a danger that the measurements might be made inadvertently in an antenna's near field. If this occurs, the results would not correctly represent the antenna's normal behavior. For antennas that are small compared to a wavelength, it is usually sufficient to separate the antennas by at least 10 wavelengths. For antennas that are large compared to a wavelength, such as microwave antennas or antenna arrays, the separation should be at least $2L^2/\lambda$, where L in this case would represent the largest dimension of the antenna.

If we eliminate the near-field components from equations (15-6) and (15-7), the description of the far field becomes

$$E = \frac{j\eta IL \sin\theta}{2\lambda R} e^{j(2\pi ft - \beta R)} = \left(\frac{j377}{2}\right)\left(\frac{IL}{\lambda}\right)(\sin\theta)\,\frac{1}{R}\, e^{j(2\pi ft - \beta R)} \tag{15-8}$$

and is polarized in the Z-direction. Now let us see if we can make some sense of this equation. The first term in parentheses is simply a constant. The j indicates a 90° phase shift which we can usually ignore. The I in the next term predicts that the field strength is proportional to the exciting current, and L/λ predicts that the field strength is proportional to the length of the antenna relative to the wavelength. The sin θ term describes how the field strength varies with the orientation of the measuring point relative to the antenna. If this function is plotted, it will produce a toroidal (doughnut) shape as shown in Fig. 15-5. The figure-eight pattern of Fig. 15-5(b) means that as the observation point moves, changing the angle θ, the field strength will vary from a maximum at $\theta = 90°$ down to zero at $\theta = 0°$ or $180°$. The circular pattern of Fig. 15-5(c) means that an observer moving around the antenna at a fixed radius in the X-Y plane will measure an unchanging value for the field strength. That is, there is no dependence on the angle ϕ. This means that the antenna pattern is directional in the Z-Y and Z-X planes but omnidirectional in the X-Y plane.

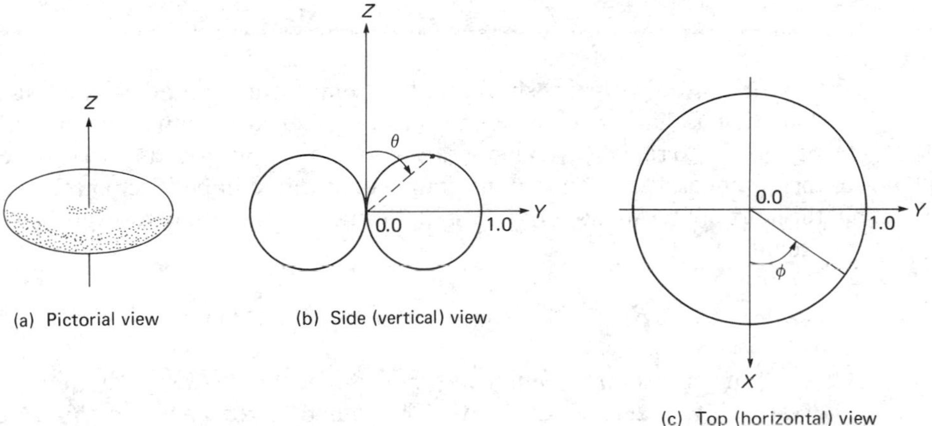

Figure 15-5 Radiation pattern of Hertzian dipole.

The $1/R$ term means that the field strength is varying in inverse proportion to the distance from the antenna. Since the power density of an EM wave is given by the square of the electric field strength divided by the intrinsic impedance, as in equation (15-4), the power density will vary as $1/R^2$. This is known as inverse-square-law behavior. Finally, the exponential term gives the instantaneous phase as a function of distance and time and identifies the field as a traveling wave. There is a sinusoidal variation with distance and time as the wave radiates outward from the source.

EXAMPLE 15-1

Find the magnitude of the electric field strength and the power density 1 km away from a Hertzian dipole that is 10 cm long if it carries a uniform current of 10 A_{rms} at a frequency of 150 MHz. Assume that the measurement is made in the X-Y plane (i.e., $\theta = 90°$).

(continued)

$$\lambda = \frac{c}{f} = \frac{3 \times 10^8 \text{ m/s}}{150 \times 10^6 \text{ Hz}} = 2 \text{ m}$$

$$|E| = \left(\frac{377}{2}\right)\frac{IL}{\lambda}(\sin\theta)\frac{1}{R}$$

$$= \left(\frac{377}{2}\right)(10)\left(\frac{0.1}{2}\right)(\sin 90°)\left(\frac{1}{1000}\right)$$

$$= 0.094 \text{ V/m} = 94.3 \text{ mV/m}$$

$$\text{power density} = \frac{|E|^2}{377} = \frac{(0.094)^2}{377} = 23.6\ \mu\text{W/m}^2$$

One reason why very short dipoles are not usually used as antenna elements is that their radiation resistance is very low; hence antenna currents have to be large in order to radiate even moderate amounts of power. This lowers the efficiency of the antennas and often requires more complicated impedance-matching devices. If the current on a short dipole is uniform, the radiation resistance will be approximately

$$R_{\text{rad}} = 80\pi^2\left(\frac{L}{\lambda}\right)^2 \quad \Omega \qquad \text{(Hertzian dipole)} \tag{15-9}$$

Usually, the current on a center-fed short dipole will be a maximum in the middle and taper off to zero at the ends. The radiation resistance in this case will be approximately

$$R_{\text{rad}} = 20\pi^2\left(\frac{L}{\lambda}\right)^2 \qquad \text{(short dipole, tapered current)} \tag{15-10}$$

The total antenna impedance in these cases would consist of the radiation resistance in series with a small capacitance.

EXAMPLE 15-2

Find the radiation resistance of a short dipole that has a length of $\lambda/20$. Calculate the required input current if the antenna is to radiate 100 W, assuming that the antenna has no losses.

$$R_{\text{rad}} = 20\pi^2\left(\frac{L}{\lambda}\right)^2 = 20\pi^2\left(\frac{1}{20}\right)^2 = 0.494\ \Omega$$

$P = I^2R$, so

$$I = \sqrt{\frac{P}{R}} = \sqrt{\frac{100}{0.494}} = 14.23\ \text{A}_{\text{rms}}$$

15-5 RESONANT ANTENNAS

If the length of the dipole antenna is increased to a half-wavelength, or multiple thereof, a standing-wave current pattern will be able to exist on the antenna and the input impedance will be totally real. This is an advantage because it is then much easier to match the antenna input impedance to that of a transmission line.

15-5.1 Half-Wavelength Dipole

By far the most common and useful resonant antenna is the half-wavelength dipole. When a dipole that is one-half-wavelength long is excited by a transmission line connected to its center, a sinusoidal standing-wave pattern is set up in which there is a current maximum at the center of the antenna. The current tapers down to zero at the ends. The radiation resistance for a very thin half-wavelength dipole is about 73 Ω but is slightly inductive due to the electric field lines extending past the antenna ends (so-called "end effects"). The input impedance will be purely resistive at about 67 Ω if the antenna is shortened by about 5%. The input impedance will also change if the diameter of the conductors used to make the antenna is increased. This also lowers the Q and broadens the resonance, so that the antenna approximates a real resistance over a larger frequency range.

The radiation pattern of a half-wavelength dipole is similar to that of the Hertzian dipole except that the variation with θ is no longer circular but is more elongated. As before, the preferred radiation direction is θ = 90 and the antenna is still omnidirectional with respect to φ. This behavior is sketched in Fig. 15-6.

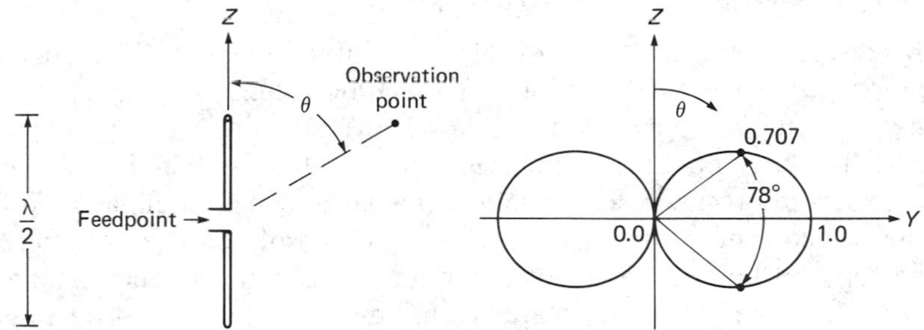

Figure 15-6 Radiation pattern of half-wavelength dipole.

The electric field strength for a half-wavelength dipole is given by

$$E = \frac{jI\eta}{2\pi R}\,\frac{\cos[(\pi/2)\cos\theta]}{\sin\theta}\,e^{j(2\pi ft - \beta R)} \tag{15-11}$$

where I is the current supplied to center terminals of the antenna by a transmission line and the other terms are as described for equation (15-6). *Note*: The arguments of the sin and cosine terms are in radians.

EXAMPLE 15-3

Find the magnitude of the electric field strength 10 km away from a half-wavelength antenna if the feedpoint current is 10 A (rms) and the angle between the antenna axis and the observer is:
(a) 90° or $\pi/2$ radians
(b) 60° or 1.05 radians
(c) 0° or 0 radians

$$|E_\theta| = \frac{I\eta \cos[(\pi/2)\cos\theta]}{2\pi R \sin\theta}$$

(a) $$|E_\theta| = \frac{(10)(377)\cos[(\pi/2)\cos(\pi/2)]}{2\pi(10{,}000)\sin(\pi/2)} = 0.06 \text{ V/m} = 60 \text{ mV/m}$$

(b) $$|E| = \frac{(10)(377)\cos[(\pi/2)\cos(1.05)]}{2\pi(10{,}000)\sin(1.05)} = 0.049 \text{ V/m} = 49 \text{ mV/m}$$

(c) $$|E| = \frac{(10)(377)\cos[(\pi/2)\cos(0)]}{2\pi(10{,}000)\sin(0)} = 0 \text{ V/m}$$

Obviously, the maximum radiation occurs in the plane where $\theta = 90°$. The electric field is oriented in the Z-direction in this case.

15-5.2 Longer Resonant Dipoles

Increasing the length of a dipole above a half-wavelength will change the feedpoint impedance, current distribution, and radiation pattern. For example, a wire antenna that is a full wavelength long will allow two standing-wave current maximums to exist. The center of the antenna will correspond to a null in the current distribution, so this would result in a high input impedance if the antenna is fed in the center. The two current maximums are always 180° out of phase, so the fields radiated by the top and bottom halves of the antenna will always cancel along the $\theta = 90°$ direction. Four strong lobes are created by this antenna in directions along which the fields add in phase. This occurs at approximately $\theta = 54°$. Several multilobed resonant linear antennas are shown in Fig. 15-7. The number of lobes is equal to twice the number of half-wavelengths in the antenna length. The antenna that is three wavelengths long, for example, has 12 lobes, four major lobes, and eight minor lobes. The radiation patterns of the longer linear antennas do not match the requirements of most applications; therefore, they are used far less commonly than the half-wavelength dipole.

The radiation pattern and input impedance of a linear antenna can also be changed by offsetting the feedpoint away from the center of the antenna. Changes in impedance and radiation pattern can also be produced by the presence of a large reflecting ground plane.

TABLE 15-1 Resonant Linear Antennas

Length	Current Distribution	Input Impedance	Radiation Pattern	Power Gain
$L = \frac{\lambda}{2}$	L, 0, I	Low $\simeq 73\Omega$	θ	1.64 (2.15 dB)
$L = \lambda$		High	$\theta = 0°$, $\theta = 54°$, $\theta = 90°$	1.8 (2.55 dB)
$L = \frac{3\lambda}{2}$		Low		2.0 (3.01 dB)
$L = 2\lambda$		High		2.3 (3.62 dB)
$L = 3\lambda$		High	Major lobe, Minor lobes	2.8 (4.47 dB)

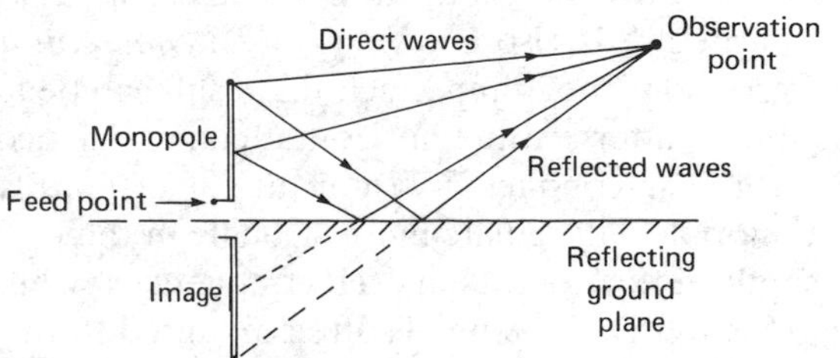

Figure 15-7 Vertical monopole over a ground plane.

15-5.3 Vertical Monopoles over a Ground Plane

A monopole is half of a dipole. When mounted over a good conducting surface, reflections from this surface will image the other side of the dipole, as shown in Fig. 15-7. This phenomenon is often used at low and medium frequencies where a half-wavelength dipole would be very large. For example, a half-wavelength antenna to be used at 1000 kHz, in the middle of the AM broadcast band, would have a length of 150 m. The corresponding monopole would require a tower only 75 m high. The tower and its supporting guy wires have to be insulated from the ground and the incoming feedline would be connected at the base of the tower. Antennas of this type, often called *Marconi antennas*, can be used to radiate many thousands of kilowatts of radio wave energy for broadcast-band stations. The radiated fields will be polarized vertically and the radiation pattern will be the same as the upper half of the corresponding dipole pattern, as shown in Fig. 15-8. Since the radiation pattern is omnidirectional in the horizontal plane, it is ideal for broadcast stations.

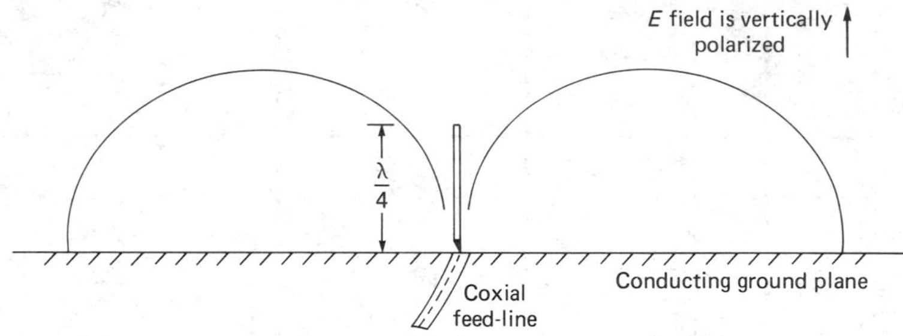

Figure 15-8 Radiation pattern of a quarter-wavelength monopole.

This type of antenna requires a good ground plane to work properly. The moisture in the earth makes the ground a reasonably good conductor at medium frequencies, although marshy ground or open water work best. Often, an artificial ground plane is used near the base of monopole antennas to improve performance. This usually consists of a large number of radial wires extending outward from near the base of the antenna. If the wires are made a quarter-wavelength long and are grounded at the inner end, they will make a very effective resonant ground plane. This structure is also known as an *antenna counterpoise*.

If the ground plane is imperfect, the result can be a distortion of the otherwise omnidirectional pattern in the horizontal plane. This could occur, for example, if the ground varies in conductivity (moisture level) in different directions. The field from the monopole will radiate more strongly in the direction of the better ground plane. A similar effect occurs at VHF frequencies when monopoles are mounted on motor vehicles. The antenna is often mounted in the middle of the roof, which serves as a reasonably good ground plane, especially if it is several wavelengths in

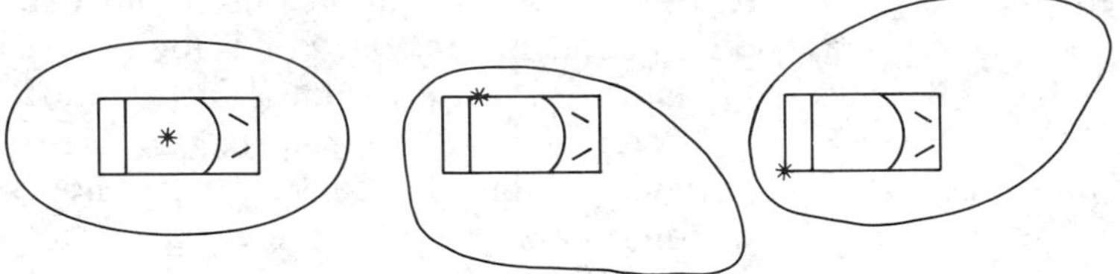

Figure 15-9 Radiation patterns of vertical monopoles (marked ∗) on a car.

extent. Figure 15-9 shows the possible pattern distortion effects of mounting a VHF monopole at different locations on a car.

If the length of the monopole is made longer than a quarter-wavelength, the direction of maximum radiation will lift off the horizontal a bit, which may be useful in some applications. Figure 15-10 illustrates this behavior. Multilobing will start to occur if the monopole length exceeds a half-wavelength.

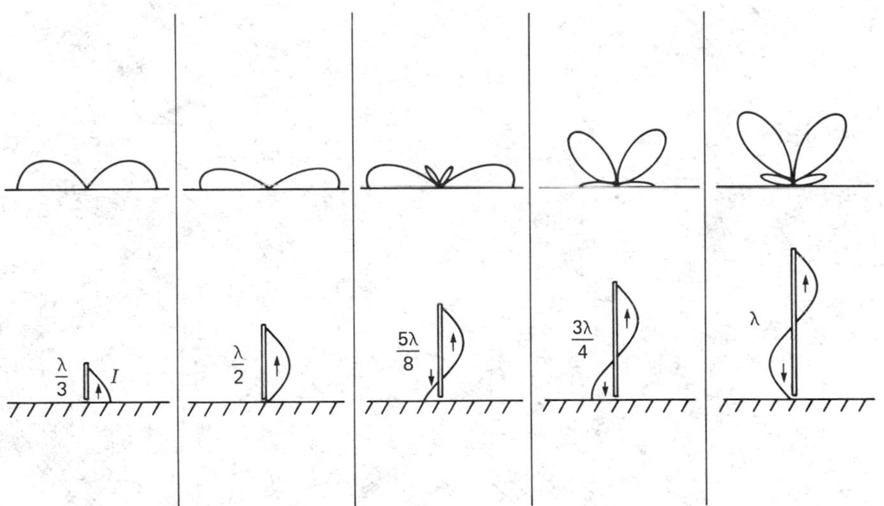

Figure 15-10 Current distributions and radiation pattern for vertical monopoles of different lengths over a perfect ground plane.

The input impedance of a monopole antenna is about half that of the corresponding dipole. A thin quarter-wavelength monopole over a perfect ground plane will have an input impedance of about $37 + j21\ \Omega$. At frequencies for which the length is not a quarter-wavelength, the input impedence will vary in a complex manner. Other factors, such as antenna diameter and capacitance, between the antenna and the ground plane will also have an effect. Although difficult to calculate, the input impedance of a monopole can easily be measured over any desired frequency range. Figure 15-11 shows typical measured impedance plots for two-monopole antennas of different diameters. One plot is for a "thin" antenna, whose length is 236 times the diameter. The other plot is for a "thick" antenna whose length is 10 times the diameter. As the frequency is varied, the electrical length of the antenna varies also, although the physical length does not change, of course. The numbers at different points on the curves represent the actual length expressed in wavelengths. At the lowest measured frequency, the monopole is about a tenth of a wavelength long, and at the highest frequency it is about one wavelength long. Remember that the radiation pattern would be changing as well as the input impedance.

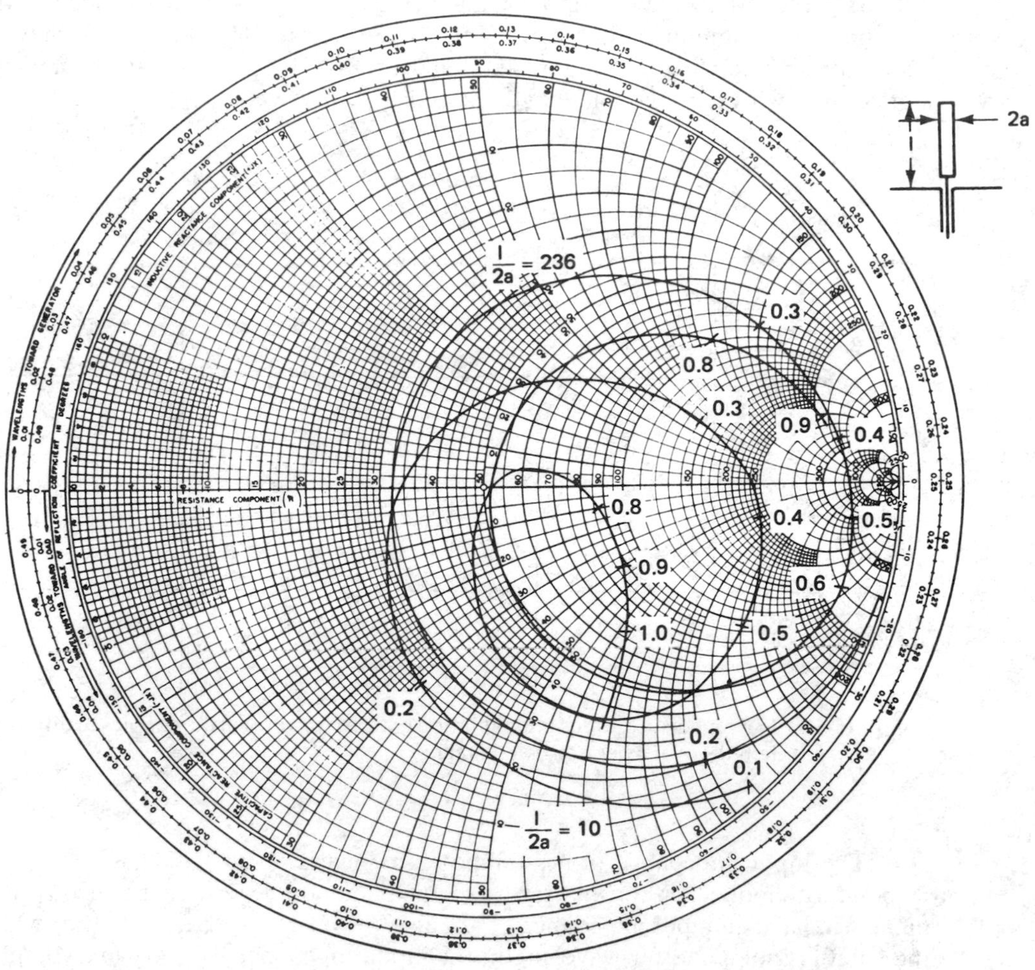

Figure 15-11 Typical input impedance of monopole antennas with length/diameter ratios of 236 and 10. (Adapted from Sinnema, *Electronic Transmission Technology*, 2nd ed., p. 283, Figure 10-35, Prentice-Hall, Inc., Englewood Cliffs, N.J., 1988.)

15-5.4 Coil and Top-Hat Loading

Another factor that can affect antenna performance is antenna loading, which is generally used to make an antenna appear electrically longer than its actual length. This permits the antenna to be more compact physically. Adding an inductor near the base of a monopole will have this effect. Adding a capacitor at the top will also work. Figure 15-12 illustrates the principle of "top-hat" loading. The effect of top-hat loading on the input impedance is to shift the impedance curves more into the capacitive side of the Smith chart, as shown in Fig. 15-13.

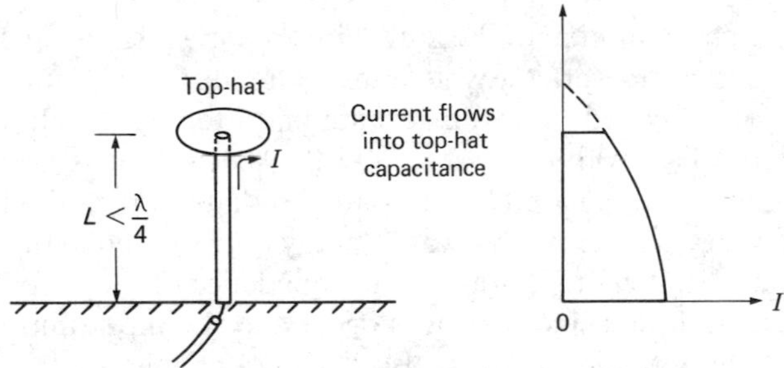

Figure 15-12 Top-hat loading of a monopole.

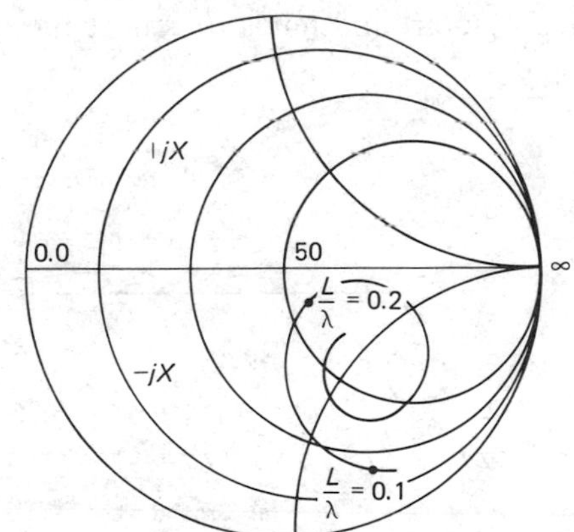

Figure 15-13 Effect of capacitive loading on impedance (Smith chart representation).

15-6 NONRESONANT LINEAR ANTENNAS

One problem with resonant dipoles (or monopoles) is that they work well only near the frequency for which they were designed. As Fig. 15-11 illustrates, the input impedance of a linear antenna changes drastically with frequency. The operating bandwidth of such antennas is therefore quite restricted, especially when high-Q matching networks are used to match their input impedance to transmission lines. Another problem is that their radiation pattern tends to change with frequency as well, as depicted in Fig. 15-10. Two antennas that work reasonable well over a very wide frequency range are the long-wire (Beverage) antenna and the rhombic antenna, which are examples of the traveling-wave class of antenna. Another example of this type is the *double Archimedes spiral*, used as an omnidirectional antenna at high frequencies.

15-6.1 Nonresonant Long-wire (Beverage) Antenna; Wave Antennas

The simple configuration of the long-wire antenna is shown in Fig. 15-14. Essentially it is just a long (several wavelengths) wire terminated in a resistive load. This antenna, often called a Beverage antenna after its developer, can operate over quite a large bandwidth, say 3 to 30 Mhz, because it is a nonresonant structure. Traveling waves stimulated by the source radiate part of their energy as they move down the wire. The remaining wave energy is dissipated in the resistive termination, which should match the characteristic impedance of the wire. Since reflected waves are not created, no standing waves occur and the input impedance stays relatively constant. The radiation pattern also does not change significantly with frequency. Unfortunately, the pattern is not very directional. As Figs. 15-14(b) and (c) show, the radiation pattern in each plane consists of a strong set of lobes to either side of the wire axis. Actually, the lobe is three-dimensionally symmetrical around the wire axis, although the ground tends to reflect the downward lobe to strengthen the skyward radiation.

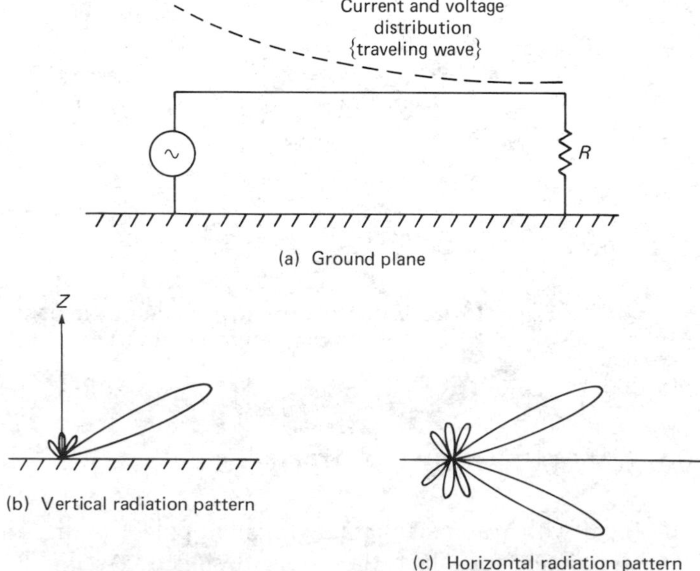

Figure 15-14 Nonresonant long-wire (Beverage) antenna.

15-6.2 Rhombic Antenna

The rhombic antenna is so named because it is formed in the shape of a rhomboid (diamond). Each side is a long-wire antenna which may be from 1 to 10 wavelengths long. If the angle θ is chosen correctly for the length, the radiation pattern will have a strong lobe along the axis of symmetry. Each of the arms of the antenna has a radiation pattern like the simple Beverage antenna, but if the angle θ is chosen correctly, the sideways lobes will cancel out, leaving the forward-directed lobes to add together, as shown in Figure 15-15. In the vertical plane, the forward lobe is elevated above the horizontal. The amount of lift can be controlled to some extent by varying the height of the antenna structure above the ground. The input impedance is generally in the range 600 to 800 Ω, which is in the range of characteristic impedances of balanced parallel-wire transmission lines. This means that a direct connection to a transmission line can be made, eliminating bandlimiting matching networks. This allows the antenna to function over a large frequency range, almost 10:1. The terminating resistor should be the same as the input impedance. Unfortunately, this antenna tends to be rather larger at low to medium frequencies.

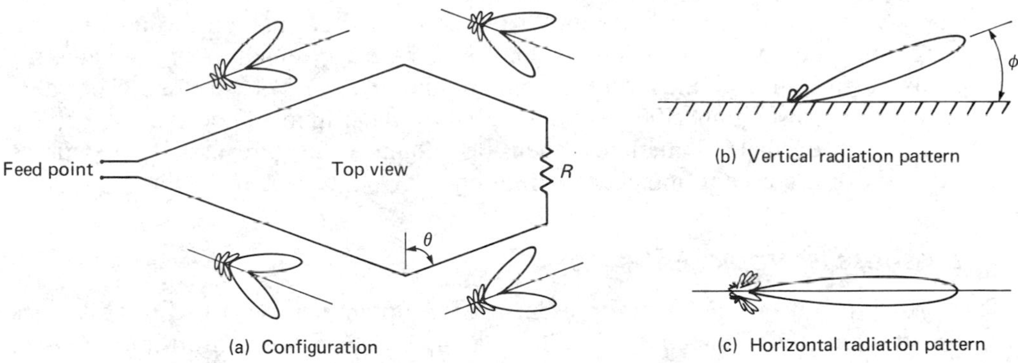

Figure 15-15 Rhombic dipole.

15-7 TERMS AND DEFINITIONS

There are quite a number of specialized terms that are used to describe antenna behavior. In this section we deal with some of the most common terms. For a more complete list of standard antenna terms, and more rigorous definitions, the reader should consult IEEE Std 145-1983.[1]

1. *IEEE Standard Definitions of Terms for Antennas*, IEEE Std 145-1983, published by The Institute of Electrical and Electronics Engineers, Inc., 345 East 47th Street, New York, NY 10017 (ISBN 0018-926X).

Relative Gain

The relative gain of an antenna is defined as the ratio of the maximum radiation power density from an antenna (S_{max}) to the maximum radiation power density from a reference antenna ($S_{max[ref]}$), where the reference antenna is fed with the same input power.

$$g_r = \frac{S_{max}}{S_{max[ref]}} \tag{15-12}$$

where S = power density in W/m^2, or

$$G_r = 10 \log(g_r) \qquad \text{dB} \tag{15-13}$$

Since antennas are passive devices, they cannot increase the total power, so the use of the word "gain" implies a different meaning than when it is used with regard to other electronic circuits. With respect to antennas, *gain* relates to the amount by which one antenna will concentrate radiated energy in some direction, compared to another reference antenna. Two reference antennas are commonly used: (1) the loss-less half-wavelength dipole (relates to "gain over a dipole" a common practical reference), and (2) the loss-less isotropic antenna (relates to "gain over isotropic" or just "gain"; used for theoretical work).

A dipole is easy to build and often convenient to use on an antenna test range, but the point source is mathematically more convenient. For the value for gain to be meaningful, the reference antenna should always be specified. The value for gain may also be specified as a simple ratio (linear number) or as a logarithmic quantity in decibels. Sometimes when logarithmic values are used, the symbols dBd or dBi are used to indicate the reference.

Loss-less Isotropic Antenna

A loss-less isotropic antenna represents an ideal point source of EM waves. For a loss-less antenna, radiated power will equal input power. For an isotropic antenna, the radiation intensity is the same in all directions. Thus for a loss-less isotropic antenna,

$$\text{radiation power density } S_0 = \frac{P_{in}}{4\pi R^2} \qquad \text{W/m}^2 \tag{15-14}$$

That is, the radiated power is assumed to be spread evenly over the surface of a sphere, which has an area of $4\pi R^2$.

Gain; Gain over Isotropic; Absolute Gain

The terms "gain," "gain over isotropic," and "absolute gain" all relate to the gain of an antenna over a loss-less isotropic radiator. Antenna gain therefore is

$$g = \frac{S_{max}}{S_0} \tag{15-15}$$

Directivity

A related antenna parameter is directivity (D) (also called directive gain). This is the ratio of the power density from the antenna to the power density averaged over all directions. The average power density (in watts per square meter) is taken as the radiated power divided by the area of a sphere, $4\pi R^2$:

$$D = \frac{S_{max}}{S_{av}} \qquad \text{where } S_{av} \text{ is } \frac{P_{radiated}}{4\pi R^2} = (S_0)(\text{efficiency}) \tag{15-16}$$

where efficiency $= K = P_{rad}/P_{in}$.

This term is very similar to that of antenna gain, but differs from it by the amount of the efficiency term. In fact, the antenna gain is equal to the directivity times the efficiency:

$$g = D \times K \tag{15-17}$$

For a loss-less antenna the gain will equal the directivity, but otherwise the gain will be less than the directivity.

EXAMPLE 15-4

Suppose that an antenna has a power input of 40 W and an efficiency of 98% (i.e., K $= P_{rad}/P_{in} = 0.98$). Also, suppose that the radiation power density has been found, by measurement, to have a maximum value of 20 mW/m^2 at 100 m from the antenna. Find the directivity and the gain over isotropic of this antenna.

Solution

$$S_{av} = \frac{P_{rad}}{4\pi R^2} = \frac{(0.98)(40\text{ W})}{4\pi(100\text{ m})^2} = 3.12 \times 10^{-4}\text{ W/m}^2$$

$$D = \frac{S_{max}}{S_{av}} = \frac{20\text{ mW/m}^2}{0.312\text{ mW/m}^2} = 64.1 \quad \text{or} \quad 18.07\text{ dB}$$

$$S_0 = \frac{P_{in}}{4\pi R^2} = \frac{40\text{ W}}{4\pi(100)^2} = 3.18 \times 10^{-4}\text{ W/m}^2$$

$$g = \frac{S_{max}}{S_0} = \frac{20\text{ mW/m}^2}{0.318\text{ mW/m}^2} = 62.83 \quad \text{or} \quad 17.98\text{ dBi}$$

N.B.: $g < D$; in fact, $g/D = P_{rad}/P_{in}$ = efficiency.

The values given for the formulas above relate to the maximum values that would be plotted on polar plots. Other values on the plots, off the peak of the main lobe, would be referred to as the "partial gain" or "partial directivity" in some direction.

EXAMPLE 15-5

Using the formula given for the electric field strength of a dipole [equation (15-11)], find the directivity of the dipole and its theoretical beamwidth.

Solution Since the power density in a wave is equal to the square of its electric field strength divided by the intrinsic impedance:

$$D = \frac{\text{maximum radiated power density}}{\text{average power density}} = \frac{|E_{\theta\max}|^2/\eta}{P_{rad}/4\pi R^2}$$

but $|E_\theta|$ is maximum when $\theta = \pi/2$ (90°), in which case

$$\frac{\cos[\pi/2 \cos(\pi/2)]}{\sin(\pi/2)} = 1$$

Also, $P_{rad} = I^2 Z_{in} \times$ efficiency; $Z_{in} \simeq 73\ \Omega$ for a $\lambda/2$ dipole, and $\eta = 120\ \pi = 377\ \Omega$. Therefore,

$$D = \frac{(\eta I/2\pi)^2\ (1/R)^2\ (1/\eta)}{I^2 Z_{in}/4\pi R^2} = \frac{I^2(120\pi)/\pi}{I^2(73)} = \frac{120}{73} = 1.64\ (2.15\ \text{dB}) \qquad (15\text{-}18)$$

If the antenna is loss-less, the gain over isotropic of the dipole would be equal to its directivity, 2.15 dB.

From the result above, and the equation for the dipole field strength, it is apparent that the expression for the partial directivity of the dipole is

$$D(\theta) = 1.64\left[\frac{\cos(\pi/2 \cos\theta)}{\sin\theta}\right]^2 \qquad (15\text{-}19)$$

The beamwidth of the dipole can be found from this expression. The beamwidth is defined by the half-power points on the partial directivity pattern. These points occur when the squared term in equation (15-19) is equal to one-half. This occurs for $\theta = 51°$, so the beamwidth is equal to

$$\text{beamwidth} = 2(90° - 51°) = 2 \times 39° = 78°$$

This agrees with the plot of Fig. 15-16.

Bandwidth

The bandwidth of an antenna may be defined as the range of frequencies within which the performance of the antenna conforms to a specified standard. The standard might be a stated permissible range of: input impedance, radiation efficiency, beamwidth, gain, sidelobe level, polarization, or any other antenna parameter. The specification would depend on the intended application of the antenna. For example, the bandwidth in one application may be defined by the frequency range over which the radiated power is within 3 dB of the optimum value. Or sometimes a manufacturer will quote a bandwidth as being a frequency range over which the

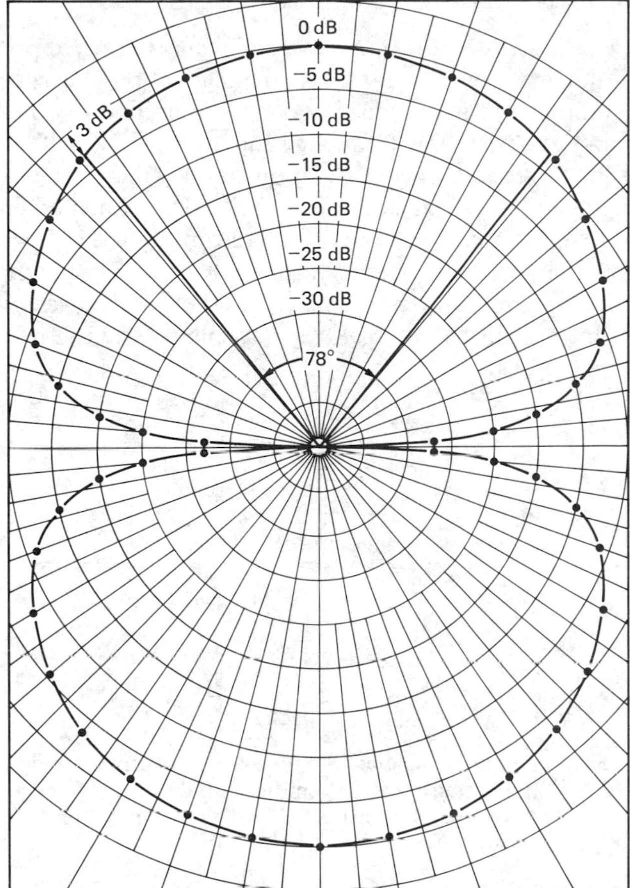

Figure 15-16 Polar Plot for λ/2 dipole.

input VSWR is acceptably close to the ideal value of 1.0 (such as, say, 1.8 maximum). This kind of specification relates well to impedance plots such as Fig. 15-12 or 15-14. If a VSWR circle is drawn on such a plot, the part of the impedance curve that falls within the VSWR circle defines the bandwidth. It should be remembered, though, that the radiation pattern changes with frequency as well and in some applications might be the limiting factor.

Linear Polarization

Linear polarization refers to the case where the electric field vector of an EM wave is always in a specific direction (as in vertical or horizontal polarization). The antennas discussed so far produce linearly polarized waves.

Circular Polarization; Elliptical Polarization

If the electric field vector of a wave rotates as the wave propagates, elliptical (or circular) polarization results. Circular polarization is a specific case of the more general elliptical form. Circular polarization occurs if two orthogonal linearly polarized waves, of the same frequency and intensity but with a 90° phase shift, travel in the same direction. Such a situation can be created by a set of dipoles, each at right angles to the other, as shown in Fig. 15-17. Each antenna is driven from the same source, but the currents in one dipole are shifted by 90° relative to the currents in the other by loss-less phase-shifting networks. This configuration is sometimes known as a *turnstile antenna*. Figure 15-18 sketches the vector relationship between the two antenna fields (vertical and horizontal) and the resultant electric field.

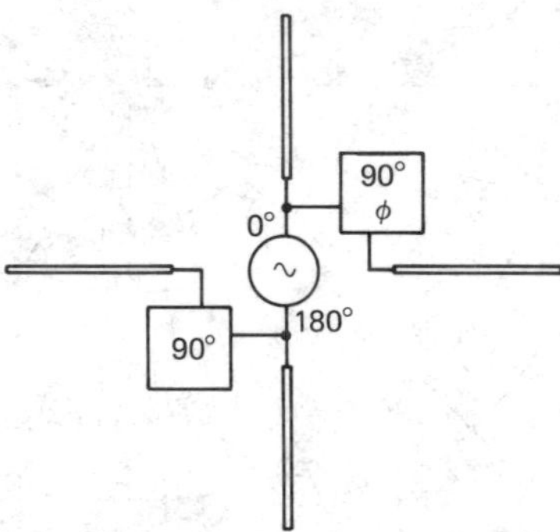

Figure 15-17 Dipole arrangement to produce circular polarization. Two dipole antennas, fed by sources of the same amplitude and frequency, but 90° out of phase. If the phase relationship between the oscillator feeds is not exactly 90°, elliptical polarization results (rarely used).

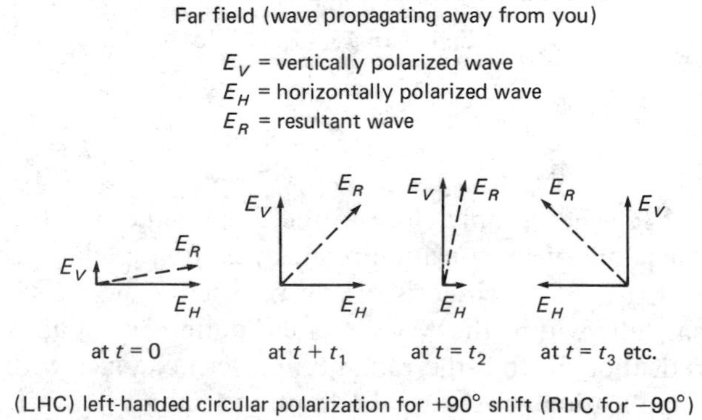

Far field (wave propagating away from you)

E_V = vertically polarized wave
E_H = horizontally polarized wave
E_R = resultant wave

E_V E_R E_H at $t = 0$ | E_V E_R E_H at $t + t_1$ | E_V E_R E_H at $t = t_2$ | E_R E_V E_H at $t = t_3$ etc.

(LHC) left-handed circular polarization for +90° shift (RHC for −90°)

Figure 15-18 Electric field vectors for a circularly polarized wave.

At a particular point in space, the electric field appears to rotate with time (see Fig. 15-19). If the electric field vector has an anticlockwise rotation when receding from the observer, it is called *left-handed circular polarization* (LHC). If the electric field is rotating in a clockwise direction, it is called *right-handed circular*

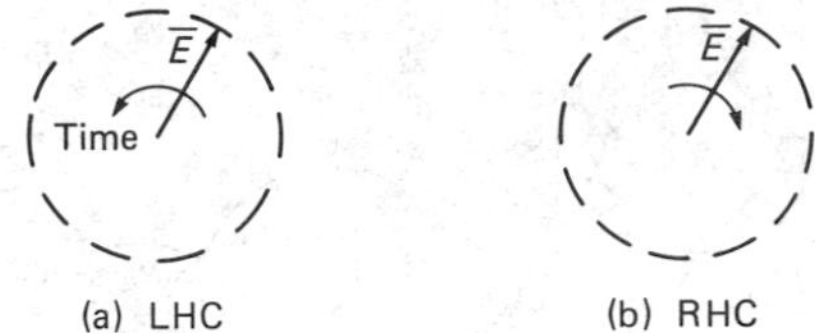

Figure 15-19 Rotating polarization vector: (a) LHC, anticlockwise; (b) RHC, clockwise.

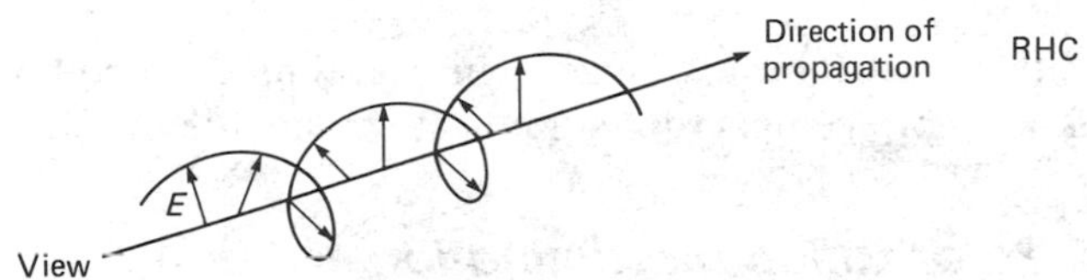

Figure 15-20 Spiraling E field in a RHC wave.

polarization (RHC). This is illustrated in Fig. 15-20. At a particular instant of time, the peak of the electric field vector seems to trace a spiral through space.

Uses of circular polarization include:

1. *Nonstabilized platforms*. In this case the orientation of the sender and receiver may be changing. Good examples of this are some satellites and missiles. Although some communications satellites are stabilized, and hold a fixed orientation with the earth, many others are not. If the satellite is spinning with respect to the earth, linear polarization would not be satisfactory, but circular polarization would be fine.
2. *Terrestrial communications*. Where obstacles or weather conditions (rain) cause depolarization of linear waves (transfer of some wave energy into orthogonal polarization), better performance can sometimes be obtained by using circular polarization.
3. *Radar target discrimination*. A plane, linearly polarized wave reflecting from a curved object will produce an elliptically polarized wave. The extent of depolarization can sometimes be used to deduce some information about the shape of the scattering object. This is used primarily in military radar systems.

Helical Antenna

Circular polarization can also be produced from a helix-shaped antenna. The diameter, sense, pitch, and number of turns relative to the wavelength determine the polarization state and gain of the antenna. A typical configuration is shown in Fig. 15-21. If the helix diameter is made less than a wavelength and the overall length is several wavelengths, the antenna will radiate a narrow beam of circularly polarized waves off the end of the spiral. This type of antenna often has a reflecting ground plane at the base to help prevent radiation in the reverse direction. Another example of a helical antenna is the "rubber duckie" miniature flexible antenna

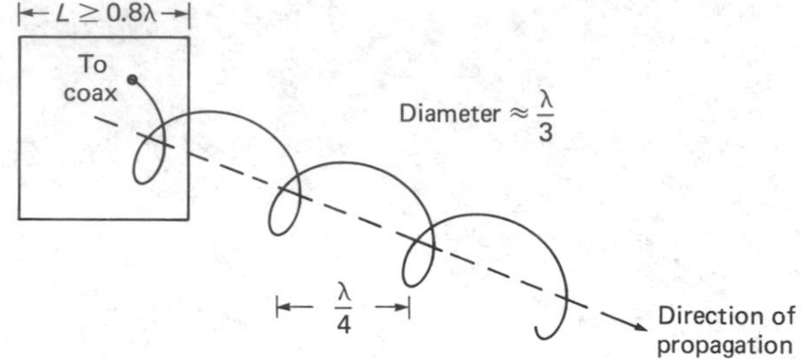

Figure 15-21 Helical antenna with a backplane reflector (LHC).

used with hand-held transceivers. This is essentially a coil of wire wound in a helix to make it more compact than a normal monopole.

Cross-Polarization Discrimination

Cross-polarization discrimination refers to the ability of an antenna to reject an incoming wave which has a polarization orthogonal to that for which the antenna was designed. This can be particularily important in some microwave communications systems, such as satellite and terrestrial relays, where orthogonal polarizations are sometimes used to double the amount of information that can be carried by the link, within a given frequency range.

Front-to-Back Ratio

The front-to-back ratio is the ratio of the wave energy sent (or received) in the forward direction of an antenna compared to the energy sent (or received) from the opposite direction. This term is relevant to unidirectional antennas only, such as the helix of Fig. 15-21 and some of the arrays and microwave antennas to be discussed later.

Effective Length of a Linearly Polarized Antenna

The effective length of a linearly polarized antenna is the ratio of the magnitude of the open-circuit voltage developed at the terminals of the antenna to the magnitude of the electric field strength of a plane wave that is oriented in the direction of the antenna polarization.

Effective Area of an Antenna

The effective area of an antenna is the ratio, in a given direction, of the available power at the terminals of a receiving antenna to the power flux density of a plane wave incident on the antenna from that direction. It is assumed that the wave polarization matches that of the antenna. The effective area of an antenna is equal to the square of the operating wavelength times its gain in that direction divided by 4π:

$$\text{effective area} \qquad A = \frac{\lambda^2 g}{4\pi} \tag{15-20}$$

EXAMPLE 15-6

Find the effective area of a half-wavelength loss-less dipole antenna being operated at a frequency of 30 MHz.

Solution First find the wavelength $\lambda = c/f = (3 \times 10^8)/30 \text{ MHz} = 10$ m.

$$A = \frac{(10)^2(1.64)}{4\pi} = 13.05 \text{ m}^2$$

Note: This would probably be much larger than the physical cross-sectional area.

15-8 BROADBAND AND LOOP ANTENNAS

The long-wire and rhombic antennas examined previously are examples of broadband antennas. Other examples are the folded dipole, the discone, and the biconical antenna.

15-8.1 Folded Dipole

A folded dipole consists of a half-wavelength dipole that has a second dipole, usually of the same diameter, mounted above it and connected at the ends, as depicted in Fig. 15-22. The separation between the top and bottom dipoles does not matter very much provided that it is quite small compared to a wavelength. The radiation resistance is four times that of a half-wavelength dipole, or about 290 Ω if equal-diameter conductors are used, so it is a good match to a 300-Ω parallel wire line. Many variations are possible for this kind of antenna. Changing the diameters of the conductors, or adding extra elements, or increasing the separation, or moving the feedpoint off-center will change the radiation resistance and bandwidth. The radiation pattern is the same as for a normal half-wavelength dipole. The beam width is also the same. The bandwidth, however, is much greater, typically 10% of the center frequency, because the top element tends to compoensate for the reactance variation of the input impedance with frequency. That is, below the resonant frequency where the dipole's reactance is capacitive, the top section is inductive and hence the input impedance is shifted closer to a pure

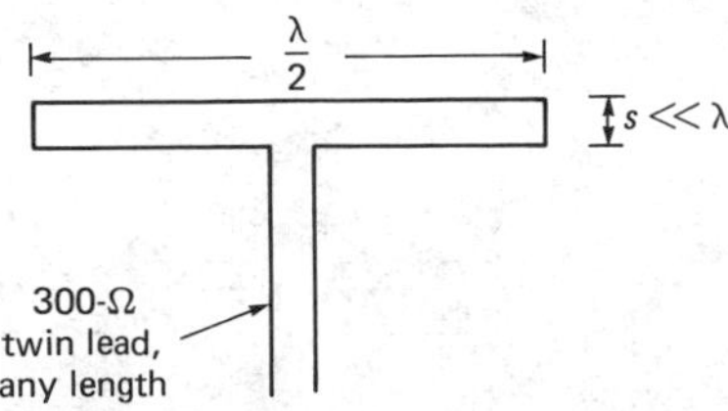

Figure 15-22 Folded dipole.

resistance. Above the resonant frequency the top is capacitive and the lower dipole is inductive. As a result, the input impedance stays closer to 300 Ω over a wider frequency range than that for a normal dipole. This wider bandwidth makes the folded dipole very useful for FM receiver antennas and for television antennas. A dipole cut for channel 4, 69 MHz, would have a bandwidth of about 7 MHz, more than adequate for the 6-MHz-wide TV channel.

The electrical length of the folded dipole should be half-wavelength. The physical length may be different due to two factors:

1. Fringing flux at the ends may make the antenna appear longer than it is. Typically, the actual length should be 0.95 of the electrical length to compensate for this.
2. The velocity of propagation along the antenna itself may be less than the speed of waves in air if the antenna elements are surrounded by a dielectric.

EXAMPLE 15-7

Using 300-Ω TV-type twin lead, which has a velocity of propagation of 80% of the speed of light, design a folded dipole to receive channel 4, which has a center frequency of 69 MHz.

$$\text{Length} = \frac{(0.95)0.8c}{2f} = \frac{(0.95)(0.8)(3 \times 10^8)}{2 \times 69 \text{ MHz}} = 1.65 \text{ m}$$

15-8.2 Discone Antenna

The discone antenna is actually a variation of the half-wavelength dipole. As Fig. 15-23 shows, the top radiator is formed into a disk and the bottom radiator is widened out into a cone. This antenna has a radiation pattern similar to that of a half-wavelength dipole. Its gain is less than that of a dipole, but its bandwidth is much larger, typically a 7:1 frequency range, for example, 10 to 70 MHz, for an input VSWR of less than 1.5. The dimension D is normally chosen to be a quarter-wavelength at the lowest frequency of desired operation.

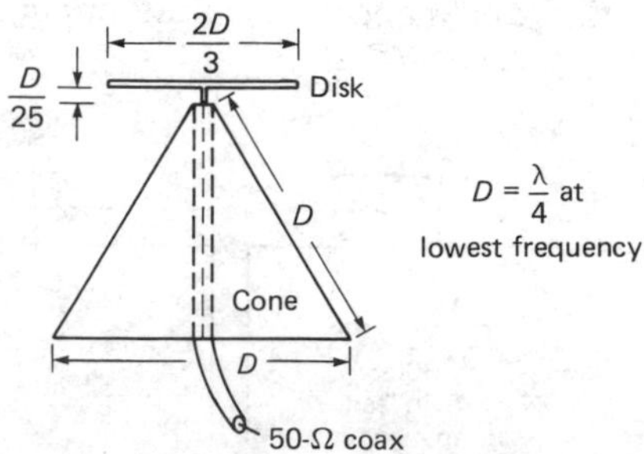

Figure 15-23 Discone antenna.

15-8.3 Biconical Antenna

This is another variation of the dipole in which two conical radiators are used instead of linear elements, as in Fig. 15-24. The dimension D should be a quarter of the free-space wavelength at the lowest desired operating frequency. Like the discone, the gain is low, but it is omnidirectional in the horizontal plane and can be used at frequencies much above the design frequency (in which case the dimension D becomes several wavelengths long). To reduce the weight, the cones are sometimes made out of wire spokes instead of sheet metal.

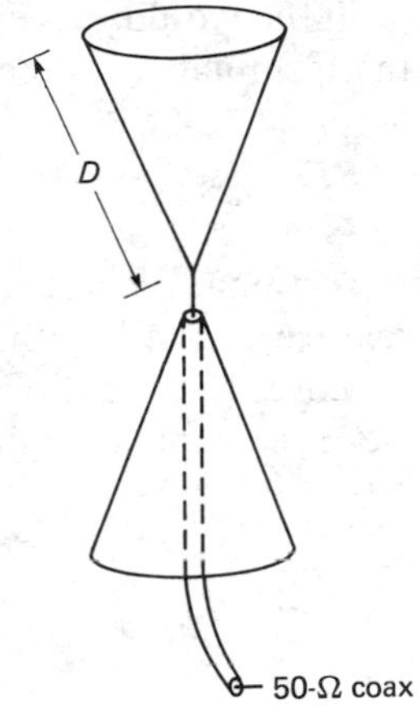

Figure 15-24 Biconical antenna.

15-8.4 Loop Antenna

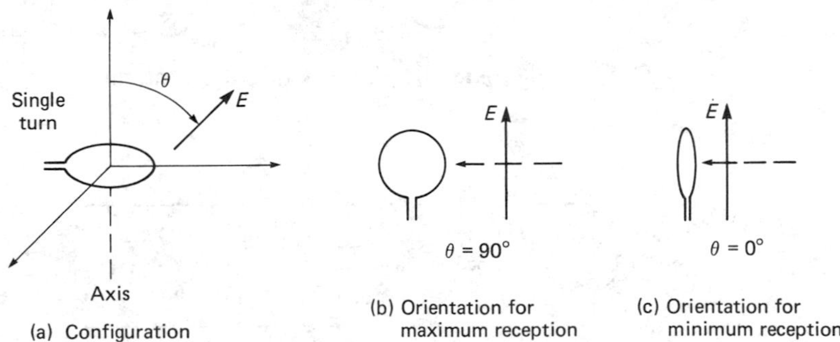

Figure 15-25 Loop antenna.

A loop antenna consists of a single loop of wire, or a number of loops, whose diameter is much smaller than a wavelength. This arrangement, shown in Fig. 15-25, simulates a magnetic dipole. When the loop is placed in the field of an incoming, linearly polarized plane wave, the open-circuit voltage induced across its terminals is given by

$$V = \frac{2\pi}{\lambda} NAE \sin\theta \qquad (15\text{-}21)$$

where N is the number of turns, A the actual area of the loop, E the electric field strength, and θ an angle relative to the axis of the loop and the incoming propagation direction. If the direction of propagation of the incoming wave coincides with the axis of the loop, there will be no reception. This strong nulling action has allowed the loop to be used in direction-finding applications.

A loop can also be used when very compact antennas are needed. Since the diameter of the loop is much smaller than a wavelength, in general, the overall size of the antenna can be much less than a comparable dipole. In this kind of application, a multiturn loop is often wound on a high-permeability core as shown in Fig. 15-26. This type of antenna is commonly used in portable AM broadcast-band radio receivers because the equivalent dipole would be very large. The maximum open-circuit voltage available from the terminals of the coil is given by

$$V = \frac{2\pi}{\lambda} NAEFu_r \qquad (15\text{-}22)$$

where N is the number of turns, A the cross-sectional area of the ferrite core, F a term dependent on the length of the coil relative to the length of the ferrite rod, and u_r the relative permeability of the ferrite core. In operation, the coil is placed in parallel with a tunable capacitor to make a resonant circuit, as shown in Fig. 15-27.

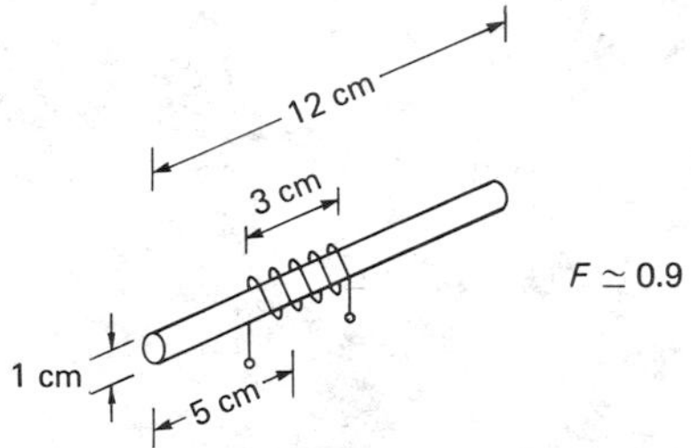

Figure 15-26 Ferrite (stick) loop antenna.

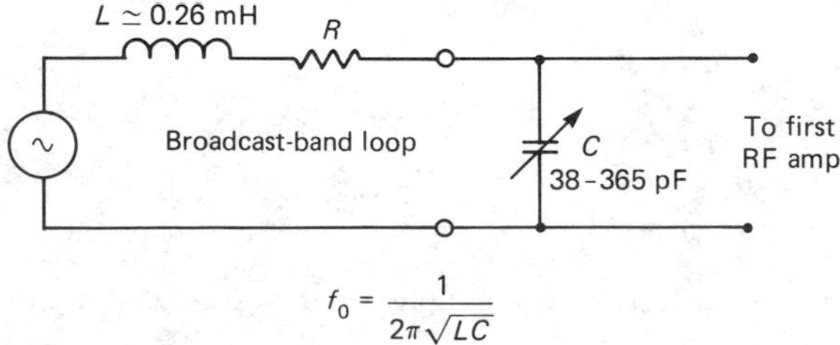

Figure 15-27 Equivalent circuit for the ferrite loop antenna.

The practical frequency range for loop antennas depends a lot on the characteristics of available ferrite materials. Ferrite materials have been developed that have low losses at high frequencies, so this type of antenna will probably find more applications in the VHF range in the future.

15-9 ANTENNA ARRAYS

It is often required that the radiation pattern of an antenna have a very narrow lobe or some other feature that is difficult to achieve with a single antenna element. In such situations it is often useful to combine several antennas into an array. The number of possible configurations of antenna arrays is without limit because a designer has many variables to consider, including number of antenna elements, types of elements, configuration (orientation and spacing of elements), phase of antenna currents, and use of reflectors or parasitic elements. The synthesis of antenna arrays is beyond the scope of this book, so we simply introduce some common array types in this section.

A convenient way of examining arrays is to model the radiation pattern of the entire array as the product of an array pattern and an antenna element pattern. The array pattern, or array factor, is the radiation pattern that would result if the antenna elements were isotropic, point-source antennas. To simplify things, we will consider an array of only two elements, as in Fig. 15-28. The two isotropic antennas are separated by a distance d and the currents induced in them have a phase difference of α radians. The total field at some distant observation point will depend on the phase of the two signals at that point. This relative phase will depend on α and d and on the location of the observation point. It is assumed that the amplitudes of the two signals will be the same. This is a reasonable assumption if the distance d is much smaller than the distance to the observation point. The phase difference between the two waves will be

$$\text{phase difference} = \beta d \cos \theta + \alpha \tag{15-23}$$

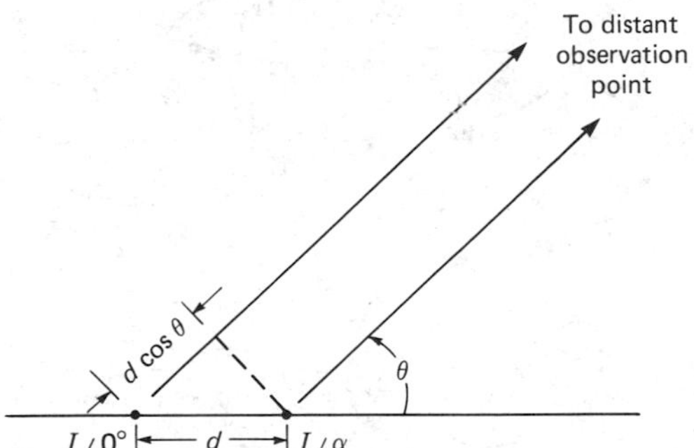

Figure 15-28 Antenna array of two isotropic antennas.

The magnitude of the total field will be given by

$$|E| = 2|Ei| \cos \left(\frac{\pi d \cos \theta}{\lambda} + \frac{\alpha}{2} \right) \tag{15-24}$$

where $|Ei|$ represents the magnitude of the electric field strength due to one radiating point source. To illustrate the effect of changing the spacing and phase angle in this equation, Fig. 15-29 depicts several array patterns for selected values of d and α. Next we will replace the isotropic antennas with dipoles to see the resultant radiation pattern.

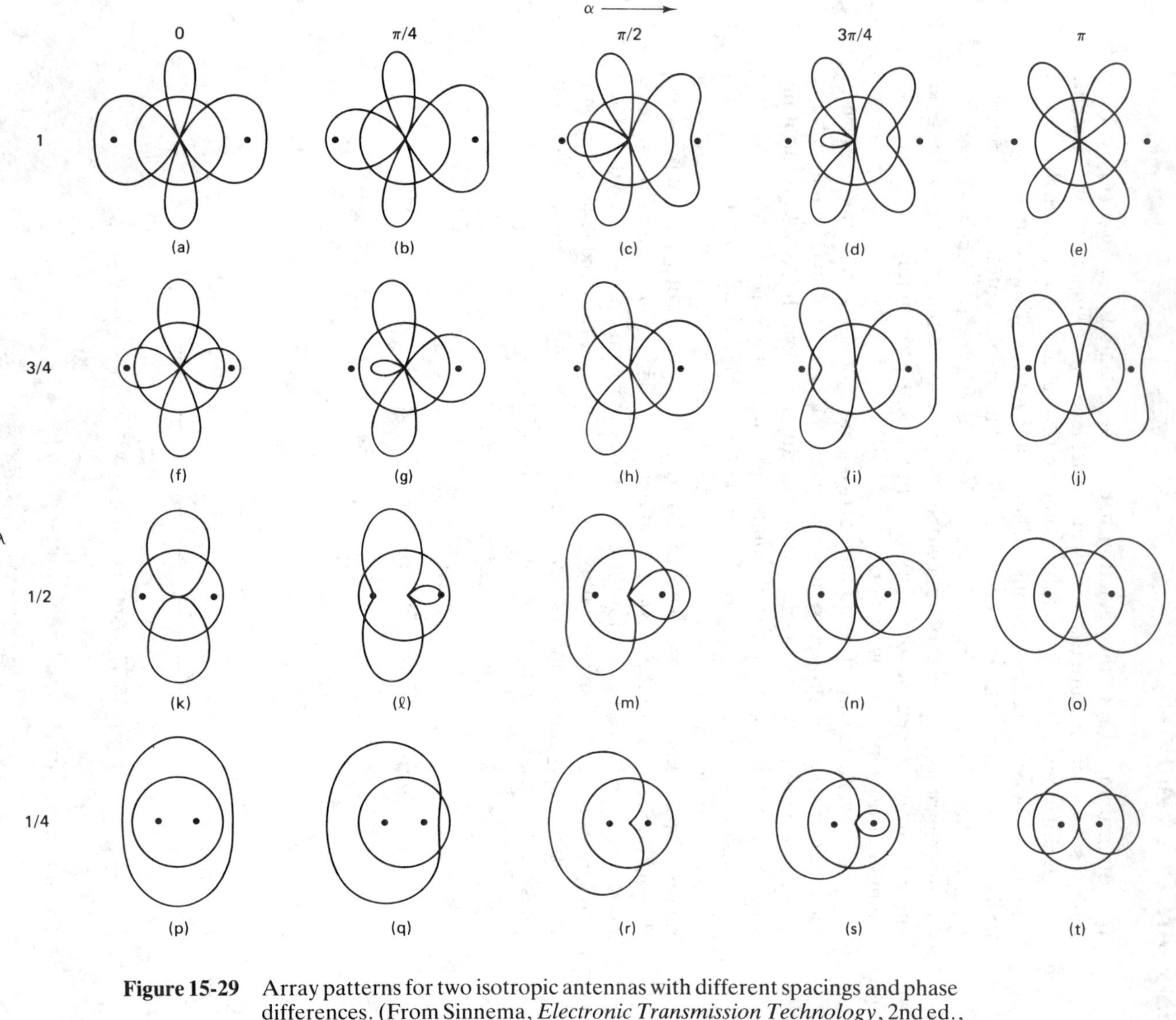

Figure 15-29 Array patterns for two isotropic antennas with different spacings and phase differences. (From Sinnema, *Electronic Transmission Technology*, 2nd ed., p. 290, Figure 10-45, Prentice-Hall, Inc., Englewood Cliffs, N.J., 1988.)

15-9.1 Two-Element Broadside Array

The two-element broadside array consists of two dipole antennas stacked one on top of the other, separated by a half-wavelength and driven in phase ($\alpha = 0$). This corresponds to the array pattern (k) of Fig. 15-29. When two antennas A and B, a half-wavelength apart, are excited in phase, their radiation is concentrated along a line at right angles to their plane. Since the currents in A and B of Fig. 15-30 are equal and in phase, the fields radiated at any instant are identical in polarity, phase, and amplitude. However, by the time the energy from A reaches B, the phase of the signal in the field at B has changed by $\lambda/2$, or π radians, and the two fields effectively cancel. The result is that radiation is prevented from going either up or down. At the same time, radiation fields from the two antennas arriving at any point at right angles to the plane of the antennas are in phase and therefore add, producing a stronger radiation field at that point. This array has a bidirectional radiation pattern. The effect of combining the two dipoles in this way is to make the antennas directional in both the vertical and horizontal planes.

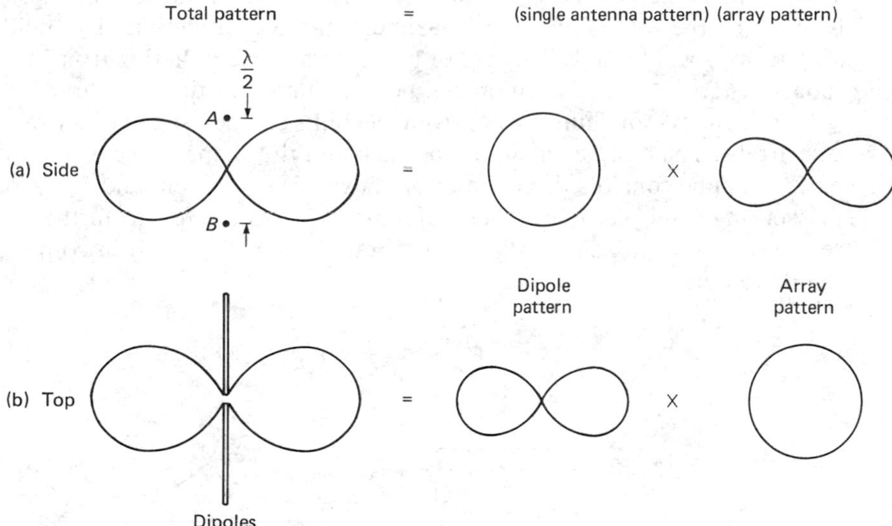

Figure 15-30 Two-element broadside array.

15-9.2 Two-Element End-Fire Array

The two-element end-fire array is designed to send energy in only one direction instead of bidirectionally like the broadside. Figure 15-31 shows the configuration and the radiation pattern. Once again, the total pattern can be obtained by multiplying the array pattern [pattern (r) of Fig. 15-29] by the radiation pattern of the individual dipoles.

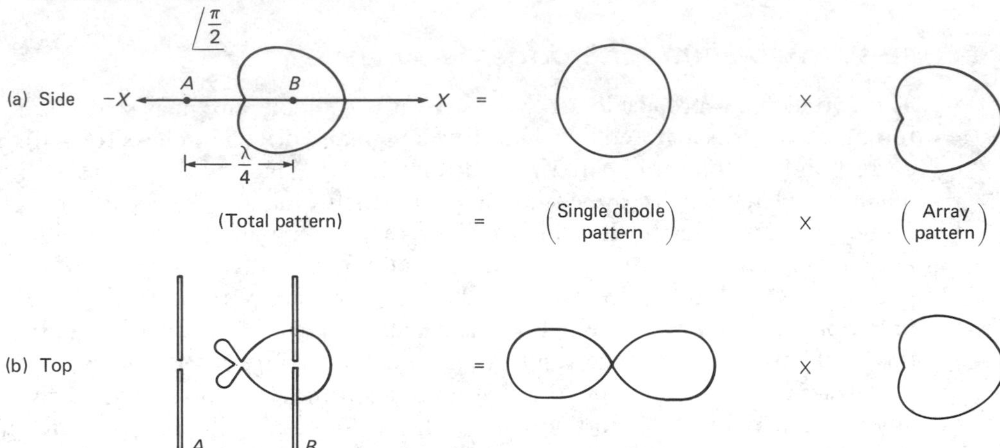

Figure 15-31 Two-element end-fire array.

The phase of the current in antenna A is leading that of antenna B by the amount $\pi/2$. The fields arriving at B from A will be in phase with those radiating from B, so the radiation will be maximum in the X direction. The fields arriving at A from B will be π radians out of phase with those radiating from A and so will cancel. Therefore, the radiation will be a minimum in the $-X$ direction.

In general, any number of driven elements can be used in end-fire and broadside arrays. The more elements used, the better the shape of the resultant radiation pattern can be controlled. Figure 15-32 shows a five-element end-fire array. Often, the spacings and phasings of the antennas are also altered to further control the response. Also, parasitic radiators and reflectors can be used in conjunction with the driven elements.

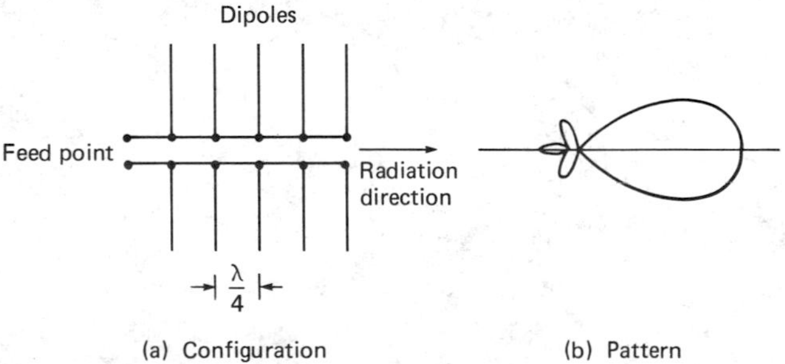

Figure 15-32 Five-element end-fire array.

15-9.3 Log-Periodic Array

Another array type that deserves mention is the log-periodic, which is shown in Table 15-2. This is a special kind of end-fire array in which the element lengths and spacings have been varied to achieve a very wide bandwidth. Its gain is not as great as that of a normal end-fire array, but its bandwidth is much larger, which makes it very useful as a television receiving antenna.

TABLE 15-2 Common Antennas

Name	*Radiation pattern*	*Polarization*	*Impedance*	*Gain*	*Bandwidth*
Log-periodic dipole array 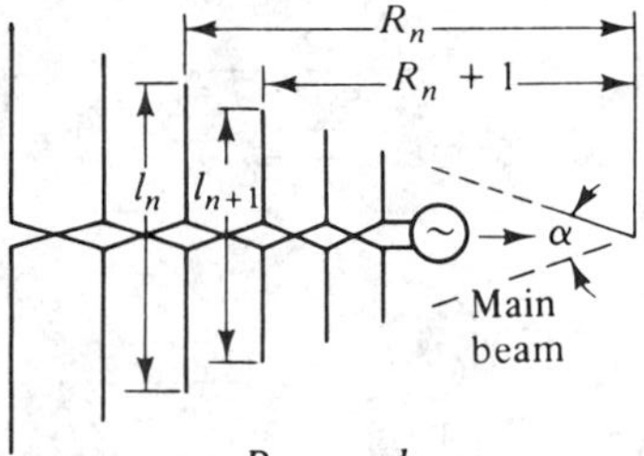 R_n; $R_n + 1$; l_n; l_{n+1}; α; Main beam $t = \frac{R_{n+1}}{R_n} = \frac{l_{n+1}}{l_n}$	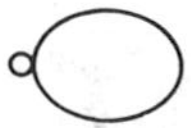 E plane (plane containing elements)  H plane (⊥'r to E plane) for $\alpha = 70°$ $\tau = 0.89$	Linear	$R = 60\ \Omega$ VSWR over period <1.6	6.5 dB	10:1 Depends upon length of shortest and longest elements
Pyramidal horn 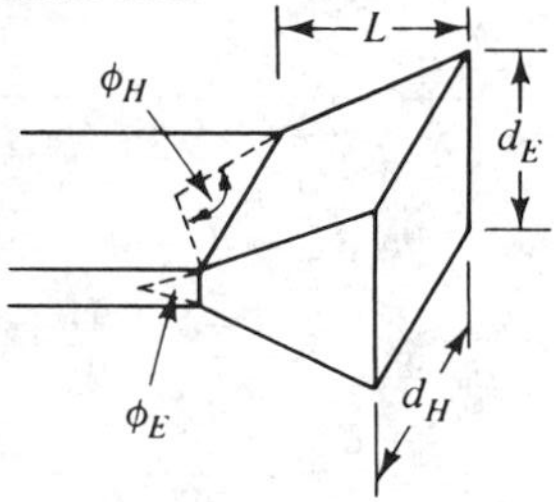 L; d_E; d_H; ϕ_H; ϕ_E ϕ's $< 40°$ Optimum horn $\text{Cos}\ \phi/2 = \frac{L/\lambda}{S + L/\lambda}$ where $S_H = 0.4$ $S_E = 0.25$	 Horizontal 3 dB beam width $= \frac{70\lambda°}{d_H}$  Vertical 3 dB beam width $= \frac{56\lambda°}{d_E}$	Linear	VSWR < 2	$\frac{7.5\ d_E\ d_H}{\lambda^2}$	2:1

(continued)

TABLE 15-2 (Continued)

Name	*Radiation pattern*	*Polarization*	*Impedance*	*Gain*	*Bandwidth*
Dielectric rod d_{max}, d_{min}, l $l \approx 4\lambda°$ for $\epsilon_r = 2.56$ $\frac{d_{min}}{\lambda°} \approx 0.23$ $\frac{d_{max}}{d_{min}} \approx 1.6$	3 dB beamwidth $55\sqrt{\frac{\lambda_0°}{1}}$	Linear	VSWR $\approx$ 1.5	$7\frac{1}{\lambda}$	10%
Wire trapezoidal tooth log periodic ψ, Feed point, Main beam α, R_3, R_2 Period = $\ln \frac{1}{\tau}$ $\tau = \frac{R_{n+1}}{R_n}$	Vertical 3 dB beamwidth $\approx 70°$ Horizontal 3 dB beamwidth $\approx 100°$ for $\alpha = 60°$ $\tau = 0.6$ $\psi = 35°$	Linear	$R \approx 110\ \Omega$ VSWR over period < 3	$\approx$ 6 dB	10 : 1 Depends upon length of shortest and longest elements

TABLE 15-2 (Continued)

Name	*Radiation pattern*	*Polarization*	*Impedance*	*Gain*	*Bandwidth*
Helical 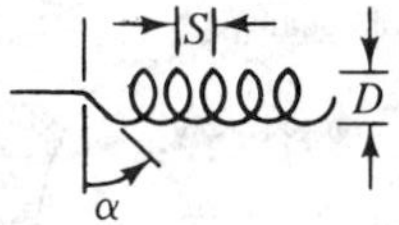 S, D, α $S \approx \lambda/4$ $\pi D \approx \lambda$ $\alpha \approx 14°$ No. of turns (N) > 3	 3 dB beamwidth $\frac{52\lambda}{\lambda D}\sqrt{\frac{\lambda}{NS}}$	Approximately circular	$R \approx 140 \frac{\pi D}{\lambda}$ $X \approx 0$	$\frac{15\, NS\,(\pi D)^2}{\lambda^3}$	1.8 to 1
3 Element Yagi 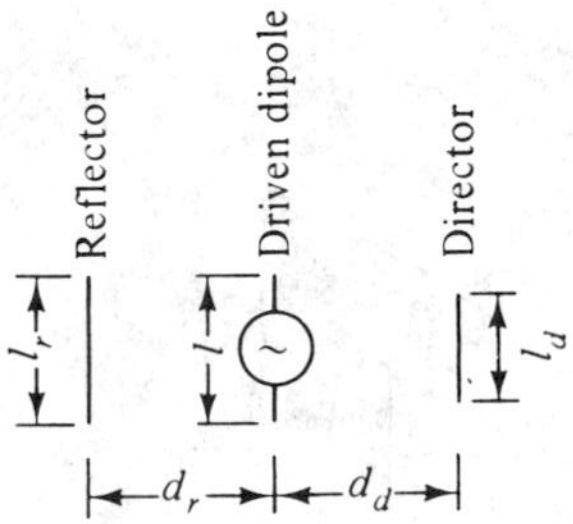 Reflector, Driven dipole, Director l_r, l, l_d, d_r, d_d $d_r = \lambda/4$ $d_d = 0.2\lambda$ $l_r = \lambda/2$ $l = \lambda/2$ $l_d = 0.45\lambda$	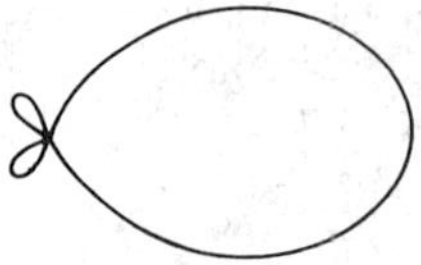	Linear	$30 + j60$ May approach 50 Ω with larger array	≈ 7 dB Gains may approach 16 dB with larger array	10% Reduced BW with larger array

TABLE 15-2 (Continued)

Name	*Radiation pattern*	*Polarization*	*Impedance*	*Gain*	*Bandwidth*
Slot 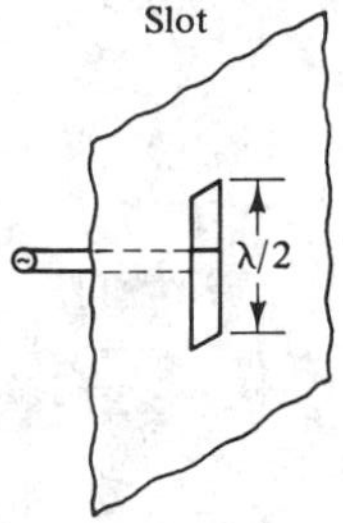 $\lambda/2$	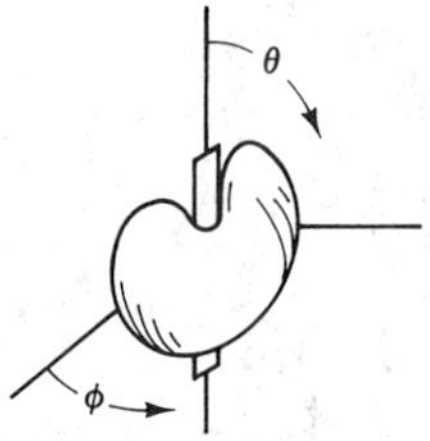 θ, ϕ Same as dipole except that E and H field directions are interchanged	Linear E_ϕ, H_θ	$Z_{slot} = \dfrac{\eta_o^2}{4Z_{dipole}}$ Thin $\lambda/2$ slot: $Z_{dipole} = 73 + j43\ \Omega$ $Z_{slot} = 361 - j213\ \Omega$ Thin resonant slot: $Z_{dipole} = 67\ \Omega$ $Z_{slot} = 530\ \Omega$	Same as dipole	Same as dipole
Microwave parabolic dish 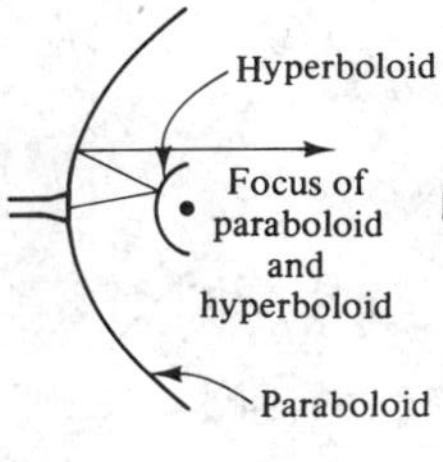 Hyperboloid Focus of paraboloid and hyperboloid Paraboloid D Cassegrain feed 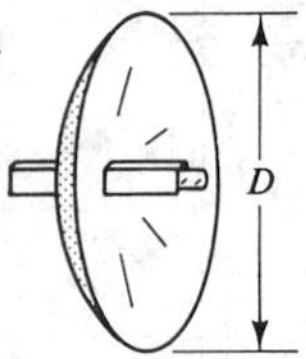 Waveguide/dipole feed	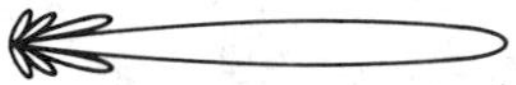 For uniformly illuminated aperture 3-dB beamwidth = $\dfrac{70\lambda°}{D}$ 3-dB beamwidth between first nulls $\dfrac{149\lambda°}{D}$	Depends on feed antenna	Depends on feed antenna	$\dfrac{K\pi^2 D^2}{\lambda^2}$ K is an efficiency constant, around 0.55 for most horn-fed antennas	Depends on feed antenna

TABLE 15-2 (Continued)

Name	*Radiation pattern*	*Polarization*	*Gain*	*Bandwidth*
Resonant waveguide slotted array a N slots $\lambda_g/2$	a 2nd null 1st null	Horizontal	20 log N (dB)	Small
For TE_{10} air-filled waveguide $\lambda = \dfrac{\lambda_g \lambda_c}{\sqrt{\lambda_g^2 + \lambda_c^2}}$ All slots radiate in phase, achieved by reversing the phase of adjacent slots	Nulls occur at $\dfrac{2n\lambda}{N\lambda_g}$ n = number of null N = number of slots λ = free-space wavelength λ_g = waveguide wavelength Amplitude of null relative to amplitude of main lobe $\dfrac{1}{N \sin\left\{\dfrac{(n + \frac{1}{2})\pi}{N}\right\}}$		The direction of the lobe can be scanned without physical motion of the antenna by changing the phase between various slots. This can be accomplished by a variation in frequency.	

15-9.4 Parasitic Antenna Elements; Yagi–Uda Antenna

In an antenna array, all the elements need not be driven directly from the source; one or more of them may receive their energy by electromagnetic coupling from a driven element. Elements excited in this manner are called *parasitic* antenna elements. The current induced in a parasitic element by a driven element itself produces an electromagnetic field. The system radiation pattern is the sum of the radiation pattern of all the elements, both driven and parasitic. Two factors that determine the phase relationship between the currents in the parasitic and driven elements are the length of each parasitic element and the physical separation between the parasitic and driven elements. A variation of either of these parameters will change the radiation pattern.

One of the simplest parasitic arrays is the half-wavelength dipole used with a reflector wire (parasitic element), shown in Fig. 15-33. For a separation of a quarter-wavelength, both the parasitic element and the driven element are the same length (about $\lambda/2$). More gain can be obtained if the reflector is moved closer and one or more director elements are added. This configuration, shown in Fig. 15-34 is known as a Yagi–Uda antenna (named for its developers). If the spacing is not a quarter-wavelength, the lengths of the parasitic elements must be modified to compensate. Making the reflector longer than a half-wavelength adds an inductive, lagging phase shift, while making the directors shorter will give them a capacitive, leading phase shift. The directors act as a kind of lens to help focus the radiated signal. The radiation pattern of the Yagi–Uda is similar to that of the normal end-fire array. The gain varies from about 7dB for a three-element array to over 15 dB for a five-element array. The Yagi–Uda array of Fig. 15-34 is shown with a folded dipole driven element. Adding additional directors gives more gain but narrows the effective bandwidth and complicates the matching problem. Many different variations are commercially available. This is an antenna commonly used for television reception because of the good match to the 300-Ω line and because of the wide bandwidth.

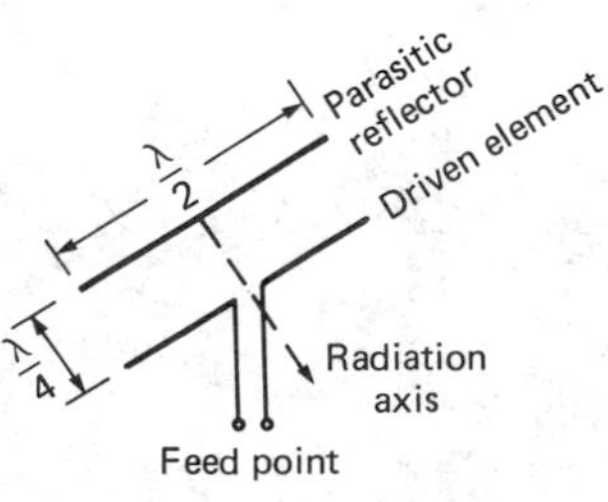

Figure 15-33 Simple dipole with parasitic reflector.

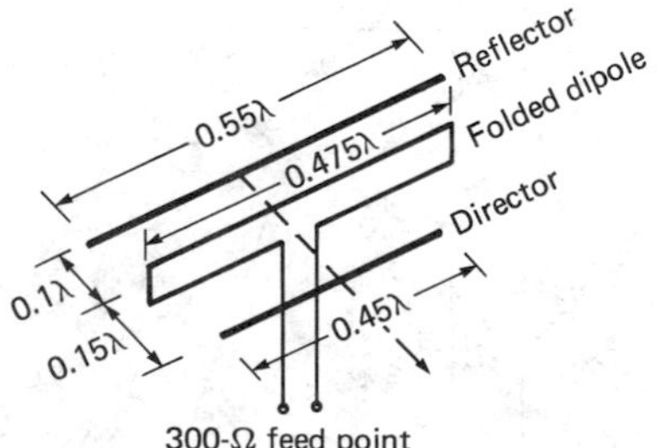

Figure 15-34 Yagi–Uda antenna, folded dipole with parasitic reflector and single director.

15-10 MICROWAVE ANTENNAS; APERTURE ANTENNAS

At microwave frequencies the free-space wavelengths of signals are physically small. This means that antennas with reasonably small dimensions may be very large compared to the wavelength of the operating signal. Because of this it is possible to build microwave antennas that have very large gains and very narrow beamwidths. The radiation pattern and beamwidth of this class of antennas depend mainly on their overall dimensions rather than their exact configuration. Microwave antennas are often thought of as aperture antennas, since their characteristics are primarily a function of the area of the aperture and the field distribution within the aperture. For example, the parabolic antenna and back-illuminated "hole-in-the-wall" aperture of Fig. 15-35 would both produce a plane wave of width D in front of themselves. The far fields of the two antennas would have to be the same as well.

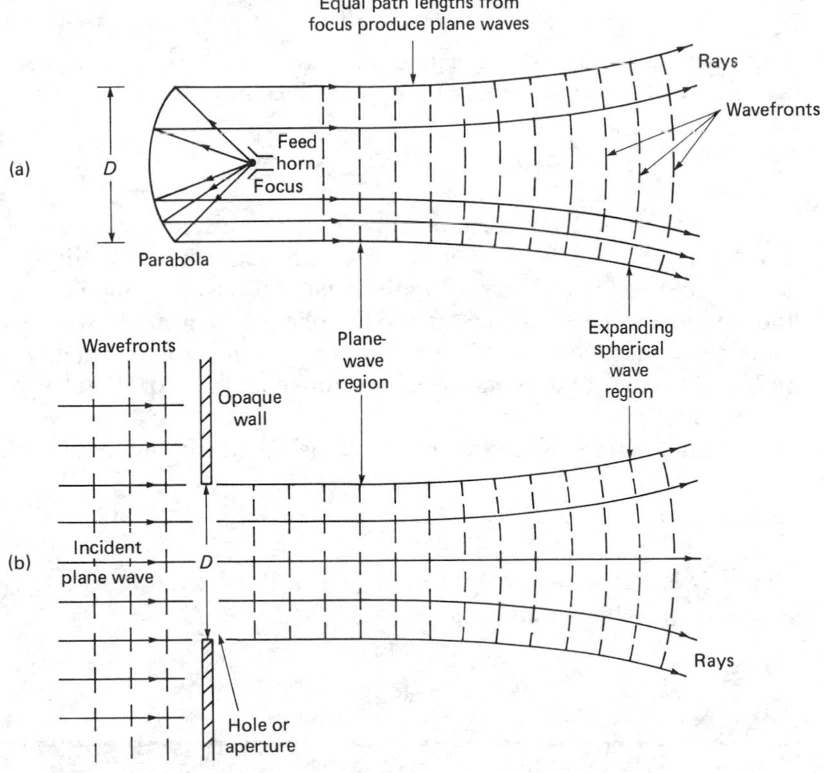

Figure 15-35 (a) Parabolic antenna; (b) illuminated circular aperture.

15-10.1 Gain of an Aperture Antenna (Parabolic Reflector)

The wave behavior introduced above will allow us to devise a formula for a generalized aperture antenna. This formula can be used with antennas such as parabolic reflectors. From the formula given previously for the effective area of an antenna [formula (15-20)] the gain (over isotropic) of an antenna can be stated in terms of its effective area as

$$g = \frac{4\pi}{\lambda^2} A \tag{15-25}$$

The cross-sectional area of a round aperture, such as a parabolic reflector, is

$$A = \frac{\lambda}{4} D^2 \tag{15-26}$$

where D is the diameter of the opening, or antenna. If we substitute formula (15-26) into (15-25), the gain of the aperture antenna becomes

$$g = \frac{4\pi}{\lambda^2}\left(\frac{\pi}{4} D^2\right) = \frac{\pi^2 D^2}{\lambda^2} \tag{15-27}$$

This formula assumes that the antenna is loss-less. If we want to account for losses, we simply multiply this equation by the efficiency factor (K). This yields

$$g = K \frac{\pi^2 D^2}{\lambda^2} \tag{15-28}$$

as a general expression for the gain of an aperture antenna (over isotropic). The value for the efficiency depends on factors such as the quality of the reflecting surface and feedhorn (in a parabolic) and the degree of illumination taper within the aperture. The efficiency of a parabolic antenna made with a mesh reflector and a simple feedhorn might be as low as 50%, while a good-quality spun-aluminum reflector and a well-designed feed structure could be expected to give an efficiency closer to 90%.

Equation (15-28) can be rewritten using logarithmic notation as

$$G = 20 \log\left(\frac{\pi}{\lambda} D\right) + 10 \log K \quad \text{dB} \tag{15-29}$$

where D is the diameter of the antenna, λ the operating free-space wavelength, and K is the antenna efficiency.

EXAMPLE 15-8

Find the gain of a 4-ft (1.22-m)-diameter dish antenna at 4GHz and at 12 GHz with both 100% and 60% efficiency.

Solution First calculate the wavelengths.

$$\lambda_{4GHZ} = \frac{c}{4GHz} = 0.075 \text{ m} = 7.5 \text{ cm}$$

$$\lambda_{12GHZ} = \frac{c}{12 \text{ GHz}} = 0.025 \text{ m} = 2.5 \text{ cm}$$

Note: $10 \log K = 10 \log(1.0) = 0$ dB for 100% efficiency
$= 10 \log(0.6) = -2.22$ dB for 60% efficiency

f (GHz)	(cm)	g (K = 1)	G dB (K = 1)	g (K = 0.6)	G dB (K = 0.6)
4	7.5	2608	34.16	1564.8	31.94
12	2.5	23,473	43.70	14,083.8	41.48

15-10.2 Beamwidth of an Aperture (Parabolic) Antenna

The larger an aperture antenna is, the larger is its gain and the smaller is its beamwidth. An approximate formula for the 3-dB beamwidth of a fully illuminated aperture antenna is

$$\theta \simeq \frac{70\lambda}{D} \quad \text{degrees} \qquad (15\text{-}30)$$

where D is the diameter of the antenna and λ is the operating wavelength. Often the feedhorn of a parabolic reflector does not fully illuminate the reflector surface. In this case, if an accurate result is to be obtained, the effective diameter must be used instead of the actual diameter.

EXAMPLE 15-9

Find the beamwidths of the 4-ft (1.22-m) parabolic antenna of Example 15-8, at 4 and 12 GHz, assuming both full and 75% illumination.

Solution

f (GHz)	Beamwidth (full illumination)	Beamwidth (75% illumination)
	(deg)	(deg)
4	4.30	5.74
12	1.43	1.91

EXAMPLE 15-10

To demonstrate that equation (15-29) relates to all electromagnetic waves, not just to antenna fields, consider the case of a visible-light laser. One manufacturer's typical low-power HeNe laser produces light with a wavelength of 632.8 nm and the diameter of the exit beam is about 0.7 mm. Calculate the minimum expected divergence angle of the laser beam.

Solution This is simply a case of using formula (15-29):

$$\text{Expected beamwidth} = \frac{70\lambda}{D} = \frac{(70)(632.8\ \text{nm})}{0.7\ \text{mm}} = 0.0633°$$

Light waves, millimeter waves, and radio waves all obey the same physical rules since they are all electromagnetic waves. This enables us, among other things, to make use of optical ray-tracing techniques in predicting the performance of antennas and arrays over a wide frequency range.

15-10.3 Parabolic Antennas

Table 15-3 shows some variations on the configuration of parabolic reflectors. They differ mainly in the design of the feed structure, and generally, equations (15-28) and (15-29) will apply. The Cassegrain configuration has slightly better gain than that of the prime focus antenna, due to better control of the illumination taper. The sidelobe levels are also lower, for the same reason. The Gregorian configuration can give slightly improved sidelobe performance as well. The hog horn configuration has good cross-polarization discrimination and is sometimes used for broadband communications systems. Some specialized applications may call for truncated reflectors. For example, an airport radar should have a narrow beamwidth in the horizontal (azimuth) plane, but it should have a wider beamwidth in the vertical plane (elevation) in order to detect aircraft at different heights. The appropriate shape for the main reflector, therefore, would be a roughly rectangular shape that is wide horizontally but short vertically.

Table 15-2 shows characteristics of various antenna types. It is interesting to note that the gain and beamwidth of the pyramidal horn antenna is quite close to that of the parabolic antenna.

TABLE 15-3 Variations on the Parabolic Reflector

Type	Comments
$x^2 = 4fy$; y; D; h; x; Focus; θ_A; f Prime focus parabolic	To find the foral length, knowing the depth (h) and diameter (D), $f = D_2/16\,h$ Aperture angle $\Theta_A = 2\tan^{-1}\dfrac{1}{2\,(f/D) - 1/16(f/D)}$

(continued)

<table>
<tr><th>Type</th><th>Comments</th></tr>
<tr><td>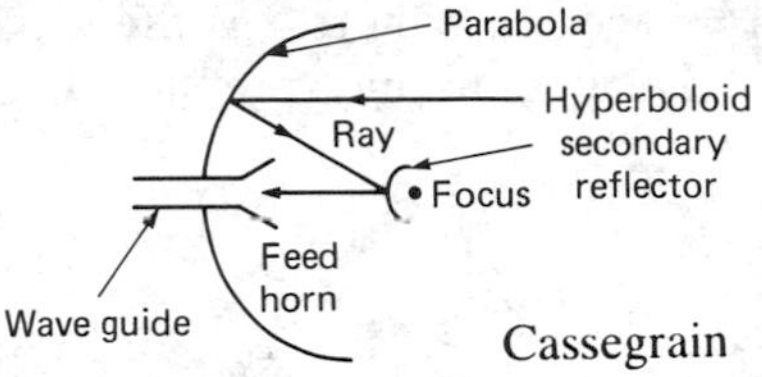

Cassegrain</td><td>Better performance than prime focus; lower sidelobes; blockage of the main reflector by the secondary reflector may be a problem; offset focus configurations are possible</td></tr>
<tr><td>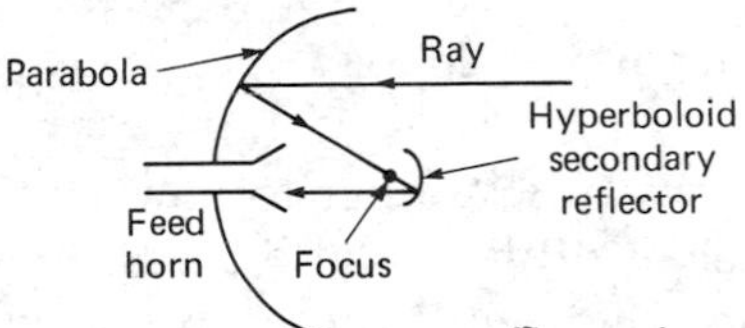

Gregorian</td><td>Good sidelobe performance; secondary reflector should be made small to reduce blockage</td></tr>
<tr><td>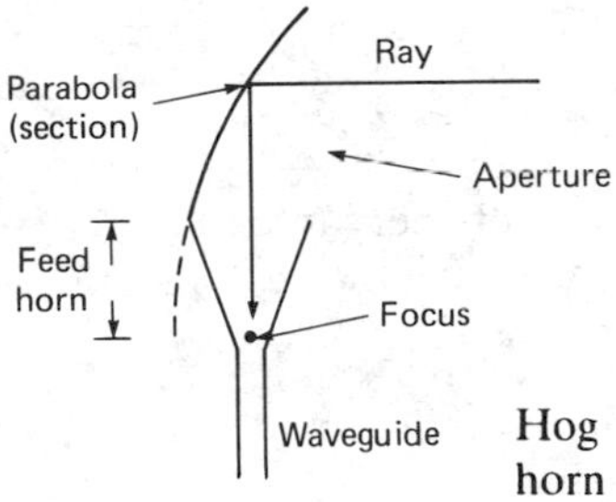

Hog horn</td><td>Good cross-polarization discrimination; good sidelobe performance; no aperture blockage, no subreflector</td></tr>
</table>

15-11 COUPLING BETWEEN ANTENNAS

It is very useful to be able to predict the amount of signal power that a receiving antenna will be able to provide for a receiver. This calculation can be made, assuming free-space propagation, if one knows the transmitted power (P), the gain over isotropic of the transmitting (g_t) and receiving (g_r) antennas, the distance between the antennas (R), and the frequency (f) [or wavelength (λ)].

An isotropic transmitting antenna will produce a radiation power density of $P_t/4\pi R^2$ W/m^2 at some distance R away from the antenna, if it radiates P_t watts. If the transmitting antenna is not in fact a point source, but instead concentrates the radiated signal in some direction, it will have a gain over isotropic (g_t) greater than 1. The radiation power density in front of this antenna will be greater than that from the isotropic antenna by the amount of the gain; that is, the radiation power density in front of the antenna will be

$$\frac{P_t}{4\pi R^2} g_t \qquad \text{W/m}^2 \tag{15-31}$$

If we multiply this power density by the effective aperture (A) of the receiving antenna, we will have the amount of signal power collected by the antenna (P_r). The effective area (aperture) of an antenna (or receiving cross section) is related to its gain over isotropic by the relation $A = \lambda^2 g_r/4\pi$, where λ is the wavelength; $\lambda = c/f$ as usual. Combining the relations above gives us an expression for received power:

$$P_r = \frac{P_t g_t \lambda^2 g_r}{(4\pi R^2)(4\pi)} = P_r = P_t g_t g_r \left(\frac{\lambda}{4\pi R}\right)^2 \qquad (15\text{-}32)$$

This can be written using logarithmic units as

$$P_r\,(\text{dBW}) = P_t\,(\text{dBW}) + G_t\,(\text{dB}) + G_r\,(\text{dB}) - 20\log\frac{4\pi R}{\lambda} \quad \text{dB} \qquad (15\text{-}33)$$

The units of power used in the relation above (dBW) refer to power referenced to 1 W. Other logarithmic power units, such as dBm, can also be used.

The last term in equation (15-33), and the squared term in the one above that, represent the *free-space loss* (FSL), the loss that occurs when the signal travels from one antenna to the other:

$$\text{FSL} = 20\log\frac{4\pi R}{\lambda} \quad \text{dB} \qquad (15\text{-}34)$$

where λ is the wavelength and R is the distance between antennas.

EXAMPLE 15-11

Calculate the received power when: $f = 4$ GHz, $P_t = 37$ dBm (about 5 W), $G_t = 30$ dB (gain over isotropic, dBi), $G_r = 35$ dB (gain over isotropic, dBi), $R = 20$ km, and

$$\lambda = \frac{3 \times 10^8\ \text{m/s}}{4 \times 10^9/\text{s}} = 0.075\ \text{m}$$

[These are typical terrestrial microwave link values (e.g., the C-band 3.7 to 4.2-GHz TD-2 system).]

Solution

$$P_r = P_t g_t g_t \left(\frac{\lambda}{4\pi R}\right)^2$$

$$P_r\,(\text{dBm}) = P_t\,(\text{dBm}) + G_t\,(\text{dB}) + G_r\,(\text{dB}) - 20\log\frac{4\pi R}{\lambda} \quad \text{dB}$$

$$= 37\ \text{dBm} + 30\ \text{dB} + 35\ \text{dB} - 130.5\ \text{dB}$$

$$= -28.5\ \text{dBm}\ (1.4 \times 10^{-3}\ \text{mW or } 1.4\ \mu\text{W})$$

EXAMPLE 15-12

Let f, P, and R be the same as before but presume 20 dB of additional losses and find the required gain of identical antennas if the received signal power must be at least 1 μW (−30 dBm).

$$P_r\ (\text{dBm}) = P_t + 2G - 20\log\frac{4\pi R}{\lambda} - \text{additional losses}$$

$$G = \left(\frac{1}{2}\right)\left\{P_r - P_t + 20\log\left(\frac{4\pi R}{\lambda}\right) + \text{additional losses}\right\}$$

$$= \left(\frac{1}{2}\right)(-\ 30\ \text{dBm} - 37\ \text{dBm} + 130.5\ \text{dB} + 20\ \text{dB})$$

$$= 41.8\ \text{dBi} \qquad \text{(a linear ratio of 15,135.6)}$$

N.B.: (1) this would only have been 31.8 dB if not for the additional losses; and (2) other combinations are possible as long as $G_t + G_r = 83.6$ dB.

EXAMPLE 15-13

Two λ/2 dipole antennas are set up as shown in Fig. 15-36. When a dipole antenna (a) is replaced by a properly oriented Yagi–Uda antenna (b), the reading on the VSWR meter goes up by 10 dB. What is the gain over isotropic of the Yagi antenna?

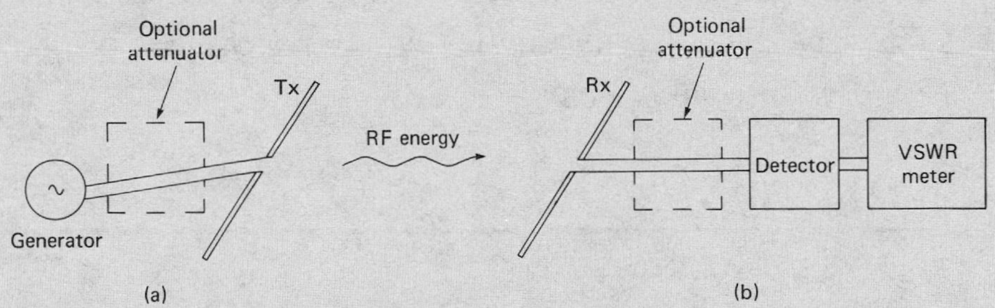

Figure 15-36 Antenna gain measurement.

$$\text{Gain over a dipole} = 10\ \text{dB, gain over isotropic} = G\ \text{dBi (unknown)}$$

$$G = \text{gain over isotropic of the dipole (dB)}$$
$$+\ \text{gain over a dipole of the Yagi}$$
$$= 2.15\ \text{dB} + 10\ \text{dB} = 12.15\ \text{dB}$$

(this assumes that the test dipoles are loss-less)

(continued)

Note: separation between the antennas; operating frequency or wavelength; transmitted power; and detector sensitivity were not specified (not needed). But the detector must indicate relative power accurately or an accurately calibrated attenuator can be used.

If the test dipole antennas are known to have an efficiency of 91.2%, what is the gain over isotropic of the Yagi antenna?

Now

$$\begin{aligned}\text{gain over isotropic of dipole} &= D(\text{dB}) + \text{efficiency (dB)} \\ &= 2.15\text{ dB} + 10\log(0.912)\text{ dB} \\ &= 2.15\text{ dB} + (-0.4\text{ dB}) = 1.75\text{ dB}\end{aligned}$$

Therefore, gain of the Yagi = 1.75 dB + 10 dB = 11.75 dB.

15-12 EFFECTIVE ISOTROPIC RADIATED POWER

Effective isotropic radiated power (EIRP) is an effective power level that combines the information of the transmitted power and the gain over isotropic of the transmitting antenna. It is often specified for satellites as a convenience instead of giving radiated power and partial gain over isotropic (in the direction of the downstation) of a satellite's transmitting antenna. EIRP is usually given in dBW, although dBm or other logarithmic power units could be used at the discretion of the designer.

$$\text{EIRP (dBW)} = P_t\text{ (dBW)} + G_t\text{ (dB)} \tag{15-35}$$

NB: 1 W = 1000 mW; 0 dBW = 30 dBm.

EXAMPLE 15-14

For a certain satellite

$P_t = 5$ dBW (+35 dBm)(3.162 W) radiated power

$G_t = 30$ dBi gain of transmitting antenna

Therefore, EIRP $= P_t + G_t = 5$ dBW + 30 dB = 35 dBW (65 dBm).

Find P (received power) if $G = 41$ dBi, $f = 4$ GHz, and slant range from satellite = 37,700 km. Assume 0.5 dB of additional losses in the atmosphere.

$$\begin{aligned}P(\text{dBW}) &= \text{EIRP (dBW)} + \text{G (dB)} - \text{FSL (dB)} - \text{additional loss (dB)} \\ &= 35\text{ dBW} + 41\text{ dB} - 196\text{ dB} - 0.5\text{ dB} = -120.5\text{ dBW (0.891 pW)}\end{aligned}$$

Example 15-14 involved an extra loss factor for atmospheric attenuation. When using equations (15-32) or (15-33), it should be remembered that they do not take into account extra losses due to weather conditions. Rain or fog or even blowing dust can increase the losses of propagating waves. These factors should be taken into consideration by allowing an extra loss margin, as in Example 15-12, when designing a communication link. Other loss factors, such as misaligned equipment, aging components, or extra transmission line losses, contribute to the loss margin as well.

Problems

15-1. **(a)** Find the magnitude of the electric field strength at a distance of 14 km from a half-wavelength dipole whose feedpoint current is 4 A (rms) if the angle (θ) between the antenna axis is 70°.
(b) Find the radiation power density at this point.

15-2. An antenna has an input current of 3.8 A (rms) and is radiating 400 W of power. Find its radiation resistance.

15-3. An antenna array operated at 100 MHz has a maximum dimension of 9 m. How far away do we have to get from this antenna to be in the far-field region?

15-4. The 3-dB beamwidth of the half-wavelength dipole of Fig. 15-16 is 78°. What is the 6-dB beamwidth of this antenna?

15-5. An antenna that radiates a total power of 80 W causes a maximum electric field strength of 8 mV/m at a distance of 24 km from the antenna.
(a) What is the directivity of this antenna (in dB)?
(b) If the antenna has a maximum gain of 95%, what is its maximum gain (in dB)?

15-6. The mathematical description of the electric field pattern of a certain antenna contains the term $\cos[(\pi/4)(\cos\theta - 1)]$ (angle in radians).
(a) On polar plot paper, plot the pattern versus the angle θ.
(b) From the plot determine the 3-dB beamwidth of this antenna.

15-7. Which type of antenna can produce circular polarization?
(1) Turnstile
(2) Loop
(3) Discone
(4) Hertzian dipole

15-8. **(a)** What, theoretically, is the front-to-back ratio of the antenna of Problem 15-6 (in dB)?
(b) What is the front-to-back ratio of the dipole antenna of Fig. 15-16 (in dB)?

15-9. Two isotropic antennas are placed as shown in Fig. P15-9. The two antennas are fed currents of equal magnitude (I), but their phase differs by 45°. The field strength due to each antenna at a distance of 30 km is 20 μV/m. Calculate the magnitude of the field strength along the array line due to both antennas at a distance of 30 km. [*Hint:* Use equation (15-23).]

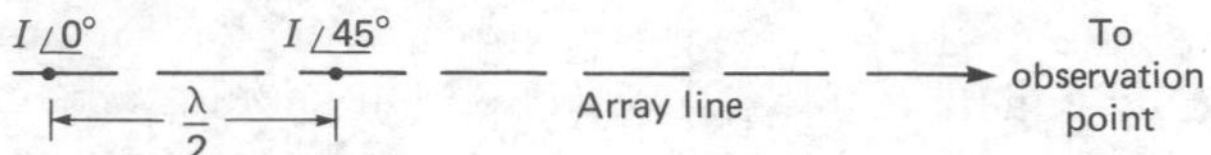

Figure P15-9 Array of isotropic antennas for Problem 15-9.

15-10. An antenna has a gain over isotropic of 7 dB and is used at 300 MHz.
(a) What is the effective area of this antenna (in meters)?
(b) If a suitably polarized plane wave with a power density of 0.2 $\mu W/m^2$ is incident on the antenna, what would be the received power?

15-11. An antenna that is known to have a gain over isotropic of 12 dB is connected to a power meter that reads 40 mW for an incident 150-MHz signal.
(a) What is the power density of the incoming wave (in $\mu W/m^2$)?
(b) What is the magnitude of the incoming electric field strength (in V/m)?

15-12. An experimental antenna operating at 365 MHz is radiating 404 W of RF power into free space. A calibrated dipole located 10 km from the source detects an electric field strength of 22 mV/m. What is the gain of this antenna in the direction of the calibrated dipole (in dBi)?

15-13. A microwave antenna operating at 2.354 GHz radiates 90% of the power delivered to its terminals. The antenna has a directivity (D) of 10 dB. When oriented for maximum gain to a remote receiving antenna, the signal level received indicates that the transmitting antenna has an effective isotropic radiated power (EIRP) of 1 kW. Determine the power level of the 2.354-GHz signal being delivered to the terminals of the transmitting antenna.

15-14. The power level (EIRP) from a particular satellite transponder in the direction of a TVRO site is 32 dBW. The slant range to the satellite is 37,800 km. If the transponder output frequency is 3840 MHz, the receiving antenna gain is 40.8 dBi and we allow 0.4 dB for extra atmospheric losses. Calculate the received signal power.

15-15. (a) Calculate the gain over isotropic of a parabolic antenna with a diameter of 1.8 m used at 11.7 GHz if its efficiency is 64%.
(b) Calculate the 3-dB beamwidth of this antenna if the feedhorn effectively collects energy from only the inner 80% of the antenna's diameter.

15-16. An antenna test range uses two half-wavelength dipole antennas as in Fig. 15-37.
(a) If the efficiency of these dipoles is 85%, what is their gain over isotropic (in dB)?
(b) When one of the dipoles is replaced by a test antenna, the received signal power is seen to go up by 5 dB. What is the gain over isotropic of the test antenna (in dB)?

15-17. Calculate the gain over isotropic of a parabolic antenna with a diameter of 1.6 m used at 11.7 GHz if its efficiency is 50%.

15-18. A certain terrestrial CATV link has the following parameters: frequency, = 12,646.665 MHz, transmitted power, = 1 W, gain of transmitting antenna = 44 dBi, gain of receiving antenna = 40 dBi, and distance between antennas, = 20 km. If we want to allow for an extra 6 dB of losses, calculate the expected received power in dBW.

An Introduction to Fiber Optic Communications

Robert McPherson

16-1 INTRODUCTION

An increasing portion of the information to be transferred from point to point is presented to the communication system in a digital rather than analog format. Even in cases where the information to be sent is analog in nature it is not unusual to sample the analog signal and then transmit a digital representation of the sample amplitude, rather than directly transmitting the analog signal. The telephone industry, for example, is routing an increasing portion of its analog voice signals through pulse code modulation (PCM) systems that perform these conversions. In terms of maintaining the quality of the recovered analog signal and the ease of directing the information to a particular sink, there are a number of distinct advantages to transmitting a digital representation of an analog signal rather than the analog information itself. The reader is directed to Chapter 11 for a more detailed discussion of the relative merits of digital transmission.

When a number of different digital information channels are to be connected between two relatively distant points, it is generally more economical to time-division multiplex (TDM) the channels into a single higher data rate channel and install a single transmission facility between the two points, rather than installing multiple parallel transmission facilities. The data rate of the single multiplexed channel varies with the particular application, but currently available systems run with data rates as high as a few hundred megabits per second.

The transmission facility used to carry a signal of this type must have a very wide bandwidth capability in order to carry a high-rate digital signal such as this without introducing excessive distortion on the signal. The transmission facilities typically considered are coaxial cable, waveguide, and fiber optic systems. The fiber optic system's size, weight, and low attenuation characteristics result in it often being the most economical method of transmitting the data, particularly in those applications that require very high data rates or long-distance runs.

16-2 BASIC FIBER OPTIC COMMUNICATION FACILITY

The basic components of a simple fiber optic communication system are shown in Fig. 16-1. The binary information to be transmitted is used to modulate the intensity of a light source, the most common light sources being light-emitting diodes and laser diodes. A portion of the light generated is captured in the fiber optic cable and the modulated light is guided to the receiving end by the fiber optic cable. At the receiving end a photo detector is used to translate the intensity modulated light beam back into a binary signal.

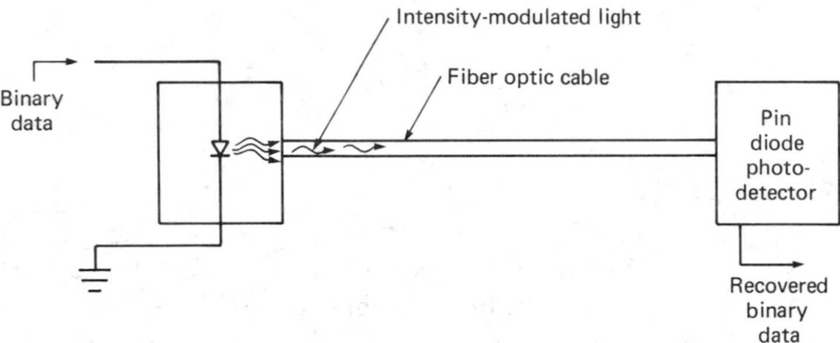

Figure 16-1 Optical communication link.

In order to examine more closely the capabilities and limitations of this form of signal transfer, it is necessary to examine in more detail how the light impinging on the optical fiber is captured and guided to the receiving end. The wavelength of the electromagnetic radiation emitted by the light sources used is typically quite small compared to the size of the physical boundaries in an optical fiber. As a result an examination of the energy propagation in an optical fiber may be undertaken using optical ray theory (see Chapter 14) rather than the more involved boundary value techniques used to describe modes in waveguides that are not much larger than a wavelength in extent (see Chapter 13). The one exception to this occurs when what is termed single-mode fiber is used; this exception will be examined later in the chapter.

A fiber optic cable is made out of an optically clear material into which small amounts of specific contaminants have been added to change the relative dielectric (ε_r) property of the material. The result of this change is a change in the velocity of propagation (v) of the light in the material. For nonmagnetic dielectric materials,

$$v = \frac{C}{\sqrt{\varepsilon_r}} \tag{16-1}$$

where $C = 3 \times 10^8$ m/s is the velocity of light in free space. In ray optic analysis the degree by which wave velocity in a material is reduced from the free-space velocity (C) is referred to as the index of refraction (n) of the material.

$$n = \frac{C}{v} = \frac{C}{C/\sqrt{\varepsilon_r}} = \sqrt{\varepsilon_r} \tag{16-2}$$

For example, in a material with an index of refraction of 3, light will travel at one-third the speed of light in free space.

The contaminants in a fiber optic cable are carefully added in the manufacturing process in a nonuniform distribution such that the index of refraction is not the same everywhere in the fiber. An example of what is referred to as multimode step-index fiber is shown in Fig. 16-2. The fiber is essentially just a very thin optically clear glass (or plastic) rod.

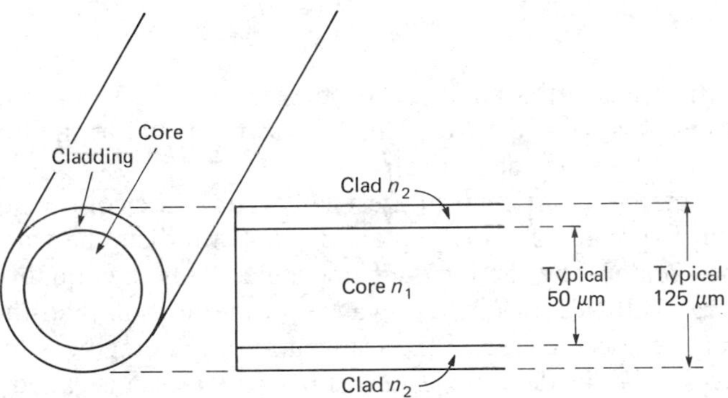

Figure 16-2 Multimode step-index fiber.

The glass cladding the outside of the rod has a different concentration of contaminants than the core area of the glass. As a result, the index of refraction of the glass in the core area is different from the index of refraction of the cladding area. The "step-index" part of the name refers to the index of refraction undergoing a sharp transition at the core–cladding glass interface. The fiber will capture and guide a portion of the light directed into it if the core glass index of refraction (n_1) is greater than the cladding glass index of refraction (n_2).

To examine the light-capturing effect, let us consider a light ray impinging on the end of an optical fiber as shown in Fig. 16-3. A portion of the incident ray energy at the air–core interface is reflected away at an angle equal to the incident angle. The rest of the ray energy is transmitted into the core glass at angle θ_2. The change in wave velocity between the two mediums at the interface results in the transmitted ray undergoing a directional shift. The amount of shift may be determined using Snell's law. Snell's law requires that the product of the index of

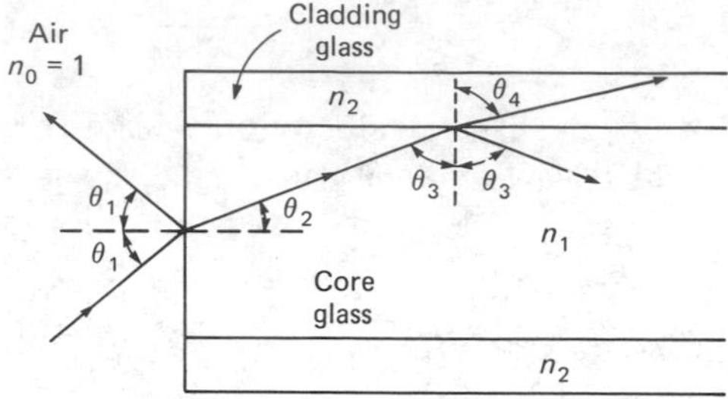

Figure 16-3 Optical ray paths in a fiber.

refraction and the sine of the ray angle be the same on both sides of the interface. In the current case, then, this requires that

$$n_0 \sin \theta_1 = n_1 \sin \theta_2 \tag{16-3}$$

Similarly, at the core–cladding interface, the requirement

$$n_1 \sin \theta_3 = n_2 \sin \theta_4 \tag{16-4}$$

allows the transmitted angle θ_4 to be determined. *Please note* the convention of measuring all angles relative to a line drawn normal to the boundary between the regions of different index of refraction.

An effect of particular interest can occur at the core–cladding interface. Note that if n_1 is greater than n_2, equation (16-4) predicts the transmitted ray angle θ_4 must be greater than the incident ray angle θ_3 for the equality to hold. The transmitted ray is thus bent further away from the normal than the incident ray. If we consider the specific case of the transmitted ray being bent to follow the interface ($\theta_4 = 90°$), the incident angle in this particular case is called the critical angle ($\theta_3 = \theta_c$). Substituting into equation (16-4) the critical angle is

$$\theta_2 = \arcsin \frac{n_2}{n_1} \tag{16-5}$$

It is of interest to note that if the incident angle (θ_3) exceeds the critical angle, no transmitted angle θ_4 exists that will satisfy equation (16-4). The physical interpretation for this is simply that no transmitted ray exists in the cladding. The incident ray thus generates only a reflected ray and no transmitted ray. The ray undergoes *total internal reflection* if the incident angle θ_3 exceeds θ_c. A ray that experiences total internal reflection is therefore "captured" in core glass and will eventually arrive at the photodetector placed across the core glass face at the receiving end of the fiber optic cable.

Rays that impinge on the core–cladding interface at less than the critical angle will be only partially reflected. As a result, a portion on the ray energy is lost to the transmitted ray, leaving the core with each reflection. Rays impinging in this manner are quite rapidly attenuated and do not reach the photodetector at the receiving end of the fiber.

If a ray at the critical angle ($\theta_3 = \theta_c$) is traced back to the original air–core interface, the input angle (θ_1) in this situation is called the *acceptance angle* (θ_a). Using equations (16-5) and (16-3) it can be determined that

$$\sin \theta_a = \sqrt{n_1^2 - n_2^2} \tag{16-6}$$

Rays impinging on the end of the fiber at angles less than the acceptance angle will be captured in the core glass by total internal reflection. Rays impinging on the end of the fiber optic cable outside the acceptance cone (see Fig. 16-4) established may initially enter the fiber but will be quickly attenuated due to reflection losses.

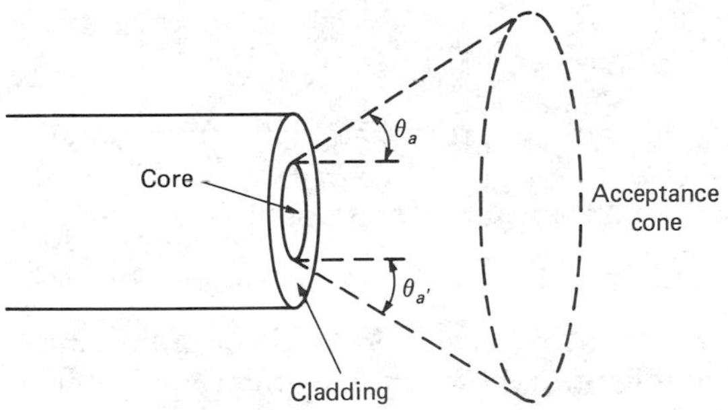

Figure 16-4 Acceptance cone.

A figure of merit often used to relate how much of the light incident on a fiber is captured is the fiber's *numerical aperture* (NA):

$$\text{NA} = \sin \theta_a = \sqrt{n_1^2 - n_2^2} \tag{16-7}$$

Note that if all the light impinging on the core is captured ($\theta_a = 90°$), the NA is equal to 1. The NA can thus vary between zero and 1. Its value is determined in the manufacturing stage when the n_1 and n_2 values are selected.

16-3 ATTENUATION AND DISPERSION IN FIBER OPTIC SYSTEMS

The transmission of information through a fiber optic system is limited by two principal effects. To examine these let us consider the transmission of a 101-bit pattern through the system shown in Fig. 16-5.

In addition to an end-to-end time delay through the system, we may note that the bit pulses have been reduced in amplitude (attenuated) and the pulse shape has been distorted by being spread out in time (dispersed). In a digital communication system the signal degradation indicated is not overly critical. As long as the ones and zeros of the bit stream can still be differentiated, no information is lost. Regenerative repeaters can be placed at regular intervals along the length of the fiber optic system to detect the digital signal and regenerate an undistorted

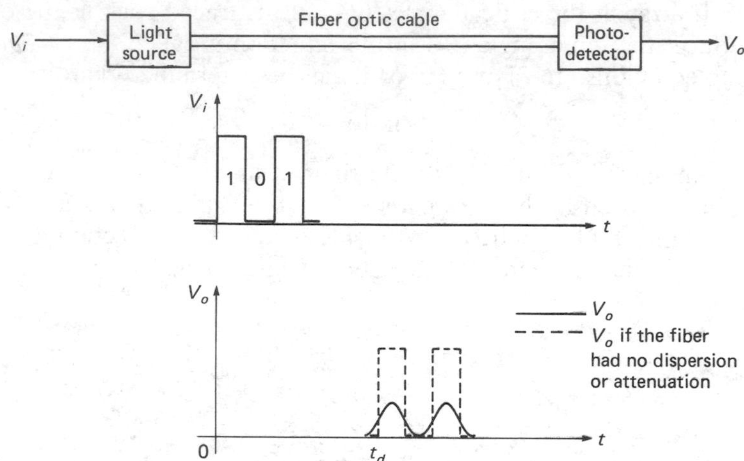

Figure 16-5 Transmission in a fiber optic system.

pulse stream for further transmission. Pulse attenuation and distortion effects do, however, set a limit on how far apart the repeaters may be placed.

The fiber optic communication system described could fail if the intensity of the light transmitted is allowed to fall so low that the photodetector used cannot differentiate between the light pulse and the ambient dark level. The repeaters must be spaced close enough together to avoid this and their maximum separation is thus set by this *attenuation limit*. Consider, by way of example, the case where a light source injects −5 dBm of optical light power into a cable that has an attenuation rate of 10 dBm/km. If the detector in the system can reliably differentiate a −35 dBm light burst from the ambient dark level then up to 30 dB of loss [−5 dBm − (−35 dBm)] is acceptable. A cable loss of 10 dB/km would imply an attenuation limit of 3 km for the repeater spacing.

The system may also fail to recover the digital information as a result of the dispersion effect on the bit stream pulses. The dispersion effect spreads the pulses out in time. If this occurs excessively, the individual bit pulse spreads into the preceding and subsequent bit time slots and the detection of any individual bit contaminated with multiple overlapping bit signals becomes impossible. The overlapping problem may be reduced by using wider time slots for each bit pulse. The wider time slots, however, imply a reduced bit rate and a corresponding reduction in the bandwidth of the transmitted signal. The *dispersion limit* to fiber optic information transmission is normally described by attributing a distance–bandwidth product to the fiber optic cable. By way of example, consider a cable that has a distance–bandwidth product of 500 MHz-km. If a digital signal operating at a data rate that required a 100-MHz bandwidth was transmitted on this cable, the repeaters on this system would have to be placed at least every 5 km to maintain a distance–bandwidth product less than the specified maximum A fiber optic communication system must be operated within both the attenuation and dispersion limits, and therefore the limit resulting in the shorter repeater spacing must be used.

16-4 SOURCES OF DISPERSION

The pulse-spreading dispersion effects observed in a fiber optic system may be attributed to two principal mechanisms. The first is referred to as *modal dispersion* and the second is called *chromatic dispersion* (or material dispersion or wavelength dispersion). To examine the mechanism that results in modal dispersion, consider a pulse applied to the fiber optic system depicted in Fig. 16-6. The pulse produces a burst of light rays, some portion of which will be within the acceptance cone of the fiber. Those rays in the acceptance cone will be captured in the fiber core and guided to the detector. For simplicity only two of the captured rays are shown on the figure. At the detector the net composite of light arriving due to all the different captured rays will establish the output signal level. Unfortunately, all the rays of the original burst do not arrive at the detector at the same time. The portion of the burst carried by ray 2, for example, has to travel a longer distance than ray 1 and as a result will arrive at the detector later in time, the net result being that the light arriving at the detector is spread out in time (dispersed) due to the multiple different modes by which the light energy is transferred. The amount of modal dispersion introduced by a fiber is largely a function of how the fiber is constructed. In a later section of this chapter we describe different fiber construction formats and the relative merits of each fiber type in terms of modal dispersion.

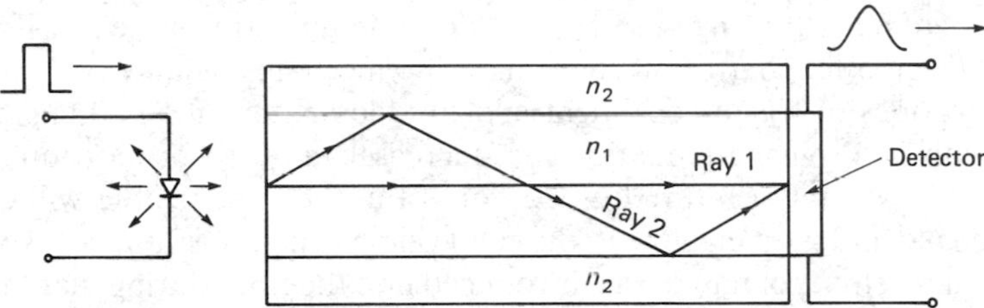

Figure 16-6 Modal dispersion in a fiber optic cable.

Chromatic dispersion occurs by a different mechanism than modal dispersion and the total dispersion occurring on a system will be due to the combination of both effects. The basic mechanism that results in chromatic dispersion can be observed by allowing white light to strike a glass prism (see Fig. 16-7). We may note two effects: the first is that white light is incoherent light, by which we mean

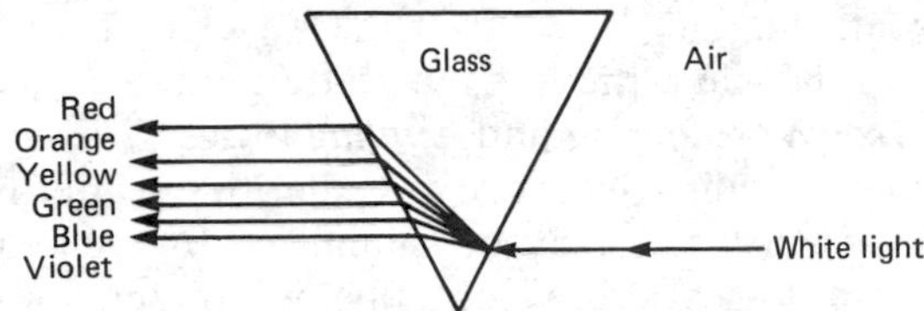

Figure 16-7 Effect of a prism on white light.

that it is made up of light of many different wavelengths (colors). We may note further that each wavelength component is refracted at a slightly different angle. Given that the incident angle is the same for all wavelengths, Snell's law would suggest that the glass is presenting a different index of refraction to each wavelength of light. A different index of refraction would suggest that each wavelength of light travels at a slightly different velocity in the material of the glass.

Consider what happens in a fiber optic cable when a bit pulse generates a burst of incoherent light to be guided to the load detector. The energy of light burst will be distributed among the range of wavelengths emitted by the light source. Each wavelength of light, however, travels at a different velocity and will therefore arrive at the load end detector at a different time, even if the same ray path is used. The result at the receiving end is a detector output that is spread in time (dispersed) relative to the input pulse.

Chromatic dispersion (material dispersion, wavelength dispersion) can be avoided if a coherent light source such as a laser diode is used, since it will emit light at only one wavelength. The relative merits of incoherent light sources such as light-emitting diodes and coherent sources such as laser diodes are presented in more detail in a later section of this chapter.

16-5 SOURCES OF ATTENUATION

A light pulse traveling in a fiber optic communication system will reduce in intensity through a number of mechanisms. The mechanisms may be broadly split into the categories of length-dependent and fixed-loss attenuations. Length-dependent losses result in a signal attenuation that increases in direct proportion to the fiber length (dB/km). Light captured in the core of a fiber optic cable will experience length-dependent losses mainly via the following mechanisms. *Material absorption* is caused by undesired contaminants introduced into the fiber during manufacture. The losses are strongly dependent on the purity of the glass and somewhat dependent on the wavelength of light used. If extremely pure materials are used in the fiber material, absorption losses can be minimized and the attenuation rate reduced to approach the *Rayleigh scattering* limit. Rayleigh scattering occurs as a result of the effect the microscopic structure of the glass has on the electromagnetic wave (light ray) propagating through it. Macroscopically, it appears that a small portion of a ray's energy is scattered into light rays traveling in different directions than the main ray. A portion of the scattered light rays will impinge on the core–cladding interface at less than the critical angle and their energy will eventually be lost. The Rayleigh scattering loss is wavelength dependent ($1/\lambda^4$) inasmuch as longer-wavelength light suffers less attenuation. The longer the wavelength compared to the microscopic fluctuation in the glass, the less the scattering effect. Additional loss mechanisms include *waveguide scattering* and *bending* losses. Waveguide scattering occurs as a result of core–cladding interface irregularities scattering light energy out of the core region. Such losses may be minimized by well-controlled fabrication techniques. Bending losses occur when a fiber is redirected. Consider a ray just captured in the core of a fiber as shown in Fig. 16-8. Bending the fiber will allow some of the light rays in the core to impinge on the core–cladding interface at less than

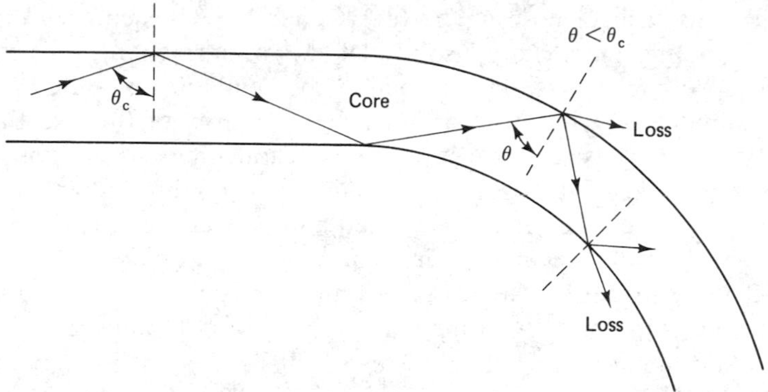

Figure 16-8 Bending loss.

the critical angle, and as a result, some of the light energy will escape the core region. Bending losses may be minimized by physically restricting the minimum radius to which the fiber may be bent. Typically, the fiber optic cable will be laid in a plastic conduit designed to restrict the bending radius or will have a polymer coating applied to restrict fiber bending.

A very pure low loss fiber with minimal waveguide and bending losses will have attenuation rates akin to those shown in Fig. 16-9. Note that the attenuation decreases as the wavelength of the applied light is increased due to a reduction in Rayleigh scattering. This continues up until about the 1.6-μm point, beyond which infrared absorption in the glass becomes progressively more significant. The attenuation curve shows why it is desirable to use the newer long-wavelength (1.3 to 1.6 μm) light sources rather than the older short-wavelength (0.65 to 0.9 μm) light sources.

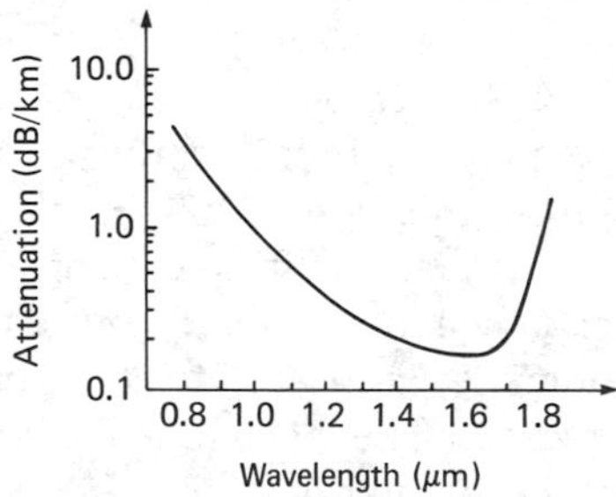

Figure 16-9 Attenuation in a low-loss fiber.

A fixed (length-independent) loss occurs whenever a fiber optic cable is interfaced. An example of this is where light is coupled into or out of the fiber and at points where lengths of fiber must be spliced together. Splicing losses are of particular interest since one bad splice can introduce as much signal attenuation as many kilometers of low-loss cable. Permanent splices are obtained by preparing the fiber ends and then bonding them together. Most commonly the ends are fused together using an electric arc; however, a number of splicing packs utilizing optically clear bonding cements are also available. Nonpermanent joins can be made using screw-type connectors. The prepared fiber ends are mounted in the connectors. The connectors, when joined, align the core area of the two fibers and position the two fiber ends close together, but not touching (so the polished ends are not damaged). The loss across a splice joining two lengths of identical fiber is a function of how well the core areas are aligned and how closely the fiber ends are brought together. A well-executed permanent splice could have an attenuation as low as about 0.2 to 0.5 dB. Removable connectors typically show much higher losses because of the difficulty of maintaining precise core alignment.

Additional loss mechanisms exist when the two fibers being joined are not identical. It is, for example, common for the light sources used in fiber optic communication to have a short length of fiber (a pigtail) factory-mounted on the light emitter. In the field installation, the pigtail is then spliced onto the fiber in use. Unfortunately, the fiber optic cable selected for the system is unlikely to be the same as the one used for the pigtail. If the two fibers have different core areas or different numerical apertures, additional losses *may* occur. If the NA of the system fiber is larger than the NA of the pigtail, all the rays in the pigtail will be within the acceptance cone of the system fiber and no additional loss will occur. If, however, the NA of the system fiber is less than the pigtail, some of the rays in the pigtail fiber will be outside the acceptance cone of the system fiber, and therefore not captured, resulting in NA loss.

A similar effect occurs if the core areas of the two fibers are different. When light travels from a smaller core area fiber to a larger core fiber, there is no additional loss. There will, however, be a loss if the light travels from a larger-core-area fiber to a smaller-core-area fiber.

16-6 TYPES OF FIBER OPTIC CABLE

While a host of different fiber optic cable designs have been proposed, the vast majority of production can be broadly grouped into three main types: *step-index multimode*, *graded-index multimode*, and *single mode*. In this section we briefly outline the relative merits and deficiencies of each of these fiber types. Typical parameters will be given where possible, but the reader is forewarned that ongoing research and development will result in the "typical" values changing.

The first fiber type to be examined is the *step-index multimode*. Our descriptions to this point in the chapter have, by and large, focused on this type of fiber. A diagram of the variation of the index of refraction across the face of the fiber is shown in Fig. 16-10. The refractive index undergoes a step change at the core–cladding interface. Compared to the wavelength of the typical light sources used (0.65 to 1.6 μm) the interface is large and smooth. Light-wave propagation in the core can thus be described using optical ray theory. Light rays are trapped in the core by total internal reflection at the interface. Energy is carried in the fiber by multiple modes, and as a result the fiber will demonstrate modal dispersion. A

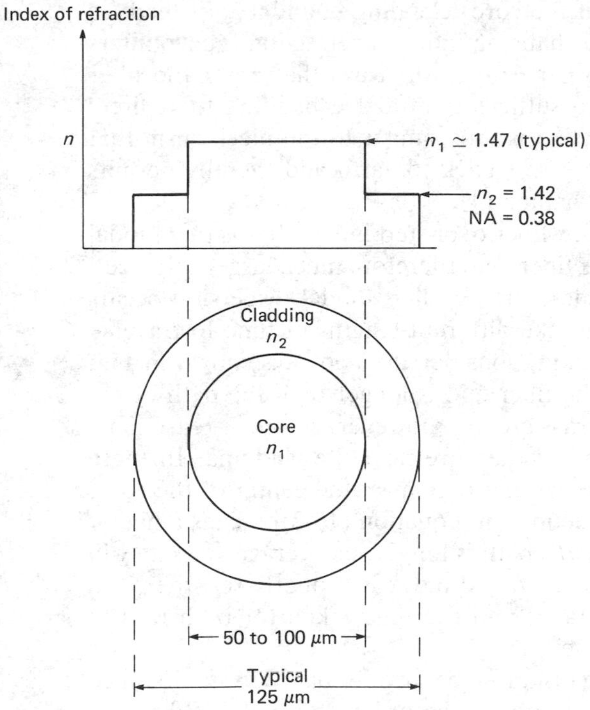

Index of refraction
n
$n_1 \simeq 1.47$ (typical)
$n_2 = 1.42$
NA = 0.38
Cladding
n_2
Core
n_1
50 to 100 μm
Typical
125 μm

Figure 16-10
Refractive index of step-index multimode fiber.

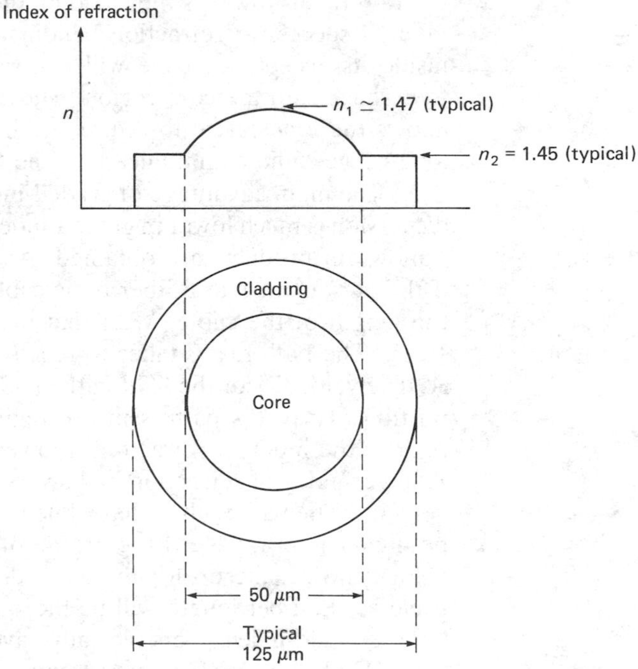

Index of refraction
n
$n_1 \simeq 1.47$ (typical)
$n_2 = 1.45$ (typical)
Cladding
Core
50 μm
Typical
125 μm

Figure 16-11
Refractive index of graded-index multimode fiber.

typical range of properties for this type of fiber follows. Numerical apertures are in the range of about 0.2 to 0.5. Distance–bandwidth products are in the range 20 to 200 MHz-km. Attenuation rates vary substantially but are typically greater than 10 dB/km.

The second type of fiber to be examined is the *graded-index multimode* fiber. Figure 16-11 shows how the index of refraction varies across the face of this fiber type. In this fiber type the core's index of refraction makes a gradual transition to the cladding value. Light rays will still be captured in the core area with this configuration, although the guiding mechanism is slightly different. Figure 16-12 is a sketch of the paths taken by two different light rays captured in a graded-index fiber.

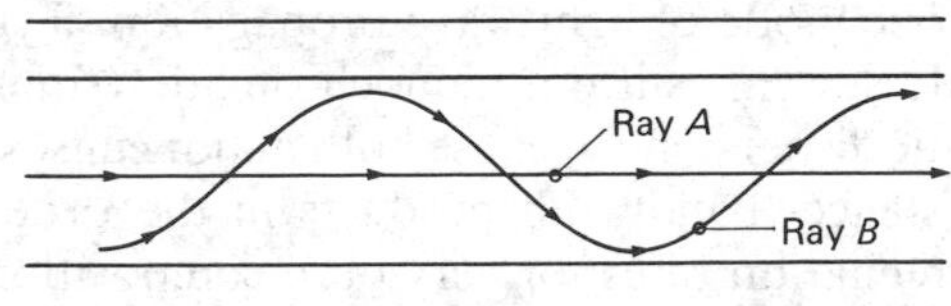

Figure 16-12 Light rays in a graded-index multimode fiber.

Note that rays are not reflected at the core–cladding boundary. Instead, as the ray travels towards the cladding the changing index of refraction it encounters induces successive refraction bending of the ray path. Rays that enter the fiber inside its acceptance cone will undergo sufficient refractive bending to redirect them back into the core region. The action is thus similar to the mechanism that allows the ionosphere to redirect radio waves back to earth and thereby permits long-range radio communication (see Chapter 14).

The main advantage of graded-index fiber over step-index fiber is that modal dispersion is much lower in graded-index fiber, and therefore much larger distance–bandwidth products are obtained. As described earlier, modal dispersion occurs if the various rays in a fiber optic cable take different lengths of time to traverse the length of the cable. With that in mind, consider the two rays shown in Fig. 16-12. The path ray B takes to reach the fiber end is longer than the path of ray A. The velocities of the light in the two rays are *not*, however, equal. A reasonable portion of ray B's path is in the region of the core near the cladding. In these regions the index of refraction is lower than the regions at the center of the core. A lower index of refraction by its very definition [equation (16-2)] means a higher velocity. The net result is that while ray B's path is longer, its average velocity will be higher than ray A. If the profile of the graded index is properly set, the extra velocity will make up for the extra distance and the time taken for both rays to traverse the fiber length will be the same.

Graded-index fibers typically have distance–bandwidth products in the range 200 MHz-km to 3 GHz-km. Attenuation rates in the order of 2 to 10 dB/km and numerical apertures in the range 0.15 to 0.2.

The final fiber type to be examined is *single-mode fiber*. Figure 16-13 shows how the index of refraction varies across the face of a single-mode fiber. The principal difference between step-index multimode fiber and single-mode fiber is the size of the core area. The core–cladding interface in single-mode fiber is *not* large or smooth compared to the wavelength of the light to be transmitted. In this situation optical ray theory cannot be used to predict the propagation of an electromagnetic (light) wave. The fiber must be treated as an optical waveguide and the various forms of electromagnetic field configurations (modes) determined. The sequence is similar to that used to determine the modes in a rectangular waveguide (see Chapter 13). In single-mode fiber, light traveling on the fiber has a single electromagnetic field configuration (a single mode). The fields are confined largely to the core but do have some components extending into the cladding area. If we wish to revert to an optical ray theory representation, we may represent the propagation as a single ray of light going straight down the fiber core. One obvious advantage of single-mode fiber is that no modal dispersion can be expected, as there is only one mode of light-wave propagation. Typically, laser diodes are used in single-mode systems since a monochromatic (single-wavelength) source eliminates chromatic dispersion. Because both major causes of dispersion are minimized, very large distance–bandwidth products in the order of 20 to 200 GHz-km are obtained. Attenuation rates for very pure composition single-mode fibers are typically in the range 0.2 to 2 dB/km and the numerical apertures are usually less than 0.12.

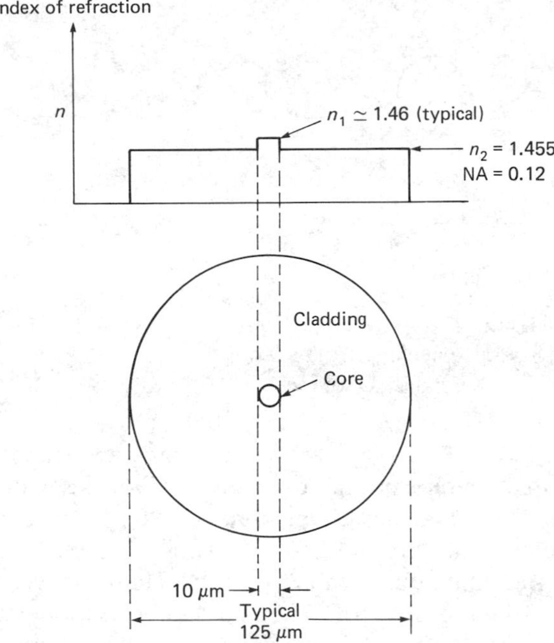

Figure 16-13 Refractive index of single-node fiber.

Single-mode fiber, because of its large bandwidth capability and low attenuation rates, is far superior to the multimode fibers. As a result, it is widely used in systems which operate at high data rates (>100 Mbits/s) and/or operate over large distances (>10 km). Single-mode fiber is not without its difficulties. The core size is very small, and very fine tolerances are required to align the cores for splicing. The small core area and the small numerical aperture of the fiber makes it very difficult to couple light efficiently into the fiber.

In applications operating over shorter distances and/or lower data rates, the step- and graded-index multimode fibers will often prove a more economic choice. The much larger core sizes involved result in the light coupling and splicing techniques being much less critical and costly.

16-7 LIGHT SOURCES USED IN FIBER OPTIC COMMUNICATION

As has been pointed out earlier, the two principal light sources used in fiber optic communication links are light-emitting diodes (LEDs) and laser diodes (LDs). LEDs are incoherent light sources that emit light over a range of wavelengths (see Fig. 16-14). LDs are constructed such that the active region which emits light is contained within an optical resonant cavity. The cavity supports and reinforces stimulated emission only at the resonant frequency of the cavity. As a result, light

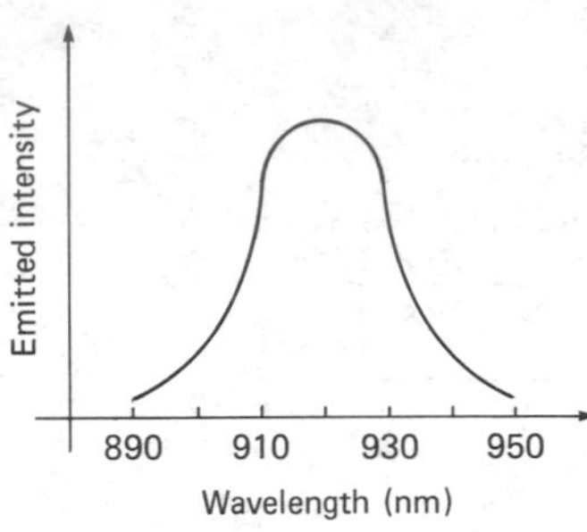

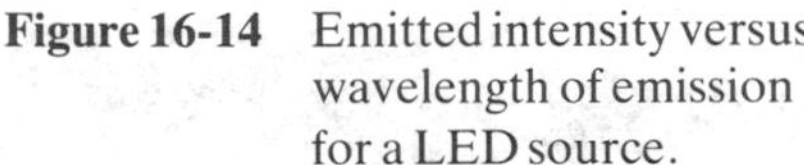

Figure 16-14 Emitted intensity versus wavelength of emission for a LED source.

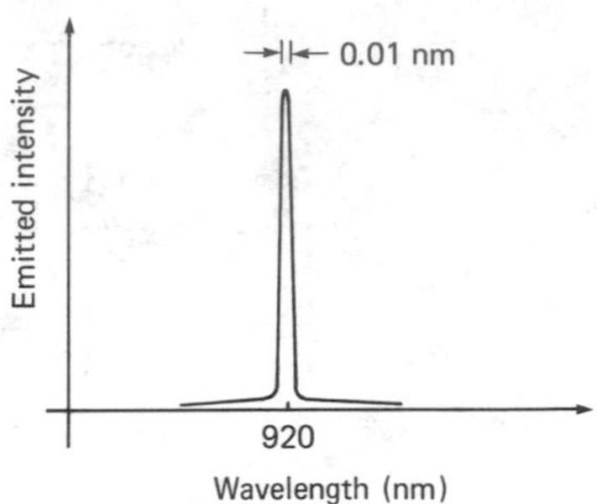

Figure 16-15 Emitted intensity versus wavelength of emission for a LD source.

is emitted only over a very narrow range of wavelengths, as shown in Fig. 16-15. The essentially monochromatic emission of the laser diode eliminates chromatic dispersion effects and allows single-mode fibers to have distance–bandwidth products that extend into the hundreds of GHz-km range. The laser diode light emission level responds more quickly to modulations in the drive current than the emissions of an LED. As a result, a LED source cannot be modulated at as high a frequency as an LD (a few hundred megahertz versus several gigahertz). An additional advantage of LDs is that they radiate their light in a much more directional pattern than an LED. As a result, substantially more of the light emitted can be coupled into the core of the fiber optic cable. The injection of higher optical power levels allows greater repeater spacing in attenuation limited systems.

The superior characteristics of the laser diode make it, in combination with single-mode fiber, the natural selection for fiber optic communication systems operating at high data rates and/or over long distances. In less critical applications, however, the LED has a number of operational advantages which may result in its selection as a light source. Consider, for example, the light emission characteristics shown in Fig. 16-16. The LED emits incoherent light in essentially direct proportion to the drive current. The laser diode emits small amounts of incoherent light below a threshold current level and then begins lasing above the threshold. The laser diode is normally biased to operate slightly above the threshold point and the information signal modulates the drive current in the *a-b* region shown in the figure. Unfortunately, the threshold current level shifts with temperature and device age. As a result, special compensating circuits are required and the provision of thermal cooling for the LD is not unusual. The power intensity in the optical cavity of a laser diode is very high and as a result, LDs degrade with time and eventually fail. Early versions of LDs were plagued by relatively short operational lifetimes; however, substantial fabrication improvements have been obtained in this area and newer laser diodes now have operating life spans on the order of 1 million hours. The specialized support circuitry and higher initial cost of laser diodes sources result in the relatively simple and robust light emitter diode being the more reasonable selection in short-range and/or low-data-rate fiber optic communication systems.

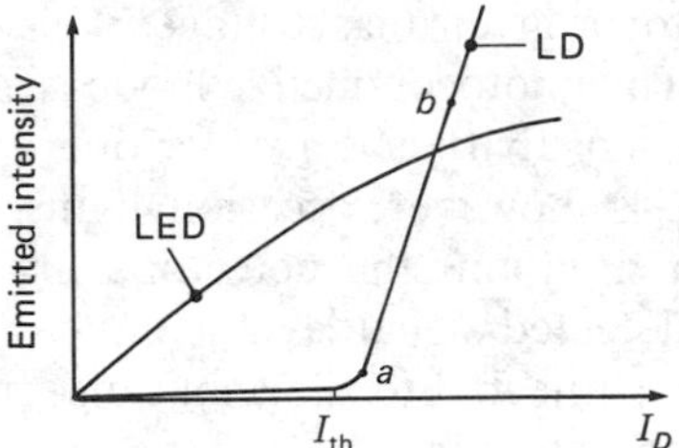

Figure 16-16 Emission intensity versus drive current.

16-8 LIGHT DETECTORS USED IN FIBER OPTIC COMMUNICATION

The detector in a fiber optic communication system must convert the fluctuations in the light amplitude delivered by the fiber optic cable into an electrical signal. There exist a wide range of devices that can perform this task, but the two devices most commonly used in fiber optic communication are the PIN diode and the avalanche photodiode (APD). The basic physical arrangement of a PIN diode is shown in Fig. 16-17.

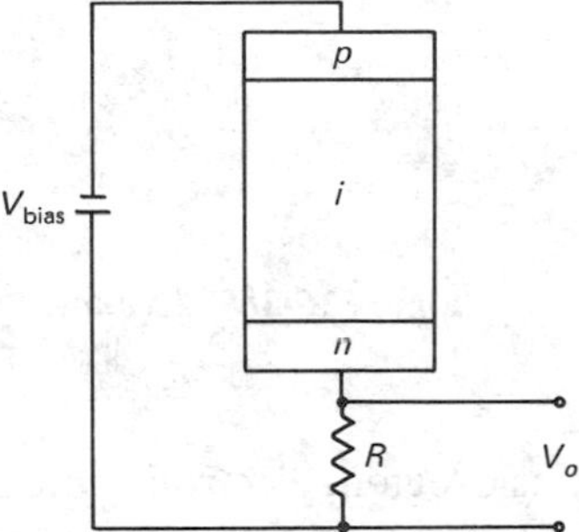

Figure 16-17 PIN Diode.

Heavily doped P and N semiconductor material is placed on either side of an essentially undoped (intrinsic) semiconductor region. The lettering associated with the layers (PIN) is the basis for the device's name. The intrinsic region, lacking the charge carriers provided by the doping, demonstrates a relatively large resistance. If, however, the intrinsic region is exposed to light from a fiber optic cable, the impinging light generates electron–hole pairs (charge carriers) in the otherwise largely insular intrinsic region. This effect can be used to generate a detector output signal in either of two ways. Relatively small voltages will be generated across the PIN device by the impinging light creating the charge carriers. Providing an electrical output in response to light input is called *photovoltaic* operation. Alternatively, the PIN device may be biased from an external supply and the current level monitored. The extra charge carriers induced by the impinging light will increase the current flow for a given bias voltage. This operating mode is referred to as *photoconductive*. The photoconductive mode is generally the preferred operational mode since better sensitivity and a wider operating bandwidth are obtained. The photovoltaic mode is, however, occasionally used when the detected light level is very low. In such situations the detector's intrinsic noise level (see Chapter 9) compared to the detected signal level is quite critical. Zero-bias photovoltaic operation will not be subjected to the shot noise associated with the bias current of photoconductive operation.

The basic structure of an avalanche photodiode (APD) is shown in Fig. 16-18. The use of an APD is similar to a PIN diode operating in the photoconductive mode. Two major differences exist, however. The supply voltage used with an APD is much higher, typically into the few hundred-volt range. The second difference is the APD is constructed with an additional *p*-doped region between the *n*-doped and intrinsic regions. The diode junction formed at this point is reverse biased by the bias supply voltage and the great majority of the bias supply potential is dropped across this junction. The very large voltage levels used places the junction near breakdown voltage levels.

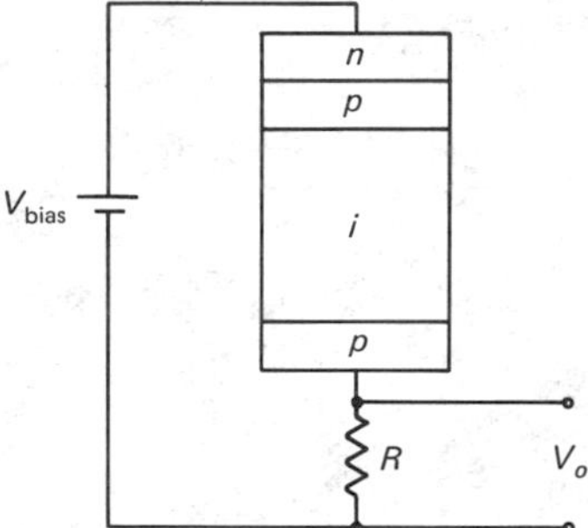

Figure 16-18 Avalanche photodiode.

When light impinges on the intrinsic region, electron–hole pair charge carriers are generated which then drift in opposite directions in response to the applied potential. The charge carriers reaching the P-N junction will be highly accelerated by the intense junction differential voltage, and if the potential is high enough, initiate an avalanche breakdown in the junction by impacting and freeing successively larger numbers of charge carriers. Each electron–hole pair liberated by incident light results in a large number of charge carriers being liberated in the avalanche breakdown process. As a result of this amplification process the APD has a much higher light sensitivity than a PIN diode.

The APD must be operated with a high-voltage bias supply adjusted to the verge of avalanche breakdown. As the breakdown effect is temperature and voltage dependent, the bias circuits must include temperature compensation and be selected carefully so that the breakdown process does not lead to irreversible damage in the device. The avalanche breakdown process also introduces additional noise above the shot noise level of a PIN diode. PIN diodes require (at most) a low-voltage supply and their operation is not critically affected by temperature variations. They generally have wider operating bandwidths and have lower internal noise levels than APDs. The greater sensitivity of APDs can, however, be used to extend the maximum repeater spacing allowed in a system. The selection of detector is thus largely dependent on the requirements of the specific fiber optic system it is to be placed in.

Problems

16-1. A light ray is split by passing it through a semicircular slab of glass as shown in Fig. P16-1. Given the angles measured, determine the index of refraction of the glass.

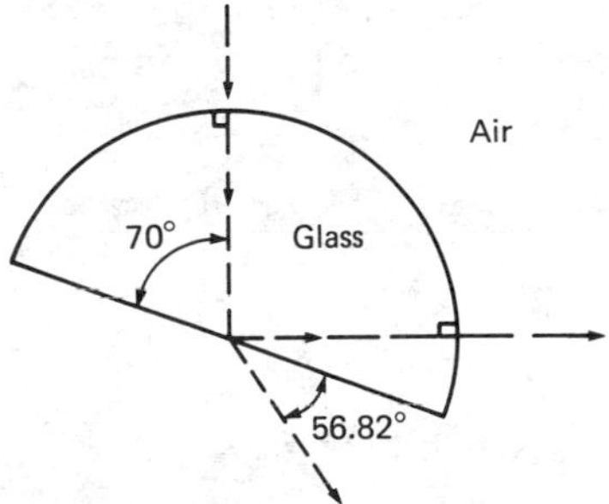

Figure P16-1.

16-2. Given the definition of index of refraction, one would expect the smallest possible value of n a material could possess would be
(1) $-\infty$
(2) 0
(3) 1
(4) 1.414

16-3. A laser diode source injects a -3 dBm of optical power into a single-mode fiber. The fiber has an attenuation rate of 0.3 dB/km and a distance–bandwidth product of 20 GHz-km. The signal to be transmitted requires a bandwidth of 500 MHz. The fiber comes in 2-km reels that must be spliced together; each splice introduces an additional 0.5 dB of loss. The PIN detectors used can reliably recover a digital signal from received optical signals as low as -36 dBm. In the course of the system's operating life, backhoe operators and rabid gophers can be expected to damage the cable and require up to five more splices to be required between repeaters. The maximum allowable repeated spacing will be:
(1) 110 km
(2) 55.5 km
(3) 40 km
(4) 30 km
(5) 27.7 km
(6) 41.9 km
(7) 19.5 km

16-4. A long-wavelength laser diode is used as the light source in a system rather than a short-wavelength LED. This selection leads to:
(1) Less Rayleigh scattering in the fiber
(2) Less chromatic dispersion in the fiber
(3) Less modal dispersion in the fiber
(4) A reduction in the fiber's NA
(5) Less reflection at the core–cladding interface
(6) Both (1) and (2) above

16-5. The index of refraction of a multimode fiber's core is adjusted to create a gradual transition to the level of the cladding's index of refraction. The advantage of this over a fiber that has an abrupt change of refractive index at the core–cladding boundary is
(1) A reduction in the Rayleigh scattering dispersion
(2) A reduction in the chromatic dispersion
(3) A reduction in the modal dispersion
(4) A reduction in the bending losses
(5) None of the above

16-6. A beam of light is focused on the end of an optical fiber cable. A ray of light that enters at the highest possible angle and still experiences total internal reflection in the fiber is shown in Fig. P16-6.

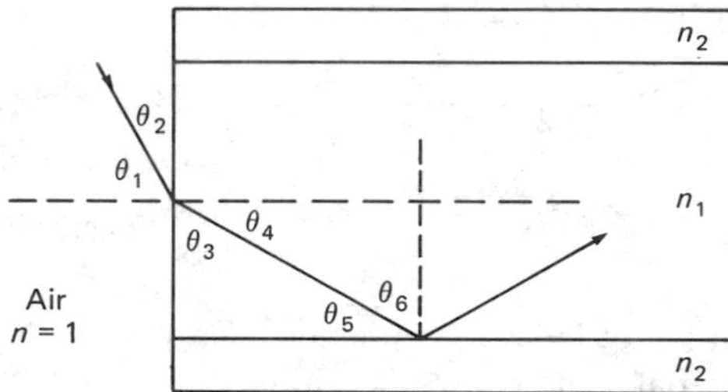

Figure P16-6.

(a) Which angle is the acceptance angle?
(1) θ_1
(2) θ_2
(3) θ_3
(4) θ_4
(5) θ_5
(6) θ_6

(b) Which angle is the critical angle?
(1) θ_1
(2) θ_2
(3) θ_3
(4) θ_4
(5) θ_5
(6) θ_6

16-7. The cable in Fig. P16-6 is step-index multimode cable with a core index of refraction of 1.4. It has a numerical aperture of 0.4. Determine the index of refraction of the cladding.

16-8. An application requires the transmission of very high data rates over very long distances. What type of light emitter and what type of fiber optic cable should be used? Justify your selection in terms of the limitations that can be encountered on a fiber optic link.

16-9. A multimode-step index fiber optic cable has a range–bandwidth product of 200 MHz-km due to a modal dispersion. It is desired to send a digital signal having a 400-MHz bandwidth down this line.
(1) This is not possible on this type of cable.
(2) This is possible if repeaters are placed at least every 2 km.
(3) This is possible if repeaters are placed at least every 1/2 km.
(4) This is possible if the numerical aperture of the cable is large enough.

16-10. Two different types of fiber optic cable are fusion-spliced together and the resulting splice has a loss of 1.0 dB. The loss could be due to:
(1) Fresnel loss
(2) Different numerical apertures
(3) Different core areas
(4) Misalignment of the cores
(5) All of the above

16-11. When splicing fiber optic cable, which of the following items is the most important to have?
(1) Hardhat
(2) Safety glasses
(3) Steel-toed boots
(4) Earplugs
(5) Lead-lined underwear

16-12. A graded-index multimode fiber optic cable is superior to a step-index multimode-fiber optic cable because:
(1) It has less modal dispersion.
(2) It has more wavelength dispersion.
(3) It is easier to manufacture.
(4) It does not experience bending losses.
(5) All of the above.

16-13. A glass ($n = 2$) bar rather than a mirror is used to redirect light in the camera as shown in Fig. P16-13. If the glass bar is formed with too small an angle (θ), a large amount of light is lost in the process.

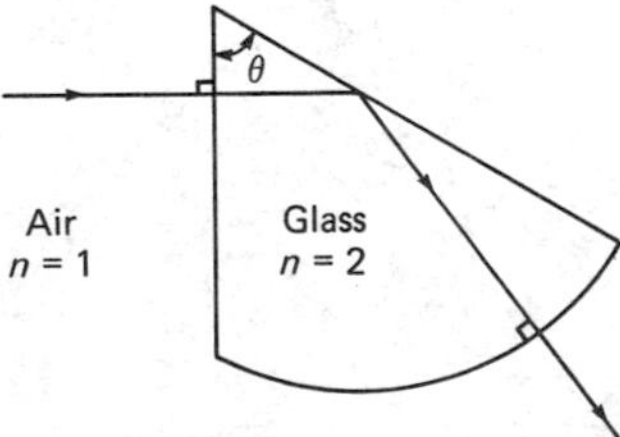

Figure P16-13.

(a) Find the minimum angle that θ can be before this occurs.
(b) The light travels a total of 3 cm in the glass bar. How long does it take the light to travel through the glass?

16-14. The hardest fiber optic cable to splice is:
(1) Step-index multimode
(2) Step-index monomode
(3) Graded-index multimode
(4) Beldon RG8/U

16-15. The numerical aperture of a step-index multimode cable is normally kept reasonably small in order to:
(1) Maximize the amount of light injected by the light source
(2) Minimize the amount of modal dispersion on the cable
(3) Permit the injection of light from a light-emitting diode source
(4) Reduce the skin depth of the cladding
(5) Reduce the amount of wavelength dispersion on the cable

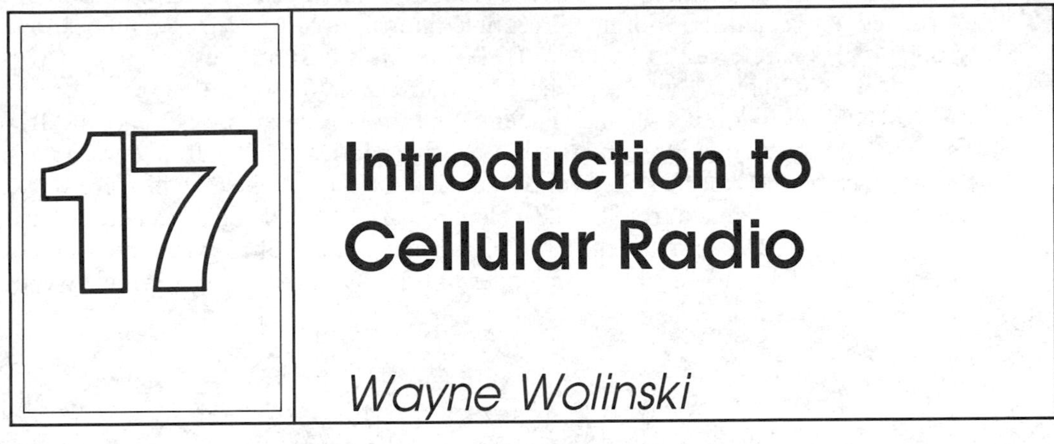

17 Introduction to Cellular Radio

Wayne Wolinski

17-1 GENERAL CELLULAR RADIO DESCRIPTION

Each system consists of a pattern of more-or-less hexagonal coverage areas (generally, 5 to 10 km in diameter) referred to as cells, similar to that illustrated in Fig. 17-1. Each cell contains base stations, antennas, and control equipment providing communications with subscriber radios (such as mobiles or portables) similar in quality to that of the average home-based telephone. While the coverage areas shown in Fig. 17-1 are somewhat ideal, it is possible to approximate this structure by utilizing low-powered UHF transmitters and carefully selected/placed antennas within each cell. Restricting the coverage area of each cell enables frequency reuse; this is demonstrated in Fig. 17-1 with numbers shown within each cell which represent a group of channels (note the distance between two 1 cells, two 2 cells, etc.).

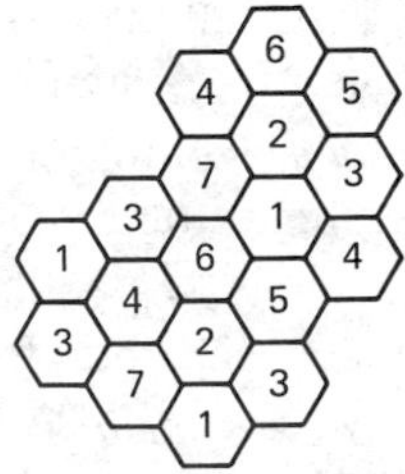

Figure 17-1 Example of a cellular system. Numbers within each cell represented the group of frequencies used.

A mobile or portable moving between coverage areas will automatically be switched between cells, with no perceptible interruption in voice communications (this process is referred to as *handoff*). Figure 17-2 illustrates an example of two practical cells, showing the area where handoff occurs between them.

The 5 to 10-km cell diameter mentioned previously applies to an 800-MHz system, the most common band used in North America. Lower-frequency operations, such as the Nordic system in Scandinavia or the province-wide Aurora 400 used in Alberta, Canada (roughly 120 cells covering 1,843,200 km^2), generally have larger-diameter cells, but like the 800-MHz system, actual coverage depends largely on terrain and obstacles. In the remainder of this chapter we deal specifically with 800-MHz systems.

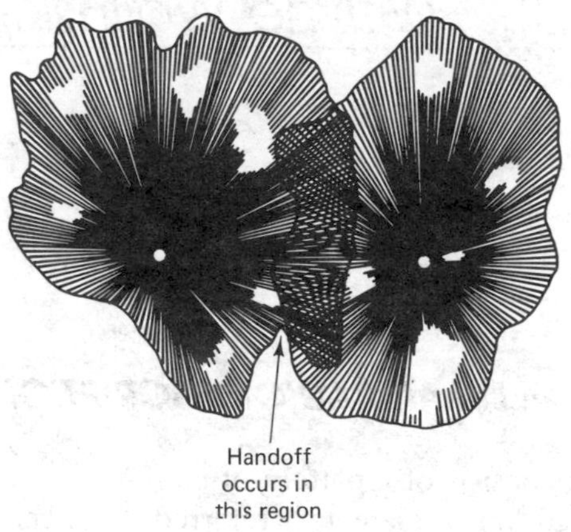

Figure 17-2 Example of two practical cells.

17-2 SYSTEM COMPONENTS

Excluding subscriber cellular radio transceivers, there are five major components of a cellular system:

1. *Radio control subsystem:* consists of cell site radio equipment required to interface subscribers with call processing and switching subsystems, including towers, antennas, transceivers, and so on.
2. *Call processing subsystem:* computer hardwawre/software required to set up, connect, monitor, terminate, and hand off calls.
3. *Switching subsystem:* interfaces cellular equipment with existing telephone equipment (interconnects cell site "trunks" with the phone network).
4. *Administrative support subsystem:* computer hardware/software required for tasks such as customer billing and maintenance records.
5. *Network Facilities:* links between subsystems in different locations (i.e., such as wire connecting a cell site to the rest of the system).

17-3 SYSTEM ARCHITECTURE

Regardless of the system, the cell site must contain the radio control subsystem (item 1 in Section 17-2). The configuration of the other subsystems depends on the equipment manufacturer, but basically falls between two extremes: (1) fully centralized: call processing, switching, and administrative support subsystems located at the same place; and (2) fully decentralized: the aforementioned subsystems are physically separated.

17-3.1 Centralized System

Figure 17-3 illustrates a centralized system. The mobile telephone switching office (MTSO) is the central coordinating element, and is computer controlled to allow it to process all mobile phone calls under programmed control. When distances between the cells and the MTSO are short, and good access to the public switched telephone network (PSTN) is available, this approach can be quite cost-effective.

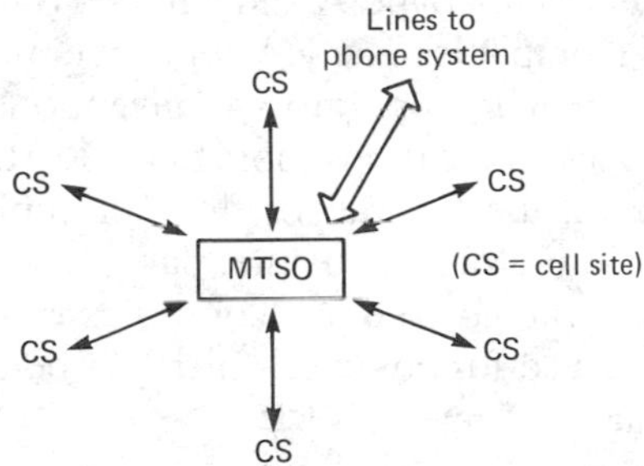

Figure 17-3 Example of a centralized system.

17-3.2 Decentralized Approach

A decentralized approach distributes the processing among several smaller control centers, as illustrated in the example system in Fig. 17-4. The cell sites (marked CS) typically contain an intelligent radio subsystem capable of communicating with subscriber radios. The A blocks represent control centers capable of managing several cell sites, handling functions such as paging (locating a subscriber), cell handoff (when another cell will communicate with the subscriber better than the currently used one), call termination, and maintenance of administrative and/or status records regarding subscribers in the area. In addition, interfacing to the PSTN can occur here. The B blocks generally are data processing centers capable of providing administrative support regarding such items as customer records, billing information, and traffic analysis (i.e., how busy the system is, and where). Human–machine interfacing is also likely to occur here.

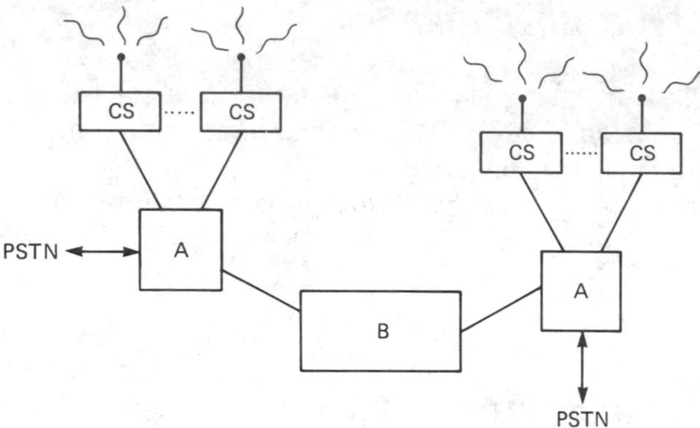

Figure 17-4 Decentralized example.

17-4 FREQUENCY ALLOCATION

Presently, 666 channel pairs are allocated, being further subdivided into two groups of 333 channel pairs called band A and band B. Band A is reserved for *non-wireline* system operators (companies that are not conventional telephone companies). This leaves band B for *wireline* operators (conventional telephone companies). Table 17-1 lists the frequency allocations (note that TX, RX refer to the cell site; therefore, subscriber radios will be the reverse for full-duplex operation). Each channel is allocated 30 kHz; therefore, a channel pair (required for full duplex) occupies 60 kHz. Channels are categorized as "control" or "voice." Twenty-one channels in each band are reserved for control, which can be further subdivided into "setup," "paging," and "access" channels.

TABLE 17-1 Cell Site Frequency Allocations

Band A:	RX	825.03–834.99 MHz
	TX	870.03–879.99 MHz
Band B:	RX	835.02–844.98 MHz
	TX	880.02–889.98 MHz

Setup Channels

One or more of these channels transmits continuous data containing system parameters required by a cellular transceiver going into operation, to enable it to access the system (if authorized to do so). Paging and access channel numbers are obtained from data transmitted on the setup channel.

Paging Channels

A mobile monitors a paging channel when not in use, in case it receives a call from the base, at which time it will switch to an access channel (paging a mobile determines if it can take a call).

Access Channel

Access channels are used to respond to a page or to originate a call (from the mobile or portable).

Voice Channels

A voice channel is used for user information transfer, whether that be an ordinary person–person conversation or a low-speed data transfer between a remote terminal and a mainframe (for example). It should be noted here that bit error rates for data communications via cellular are typically no worse than those encountered using the dial-up PSTN; hence we can expect an onslaught of voice and/or data terminals/peripherals for cellular usage in the near future. The only drawback of data transmission via cellular is that it is best done from a stationary cellular unit, since cell handoff causes a short interruption in the voice channel (not perceptible during conversation).

Frequency Reuse

The cellular concept allows for frequency reuse, provided that the distance between cells using the same channels is adequate. This distance will depend largely on the antenna transmission pattern(s) within the cells. Figure 17-5 illustrates a 21-channel group/7 cell repeat pattern, which can be used for 120° sectored cell transmission patterns. If an omnidirectional transmission pattern is used (antenna placement in cell centers), a 24 frequency group/12 cell repeat pattern is recommended, and if a 60° sector transmission pattern is used, a 24 group/4 cell repeat pattern is recommended.

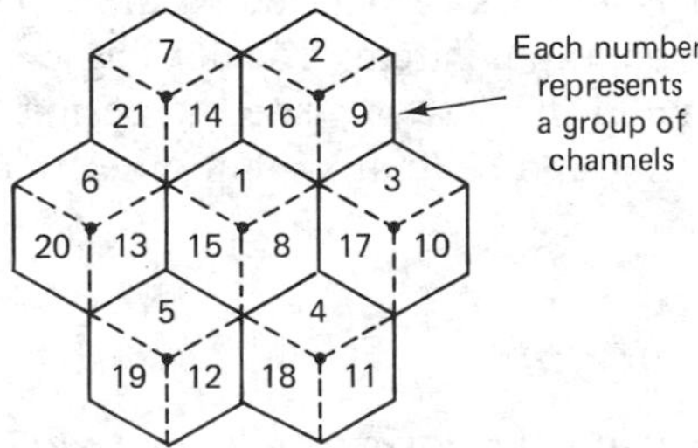

Figure 17-5 120° transmission pattern, channel reusage.

17-5 CELL SPLITTING

When system congestion becomes a problem and the grade of service is no longer acceptable, individual cells may be subdivided into smaller cells as illustrated in Fig. 17-6. Of course, this requires system redesign to reflect the new coverage areas (i.e., adjustment of cell site power levels, antennas, etc.); therefore, it is wise to design the initial system carefully to avoid changes such as this soon after system startup.

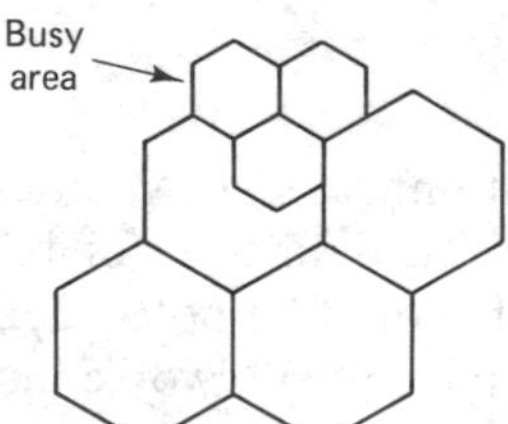

Figure 17-6 Concept of cell splitting.

17-6 SUBSCRIBER RADIOS

A wide range of mobile and portable units is currently available from manufacturers such as General Electric, Motorola, Novatel, Ericsson, Advanced Communications, OKI, and Tandy, just to name a few. The price ranges currently from approximately $500 (Cdn) for a bottom-of-the-line mobile, to approximately $4000 (Cdn) for a top-of-the-line hand-held portable. It is likely that units will be available for under $350 (Cdn) within two years of the time of writing (1990).

Compatibility

Unfortunately, a worldwide specification for cellular systems does not exist; however, in North America, cellular units conforming to EIA/FCC specifications will operate on any system, providing they have been "configured" properly (by maintenance personnel) and are recognized by the system to which access is desired. If visiting another geographic area, "roamer" agreements must be in place between the home system and the system visited to permit access (a *roamer* is defined as a subscriber operating in a cellular system other than the one from which service is subscribed).

General Description

Both portables and mobiles cover 666 channel pairs (825.03 to 844.89 MHz transmit, and 870.03 to 889.98 MHz receive), allowing for full-duplex operation on any of the voice channels used (on the 800-MHz band, a duplexer is quite small, permitting the same antenna to be used for receive and transmit simultaneously). State-of-the-art technology, including surface-mount devices, custom LSI, modular construction, and low-loss duplexers help keep the size, weight, and power consumption of these units to a minimum.

Operational Features

Operational features too numerous to list are provided by many of the subscriber radios; however, some of the more commonly utilized features are such things as hands-free operation, last-number recall, function/status display, pushbutton dialing, electronic lock (preventing unauthorized usage), and volume. For a more complete list, it is recommended that the reader obtain a sales brochure from a local cellular sales office.

Functional Features

Following is a sample of typical functional features found in most cellular units:

1. Automatic leveling of transmitter power output (as directed by cell site)
2. Temperature-compensated crystal oscillators resulting in good frequency stability
3. Dual-conversion superheterodyne receiver
4. Transmitter keying prevented on hardware/software error detection
5. VSWR-protected power amplifiers in the transmitter
6. High/low supply voltage protection
7. Thermal shutdown in the event of abnormally high unit temperatures
8. Battery backup of memory data during receiver power-down

This is only a sample, and for more complete information on any particular unit, contact the manufacturer.

Transceiver Architecture

With such a variety of manufacturers producing cellular radios, it is reasonable to expect the exact designs to vary somewhat; however, regardless of manufacturer, all subscriber units can be subdivided into two main sections: the control section and the RF section. The control section will be microprocessor-based (at least one), with provision for input–output with the user (i.e., such as through the handset), while the RF section will consist of a state-of-the-art transmitter–receiver (often containing surface-mounted devices for minimal size/weight) and duplexer (to enable reception/transmission simultaneously using only one antenna).

Figure 17-7 is a block diagram representing a typical cellular transceiver. Examining the receive path first, the incoming signal is amplified, heterodyned down to an intermediate frequency, and then amplified again before being detected (the reader unfamiliar with this technique may refer to Section 1-7.2). Note that the detector block has three outputs: the top output (Fig. 17-7) will normally be audio intended for the earpiece of the handset (voice). The use of an expansion block restores the previously compressed (at the transmitter) audio to normal (this is used to improve system noise immunity). The middle detector output provides

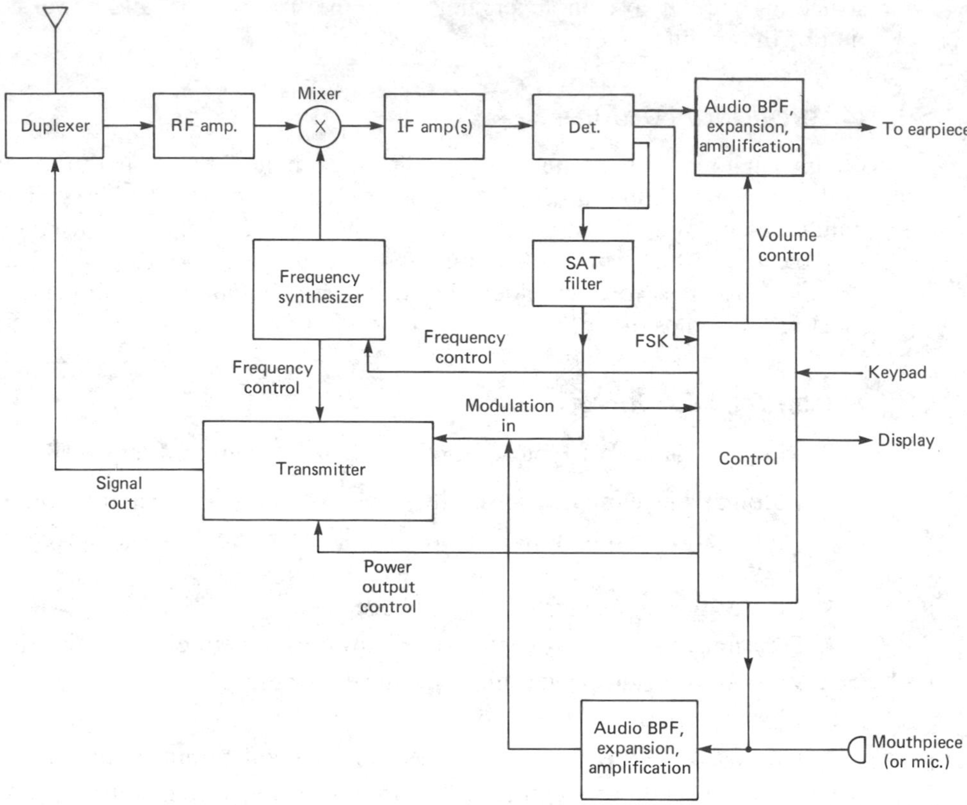

Figure 17-7 Representative block diagram of typical cellular unit.

for reception of FSK (frequency-shift keying—digital "ones" and "zeros" corresponding to 8-kHz deviations above or below the carrier frequency), used for data reception of important information sent from the cell site. Note that this information is passed to the control section of the transceiver, where information transmitted by the cell site will be interpreted (and acted upon if necessary). The last (bottom) detector output is used for reception of a 5.97-, 6.0-, or 6.03-kHz SAT (supervisory audio tone). In addition to feeding the control section (so it can tell when SAT is present), SAT is also fed back to the transmitter. By looping back SAT to the cell site (during a call-in progress) the cell site can tell if contact with the subscriber is lost for some reason (SAT is not audible to the user).

As Fig. 17-7 shows, a frequency synthesizer under direction of the control section determines the receive/transmit frequencies being used. This is reasonable, as it is desirable for the cell site to be capable of directing the unit to a new frequency (such as during a handoff, while the user is crossing between coverage areas of two cell sites), or for the user (under certain conditions such as during maintenance) to be able to select frequencies via the handset. Alternatively to SAT modulating the transmitter, two other paths can be seen: one from the control section (so digital messages can be sent to the cell site from the subscriber unit); the other being, of course, the audio coming in from the handset (which is compressed to improve system noise immunity, and bandlimited to restrict channel bandwidth requirements).

The transmit power output can be controlled in discrete steps by the cell site (it can send a message to the subscriber control unit, which in turn, can direct the power amplifier) since it is desirable to restrict transmit patterns geographically (as ultimately in a large system frequencies are reused) and to minimize power consumption (especially for hand-held units). Since these units operate at a fairly high frequency, it is possible to incorporate small state-of-the-art, low-loss duplexers to facilitate full-duplex operation (simultaneous transmission/reception) using a single antenna.

Maintenance and Repair Requirements

As the cellular unit is considerably more complex than the typical mobile radio of only a few years ago, service personnel troubleshooting to the component level must be well versed in both analog and digital theory (unless of course, a board-replacement approach is taken). Also, as may be expected, a new generation of complex test equipment has surfaced on the marketplace for servicing of cellular transceivers. Test units capable of communicating with cellular transceivers are currently offered by several manufacturers (ie; Motorola, Marconi, IFR), and they offer the service technologist a vehicle for troubleshooting in seemingly hopeless areas—areas difficult to service otherwise, even when armed with logic analyzer, scope, and proper technical documentation.

Problems

All problems deal with 800 to 900-MHz systems only.

17-1. What is the diameter of a typical coverage area as specified in the text?

17-2. Excluding the subscriber radio, what are the five major components of a cellular system?

17-3. List one advantage of a decentralized system over a centralized one.

17-4. Generally, under what conditions will a handoff occur?

17-5. What frequency band has been allocated for band A, subscriber radio transmitters?

17-6. What bandwidth is occupied by a channel pair?

17-7. What kind of information is passed on a voice channel?

17-8. Under what general conditions should cell splitting be considered?

The following problems pertain to subscriber radio units.

17-9. What function is performed by the duplexer?

17-10. Why is compression/expansion used?

17-11. Can a cell site communicating with a mobile direct the mobile to change transmit output power? If so, is there any advantage to be gained?

17-12. What is the purpose of the FSK line out of the detector in Fig. 17-7?

17-13. What are three possible frequencies corresponding to SAT tones?

Satellite Communications

Walter Kalin

18-1 INTRODUCTION

Satellite system communications are well visible to the over 1,000,000 homes in North America equipped with antenna dishes for the reception of satellite television. What may not be so well known is that satellites handle over 60% of the world's international communications traffic.

Satellites also service many regional and domestic needs, such as gathering and transmitting weather information, data communications, remote sensing, military reconnaissance, and navigation, together with a host of other applications.

18-2 SATELLITE ORBITS

From early observations about planetary motion we know that any body placed in an orbit in space will trace a path around the larger body it is orbiting. In addition, once a body is placed in orbit, its orbit will remain virtually unchanged. Satellites that orbit the earth follow the same laws that govern the motion of the planets. Johannes Kepler empirically derived a few laws describing planetary motion that can be applied to satellite orbits.

18-2.1 Kepler's First Law

Kepler's first law states that the path followed by a satellite around the earth will be an ellipse. An ellipse can be described by two focal points, f_1 and f_2, as shown in Fig. 18-1. The larger mass, being earth, must be located at one of the two foci.

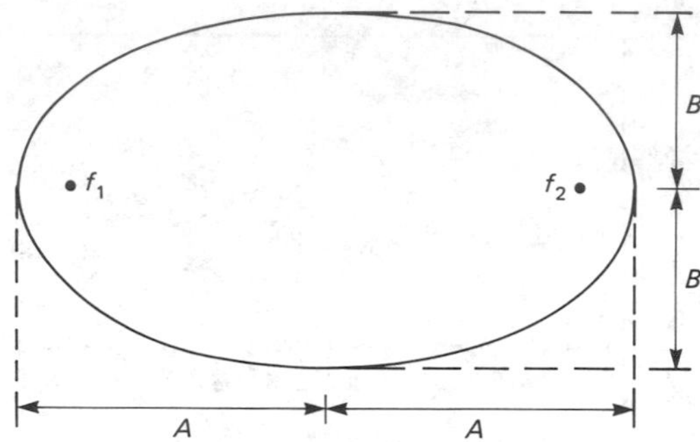

Figure 18-1 Ellipse with two foci f_1 and f_2.

Because of the enormous size of the earth, it is the center of the earth that lies at one of the foci. The eccentricity e of an ellipse is given by

$$e = \frac{\sqrt{A^2 - B^2}}{A} \tag{18-1}$$

where A is the length of the semimajor axis and B is the length of the semiminor axis. An elliptical orbit has $0 < e < 1$. When $e = 0$, $A = B$, and the orbit becomes circular. Most satellites transmitting television signals travel in circular orbit about one focal point, the earth's center.

18-2.2 Kepler's Second Law

Kepler's second law states that for a given time interval, a satellite will sweep out equal areas when focused from the center of the earth. Referring to Fig. 18-2, area A_1 will equal area A_2 if it takes the same time for a satellite to travel distances d_1 and d_2. From this it follows that the velocity along path d_1 must be greater than

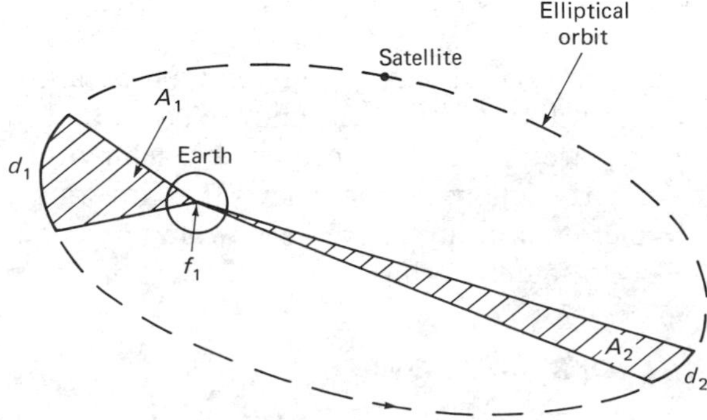

Figure 18-2 Kepler's second law: For a given time period, a satellite will sweep out equal areas $A_1 = A_2$.

along path d_2. Thus a satellite takes longer to travel a given distance when it is farther away from the earth's center. Certain types of satellites make use of this property to increase the time a satellite lies over a particular geographic region of the planet.

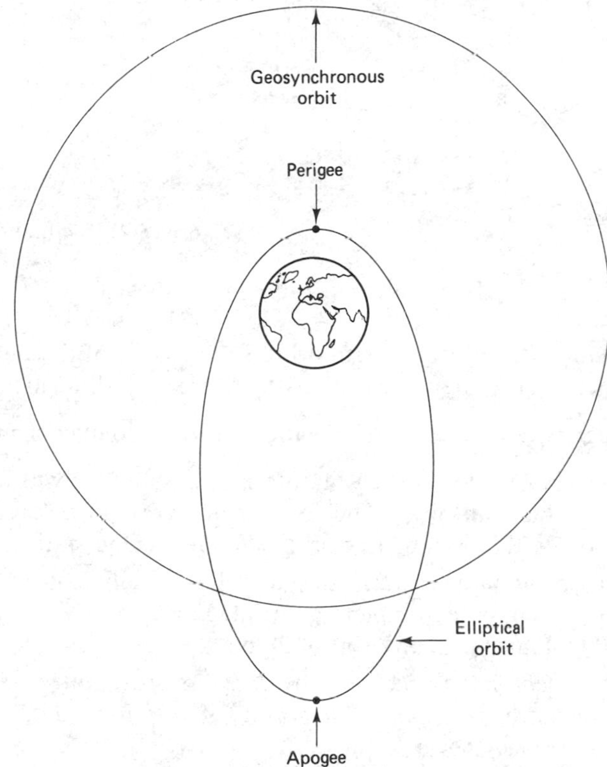

Figure 18-3 Apogee and perigee of an orbit.

Two parameters of interest for elliptical orbits are the apogee and perigee. *Apogee* is the point on the orbit farthest from the earth's center, as shown in Fig. 18-3. *Perigee* is the point on the orbit closest to the earth's center. When apogee = perigee, the orbit is circular as shown in Fig. 18-3. The two principal parameters of any orbit are:

1. *Inclination:* the inclination of an orbit is a measure of angle between the earth's equator and the satellite orbit plane. For example, an orbit around the earth's equator has an inclination of 0°, while an orbit around the earth's poles has an inclination of 90°, as shown in Fig. 18-4. Many satellites follow a polar orbit, such as Soviet communications satellites, which fly in high elliptical orbits to cover the northern regions of the Soviet Union, and the American weather satellites, which fly in low orbits.

Figure 18-4 Inclination angles.

2. *Altitude:* the altitude of an orbit is a measure of the distance from the earth's center to a point on the orbit path. A circular orbit will have a constant altitude.

There are four altitude classifications for circular orbits:

1. *Low orbit:* below 22,279 statute miles. Satellites will take less than 24 hours to complete an orbit. Thus earth stations require complex tracking antennas to follow the path of low-orbit satellites. (One statute miles = 5280 ft.)
2. *High orbit:* above 22,279 statute miles. Satellites in high orbit will take more than 24 hours to complete an orbit. For example, the moon at an altitude of 240,000 miles has an orbit of 28 days.
3. *Geosynchronous orbit:* exactly 22,279 statute miles. By placing a satellite in geosynchronous orbit, the period of rotation equals that of the earth (24 hours).
4. *Geostationary orbit:* exactly at 22,279 statute miles and at an inclination of 0°. The geostationary orbit is the orbit in which a satellite appears stationary relative to the earth.

The geostationary orbit is often called the *Clark belt* and is sometimes improperly called a *geosynchronous orbit.* Both geostationary and geosynchronous orbits have the same altitude; however, geosynchronous orbits can have any inclination angle, whereas a geostationary orbit always has an inclination of 0°. Thus there are an infinite number of geosynchronous orbits, but only one geostationary orbit.

Geostationary orbit is the most commonly used of all orbits since an earth station pointed at a geostationary satellite will follow it automatically. Thus elaborate satellite tracking systems are not required. For an orbit to be geostationary, the satellite must orbit the earth's equator in the same direction as the earth spins and at a constant speed. For constant speed, Kepler's law requires that the orbit be circular.

18-2.3 Major Forces Acting on a Satellite

When a satellite travels in a circular orbit around earth, there are two major forces acting on the satellite. According to Newton, a gravitational force of attraction exists between two bodies. This force is directed radially toward the center of the earth as shown in Fig. 18-5.

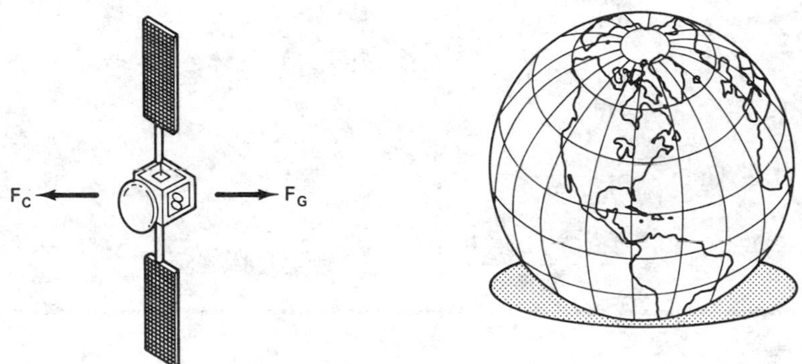

Figure 18-5 Major forces acting on a satellite.

$$F_G = \frac{Gm_1m_2}{r^2} \tag{18-2}$$

where F_G is the force of gravity (N), G is a constant $= 6.673 \times 10^{-11}$ N·m²/kg², m_1 the mass of the satellite (kg), m_2 the mass of the earth $= 5.975 \times 10^{24}$ kg, and r the distance from the center of the earth to the satellite (m) (the radius of the earth $=$ 6400 km).

The other major force acting on the satellite is the centrifugal force resulting from satellite motion balancing the earth's gravitational pull. The centrifugal force is directly radially outward, as shown in Fig. 18-5.

$$F_c = m_1a = \frac{m_1v^2}{r} \tag{18-3}$$

where F_c is the centrifugal force (N) and v is the velocity of the satellite in space (m/s). Since a satellite remains at the same distance above the equator when in orbit, the two forces must be equal. By setting $F_c = F_G$ and solving for v, we get the following:

$$v = \sqrt{\frac{Gm_2}{r}} \tag{18-4}$$

[*Note:* The velocity of a satellite is independent of its mass (m_1).]

EXAMPLE 18-1

A satellite is orbiting the earth at an altitude of 500 km above the surface. Calculate the **(a)** velocity of the satellite and **(b)** time to complete one orbit.

Solution

(a) r = km + radius of earth = 500 km + 6400 km = 6900 km

$$v = \sqrt{\frac{Gm_2}{r}} = \sqrt{\frac{(6.673 \times 10^{-11})(5.975 \times 10^{24})}{6900 \times 10^3}}$$

$$= 7602 \text{ m/s}$$

(b) $$\text{Time} = \frac{\text{distance}}{\text{velocity}} = \frac{\text{circumference}}{\text{velocity}} = \frac{2\pi r}{v}$$

$$= \frac{2\pi(6900 \times 10^3\text{m})}{7602 \text{ m/s}} \simeq 5703 \text{ s} \simeq 1.58 \text{ h}$$

Satellites would remain in a fixed position relative to the earth if only the earth's gravity and the centrifugal force acted on the satellite. Unfortunately, other forces that can be significant also act on the satellite. These forces along with meteorite collisions cause the satellites to drift away from their assigned locations. Ground control stations periodically adjust satellite positions to counteract the drifting.

The other significant forces acting on a satellite include the gravitational forces of the moon and sun, and atmospheric drag. Atmospheric drag mostly affects low-orbiting satellites (below 1000 km) and has a negligible effect on geostationary satellites. The gravitational forces of the moon and sun, on the other hand, have a negligible effect on low-orbiting satellites, but they do affect geostationary satellites. In geostationary orbit there are two "zero-movement" locations where all forces balance out so as to keep an object stationary. The two zero-movement locations are at approximately 75.5°E and 104.5°W longitude. Unfortunately, no satellites are located at these zero-movement locations.

18-2.4 Satellite Launch

Satellites are launched into space by using either a launch vehicle or by using an expendable launcher such as the American Atlas-Centaur and Delta rockets. When a satellite is to be orbiting at less than 200 km above the earth's surface, it is economical to launch the satellite directly into its final orbit. However, if the satellite is to be orbiting at greater than 200 km above the surface of the earth, such as geostationary orbit, it is not economical to inject the satellite into its final high-altitude orbit. In these cases a transfer orbit is used between the initial low earth orbit and the final high earth orbit. If the transfer orbit is one that minimizes

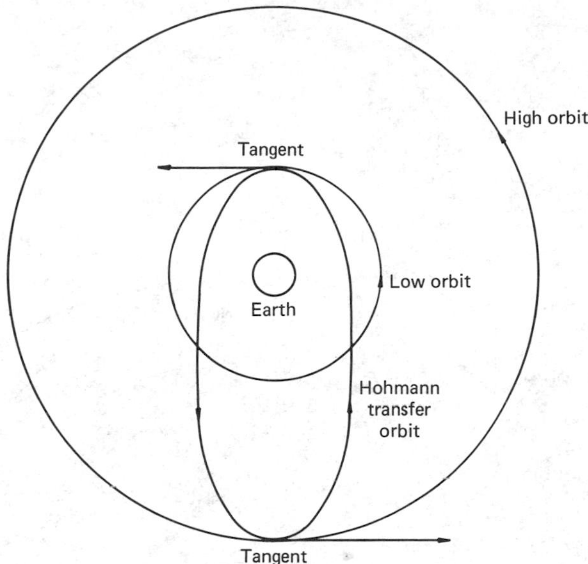

Figure 18-6 Hohmann transfer orbit.

energy costs, it is called a Hohmann transfer orbit. A Hohmann transfer orbit is tangent to the low-altitude orbit at its perigee and tangent to the high-altitude orbit at its apogee as illustrated in Fig. 18-6. It is at these points that the satellite would fire its thruster rockets to change into the new orbits. Also, to keep launch fuel costs down it is best to launch geostationary satellites as close to the equator as possible to minimize the adjustment for inclination that is necessary. This is one reason why a southern state such as Florida is used in the United States as a launch site.

18-3 GEOSTATIONARY SATELLITE SYSTEMS

The most basic satellite communications system consists of an earth station transmitter (uplink), a satellite receiver/transmitter, and an earth station receiver (downlink) as shown in Fig. 18-7.

18-3.1 The Uplink

The uplink must generate and transmit the RF signal with enough power so that the signal can be recovered by the satellite located far away in space. The transmitters often use very large antennas which may weigh in excess of 250 tons. It was mentioned earlier that some drift in satellite position does occur, and this means that a provision must be made for a limited degree of tracking or movement of the large antenna. In colder climates where snow and ice conditions are likely, built-in heaters are required for the antenna and assembly.

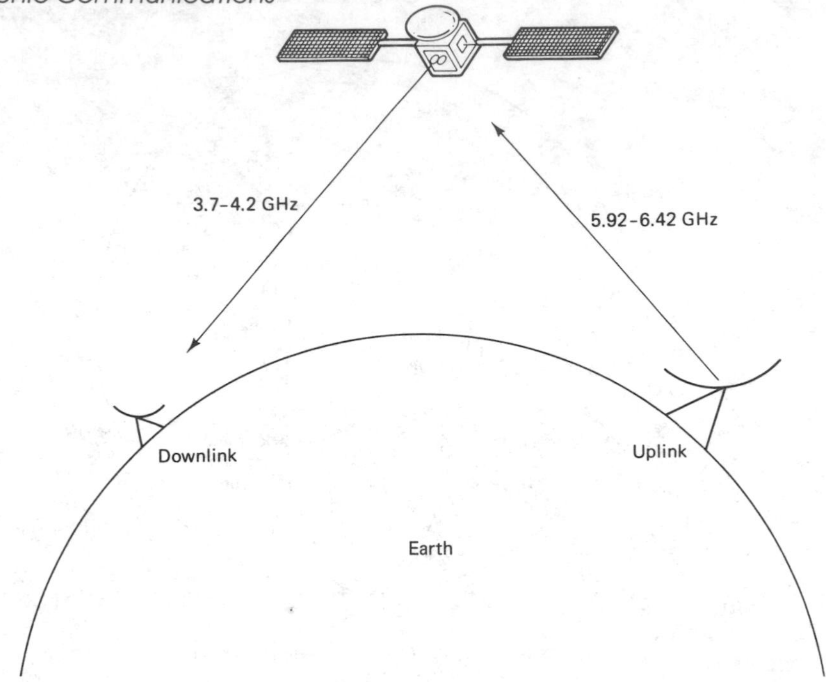

Figure 18-7 Basic satellite links.

Earth stations must also make provisions for continuity of power supply. Power hookup in the form of batteries and generators is available on most uplinks. Should commercial power fail, batteries will immediately take over with no interruption. Generators will also start up and then automatically take over battery-supplied power when their motors are up to speed.

Uplink transmission for geostationary satellites uses two frequency bands. The most common uplink band is 5925 to 6425 MHz, known as the C band. To allow for more satellites, a new band of frequencies from 13.7 to 14.2 GHz, known as the Ku band, is now also being used.

18-3.2 The Satellite

The satellite receives the C- or Ku-band signals, amplifies the signals, and then retransmits the signals at a down-converted frequency range. In the case of C-band signals, the downlink frequency band is 3700 to 4200 MHz. The Ku band is retransmitted down to earth using the frequency band of 11.7 to 12.2 GHz.

Satellite equipment is functionally classified into a payload and a bus. The *payload* refers to the equipment used for communicating the signals, such as the antenna subsystem and transponders. The *bus* refers to the various subsystems that support the payload, such as power supply, satellite position control, telemetry, and thermal stabilization control.

Satellites must be equipped with antennas to receive and transmit signals. Antennas range from a dipole type to a highly directional type. These antennas are retracted for satellite launch and then positioned once the satellite is in space.

The satellite transponder consists of the electronics needed to receive, amplify, down-convert, and retransmit a single communications channel. One transponder might be a single TV signal or a number of separate carriers. A satellite will have anywhere from 12 to 32 primary transponders. A set of backup transponders is also aboard satellites, so that if one primary fails, the backup switches in automatically.

Satellite transponders contain the same general blocks as terrestrial communications systems. A wideband solid-state receiver is used to amplify all the received channels. Next, a demultiplexer separates the broadband input into separate channels. Each channel is then amplified. The final output power boost is provided by traveling-wave tubes. Finally, all the channels are recombined using a multiplexer.

The power for operating the electronics aboard a satellite comes from solar cell arrays. Solar cell arrays come in two forms: cylindrical and flat rectangular. Cylindrical solar arrays are used with spinning satellites such as the Canadian Anik C and D series. Cylindrical solar cells use the gyroscopic effect to provide mechanical stability; however, only part of the cell array is sunshine at any given time.

Flat rectangular solar arrays were used by earlier satellites, such as the Canadian Anik B series. These arrays must be folded during satellite launch and then extended into position once in orbit. Rectangular shape arrays have the advantage that the complete array is sunshine nearly all the time.

Satellite solar cell arrays in geostationary orbit must contend with eclipses that occur twice a year, during the spring and summer equinoxes, as shown in Fig. 18-8. An eclipse occurs when the earth is positioned between the satellite and the sun, blocking sunshine to the solar cells. Eclipses begin 23 days before the equinox and end 23 days after the equinox. An eclipse can last from 10 to 72 minutes per day in this 47-day period. During an eclipse the solar cells do not function and primary power must be supplied from batteries. Nickel–cadmium batteries were

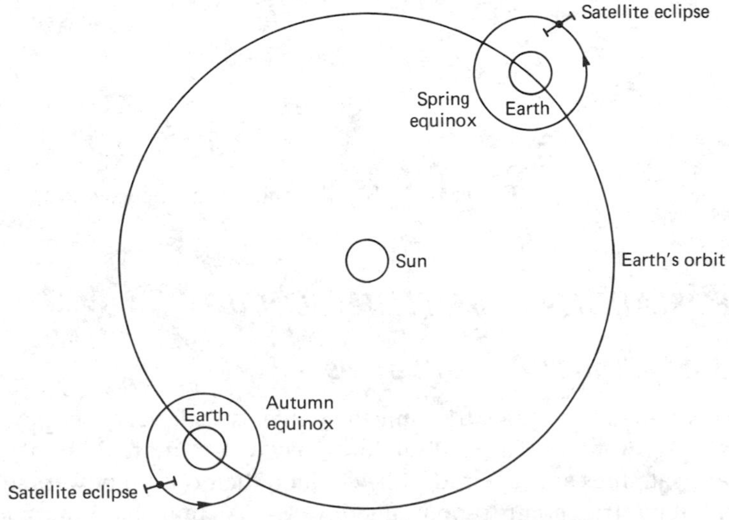

Figure 18-8 Spring and autumn eclipse.

used on older satellites, but lighter nickel–hydrogen batteries are now being used instead.

Satellite position control is another bus function. Should the satellite drift from its position, control jets will be fired to move the satellite back into position. Ground control monitors the satellite's position and operation and uses telemetry to keep the satellite in a correct orbit and orientation. A satellite must be positioned correctly in three axes. These axes are defined as roll, pitch, and yaw axes, as shown in Fig. 18-9. The yaw axis is directed toward the earth's center. The pitch axis is perpendicular to the geostationary orbit, and the roll axis is tangent to the orbit.

The final bus function is thermal stability control. Satellites are subject to larger thermal gradients; however, the equipment must operate in a stable temperature environment. Thermal blankets, shields, and heaters are used to make up for heat reduction. Radiation mirrors, on the other hand, are used to remove heat from the communications payload.

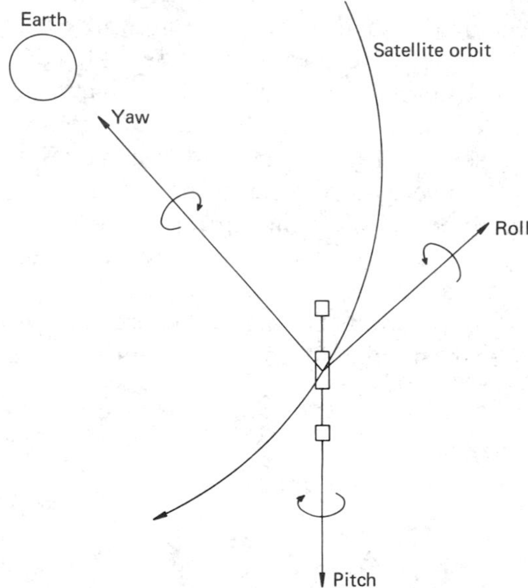

Figure 18-9 Satellite roll, pitch, and yaw axes.

18-4 GEOSTATIONARY SATELLITE SIGNALS

18-4.1 Signal Power Levels

Most C-band satellites transmit their signals with very low power levels of only 5 to 24 W, whereas Ku-band satellites may output up to 200 watts of power. The C-band satellites are required by law to limit their output powers so as not to interfere with terrestrial point-to-point microwave links that share the same frequency band as C-band satellites. There are no terrestrial links using the same frequencies as

Ku-band satellites, and as a result these satellites may transmit with higher output power levels.

The C-band satellites also reduce possible interference with terrestrial links by frequency dithering the uplink signal. Frequency dithering is the adding of a dispersal waveform to the signal so as to spread out the energy content of the signal more evenly. This will result in less interference to terrestrial links. The dispersal waveform is usually a triangular (sawtooth) of 30 Hz for North American television signals, or 25 Hz for other television format signals. The downlink receiving antenna also faces possible interference from terrestrial point-to-point links. This interference is best avoided by placing the receiving antenna outside the point-to-point link transmission path.

18-4.2 Signal Spectrum

The downlink frequency spectrum for C-band satellites varies considerably depending on such things as intended use, country of origin, and design. Figure 18-10 shows a detailed spectral diagram for typical North American C band satellites. The total spectrum bandwidth available is 500 MHz (4200 to 3700 MHz). For normal video use, the bandwidth requirement is 40 MHz (36 MHz plus a 4-MHz guard band). With a total of 500 MHz bandwidth we would expect to have up to 500/40 = 12.5 or 12 transponders. Early Western Union satellites such as Westar I and II, along with early Canadian Anik satellites, relayed 12 television programs (transponders) via satellite.

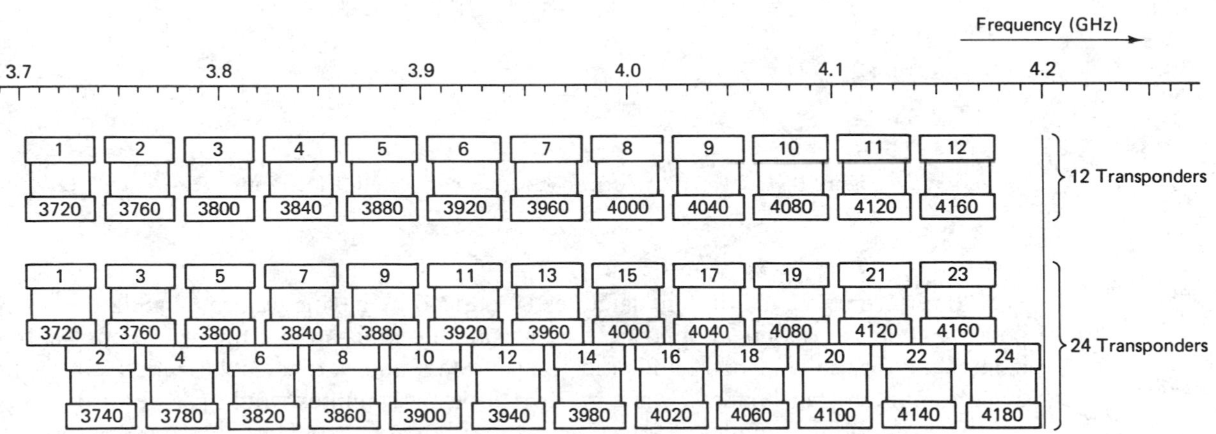

Figure 18-10 North American C-band transponder spectrum.

Today, most North American satellites, such as the RCA Satcom series, the Comstar series, and the new Anik C series, each handle 24 channels. This is possible through signal polarization. Signal polarization doubles the maximum number of transponders from 12 to 24. Twelve transponders per polarity use 12 × 40 MHz = 480 MHz of bandwidth. The remaining 500 − 480 = 20 MHz is used for satellite telemetry. The North American Ku-band satellites have as many as 32 transponders, while older satellites have as few as six transponders.

18-4.3 Signal Polarization

Radio waves are made up of an electric field and a magnetic field. These two fields are always perpendicular to each other. Signal polarization refers to the direction of the electric field for a given wave. North American satellites use horizontal or vertical plane polarization, meaning that the electric field lies in a horizontal or vertical plane with respect to the earth's surface. For example, the odd-numbered channels might be transmitted with horizontal polarization, while the even-numbered channels transmitted with vertical polarization. This provides a 20- to 25-dB isolation between odd- and even-numbered transponders. Further isolation is gained by offsetting the center frequencies of the even-numbered transponders by 20 MHz from the odd-numbered transponders, as shown in Fig. 18-10.

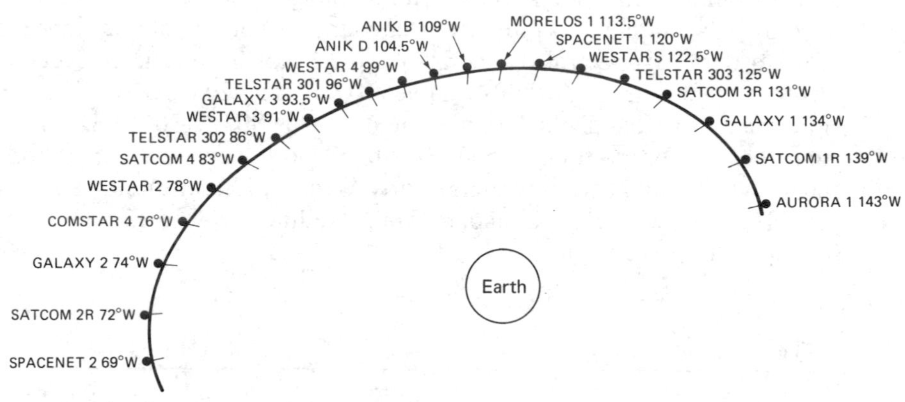

Figure 18-11 North American geostationary satellite locations.

Interference from adjacent satellites is reduced by having adjacent satellites polarize their transponders oppositely. For example, one satellite might use horizontal polarization on odd-numbered transponders and vertical polarization on even-numbered transponders, while the two adjacent satellites would use vertical polarization on odd-numbered transponders and horizontal polarization on even-numbered transponders. Alternating polarization between adjacent satellites is known as cross-polarization discrimination. It provides up to 20 dB of protection against interference from adjacent satellites. Whereas (most) North American satellites use linear polarization (horizontal or vertical), most international satellites use right-hand or left-hand circular polarization. Intelsat international satellites use right-hand circular polarization for transmission of video information. Circular polarization is also being used on Ku-band satellites. Circular polarization is like a corkscrew, where as the signal travels through space, the electric field rotates at a rate of 90° for every quarter wavelength of forward movement.

18-4.4 Satellite Positions

There are many satellites located in geostationary orbit, as shown in Fig. 18-11. Since there is only one geostationary orbit, the placing of satellites in this orbit is heavily regulated. The legal body regulating the spacing of satellites in geostationary orbit has reduced the spacing minimum between satellites from 4° to 2°. This will allow us to place more satellites into the valuable geostationary orbit. However, this reduction in spacing leads to a problem of beamwidth interference, which will be looked at later in more detail.

The locations of satellites are constantly changing as new satellites are added every year and as old satellites wear out. Satellites have a life expectancy of between 7 to 10 years and this generally depends on factors such as how long the hydrazine fuel supply for its rocket boosters lasts, or how long the electronics last without failure. Occasionally, a satellite is struck by a meteorite, making it inoperable. Some day, space vehicles such as the American space shuttle may be used to recover, repair, and relaunch old satellites.

18-5 TELEVISION RECEIVE-ONLY SYSTEMS

The downlink side to satellite television is often called television receive only (TVRO). To prevent confusion we will only examine C-band TVRO. To view satellite television, the following five major pieces of equipment are required: (1) dish (reflector), (2) feedhorn/polarizer (antenna), (3) low-noise amplifier (LNA), (4) downconverter/receiver, and (5) RF modulator/television (monitor). All parts of the system are equally important as the final picture quality is determined by the weakest link in the system. Now let us examine the individual parts of a TVRO system.

18-5.1 Dish (Reflector)

C-band satellites transmit television signals at an output power of only 5 to 24 W. By the time the signal reaches the earth it has suffered a space loss of almost 200 dB. The dish acts as a passive signal reflector which gathers the very weak signal and focuses the energy spread over the dish onto the entrance of the feedhorn. TVRO dishes comes in two shapes: a parabolic or a section of a sphere, with parabolic dishes being more popular. Spherical dishes are good should one want to watch mainly one satellite. Parabolic dishes are more desirable and much more popular because they will generally be smaller relative to a spherical dish. It is also easier to move a parabolic dish from one satellite to another should the dish be mounted on a polar mount.

There are three principal types of dishes available for TVRO use: spun, mesh, and fiberglass dishes. Spun steel or aluminum dishes usually have the most accurate surface and are also usually the most expensive. Spun dishes come in two forms: sectional and one-piece. The sectional spun dish is made by rolling metal pieces into sections of a dish. The sections are then assembled to form a parabolic shape.

The one-piece spun dish is made from a large sheet of metal which is pressed into shape by a large hydraulic press. The one-piece dish is more difficult to transport, but it has a more accurate surface.

Mesh dishes are made from parabolic sections of mesh wire and are the most popular. A mesh dish is very light and allows winds to pass through. This may be significant in windy areas, where a 10-ft solid dish must be able to withstand wind loads of over 1 ton. The slight amount of gain lost by the mesh surface over a solid surface is negligible. However, the mesh surface can easily be damaged should objects accidently come into contact with the dish.

Fiberglass dishes are the most heavy and generally the least expensive of the three. Fiberglass dishes consist of a reflector surface covered with layers of fiberglass. Three types of reflector surfaces are used: a fine mesh screen, a flame-sprayed metallic coating, and a fine aluminum foil. One drawback to fiberglass dishes is that the individual layers of fiberglass may separate over time, causing warping of the reflective surface.

The three types of dishes come in four common sizes: 6, 8, 10, and 12 feet in diameter. These sizes are approximate as an 8-ft dish may actually measure $8\frac{1}{2}$ ft in diameter. The larger the dish size, the more signal that is gathered and the sharper the final picture.

The mount that a dish rests on is just as important as the dish itself. It must be rugged enough to hold the dish stable under all weather conditions. Many TVRO systems are not operable after a bad storm if the mount is of poor quality. Mounts come in two forms, either an azimuth-elevation mount or a polar mount. Azimuth-elevation mounts require two adjustments, rotation (azimuth) and elevation, when changing from one satellite to another. A polar mount only requires that the angle of azimuth be changed to watch another satellite, as the adjustment in elevation is inherent in the design. The most common type of mount is a polar mount.

Actuators are often installed on polar mounts to move the dish from satellite to satellite. An actuator is a motor that rotates a big screw attached to the dish. As the motor turns it rotates the screw and moves the dish about a pivot point. Many actuators come with remote controls for operation from indoors. Some actuators are programmable and store the exact location of the various satellites for easy selection of another satellite.

18-5.2 Feedhorn/Polarizer

The dish gathers the weak signal from space and focuses it into one point, called the *focal point.* The focal point is located a distance equal to the focal length above the center of the dish as shown in Fig. 18-12. The entrance of the feedhorn, called the *throat*, should be located at the focal point of the dish. The feedhorn throat gathers the microwaves in the general area of the focal point, as shown in Fig. 18-13.

The feedhorn throat is actually a circular waveguide which directs the satellite signals toward the feedhorn antenna. A circular waveguide, being nonpolarized, is used rather than a rectangular waveguide, which is polarized. Signals from space come in two polarities: horizontal and vertical. The real antenna in a TVRO system is located near the end of the circular waveguide, as shown in Fig. 18-14. The waveguide and antenna lengths/distances are matched to the C frequency band. There are feedhorns called "dual feeds" which propagate both C- and Ku-band satellite signals. The antenna in the feedhorn rotates to allow for selection of horizontally or vertically polarized signals. This is done by a servomotor that is controlled by a switch located indoors.

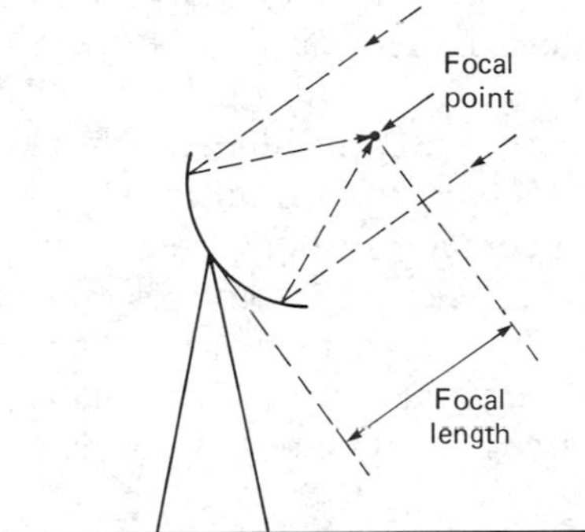

Figure 18-12 Focal length and focal point of a parabolic dish.

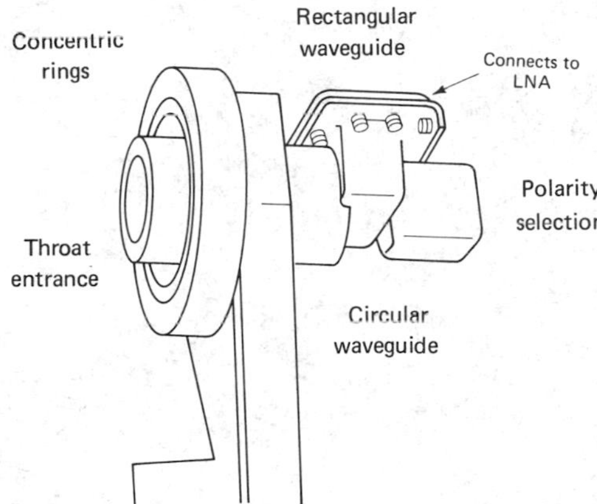

Figure 18-13 Typical feedhorn.

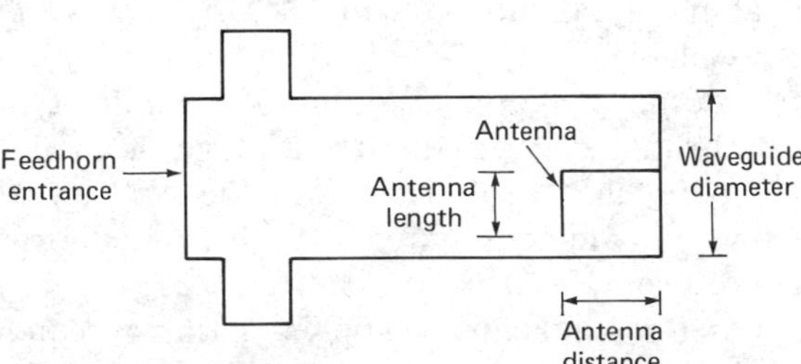

Figure 18-14 Feedhorn dimensions.

Two other methods are used in some feedhorns to perform polarity selection. One method uses two pin diodes to provide much faster polarity selection. The other method involves using a ferrite rotator for polarity selection. Neither method has become very popular, partly because they introduce 0.25- to 0.5-dB losses, while the servomotor mechanical probe introduces almost no loss.

The single polarity signal from the antenna is sent to a rectangular waveguide which is polarized. The low-noise amplifier connects to the rectangular waveguide flange as shown in Fig. 18-13. The feedhorn flange must make metal-to-metal contact with the low-noise amplifier flange to ensure maximum signal transfer. There is also a groove around the flange that is used to hold a gasket that helps to keep water out of the waveguide. Water absorbs the microwave frequency satellite signals and also corrodes the metal contacts of the waveguide.

Feedhorns are designed to pick up the most energy from the center of a dish and the least energy from the outside parts. This is done to avoid picking up ground noise, which flows from around the dish as shown in Fig. 18-15. An ideal feedhorn will illuminate 14 dB less signal at the dish edges compared to the dish center to maintain good carrier-to-noise (C/N) ratios.

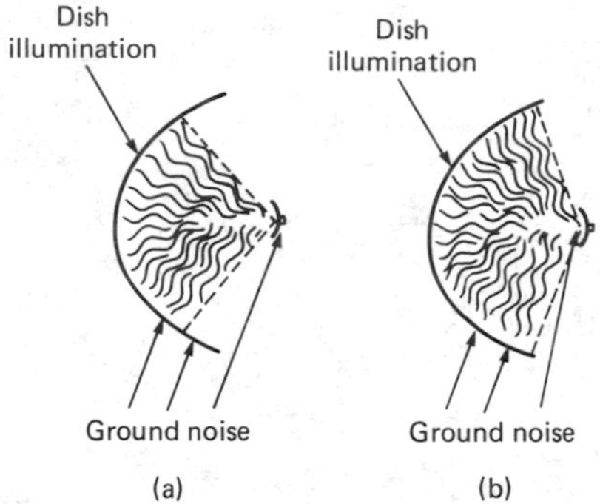

Figure 18-15 (a) Properly designed feedhorn picks up less ground noise; (b) improperly designed feedhorn will pick up too much unwanted noise.

Dishes are made with different focal lengths. A common dish specification is the *f/D* ratio, which represents the ratio of focal length over dish diameter. A dish with an *f/D* ratio of from 0.27 to 0.33 is called a "deep" dish, while a dish with an *f/D* ratio greater than 0.33 and up to 0.45 is called a "shallow" dish. This means that a deep dish will require a wider angle of illumination than a shallow dish to obtain the same signal energy as shown in Fig. 18-16. A 0.375 *f/D* is the most common feedhorn size available.

Feedhorn manufacturers do not make many different feedhorn types to accommodate all the possible *f/D* ratios because there is a range of about 0.04 *f/D* ratio that can be used for a particular feedhorn before the *C/N* level changes significantly. Thus some feedhorn manufacturers provide three feedhorn types to accommodate 3 *f/D* ranges. For example, 0.33 to 0.36 *f/D*, 0.37 to 0.41 *f/D*, and 0.42 to 0.45 *f/D*. Another major manufacturer supplies a ring to convert their

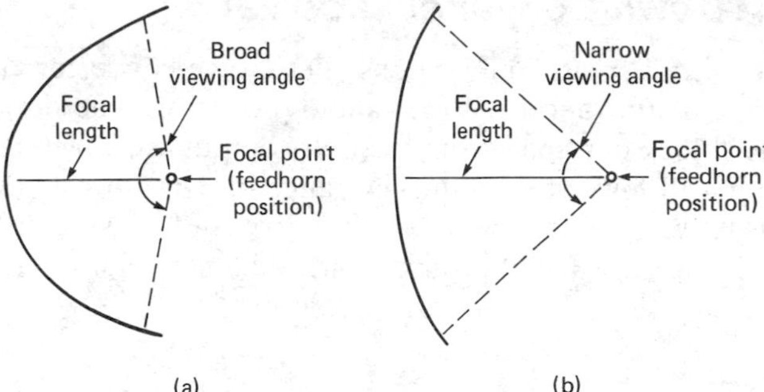

Figure 18-16 (a) Deep dish versus (b) shallow dish.

standard 0.375-f/D feedhorn into an equivalent 0.3-f/D feedhorn. The ring is attached to the feedhorn throat. The ring will improve C/N of a 0.3 f/D by 0.9 dB.

The feedhorn rings shown in Fig. 18-13 act as quarter-wavelength chokes to reduce fringing caused by a mismatch of the E and H fields. The E and H fields have different characteristics, resulting in the E field wanting to spill over the walls of the feedhorn. Placing a series of quarter-wave chokes around the feedhorn opening will help cancel the E-field fringing and produce a nearly flat C/N response across the entire C band of frequencies.

18-5.3 Low-Noise Amplifier

A low-noise amplifier (LNA) is a broadband nontunable amplifier. The purpose of an LNA is to amplify the signal while adding as little noise as possible. Basically, the LNA is used to increase the carrier-to-noise level of the system. A LNA must be located at the dish, as it would be of little use inside the receiver, as it would also amplify cable noise. Commercial LNAs provide gains of 45 to 55 dB over the complete satellite band of frequencies. The LNA gain is not an important specification. As long as the LNA gain exceeds about 45 dB, adding more gain at this stage will not improve overall system performance.

TVRO LNAs are manufactured using GaAs FETs which are very low noise transistors. The noise level of an LNA is expressed in kelvin (K), with a lower rating corresponding to a lower noise level. LNAs have noise temperatures from 120 K down to 45 K. A smaller dish should have a lower noise temperature LNA to obtain a good picture. Many older TVRO systems with 120-K LNAs are being replaced with 45-K LNAs to improve the picture quality of a system. For military and industrial applications, LNAs are constructed using a cryogenically cooled parametric amplifier which has a noise level approaching 0 K.

An ideal feedhorn will present a 50-Ω impedance to the LNA input. Since it is hard to match impedances, an isolator is used at the LNA input. The LNA output is an N-type connector used to connect coaxial cable to the receiver's down-converter. Often the LNA and down-converter are integrated into a single unit known as a low-noise converter (LNC).

18-5.4 Down-Converter/Receiver

Receivers consist of four sections: (1) Down-converter, (2) IF amplifier/filter, (3) Demodulator, and (4) Video/audio processor. The down-converter receives a 500-MHz block of signals from the LNA output as shown in Fig. 18-17. The down-converter will lower the frequency range of 3700 to 4200 MHz to a more usable IF of usually 70 MHz. There are three types of down-conversion methods in use today: single stage (single conversion), dual conversion, and block down-conversion.

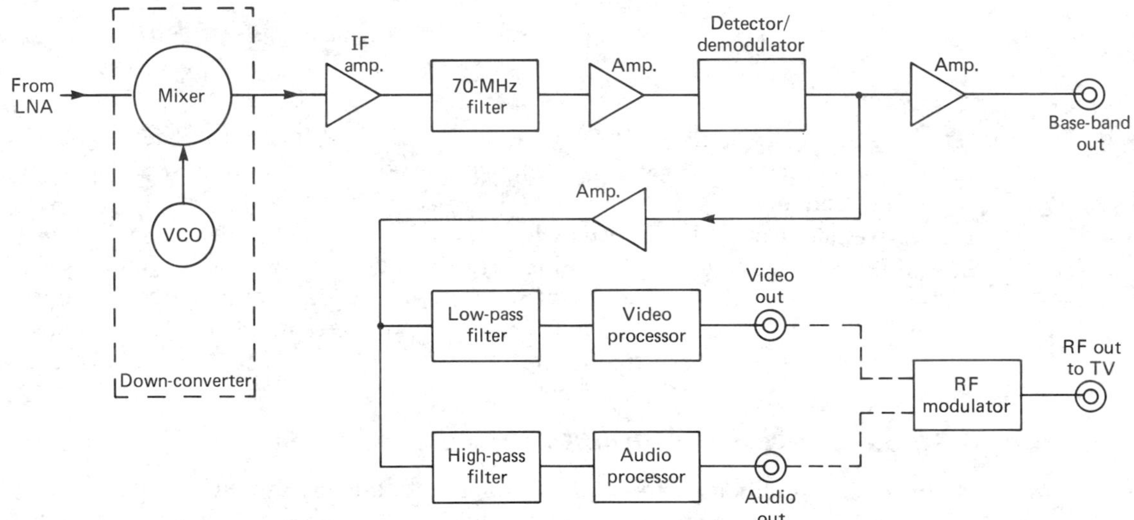

Figure 18-17 Receiver block diagram.

The IF of 70 MHz became an arbitrary standard for TVRO use. It was probably chosen for a few reasons. First, the phone companies used a 70-MHz IF for many of their microwave down-converters and the technology was well developed. Second, early receivers used phase-locked-loop (PLL) integrated circuits capable of operation only up to about 40 MHz. An ECL flip-flop configured to divide frequency by 2 would reduce the 70-MHz IF to 35 MHz, and then be sent to the PLL. Today, PLLs are available with maximum operating frequencies well above 500 MHz; however, most manufacturers still use 70 MHz as the IF.

Figure 18-18 shows a single-conversion down-converter. Single conversion is the simplest and least expensive method of down-conversion. With this method the down-converter is located outdoors separate from the rest of the receiver. Thus the unit must be designed to withstand local weather conditions. Selecting a channel from indoors will send a voltage to the voltage-controlled oscillator (VCO) in the down-converter. The VCO will produce a signal that is 70 MHz above or below the channel center frequency. The channel center frequencies are listed in Fig. 18-10 for North American C-band satellites. This signal is mixed with the C-band block to produce the channel signal centered at 70 MHz. For example, to

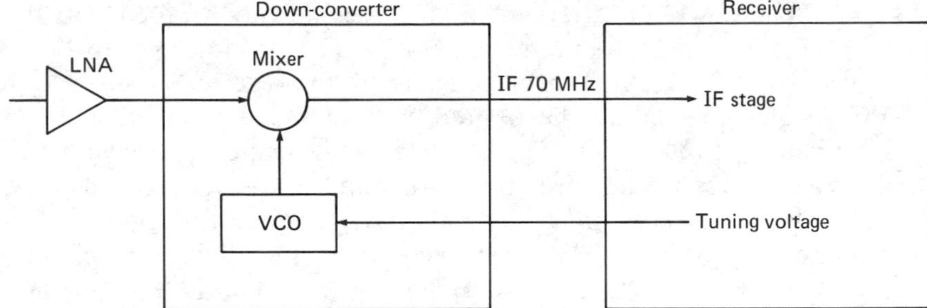

Figure 18-18 Single-conversion down-converter.

select channel 18 having a center frequency of 4060 MHz, the mixing frequency would be set at either 3990 or 4130 MHz.

One problem with single conversion is that image signals 70 MHz above or below a desired channel center frequency are produced and they can leak back into the system. If the image signal is less than 12 dB weaker than the desired signal, the image will become visible in the background of the television picture. Single-conversion systems are equipped with an image-reject mixer which typically reduces the image signal to a level of 15 to 20 dB below the desired signal. Ferrite isolators are also installed on single-conversion system inputs to provide additional protection against the unwanted image signal.

Dual-conversion down-converters do not have the image signal problem. Dual conversion uses two stages of frequency conversion to obtain the final IF as in Fig. 18-19. Part of the frequency conversion takes place outdoors and part takes place indoors. In the first stage, the 3700- to 4200-MHz signal is reduced with a VCO to an intermediate frequency, typically 810 MHz. This signal is carried into the receiver via coaxial cable. Inside, the signal is then reduced with a fixed oscillator to the final IF, usually 70 MHz. The main disadvantage of dual conversion is that the microwave signal at 810 MHz must be carried down a lossy coaxial cable resulting in signal loss. More bulky and expensive heliax cable can be used to reduce the signal loss to ensure that minimum sensitivity requirements of the receiver are met.

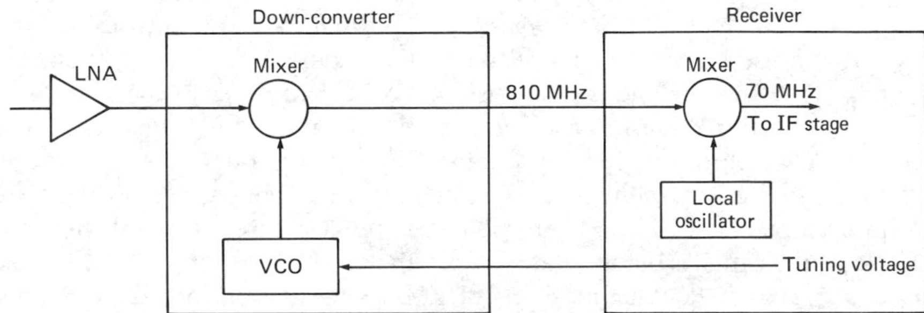

Figure 18-19 Dual-conversion down-converter.

Block down-converters use a local oscillator to down-convert the entire 500-MHz signal band to an intermediate range, usually either 950 to 1450 MHz or 450 to 950 MHz. This 500-MHz block is sent to the receiver, where it is mixed with a VCO signal to produce the desired channel signal at the final IF (70 MHz) as shown in Fig. 18-20. The main advantage of block down-converters is that many users can share the same dish, feedhorn, and LNA system, and also watch different channels at the same time. All users have their own receiver and the complete satellite band of frequencies at their disposal. Sometimes a block down-converter is combined with an LNA into a single unit called an LNB (low-noise block converter). Today, most of the TVRO market uses block down-converters or LNBs.

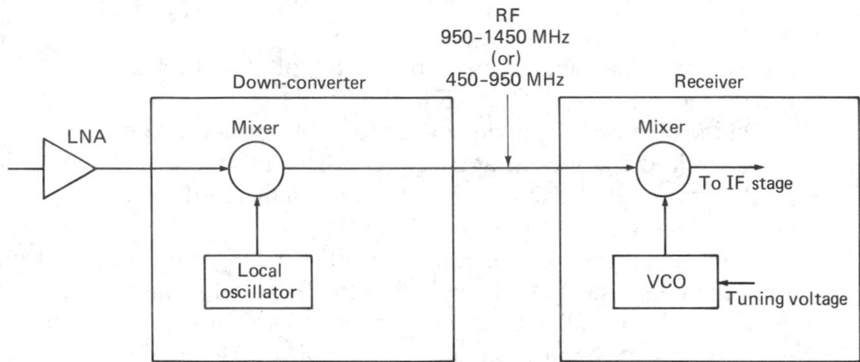

Figure 18-20 Block conversion down-converter.

The receiver's second stage is the IF amplifier and filter. This stage amplifies the signal further and then sets the bandwidth of the receiver. There are several devices that are typically used for 70-MHz wideband amplification. Devices include transistors, integrated circuits, and hybrid modules, with all three providing the same actual end performance. After amplification the signal is bandpass-filtered to remove any out-of-band signals. A transponder transmits 36 MHz of bandwidth. However, a receiver need not have a 36-MHz bandwidth to recover the original signal. A watchable picture can be reproduced using an IF bandwidth of only 15 MHz, and it has been found that the picture quality of a 24-MHz IF bandwidth receiver is virtually indistinguishable from that of a 36-MHz receiver. Most receivers use an IF bandwidth between 22 and 28 MHz. As IF bandwidth drops below 24 MHz, noise power drops, and unwanted "sparkles" common to satellite TV disappear. However, picture details and color shades become worse.

The third stage in a receiver is the demodulator stage. This stage converts the FM satellite TV signal into a baseband signal. First noise is removed from the FM signal using a limiter. A limiter may be as simple as two diodes or an ECL integrated circuit. The output will be a clipped square wave with varying frequency. Next, the signal is demodulated. Here the carrier is separated from the information it is carrying. Popular methods of FM demodulation include PLL, coaxial delay line, IC balanced demodulator, and the quadrature detector. Most satellite receivers use PLL demodulators since they are better equipped to detect weak signals. However, a poorly designed PLL demodulator will produce a fuzzy picture. The output of the demodulator is a baseband signal of video plus audio subcarriers, which is about 10 MHz wide. The baseband output on a receiver is often used as an input to stereo decoders, descramblers, and other special applications.

The fourth stage in a receiver is the video and audio signal processors. The video processing to be done includes amplification, video deemphasis, and clamping. First, the video signal must be amplified to 1-V peak-to-peak levels of standard video. Second, the high-frequency boost provided to signals for transmission must be de-emphasized. Third, the 30-Hz triangular dithering signal must be removed with a video clamp. A 30-Hz signal was added on the uplink to comply with internationally agreed maximum power flux density levels. These power flux density levels were set so that satellite signals transmitted down to earth would not interfere with terrestrial microwave links. If this 30-Hz signal is not removed, picture jittering or flickering will result. The video processor will produce a 0- to 4.2-MHz standard signal at the video output.

Audio subcarriers will be located in the signal baseband at frequencies from 5 to 8 MHz. The audio signal that belongs to the video signal is usually located at 6.8 MHz. The input signal to the audio demodulation circuitry is the baseband output. The 10-MHz bandwidth is high-pass filtered above 4.5 MHz to remove the audio subcarriers. A PLL is used to demodulate the audio signal from its carrier. Once the audio and video are properly processed, they can be viewed on a monitor. If a television is to be used rather than a monitor, an RF modulator must be used to combine the video and audio signals into a standard terrestrial television format. Most modulators are designed to output the television signal to either channel 2, 3, or 4.

18-6 SATELLITE DISH CALCULATIONS

18-6.1 Azimuth-Elevation Mount

Dishes are mounted using either azimuth-elevation or polar mounts. Polar mounts make it easy to locate another geostationary satellite; however, they cannot be used to track nongeostationary satellites. Azimuth-elevation mounts allow for tracking of all satellites, but are difficult to use since two adjustments (elevation and azimuth) must be made to locate any satellite.

Dish elevation is a measure of the angle between the earth's surface and an imaginary line pointing from the dish center to the satellite known as a *boresight.* This angle is shown in Fig. 18-21 and is measured in the north–south direction. If a dish is located on the equator and it is pointed toward a geostationary satellite, the angle of elevation is always 90°. Thus a dish located on the equator requires no elevation adjustment to track the geostationary satellites. However, most dish site locations are not on the equator and do require an elevation adjustment as shown in Fig. 18-22.

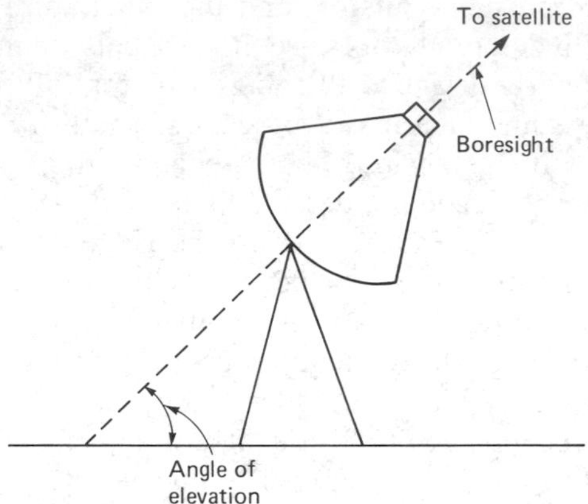

Figure 18-21 Dish elevation.

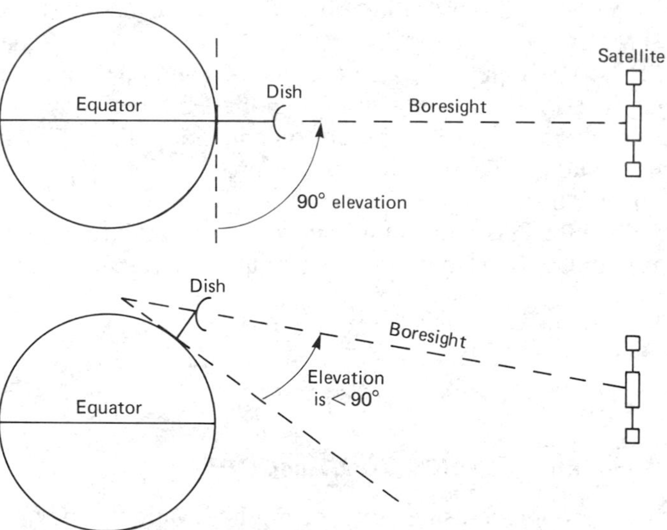

Figure 18-22 Elevation angles.

An azimuth-elevation mount also requires an azimuth adjustment. This angle is measured in the east-west direction as shown in Fig. 18-23. If a satellite has the same azimuth location as the dish location the azimuth is set to 180° (i.e., pointing south). If the satellite is east of the dish site, the azimuth angle is less than 180°. If the satellite is west of the dish site the azimuth angle is greater than 180°. Using trigonometry it is possible to derive formulas to determine the azimuth and elevation setting for given earth station coordinates. The formulas are:

$$\text{azimuth} = 180° + \arcsin\frac{\sin(Z - Y)}{\sin\theta} \qquad \text{where } \theta = \arccos[(\cos X)(\cos(Z - Y))] \tag{18-5}$$

$$\text{elevation} = \arctan\left(\frac{[\cos(Z - Y)](\cos X) - 0.15126}{\sqrt{[\sin^2(Z - Y)] + [\cos^2(Z - Y)](\sin^2 X)}}\right) \tag{18-6}$$

where X is the earth-site latitude (degrees), Y the earth-site longitude (degrees), and Z the satellite longitude (degrees)(may be found from Fig. 18-11).

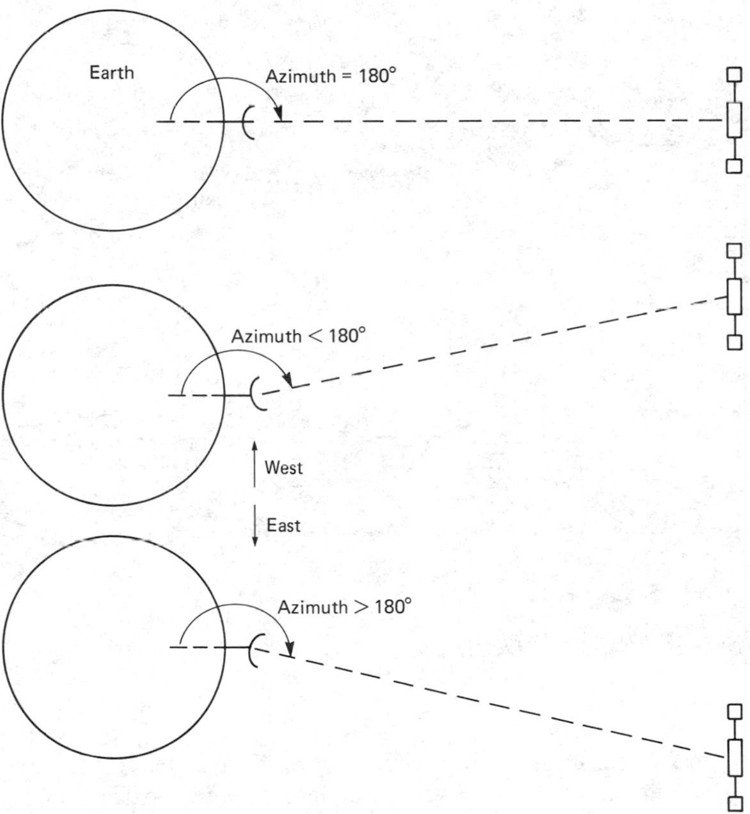

Figure 18-23 Azimuth angles.

EXAMPLE 18-2

Calculate the azimuth and elevation angles for a dish located at 45°N and 105°W looking at the Galaxy 1 satellite.

Solution

$$X = 45.00°\text{N}$$

$$Y = 105.00°\text{W}$$

$$Z = 134.00°\text{W (from Fig. 18-11)}$$

$$\begin{aligned}\theta &= \arccos[(\cos X)(\cos(Z - Y))]\\ &= \arccos[(\cos 45°)(\cos(134 - 105))]\\ &= 51.797°\end{aligned}$$

$$\begin{aligned}\text{azimuth} &= 180° + \arcsin\frac{\sin(Z - Y)}{\sin\theta}\\ &= 180° = \arcsin\frac{\sin(134 - 105)}{\sin(51.797)}\\ &= 218.09°\end{aligned}$$

$$\begin{aligned}\text{elevation} &= \arctan\left(\frac{[\cos(Z - Y)](\cos X) - 0.15126}{\sqrt{[\sin^2(Z - Y)] + [\cos^2(Z - Y)](\sin^2 X)}}\right)\\ &= \arctan\left(\frac{[\cos(134 - 105)](\cos 45) - 0.15126}{\sqrt{(\sin^2(134 - 105)] + [\cos^2(134 - 105)](\sin^2 45)}}\right)\\ &= 30.73°\end{aligned}$$

18-6.2 Polar Mount

A polar mount is the most popular type of mount used for tracking satellites in geostationary orbit. Once a polar mount has been properly set, the angle of elevation will be adjusted automatically for all satellites in geostationary orbit. Thus only a change in azimuth setting is necessary to locate a different geostationary satellite. To track all satellites in geostationary orbit, a polar mount must be set up properly by setting an elevation angle and one or two declination angles, as shown in Fig. 18-24a. The elevation angle ϕ is set equal to the earth-site latitude. A polar mount can have one or two declination angles. A true polar mount will have two declination angles and thus track geostationary satellites more accurately than a polar mount with only one declination angle. The correct settings for declination angles depend on the mount design and are usually supplied by the manufacturer. Figure 18-24b shows typical declination angle settings for a polar mount. The declination angles range from 0° at the equator to about 8.5° for dish sites at the north or south poles.

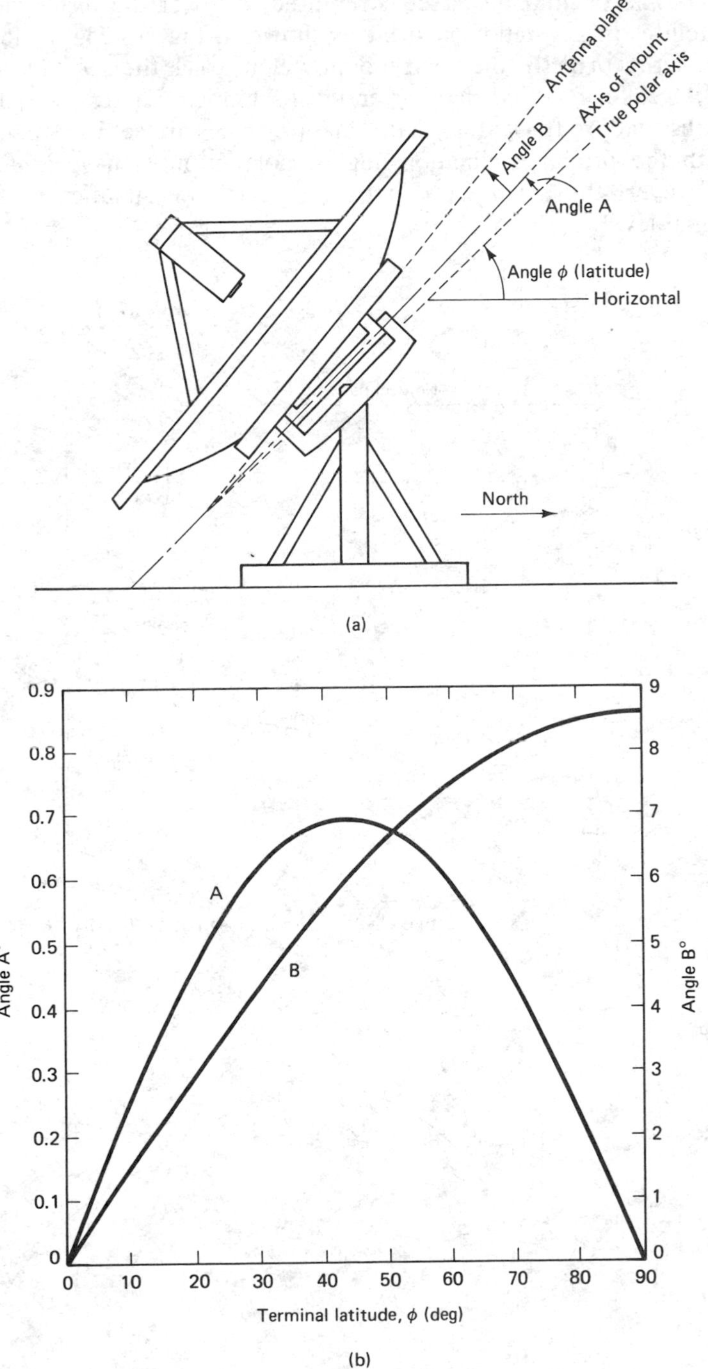

Figure 18-24 (a) Polar mount; (b) declination angles.

The declination offset is required to lower the dish's view onto the arc of satellites in geostationary orbit as shown in Fig. 18-25. If the declination angles are set incorrectly, the polar mount will mistrack the geostationary orbit, resulting in a weaker received signal from most satellites. Figure 18-26 illustrates the actual paths that are tracked if the declination angles are set too small or too large. Even with the proper declination angles, polar mounts have a small tracking error. However, this small error in tracking does not significantly affect the received signal level.

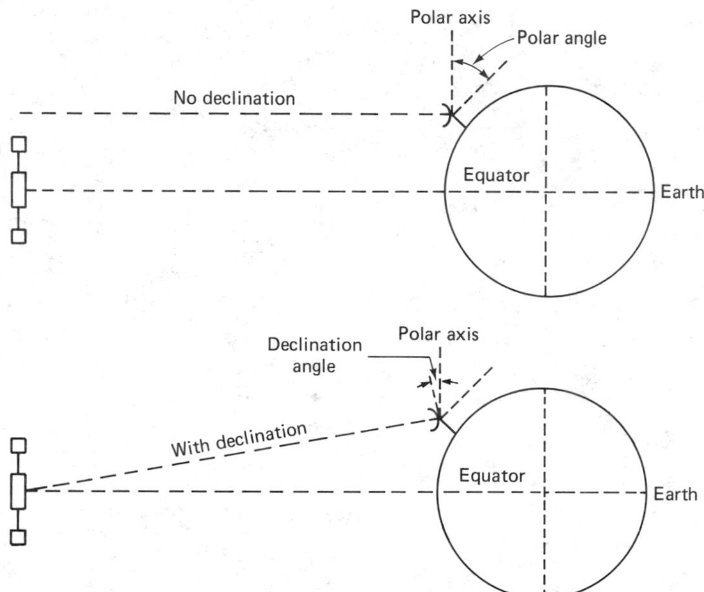

Figure 18-25 Declination angle(s) offset.

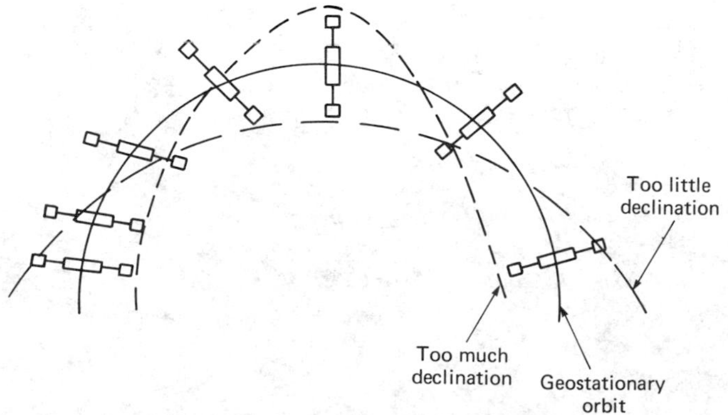

Figure 18-26 Incorrect declination angle setting results.

EXAMPLE 18-3

Determine the elevation angle ϕ and the two declination angles for a polar mount located at 35°N and 90°W.

Solution The elevation angle = site elevation = 35°. Using Fig. 18-24b and 35° elevation, we obtain

$$\text{angle } A = 0.67° \qquad \text{angle } B = 5°$$

18-6.3 Dish Radiation Patterns and Beamwidth

Every satellite dish has an antenna radiation pattern. A parabolic dish has a strong main lobe along its boresight, together with weaker sidelobes, as shown in Fig. 18-27. The radiation pattern is strongest along the boresight and becomes progressively weaker when moving away from the boresight. The sidelobes also have peak radiation levels. These peak levels of the sidelobes should be at least 20 dB below the main-lobe boresight level if the dish is to avoid interference from other satellites, which may be located on a line from the dish center through a sidelobe peak. Sidelobes also extend behind the dish as shown in Fig. 18-27. This causes a "spillover" effect, making dishes detect signals and noise from behind the dish. A well-made dish will have minimal sidelobe gain, thus avoiding detection of unwanted noise.

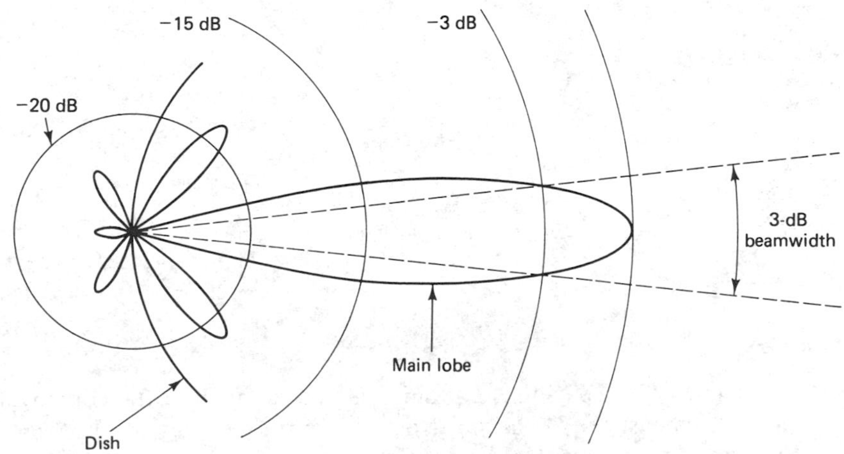

Figure 18-27 Parabolic dish radiation pattern.

The main parameter of interest for a dish's radiation parameter is its beamwidth. The beamwidth of a dish is simply a measure of the angle in degrees from the center of the dish outward to a point where the beam strength is 3 dB less than along its boresight. The beamwidth is illustrated in Fig. 18-27. The beamwidth for a parabolic dish can also be calculated using the following approximate formula:

$$\theta \simeq \frac{70\lambda}{D} \tag{18-7}$$

where θ = 3-dB beamwidth (degrees), λ is the wavelength (meters), and D is the diameter of the dish (meters). From formula (18-7) it can be seen that larger-diameter dishes have smaller beamwidth angles. This is important, as the Federal Communications Commission has approved 2° spacing between satellites in geostationary orbit to make room for more satellites. Smaller dishes, with a wider 3-dB beamwidth, may have problems rejecting signals from adjacent satellites resulting in interference. This is illustrated in Fig. 18-28.

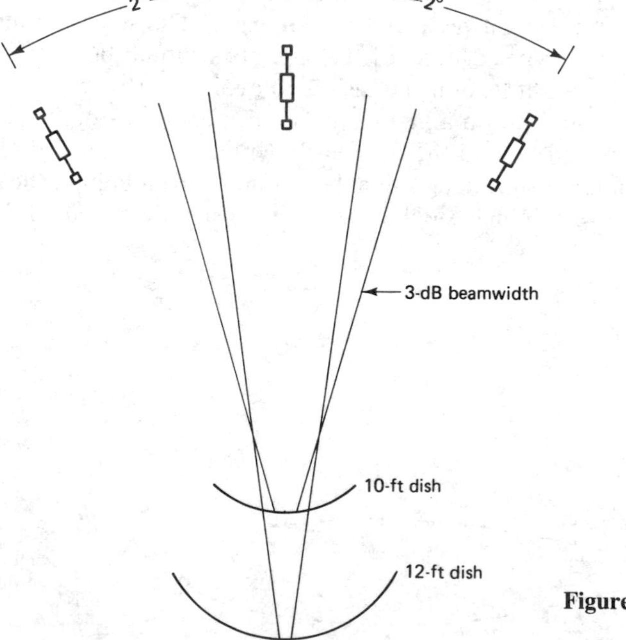

Figure 18-28 Dish beamwidth versus dish diameter.

The FCC 2° spacing is not quite as bad as much of the controversy suggests. The FCC measures 2° spacing from the earth's center as shown in Fig. 18-29. However, dishes are not located at the earth's center, but somewhere on its surface. Thus 2° spacing actually means slightly greater than 2° spacing when viewed by a dish on the earth's surface.

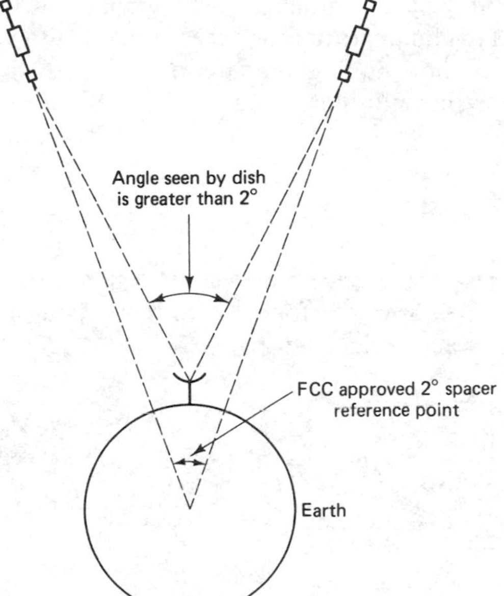

Figure 18-29 How is 2° spacing measured?

18-6.4 Parabolic Dish Gain

The gain of a parabolic dish is given by:

$$g = \frac{K4\pi A}{\lambda^2} \tag{18-8}$$

where g is the dish gain, A the dish area (m^2), λ the wavelength (m), and K the efficiency of the dish (decimal).

From equation (18-8) we see that dish gain is dependent on three variables: dish size, signal frequency, and efficiency. First, as the dish area or dish diameter increases, more signal can be intercepted and reflected to the focal point, resulting in more gain.

Second, gain increases as wavelength decreases or frequency increases. Signals traveling at higher frequencies travel in a more concentrated area than signals at lower frequencies which spread out the signal energy over a larger area. Thus more signal energy per given area is received on the dish surface at higher frequencies.

Third, gain increases as the efficiency increases. Dish efficiency is basically dependent on three factors: the parabolic surface accuracy, the dish alignment accuracy, and the match of the feedhorn to the dish. The parabolic surface must be smooth and have its shape described by a parabola to focus the signals onto its focal point. The dish boresight should also point directly at the satellite. A slight misalignment will result in lost signal gain. Finally, the feedhorn must match the dish f/D ratio and be placed at the dish's focal point. A 100% efficient dish would require all the signals striking the dish to be reflected into the feedhorn. A practical dish will have an efficiency of 50 to 75%. The main reason for the low efficiency levels is because feedhorns are set to illuminate only about 75% of the dish's surface, as shown in Fig. 18-15(a). If a feedhorn fully illuminated the dish, it would also collect unwanted ground noise, as shown in Fig. 18-15(b). Thus an efficiency loss is traded off for less ground noise.

EXAMPLE 18-4

Calculate the gain and 3-dB beamwidth angle for a 3-m dish operating at 4060 MHz. Assume that the dish has an efficiency of 65%.

Solution

$$\lambda = \frac{v}{f} = \frac{3 \times 10^8 \text{ m/s}}{4060 \times 10^6 \text{ Hz}} = 0.07389 \text{ m}$$

$$g = \frac{K \cdot 4\pi A}{\lambda^2} = \frac{K \cdot 4\pi(\pi r^2)}{\lambda^2}$$

$$= \frac{(0.65)(4\pi)(\pi)(1.5 \text{ m})^2}{(0.07389 \text{ m})^2} = 10{,}575$$

$$G(\text{dB}) = 10 \log g = 10 \log(10{,}575) = 40.2 \text{ dB}$$

$$\theta = \frac{70\lambda}{D} = \frac{70(0.07389)}{3 \text{ m}} = 1.72°$$

18-6.5 Dish Noise

A feedhorn will pick up ground noise in addition to the satellite signal, as shown in Fig. 18-15. The amount of noise that enters the feedhorn depends not only on the illumination angle of the dish, but also on the elevation of the dish. If a dish is pointing vertically upward, the feedhorn will pick up the least amount of ground noise. However, if the dish is elevated so that it points into the horizon, the feedhorn will pick up the most amount of ground noise. Figure 18-30 graphically displays the amount of noise that reaches a feedhorn for a given dish size and elevation. Dish noise also depends on the dish's radiation pattern, which is dependent on the surface quality of the dish. Larger sidelobes will cause more ground noise to be detected.

A dish is also subject to sun transit outage when the sun comes within the beamwidth of the dish. A sun transit outage occurs during equinoxes and causes the sun to appear as an extremely noisy source that completely blanks out the satellite signal. A sun transit outage lasts about 10 minutes at most.

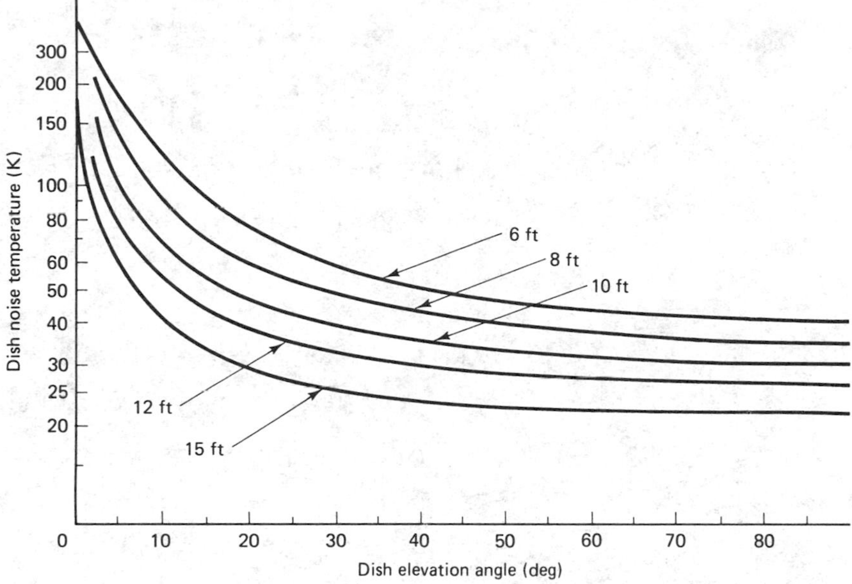

Figure 18-30 Noise temperature of a dish. (From Steve J. Birkill, On Experimental Earth Terminals, *CATJ*, April 1980. © Communications Technology Publications Corp.)

18-6.6 G/T Figure of Merit

A G/T figure of merit is a specification used to determine whether a dish and LNA system will be adequate enough to receive good television signals. It is described by

$$G/T(\text{dB}) = G(\text{dB}) - 10\log(N_D + N_{\text{LNA}}) \qquad (18\text{-}9)$$

where G/T(dB) is a figure of merit (dB), G the dish gain (dB), N_D the dish noise (kelvin) (from Fig. 18-30), and N_{LNA} the LNA noise (kelvin). The larger the value of G/T, the better. Typical G/T values range from 16 to 23 for TVRO residential systems.

EXAMPLE 18-5

Determine the G/T figure of merit for a 12-ft dish operating at 4000 MHz with a 60-K LNA. Assume that the dish elevation is 40° and the dish efficiency is 55%.

(continued)

Solution

$$\lambda = \frac{c}{f} = \frac{3 \times 10^8 \text{ m/s}}{4000 \times 10^6 \text{ Hz}} = 0.075 \text{ m}$$

$$12 \text{ ft} = 12 \text{ ft } (0.3048 \text{ m/ft}) = 3.658 \text{ m}$$

$$g = \frac{K(4\pi)(\pi r^2)}{\lambda^2} = \frac{(0.55)(4\pi)(\pi)(3.658/2)^2}{(0.075)^2}$$

$$= 12{,}913$$

$$G(\text{dB}) = 10 \log g = 10 \log(12{,}913) = 41.1 \text{ dB}$$

$N_D = 32\text{K}$ from Fig. 18-30 at 40° elevation and using a 12-ft dish.

$$N_{\text{LNA}} = 60 \text{ K}$$

Finally,

$$G/T = G(\text{dB}) - 10 \log(N_D + N_{\text{LNA}})$$

$$= 41.1 - 10 \log(32 + 60)$$

$$= 41.1 - 19.6$$

$$= 21.5$$

18-7 SIGNAL TRANSMISSION LOSSES

18-7.1 Effective Isotropic Radiated Power

A signal transmitted from a satellite could theoretically reach one-half of the earth's surface. However, the antenna radiation patterns of satellites concentrate the signal power to cover a selected area on the earth's surface. This distribution of satellite signal energy is shown on a map of the earth called a "footprint" map. Figure 18-31 is a typical footprint map for an Anik satellite transponder. Each transponder of a satellite will have a unique footprint depending on the desired area of coverage. A footprint map will have one or more points where the signal strength is the strongest, known as the *footprint boresight.* Signal strength decreases as the signal moves away from these boresights as is shown by the contour lines of signal strength. The signal strength levels are termed *effective isotropic radiated power* (EIRP). EIRP can be calculated for a given point as

$$\text{EIRP(dBW)} = P(\text{dBW}) + G(\text{dB}) \tag{18-10}$$

where P is the satellite transponder output power and G is the gain of the satellite antenna. If the satellite uses a parabolic antenna to transmit the signal, formula (18-8) can be used to calculate G.

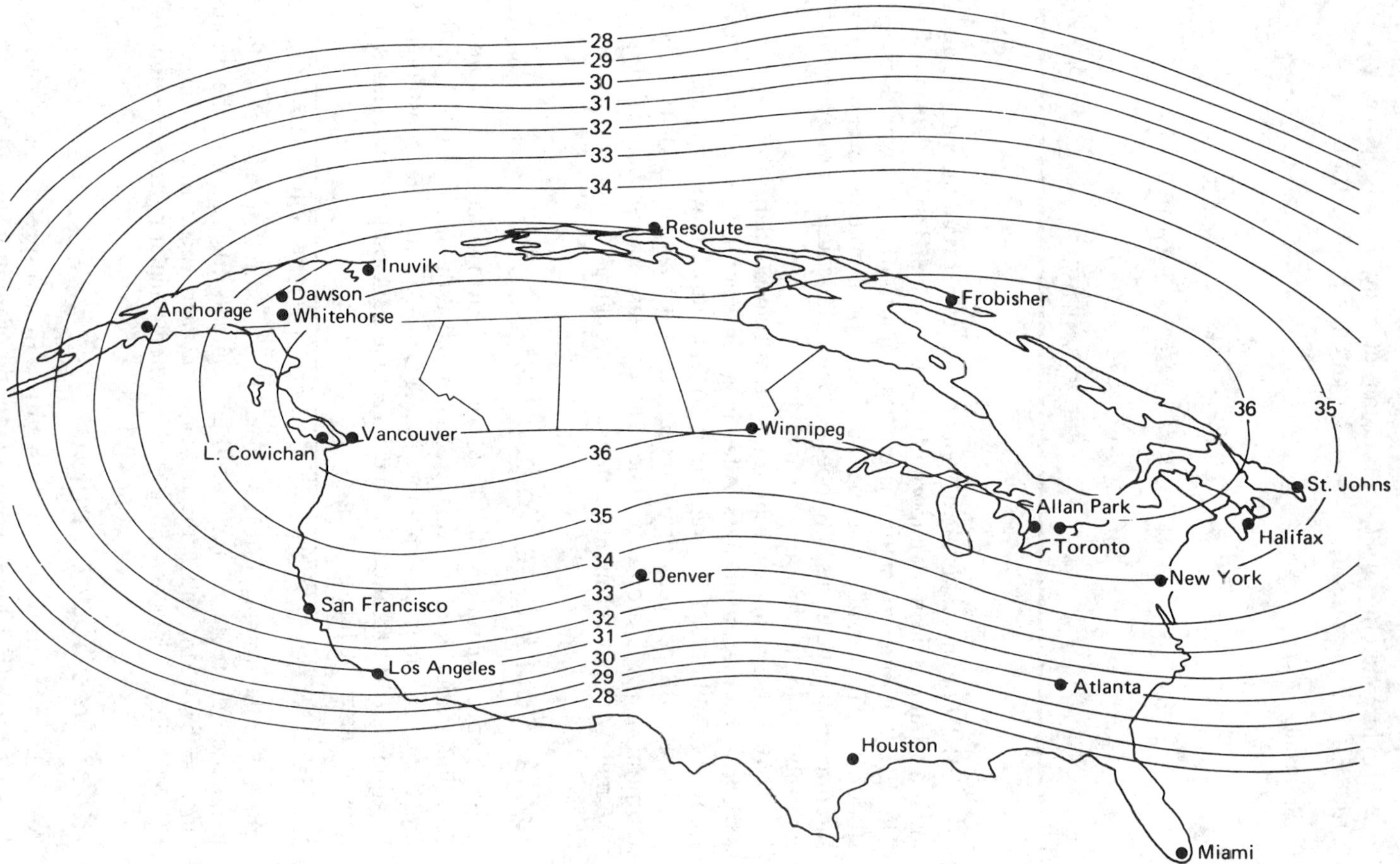

Figure 18-31 Typical Anik EIRP contours (in dBW). (Courtesy of Telesat Canada.)

EXAMPLE 18-6

A satellite operating at 4 GHz transmits at an output power of 6 W. Calculate the EIRP if the satellite transponder antenna gain is 30 dB.

Solution

$$P(\text{dBW}) = 10 \log \left(\frac{6\text{ W}}{1\text{ W}}\right) = 7.78\text{ dBW}$$

$$\begin{aligned}\text{EIRP(dBW)} &= P(\text{dBW}) + G(\text{dB}) \\ &= 7.78 + 30 \\ &= 37.78\text{ dBW}\end{aligned}$$

EIRP does not take into account the losses between the satellite and receiving antenna, but it is the most important indicator of signal strength available on Earth. The EIRP at an earth station is the maximum power that would be seen by the station if the transponder was operating at full or saturated power output. However, satellite transponders usually operate below full output power levels to reduce intermodulation distortion and crosstalk.

Note when using a footprint map that transponder output power levels tend to decrease as the satellite ages. A few magazines publish annual power-level reports for many satellites. On average, transponder output power levels drop by approximately 0.5 dB per year of operation. Thus the actual EIRP levels shown on a footprint map will drop over time. The decrease in transponder output power is due mainly to power amplifier and solar cell aging.

18-7.2 Slant Range Calculation

The distance from a satellite to a point on earth is called the *slant* range. If a downlink station is located on the equator and at the same longitude as the satellite under investigation, the slant range is 22,279 miles. Any other geostationary satellite location or earth location will have a larger slant range. The slant range for any geostationary satellite is given by

$$S = \sqrt{(r + h)^2 + r^2 - 2(r + h)(r)(\cos X)[\cos(Z - Y)]} \qquad (18\text{-}11)$$

where S is the distance (km), X the earth-site latitude (degrees), Y the earth-site longitude (degrees), Z the satellite longitude (degrees), r the radius of the earth = 6370 km, and h the radius of the geosynchronous orbit = 35,840 km (22,279 miles). The slant range is used when calculating transmission losses.

18-7.3 Free-Space Path Loss

A satellite signal traveling to earth suffers from a spreading or free-space loss. A *free-space loss* (FSL) is defined as the ratio of the transmitted power to the power obtainable at the receiving downlink, assuming ideal isotropic antennas. This loss is given by

$$\text{FSP} = \left(\frac{4\pi S}{\lambda}\right)^2 \tag{18-12}$$

where S is the slant range (m) and λ is the wavelength (m). The free-space loss is due to the spreading of energy in the signal wavefront as it travels through space and is proportional to the square of the distance. Only a small portion of the energy radiated from the satellite reaches the receiving antenna since most of the energy is spread over the area outside the area of the receiving antenna.

The free-space loss in decibels is given by

$$\begin{aligned} \text{FSL(dB)} &= 10 \log\left(\frac{4\pi S}{\lambda}\right)^2 \\ &= 20 \log\left(\frac{4\pi S}{\lambda}\right) \end{aligned} \tag{18-13}$$

If S is in miles and frequency is in megahertz, then equation (18-13) reduces to

$$\text{FSL(dB)} = 36.6 + 20 \log S + 20 \log F \tag{18-14}$$

If S is in kilometers and frequency is in megahertz, equation (18-13) reduces to

$$\text{FSL(dB)} = 32.4 + 20 \log S + 20 \log F \tag{18-15}$$

EXAMPLE 18-7

Calculate the slant range and free-space loss from Galaxy 3 to a location on earth having coordinates of 38°N latitude and 90°W longitude. Assume that transponder 21 (4120 MHz) is being received.

Solution From Fig. 18-11, Galaxy 3 location is 93.5°W.

$$S = \sqrt{(r+h)^2 + r^2 - 2(r+h)(r)(\cos X)\cdot\cos(Z-Y)]}$$

$$= \sqrt{(6370 + 35{,}840)(6370)^2 - 2(6370 + 35{,}840)(6370)(\cos 38°)[\cos(93.5 - 90)]}$$

$$= 37{,}407 \text{ km}$$

$$\begin{aligned} \text{FSL} &= 32.4 + 20 \log S + 20 \log F \\ &= 32.4 + 20 \log(37{,}407) + 20 \log(4020) \\ &= 32.4 + 91.46 + 72.09 \simeq 196 \text{ dB} \end{aligned}$$

This is a very large loss. If the EIRP is 37.8 dBW as in Example 18-6 and the downlink antenna gain is 41 dB, the received power level is 37.8 dBW − 196 dB + 41 dB = −117.2 dBW. This is only 1.9 picowatts!

18-7.4 Total Path Loss and Other Losses

The total path loss for a signal traveling to earth is made up of two components: free-space loss (spreading loss) as discussed previously, and various atmospheric losses. Free-space loss includes only losses due to a signal traveling through space. However, a signal also travels through various levels of the earth's atmosphere. Figure 18-32 shows that atmospheric losses are greater at higher earth-site latitudes. This is due to the signal having to travel a longer distance through earth's atmosphere at a higher elevation, as shown in Fig. 18-33. Atmospheric attenuation is typically quite small and often neglected for C-band satellites, except at very high latitudes. However, Ku-band satellites operating at a much higher frequency are

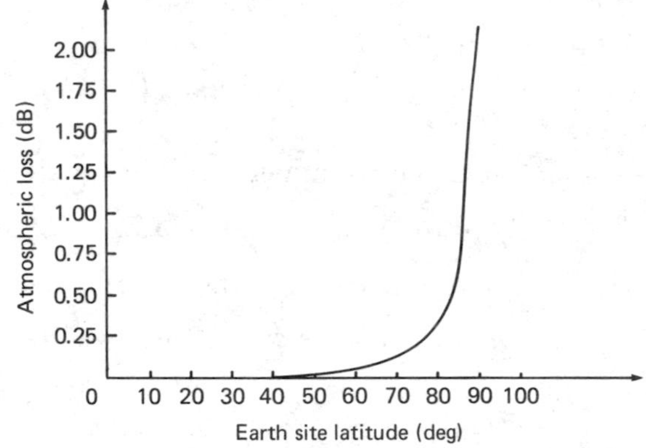

Figure 18-32 Typical atmospheric loss of 4 GHz.

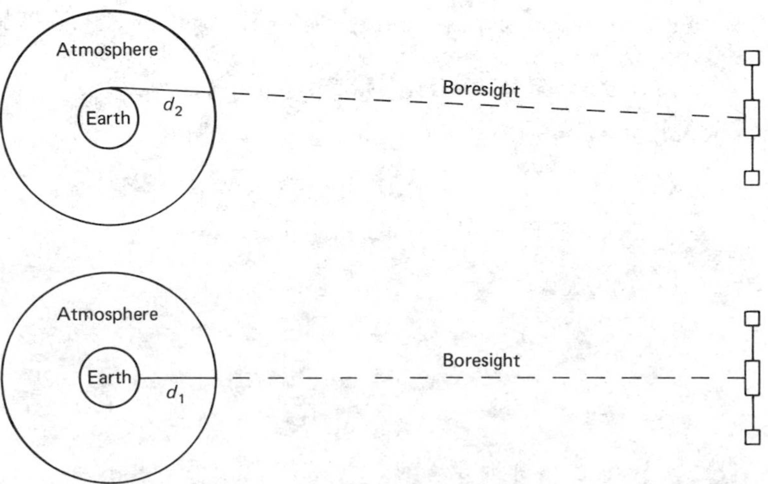

Figure 18-33 Atmosphere path distance at different earth latitudes ($d_2 > d_1$ at higher earth latitude).

susceptible to a large atmospheric attenuation under medium-to-heavy rain conditions. This can be a problem should the downlink be located in an area having large rainfall figures.

18-7.5 Carrier-to-Noise Levels

The carrier-to-noise (C/N) level at the receiver's input indicates the quality of the final television signal. For TVRO applications typical values of C/N range from 9 to 12 dB for a "good"-quality picture. Table 18-1 describes how video quality varies with various C/N levels. The C/N level at a receiver input can be calculated using

$$C/N(\text{dB}) = C(\text{dBW}) - N(\text{dBW}) \tag{18-16}$$

$$C(\text{dBW}) = \text{EIRP} + G - \text{TPL} \tag{18-17}$$

$$N(\text{dBW}) = 10 \log K + 10 \log(T_D + T_{\text{LNA}}) + 10 \log B \tag{18-18}$$

where EIRP is the effective isotropic radiated power (dBW), G the dish gain (dB), TPL the total path loss (dB), K is Boltzmann's constant $= 1.38 \times 10^{-23}$ J/K, T_D is the dish temperature (K), T_{LNA} the LNA temperature (K), B the receiver IF bandwidth (Hz), C the carrier level (dBW), N the noise level (dBW), and C/N the carrier-to-noise ratio at the receiver IF input (dB).

TABLE 18-1 *C/N* versus Television Picture Quality

C/N (dB)	Picture Description
4	Heavy noise and sparkles, little or no color, very poor quality video
5–6	Marginal quality with noise, tearing, and audio noise
7–8	Medium quality with some noise and sparkles, good color, good audio
9–10	Good quality with very good picture, sparkles only on saturated colors
11–12	Excellent quality with no noise, videotape quality
13	Cable TV head-end performance

EXAMPLE 18-8

Find the C/N level for a TVRO system watching Galaxy 3 on transponder 21 (4120 MHz). Assume that the downlink is located at 38°N latitude and 90°W longitude. The TVRO system consists of a 3-m dish with 70% efficiency, an LNA with an 80-K noise level, and a 26-MHz receiver bandwidth (assume that the EIRP level is 32 dBW).

(continued)

Solution

$$\lambda = \frac{c}{f} = \frac{3 \times 10^8 \text{ m/s}}{4120 \times 10^6 \text{ Hz}} = 0.07282 \text{ m}$$

Dish gain:

$$g = \frac{K \cdot 4\pi A}{\lambda^2} = \frac{K \cdot 4\pi(\pi r^2)}{\lambda^2}$$

$$= \frac{(0.70)(4\pi)(\pi)(1.5 \text{ m})^2}{(0.07282 \text{ m})^2} = 11{,}727$$

$$G(\text{dB}) = 10 \log g = 10 \log(11{,}727) = 40.69 \text{ dB}$$

Total path loss: TPL $\simeq$ FSL at 38°N latitude. From Example 18-7, FSL = 196.0 dB.
Carrier level:

$$C = \text{EIRP} + G - \text{TPL}$$

$$= 32 \text{ dBW} + 40.69 \text{ dB} - 196 \text{ dB}$$

$$= -123.31 \text{ dBW}$$

Dish noise: From Fig. 18-30 at 38° latitude with $T_D \simeq 42$ K.
Total noise:

$$N(\text{dBW}) = 10 \log K + 10 \log(T_D + T_{\text{LNA}}) + 10 \log B$$

$$= 10 \log(1.38 \times 10^{-23}) + 10 \log(42 + 80) + 10 \log(926 \text{ MHz})$$

$$= -228.6 + 20.86 + 74.15$$

$$= -133.59 \text{ dBW}$$

C/N:

$$C/N = C - N$$

$$= -123.31 - (-133.59)$$

$$= 10.3 \text{ dB}$$

18-8 TVRO RECEIVER CALCULATIONS

The satellite baseband signal, one of the receiver's outputs, spans from 0 to 10 MHz. The video signal only occupies part of the total baseband from 0 to 4.2 MHz. Satellite video is frequency modulated, unlike terrestrial television, which uses vestigial sideband amplitude modulation. The remaining baseband from 4.2 to 10 MHz is used for television audio subcarriers, broadcast radio subcarriers, data transmission, and a host of other services. Audio subcarriers on satellite transmission are frequency modulated and usually occupy between 5.0 and 7.4 MHz of the baseband signal. The audio carrier for the video signal is usually located at either 5.8, 6.2, 6.8, or 7.4 MHz, with 6.8 MHz being the most commonly used carrier frequency. Figure 18-34 shows a typical baseband for a satellite transponder.

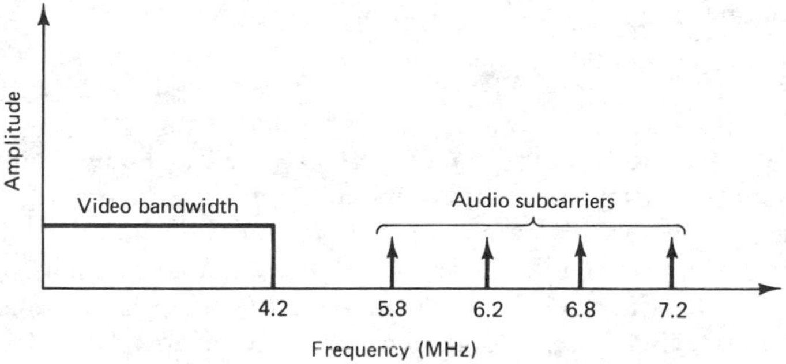

Figure 18-34 Typical transponder baseband.

The satellite baseband often has radio subcarriers and the audio for some television programs in stereo. Three different stereo formats are presently in use:

1. *Discrete stereo.* Two separate subcarriers are used: one for the left channel and one for the right channel.
2. *Matrix stereo.* Two separate subcarriers are used: one for a left plus right (L + R) signal and one for a left minus right (L − R) signal. A receiver with a stereo processor will algebraically add the two signals to produce an R signal and subtract the two signals to produce an L signal. A mono receiver will use the (L + R) signal alone.
3. *Multiplex stereo.* Both channels are transmitted using a single pilot subcarrier similar to broadcast FM radio. It uses an FM subcarrier for the (L + R) signal, a double-sideband suppressed carrier for the (L − R) signal, and a 10-kHz synchronizing pilot subcarrier signal as a reference to aid in recovering the original L and R signals.

The audio that is broadcast by satellite is transmitted with either wideband or narrowband deviation. Wideband deviation of 200 kHz is used for the audio that belongs to the video. The other subcarriers normally use a narrowband deviation of either 25 kHz or 50 kHz Some TVRO receivers have built-in circuits which determine the type of deviation and stereo formats being used and automatically process the audio. Stereo processors can also be added to the unfiltered baseband output of a nonstereo receiver to obtain stereo sound.

18-8.1 Bandwidth Requirements

In theory, a frequency-modulated signal requires a bandwidth of infinity to be totally recovered. However, satellite systems use a finite bandwidth of 36 MHz plus a 4-MHz guardband to transmit an FM signal. The required bandwidth of an FM system is usually estimated by Carson's bandwidth rule. Carson's rule states that the significant bandwidth occupied by the spectrum of a frequency-modulated signal is equal to twice the sum of the maximum frequency deviation and the modulating frequency and is given by

$$B = 2(\Delta f + f_m) \tag{18-19}$$

where B is Carson's bandwidth (Hz), Δf is the maximum frequency deviation (Hz), and f_m is the modulating frequency (Hz). The maximum frequency deviation Δf for satellite transmission varies depending on the source and type of information it carries. Video signals use a maximum frequency deviation, while other subcarriers may use even less frequency deviation than the audio signals.

If we consider a typical North American television signal that is broadcast using the NTSC system containing frequencies up to 4.2 MHz, and frequency modulation with a maximum deviation of 10.75 MHz, the required Carson's bandwidth would be

$$B = 2(10.75 \times 10^6 + 4.2 \times 10^6) = 29.9 \text{ MHz}$$

Carson's rule provides a good estimate of the significant bandwidth required only when the deviation ratio $n < 2$.

$$n = \frac{\Delta f}{f_m} \tag{18-20}$$

where n is the deviation ratio. For values of $n > 2$, Carson's rule underestimates the bandwidth required.

In the case of satellite video $n = 10.75/4.2 = 2.56$ and 29.9 MHz is an underestimate of the required bandwidth. A better estimate is given by

$$B = 2(\Delta f + 2f_m) \tag{18-21}$$

Now the required bandwidth for satellite video is

$$B = 2[10.75 \times 10^6 + (2 \times 4.2 \times 10^6)] = 38.3 \text{ MHz}$$

Formula (18-21) tends to overestimate the required bandwidth. The two estimates of bandwidth for video are 29.9 and 38.3 MHz. In addition, audio subcarriers may use up to 1 MHz of bandwidth. Thus a standard bandwidth of 36 MHz is used to reasonably recover the original audio and video signals.

18-8.2 FM Receiver Threshold and FM Improvement Factor

A major advantage of frequency modulation is that an improvement can be achieved in the signal-to-noise ratio (S/N) at the output of the receiver over the input carrier-to-noise (C/N). If the C/N at the input is plotted against the output S/N, a linear relationship exists as shown in Fig. 18-35 for normal operation. Under normal operation the output S/N is higher than the input C/N by a factor known as the FM improvement factor.

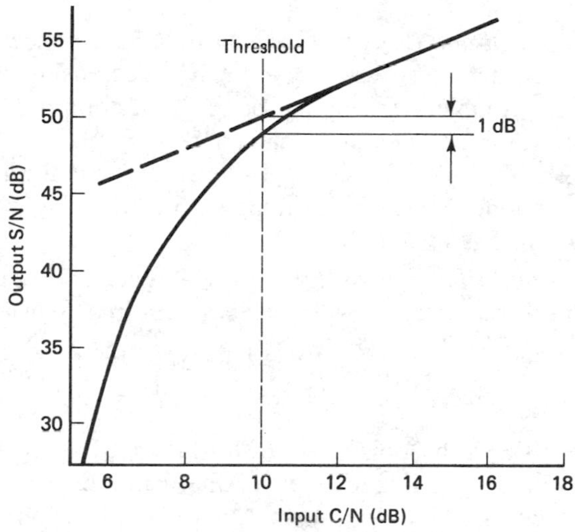

Figure 18-35 FM threshold and FM improvement factor.

The FM threshold is that point where the deviation from a linear (normal operation) is 1 dB. The nominal threshold in satellite receivers is typically at a C/N level of 10 dB. If the C/N level of a signal is below the threshold level, the television video will show noise in the form of "sparkles." Thus the threshold of a video receiver determines how weak an input signal can be before a picture is judged as unacceptable. This judgment is, of course, subjective in nature.

The output S/N figure is important because it measures the quality of the picture on a television set. Quantitative research has been done to describe various reactions to different S/N ratios by "average viewers." The average viewer can perceive some signal noise at or below an output S/N ratio of 47 dB. This can be compared to an average VCR with an S/N of 45 dB or to an average NTSC television broadcast with an S/N of 50 dB. At the threshold point of most receivers we can expect about 47 dB of S/N, which will result in a "good" picture. Thus the basic requirement for a TVRO system is to have enough C/N at the input to reach the receiver threshold point.

Threshold extension circuits are often used which reduce the threshold level by 3 dB. The receiver threshold is lowered by using low-noise components and/or by reducing video bandwidth. Some threshold extension circuits automatically reduce the receiver bandwidth when the input signal is too weak. Figure 11-35 illustrates that an apparent gain of about 40 dB is obtained in output *S*/*N* over input *C*/*N* by using frequency modulation. This 40-dB gain is referred to as an *FM advantage* or an *FM improvement factor*. This FM improvement factor is made up of several components:

1. *Noise Weighting Factor (NWF).* Noise has less affect on video quality at certain frequencies relative to others. This results in a 10.2-dB gain in *S*/*N* over *C*/*N*.
2. *Pre-emphasis and Conversion Factor (PCF).* Noise power density increases as frequency increases, resulting in the *S*/*N* decreasing. To equalize the *S*/*N* level over the complete frequency band, a pre-emphasis network is introduced before transmission and a matching de-emphasis network is placed in the receiver. The overall result is to leave the signal unchanged and reduce the high-frequency noise. This, along with a conversion factor, results in an 8.5-dB gain in *S*/*N* over *C*/*N*.
3. *Noise Bandwidth Factor (NBF).* If a receiver is operating at or above its threshold, the receiver will realize a gain in *S*/*N* over *C*/*N* given by

$$\text{NBF} = 10 \log\left[\frac{3}{2}\left(\frac{\Delta f}{f_m}\right)^2 \frac{B_{\text{IF}}}{f_m}\right] \tag{18-22}$$

where NBF is the noise bandwidth factor (dB), Δf the maximum frequency deviation (10.75 MHz for North American television), f_m the maximum video frequency (4.2 MHz for North American television), and B_{IF} the receiver IF bandwidth (Hz). Thus *S*/*N* and *C*/*N* are related as

$$S/N = C/N + \text{NWF} + \text{PCF} + \text{NBF} \tag{18-23}$$

$$S/N = C/N + 18.7 \text{ dB} + \text{NBF} \tag{18-24}$$

EXAMPLE 18-9

Calculate the output *S*/*N* for a 28-MHz satellite television receiver. Assume the input *C*/*N* = 10 dB, which is above the receiver threshold.

Solution

$$\begin{aligned}\text{NBF(dB)} &= 10 \log\left[\frac{3}{2}\left(\frac{\Delta f}{f_m}\right)^2 \frac{B_{\text{IF}}}{f_m}\right] \\ &= 10 \log\left[\frac{3}{2}\left(\frac{10.75}{4.2}\right)^2 \left(\frac{28}{4.2}\right)\right] \\ &= 18.2 \text{ dB}\end{aligned}$$

(continued)

$$S/N(\text{dB}) = C/N + 18.7 + \text{NBF}$$
$$= 10 \text{ dB} + 18.7 + 18.2$$
$$= 46.9 \text{ dB}$$

From a receiver design viewpoint, the only parameter that we can change to improve the output S/N is the IF bandwidth. At or above the threshold, increasing the IF bandwidth will result in a slight S/N gain. However, below the threshold, equation (18-24) no longer applies, and by decreasing the IF bandwidth we realize a significant improvement in output S/N. As the IF bandwidth drops below 24 MHz, noise in the form of sparkles disappears. However, there is a limit on how far the IF bandwidth can be reduced, as picture details and color shades also worsen as the IF bandwidth drops.

18-8.3 NTSC Television Format

The video or baseband output from a TVRO receiver cannot be fed directly into a television. The signal must be converted into a standard NTSC (National Television System Committee) format with an RF modulator. However, the baseband output containing the video and audio information can be directly input into a monitor, without an RF modulator. An RF modulator is a device that converts the satellite baseband signal into an amplitude-modulated form used by television sets. Modulators are available to output the baseband signals onto any channel. However, channel 2, 3, or 4 is most commonly used.

A channel 3 modulator will produce an NTSC signal format as shown in Fig. 18-36. Each NTSC signal has a bandwidth of 6 MHz. The channel 3 modulator amplitude modulates the video onto a carrier centered at 61.25 MHz. The lower sideband of the modulated signal is filtered off at 60 MHz by a process known as vestigial sideband amplitude modulation. The upper sideband of the modulated signal contains all 4.2 MHz of video information and is positioned from 61.25 MHz

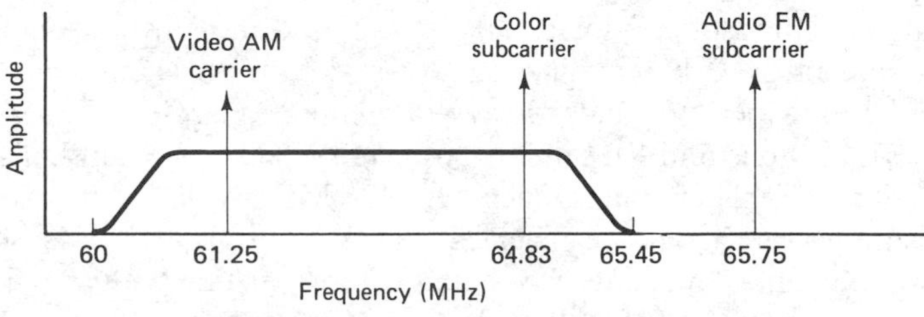

Figure 18-36 NTSC channel 3 spectrum.

up to 61.25 + 4.2 = 65.45 MHz. The upper sideband is a composite of the visual information signals and synchronization signals and is compatible with both monochrome and color formats. The modulator also frequency modulates the audio signal onto an FM subcarrier located at 65.75 MHz. Channel 4 starts at 66 MHz, and thus the space between 65.75 MHz and 66 MHz serves as a guardband. The NTSC system is used by Canada, Japan, and the United States. Other countries use different broadcast formats requiring a different modulator and different television set. France and a few other countries use the SECAM (Sequential Couleur a Memoire) broadcast format. Many other countries use a system called PAL (Phase Alternation Line) format.

18-9 DIRECT BROADCAST SATELLITE

The broadcasting of television signals directly to the home using the Ku band is known as direct broadcast satellite service (DBS). Ironically, DBS is intended for home use, but very little in the way of programming is available at the moment, while C-band signals, which were never intended for home reception, are very popular in this market. The main goal of DBS is to provide the home market with many television programs using a relatively inexpensive TVRO downlink station.

The major cost of a TVRO system is a dish. The dish cost for Ku-band reception is dramatically reduced, for two reasons. First, the Ku band uses a 11.7- to 12.2-GHz downlink, which is approximately three times higher in frequency than the C-band downlink. Thus the wavelength is one-third shorter, requiring a smaller dish.

Second, Ku-band satellites transmit signals down to earth using a larger output power level than C-band satellites. C-band satellites are required by law to limit output power levels to values of 5 to 24 W, while Ku-band satellites may use power levels of up to 200 W. This means that the received signal will be much stronger, requiring less gain and thus a smaller dish can be used.

Ku-band dishes are typically 1 m in diameter. The dishes are usually a solid spun type rather than a mesh type, as a mesh dish will not perform well at the higher Ku band. At the moment smaller dishes perform well at the Ku-band frequencies. However, if more satellites operating at Ku frequencies are placed into orbit, smaller dishes will suffer from 3-dB beamwidth interference.

The Ku-band signals are not subject to terrestrial interference, as there are no terrestrial links using the same frequencies. However, Ku-band signals suffer significant atmospheric losses due to rain and moisture attenuation, over what is experienced by C-band signals. As a result, DBS signals will be more noisy or completely wiped out under heavy rain conditions.

Both the C and Ku satellites transmit a 500-MHz band. The C-band satellites transmit up to 24 channels, each being 36 MHz wide, while the Ku-band satellites transmit up to 32 channels, each 24 MHz wide. In both bands, some overlap occurs between chanels and thus the signals are polarized. Unlike C-band signals, which are vertically and horizontally polarized, Ku-band channels are alternately polarized left-hand-circular (LHC) and right-hand-circular (RHC). There are feedhorns available that can receive both C- and Ku-band signals, called dual-feeds, and there are also receivers that handle both bands.

18-10 SIGNAL SCRAMBLING/DESCRAMBLING

Recently, many of the television signals carried by C-band satellites have become scrambled. Scrambling systems have developed to prevent unauthorized use of the television signals by individuals owning TVRO systems. Television scrambling involves altering the audio and/or video components of a signal so that a television set cannot reproduce the original signal.

There are several methods of scrambling the audio signal. Simpler systems will remodulate the original audio carrier to a higher frequency or use a second signal to interfere with the audio information. More complex systems convert the audio signal into a digital signal. The digital signal is then encrypted and mixed with control and addressing information.

There are also several methods of video signal scrambling. Simpler video scrambling may consist of adding an interfering signal between the video and audio carriers, or reversing the voltage pattern of the video signal known as video inversion, or inserting a notch filter that attenuates a range of frequencies from the signal. Slightly more complex video scrambling ranges from using notch filters, video inversion, or interfering signals that are addressable, to suppressing, removing, or shifting the horizontal synchronization pulses. The more complex video descramblers will vary from one method of scrambling to another based on a control code that is hidden in the video signal.

With all the possibilities of scrambling available, two main scrambling systems have become the most popular for satellite television. The two systems are the Canadian Cancom system and the American Video Cypher II system. Both systems require that you purchase a descrambling circuit which connects to your receiver. Both descrambling circuits are microprocessor-controlled descrambling devices, each containing a unique address. In both systems, the unique address for a unit is stored in a nonvolatile memory integrated circuit.

The American Video Cypher II system uses 13 combinations of video scrambling along with digital encryption of the audio signal. Digital control information is inserted into the horizontal sync pulse in each scan line as three 8-bit words. The control information allows one central computer to manage almost 100 million decoders. The control information allows any decoder to descramble from one to as many as 56 scrambled programs. These can be packaged into any combination for a particular unit address. The owner of a Video Cypher II unit simply calls up a number and pays a fee to have their descrambler unit turned on to certain channels. A computer operator then enters the descrambler serial number and channels to be descrambled into a computer. Control data are then transmitted with the horizontal sync pulse and received by the decoder turning it on for only those services that were paid for. The system can handle 600,000 subscribers per hour. The system also sends periodic turn-off codes to Video Cypher units which are owned by people who have not paid their bills.

Some Canadian satellite signals are scrambled using the Oak Orion scrambling system. A Cancom descrambler unit must be purchased and a fee paid to descramble the programming. The main purpose of the Cancom system is to provide program services to remote parts of the country that are out of reach of cable services and prevent unauthorized use of the programs by others.

18-11 OTHER SATELLITE SERVICES

Satellite downlinks handle many types of signals other than television and radio broadcasts. Some of these services are hidden in the television signals, whereas others use a complete transponder or satellite. The Telestar 300 series of satellites is used to relay long-distance telephone traffic. Both digital and analog transmission techniques are used, although the majority of transmissions still employ analog schemes. Digital schemes involve time-division multiplexing (TDM) of pulse-code-modulated (PCM) telephone signals. Analog schemes use either single channel per carrier (SCPC) or single-sideband (SSB) frequency-division multiplexing (FDM), with the latter being more popular.

SSB FDM involves modulating a telephone channel bandwidth of 4 kHz with other telephone channels to form a wideband channel. This wideband channel may accommodate up to 2000 telephone conversations onto one transponder. Decoding of the SSB FDM signals can be done at a satellite receiver's unfiltered baseband output using a SSB shortwave receiver. A good shortwave receiver capable of continuous coverage from 150 kHz to 10.7555 MHz is required to properly demodulate the telephone conversations. Often the upper sideband carries one side of the conversation, while the lower sideband of a channel carries the other side of the conversation.

The other analog technique SCPC uses a separate subcarrier for each channel to carry telephone traffic. SCPC does provide a higher signal power but is not as frequency efficient as SSB FDM. However, SCPC does allow the bandwidth of a channel to be altered for special transmissions of modem data. Modem data carried by satellites are usually transmitted in three formats: frequency shift keyed (FSK), binary phase-shift keyed (BPSK), and quadrature phase-shift keyed (QPSK). The data being transmitted are usually news, sports, or stock quotations carried by wire services.

The vertical blanking interval of a television signal is also used to carry information. The NTSC format has 525 lines scanned 30 times each second. The field lines start at the top of a screen and work downward. When one screen of lines has been displayed, the electron beam is positioned back at the top of the screen. The vertical blanking interval is the time used to move the electron beam from the bottom to the top of the screen. The vertical blanking interval lasts the equivalent of 21 horizontal blanking intervals, as shown in Fig. 18-37. The first nine horizontal line periods are used for synchronization purposes. The last five horizontal line periods are used for test signals known as vertical interval test signals. The remaining seven horizontal line periods, labeled 10 to 16 in Fig. 18-37, are sometimes used to transmit other digital data. The data services carried include news wires, closed-caption data feeds, stock quotations, financial and sports wires, charts and graphs, and program listings. There are many different protocol standards used to transmit the data. A data communications tester and small microcomputer are required to view these services.

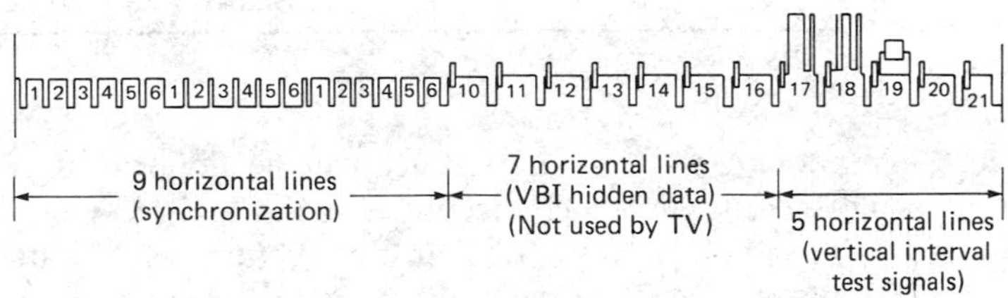

Figure 18-37 Vertical blanking interval.

It is also possible to see the vertical blanking pulses on a monitor by rolling down the television picture using the vertical hold control. The blank line between frames will show some sparkling dots. These sparkling dots indicate the presence of digital information being hidden in the video signal. The dots do not exist on all stations, only those stations carrying information in the vertical blanking interval.

18-12 DOWNLINK TVRO INSTALLATION

18-12.1 Site Survey

The first major task in installing a TVRO system is finding a good location for the dish. This is perhaps one of the most important tasks and unfortunately, the most often neglected. Two steps are involved in finding a good site: (1) ensuring that the dish's view is free from physical obstructions, and (2) testing for terrestrial interference.

First, the space between the dish and the satellites in geostationary orbit should be free from physical obstructions such as trees and buildings. Obstructions totally or partially blocking the view of a dish will seriously affect signal reception. Trees are usually worse than buildings, due to their high water content, which absorbs microwaves.

Second, a dish is best located in an area free of terrestrial interference. Terrestrial interference results when terrestrial signals, sharing the same band of frequencies as the TVRO downlink, cross the path of the dish. Terrestrial interference will have a different effect on a TVRO system, depending on its intensity, as shown in Table 18-2.

TABLE 18-2 Terrestrial Interference versus Picture Quality

Terrestrial Interference Relative to Carrier Level (dB)	Picture Quality
< −18	No visible effect
−18 to −10	Light sparkles
−10 to −5	Medium sparkles
−5 to 0	Heavy sparkles and lines
0 to 10	No picture
> 10	Blackout of picture

A general site test for interference can be done using a portable satellite system or more easily just by connecting an LNA and receiver to a TV set. At the proposed dish location point the LNA in all directions while watching the television set. Interference will cause horizontal bars to appear on the set. A snowy pattern on the screen indicates that no interference is present. All channels should be tested as some types of terrestrial interference have a limited bandwidth. If horizontal bars appear on many channels, a portable satellite system should be used to determine the severity of the interference. If the terrestrial interference cannot be avoided by moving the dish location, it can be reduced. Filters and artificial shields will reduce terrestrial interference. In addition, properly grounding the system and mounting the dish as near to ground level as possible will also help combat terrestrial interference.

18-12.2 Dish Foundation

Once the best possible location is known, the next major task is to build a foundation for the dish. Depending on the terrain, either a concrete slab or a concrete post foundation is used. If the downlink location is a rocky area, it is easier to construct a slab foundation. In soft ground conditions, a post foundation is better, as it requires less cement. If a polar mount is used, a 2- to 3-in. diameter pole must be set in a vertical position as shown in Fig. 18-38. The pole must be vertical so that the dish will properly track the geostationary satellites. One can ensure that the pipe is vertical by placing a level on the top and the side of the pipe, as shown in Fig. 18-38. The pipe should extend below the frostline so that freezing will not cause any movement of the dish assembly. Roof installations often require that

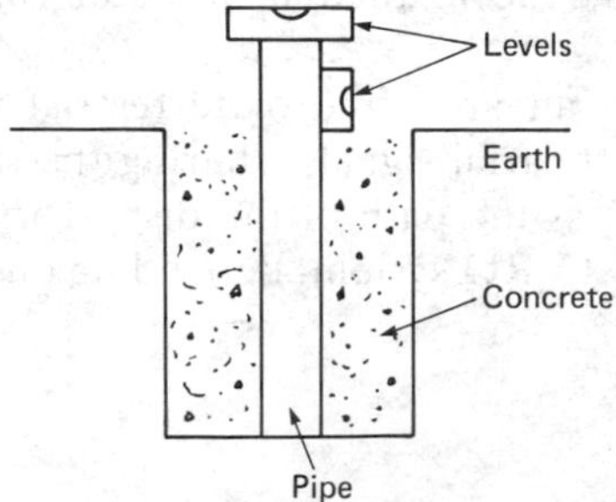

Figure 18-38 Concrete pole foundation.

special supports be used to keep the dish stationary. Winds create large forces when they blow on a dish, and a support must be able to withstand them.

18-12.3 Dish Assembly and Alignment

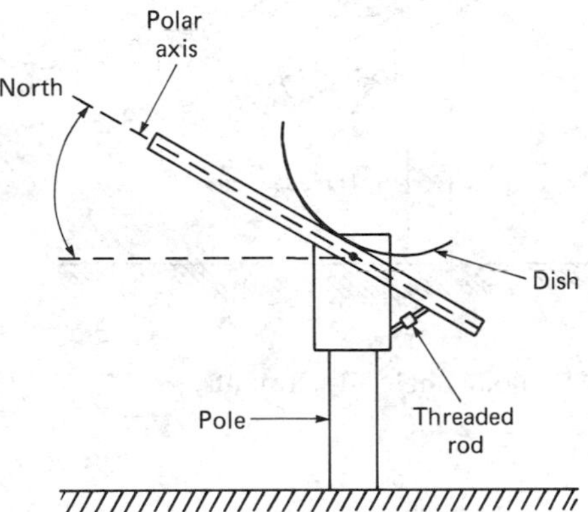

Figure 18-39 Polar axis alignment.

The dish and mount should be carefully assembled according to the manufacturer's instructions. If the dish is a sectional type, its surface should be checked with a template pattern to ensure that the surface is smooth and even. Once the dish is set up it must be aligned. Most dishes come with a polar mount, which requires the setting of three angles: (1) N-S (polar) axis, (2) polar (elevation) angle, and (3) offset (declination) angle(s). First, the polar axis must be aligned with true north–south as shown in Fig. 18-39. Finding true north–south can be done in four ways:

1. Using the North Star, which points to true north (see Fig. 18-40).
2. Using the wall of a building that runs in the north–south axis.
3. Using the sun's shadow at midday.
4. Using a compass, and then correcting for magnetic variation. Magnetic north varies from true north by a correction angle that can be obtained from a local airport.

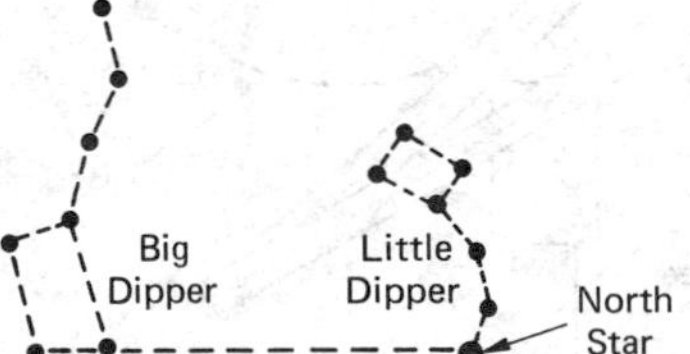

Figure 18-40 North Star location.

The second angle, the polar angle, should be set equal to the latitude of the site. The polar angle is the angle between the ground plane and the polar axis, which points to true north. The polar angle is adjusted by rotating a threaded rod on the polar mount as shown in Fig. 18-41. An inclinometer can be used to set the polar angle.

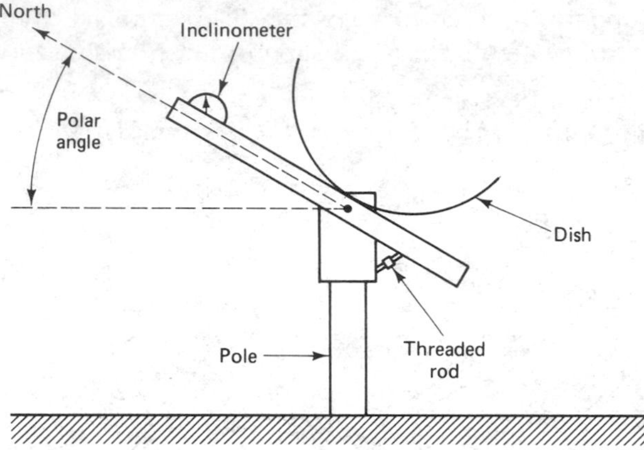

Figure 18-41 Polar angle adjustments.

The final adjustment is setting the offset or declination angle(s). Declination angles for a site latitude can be found in graphs or charts supplied by the dish mount manufacturer. Once again a threaded rod is rotated until an inclinometer reads the proper declination, as shown in Fig. 18-42. Some polar mounts do not have adjustment offset angles. These mounts have a built-in offset angle that is set for the middle latitude of the United States, which will result in a slight tracking error when used at extreme latitudes. Once all three adjustments have been made, the dish should track all visible geostationary satellites. Some fine tuning may be required after the complete downlink is set up.

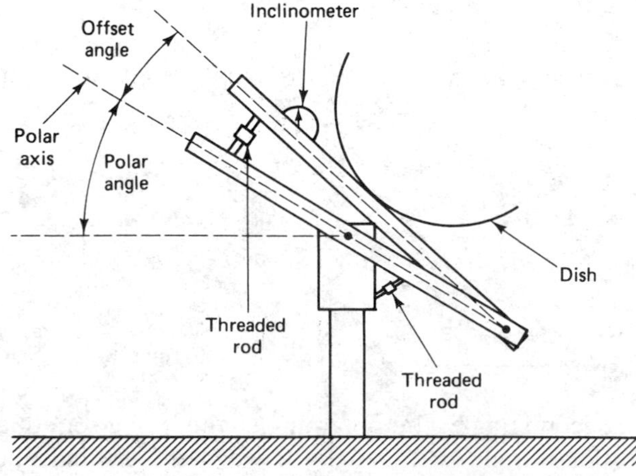

Figure 18-42 Declination angle adjustment.

18-12.4 Feedhorn Installation

The system performance can be peaked by correctly positioning the feedhorn assembly. The feedhorn should be centered and located at the focal point of the dish (Fig. 18-43). The focal point is measured from the center of the dish to the front edge of the feedhorn. Some feedhorns require that this distance extend $\frac{1}{4}$ in. inside the horn throat. The feedhorn can be centered above the dish by measuring the distance from the outer rim of the dish to the edge of the feedhorn in three different locations. The feedhorn is centered when all three measurements are equal (Fig. 18-44).

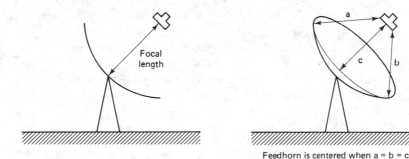

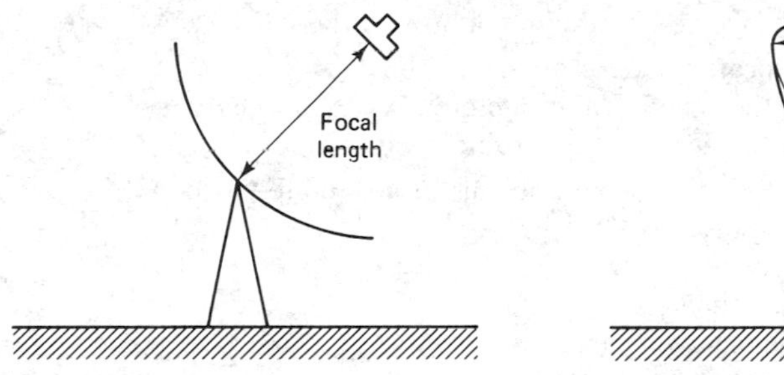

Figure 18-43 Focal length.

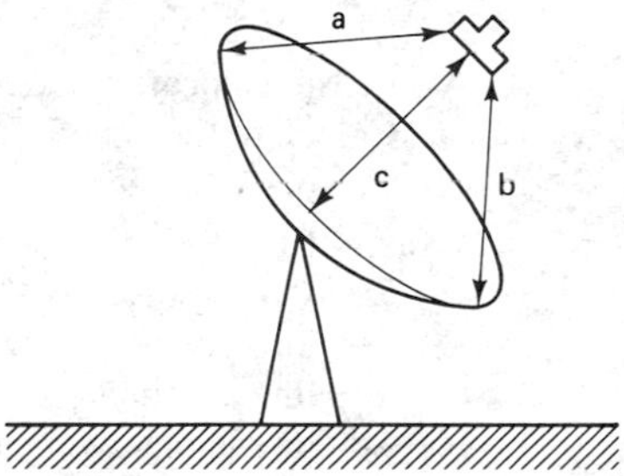

a
b
c
Feedhorn is centered when a = b = c

Figure 18-44 Feedhorn centering.

18-12.5 Installing Cables and Lightning Rods

Cables from the dish to the receiver should be run through electrical piping and buried underground. Cables required for a TVRO system include one coaxial cable to the LNC, three- or five-conductor cable for polarity adjustment to the feedhorn, six-conductor cable for motorized dish drives, and any spare cables that may be required in the future. A lightning rod should also be used to provide some security against nearby lightning strikes, which may cause large surges in the system. A lightning rod consists of a 6-ft metal rod driven into the ground and a cable that connects the rod to the dish assembly.

18-12.6 Waterproofing and Rustproofing

After the downlink is aligned and fine tuned, all connectors and mechanical assemblies should be sealed. Water may corrode connectors, eventually leading to a degradation in picture quality. The N-type coaxial connectors between the LNA and down-converter are particularly important when it comes to waterproofing. The electrical pipe carrying the cables should have its ends sealed with silicon to keep water out. Mechanical assemblies such as polarization rotators and dish-drive actuators may become damaged if water collects inside this housing. If water collects and freezes, it may expand and damage the actuator. It is common practice to drill a small hole on the actuator bottom so that water can escape from the actuator housing.

Problems

18-1. What is known about the speed of a satellite in geosynchronous orbit based on Kepler's second law?

18-2. A geostationary orbit is:
(1) A geosynchronous orbit having an inclination of 0°
(2) An orbit located at above 22,279 statute miles
(3) The least common of all orbits since it is difficult to track satellites in this type of orbit
(4) The orbit in which a satellite does not appear stationary relative to the earth
(5) The orbit followed by the moon around the earth

18-3. A satellite is located in geostationary orbit.
(a) Calculate the speed of the satellite (express in m/s).
(b) Calculate the period of one orbit (express in hours).

18-4. Which of the following would be part of a satellite payload?
(1) Solar cell arrays
(2) Satellite position control
(3) Satellite antenna
(4) Satellite launch vehicle
(5) Thermal stability control

18-5. C-band satellites use an uplink frequency band of ____________ while Ku-band satellites use an uplink frequency band of ____________.
(1) 5.925 to 6.415 MHz, 13.7 to 14.2 GHz
(2) 3.7 to 4.2 GHz, 11.7 to 12.2 GHz
(3) 5925 to 6425 MHz, 13.7 to 14.2 GHz
(4) 5925 to 6425 MHz, 13.7 to 14.2 GHz
(5) 3.7 to 4.2 MHz, 11.7 to 12.2 MHz

18-6. The total C-band spectrum bandwidth for 24 transponder North American satellites is:
(1) 960 MHz
(2) 432 MHz
(3) 40 MHz
(4) 500 MHz
(5) 4200 MHz

18-7. How are C-band satellites able to transmit 24 transponders each having a bandwidth of 40 MHz using less than 960 MHz of total bandwidth?

18-8. Interference between transponder signals is reduced by:
(1) Cross-polarization discrimination
(2) Separating satellites by a larger distance
(3) Using signal polarization
(4) Offsetting the center frequencies of odd-numbered transponders from the even-numbered transponders
(5) All of the above

18-9. The most important part of a TVRO system is:
(1) The dish
(2) The television

(3) The feedhorn/polarizer
(4) The receiver/down-converter
(5) They are all equally important.

18-10. Which of the following statements is false?
(1) The dish reflector is made of metal.
(2) Mesh dishes allow winds to pass through but suffer from a large gain loss as C-band microwaves pass through the mesh spacing.
(3) Most TVRO dishes come in basically two shapes: a parabola or a spherical section.
(4) Solid-spun dishes usually have the most accurate surface.
(5) Fiberglass dish surfaces may deform over time.

18-11. Which of the following statements is true?
(1) A deep dish requires a wider angle of illumination than a shallow dish to reflect the same amount of signal energy.
(2) Feedhorn rings act as quarter-wavelength chokes to increase fringing caused by E and H fields.
(3) Feedhorns pick up the most energy concentration from the edges of a dish since the dish surface is most accurate at the edges.
(4) Polarity selection is performed by rotating the dish by 90°.
(5) The feedhorn throat is a rectangular waveguide.

18-12. Explain the advantages and disadvantages of single-conversion down-converters versus double-conversion down-converters.

18-13. The image signal produced by a single-conversion down-converter can be reduced by:
(1) Using low-loss heliax cable
(2) Filtering the terrestrial interference to a level of 18 dB below the carrier level
(3) Using a high-speed image gun eliminator
(4) Using an image-reject mixer and ferrite isolators
(5) Decreasing the IF bandwidth of the TVRO receiver

18-14. Calculate the azimuth and elevation angles to point a dish at Galaxy 3 from a location on the earth having coordinates of 43°N latitude and 101°W longitude.

18-15. Determine the declination angles to point a dish at Space Net 1 from a location on the earth having coordinates of 39°N latitude and 94°W longitude.

18-16. Which of the following statements is true?
(1) Two-degree spacing as defined by the FCC really means slightly less than 2° spacing when measured on the earth's surface.
(2) A properly set up polar mount will require movements of azimuth and elevation to track a geostationary satellite.
(3) Polar mounts always have some small tracking error even with correctly set declination angles.
(4) The correct declination angle setting on a polar mount should not vary throughout North America.
(5) A parabolic dish will not detect any noise located behind it.

18-17. Determine the minimum spacing between two C-band satellites to avoid 3-dB beamwidth interference using a 10-ft mesh dish.

18-18. Calculate the gain (in dB) for a 70% efficient 10-ft parabolic dish. Assume that transponder 21 (4120 MHz) is being watched.

18-19. The gain of a parabolic dish decreases if:
(1) The dish diameter decreases.
(2) The dish surface accuracy decreases.
(3) The satellite signal frequency decreases.
(4) Both (1) and (2)
(5) All of the above

18-20. Determine the noise temperature of a 10-ft parabolic dish having coordinates of 43°N latitude and 101°W longitude. Assume that the dish is pointed at Galaxy 3.

18-21. Find the G/T figure of merit for the following TVRO system: a 70% efficient 10-ft dish, dish pointed at Galaxy 3 (transponder 21), 60-K LNA, and earth coordinates of 43°N latitude and 101°W longitude.

18-22. A TVRO downlink located at 110°W longitude and 35°N latitude consists of a 65% efficient 12-ft dish, a dish pointed at Satcom 3R (transponder 13), a 80-K LNA, a 28-MHz receiver bandwidth, and EIRP = 31 dBW.
(a) Find the slant range.
(b) Find the free-space loss.
(c) Calculate the receiver's input C/N ratio (assume that the total path loss = free-space loss + 0.5 dB).
(d) Calculate the receiver's output S/N ratio.

18-23. Satellite transmission of North American audio/video television signals is accomplished by:
(1) FM for audio and FM for video
(2) FM for audio and vestigial sideband AM for video
(3) AM for audio and AM for video
(4) FSK for audio and FM for video
(5) Vestigial sideband AM for audio and FM for video

18-24. Terrestrial transmission of North American audio/video television signals is accomplished by:
(1) FM for audio and FM for video
(2) FM for audio and vestigial sideband AM for video
(3) AM for audio and AM for video
(4) FSK for audio and FM for video
(5) Vestigial sideband AM for audio and FM for video

18-25. Give two reasons why Ku-band TVRO downlink is less expensive than a C-band TVRO downlink.

18-26. Which of the following statements is true?
(1) The Video Cypher II system uses 20 combinations of video scrambling.
(2) Scrambling systems have developed to improve the audio quality of TVRO downlink.
(3) The Video Cypher II system inserts three 8-bit words of control information into the horizontal sync pulse of the video.
(4) Only the American satellite signals are scrambled.
(5) Both the Canadian Cancom system and the American Video Cypher II system are microprocessor-controlled systems, but only the Video Cypher II system units come with a unique address stored in nonvolatile memory.

18-27. List three services other than television and radio that are carried by satellites.

18-28. How can one reduce or prevent terrestrial interference?

18-29. Terrestrial interference:
(1) Reduces the noise weighting factor, causing an increase in the FM improvement factor

(2) Can be avoided only in rural areas
(3) Is due to obstructions such as buildings and trees
(4) Is often frequency-specific and directional
(5) All of the above

18-30. True north is given by
(1) The North Star
(2) A reading of 0° on an inclinometer
(3) The North Star plus an adjustment angle to correct for star field magnetic variation
(4) A reading of north on a compass
(5) All of the above

18-31. Calculate the azimuth and elevation angles to point a parabolic dish at Satcom 3R from a location on the earth having coordinates of 100°W longitude and 40°N latitude.

18-32. Calculate the two declination angles to point a parabolic dish at Anik D1 from a location on the earth having coordinates of 120°W longitude and 30°N latitude.

18-33. Determine the minimum spacing between satellites before 3-dB beamwidth interference occurs on a 12-ft parabolic dish downlink when used for C-band reception. Assume that the range of received frequencies is 3700 to 4200 MHz.

18-34. Calculate the gain (in decibels) for a 12-ft parabolic dish having an efficiency of 55%. Assume that the received frequency is 3700 MHz.

18-35. Determine the noise temperature of a 12-ft-diameter parabolic dish pointed at Satcom 3R from a location on the earth having coordinates of 100°W longitude and 40°N latitude.

18-36. Determine the G/T figure of merit for a 12-ft-diameter parabolic dish pointed at Satcom 3R from a location on the earth having coordinates of 100°W longitude and 40°N latitude. The dish is 55% efficient and connected to an 80-K LNA. Assume that the frequency of operation is 3700 MHz.

18-37. A downlink located at 100°W longitude and 40°N latitude consists of the following: a 10-ft parabolic dish with an efficiency of 70%, a 24-MHz receiver bandwidth, a 100-K LNA, and an EIRP of 30 dBW. The downlink is pointed at Westar 4 and receiving at 3800 MHz.
(a) Calculate the slant range.
(b) Determine the free-space loss.
(c) Determine the receiver input C/N ratio. (Assume that the total path loss = free-space loss + 0.5 dB.)
(d) Calculate the output S/N ratio.

18-38. Calculate the azimuth and elevation angles to point a parabolic dish at Comstar 4 from a location on the earth having coordinates of 110°W longitude and 30°N latitude.

18-39. A downlink located at 110°W longitude and 30°N latitude consists of the following: a 10-ft parabolic dish with an efficiency of 65%, a 30-MHz receiver bandwidth, a 55-K LNA, and an EIRP of 32 dBW. The downlink is pointed at Comstar 4 and receiving at 3800 MHz.
(a) Calculate the slant range.
(b) Determine the free-space loss.
(c) Determine the receiver input C/N ratio. (Assume that the total path loss = free-space loss + 0.5 dB.)
(d) Calculate the output S/N ratio.

APPENDIX A Motorola C-Quam AM Stereo Encoding

Wayne Wolinski

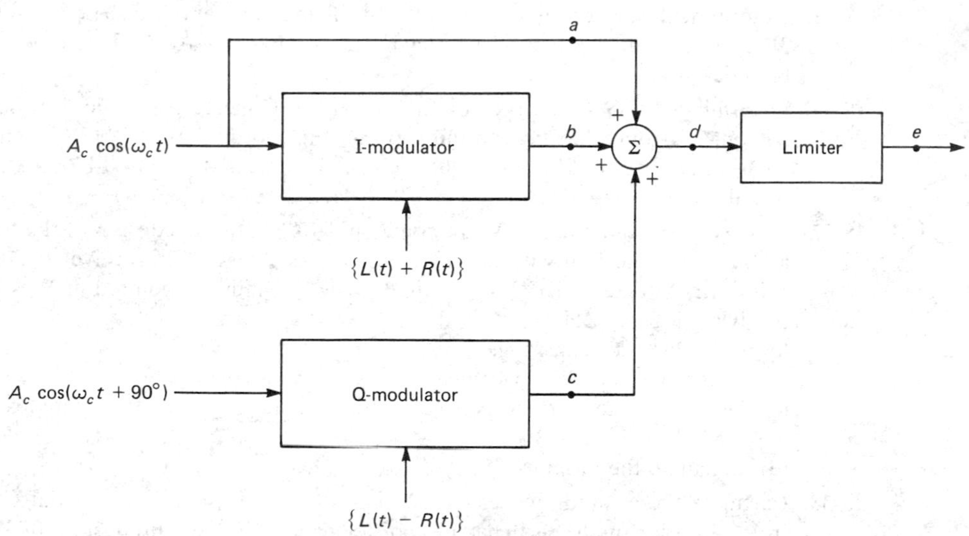

Figure A-1 C–QUAM encoding.

A partial reproduction of Fig. 6-26 is given in Fig. A-1. As given in Sec. 6-7.3, the output from the I-modulator (signal point b in Fig. A-1) can be expressed as

$$e_b(t) = M_s\,[L(t) + R(t)]A_c \cos \omega_c t \tag{A-1}$$

and the output from the Q-modulator as

$$e_c(t) = M_d\,[L(t) - R(t)]A_c \cos(\omega_c t + 90°) \tag{A-2}$$

where M_s is the index of modulation for sum information $L + R$, and M_d is the index of modulation for difference information $L - R$. Summing the signals at points a, b, and c yields the result at point d:

$$\begin{aligned} e_d(t) &= A_c \cos \omega_c t + M_s[L(t) + R(t)]A_c \cos \omega_c t + \cdots \\ &\quad + M_d[L(t) - R(t)]A_c \cos(\omega_c t + 90°) \\ &= A_c \cos \omega_c t + M_s A_c L(t) \cos \omega_c t + M_s A_c R(t) \cos \omega_c t + \cdots \\ &\quad + M_d A_c L(t) \cos(\omega_c t + 90°) - M_d A_c R(t) \cos(\omega_c t + 90°) \end{aligned} \tag{A-3}$$

A phasor diagram for equation (A-3) can now be constructed, as given in Fig. A-2. Phasor $\overline{R_2}$ represents the signal at point d. From the phasor diagram,

$$|\overline{R_1}| = A_c M_d[L(t) - R(t)]$$

and

$$|\overline{R_2}| = \sqrt{\{A_c M_d[L(t) - R(t)]\}^2 + \{A_c + A_c M_s[L(t) + R(t)]\}^2}$$

$$\begin{aligned} \angle \overline{R_2} &= \tan^{-1} \frac{A_c M_d[L(t) - R(t)]}{A_c + A_c M_s[L(t) + R(t)]} \\ &= \tan^{-1} \frac{A_c\{M_d[L(t) - R(t)]\}}{A_c\{1 + M_s[L(t) + R(t)]\}} \\ &= \tan^{-1} \frac{M_d[L(t) - R(t)]}{1 + M_s[L(t) + R(t)]} \end{aligned} \tag{A-4}$$

Translating phasor $\overline{R_2}$ into the time-domain and limiting the amplitude to value A_c yields, at point e,

$$e_e(t) = A_c \cos \left\{ \omega_c t + \tan^{-1} \frac{M_d[L(t) - R(t)]}{1 + M_s[L(t) + R(t)]} \right\} \tag{A-5}$$

as given in equation (6-33).

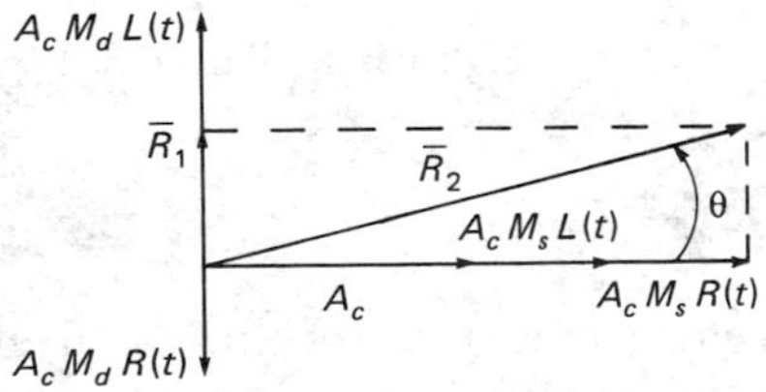

Figure A-2 Phasor diagram for $e_d(t)$.

Signal-to-Noise Ratios for FM

Wayne Wolinski

Consider the partial FM receiver given in Fig. B-1 (repeat of Fig. 8-29, for convenience). The signal-to-noise (S/N) ratio at any point in the receiver can be defined as

$$\frac{S}{N} = \frac{\text{signal power without noise}}{\text{noise power in the presence of a carrier (unmodulated)}}$$

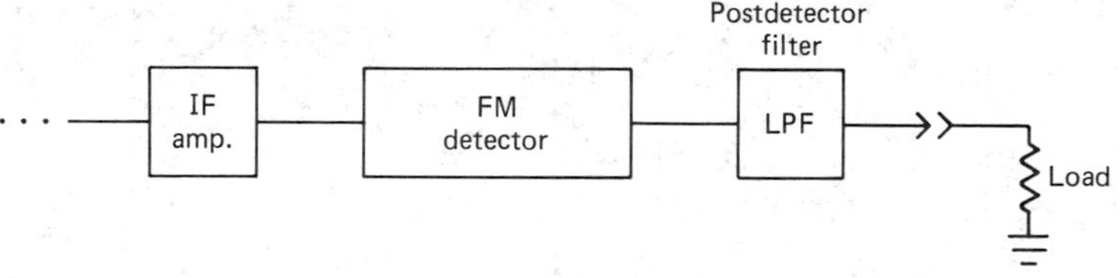

Figure B-1 Partial FM receiver.

B-1 SIGNAL POWER

For an FM detector (frequency-to-voltage converter), the instantaneous detector output voltage is proportional to frequency departure; therefore, for a single-tone-modulated FM input, the detector output can be expressed as

$$V_{0s}(t) = K_1 \, \Delta f \cos \omega_m t \tag{B-1}$$

where K_1 = constant depending on detector sensitivity
Δf = deviation
ω_m = modulating signal angular frequency

The resulting power dissipated into a 1-Ω load placed at the detector output would then be

$$\frac{K_1^2\,\Delta f^2}{2} \tag{B-2}$$

B-2 NOISE POWER

Consider the noise spectrum to be flat and continuous for analytical purposes (i.e., an infinite number of equal-amplitude frequency components) as illustrated in Fig. B-2.

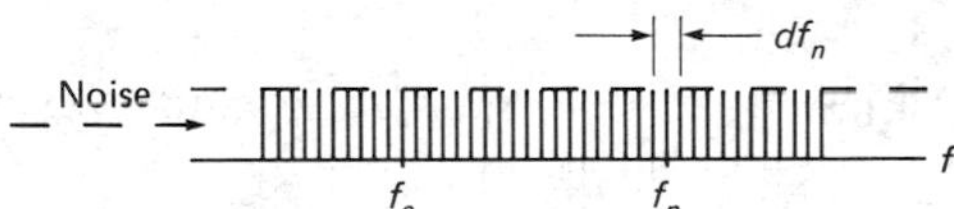

Figure B-2 Noise spectrom at detector input.

Although noise is random in nature, and spectrally distributed according to source (see Chapter 9), the FM receiver IF stages typically have sharp bandpass filtering; therefore, over the relatively small IF bandwidth passed, the contribution to total input noise at the detector by each source can be assumed to be constant over the bandwidth in question.

For the spectrum of Fig. B-2, a spectral noise density of η watts per hertz will be assumed at the input to the detector, and since η is normally specified for a 1-Ω load, assume the detector of Fig. B-1 to have an input impedance of 1 Ω. If a very narrow bandwidth represented by df_n in Fig. B-2 is now selected, the noise power contained in a signal at $f \approx f_n$ having amplitude E_n volts peak may now be determined:

$$\frac{E_n^2}{2} = \eta\, df_n \tag{B-3}$$

Recall now that we are trying to find the noise power in the presence of a carrier; therefore, the detector input can be assumed to be of the form

$$e_{\text{IF}}(t) = E_c \cos \omega_c t + E_n \cos\left[\omega_c t + (\omega_n - \omega_c)t\right] \tag{B-4}$$

Letting $\omega_n > \omega_c$ (for example), a phasor diagram can be constructed, as illustrated in Fig. B-3. The resultant (R) of Fig. B-3 represents a signal having instantaneous amplitude and phase variations (the phase of the resultant varies with respect to the carrier as the vectors in Fig. B-3 are imagined to rotate about the origin). The amplitude variations may be removed by preceding the detector with a limiter stage, or using a type of detector that does not respond to amplitude variation; however, the phase variations remain. As instantaneous frequency variations accompany instantaneous phase variations, a noise power out of the detector will result, and it is this that we now seek.

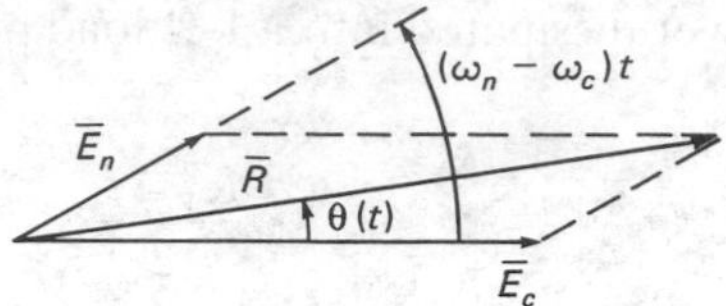

Figure B-3 Phasor diagram of detector input.

The first step is to obtain an expression for the instantaneous frequency variations at the detector input. Start by considering the phase angle between the carrier and resultant for various instants in time, as illustrated in Fig. B-4. The variation of $\theta(t)$ with time will be sinusoidal, as illustrated in Fig. B-5, and the peak value will be approximately E_n/E_c, provided that the amplitude of the noise signal is much less than that of the carrier (as assumed for Fig. B-4).

Based on Figs. B-4 and B-5, an approximate expression for $\theta(t)$ results:

$$\theta(t) \approx \frac{E_n}{E_c} \sin(\omega_d t + \theta_0), \qquad E_n << E_c \tag{B-5}$$

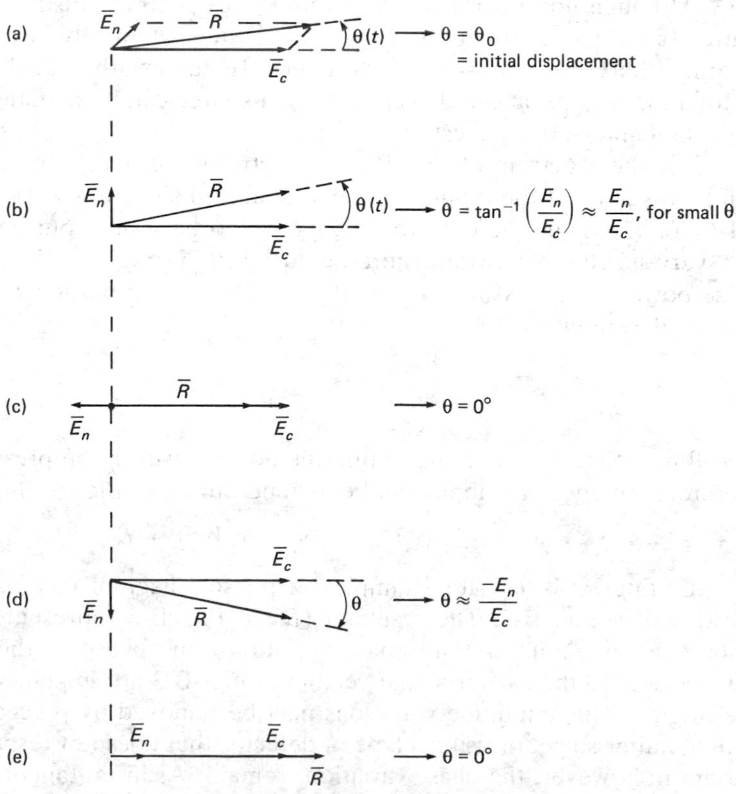

Figure B-4 Phase angle $[\theta(t)]$ variation.

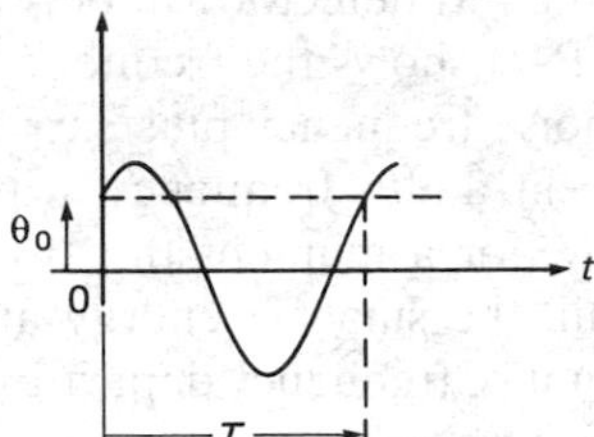

Figure B-5 $\theta(t)$.

The instantaneous frequency accompanying this phase variation may now be determined:

$$\omega_i(t) = \frac{d\theta(t)}{dt} \approx \frac{E_n}{E_c} [\cos(\omega_d t + \theta_0)] (\omega_d)$$

Therefore,

$$f_i \approx \frac{E_n}{E_c} f_d [\cos(\omega_d t + \theta_0)] \tag{B-6}$$

where $f_d \triangleq f_n - f_c$, $\omega_d \triangleq \omega_n - \omega_c$. Again detector output is a function of the instantaneous frequency at the input; therefore, the output noise voltage due to input specified by (B-6) is

$$v_{on}(t) \approx \frac{K_1 E_n f_d}{E_c} [\cos(\omega_d t + \theta_0)] \tag{B-7}$$

The power dissipated into a 1-Ω load by this noise voltage will be

$$N_{o1}\Big|_{R_L = 1\Omega} \approx \frac{K_1^2 E_n^2 f_d^2}{2E_c^2} \tag{B-8}$$

Equation (B-3) stated:

$$\frac{E_n^2}{2} = \eta\, df_n$$

Therefore, (B-8) may be expressed as

$$N_{o1}\Big|_{R_L = 1\Omega} \approx \frac{K_1^2 f_d^2 \eta\, df_n}{E_c^2} \tag{B-9}$$

This result demonstrates a parabolic increase in noise power out as the magnitude of $f_n - f_c$ increases, as illustrated in Fig. B-6.

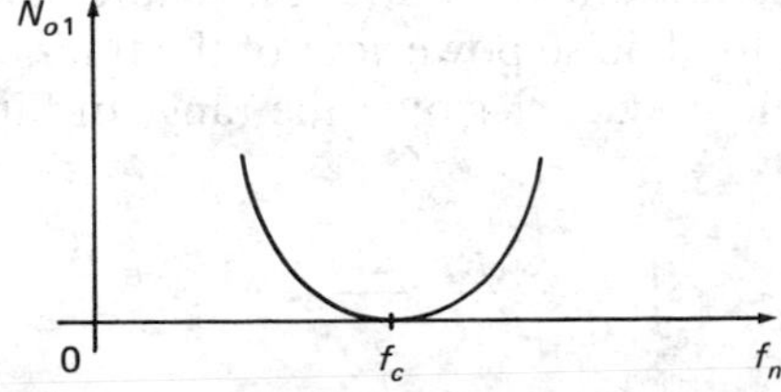

Figure B-6 Noise power out versus input noise frequency.

Recall now that following the FM detector will be some sort of low-pass filter (postdetector filter). Equation (B-7) shows the frequency of the noise voltage out of the detector due to a single noise frequency plus carrier input to be f_d (or $f_n - f_c$); therefore, f_d can easily be an audio frequency within the bandpass of the postdetector filter. Consider now for a moment the situation at the transmitter; normal speech and music contain the smallest energy at high audio frequencies, thereby producing the least amount of frequency departure out of an FM modulator (if no modification to the audio baseband were permitted). At the receiver, the detector output voltage is proportional to frequency departure present at the input; therefore, it would appear that the demodulated signal will contain the smallest amplitudes, and hence power, for the highest audio frequencies.

Figure B-6 has illustrated the noise power out of the detector to be the greatest at high audio frequencies; therefore, the potential for very poor signal-to-noise performance at high audio frequencies exists were it not for "pre-" and "deemphasis." *Preemphasis* involves boosting high audio frequencies in the baseband before passage to the modulator section of the transmitter. At the receiver, *deemphasis* is used, whereby the high audio frequencies present at the detector output are attenuated (to restore the proper amplitude–frequency relationship to the original baseband signal). Pre-/deemphasis is normally done according to standard curves, as illustrated in Fig. P8-18b, and use of this technique results in a greatly improved S/N ratio at the output of the receiver, since no net change to signal power out has occurred (compared to the case for no pre-/deemphasis), while the noise spectral density has been altered considerably, as illustrated in Fig. B-7.

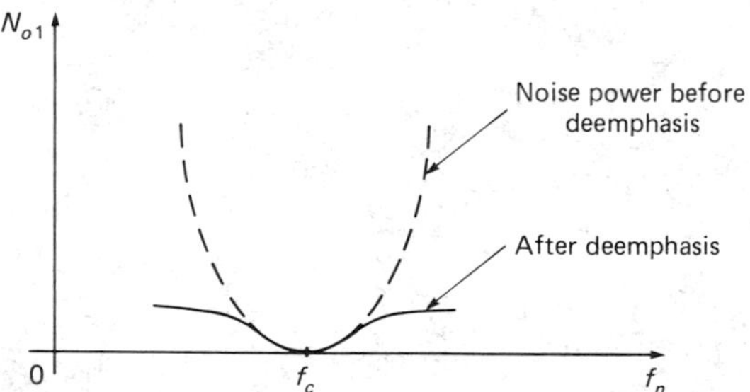

Figure B-7 Effect of deemphasis on noise output.

Having discussed pre-/deemphasis, let us now continue our analysis for the case with *no pre-/deemphasis*. For the following, assume the postdetector filter to be an ideal low-pass filter having bandwidth W, and a 1-Ω load connected to the output; under these conditions, the total noise power out of the filter can be found by summing the effects of all possible frequencies over the range of filter bandpass (i.e., by integration):

$$N_o \Big|_{RL=1\Omega} \approx \int_{-W}^{+W} \frac{K_1^2 f_d^2 \eta \, df_n}{E_c^2} \tag{B-10}$$

Note the limits of integration; these are due to the fact that noise frequencies at detector input both above and below f_c will produce audio out of the filter (since $f_d = f_n - f_c$), provided of course that f_d falls within the postdetector passband. Equation (B-10) yields

$$N_o \bigg|_{RL=1\Omega} \approx \frac{2K_1^2 \eta W^3}{3E_c^2} \tag{B-11}$$

B-3 SIGNAL-TO-NOISE RATIO OUT

Thus the S/N ratio at the output of the postdetector filter may now be expressed as

$$\left(\frac{S_o}{N_o}\right)_{\text{FM}} \approx \frac{3\,\Delta f^2 E_c^2}{4\eta W^3} \tag{B-12}$$

Equation (B-12) reveals that best S/N performance will be obtained when the frequency deviation is as large as possible (i.e., use as much of the maximum as possible).

B-4 COMPARISON TO AM

Theoretically, the optimum S/N ratio present at the output of an AM detector will occur for 100% modulation, and under these conditions

$$\left(\frac{S_o}{N_o}\right)_{\text{AM}} = \left(\frac{C_i}{N_i}\right)_{\text{AM}} \tag{B-13}$$

for the AM detector,

where C_i = carrier input power
N_i = noise input power
S_o = signal output power
N_o = noise output power

To relate FM performance (with no pre-/deemphasis) to AM performance, an expression for C/N at the input to the FM detector must first be found. The carrier power at the input to a detector having a 1-Ω input impedance is

$$C_i = \frac{(E_c/\sqrt{2})^2}{1\ \Omega} = \frac{E_c^2}{2} \tag{B-14}$$

The noise power input to the detector can be expressed by

$$N_i = \eta B \tag{B-15}$$

where B = IF bandwidth. Now an expression for C/N at detector input results:

$$\left(\frac{C_i}{N_i}\right)_{\text{FM}} = \frac{E_c^2}{2\eta B} \tag{B-16}$$

To make a fair comparison, the IF bandwidth should be limited to that for the AM case (i.e., $2W$, since W = highest baseband frequency). Restricting the bandwidth as such yields

$$\left(\frac{C_i}{N_i}\right)_{\text{FM}} = \frac{E_c^2}{2\eta(2W)} = \frac{E_c^2}{4\eta W} \tag{B-17}$$

Now using (B-12) and (B-17), we obtain

$$\left(\frac{S_o}{N_o}\right)_{\text{FM}} = \frac{3\,\Delta f^2}{W^2}\left(\frac{C_i}{N_i}\right)_{\text{FM}} \tag{B-18}$$

For a single-tone modulated detector input, W becomes f_m (modulating frequency); therefore,

$$\left(\frac{S_o}{N_o}\right)_{\text{FM}} = 3\left(\frac{\Delta f}{f_m}\right)^2\left(\frac{C_i}{N_i}\right)_{\text{FM}} = 3(M_f)^2\left(\frac{C_i}{N_i}\right)_{\text{FM}} \tag{B-19}$$

Now to compare AM to FM (no pre-/deemphasis), let the C/N ratio be the same at the inputs to the respective detectors; that is,

$$\left(\frac{C_i}{N_i}\right)_{\text{FM}} = \left(\frac{C_i}{N_i}\right)_{\text{AM}} \tag{B-20}$$

Now using (B-13), (B-17), and (B-18) yields

$$\left(\frac{S_o}{N_o}\right)_{\text{FM}} = 3(M_f)^2\left(\frac{S_o}{N_o}\right)_{\text{AM}} \tag{B-21}$$

This result reveals that the S/N ratio out of an FM receiver can be made considerably higher than for an AM receiver simply by increasing M_f; however, the price to be paid is increased bandwidth.

Index

Credits

Key to abbreviations: DCCA Tom McGovern, *Data Communications: Concepts and Applications*, © 1988. Reprinted by permission of Prentice-Hall Canada Inc., Scarborough, Ontario; ETT William Sinnema, *Electronic Transmission Technology: Lines, Waves, and Antennas*, 2e, © 1988. Reprinted by permission of Prentice Hall, Inc., Englewood Cliffs, New Jersey.

Figure 2-16 William Sinnema/Tom McGovern, *Digital, Analog, and Data Communication*, 2/e, © 1986, p. 287. Reprinted by permission of Prentice Hall Inc., Englewood Cliffs, New Jersey; Table 3-1 R. W. Johnson, "Response of pi-pi-L and tandem quarter-wave-line matching networks," *Ham Radio Magazine*, February 1982, p. 13. Reprinted with permission; Figure 3-23 Reprinted with permission from *Electronics* magazine, copyright © August 16, 1973, VNU Business Publications Inc.; Figure 5-13(d) Courtesy of GE Solid State; Figure 5-38 Copyright of Motorola, Inc. Used by permission; Figure 5-40 Courtesy of Signetics Corp.; Figure 5-48 Courtesy of Signetics Corp.; Figure 5-49 Courtesy of GE Solid State; Figure 5-50 Copyright of Motorola, Inc. Used by permission; Figure P5-16 Copyright of Motorola, Inc.; Figure 6-6 Courtesy of Midland International; Figure 6-8 Courtesy of Midland International; Figure 6-19 Courtesy of GE Solid State; Figure 6-26 Copyright of Motorola, Inc. Used by permission; Figure 6-28 Copyright of Motorola, Inc. Used by permission; Figure P6-9 Courtesy of Midland International; Figure 7-4 A.I.H. Wade, "Design criteria for SSB phase-shift networks," *Ham Radio Magazine*, June 1970, figures 4 and 5. Reprinted with permission; Table 7-1 Courtesy Bird Electronic Corporation; Figure 7-14, Figure 7-15, Figure 7-17 Courtesy Resdel Industries; Figure 8-2 Wayne Tomasi, *Electronic Communications Systems: Fundamentals Through Advanced*, © 1988, p. 276. Reprinted by permission of Prentice Hall, Inc., Englewood Cliffs, New Jersey; Figure 8-29, Table 8-3, Figure 8-31, Figure 8-32, Figure 8-33, Figure 8-34, Figure 8-36, Figure 8-38, Table 8-4 Courtesy of GE Mobile Communications Canada; Figure 9-3 Reproduced with permission of the publisher, Howard W. Sams & Co., Indianapolis, *Reference Data for Radio Engineers*, 5th ed., by Howard W. Sams & Co., copyright 1968; Figure 10-62, Figure 10-63 Courtesy of Mitel Corporation; Figure 12-1 DCCA, p. 2; Figure 12-2 DCCA, p. 7; Figure 12-3, Figure 12-4 DCCA, p. 15; Figure 12-5 DCCA, p. 17; Figure 12-6, Figure 12-7 DCCA, p. 18; Figure 12-8 DCCA, p. 19; Figure 12-9, Figure 12-10 DCCA, p. 20; Table 12-1 DCCA, p. 16; Figure 12-11 DCCA, p. 39; Figure 12-12 DCCA, p. 48; Figure 12-13 DCCA, p. 68; Figure 12-14, Figure 12-15 DCCA, p. 69; Figure 12-16, Figure 12-17 DCCA, p. 71; Figure 12-18 DCCA, p. 72; Figure 12-19, Figure 12-20 DCCA, p. 73; Figure 12-21 DCCA, p. 75; Figure 12-22, Figure 12-23 DCCA, p. 76; Figure 12-24, Figure 12-25 DCCA, p. 77; Table 12-2 DCCA,